Lecture Notes in Artificial Intelligence 9773

Subseries of Lecture Notes in Computer Science

More information about this series at http://www.springer.com/series/1244

De-Shuang Huang · Kyungsook Han
Abir Hussain (Eds.)

Intelligent Computing Methodologies

12th International Conference, ICIC 2016
Lanzhou, China, August 2–5, 2016
Proceedings, Part III

 Springer

Editors
De-Shuang Huang
Tongji University
Shanghai
China

Abir Hussain
Liverpool John Moores University
Liverpool
UK

Kyungsook Han
Inha University
Incheon
Korea (Republic of)

ISSN 0302-9743   ISSN 1611-3349 (electronic)
Lecture Notes in Artificial Intelligence
ISBN 978-3-319-42296-1   ISBN 978-3-319-42297-8 (eBook)
DOI 10.1007/978-3-319-42297-8

Library of Congress Control Number: 2016943870

LNCS Sublibrary: SL7 – Artificial Intelligence

Printed on acid-free paper

This Springer imprint is published by Springer Nature
The registered company is Springer International Publishing AG Switzerland

Preface

The International Conference on Intelligent Computing (ICIC) was started to provide an annual forum dedicated to the emerging and challenging topics in artificial intelligence, machine learning, pattern recognition, bioinformatics, and computational biology. It aims to bring together researchers and practitioners from both academia and industry to share ideas, problems, and solutions related to the multifaceted aspects of intelligent computing.

ICIC 2016, held in Lanzhou, China, August 2–5, 2016, constituted the 12th International Conference on Intelligent Computing. It built upon the success of ICIC 2015, ICIC 2014, ICIC 2013, ICIC 2012, ICIC 2011, ICIC 2010, ICIC 2009, ICIC 2008, ICIC 2007, ICIC 2006, and ICIC 2005 that were held in Fuzhou, Taiyuan, Nanning, Huangshan, Zhengzhou, Changsha, China, Ulsan, Korea, Shanghai, Qingdao, Kunming, and Hefei, China, respectively.

This year, the conference concentrated mainly on the theories and methodologies as well as the emerging applications of intelligent computing. Its aim was to unify the picture of contemporary intelligent computing techniques as an integral concept that highlights the trends in advanced computational intelligence and bridges theoretical research with applications. Therefore, the theme for this conference was "Advanced Intelligent Computing Technology and Applications." Papers focused on this theme were solicited, addressing theories, methodologies, and applications in science and technology.

ICIC 2016 received 639 submissions from 22 countries and regions. All papers went through a rigorous peer-review procedure and each paper received at least three review reports. Based on the review reports, the Program Committee finally selected 236 high-quality papers for presentation at ICIC 2016, included in three volumes of proceedings published by Springer: two volumes of *Lecture Notes in Computer Science* (LNCS), and one volume of *Lecture Notes in Artificial Intelligence* (LNAI).

This volume of *Lecture Notes in Artificial Intelligence* (LNAI) includes 80 papers.

The organizers of ICIC 2016, including Tongji University and Lanzhou University of Technology, China, made an enormous effort to ensure the success of the conference. We hereby would like to thank the members of the Program Committee and the referees for their collective effort in reviewing and soliciting the papers. We would like to thank Alfred Hofmann, of Springer, for his frank and helpful advice and guidance throughout and for his continuous support in publishing the proceedings. Moreover, we would like to thank all the authors in particular for contributing their papers. Without the high-quality submissions from the authors, the success of the conference would not have been possible. Finally, we are especially grateful to the IEEE Computational Intelligence Society, the International Neural Network Society, and the National Science Foundation of China for their sponsorship.

May 2016

De-Shuang Huang
Kyungsook Han
Abir Hussain

Organization

General Co-chairs

De-Shuang Huang	China
Cesare Alippi	Italy
Jie Cao	China

Program Committee Co-chairs

Kang-Hyun Jo	Korea
Vitoantonio Bevilacqua	Italy
Jinyan Li	Australia

Organizing Committee Co-chairs

Aihua Zhang	China
Ce Li	China

Organizing Committee Members

Weirong Liu	China
Erchao Li	China
Xiaolei Chen	China
Hui Chen	China
Suping Deng	China
Lin Zhu	China
Gang Wang	China

Award Committee Chair

Kyungsook Han	Korea

Tutorial Co-chairs

Laurent Heutte	France
Abir Hussain	UK

Publication Co-chairs

M. Michael Gromiha	India
Valeriya Gribova	Russia
Juan Carlos Figueroa	Colombia

Workshop/Special Session Chair

Ling Wang   China

Special Issue Co-chairs

Henry Han   USA
Phalguni Gupta   India

International Liaison Chair

Prashan Premaratne   Australia

Publicity Co-chairs

Evi Syukur   Australia
Chun-Hou Zheng   China
Jair Cervantes Canales   Mexico

Exhibition Chair

Lin Zhu   China

Program Committee

Andrea F. Abate	Fengfeng Zhou	Ming Jiang
Akhil Garg	Francesco Pappalardo	Jijun Tang
Vangalur Alagar	Shan Gao	Joaquin Torres
Angel Sappa	Liang Gao	Jun Zhang
Angelo Ciaramella	Kayhan Gulez	Kang Li
Bingqiang Liu	Hongei He	Ka-Chun Wong
Shuhui Bi	Huiyu Zhou	Seeja K.R
Bin Liu	Fei Han	Kui Liu
Cheen Sean Oon	Huanhuan Chen	Min Li
Chen Chen	Mohd Helmy Abd Wahab	Jianhua Liu
Wen-Sheng Chen	Hongjie Wu	Juan Liu
Michal Choras	Indrajit Saha	Yunxia Liu
Xiyuan Chen	Ivan Vladimir Meza Ruiz	Haiying Ma
Chunmei Liu	John Goulermas	Maurizio Fiasche
Costin Badica	Jianbo Fan	Marzio Pennisi
Dah-Jing Jwo	Jiancheng Zhong	Peter Hung
Daming Zhu	Junfeng Xia	Qiaotian Li
Dongbin Zhao	Jiangning Song	Qinmin Hu
Ben Niu	Jian Yu	Robin He
Dunwei Gong	Jim Jing-Yan Wang	Wei-Chiang Hong

Emanuele Lindo Secco
Shuigeng Zhou
Shuai Li
Shihong Yue
Saiful Islam
Jiatao Song
Shuo Liu
Shunren Xia
Surya Prakash
Shaoyuan Li
Tingwen Huang
Vasily Aristarkhov
Fei Wang
Xuesong Wang
Weihua Sun

Weidong Chen
Wei Wei
Zhi Wei
Shih-Hsin Chen
Wu Chen Su
Shitong Wang
Xiufen Zou
Xiandong Meng
Xiaoguang Zhao
Minzhu Xie
Xin Yin
Xinjian Chen
Xiaoju Dong
Xingsheng Gu
Xiwei Liu

Yingqin Luo
Yongquan Zhou
Yun Xiong
Yong Wang
Yuexian Hou
Chenghui Zhang
Weiming Zeng
Zhigang Luo
Fa Zhang
Liang Zhao
Zhenyu Xuan
Shanfeng Zhu
Quan Zou
Zhenran Jiang

Additional Reviewers

Yong Chen
Peng Xie
Yunfei Wang
Selin Ozcira
Stephen Tang
Badr Abdullah
Xuefeng Cui
Lumin Zhang
Chunye Wang
Qian Chen
Kan Qiao
Yingji Zhong
Wei Gao
Tao Yi
Liuhua Chen
Faliu Yi
Xiaoming Liu
Sheng Ding
Xin Xu
Zhebin Zhang
Shankai Yan
Yueming Lyu
Giulia Russo
Marzio Pennisi
Zhile Yang
Enting Gao
Min Sun

Lingjiao Pan
Ying Bi
Chao Jin
Shiwei Sun
Mohd Shamrie Sainin
Xing He
Xue Zhang
Junqing Li
Chen Chen
Wei-Shi Zheng
Chao Wu
Tingli Cheng
Francesco Pappalardo
Neil Buckley
Bolin Chen
Pengbo Wen
Long Wen
Bogdan Czejdo
Jing Wu
Weiwei Shen
Ximo Torres
Lan Huang
Jingchuan Wang
Savannah Bell
Alexandria Spradlin
Christina Spradlin
Li Liu

Geethan Mendiz
Jingsong Shi
Lun Li
Cheng Lian
Jin-Xing Liu
Obinna Anya
Lai Wei
Yan Cui
Peng Xiaoqing
Vivek Kanhangad
Yong Xu
Morihiro Hayashida
Yaqiang Yao
Chang Li
Jiang Bingbing
Haitao Li
Wei Peng
Jerico Revote
Xiaoyu Shi
Jia Meng
Jiawei Wang
Jing Jin
Yong Zhang
Biao Xu
Vangalur Alagar
Kaiyu Wan
Surya Prakash

Yongjin Li
Changning Liu
Xionghui Zhou
Hong Wang
Gongjing Chen
Yuntao Wei
Fangfang Zhang
Jia Liu
Jing Liu
Jnanendra Sarkar
Sayantan Singha Roy
Puneet Gupta
Shaohua Li
Zhicheng Liao
Adrian Lancucki
Julian Zubek
Srinka Basu
Xu Huang
Liangxu Liu
Qingfeng Li
Cristina Oyarzun Laura
Rina Su
Xiaojing Gu
Peng Zhou
Zewen Sun
Xin Liu
Yansheng Wang
Xiaoguang Zhao
Qing Lei
Yang Li
Wentao Fan
Hongbo Zhang
Minghai Xin
Yijun Bian
Yao Yu
Vasily Aristarkhov
Qi Liu
Vibha Patel
Jun Fan
Bojun Xie
Jie Zhu
Long Lan
Phan Cong Vinh
Zhichen Gong
Jingbin Wang
Akhil Garg

Wei Liao
Tian Tian
Xiangjuan Yao
Chenyan Bai
Guohui Li
Zheheng Jiang
Li Hailin
Huiyu Zhou
Baohua Wang
Kasi Periyasamy
Li Nie
Zhurong Wang
Ella Pereira
Danilo Caceres
Meng Lei
Changbin Du
Shaojun Gan
Yuan Xu
Chen Jianfeng
Chuanye Tang
Bo Liu
Bin Qian
Xuefen Zhu
Haoqian Huang
Fei Guo
Jiayin Zhou
Raul Montoliu
Oscar Belmonte
Farid Garcia-Lamont
Alfonso Zarco
Yi Gu
Ning Zhang
Jingli Wu
Xing Wei
Shenshen Liang
Nooraini Yusoff
Yanhui Guo
Nureize Arbaiy
Wan Hussain Wan Ishak
Yizhang Jiang
Pengjiang Qian
Si Liu
Chen Aiguo
Yunfei Yi
Rui Wang
Jiefang Liu

Aijia Ouyang
Hongjie Wu
Andrei Velichko
Wenlong Hang
Lijun Quan
Min Jiang
Tomasz Andrysiak
Faguang Wang
Liangxiu Han
Leonid Fedorischev
Wei Dai
Yifan Zhao
Xiaoyan Sun
Yiping Liu
Hui Li
Yinglei Song
Elisa Capecci
Tinting Mu
Francesco Giovanni Sisca
Austin Brockmeier
Cheng Wang
Juntao Liu
Mingyuan Xin
Chuang Ma
Marco Gianfico
Davide Nardone
Francesco Camastra
Antonino Staiano
Antonio Maratea
Pavan Kumar Gorthi
Antonio Brunetti
Fabio Cassano
Xin Chen
Fei Wang
Chen Xu
Gianpaolo Francesco
   Trotta
Alberto Cano
Xiuyang Zhao
Zhenxiang Chen
Lizhi Peng
Nagarajan Raju
e Wang
Yehu Shen
Liya Ding
Tiantai Guo

Contents – Part III

Machine Learning

Knowledge Discovery and Natural Language Processing

Nature Inspired Computing and Optimization

Intelligent Control and Automation

Intelligent Data Analysis and Prediction

Computer Vision

Knowledge Representation and Expert System

Bioinformatic

Optimization, Neural Network
and Signal Processing

A Method of Reducing Output Waveform Distortion in Photovoltaic Converter System

Huixiang Xu and Nianqiang Li[✉]

School of Information Science and Engineering,
University of Jinan, Jinan 250022, China
jd_chuying1990@163.com

Abstract. For the problem of leak inductance energy and voltage spike, given in the fly-back photovoltaic converter system (hereinafter abbreviated as PV), a new physical circuit of RCD (resistor-capacitor-diode) is proposed which based on the Boost circuit topology to recovering leak inductance energy and depressing main switch voltage spike during the commutation of power MOSFET. The circuit can recycle the leak inductance energy effectively [1–2], and suppress output capacitance (hereinafter abbreviated as Coss) electric discharge which caused the voltage spike during MOSFET switch-on. In addition, after detailed analysis, the new physical circuit can solve the problem of the energy loss generate from the Coss electric discharge which is not pay more attention by the other papers. The new design can clamp the drain-source voltage and implement the control policy of soft-switch. In this paper, we give the theoretical derivation of the circuit parameter and the implemented physically experimental curve of the physical circuit by analyzing the causes of voltage spike generated. Compared with the traditional method, improved the converter efficiency from 94.2 % to 96.2 %, and the MPPT efficiency to 99.4 %.

Keywords: Leak inductance energy recovery · Voltage spike depress · Coss capacity discharge · Soft-switch · RCD · Boost topology · Converter

1 Introduction

The PV converter is a controller that input DC and output AC, which include two ways to implement it, converter-booster(firstly, inverse the DC to AC, then boost the voltage up to 230 V) and booster-converter(firstly, boost the DC voltage up to 230 V, then inverse the DC to AC). The former need using large transformer, so it's more used in high power inverter design. Traditional design prefer to use converter-booster to conversion the voltage, because of its directness. However, in order to protect the electronic component from being damaged, the cost of this method is very high. For the distributed converter with high frequency and miniature, its more prefer to used the method of booster-converter [3], which is cheaper than traditional method. The fly-back is the main transformer in the secondary side which include the half-wave diode rectifier to output the half-wave with the 100 Hz frequency and 330 V peak value, then conversion 50 Hz sine wave through the H-bridge converter circuit.

© Springer International Publishing Switzerland 2016
D.-S. Huang et al. (Eds.): ICIC 2016, Part III, LNAI 9773, pp. 3–10, 2016.
DOI: 10.1007/978-3-319-42297-8_1

Experiments show that, during an DC-AC conversion, producing a good or bad half wave is the key to affect the wave shape and the conversion efficiency. In addition the quality of half wave affected by the leak inductance energy and the power MOSFET parasitic capacitor discharge more seriously. if its magnitude is excessive, the problem will cause the voltage spike, energy loss and the converter efficiency greatly reduced [4]. For this problem, designer usual use the circuit of leak inductance energy recovery based on the Buck topology to absorb the leak inductance energy, just like mentioned in the Ref. [2]. But if the leak inductance energy is too much or unstable, the output voltage of buck circuit will be influenced seriously. Moreover, in the circuit, the traditional method do not consideration the effect of capacitor electric discharge, and no theory analysis about the circuit parameter selection. In this paper, the control policy of MOSFET used the sinusoidal pulse width modulation (hereinafter abbreviated as SPWM) to achieve zero voltage switching (ZVS) [5]. The new design has improved the ability of physical circuit energy absorption, and clamped the drain-source voltage. In addition, inversion curve and converter efficiency curve are measured by circuit parameter selection with theory analysis.

2   The Cause of the Fly-Back Leak Inductance and Voltage Spike

Compare to the traditional one converter power system for multiple PV modules, the fly-back converter suitable for medium and small power transformation, can be used to one converter and one PV modules connected. It is more easily to work in the Maximum Power Point Tracking (MPPT) [2, 6, 7]. The working principle of the fly-back transformer is that the secondary coil do not output power to the load, until the primary coil incentive cut off. Because of the fly-back have an air gap when its designed, so the original boundary energy can not be 100 % transformation to the secondary side, thus cause the leak inductance. The existence of leak inductance will necessarily produces energy loss, and may cause counter assault if its magnitude is excessive [8–10]. The output power can not meet the requirements if too much leak inductance energy retention during input. Reflected in the waveform has been shown in Fig. 1. The energy loss are generated during each half wave connect. Besides, the leak inductance not only affect the power conversion efficiency, also can cause the voltage spikes of the power switches in turn off instant [11].

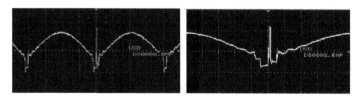

Fig. 1. Leak inductance energy loss

When the grid-source electrode of the power MOSFET be shorted, the capacitance value during the drain-source electrode is called Coss output capacitance. The principle diagram of the MOSFET has been shown in Fig. 2.

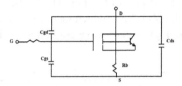

Fig. 2. Principle diagram of MOSFET with the parasitic capacitance

Let Cgd means the terminal capacoty between Gate (G) and Drain (D), and Cds means the terminal capacoty between Drain and Source (S). Cgd added Cds is equal to coss. The Coss capacitor charged when the power tube turn off, during this time, most of the electricity flowing through the MOSFET, and relative small current through the Coss capacitor. Therefore, the voltage between drain and source maintain in a low voltage value. Before the gate-source voltage value (VGS) from zero to gate threshold voltage value (VTH), the MOSFET do not load inductance current. The Coss capacitor will discharge after the VGS value reaching the power MOSFET switch-on voltage value [12–14]. If the system is not set reasonable discharged circuit, the voltage of Coss capacitor will be added to D-S side. That cause the high voltage spike, especially in the converter design. The fly-back rectifying half wave waveform with approximately 70 V voltage spike has been show in Fig. 3. From the theoretical analysis, the voltage spike are generated by the leak inductance retention and the parasitic capacitor discharged. If these condition are not suppressed effectively, it will produce a high current stress during the commutation of the power, even breakdown the MOSFET. In addition, because of its instability, PWM controller and zero synchronous detection measurement will be disturbed, which can lead to reducing the converter efficiency.

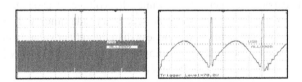

Fig. 3. Half wave with voltage spike(10 V/lattice)

3   Analysis for the Improved Circuit

3.1   The Working Principle of the Circuit

According to the principle of fly-back transformer and the generate cause of voltage spike, the energy stored in the transformer primary side during the power turn on, and output to the secondary side when the power turn off. For the issue, the circuit needs to increase the discharge absorbing loop. For the problem of output voltage instability in buck circuit, utilize the parasitic capacitance of MOSFET and the parasitic inductance of fly-back transformer, we design the circuit of RCD which based on Boost circuit topology to recovering leak inductance energy and depressing main switch voltage spike during switch-on (the dotted line part of the Fig. 4). The new design removes the high frequency transformer and the MOSFET, which mentioned in ref. 2 to reduce the energy consumption. The main topology of half wave converter has been shown in Fig. 4.

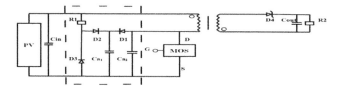

Fig. 4. Topology of half wave converter

When the MOSFET turns off, the leak inductance energy back to the clamp capacitance Ca_1 by rapid conduction diode D1, absorbing the excess energy. The clamp capacitance Ca_1 will suppress the Coss capacitance discharge voltage, and depress the voltage spike when the MOSFET switch-on. After the MOSFET turn on, the D1 cutoff and the absorbed energy recycle to the DC input side through resistance R1. The drain-source voltage of the main switch is clamped with D1 and D2 to prevent the voltage spike caused by Coss capacitance. The Ca_2 protect the MOSFET, and partic-ipates in the implementation of soft-switch control policy.

3.2   The Circuit Parameter Design

For the RC parameters designed, if the resistance value is too small, before the power conduction, the capacitance value may drop to zero, then the reflect voltage energy will be consumed. If capacitance get a large value, the voltage of capacitance rise slowly and the absorbed energy can not pass to the secondary side quickly. If the capacitance value is excessive, the peak voltage will be less than the reflect value, which lead clamp resistance R1 become a dead load to consumption magnetic core energy [15]. The ideal clamp circuit voltage waveform has been shown in Fig. 5.

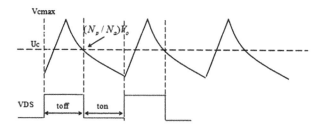

Fig. 5. The ideal clamp circuit voltage waveform

According to the geometrical relationship of clamp voltage wave and drain-source voltage of power MOSFET in Fig. 5, the following formula can be obtained.

$$\frac{\frac{Np}{Ns}Vo}{Vcmax} = \frac{ton}{T} = Dmax \tag{1}$$

$$Vcmax = \frac{Np \bullet Vo}{Ns \bullet Dmax} \tag{2}$$

To meet the capacitance value to $\frac{Np}{Ns}Vo$, when the power conduction, need to

$$Vcmax \bullet e^{\frac{-(1-Dmax)T}{Rc}} = \frac{Np}{Ns} \bullet Vo \tag{3}$$

Substitute Eqs. (2) to (3),

$$RC = (Dmax - 1)T/\ln Dmax \tag{4}$$

$$Dmax = D = \frac{Ug}{N \bullet Upv + Ug} \tag{5}$$

In the formula, the D means the duty cycle, the Upv means the value of network voltage, the N means the ratio of turns and the T means the working frequency. According to calculation, when the working frequency is 10 KHZ, then R1 = 100 K, Ca = 3.3 nf.

4   The Experimental Result of the Improved Circuit

In this paper, the converter rated capacity is 230 VA, with the input 30 V DC and output 220 V AC. The main power MOSFET was used IRF4321, in which the parasitic capacitance Coss = 390PF, the primary induction 5.4 μH, the leak inductance 0.2 ηH, and the ratio of turns is 1:6. The clamp capacitance Ca is 3.3 nF. The leak inductance recovery resistance R1 is 100 K. The clamp diode used ES1D. The switch tube in secondary side was used 17N80, which include the commutation diode C2D05120.

The experiments waveform have been shown in Fig. 6. The voltage waveform of clamp capacitance, which based on the theoretical derivation as shown in Fig. 6(a). The waveform close to the ideal curve. The voltage curve of the Ugs (the voltage between gate and source of MOSFET) and the Uds (the voltage between drain and source of MOSFET) has been shown in Fig. 6(b). From the waveform we can see that the main switch achieve zero voltage switching (ZVS). As the Fig. 6(c) shown that the grid voltage waveform with the peak value is about 330 V, and very smooth. The experimental results show that, in this paper, the correctness of the theoretical analysis and the efficiency of the improved physical circuit.

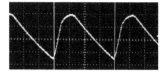

(a) Voltage waveform of clamp capacitance(t=5ms/lattice)

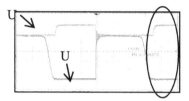

(b) The voltage of the drive of main switch and the drain-source

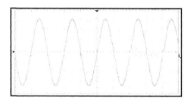

(t=10ms/lattice、 peak value=100V/lattice)

(c) The converter AC output voltage

Fig. 6. Experiment waveform

5   Conclusions

The comparison of the waveform before and after modification has been shown in Fig. 7. From the experiment waveform, we can found that the shake in half wave connection and the voltage spike which is generated by the Coss electric discharge during the commutation of power MOSFET are absorbed and suppressed.

In this paper, the test curve of converter efficiency (η) has been shown in Fig. 8, which the output converter efficiency can reach 96.2 %.p is output power. The converter can achieve higher output efficiency in the whole load range. The efficiency of MPPT has been shown in Fig. 9, which can reach 99.4 % when full load. The results show that the new circuit which mentioned in this paper, can absorb and recycle the leak inductance energy, and suppress the voltage spike caused by the Coss, and improve the converter efficiency.

The Theoretical analysis and experimental results show that, the physical circuit which is proposed in this paper has the efficient, stable and reliable advantages to deal with the problem of leak inductance and the voltage spike. The new method solves the issues of leak inductance energy recovery and voltage spike phenomenon which caused by the leak inductance and the parasitic capacitance discharged during MOSFET switch-on. The design can protect the power switch by clamping the voltage spike

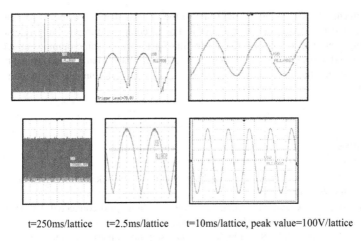

t=250ms/lattice   t=2.5ms/lattice   t=10ms/lattice, peak value=100V/lattice

Fig. 7. The comparison of the waveform

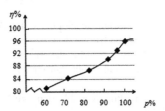

Fig. 8. The efficiency test curve

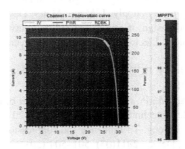

Fig. 9. The MPPT test curve

value. In this paper, compare with the *ref.* 2, we improved the domestic converter efficiency from 94.2 % to 96.2 %, while the world's highest converter efficiency up to 98 % [16].

Acknowledgement. The authors would like to the express thanks for the helpful discussions with Dr Li, and the AMETEK Programmable Power company for providing relevant photovoltaic equipment. The author also thank Mr. Sun for providing some opinions about the photovoltaic converter system design. Authors would like to thank Han Yu of University of Jinan Computational Intelligence Lab for providing relevant information about this meeting announced.

References

1. Cai, C., Qin, H.: A design of the snubber circuit for flyback converter. Chin. J. Electron Devices **04**, 469–472 (2013)
2. Gu, J., Wu, H., Chen, G., Xing, Y.: Soft-switching grid-connected PV inverter with interleaved flyback topology. Proc. CSEE **12**, 40–45 (2011)
3. Tseng, K.C., Liang, T.J.: Novel high-efficiency step-up converter. Electr. Power Appl. **2004** (03), 182–190 (2004)
4. da Rocha, J.F., dos Santos, M.B., Costa, J.M.D.: Voltage spikes in integrated CMOS buck DC-DC converters: analysis for resonant and hard switching topologies. Procedia Technol. **17**, 327–334 (2014)
5. Sicheng, W., Shijie, Y., Delin, W., et al.: Development of 3KW dispatchable grid connected inverter. Acta Energiae Solaris Sinica **01**, 17–20 (2001)
6. Loukriz, A., Haddadi, M., Messalti, S.: Simulation and experimental design of a new advanced variable step size Incremental Conductance MPPT algorithm for PV systems. ISA Trans. **08**, 6–14 (2015)
7. Ma, Y., Guo, Q., Zhou, X.: A modified variable step size MPPT algorithm. Microcomput. Appl. **34**(17), 78–80,84 (2015)
8. Zhang, H., Zhao, Y.: Design of multifunctional flyback switching power supply. Electr. Power Autom. Equipment **01**, 113–117 (2011)
9. Lv, L., Xiao, J., Zhong, Z., Shi, Yu.: Design of single-ended flyback transformer in high-frequency switching power supply. J. Magn. Mater. Devices **01**, 36–38 (2006)
10. Zhu, S.: The research and design of grid-connected microinverter based on an interleaved flyback structure. School of Electrical Engineering Shenyang University of Technology (2013)
11. Liu, G., Dong, J.: Study of RCD clamp circuit in flyback converter. Jiangsu Electr. Apparatus **01**, 20–24 (2011)
12. Liu, J., Li, J., Cui, Y., Han, M.: Power MOSFET gate driver of the high frequency resonant inverter. Trans. China Electrotechnical Soc. **05**, 113–118 (2011)
13. Bo, W., Au, H., Ming, C., Yong, T.: Influence and suppression of voltage spike in turn-off transient of IGBT. Mater. Device **2011**(07), 501–504 (2011)
14. Dai, Y., Wang, G., Guan, Y., Wu, L., Li, X.: Degradation analysis of power MOSFET parasitic capacitance intransient response of switching. Chin. J. Power Sources **04**, 661–664 (2014)
15. Chen, X., Li, G., Liu, F.: Charging Efficiency Analysis for RC Circuit. J. EEE **02**, 32–35 (2012)
16. Jian, F., Li, Z.: Technology status and future development of PV inverter. High Power Converter Technol. **03**, 5–9 (2014)

Secure and Pairing-Free Identity-Based Batch Verification Scheme in Vehicle Ad-Hoc Networks

Xiaoming Hu[1(✉)], Jian Wang[1], Huajie Xu[2], Yan Liu[1], and Xiaojun Zhang[3]

[1] College of Computer and Information Engineering,
Shanghai Polytechnic University, Shanghai 201209, China
xmhu@sspu.edu.cn
[2] School of Computer and Electronic Information,
Guangxi University, Nanning 530004, China
[3] E & A College, Hebei Normal University of Science and Technology,
Hebei 066004, China

Abstract. Identity-based batch verification (IBBV) scheme is very desirable to solve efficiency, security and privacy preservation issues for vehicular ad hoc network (VANET). In 2015, Tzeng et al. proposed an IBBV scheme which was published in IEEE Transaction on Vehicular Technology. Their scheme has superior performance than other existing similar schemes in terms of security, computation cost and transmission overhead by performance evaluations. However, one time signature verification of their scheme needs two bilinear pairing operations. As it is well known, bilinear pairing is one of the most time-consuming operation in modern cryptography. Therefore, some efforts can be made to prevent the appearance of pairing and obtain better efficiency. In this paper, we propose an improved scheme of Tzeng et al.'s IBBV. Our improved IBBV scheme needs not use bilinear pairing without the lack of security and privacy-preserving. The total computation cost for signing and verifying is the constant 1.2 *ms* for single message and *n* messages respectively, which is far better performance than Tzeng et al. scheme and other similar schemes. So our improved IBBV scheme is more suitable for practical use. Finally, we apply the recovering technology of the vehicle's real identity of Tzeng et al.'s IBBV scheme to a public key authentication scheme for mobile Ad-hoc networks to address an improved pairing-free authentication scheme.

Keywords: Information security · Ad hoc network · Vehicular ad hoc network (VANET) · Identity-based batch verification · Authentication scheme · Pairing-free Diffie-Hellman

1 Introduction

Mobile ad hoc networks (MANETs) [1–4] have attracted many researchers' interests due to the nonexistence of fixed network infrastructure, but which increases the diffi-culty of providing security for MANETs. Vehicle ad hoc network (VANET) [5, 6] is a

© Springer International Publishing Switzerland 2016
D.-S. Huang et al. (Eds.): ICIC 2016, Part III, LNAI 9773, pp. 11–20, 2016.
DOI: 10.1007/978-3-319-42297-8_2

variant of MANETs which can improve the traffic safety and efficiency. In a VANET, vehicles are equipped with on-board units (OBUs) which can be used to communicate with road side units (RSU). The vehicles can also use OBUs to communicate each other. However, wireless communication mode makes the security of VANET is complex. Many attack issues need to be considered, including intercept, replay, delete and so on. A secure VANET ought to include the following properties [6]: unforgeagility, identity privacy preservation, traceability, message authentication, non-repudiation, unlinkability and replaying resistance. Recently, a lot of research [7–12] was made on security of VANET. However, most of them involved low efficiency and expensive deployment.

In 2008, an identity-based batch (IBBV) scheme was proposed by zhang et al. [13, 14]. In their scheme, a batch of messages to be signed can be verified together, which is greatly saving time. In 2013, Lee and Lai [15] pointed that there existed some drawbacks in Zhang et at.' IBBV scheme [14]. Lee and Lai also addressed an improved IBBV with high efficiency. Unfortunately, in 2015, Tzeng et al. [6] found out that Lee and Lai's IBBV scheme was vulnerable to some attacks, including forgeability and real identity extracting. As an improvement, Tzeng et al. [6] proposed a new IBBV scheme with enhancing security. They also made a concrete simulation evaluation that showed that their scheme had a superior performance in terms of computation cost and transmission overhead compared with other similar IBBV schemes. However, Tzeng et al.'s scheme [6] needed to use bilinear pairing operation which is one of the most consuming-time operations in modern cryptography. In this paper, we make a slight modification for Tzeng et al.'s scheme to propose an improved IBBV scheme. The improved scheme removes the bilinear pairing and make the computation cost be very low, i.e., constant 1.2 *ms* for any message signatures. Therefore, our scheme obtains a better performance than Tzeng et al.'s scheme and other similar schemes.

The rest of the paper is organized as follows. In second section, we give some preliminaries. In Sect. 3, we review the Tzeng et al.'s IBBV scheme. In Sect. 4, we present our improved scheme and the evaluation for security and performance of our scheme. In Sect. 5, an improved authentication scheme is presented. In Sect. 6, we conclude this paper.

2 Preliminaries

2.1 Bilinear Map

Let G1 and G2 be two groups with the same prime order q. Assume that P is a generator of G1 and $e: G1 \times G1 \rightarrow G2$ is a bilinear map if it holds the following properties:

- Bilinearity: $e(U^x, V^y) = e(U, V)^{xy}$ for all $U, V \in G_1$ and $x, y \in Z_q^*$.
- Non-degeneracy: $e(P, P) \neq 1$.
- Computability: for all $U, V \in G_1$, $e(U, V)$ can be computed efficiently.

2.2 Discrete Logarithm (DL) Problem

Given two random elements $P_1, P_2 \in G_1$, the DL problem is to find a which satisfies $P_2 = aP_1$.

3 Review of Tzeng et al.'s IBBV Scheme

In order to state conveniently and show the difference of our improved IBBV scheme and Tzeng et al.'s scheme clearly, here we briefly review Tzeng et al.'s IBBV scheme [6] for VANET.

Tzeng et al.'s IBBV scheme includes three parts: system setup (Syssetup), identity and signature generation (ID-SIGgeneration) and signature verification (SIGverification).

3.1 System Setup(Syssetup)

In this phase, a trust authority (TA) setups and publishes some system parameters as follows.

- TA selects two groups G_1 and G_2 with the same prime order q. Then choose two random generators of $G_1 : P_1$ and P_2. $e : G_1 \times G_1 \rightarrow G_2$ is a bilinear map as defined above. Define two one-way hash functions: $H_1 : \{0,1\}^* \rightarrow G_1$ and $H_2 : \{0,1\}^* \rightarrow Z_q^*$. Then, TA chooses a random element $s \in Z_q^*$ as the master private key and computes $P_{pk} = sP_1$ as the system public key. Next, the system public parameters are $params = \{G_1, G_2, q, e, P_1, P_2, P_{pk}, H_1, H_2\}$.
- TA generates a real identity RID and a secret password PWD for each vehicle when it makes the first registration. After that, TA preloads its identity RID, the password PWD and the master private key s into each vehicle's tamper-proof device. Then, TA publishes $params$ to all RSUs and all vehicles.

3.2 Identity and Signature Generation (ID-SIGgeneration)

In this phase, the tamper-proof device of each vehicle performs the anonymous identity generation and signature generation. In order to do these works, the tamper-proof device of each vehicle consists of three modules: authentication module (AM), anonymous identity generation module (AIGM) and signature module (SM). Each module works as follows.

- Before making message signing, the vehicle C must pass the identity authentication which is performed by the authentication module of tamper-proof device. C first inputs its own RID and PWD into tamper-proof device. Then, the AM verifies the correctness of RID and PWD. If they are both not correct, the tamper-proof device ends the following operation. Otherwise, it performs the AIGM as follows.

- The AIGM picks up a random element $k \in Z_q^*$ and computes (ID_1, ID_2) where $ID_1 = kP_1$ and $ID_2 = RID \oplus H_1(kP_{pk})$. Then, (ID_1, ID_2) is an anonymous identity for vehicle C. Next, SM begins to work.
- The vehicle C constructs a message m and sends m to the SM of tamper-proof device. SM performs the following signature operations on m. First, the SM generates a current time stamp T. Then, it computes

$$V = (k + sH_2(ID_1 \,||\, m \,||\, ID_2 \,||\, T).$$

Finally, the tamper-proof device obtains $\sigma = \{(ID_1, ID_2), m, V, T\}$ and the vehicle C sends σ to the neighbouring RSU and vehicles.

3.3   Signature Verification (SIGverification)

- Single signature verification. After RSU or a vehicle receives a message $\sigma = \{(ID_1, ID_2), m, V, T\}$, the RSU or vehicle verifies the message as follows. First, it checks if the time T is fresh. If T is fresh, it checks if

$$e(V, P_1) = e(ID_1 + H_2(ID_1 \,||\, m \,||\, ID_2 \,||\, T)P_{pk}, P_2)$$

If the above equation is satisfied, the RSU or vehicle accepts the message or rejects it.
- Batch signatures verification. In order to prevent a lot of messages congesting the RSU, batch verification is used as a very efficient method to improve the verification speed. Assume that the RSU receives l messages, such as $\{(ID_{1,1}, ID_{1,2}), m_1, V_1, T_1\}$, $\{(ID_{2,1}, ID_{2,2}), m_2, V_2, T_2\}, \ldots\ldots, \{(ID_{l,1}, ID_{l,2}), m_l, V_l, T_l\}$. RSU first checks if all time T_i is fresh for $1 \leq i \leq l$. Then, RSU uses the concept of small exponent test [6, 9] to check the following equation

$$e\left(\sum_{i=1}^{l} t_i V_i, P_1\right) = e\left(\sum_{i=1}^{l} t_i ID_{i,1} + \sum_{i=1}^{l} t_i H_2(ID_{i,1} \,||\, m_i \,||\, ID_{i,2} \,||\, T_i)P_{pk}, P_2\right)$$

where $t_i (1 \leq i \leq l)$ is a random l-vector referring to [6, 9] for more information on t_i.

4   The Improved IBBV Scheme

From the above description of Tzeng et al.'s IBBV scheme, it can see that Tzeng et al.'s IBBV scheme needs two bilinear pairing operations for single signature verification. In this section, we make a slight modification for Tzeng et al.'s IBBV scheme to present an improved IBBV scheme which is pairing-free.

4.1   The Scheme

Our improved scheme consists of three phases as Tzeng et al.'s IBBV scheme: system setup (Syssetup), identity and signature generation (ID-SIGgeneration) and signature verification (SIGverification).

The Syssetup phase of our improved scheme is the same to that of Tzeng et al.'s scheme. The main difference is the ID-SIGgeneration and SIGverification phases.

ID-SIGgeneration

– After RID and PWD pass the authentication of the AM, the AIGM picks up a random element $k \in Z_q^*$ and computes (ID_1, ID_2) where

$$ID_1 = kP_1$$
$$ID_2 = RID \oplus H_1(kP_{pk}).$$

Then, (ID_1, ID_2) is an anonymous identity for vehicle C.
We can reduce the computation cost of anonymous identity by the following method. The AIGM generates a current time stamp T and computes

$$ID = RID \oplus H_1(T, s).$$

Then, ID is the anonymous identity which can be used to generate signature lately. It can be saw that a point multiplication operation is reduced than the previous method. When the real identity of vehicle needs to be recovered, TA computes

$$RID = ID \oplus H_1(T, s).$$

Then TA obtains the vehicle's real identity RID.
– After the anonymous identity generation, the vehicle C constructs a message m and sends m to the SM of tamper-proof device. The SM generates a current time stamp T. Then, it computes

$$r = k + sH_2(ID_1 \,||\, m \,||\, ID_2 \,||\, T).$$

Finally, the tamper-proof device gets the signature $\{(ID_1, ID_2), m, r, T\}$ and the vehicle C sends $\{(ID_1, ID_2), m, r, T\}$ to the neighbouring RSU and vehicles.

Signature Verification (SIGverification)

– Single signature verification. After RSU or a vehicle receives a signature $\{(ID_1, ID_2), m, r, T\}$, the RSU or vehicle verifies the message as follows. First, it checks if the time T is fresh. If T is fresh, it checks if

$$rP_1 = ID_1 + H_2(ID_1 \,||\, m \,||\, ID_2 \,||\, T)P_{pk}.$$

If the above equation is satisfied, the RSU or vehicle accepts the message or rejects it. The correctness of the above equation can be verified by the following method.

$$rP_1$$
$$= (k + sH_2(ID_1 \| m \| ID_2 \| T))P_1$$
$$= kP_1 + sH_2(ID_1 \| m \| ID_2 \| T)P_1$$
$$= ID_1 + H_2(ID_1 \| m \| ID_2 \| T)P_{pk}.$$

– Batch signatures verification adopts the same technology as Tzeng et al.'s scheme. Assume that the RSU receives l messages, such as

$$\{(ID_{1,1}, ID_{1,2}), m_1, r_1, T_1\},$$
$$\{(ID_{2,1}, ID_{2,2}), m_2, r_2, T_2\},$$
$$\ldots\ldots$$
$$\{(ID_{l,1}, ID_{l,2}), m_l, r_l, T_l\}.$$

RSU first checks if all time T_i is fresh for $1 \le i \le l$. Then, RSU checks the following equation

$$\sum_{i=1}^{l} t_i r_i P_1 = \sum_{i=1}^{l} t_i ID_{i,1} + \sum_{i=1}^{l} t_i H_2(ID_{i,1} \| m_i \| ID_{i,2} \| T_i)P_{pk}.$$

where $t_i (1 \le i \le l$ is a random l-vector as that of Tzeng et al.'s scheme.

4.2 Performance and Security Evaluation

Performance Analysis

In order to facilitate comparison with other IBBV schemes, we adopt the same items to Tzeng et al.'s scheme and only consider the dominated operations, such as $T_{pairing}$ as the time for one bilinear pairing operation, $T_{multiple}$ as the time for one point multiplication operation on G_1. Then, from the Table 1, we can get that Tzeng et al.'s scheme needs $2T_{multiple} + 2T_{pairing}$ for single message signing and verifying, and at the same needs $(n + 1)T_{multiple} + 2T_{pairing}$ for n messages signing and verifying. According to the simulation result from Tzeng et al.'s literature [6], $T_{pairing}$ is 4.5 ms and $T_{multiple}$ is 0.6 ms. So the total time for Tzeng et al.'s scheme is $4.5\,ms \times 2 + 0.6\,ms \times 2 = 10.2\,ms$ for single message and $4.5\,ms \times 2 + 0.6\,ms \times (n + 1) = (9.6 + 0.6n)\,ms$ for n messages. Comparison with other schemes [14–18], Tzeng et al.'s scheme has the best efficiency.

However, from the Table 1, we can get that our improved IBBV scheme needs $2T_{multiple}$ for single message signing and verifying and $2T_{multiple}$ for n messages signing and verifying, and the total time is the constant $0.6\,ms \times 2 = 1.2\,ms$ for single message and n messages respectively. Therefore, our scheme has the constant time cost which does not increase with the number of message. So, by the above analysis, we can find that our improved scheme has better efficient than other schemes, including Tzeng et al.'s scheme [6, 14–18]. We will present the practical experiment data of our improved scheme in the further work.

Table 1. Computational comparison with scheme [6]

Scheme	Single signing + verifying	n signing + verifying
[6]	$2(T_{pairing} + T_{multiple})$	$(n+1)(T_{multiple} + 2T_{pairing})$
Our	$2T_{multiple}$	$2T_{multiple}$

Security Analysis

Our improved IBBV scheme is slight change of Tzeng et al.'s IBBV scheme. So, we can adopt the similar proof technology to prove the security of our improved scheme. Next we give a simple analysis on the unforgeability of our scheme, which is the main security property of IBBV scheme.

The unforgeability of our improved IBBV scheme can be deduced to the DL problem on G_1 in the random oracle model. The main idea on the unforgeability is as follows. Given a random instance of DL problem (P_1, Q) where $Q = aP_1$ and $P_1, Q \in G_1, a \in Z_q^*$. The aim of the challenger CH is to obtain a by the adversary AD by the following running.

Setup: In order to get a, CH sets

$$P_{pk} = aP_1$$

as the system public key and a as the master private key. The other system parameters are set as Tzeng et al.'s scheme.

Query: In this phase, AD can make the random oracle query and signature query. When AD submits a $U_i \in G_1$ for a H_1 random oracle query, CH chooses a random number $b_i \in G_1$ and returns b_i to AD as the value of $H_1(U_i)$. CH records the tuple (U_i, b_i). Namely,

$$b_i = H_1(U_i).$$

When AD submits a $\{(ID_{i,1}, ID_{i,2}), m_i, T_i\}$ for a H_2 query, CH selects randomly $c_i \in Z_q^*$ and returns c_i to AD as the value of $H_2(ID_{i,1} \| m_i \| ID_{i,2} \| T_i)$. Namely,

$$c_i = H_2(ID_{i,1} \| m_i \| ID_{i,2} \| T_i).$$

When AD submits a message m_i for a signature query, CH chooses randomly $k_i. c_i \in Z_q^*, b_i \in G_1$ and sets

$$r_i = k_i,$$
$$ID_{i,2} = RID \oplus b_i.$$

Then, CH computes

$$ID_{i,1} = k_i P_1 - c_i P_{pk}.$$

$\{(ID_{i,1}, ID_{i,2}), m_i, r_i, T_i\}$ is a valid signature because

$$r_i P_1$$
$$= ID_{i,1} + c_i P_{pk}$$
$$= k_i P_1 - c_i P_{pk} + c_i P_{pk}$$
$$= k_i P_1 = r_i P_1.$$

Output: AD finally outputs a forged signature $\{(ID_1^*, ID_2^*),\ m^*,\ r^*,\ T^*\}$. Writing

$$h^* = H_2(ID_1^* \| m^* \| ID_2^* \| T^*).$$

Using the forking lemma [19], CH can obtain another signature $\{(ID_1^*, ID_2^*),\ m^*,\ r'^*,\ T^*\}$. The two signatures satisfy

$$r^* = k^* + ah^*$$
$$r'^* = k^* + ah'^*$$

Thus, CH can solve the given DL problem and obtain

$$a = (r^* - r'^*) \cdot (h^* - h'^*)^{-1}.$$

The other security properties (i.e., message authentication, identity privacy preservation, traceability and so on) for our IBBV scheme can be proved using the same analysis to Tzeng et al.'s scheme [6]. So we omit these descriptions. Therefore, our improvement scheme does not change the security properties of the original scheme, namely, the improved scheme still keeps the original security properties.

5   A Pairing-Free Authentication Scheme

In 2012, Tameem Eissa et al. [2] proposed an identity-based RSA authentication scheme for mobile ad hoc network (MANET). In their authentication scheme, the messages transmitted between mobile nodes were encrypted by RSA. In order to avoid RSA attacks, the public keys of RSA are secured and only the trust nodes can access. Due to the use of RSA, the efficiency of their authentication scheme is high. However, their scheme used bilinear pairing operation, so the computation cost still is high. Here, we use the recovering technology of the real identity of Tzeng et al.'s IBBV scheme (i.e., $RID = ID_2 \oplus H_1(kP_{pk})$) present an improved authentication scheme which is a very slight change of Tameem Eissa et al.' scheme. Due to the space, we only provide the main idea.

Assume that d_i and (e_i, N_i) is the private key and public keys on RSA for node $i \in \{A, B, CRSA\}$. When a node A submits (ID_B, P_A) to a coalition $CRSA$ for the public keys of B node where ID_B is the identity of B and

$$P_A = d_A P,$$

CRSA does as follows. *CRSA* computes

$$C = e_B \oplus H_2(d_{CRSA} P_A \,\|\, ID_A \,\|\, ID_B),$$
$$U = P_{CRSA} = d_{CRSA} P,$$
$$W = e_B P,$$
$$Y = N_B \oplus H_3(e_B).$$

After *A* gets (*U*, *C*, *W*, *Y*), it can obtain the public key (e_B, N_B) by the following computing

$$e_B = C \oplus H_d(d_A P_{CRSA} \,\|\, ID_A \,\|\, ID_B),$$
$$N_B = Y \oplus H_3(e_B).$$

6 Conclusion

In this study, we review Tzeng et al.'s identity-based batch scheme for VANET which has better efficiency than other similar schemes. As an improvement of Tzeng et al.'s scheme, we present a new IBBV scheme with pairing-free. We also make a performance evaluation between our scheme with other IBBV schemes. The comparison shows that our scheme is not only secure in the random oracle but also has constant computation cost which is independent of the number of messages. Therefore, our scheme has better performance and more practical than others.

Acknowledgements. This work was supported by the Innovation Program of Shanghai Municipal Education Commission (No.14ZZ167), the National Natural Science Foundation of China (No.61103213), the Guangxi Natural Science Foundation (No.2014GXNSFAA11838-2) and the Key Disciplines of Computer Science and Technology of Shanghai Polytechnic University under Grant No. XXKZD1604.

References

1. Ying, L., Yang, S., Srikant, R.: Optimal delay-throughput tradeoffs in mobile ad hoc networks. IEEE Trans. Inf. Theor. **54**(9), 4119–4143 (2008)
2. Eissa, T., Razak, S., Ngadi, M.: A novel lightweight authentication scheme for mobile ad hoc networks. Arab. J. Sci. Eng. **37**(8), 2179–2192 (2012)
3. Zhang, Y., Liu, W., Wei, W., et al.: Location-based compromise-tolerant security mechanisms for wireless sensor networks. IEEE J. Sel. Areas Commun. **24**(2), 247–260 (2013)

4. Shabnam, K., Mazleena, S.: A novel authentication scheme for mobile environments in the context of elliptic curve cryptography. In: Proceeding I4CT 2015, pp. 506–510. IEEE press, New York (2015)

5. Lin, X., Lu, R., Zhang, C., et al.: Security in vehicular ad hoc networks. IEEE Commun. Mag. **46**(4), 88–95 (2008)

6. Tzeng, S., Horng, S., Li, T., et al.: Enhancing security and privacy for identity-based batch verification scheme in VANET. IEEE Trans. Veh. Technol. **99**, 1–12 (2015)

7. Papadimitratos, P., Hubaux, J.: Report on the secure vehicular communications: results and challenges ahead workshop. ACM SIGMOBILE Mobile Comput. Commun. Rev. **12**(2), 53–64 (2008)

8. Chim, T., Yiu, S., Hui, C., et al.: SPECS: secure and privacy enhancing communications schemes for VANETs. Ad Hoc Netw. **9**(2), 189–203 (2011)

9. Horng, S., Tzeng, S., Pan, Y., et al.: b-SPECS+: batch verification for secure pseudonymous authentication in VANET. IEEE Trans. Inf. Forensics Secur. **8**(11), 1860–1875 (2013)

10. Nithya, D., Kumari, N.: A survey of routing protocols for VANET in Urban scenarios. In: Proceeding Pattern Recognition, Informatics and Mobile Engineering PRIME 2013, pp. 464–467. IEEE Press, New York (2013)

11. Mokhtar, B., Azab, M.: Survey on security issues in vehicular ad hoc networks. Alexandria Eng. J. **54**, 1115–1126 (2015)

12. Thenmozhi, T., Somasundaram, R.: Towards modelling a trusted and secured centralised reputation system for VANET's. Wireless Pers. Commun. **88**(2), 357–370 (2016)

13. Zhang, C., Lu, R., Lin, X., et al.: An efficient identity-based batch verification scheme for vehicular sensor networks. In: Proceeding INFOCOM 2008, pp. 816–824. IEEE Press, New York (2008)

14. Zhang, C., Ho, P., Tapolcai, J.: On batch verification with group testing for vehicular communications. Wireless Netw. **17**(8), 1851–1865 (2011)

15. Lee, C., Lai, Y.: Toward a secure batch verification with group testing for VANET. Wireless Netw. **19**(6), 1441–1449 (2013)

16. IEEE Trial-Use Standard for Wireless Access in Vehicular Environment-Security Services for Applications and Management Messages. In: IEEE standard 1609, 2 July 2006

17. Huang, J., Yeh, L., Chien, H.: ABAKA: an anonymous batch authenticated and key agreement scheme for value-add services in vehicular ad hoc networks. IEEE Trans. Veh. Technol. **60**(1), 248–262 (2011)

18. Boneh, D., Lynn, B., Shacham, H.: Short signatures from the weil pairing. J. Cryptology **17**(4), 297–319 (2004)

19. Pointcheval, D., Stern, J.: Security arguments for digital signatures and blind signatures. J. Cryptology. **13**(3), 361–396 (2000)

An Improved Ant Colony Optimization Algorithm for the Detection of SNP-SNP Interactions

Yingxia Sun[1], Junliang Shang[1,2(✉)], JinXing Liu[1], and Shengjun Li[1]

[1] School of Information Science and Engineering,
Qufu Normal University, Rizhao 276826, China
{sunyingxia026,shangjunliang110,qfnulsj}@163.com,
sdcavell@126.com
[2] Institute of Network Computing, Qufu Normal University,
Rizhao 276826, China

Abstract. An increasing number of studies have found that one of the most important factors for emergence and development of complex diseases is the interactions between SNPs, that is to say, epistasis or epistatic interactions. Though many efforts have been made for the detection of SNP-SNP interactions, the algorithm of such studies is still ongoing due to the computational and statistical complexities. In this work, we proposed an algorithm IACO based on ant colony optimization and a novel introduced fitness function *Svalue*, which combined both Bayesian networks and mutual information, for detecting SNP-SNP interactions. Furthermore, a memory based strategy is also employed to improve the performance of IACO, which effectively avoids ignoring the optimal solutions that have already been identified. Experiments of IACO are performed on both simulation data sets and a real data set of age-related macular degeneration (AMD). Results show that IACO is promising in detecting SNP-SNP interactions, and might be an alternative to existing methods for inferring epistatic interactions. The software package is available online at http://www.bdmb-web.cn/index.php?m=content&c=index&a=show&catid=37&id=98.

Keywords: SNP-SNP interaction · Bayesian network · Mutual information · Ant colony · Optimization

1 Introduction

With the development of high-throughput sequencing technologies, it is universally acknowledged that SNP (single nucleotide polymorphism) is one of the most common forms of genetic variants in human genome, which usually affects complex diseases by their nonlinear interactions, namely, epistatic interactions or epistasis [1]. Currently, epistasis has caused the extensive concerns in exploring the pathogenic mechanism of non-Mendelian diseases, such as hypertension, diabetes, Alzheimer's disease and many others [2]. Although many efforts have been made for the detection of SNP-SNP interactions, the algorithm of such studies is still ongoing due to their computational

© Springer International Publishing Switzerland 2016
D.-S. Huang et al. (Eds.): ICIC 2016, Part III, LNAI 9773, pp. 21–32, 2016.
DOI: 10.1007/978-3-319-42297-8_3

and statistical complexities, including the complexity of pathogenesis, the complexity of genetic models and the influence of environment factors.

Recently, a number of generic ant colony optimization (ACO) based methods have been proposed [1, 3–8] to detect the epistatic interactions. For instance, Christmas *et al.* [6] used generic ACO algorithm to identify the epistatic interactions in type 2 diabetes data. Results indicate that ACO algorithm is able to find statistically significant epistatic interactions. Wang *et al.* [4] proposed AntEpiSeeker based on ACO algorithm and designed two-stage optimization procedure for detection SNP-SNP interactions. Though it enhances the power of ACO algorithms, it is sensitive to main effect SNPs and it requires large amounts of ants over numerous iterations to obtain acceptable results. Shang *et al.* [1] proposed AntMiner based on ACO algorithm and incorporated heuristic information into ant colony optimization for epistasis detection, although it contributes to improving the computational efficiency and solution accuracy, it is also very time-costing. Jing *et al.* [5] proposed MACOED on the basis of ACO algorithm and designed a multi-objective method to detect SNP epistasis, which may be criticized for the complexity of multi-objective computation and the randomness of its searching strategy.

In this study, we proposed a method IACO based on ant colony optimization and a novel introduced fitness function *Svalue*, which combined both Bayesian networks and mutual information for detecting SNP-SNP interactions, effectively and efficiently evaluates how well the SNP-SNP combinations associates with the phenotype. In addition, a memory based strategy is also employed to improve the performance of IACO, which effectively avoids ignoring the optimal solutions that have already been identified and speed up the convergence of IACO algorithm. Experiments of IACO are performed on simulation data sets and a real data set of age-related macular degeneration (AMD). And we also compared IACO with some other representative methods, including AntEpiSeeker [4], AntMiner [1], and MACOED [5]. Results demonstrate that IACO outperforms others in inferring SNP-SNP interactions. Besides, the application of IACO on a real data set of age-related macular degeneration (AMD) may provide some new clues for the detection of AMD. IACO might be an alternative to existing methods for inferring epistatic interactions.

2 Methods

2.1 Mathematical Description of the Problem

In genome-wide association study (GWAS) [9], SNPs are bi-allelic labels with major and minor alleles being denoted by capital letters (e.g., *A*) and lowercase letters (e.g., *a*) respectively. There are three genotypes for each SNP: homozygous common genotype (*AA*), heterozygous genotype (*Aa*), and homozygous minor genotype (*aa*) [1]. The most common way of mapping SNPs is to collect them as a matrix, where a column represents a SNP and a row represents the genotypes of a sample. Furthermore, the elements 1, 2, 3 in the matrix on behalf of *AA*, *Aa* and *aa* respectively. In addition, we generally called the case group and the control group of each individual as a sample and used a matrix to collect them. Rows of the matrix are equivalent to the rows of SNP matrix. The label of a sample is either 1 (case) or 2 (control) [10]. In mathematics, Genome-wide SNP interaction research is described as a study of multiple SNP combinations for high-dimensional and small sample size data, which is used to greatly predict the phenotype of the sample.

2.2 Traditional Ant Colony Optimization

Swarm intelligence algorithm [1, 3–8, 10–12] is an optimization algorithm to solve the problems of large-scale data, which can handle some discrete data optimization problems through simulation of animal behavior. As a new swarm intelligence algorithm, ACO takes inspiration from the foraging behavior of some ant species [1], which has already been used to solve the problems of traveling salesman, graph coloring and others.

In the progress of foraging, ants can communicate with each other by secreting pheromones on the ground, but pheromones will gradually evaporate as time passes. The subsequent ants choose the path according to pheromones and tend to choose the paths with higher pheromones [1]. With the passage of time almost all ants choose a nearest path from a food source to their nest.

In this study, m ants are represented by $\{m_1, m_2 \cdots m_m\}$ respectively, each of which is used to detect a K-SNP interaction at each iteration, where K is a user-specified order of epistatic interactions.

The ants choose the next position by probability function (PF). PF is a probability that ant k choose the next position j from the position i at iteration t, and is denoted as $p_k^{ij}(t)$. The PF of traditional ant colony optimization is defined as

$$P_k^{ij}(t) = \begin{cases} \dfrac{\tau_{ij}(t)^{\alpha} \eta_{ij}^{\beta}}{\sum_{u \in U_k(t)} \tau_{iu}(t)^{\alpha} \eta_{iu}^{\beta}} & i \in U_k(t) \\ 0 & otherwise \end{cases} \tag{1}$$

where $\tau_{ij}(t)$ is the pheromones of position i to position j at iteration t. η_{ij} is the heuristic information of position i to position j. α and β are controlling importance of pheromones and heuristic information respectively. $U_k(t)$ is a set of positions that are not selected by ant k at iteration t.

Pheromones will evaporate in the process of movement. Pheromones of position i to position j at iteration $t + 1$ are updated according to

$$\tau_{ij}(t+1) = (1 - \rho)\tau_{ij}(t) + \Delta\tau_{ij}(t) \tag{2}$$

where ρ is a user-defined evaporation coefficient and $\Delta\tau_{ij}(t)$ is an increment of pheromones from position i to position j at iteration t. The $\Delta\tau_{ij}(t)$ is defined as

$$\Delta\tau_{ij}(t) = \sum_{k=1}^{m} \Delta\tau_{ij}^k(t) \tag{3}$$

where m is the number of ants and $\Delta \tau_{ij}^k(t)$ is the legacy of pheromones from position i to position j for ant k at iteration t, which is defined as

$$\Delta \tau_{ij}^k(t) = \begin{cases} \dfrac{Q}{S_k(t)} & f \ ant \ k \ via \ the \ path \ of \ ij \ at \ iteration \ t \\ 0 & otherwise \end{cases} \tag{4}$$

where Q is a user-defined positive constant and $S_k(t)$ is the path length for ant k at iteration t.

2.3 Evaluation Measure Svalue

2.3.1 Mutual Information

Information entropy is used to measure the characteristic of uncertainty, higher uncertainty implies the higher information entropy. Mutual information [12, 13] is a measure based on information entropy, which can measure the dependence of SNP-SNP combinations and phenotype. The IACO employs mutual information to measure how much the relevance of SNP-SNP interactions to the phenotype. The formula of mutual information can be written as

$$MI(S;Y) = H(S) + H(Y) - H(S,Y) \tag{5}$$

where $H(S)$ is the entropy of S, $H(Y)$ is the entropy of Y, and $H(S,Y)$ is the joint entropy of both S and Y, S is a position of a particle indicating a SNP combination, Y is the phenotype.

The entropy and the joint entropy are defined as

$$H(S) = -\sum_{j_1=1}^{3} \cdots \sum_{j_K=1}^{3} \left(p\left(s_{j_1}, \cdots, s_{j_K}\right) \cdot \log p\left(s_{j_1}, \cdots, s_{j_K}\right) \right) \tag{6}$$

$$H(Y) = -\sum_{j=0}^{1} \left(p\left(y_j\right) \cdot \log p\left(y_j\right) \right) \tag{7}$$

$$H(S,Y) = -\sum_{j_1=1}^{3} \cdots \sum_{j_K=1}^{3} \sum_{j=0}^{1} \left(p\left(s_{j_1}, \cdots, s_{j_K}, y_j\right) \cdot \log p\left(s_{j_1}, \cdots, s_{j_K}, y_j\right) \right) \tag{8}$$

where S is the genotype of a SNP coded as $\{1, 2, 3\}$, corresponding to homozygous common genotype, heterozygous genotype, and homozygous minor genotype, y is the label of a sample coded as $\{0, 1\}$, corresponding to control and case. $p(\cdot)$ is the probability distribution function.

Apparently, the higher the mutual information score is, the stronger the association between the SNP subset and the disease.

2.3.2 Bayesian Network

In recent years, with the rapidly development of machine learning, there are a lot of researches devoting to some statistical learning methods. For instance, Bayesian network [14, 15]. Many measures in Bayesian networks are used to evaluate the correlation between SNP-SNP combinations and disease. In this work, we adopt *BN* to measure the correlation between disease and different SNP combinations. Specifically, in the GWAS *BN*, the genotypes and phenotypes are denoted as a set of nodes, and their conditional dependences are denoted as a set of edges. On the basis of previous studies [15] we choose the *BN* score as follows:

$$BN = \sum_{i=1}^{I} \left(\sum_{b=1}^{r_i+1} \log(b) - \sum_{j=1}^{J} \sum_{d=1}^{r_{ij}} \log(d) \right) \tag{9}$$

where I is the combinatorial number of SNP nodes with different values (if l-SNP nodes are connected to disease node y, the number of SNP nodes' combinations is 3^l as the possible value of a SNP node is 0, 1, or 2), J is the state number of disease node y two for all samples), r_i is the number of cases with SNP nodes taking the i_{th} combination, and r_{ij} is the number of cases where the disease node takes the j_{th} state and its parents take the i_{th} combination [5]. In this measure, lower BN value implies greater relevance of SNP-SNP combinations and phenotype.

2.3.3 Svalue

In previous studies, most of studies used single objective evaluation method to evaluate the correlation between SNP-SNP interactions and phenotype. However, many evaluation methods have different effects on the disease in different model due to the complexity of some disease. In this work, we adopt a new evaluation measure *Svalue*, which is based on Bayesian network and mutual information. As mentioned before, the higher the mutual information score and the lower the BN score, the stronger the association between the SNP subsets and the disease. But we can't evaluate the pros and cons of different SNP-SNP combinations due to the different assessment of two evaluation measures and the different order of magnitude for two evaluation functions. So, we proposed a new evaluation measure as follows:

$$Svalue(A) = \theta \frac{MI}{BN} \tag{10}$$

where *MI* and *BN* are the mutual information score and *BN* score respectively. θ is a user-specified constant. The ability to identify between SNP-SNP combinations and phenotype will be better when the *Svalue* score is larger.

2.4 Improved Ant Colony Optimization (IACO)

2.4.1 Path Selection Strategy

IACO algorithm is based on the generic ant colony algorithm (ACO). And in this work, we use a matrix to store the information of pheromones, which number of columns is all SNPs data sets. In order to treat each SNP equally, the pheromones of all SNPs is set to 1 in the initial time of the algorithm. Then ants can independently choose the SNPs on the basis of pheromones and heuristic information. The ant-decision rule is ant k select SNP i at iteration t, and it is denoted as $P_k^i(t)$. The PF of IACO is defined as

$$P_k^i(t) = \begin{cases} R & q \leq q_0 \\ S & q > q_0 \end{cases} \tag{11}$$

where q is a number generated randomly with uniform distribution in $(0,1)$, and q_0 is a user-defined threshold to control the rate of convergence and to avoid falling into locally optimal solution. R and S are the rules for ants to select SNPs, described as

$$S = \begin{cases} 1 & i = r\,and\,(U_k(t)) \\ 0 & otherwise \end{cases} \tag{12}$$

$$R = \begin{cases} \dfrac{\tau_i(t)^\alpha \eta_i^\beta}{\displaystyle\sum_{u \in U_k(t)} \tau_u(t)^\alpha \eta_u^\beta} & i \in U_k(t) \\ 0 & otherwise \end{cases} \tag{13}$$

where $\tau_i(t)$ is the pheromones of SNP i at iteration t, and η_i is the heuristic information of SNP i.

2.4.2   Pheromone Updating Strategy

Each ant can find a K-order SNP-SNP combination at the completion of each iteration, but the connection degree of different SNP-SNP combinations and disease is distinct. Therefore we adopt a new pheromone update function to change the pheromones of different SNP-SNP combinations respectively, and described as

$$\tau_i(t+1) = (1 - \rho)\tau_i(t) + \Delta\tau_i(t) \tag{14}$$

where $\Delta\tau_i(t)$ is the variation of pheromones, which is described as

$$\Delta\tau_i(t) = \sum_{k=1}^{m} \Delta\tau_i^k(t) \tag{15}$$

$$\Delta\tau_i^k(t) = \begin{cases} Svalue(A) & k \in M_i(t) \\ 0 & otherwise \end{cases} \tag{16}$$

In those equations, $M_i(t)$ is a set of ants who select SNP i at iteration t, A is the SNP-SNP combinations which are identified by ant k at iteration i, and $Svalue(A)$ is the $Svalue$ of A.

2.4.3 Memory Based Strategy

In this work, we rank all SNP-SNP combinations that identified by ants at iteration t with descending $Svalue$, and the combinations are described as $A_1, A_2, \cdots A_m$, where their $Svalues$ are $(1, Svalue(A_1)), (2, Svalue(A_2)), \cdots (m, Svalue(A_m))$. Then we should find an inflection point f and hold the points before the inflection point as the optimal solutions of this iteration, which is defined as

$$f = arc \max_{g=3}^{m} \left(\left(Svalue(A_g) - Svalue(A_{g-1}) \right) - \left(Svalue(A_{g-1}) - Svalue(A_{g-2}) \right) \right) \quad (17)$$

In order to save the optimal solutions that have already been identified, speed up the convergence and reduce the calculation of $Svalue$ for some optimal solutions that have already been calculated, we introduce a memory strategy, which put the optimal solutions to the sets of suspected solutions. Otherwise, we design the memory strategy to remain the optimal solutions to compare with the identified solutions in the current iteration and get final suspected solutions. All suspected solutions in the last iteration will be fed into the next stage of IACO to process [5].

3 Results and Discussion

3.1 Simulation and Real Data Sets

We exemplify 4 benchmark models of SNP-SNP interactions for the experiments [14, 16–19], parameters are set as follows, penetrance is the probability of the occurrence of a disease given a particular genotype, and penetrance functions of four models is reported in Fig. 1; MAF(α)is the minor allele frequency of α. AA, Aa and aa are homozygous common genotype, heterozygous genotype, and homozygous minor genotype; and prevalence is the proportion of samples that occur a disease. Specifically, Model 1 is a model that display both marginal effects and interactive effect, the penetrance of which increases only when both SNPs have at least one minor allele [14, 16], the MAF(a), MAF(b) and the prevalence of Model 1 are 0.300, 0.200 and 0.100 respectively; Model 2 is a model showing both marginal effects and interactive effect, the additional minor allele at each locus of which does not further increase the penetrance [14], the MAF(a), MAF(b) and the prevalence of Model 2 are 0.400, 0.400 and 0.050 respectively; Model 3 is also a model displaying both marginal effects and interactive effect, which assumes that the minor allele in one SNP has the marginal effect, however, the marginal effect is inversed while minor alleles in both SNPs are present [14], the MAF(a), MAF(b) and the prevalence of Model 3 are 0.400, 0.200 and 0.010 respectively; Model 4 is a model that shows only interactive effects, which is directly cited from the reference [18]. Model 4 is exemplified here since it provides a high degree of complexity to challenge ability of a method in detecting SNP-SNP interactions, the MAF(a), MAF(b) and the prevalence of Model 4 are 0.400, 0.400 and 0.171 respectively. For each model, 50 data sets are simulated by *epi*SIM [11], each containing 4000 samples with the ratio of cases and controls being 1. For each data set, random SNPs are set with their MAFs chosen from $[0.05, 0.5]$ uniformly.

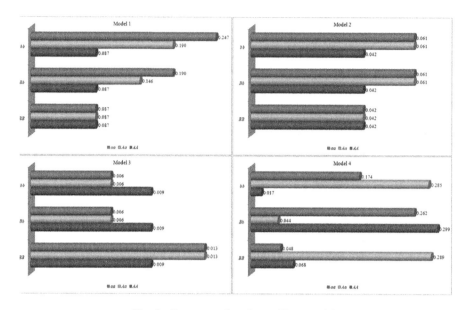

Fig. 1. Penetrance functions of four models

A real data set of age-related macular degeneration (AMD) is used for testing the practical ability of IACO. AMD, refers to pathological changes in the central area of the retina, is the most important cause of irreversible visual loss in elderly populations, and is considered as a complex disease having multiple SNP-SNP interactions [16]. The AMD data set contains 103611 SNPs genotyped with 96 cases and 50 controls, which has been widely used as a benchmark data set in the field of testing methods of detecting SNP-SNP interactions [1, 10, 12–14, 16, 19, 20].

3.2   Experiments on Simulation Data Sets

Three methods for detection SNP-SNP interactions are compared with IACO, which are AntEpiSeeker, AntMiner and MACOED. In order to equitable comparison, we set the same parameters for four methods. The number of ants m and the number of iterations T are set to 500 and 10 respectively; the dimension K for IACO is set to 2; the heuristic information η is set to 1; the controlling importance of pheromones α and heuristic information β are both set to 1; the threshold for control the rate of convergence q_0 is set to 0.6; the evaporation coefficient ρ is set to 0.2; the constant θ is set to 10000.

In order to evaluate the performance of the algorithm, we introduced power as evaluate criteria, which is defined as the proportion of 50 data sets that all connected SNPs are recognized. Power of these four compared methods on simulation data sets is reported in Fig. 2. Apparently, power of IACO is much higher than that of AntEpiSeeker; power of IACO is much higher than that of MACOED except the Model 4; power of IACO is much higher than that of AntMiner except the Model1 and Model2,

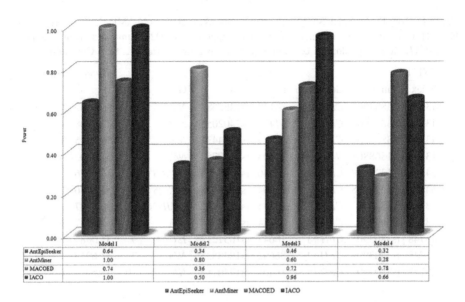

Fig. 2. Power of compared methods on simulation data sets.

on account of AntMiner incorporate heuristic information into ACO for SNP-SNP interaction, but the running time of AntMiner is significantly higher than IACO; the performance of AntEpiSeeker is worse on all models because it focus on identify the models of pathogenic SNPs; the MACOED introduce logistic regression to evaluate SNP-SNP combinations, which is unstable for different parameter settings, and the most major problem of it is the searching strategy becomes stochastic search. Specifically, the computer time of AntEpiSeeker, MACOED and IACO is on an order of magnitude, immensely lower than AntMiner. With comprehensive consideration, IACO outperforms others in detection power for large datasets and it has superb stability. In the future research, we will consider introducing heuristic information into ACO to improve the power of our methods.

3.3 Application to Real Data Set

The IACO is applied on AMD data set four times with parameter settings (m, T) being $(10000, 500)$, $(10000, 1000)$, $(20000, 500)$ and $(20000, 1000)$. In total, there are captured SNP-SNP interactions that might be associated with the AMD are reported in Table 1 with descending *Svalue*.

Obviously, all reported SNP-SNP interactions contains either rs380390 or rs1329428 except the fourth line SNPs, of which reason is that these two SNPs have strongest main effects among all tested SNPs, and these two SNPs have already proved to be significantly associated with AMD [4]. All of them in the gene of *CFH*, which is located on chromosome 1 in a region repeatedly linked to AMD in family-based studies [21].

Table 1. Top 11 captured SNP-SNP interactions associated with AMD. *CFH*: complement factor H. *CRSP8*: cofactor required for Sp1 transcriptional. *JMJD2C*: jumonji domain containing 2C. *ISCA1*: HESB like domain containing 2. *N/A*: no gene is available. Chr: Chromosome.

SNP 1				SNP 2				*Svalue*	Times
Index	Name	Gene	Chr	Index	Name	Gene	Chr		
43748	rs380390	*CFH*	1	80178	rs1363688	*N/A*	5	39.1602	4
43748	rs380390	*CFH*	1	57476	rs2402053	*N/A*	7	37.7208	4
43748	rs380390	*CFH*	1	63879	rs1374431	*N/A*	2	37.1915	3
13485	rs1368863	*N/A*	11	19405	rs1740752	*N/A*	10	37.1603	3
54108	rs1329428	*CFH*	1	79546	rs7467596	*CRSP8*	9	36.8086	1
54108	rs1329428	*CFH*	1	31604	rs9328536	*CRSP8*	9	36.8086	1
43748	rs380390	*CFH*	1	97535	rs2224762	*JMJD2C*	9	36.3767	1
43748	rs380390	*CFH*	1	75884	rs2380684	*N/A*	2	35.9957	2
43748	rs380390	*CFH*	1	10645	rs10512174	*ISCA1*	9	35.0884	1
43748	rs380390	*CFH*	1	86692	rs10488386	*N/A*	7	33.6912	3
43748	rs380390	*CFH*	1	19405	rs1740752	*N/A*	10	33.5013	1

Besides, the SNPs in the second column of the table might be identified due to they combine with rs380390 or rs1329428, since strong main effects of these two SNPs leads to their combinations with other SNPs almost displaying strong interactive effect. This also indicates that IACO is sensitive to those SNPs displaying strong main effects. It is worth noting that our methods find that the combinations of rs1368863 and rs1740752 also displaying strong interactive effect. These SNPs in the N/A gene, which are located on chromosome 11 and 10 in noncoding region, may induced AMD by gene mutation or induced diseases by changing the other disease-causing genes. The above observation may provide some new clues for the detection of AMD.

4   Conclusions

Epistasis has caused the extensive concern in exploring the mechanism underlying susceptibility to complex diseases. In this work, we proposed an algorithm IACO based on ant colony optimization and a novel introduced fitness function *Svalue*, which combined both Bayesian networks and mutual information, for detecting SNP-SNP interactions. Furthermore, a memory based strategy is also employed to improve the performance of IACO, which effectively avoids ignoring the optimal solutions that have already been identified. Experiments of IACO are performed on simulation data sets and a real data set of age-related macular degeneration (AMD). And we also compared IACO with some other representative methods, including AntEpiSeeker [4], AntMiner [1], and MACOED [5]. Results demonstrate that IACO outperforms others in detecting SNP-SNP interactions. Besides, the application of IACO on a real data set of age-related macular degeneration (AMD) may provide some new clues for the detection of AMD. The Matlab version of IACO software package is available online at http://www.bdmb-web.cn/index.php?m=content&c=index&a=show&catid=37&id=98.

Obviously, there are many advantages for IACO to identify epistatic interactions. First of all, IACO proposed a new fitness function *Svalue* based on Bayesian networks and mutual information, which unifies the different assessment of two evaluation measures, greatly improve the processing ability of the optimal solutions. In the next place, it adopted a memory based strategy to dispose the optimal solution, which rank *Svalue* of the suspected solutions with the *Svalue* of the SNP-SNP combinations identified at current iteration. One merit of memory based strategy is the optimal solutions in last iteration is used for the current iteration, which avoid ignoring the optimal solution that have already been found, speed up the convergence and reduce the calculation of *Svalue* for some optimal solutions that have already been calculated. What's more, we use a dynamic pheromone updating strategy to update the pheromones of SNPs. SNPs of bigger associated with disease will update more pheromones than that poor relevance of disease, which effectively improve the convergence speed of the algorithm.

Although IACO showed great performance for detection of epistatic interactions on both simulation and real data sets, there still remain some limitations. For instance, we set the heuristic information η to 1, which may degrade the performance of IACO. In future studies, we will consider introducing the heuristic information to IACO to deal with GWAS dataset. IACO is a continuous research project and we will continue improving in the future.

Acknowledgments. This work was in part supported by the National Natural Science Foundation of China (61502272, 61572284, 61572283), the Scientific Research Reward Foundation for Excellent Young and Middle-age Scientists of Shandong Province (BS2014DX004), the Science and Technology Planning Project of Qufu Normal University (xkj201410), the Opening Laboratory Fund of Qufu Normal University (sk201416), the Scientific Research Foundation of Qufu Normal University (BSQD20130119), The Innovation and Entrepreneurship Training Project for College Students of China (201510446044), The Innovation and Entrepreneurship Training Project for College Students of Qufu Normal University (2015A058, 2015A059).

References

1. Shang, J., Zhang, J., Lei, X., Zhang, Y., Chen, B.: Incorporating heuristic information into ant colony optimization for epistasis detection. Genes Genomics **34**(3), 321–327 (2012)
2. Maher, B.: The case of the missing heritability. Nature **456**(7218), 18–21 (2008)
3. Rekaya, R., Robbins, K.: Ant colony algorithm for analysis of gene interaction in high-dimensional association data. Revista Brasileira de Zootecnia **38**(SPE), 93–97 (2009)
4. Wang, Y., Liu, X., Robbins, K., Rekaya, R.: AntEpiSeeker: detecting epistatic interactions for case-control studies using a two-stage ant colony optimization algorithm. BMC Res. Notes **3**(1), 117 (2010)
5. Jing, P., Shen, H.: MACOED: a multi-objective ant colony optimization algorithm for SNP epistasis detection in genome-wide association studies. Bioinformatics **31**, 634–641 (2014). btu702
6. Christmas, J., Keedwell, E., Frayling, T.M., Perry, J.R.: Ant colony optimisation to identify genetic variant association with type 2 diabetes. Inf. Sci. **181**(9), 1609–1622 (2011)

7. Greene, C.S., White, B.C., Moore, J.H.: Ant colony optimization for genome-wide genetic analysis. In: Dorigo, M., Birattari, M., Blum, C., Clerc, M., Stützle, T., Winfield, A.F. (eds.) ANTS 2008. LNCS, vol. 5217, pp. 37–47. Springer, Heidelberg (2008)
8. Greene, C.S., Gilmore, J.M., Kiralis, J., Andrews, P.C., Moore, J.H.: Optimal use of expert knowledge in ant colony optimization for the analysis of epistasis in human disease. In: Pizzuti, C., Ritchie, M.D., Giacobini, M. (eds.) EvoBIO 2009. LNCS, vol. 5483, pp. 92–103. Springer, Heidelberg (2009)
9. Li, P., Guo, M., Wang, C., Liu, X., Zou, Q.: An overview of SNP interactions in genome-wide association studies. Briefings Funct. Genomics **14**, 143–145 (2014). elu036
10. Shang, J., Zhang, J., Sun, Y., Zhang, Y.: EpiMiner: a three-stage co-information based method for detecting and visualizing epistatic interactions. Digit. Sig. Process. **24**, 1–13 (2014)
11. Shang, J., Zhang, J., Lei, X., Zhao, W., Dong, Y.: EpiSIM: simulation of multiple epistasis, linkage disequilibrium patterns and haplotype blocks for genome-wide interaction analysis. Genes Genom. **35**, 1–12 (2013)
12. Ma, C., Shang, J., Li, S., Sun, Y.: Detection of SNP-SNP interaction based on the generalized particle swarm optimization algorithm. In: 2014 8th International Conference on Systems Biology (ISB), 2014, pp. 151–155. IEEE (2014)
13. Shang, J., Sun, Y., Fang, Y., Li, S., Liu, J.-X., Zhang, Y.: Hypergraph supervised search for inferring multiple epistatic interactions with different orders. In: Huang, D.-S., Jo, K.-H., Hussain, A. (eds.) ICIC 2015. LNCS, vol. 9226, pp. 623–633. Springer, Heidelberg (2015)
14. Zhang, Y., Liu, J.S.: Bayesian inference of epistatic interactions in case-control studies. Nat. Genet. **39**(9), 1167–1173 (2007)
15. Han, B., Chen, X.-w., Talebizadeh, Z., Xu, H.: Genetic studies of complex human diseases: Characterizing SNP-disease associations using Bayesian networks. BMC Syst. Biol. **6**(Suppl 3), S14 (2012)
16. Tang, W., Wu, X., Jiang, R., Li, Y.: Epistatic module detection for case-control studies: a Bayesian model with a Gibbs sampling strategy. PLoS Genet. **5**(5), e1000464 (2009)
17. Frankel, W.N., Schork, N.J.: Who's afraid of epistasis? Nat. Genet. **14**(4), 371–373 (1996)
18. Li, W., Reich, J.: A complete enumeration and classification of two-locus disease models. Hum. Hered. **50**(6), 334–349 (2000)
19. Shang, J., Zhang, J., Sun, Y., Liu, D., Ye, D., Yin, Y.: Performance analysis of novel methods for detecting epistasis. BMC Bioinform. **12**(1), 475 (2011)
20. Shang, J., Sun, Y., Li, S., Liu, J.-X., Zheng, C.-H., Zhang, J.: An improved opposition-based learning particle swarm optimization for the detection of SNP-SNP interactions. BioMed Res. Int. **2015**, 524821 (2015)
21. Klein, R.J., Zeiss, C., Chew, E.Y., Tsai, J.-Y., Sackler, R.S., Haynes, C., Henning, A.K., SanGiovanni, J.P., Mane, S.M., Mayne, S.T.: Complement factor H polymorphism in age-related macular degeneration. Science **308**(5720), 385–389 (2005)

Monaural Singing Voice Separation by Non-negative Matrix Partial Co-Factorization with Temporal Continuity and Sparsity Criteria

Ying Hu[(✉)], Liejun Wang, Hao Huang, and Gang Zhou

The Institution of Information Science and Technology, Xinjiang University,
Shengli Road, 14, 830001 Urumuqi, China
huying_75@sina.com

Abstract. Separating singing voice from music accompaniment for monaural recordings is very useful in many applications, such as lyrics recognition and singer identification. Based on non-negative matrix partial co-factorization (NMPCF), we propose an improved algorithm which restricts the activation coefficients of singing voice components to be temporal continuous and sparse in each frame. Temporal continuity is favored by using a cost term which is the sum of squared difference between the activation coefficients in adjacent frames, and sparsity is favored by penalizing nonzero values for each frame. For the separated singing voice, we quantify the performance of the system by the signal-to-noise ratio (SNR) gain and the accuracy of singer identification. The experiments show that the constraints of temporal continuity and sparsity criteria both can improve the performance of singing voice separation, especially the constraint of temporal continuity.

Keywords: Non-negative matrix partial co-factorization · Temporal continuity · Sparsity criteria · Singing voice separation

1 Introduction

Separating singing voice from music accompaniment of monaural recordings is very useful in many applications, such as lyrics recognition and alignment, singer identification, and some other music information retrieval application. Non-negative matrix factorization (NMF) is a common and effective method in singing voice separation [2, 4]. However, the standard NMF lacks separation ability for a specific source that eventually causes the consequent degradation of basis classification accuracy. To tackle this problem, non-negative matrix partial co-factorization (NMPCF) was introduced [1, 5]. In paper [1], NMPCF conducts partial co-factorization simultaneously exploiting the prior knowledge of singing voice and accompaniment. It can locate the frequency bases of each source.

On the whole, the NMPCF-based singing voice separation system presents good performance. However, when the energy ratio of singing to accompaniment is lower,

© Springer International Publishing Switzerland 2016
D.-S. Huang et al. (Eds.): ICIC 2016, Part III, LNAI 9773, pp. 33–43, 2016.
DOI: 10.1007/978-3-319-42297-8_4

it scores lower too. E.g. when the singing-to-accompaniment ratio (SAR) is -9 dB, results are even slightly lower than that of a comparative separation system.

In the field of single-channel signal separation, the researchers found that constraining the sparsity of basis vector belonging to a certain source is helpful for source separation [6, 7]. Sun and Mysore [7] focus on the speech de-noising application and propose a universal speech model based on the principle of block sparsity. Moreover, the temporal dependency of activation coefficients is also exploited to improve the performance of source separation [10, 11]. Virtanen [10] used the penalty term of temporal continuity in NMF for source separation which is the sum of squared differences between the gains in adjacent frames. This algorithm does not need training in advance but add a temporal continuity penalty term in cost function.

In this paper, we dedicate to improve the performance of singing voice separation based on NMPCF. Sparsity and temporal continuity of activation coefficients are both considered. The remainder of this paper is organized as follows. In Sect. 2, we propose a new NMPCF-based system of singing voice separation, which integrate the sparsity criteria and temporal continuity of activation coefficients. In Sect. 3, we provide quantitative evaluation of separation system from the perspectives of signal-to-noise ratio (SNR) gain and completeness of singer information, respectively. Then, we draw a conclusion in Sect. 4.

2   Proposed Method

Firstly, we describe briefly the method of NMPCF for singing voice separation and then propose a new NMPCF-based factorization model, which integrates sparsity criteria and temporal continuity of activation coefficients.

2.1   Non-negative Matrix Partial Co-factorization for Singing Voice Separation

According to the framework of NMPCF proposed in paper [1], firstly, through a pitch detection process of singing voice, a song is divided into vocal and non-vocal segments, whose magnitude spectrograms are denoted by $X \in R_+^{F \times T_x}$ and $Y \in R_+^{F \times T_y}$ respectively. T_x and T_y denote the frame lengths X of and Y, and F the number of frequency bins. A vocal segment starts at the first frame with nonzero pitch of singing voice, and ends at the last frame with non-zero pitch. The non-vocal portions are generally interludes between two vocal parts or at the first and last portions in a song. So the non-vocal portions are pure accompaniments of various intensities. Only these vocal segments are used to be separated. As another additional prior spectrogram, the spectrogram of clean singing voice covering all singers, denoting Z, also participates co-factorization. Given three input spectrogram matrices X, Y and Z, the joint decomposition approximated the matrix factorization of target mixture spectrogram matrix X and two prior spectrograms matrices Y and Z:

$$X = U_S V_S + U_A V_A \tag{1}$$

$$Y = U_A W_A \tag{2}$$

$$Z = U_S W_S \tag{3}$$

The models (1) and (2) share the basis matrix $U_A \in R_+^{F \times K_A}$ factorize the target mixture signal X and pure accompaniment Y, while the models (1) and (3) share the basis matrix $U_S \in R_+^{F \times K_S}$ to factorize the target mixture signal X and clean singing voice Z. K_A and K_S denote the number of basis vectors for accompaniment and singing voice, respectively. The matrix Y stands for prior side information about accompaniment in the target song, while the matrix Z prior side information about clean singing voice covering all singers.

In actual experiments, to reduce the amount of calculation, the basis matrix U_S was obtained in advance by the matrix decomposition of Z using most straight-forward way.

To obtain non-negative decomposition of given matrices X and Y, a objective function of NMPCF, which reflects the reconstruction errors of models (1) and (2), shown as follows:

$$L = \frac{1}{2} \|X - U_S V_S - U_A V_A\|_F^2 + \frac{\lambda}{2} \|Y - U_A W_A\|_F^2, \tag{4}$$

where K_S is a parameter reflecting the relative importance of accompaniment matrix Y in the process of joint non-negative matrix factorization.

2.2 Factorization Models with Temporal Continuity and Sparsity Criteria

As discussed in the previous section, constraining the sparsity and the temporal continuity of activation coefficients corresponding to a certain source is helpful for source separation. Figure 1 describes the pictorial illustration of NMPCF model with temporal continuity and sparsity criterion. The basis vectors U_S are first learned from the clean singing voice spectrograms covering all underlying singers. In the top portion of Fig. 1, the basis vectors of singing voice U_S are grouped into several blocks, e.g. six blocks in this figure. Meanwhile, the activation matrix V_S is learned to be block-wise sparse, except that corresponding the fourth basis block, a common block belonging to each singer.

Usually, for a vocal segment, there is only a singer at the same time. So, a constraining term of sparsity is considered to enforce $V_S \in R_+^{K_S \times T_x}$ sparsity that only fewer elements in V_S have non-zeros values at the same time. According to the suggestion [7–9], we use log $/l_1$ penalty that the sparsity penalty is set on activation coefficient V_S as follows:

$$\Omega_S([V_S]_t) = \mathbf{1} \cdot \log\left(\varepsilon + \|[V_S]_t\|_1\right) \tag{5}$$

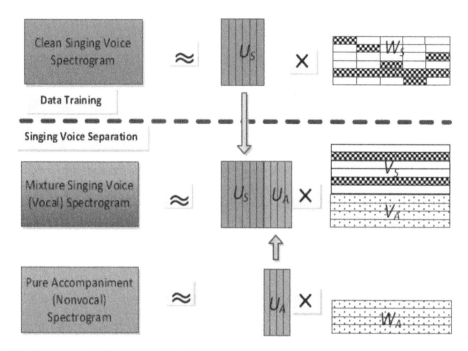

Fig. 1. A pictorial illustration of NMPCF model with temporal continuity and sparsity criterion

where the function Ω_S on $[V_S]_t$ is a penalty term that induces sparsity of V_S at each frame t. $\mathbf{1}$ is an all-one column vector with the length of T_x. ε is an arbitrarily small number and $\left\|[V_S]_t\right\|_1$ the l_1 norm of $[V_S]_t$:

$$\left\|[V_S]_t\right\|_1 = \sum_{j=1}^{J} [V_S]_{j,t} \tag{6}$$

J is the number of basis vectors in matrix U_S.

During singing, a singer often intentionally stretches the voiced sound and shrinks the unvoiced sound to match other musical instruments, thus the activation coefficients V_S tends to be temporally continuous or vary slowly. The temporal continuity term is calculated as follows [10]:

$$\Omega_T(V_S) = \sum_{k=1}^{K_S} \frac{1}{\sigma_k^2} \sum_{t=2}^{T_x} \left\{ [V_S]_{k,t} - [V_S]_{k,t-1} \right\}^2 \tag{7}$$

$$\sigma_k^2 = \frac{1}{T_x} \sum_{t=1}^{T_x} [V_S]_{k,t}^2 \tag{8}$$

The bottom portion of Fig. 1 shows a simple example where V_S is temporally continuous.

Accompaniment generally includes short-duration percussive sound produced by percussive instrument and sustained harmonic sound produced by pitched instrument. Thus, the temporal continuity constraining set on the activation coefficients of accompaniment V_A would not improve the performance of singing voice separation. This has proved by our simple experience.

2.3 Objective Function and Update Rules

Based on NMPCF, temporal continuity and sparsity criteria are considered to improve the performance of source separation. The objective function of NMPCF rewrite as follows:

$$L = \frac{1}{2}\|X - U_S V_S - U_A V_A\|_F^2$$
$$+ \frac{\lambda}{2}\|Y - U_A W_A\|_F^2 + \alpha \cdot \Omega_S([V_S]_t) + \beta \cdot \Omega_T(V_S) \tag{9}$$

Where α and β are the weights of the latter two terms, respectively. The expressions of $\Omega_S([V_S]_t)$ and $\Omega_T(V_S)$ are shown in Eqs. (5) and (7).

The basis matrix is trained in advanced thus fixed during NMPCF, while U_A, V_S, V_A and W_A should be updated iteratively. After calculating the derivation of the with respect to U_A, V_A and W_A, respectively, the multiplicative update rules to learn those matrices can be derived by taking negative terms of the partial derivative as numerator of multiplication factor, while taking positive terms as denominator:

$$U_A \leftarrow U_A \otimes \frac{XV_A^T + \lambda_A Y W_A^T}{(U_S V_S + U_A V_A)V_A^T + \lambda_A (U_A W_A) W_A^T}, \tag{10}$$

$$V_A \leftarrow V_A \otimes \frac{U_A^T X}{U_A^T (U_S V_S + U_A V_A)}, \tag{11}$$

$$W_A \leftarrow W_A \otimes \frac{Y}{U_A W_A}, \tag{12}$$

in which $\otimes$ denotes element-multiplication of matrix.

In the objective function Eq. (9), U_A, V_A and W_A affect only the first two terms, that is the reconstruction error. And therefore the update rules of U_A, V_A and W_A for objective function (9) are same as for objective function (4). While V_S in Eq. (9) affect not only the reconstruction error but also the constraints of $\Omega_S([V_S]_t)$ and $\Omega_T(V_S)$. In the derivation of the L in Eq. (9) with respect to V_S, the total negative term and positive term are calculated as follows:

$$[\nabla^-]_{k,t} = [U_S^T X]_{k,t} + \beta \cdot \frac{2T_x\left[(V_S)_{k,t-1} + (V_S)_{k,t+1}\right]}{\left[\sum_{t=1}^{T_x} (V_S)_{k,t}^2\right]_k}$$
$$+ \beta \cdot \frac{2T_x(V_S)_{k,t} \cdot \sum_{t=2}^{T_x}\left[(V_S)_{k,t} - (V_S)_{k,t-1}\right]^2}{\left[\sum_{t=1}^{T_x} (V_S)_{k,t}^2\right]_k^2} \tag{13}$$

$$[\nabla^+]_{k,t} = [U_S^T (U_S V_S + U_A V_A)]_{k,t} + \frac{\alpha}{\varepsilon + [\|V_S\|_1]_k}$$
$$+ \beta \cdot \frac{4T_x(V_S)_{k,t}}{\left[\sum_{t=1}^{T_x} (V_S)_{k,t}^2\right]_k} \tag{14}$$

the above two Eqs. (13) and (14) are represented in the form of matrix element in order to facilitate representation and understanding. Finally, the update rule of matrix element $(V_S)_{j,t}$ is:

$$(V_S)_{k,t} \leftarrow (V_S)_{k,t} \cdot \frac{[\nabla^+]_{k,t}}{[\nabla^-]_{k,t}} \tag{15}$$

After NMPCF, the activation coefficients VS and the basis matrix US of singing voice can be obtained, so the singing voice spectrograms are further obtained by a Wiener-type filtering [9]:

$$\tilde{X}_S = X \otimes \frac{U_S V_S}{U_S V_S + U_A V_A} \tag{16}$$

The singing voice waveform can be synthesized by inverse STFT and overlap-add method.

Figure 2 shows a spectrogram comparison of separated singing voice. In order to facilitate a clear understanding of the spectrogram, the amplitude values were all compressed by a cubic root operation, and only the frequency components with the frequency from 0 to 2 kHz are displayed. Comparing with the spectrogram in Fig. 2(c), it is distinct that the spectrogram after NMPCF with temporal continuity and sparsity criteria in Fig. 2(d) is closer to the spectrogram of clean singing voice in Fig. 2(b). While there are still some frequency components of about 200 Hz produced by pitch instrument around 1.1s existed in Fig. 2(c). By the constraining of temporal continuity and sparsity criteria, those frequency components of non-singing voice are removed mostly, this can be seen in Fig. 2(d).

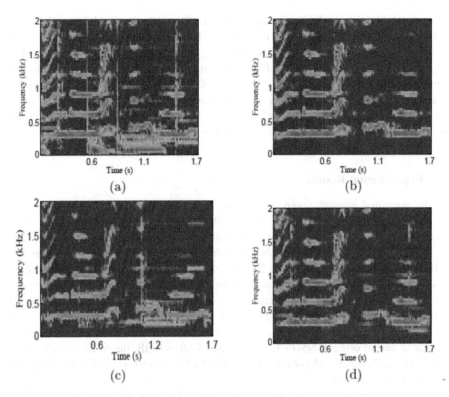

Fig. 2. Spectrogram Comparison. (a) The Mixture spectrogram of accompanied vocal signal mixed at the SAR of −6 dB. (b) The Spectrogram of clean singing voice. (c) The spectrogram of the singing voice separated by "NMPCF". (d) The spectrogram of the singing voice separated by "NMPCF + S+TC".

3 Evaluation and Comparison

3.1 Experimental Setup

The data-set is exactly the same as that used in the paper [1], that it comprises 31 songs covering 22 singers: 10 female singers and 12 male singers. The solo singing voice are served as training data set and superimposed with pure accompaniment at various singing-to-accompaniment ratios (SARs) from 0 dB to −9 dB at −3 dB intervals. Those monophonic mixture songs were first divided into vocal and non-vocal segments according the pitches of singing voice, and then those vocal segments were further segmented that each vocal segment lie within the range[1s, 4s]. There are 1568 vocal segments in all. The experiments were conducted over the segment data.

In the experiments, a frame length of 32 ms and hop length of 16 ms with the sampling frequency of 16 kHz were used. The length of DFT was 512. The numbers of basis vectors were set to 100 for U_S and 50 for U_A. λ_A was set to 12. To verify the performances of the constraints of sparsity criteria and temporal continuity, we set different combinative settings of α and β in Eq. (9) which denote different NMPCF models, respectively. These models described in Table 1. When α and β are both set to zero that it means the model corresponding original "NMPCF", while when $\alpha = 0.7$ and $\beta = 0$, it means that the "NMPCF" is with only the constraint of spasity criteria and denoted as "NMPCF + S", when $\alpha = 0$ and $\beta = 0.1$, it means that the model with only the constraint of temporal continuity is denoted as "NMPCF + TS", and when $\alpha = 0.7$ and $\beta = 0.1$, it means that the model with both the constraints of temporal continuity and of sparsity criteria is denoted as "NMPCF + S+TS".

3.2   Experimental Results

For the separated singing voice, firstly, the signal-to-noise ration (SNR) gain was selected to quantify the performance of the systems. The SNR gain, denoted as Δ SNR, is the difference between the SNR of mixture signal and that of separated singing voice [1]. Table 1 shows the average Δ SNR of 1568 samples when applying the improved NMPCF models and original NMPCF model. The last column denote the average values over the results under the conditions of four SARs. Results show that the constraints of temporal continuity and sparsity criteria are both effective. In comparison to the original NMPCF model, NMPCF + TC provides an certain improvement in all cases. For NMPCF + S model, when SAR = −3 dB, −6 dB, –9 dB, the results are superior to that of original NMPCF, while when SAR = 0 dB, the result is even worse than original NMPCF. However, NMPCF + S+TC model provides an excellent improvement in all cases.

For the quality of separated signal, different tasks have different demands. Generally, the singer identification system needs the separated singing voice of higher quality. So, we also perform the singer identification test with the separated singing voice by a singer identification system [12], which first detects the pitches of singing voice directly over the song mixed by accompaniment, then perform CASA-based segregation and finally reconstruct a complete spectrum of singing voice using missing feature method. As the first step of the singer identification system is pitch detection, so we first test the pitch detection of singing voice. The reference pitches were also detected by *Praat* [13] on clean singing voices in training data-set. A correct pitch estimate was defined to deviate less than 3 % from the reference, making it "round" to a correct musical note [14]. For pitch detection, a more appropriate measure is F-measure [15], which considers both the precision p and recall r of the test to compute the score: p is the number of correct results divided by the number of all returned results and r is the number of correct results divided by the number of results that should be returned. The F-measure is the harmonic mean of the precision and recall:

$$F - measure = \frac{2pr}{p+r} \tag{17}$$

Table 2 shows the F-measure values and the average F-measure values of four difference SARs at the frame level. Since the test samples are all pop music, the pitch range of the test samples is relatively small compared to that of the operatic song, we

Table 1. The average Δ SNR (dB) of diverse NMPCF model

SAR	0 dB	−3 dB	−6 dB	−9 dB	Avg.
NMPCF	4.86	5.34	5.51	6.05	5.43
NMPCF + S	4.78	5.52	5.84	6.08	5.55
NMPCF + TC	5.13	5.53	5.79	6.08	5.64
NMPCF + S+TC	5.68	5.74	5.92	6.69	6.05

Table 2. Average F-measure (%) of pitch estimation of singing voice

SAR	0 dB	−3 dB	−6 dB	−9 dB	Avg.
NMPCF	91.66	87.25	81.77	74.64	83.83
NMPCF + S	90.85	87.21	81.67	73.78	83.38
NMPCF + TC	91.05	87.28	81.43	74.30	83.52
NMPCF + S+TC	90.71	87.07	81.75	74.21	83.43

set [100, 1000] Hz as the plausible pitch range for all the test. We can see from the Table 3 that the NMPCF model with the constraints of sparsity criteria or temporal continuity do not improve the performance of pitch estimation of singing voice. While they do improve Δ SNR in Table 1. This maybe states that, with the constraints of sparsity criteria or temporal continuity, the NMPCF-based singing voice separation system although can further reduces the interference of accompaniment but lead to reduces the smoothness of singing voice components. Thus, the performance of pitch estimation of singing voice is not improved.

Next, we test the singer identification over the separated singing voice with estimated binary mask (EBM), and the associated parameters were set according with [1]. Figure 3 shows the average singer identification accuracy in the cases of four diverse NMPCF-based models and at four SARs. As can be seen, the constraints of temporal continuity and sparsity criteria both can improve the accuracy of singer identification in

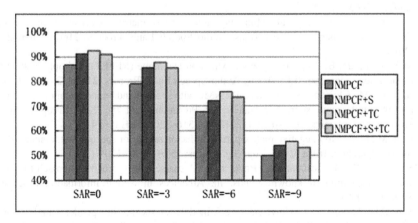

Fig. 3. Singer Identification Accuracy (%) of the separated singing voice by diverse models (Color figure online)

contrast to the original NMPCF algorithm at each SARs levels, although they can not improve the performance of pitch estimation of singing voice. However, the NMPCF + S+TC model is even inferior to the NMPCF + TC model. The results show that the NMPCF + TC model is more appropriate for singer identification by the system proposed in [12].

Based on the results of Δ SNR and singer identification accuracy, we can see that the constraints of temporal continuity and sparsity criteria both can improve the performance of singing voice separation at lower SARs, especially the constraint of temporal continuity.

4   Conclusion

In this paper, we proposed an improved NMPCF-based algorithm which restricts the activation coefficients of singing voice components to be temporal continuous and sparse in each frame. The experiments show that the model contains the constraints of both temporal continuity and sparsity criteria achieves the best Δ SNR, however, for singer identification, the model contains only the temporal continuity do the best. But the constraints of temporal continuity and sparsity can not improve the performance of pitch estimation of singing voice.

Acknowledgments.   This work is funded by the National Natural Science Foundation of China under Grants 61471311 and 61365005, and the Scientific Research Programs of the Higher Education Institution of XinJiang under Grants XJEDU2014S006.

References

1. Hu, Y., Liu, G.: Separation of singing voice using non-negative matrix partial co-factorization for singer identification. IEEE/ACM Trans. Audio Speech Lang. Process. **23**(4), 643–653 (2015)
2. Zhu, B., Li, W., Li, R., et al.: Multi-stage non-negative matrix factorization for monaural singing voice separation. IEEE Trans. Audio Speech Lang. Process. **21**(10), 2096–2107 (2013)
3. Rafii, Z., Pardo, B.: Repeating pattern extraction technique (REPET): a simple method for music/voice separation. IEEE Trans. Audio Speech Lang. Process. **21**(1), 73–84 (2013)
4. Becker, J.M., Christian, S., Christian, R.: NMF with spectral and temporal continuity criteria for monaural sound source separation. In: Proceedings of the 22nd European Signal Processing Conference (EUSIPCO), 2013, pp. 316–320. IEEE (2014)
5. Kim, M., et al.: Non negative matrix partial co-factorization for spectral and temporal drum source separation. IEEE J. Sel. Top. Sig. Process. **5**(6), 1192–1204 (2011)
6. Smaragdis, P., Madhusudana, S., Bhiksha, R.: A sparse non parametric approach for single channel separation of known sounds. In: Advances in Neural Information Processing Systems (2009)
7. Sun, D.L., Mysore, G.J.: Universal speech models for speaker independent single channel source separation. In: 2013 IEEE International Conference on Acoustics, Speech and Signal Processing (ICASSP). IEEE (2013)

8. Lefevre, A., Bach, F., Févotte, C.: Itakura-Saito nonnegative matrix factorization with group sparsity. In: 2011 IEEE International Conference on Acoustics, Speech and Signal Processing (ICASSP), IEEE (2011)

9. Kim, M., Smaragdis, P.: Mixtures of local dictionaries for unsupervised speech enhancement. IEEE Sig. Process. Lett. **22**(3), 293–297 (2015)

10. Virtanen, T.: Monaural sound source separation by nonnegative matrix factorization with temporal continuity and sparseness criteria. IEEE Trans. Audio Speech Lang. Process. **15**(3), 1066–1074 (2007)

11. Wilson, K.W., Raj, B., Smaragdis, P.: Regularized non-negative matrix factorization with temporal dependencies for speech denoising. Interspeech 2008, Brisbane Australia, 22–26 September 2008

12. Hu, Y., Liu, G.: Singer identification based on computational auditory scene analysis and missing feature methods. J. Intell. Inf. Syst. **42**(3), 333–352 (2014)

13. Boersma, P., Weenink, D.: Praat: doing phonetics by computer [Computer program], Version, vol. 5, p. 21, (2005)

14. Klapuri, A.: A perceptually motivated multiple-f0 estimation method. In: IEEE Workshop on Applications of Signal Processing to Audio and Acoustics, 2005, pp. 291–294. IEEE (2005)

15. Vincent, E., Bertin, N., Badeau, R.: Adaptive harmonic spectral decomposition for multiple pitch estimation. IEEE Trans. Audio Speech Lang. Process. **18**(3), 528–537 (2010)

An Improved Supervised Learning Algorithm Using Triplet-Based Spike-Timing-Dependent Plasticity

Xianghong Lin[✉], Guojun Chen, Xiangwen Wang, and Huifang Ma

School of Computer Science and Engineering,
Northwest Normal University, Lanzhou 730070, China
linxh@nwnu.edu.cn

Abstract. The purpose of supervised learning with temporal encoding for spiking neurons is to make the neurons emit arbitrary spike trains in response to given synaptic inputs. Recent years, the supervised learning algorithms based on synaptic plasticity have developed rapidly. As one of the most efficient supervised learning algorithms, the remote supervised method (ReSuMe) uses the conventional pair-based spike-timing-dependent plasticity rule, which depends on the precise timing of presynaptic and postsynaptic spikes. In this paper, using the triplet-based spike-timing-dependent plasticity, which is a powerful synaptic plasticity rule and acts beyond the classical rule, a novel supervised learning algorithm, dubbed T-ReSuMe, is presented to improve ReSuMe's performance. The proposed algorithm is successfully applied to various spike trains learning tasks, in which the desired spike trains are encoded by Poisson process. The experimental results show that T-ReSuMe has higher learning accuracy and fewer iteration epoches than the traditional ReSuMe, so it is effective for solving complex spatio-temporal pattern learning problems.

Keywords: Spiking neural networks · Supervised learning · Triplet-based spike-timing-dependent plasticity · Remote supervised method

1 Introduction

Artificial neural networks (ANNs) are the effective computational models to simulate biological nervous systems and reconstruct human brains. From the perceptron in the 1950s [1] to the widely-used back propagation (BP) training algorithm in the 1980s [2], ANNs have been evolving towards more powerful models based on novel formulations of the input and output spaces, neuron models, network structures, learning mechanisms, and various combinations of these. The traditional ANNs encode information by the firing rate of the biological neurons, and the outputs of neurons are generally expressed as analog variables in the given interval. However, recent neuroscience researches have shown that the transmitted information is usually represented in the brain through the precise timing of spikes (temporal encoding) [3], not only through the frequency of spikes (rate encoding). As the third generation of ANNs, spiking neural networks (SNNs) are constructed by the biological plausible spiking neuron models, they encode and process neural information through the precise timing spike trains [4].

© Springer International Publishing Switzerland 2016
D.-S. Huang et al. (Eds.): ICIC 2016, Part III, LNAI 9773, pp. 44–53, 2016.
DOI: 10.1007/978-3-319-42297-8_5

SNNs have more powerful computation capacity to simulate different neuronal signals and arbitrary continuous functions, which are very suitable for the processing of nervous signals in the brain. They have been successfully applied to various domains, such as image processing [5], robot control [6] and speech processing [7].

In recent years, many learning algorithms for spiking neural networks have been put forward [8]. Depending on whether the supervisory signals used in learning processes, learning algorithms could be divided into unsupervised learning and supervised learning. Unsupervised learning algorithms are stemmed from a synaptic plasticity hypothesis proposed by Hebb [9]. Synaptic plasticity is the ability of synaptic connections to change in strength and is believed to provide the basis for learning and memory in the brain. The strength of the connection between the presynaptic and postsynaptic neurons can vary via spike-timing [10, 11]. In fact, the spike train can not only cause persistent changes of synaptic strength, it also satisfies the spike-timing-dependent plasticity (STDP) rule [12, 13]. The conventional pair-based STDP rule changes the strength of the synapses by pairs of the precise timing of presynaptic and postsynaptic spikes. Then, many STDP variations have been proposed and got better performance [14]. The experiments have studied the detailed or specific role of spike time by triggering synaptic plasticity with triplet spikes. A special STDP learning rule, named triplet-based STDP was proposed, which uses sets of three spikes (one presynaptic spike with two postsynaptic spikes or one postsynaptic spike with two presynaptic spikes) [15]. Compared with triplet-based STDP, the experimental results indicated that classical STDP models based on pairs of spikes are not sufficient to explain synaptic changes triggered by spikes [16].

Supervised learning algorithm for spiking neurons aims to make the neurons emit arbitrary spike trains in response to given input spike trains. Based on the synaptic plasticity rule, many researchers have proposed various supervised learning algorithms suitable for SNNs. Ruf and Schmitt [17] put forward the first supervised Hebbian-type learning rule, which is based solely on single presynaptic and postsynaptic spikes. Legenstein et al. [18] presented a supervised Hebbian learning algorithm for spiking neurons, based on injecting external input current to make the learning neurons fire in a specific target spike train. Franosch et al. [19] presented a supervised STDP learning rule which trains one modality using input from another one as "supervisor". They proved that the proposed algorithm converges to a stable configuration of synaptic strengths leading to a reconstruction of primary sensory input. Ponulak and Kasiński [20] proposed a supervised learning method, named Remote Supervised Method (ReSuMe), the basic idea of the method comes from Widrow-Hoff and STDP rules. The ReSuMe algorithm adjusts the synaptic weights according to the combination of the STDP and anti-STDP processes and is suitable for various types of spiking neuron models. The results of the study show that the ReSuMe algorithm owns a very good learning performance and is widely applied to various areas [21].

In this paper, using the triplet-based STDP, an improved supervised learning algorithm for spiking neurons named triplet-based remote supervised method (T-ReSuMe) is presented to improve ReSuMe's learning performance. The structure of this paper is organized as follows: the triplet-based STDP rule and the improved supervised learning algorithm for spiking neurons are described in Sect. 2. The description of the experimental results is given in Sects. 3, and 4 provides the conclusion.

2 Improved Supervised Learning Algorithm for Spiking Neuron

2.1 Triplet-Based Spike-Timing-Dependent Plasticity Rule

The connection between a presynaptic neuron and a postsynaptic neuron is named a synapse. With the property of plasticity, synapses can change either their shapes or their functions over periods of seconds, minutes, or perhaps even the whole lifetime. The synaptic connections can be changed in the structure or function when biological nervous systems try to learn and memorize. The change of synaptic strength depends on the precise timing of presynaptic and postsynaptic spikes. In other words, synaptic strength is a function of the spike timing between the presynaptic and postsynaptic neurons.

STDP is considered as a special Hebbian rule of synaptic plasticity, which is taken spike timing into account. Considering the discovering in biological experiments, there is no doubt that the STDP rule plays an important role in learning and memory in the brain, where changes in synaptic efficacy are sensitive to the precise timing of presynaptic and postsynaptic action potentials [13]. Two types of STDP rules are considered in this paper. The conventional one is pair-based STDP, and the other one is triplet-based STDP.

The pair-based STDP is described by an exponentially decaying as a response to two separate levels of synaptic plasticity [22]. It is the classical form of STDP, used into many learning algorithms widely. The mathematical representation of the pair-based STDP rule can be expressed formally as:

$$\Delta w = \begin{cases} \Delta w^+ = +A^+ \exp(-\Delta t/\tau_+) & \text{if } \Delta t > 0 \\ \Delta w^- = -A^- \exp(\Delta t/\tau_-) & \text{if } \Delta t \leq 0 \end{cases} \quad (1)$$

where Δw is the change of synaptic weight, and $\Delta t = t_{\text{post}} - t_{\text{pre}}$ is the difference of time between a presynapstic spike and a postsynaptic spike where t_{pre} is the arrival time of presynaptic spike and t_{post} is the arrival time of postsynaptic spike. A^+ and A^- are the potentiation and depression parameters respectively, and have the positive values. The triplet-based STDP rule considers sets of three spikes [23, 24]. The triplet-based STDP is given by:

$$\Delta w = \begin{cases} \Delta w^+ = \exp(-\Delta t_1/\tau_+) & \left[A_2^+ + A_3^+ \exp\left(-\Delta t_2/\tau_x\right)\right] \\ \Delta w^- = \exp(\Delta t_1/\tau_-) & \left[A_2^- + A_3^- \exp\left(-\Delta t_3/\tau_y\right)\right] \end{cases} \quad (2)$$

In Eq. 2, the synaptic weight change is $\Delta w = \Delta w^+$ if $t = t_{\text{post}}$ and $\Delta w = \Delta w^-$ for $t = t_{\text{pre}}$. $\Delta t_1 = t_{\text{post}(n)} - t_{\text{pre}(n)}$, $\Delta t_2 = t_{\text{post}(n)} - t_{\text{post}(n-1)}$ and $\Delta t_3 = t_{\text{pre}(n)} - t_{\text{pre}(n-1)}$ denote the differences of time between presynaptic spikes and postsynaptic spikes. A_2^+, A_2^-, A_3^+ and A_3^- are the potentiation and depression parameters respectively. τ_+, τ_-, τ_x and τ_y are time constants.

2.2 Supervised Learning Algorithm Using Triplet-Based STDP

In fact, STDP as a rule of synaptic plasticity is an unsupervised learning rule. It enhances the ability of learning through adaptive weigh adjustment. We have described the triplet-based STDP rule, which uses sets of three spikes, i.e., one presynaptic spike with two postsynaptic spikes or one postsynaptic spike with two presynaptic spikes. In this section, considering the property of triplet-based STDP, the triplet-based STDP is used into a supervised learning algorithm, ReSuMe, to obtain an improved supervised learning rule.

For the input and output spike trains of spiking neurons, a spike train $S = \{t^f : f = 1, 2, \cdots, F\}$ represents the ordered sequence of spike time, the spike train can be shown as follows:

$$S(t) = \sum_{f=1}^{F} \delta(t - t^f) \tag{3}$$

where F is the number of spikes, $\delta(t)$ is the Dirac delta function, if $t = 0$, $\delta(t) = 1$, else $\delta(t) = 0$, t^f is the fth spike firing time.

ReSuMe is a supervised learning algorithm that allows the spiking neurons to adjust their synaptic weights to obtain a desired spike trains for the given synaptic inputs. In general, the ReSuMe learning rule [20] is expressed as:

$$\Delta w_{oi}(t) = S_i(t) \int_0^{\infty} a_{id}(s)[S_d(t - s) - S_o(t - s)]ds \\ + [S_d(t) - S_o(t)]\left[a_d + \int_0^{\infty} a_{di}(s)S_i(t - s)ds\right] \tag{4}$$

where $S_d(t)$ is the desired output spike train, $S_o(t)$ is the actual output spike train and $S_i(t)$ is the input spike train. a_d is the non-Hebbian term in ReSuMe which speeds up the learning process, $a_{id}(s)$ and $a_{di}(s)$ are the kernels of pair-based STDP represented by Eq. 1.

Equation 4 shows a mathematical description of ReSuMe for spike trains learning for spiking neurons. But in experiments of this rule, the kernel $a_{id}(s)$ is shown that has no effect on success. This can be explained by noting that $a_{id}(s)$ is related to the anti-causal order of spikes. The kernel $a_{id}(s)$ doesn't modify inputs that contribute to the neuron state briefly before getting the desired output spike trains. Therefore, it is selected $a_{id}(s) = 0$ for all $s \in \mathbf{R}$ to guarantee the learning speed and get better learning results [20]. Equation 4 can be reduced to the following formula:

$$\Delta w(t) = [S_d(t) - S_o(t)]\left[a_d + \int_0^{\infty} a_{di}(s)S_i(t - s)ds\right] \tag{5}$$

Now we can perform the integration of Eq. 5, the offline form of ReSuMe is expressed as:

$$\Delta w_{oi} = \eta \left\{ \sum_{t_d^g \in D} \left[a_d + \sum_{t_i^f < t_d^g} \exp\left(-\left(t_d^g - t_i^f\right)\middle/\tau_+\right) \right] - \sum_{t_o^h \in O} \left[a_d + \sum_{t_i^f < t_o^h} \exp\left(-\left(t_o^h - t_i^f\right)\middle/\tau_+\right) \right] \right\} \quad (6)$$

where η is the learning rate. In ReSuMe, $exp(-s/\tau_+)$ is a kernel that represents the pair-based STDP rule. Using the triplet-based STDP rule described in Eq. 2, an improved supervised learning algorithm for spiking neurons (T-ReSuMe) is given as:

$$\Delta w_{oi} = \eta \left\{ \sum_{t_d^g \in D} \left[a_d + \sum_{t_i^f < t_d^g} \exp\left(-\left(t_d^g - t_i^f\right)\middle/\tau_+\right)\left(A_2^+ + A_3^+ \exp\left(-\left(t_d^g - t_d^{g-1}\right)\middle/\tau_x\right)\right) \right] \right.$$
$$\left. - \sum_{t_o^h \in O} \left[a_d + \sum_{t_i^f < t_o^h} \exp\left(-\left(t_o^h - t_i^f\right)\middle/\tau_+\right)(A_2^+ + A_3^+ \exp(-(t_o^h - t_o^{h-1})/\tau_x)) \right] \right\} \quad (7)$$

3   Experimental Results

This section presents several experiments to demonstrate the learning performances of the supervised learning algorithm T-ReSuMe. According to the biological mechanism, the values of parameters of weight changes in Eq. 7 are set as: $a_d = 0.05$, $A_2^+ = 1.0$, $A_3^+ = 3.0$, $\tau_+ = 5.0$, and $\tau_x = 500$. The spiking neural networks are established using spike response model (SRM) [25]. The membrane potential of SRM neuron is expressed as:

$$u_o(t) = \sum_i \sum_f w_{oi}\, \varepsilon\left(t - t_i^f\right) + \eta\left(t - t^h\right) \quad (8)$$

In Eq. 8, w_{oi} is the weight of synapse from the input neuron i to the output neuron o, and t_i^f is the fth spike firing time of the input neuron i. $\varepsilon(t)$ denotes the postsynaptic potential that is described by $\varepsilon(t) = (t/\tau)\exp(1 - t/\tau)$ for $t > 0$, τ is a constant time. The threshold of the SRM neuron $v_{thresh} = 1$ and the absolute refractory period $t_{ref} = 1$ ms.

In order to show the learning performances of the proposed algorithm in quantity, the correlation-based measure C is used in computing the distance between the actual and desired output spike trains [20]. After the end of learning, the distance is calculated by the equation:

$$C = \frac{v_a \cdot v_d}{|v_a||v_d|} \quad (9)$$

where v_a and v_d denote the vectors of the actual and desired output spike trains with a Gaussian low-pass filter. $v_a{\cdot}v_d$ is an inner product of v_a and v_d, and $|v_a|$, $|v_d|$ are the Euclidean norms of v_a and v_d.

3.1 Learning Sequences of Spikes

In the first simulation experiment, a neuron with a number of 200 synaptic inputs is trained, and the simulation duration is 100 ms. The desired output and input spike trains are generated randomly with a Poisson process at the rate of 20 Hz. The initial weights of synapses are selected as a random distribution in the interval $(0, 0.2)$. The leaning rate is set for $\eta = 0.02$. The results of the experiment are recorded and given in Fig. 1.

Figure 1(a) shows the result of the output spikes for the spiking neuron in the learning epochs. Δ, an initial output spike train; ×, final actual output spike train; o, desired output spike train; •, actual output spike trains at some learning epochs. The spiking neurons can be seen having an ability to learn to emit the desired spike train. Figure 1(b) and (c) illustrate 200 initial weights before learning and the final weights after learning, and they can be recognized the difference obviously. Figure 1(d) shows

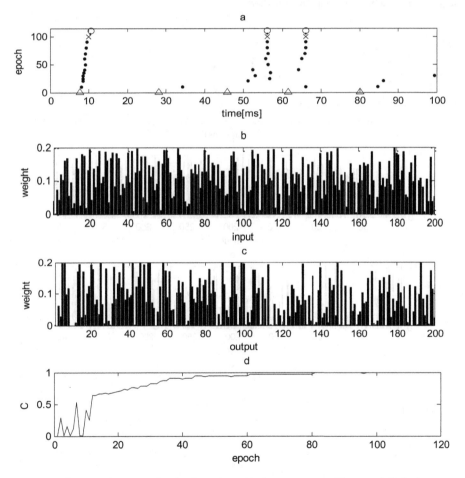

Fig. 1. Thefirst learning example of spike trains with the same rates of input and desired output. (a) The complete learning process. (b) The initial weights before learning. (c) The final weights after learning. (d) The change of the learning precision curve.

the change curve of C from around 0 at the beginning of this learning to 1 when the learning finished.

In the second experiment, a neuron with a number of 400 synaptic inputs is trained, and the duration of simulation is 100 ms. The desired output and input trains are generated randomly with Poisson process at the rates of 60 Hz and 20 Hz, and the learning rate is $\eta = 0.016$. The results of the experiment are given in Fig. 2.

Figure 2(a) shows the result of the output spikes for the spiking neuron in the learning epochs. Δ, an initial output spike train; $\times$, final actual output spike train; o, desired output spike train; •, actual output spike train at learning epochs. The proposed learning algorithm can be seen having an ability to learn the desired spike train for spiking neuron. Figure 2(b) and Fig. 2(c) illustrate 400 initial weights before learning and the final weights after learning, and they can be recognized the difference obviously. Figure 2(d) shows the changes of C from around 0 at the beginning of this learning to 1 when the learning finished.

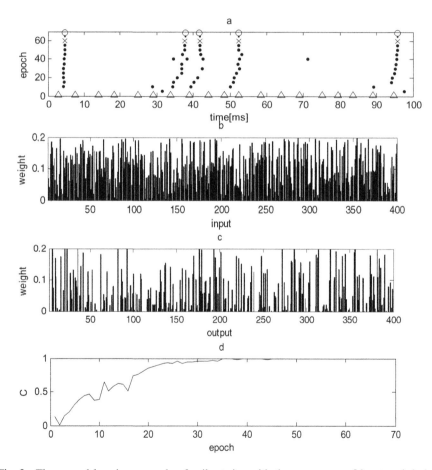

Fig. 2. The second learning example of spike trains with the same rates of input and desired output. (a) The complete learning process. (b) The initial weights before learning. (c) The final weights after learning. (d) The change of the learning precision curve.

3.2   Comparison of Two Algorithms

In addition, some comparisons are made between ReSuMe and T-ReSuMe in this section. Firstly, the learning task with 400 synaptic inputs are used for the training of sequence of spikes. The desired output and input spike trains are Poisson spike trains with the rate of 20 Hz. The initial weights of synapses are given at a random distribution in the interval (0, 0.2). The lengths of the duration are changed to 100, 200, 400, 600, 800 and 1000 ms. For two algorithms, 100 times of experiments with 2000 learning epochs are implemented in each case. Table 1 shows the average value of C in each case and the average number of learning epochs to reach the maximum C.

Table 1. Learning performance of ReSuMe and T-ReSuMe for the different length of simulation duration

		100 ms	200 ms	400 ms	600 ms	800 ms	1000 ms
	D						
	LR	0.009	0.007	0.0045	0.003	0.0023	0.0016
ReSuMe	C	1	0.994	0.901	0.753	0.603	0.482
	Epochs	244	366	638	790	952	944
T-ReSuMe	C	1	0.998	0.932	0.797	0.652	0.536
	Epochs	146	247	543	625	818	856

From Table 1, we can see that both methods can learn to get the desired output spike trains and the values of C are more than 0.9 when the lengths of the simulation duration are less than 400 ms. Besides, T-ReSuMe has better performance than ReSuMe. At each of the durations, C of our method is closer to 1, and the number of learning epochs is less than ReSuMe.

Secondly, the learning task with 400 synaptic inputs are used in the spike sequence learning. The desired output and input spike trains are Poisson spike trains with the frequency of 20, 40, 60, 80, 100, and 120 Hz. The length of the desired duration is 100 ms. For two algorithms, 100 times of experiments with 2000 learning epochs are implemented in each case of the simulation. Table 2 shows the learning rate for every case of the two algorithms, the average value of C in each case and the average number of learning epochs to reach the maximum C.

Table 2. Learning performance of ReSuMe and T-ReSuMe for the different frequency of spike trains

		20 Hz	40 Hz	60 Hz	80 Hz	100 Hz	120 Hz
	F						
	LR	0.009	0.007	0.0046	0.0028	0.0019	0.0013
ReSuMe	C	1	0.8835	0.7088	0.6025	0.5345	0.5043
	Epochs	244	553	828	844	895	858
T-ReSuMe	C	1	0.9106	0.7403	0.6496	0.5853	0.5507
	Epochs	146	467	668	776	645	689

Table 2 shows that both methods can learn to get the desired output train and the values of C is 1 when the frequency is 20 Hz. With the increment of frequency of spike trains, the average value of C decreases gradually in two methods. But for the same duration, C of our method is closer to 1, and the number of learning epochs is less than ReSuMe. So T-ReSuMe has better performance than ReSuMe for the spike trains learning problem.

4 Conclusions

This paper introduces an improved supervised learning algorithm, named T-ReSuMe, which is based on a triplet-based STDP rule. Besides, the paper further investigates the learning ability of T-ReSuMe to produce the desired output spike trains for spiking neurons. Then, the experimental results of T-ReSuMe are shown and compared with ReSuMe with the different lengths of the simulation duration and the different frequency of spike trains. In terms of the distance between the desired and actual output spike trains, it's found that the proposed supervised algorithm T-ReSuMe performed better than ReSuMe. In general, the average numbers of learning epochs to reach the maximum C of T-ReSuMe are less than ReSuMe. The reason for better performance is that the triplet-based STDP rule is closer to the actual synaptic plasticity of biological neuron.

Acknowledgement. This work is supported by the National Natural Science Foundation of China (No. 61165002, No. 61363058), the Natural Science Foundation of Gansu Province of China (No. 1506RJZA127), and Scientific Research Project of Universities of Gansu Province (No. 2015A-013).

References

1. Rosenblatt, F.: The Perceptron: A Perceiving and Recognizing Automaton Project Para. Cornell Aeronautical Laboratory (1957)
2. Rumelhart, D.E., Hinton, G.E., Williams, R.J.: Learning representations by back-propagating errors. Nature **323**(6088), 533–536 (1986)
3. Bohte, S.M.: The evidence for neural information processing with precise spike-times: a survey. Nat. Comput. **3**(2), 195–206 (2004)
4. Ghosh-Dastidar, S., Adeli, H.: Spiking neural networks. Int. J. Neural Syst. **19**, 295–308 (2009)
5. Kerr, D., McGinnity, T.M., Coleman, S., et al.: A biologically inspired spiking model of visual processing for image feature detection. Neurocomputing **158**, 268–280 (2015)
6. Cao, Z., Cheng, L., Zhou, C., et al.: Spiking neural network-based target tracking control for autonomous mobile robots. Neural Comput. Appl. **26**(8), 1839–1847 (2015)
7. Dennis, J., Tran, H.D., Li, H.: Spiking neural networks and the generalised hough transform for speech pattern detection. In: The 16th Annual Conference of the International Speech Communication Association (2015)
8. Lin, X., Wang, X., Zhang, N., et al.: Supervised learning algorithms for spiking neural networks: a review. Acta Electronica Sinica **43**(3), 577–586 (2015)

9. Hebb, D.O.: The Organization of Behavior: A Neuropsychological Theory. Wiley, New York (1949)
10. Abbott, L.F., Nelson, S.B.: Synaptic plasticity - taming the beast. Nat. Neurosci. **3**, 1178–1183 (2000)
11. Wang, H., Gerkin, R., Nauen, D., et al.: Coactivation and timing dependent integration of synaptic potentiation and depression. Nat. Neurosci. **8**(2), 187–193 (2005)
12. Markram, H., Gerstner, W., Sjöström, P.J.: Spike-timing-dependent plasticity: a comprehensive overview. Front. Synaptic Neurosci. **4**, 8 (2012)
13. Daniel, E.F.: The spike-timing dependence of plasticity. Neuron **75**(8), 558–571 (2012)
14. Serrano-Gotarredona, T., Masquelier, T., Prodromakis, T., et al.: STDP and STDP variations with memristors for spiking neuromorphic learning systems. Front Neurosci **7**(2), 1–15 (2013)
15. Pfister, J.P., Gerstner, W.: Triplets of spikes in a model of spike timing dependent plasticity. J. Neurosci. **26**(38), 9673–9682 (2006)
16. Shahim-Aeen, A., Karimi, G.: Triplet-based spike timing dependent plasticity (TSTDP) modeling using VHDL-AMS. Neurocomput. **149**, 1440–1444 (2015)
17. Ruf, B., Schmitt, M.: Learning temporally encoded patterns in networks of spiking neurons. Neural Process. Lett. **5**(1), 9–18 (1997)
18. Legenstein, R., Naeger, C., Maass, W.: What can a neuron learn with spike-timing-dependent plasticity? Neural Comput. **17**(11), 2337–2382 (2005)
19. Franosch, J.M.P., Urban, S., van Hemmen, J.L.: Supervised spike-timing-dependent plasticity: a spatiotemporal neuronal learning rule for function approximation and decisions. Neural Comput. **25**(12), 3113–3130 (2013)
20. Ponulak, F., Kasinski, A.: Supervised learning in spiking neural networks with ReSuMe: sequence learning, classification and spike shifting. Neural Comput. **22**(10), 467–510 (2010)
21. Hu, J., Tang, H., Tan, K.C., et al.: A spike-timing-based integrated model for pattern recognition. Neural Comput. **25**(2), 450–472 (2013)
22. Bi, G., Poo, M.: Synaptic modifications in cultured hippocampal neurons: dependence on spike timing, synaptic strength and postsynaptic cell type. J. Neurosci. **18**(24), 10464–10472 (1998)
23. Pfister, J.P., Gerstner,W.: Beyond pair-based STDP: a phenomenological rule for spike triplet and frequency effects. In: Advance Neural Information Process System, vol. 18, pp. 1083–1090 (2006)
24. Gjorgjieva, J., Clopath, C., Audet, J., et al.: A triplet spike-timing-dependent plasticity model generalizes the bienenstock-cooper-munro rule to higher-order spatiotemporal correlations. Proc. Natl. Acad. Sci. **108**(48), 19383–19388 (2011)
25. Gerstner, W., Kistler, W.M.: Spiking Neuron Models: Single Neurons, Populations, Plasticity. Cambridge University Press, Cambridge (2002)

Risk and Vulnerability Analysis
of Critical Infrastructure

Kaiyu Wan[1] and Vangalur Alagar[2(✉)]

[1] Xi'an Jiaotong-Liverpool University, Suzhou, People's Republic of China
Kaiyu.Wan@xjtlu.edu.cn
[2] Concordia University, Montreal, Canada
alagar@cse.concordia.ca

Abstract. A nation's Critical Infrastructures (CI) is vital to the trustworthy functioning of the economic, health care, and social sectors of the nation. Any disruption to CI will adversely affect the economy, and peaceful functioning of the government. Above all, it will adversely affect the morale and confidence of the citizens. Hence, protecting CI of a nation must be given top priority. Fundamental to protection mechanisms are risk and vulnerability analysis. Based on their outcomes protection mechanisms can be planned, designed, and implemented. In this paper we offer a concise template representation for critical assets, and explain a methodology for vulnerability assessment and risk analysis. We point out the potential role of agents, and deep learning methods in the development and commissioning of future cyber defense solutions.

Keywords: Critical infrastructure · Risk analysis · Vulnerability analysis · Agent system · Deep learning

1   Introduction

Almost every nation in the world has an abundant number of CIs on which the governance and security of the country depends, and the citizens have come to depend upon them for their everyday living. From several reports [6, 9, 12] published by governments of Canada, USA, and European Commission and the recent monograph of Lazari [16] we get an adequate understanding of "what CI is?". From them we extract the following common features to be used in this paper:

- CI can be either physical (facility) or virtual (system) or a mixture of both.
- CI provides services that are vital (essential) to the health (safety, security, governance, reliable, available at all times, resilient to attacks) of the nation.
- There exist many CI sectors, and in each sector many networks and processes exist. Consequently, CI is a System of Systems (SoS).
- Maintaining the CI services without disruption requires protecting the CI components and their interconnectedness.

© Springer International Publishing Switzerland 2016
D.-S. Huang et al. (Eds.): ICIC 2016, Part III, LNAI 9773, pp. 54–66, 2016.
DOI: 10.1007/978-3-319-42297-8_6

1.1 CI Sectors

The four common characteristics discussed above permeate through the 16 CI sectors classified by the Department of Homeland Security (DHS) [12]. To get a good understanding of the concept of SoS within each sector, and cross-sectorial interdependence we have (arbitrarily) chosen 5 of them to discuss below.

Commercial Facilities Sector: This sector is classified into Social, Residential, Business, and Military segments. Social facilities operate to fulfill the social goals of the community and on a principle of "openness", and hence it is a vulnerable segment. Access to Residential buildings might range from semi-private to strictly private. It is known [10, 24] that hackers use remote devices to infiltrate make modifications to the embedded systems in power grids as well as in automobiles. Consequently, we need better and stronger protection methods to safeguard homes, especially the "senior homes". Most of Business facilities are also centers for social get-together. Protecting such facilities is much harder, although media-based technology and closed circuit TVs are widely used. Clearly, Military facilities that store weapons, large servers with confidential information of personnel, and strategic plans of significance for homeland security face greatest threats and should by necessity secured without fail. This sector is dependent on many other sectors, including energy, transportation, water, and emergency services sectors.

Energy Sector: The three major sub-sectors are electricity, oil, and natural gas. All industries, commercial sector facilities, and emergency services rely on electric power. Some of the risks and threats to electric power grids are cyber and physical infiltration, natural disasters, incompatible changes and interoperability failure in information technology infrastructure, and human errors. Other operational hazards, such as blowouts and spills, may be caused by political instability, and terrorist activities.

Transportation Sector: This sector is vital to economy, social well-being, and strategic operations of military. The important sub-sectors are aviation, maritime transportation, mass transit of humans and products, and pipeline. The assets and services of this sector touch on almost every other sector. Most importantly it depends on energy, information technology, and water. Sectors such as healthcare and manufacturing rely on uninterrupted services of this sector.

Financial Sector: This sector is a vital CI component, because any disruption of even a small nature will disrupt social harmony and might cause economic ruin. Power outages, hacking of communication network and cyber assets, and natural disasters have an immense setback on this sector. Cyber-attacks stand out as the primary source of disruption and misuse for this sector.

Healthcare Sector: Healthcare services in many countries are offered by private agencies, public hospitals, and government controlled clinics. The stakeholders need to cooperate with Emergency Services Sector, Chemical Sector, and Government Facilities Sector in order to manage the public health component of health care. This sector is dependent on transportation, energy, communications, information technology, and emergency service sectors. Since protecting this sector is vital for the social well-being of the nation, it is necessary to protect all sectors on which it depends.

From the above discussion it is clear that a CI sector is dependent on one or more CI sectors. Identifying the dependency relation among sectors as well as among the assets within sectors is crucial to investigate the risk factors and vulnerabilities that migrate (cascade) across related sectors. Recent incident of "9/11 attacks" convinces us of the cascading effect. It is in this context that we propose an asset model using which risk and vulnerability analysis can be done effectively.

1.2   Contribution

We define the basic concepts asset, threat, risk, and vulnerability in Sect. 2. We give a concise template, table-based approach, to describe assets in Sect. 3. We explain the semantics of the template segments and emphasize how inter-related assets have a compact representation in the asset table. In Sect. 4 we use the template representation to motivate vulnerability and risk analysis. We conclude the paper in Sect. 5 by pointing out our ongoing work in investigating network of agents and robots that use "deep learning" to collectively learn, rehearse, and collaborate in protecting CI assets and processes.

2   Basic Concepts

Vulnerability and risk analysis require the following five activities, in the order stated: (1) Identification of critical assets, and their dependencies; (2) Identification and assessment of threats to each asset; (3) Vulnerability analysis, which assess the vulnerability of each asset to the threats; (4) Risk calculation, which is the "expected loss to economy, and human lives"; and (5) Risk reduction through prioritization.

A physical asset is an item of economic value that has a tangible or material existence. For most businesses and governments, physical assets usually refer to cash, equipment, inventory and properties owned by them. Virtual assets are electronic data stored on a computer or the internet. A team of CI sector experts should identify critical assets in each CI sector. In general, if economic loss or damage to the integrity of society through the loss of an asset threatens the existence of its owner then that asset must be declared critical. After selecting and ordering critical assets in all sectors, the inter-relatedness between assets within each sector, as well as the intra-relatedness (dependencies across sectors) should be identified. This analysis is essential to understand the cascading effect of asset loss in the system. The experts shall develop a list of associated Critical Cyber Assets (CCA) essential to the operation of the Physical Critical Asset (PCA).

Threat is defined [3, 4] as "any indication, circumstance or event with the potential to cause loss or damage to an asset". Threat analysis is done to recognize the plausible threats to an asset, based upon the asset features, environment characteristics, and the potential external motivations for harming the asset. Important threat types are (1) those caused by humans (insider, terrorist, military), (2) those caused by nature (hurricane, earthquake, forest fire, tornado, flood), and (3) those caused by humans using technology, cyber space, and communication network. The motive and capacity in triggering threats should be taken with threat types in assessing the likelihood of the attack, the frequency and duration of the attack, and its likely consequences. Vulnerability is defined [2] as a "weakness that can be exploited to gain access to an asset". Some of the common types of weaknesses are (1) physical (accessibility, relative location), (2) technical (eavesdropping, susceptible to cyber-attack), (3) operational (weak business and security policies), and (4) organizational (weak structuring). Risk is defined [4, 9, 12] as "the probability of loss or damage".

A systematic method to assess vulnerability starts with the identification of threats that has the potential to affect the assets in a critical system. A prerequisite is to have a model of system taxonomy, critical assets within each system, a definition of the "interrelatedness" among the assets, and an estimate of impact factor on assets. As an example, the system taxonomy in a sector may consist of the three subsystems "Mission Critical system", "Support System", and "Quality Assurance System". This taxonomy may be linked to the taxonomy of its sub-sectors, sectors that are dependent on it, and the sectors on which it depends. Once taxonomy is defined, the components within each subsystem and their assets should be identified. For example, components within "Mission Critical system" are "Control System", "Supervisory Human Experts", and "Communication System". Assets within Control System include *Instrument/Sensor/Analyzer, Controller, Control valves, and Software & Configuration*. Threat analysis for each asset within each subsystem must be conducted next. It is best based on the "interrelatedness among assets". Thus, modeling interrelatedness at sector level and at asset level within sectors is essential.

Interrelatedness property between sectors can be defined statically, with knowledge of the products and services produced by sectors. Static interrelatedness has been modeled by using matrices [18] and graph structures [13, 19]. Since assets within sectors may vary depending upon policy changes [19], cyber assets and system dynamics [23, 26] interrelatedness property may also vary dynamically. In general, an interrelatedness property is *transitive,* in the sense that if "asset A depends on asset B AND asset B depends on asset C", then asset A depends on asset C. Based on transitivity it is possible to define interrelatedness to higher orders. We say threat of asset A is *cascaded* to asset B if either "*B depends on A directly*" or "*B has a higher-order dependency on A*". Using directed graphs and higher-order dependencies an analysis of "cascading effect of risks" is given in [26]. In our research we explain how such an analysis can be done more efficiently by using the template model discussed in the following section.

3 A Template for Modeling Critical Assets

Once critical assets are identified within a sector and their distribution across the subsystems within that sector, the assets can be modeled to satisfy a set of requirements. We consider "Business/Management (BM)" and "Technical/Scientific (TS)" as the two groups who among themselves will define the system taxonomy. It is remarked in [4] that for BM group, who are involved with policy making and management of resources a top-down dynamic view is preferable, whereas for TS group a detailed bottom-up view of static and dynamic modeling is required. Both these requirements are met in our modeling technique. The key features are (1) each critical asset is modeled as a formal table in which protection mechanisms can be set for preserving information integrity and confidentiality, (2) the asset tables are linked using hyperlinks to model interrelatedness, and (3) a hierarchy of tables is defined based upon the nature of sectors. The CAS Knowledge Description Template (CASKDT), shown in Table 1, is the basic building block to create the asset model. The template CASKDT and the physical (or virtual) asset that it represents can be accessed by any authorized client in the system. Many methods discussed in [22] for protecting critical infrastructures can be adapted to secure CASKDT. The template description may be extended by adding more description elements, as and when policy changes and the criticality constraints vary. Thus, the model serves both static and dynamic requirements. The tabular format shown in Table 1 is meant for human agents in BM and TS groups. We have developed a GUI which automatically generates the XML version of the CASKDT. The XML version is used for knowledge propagation across the different system components for automatic analysis and decision making. Below we briefly discuss the structure and semantics of CASKDT components.

Table 1. Critical Asset Knowledge Description Template

Reference	<ADDRESS>, <URL>
Sector	<generic description of sector type> <link to specific sector information> <link to parent-sector> <links to subsectors>
Expert Knowledge	<link to expert opinions>
Functional Properties of Services	**Type of Use**: <service type> **Attributes**: <criticality level, access constraints> **Relations**:{depends, part-of, requires, contained-in} **Priority**:{ordered list of (asset, sector)} **Utility**:{<a_1, u_1>, <a_2, u_2>, ..., <a_k, u_k>}
Nonfunctional Aspects, Trust and Threat Types	**Cost**: cost of replacement **Availability**: <availability constraints, replacement constraints, regional constraints> **History**: attack reports vulnerability and risk metrics list of potential threats **Other Knowledge in the Context of Threat**: link to knowledge store
Policy	**Legal Rules**: for asset sharing **Context Information**: location context, availability context **Context Rule**: for asset release
Exceptional Situations	**Side Effects**: environmental impact factors, emergency procedures

The "Reference" part specifies the "ADDRESS" of the physical location of the facility in which the asset exists, and the "URL" of virtual assets related to the physical asset. The "Sector" part includes (1) the sector name, a generic natural language description of the sector, (2) a link (URL) to a description of sector characteristics, (3) a link (URL) to the CASKDT of its parent sector, and (4) links to the CASKDTs of its sub-sectors. An authorized client/expert of the CASKDT can navigate to learn and reason about its relevance (for criticality), because tacit knowledge cannot be textually described. To assist in such reasoning, the Experts Knowledge section has a link to the expert opinions on the asset's importance. In Functional Properties/Services section we list the assortment of services available through the asset, critical levels of services, access constraints, and the possible relations between this asset and other assets in the system. The "Relations" and "Priority" sections are very vital for cascade analysis and must be compiled with "experts knowledge" included in "Experts Knowledge" section. The relation "*asset A depends on Asset B*" may be used if the services of asset B are essential for one or more services of asset A. An example is "Nuclear Facility depends on Electricity". If (s, a) precedes (s', a') in "Priorities" section it means that (1) the asset b of sector S modeled in the template is related to asset a of sector s and asset a' of sector s', and (2) the impact of threat spreading from (s, a) to (S, b) is higher than the impact of threat spreading from (s', a') to (S, b). In general, we can expect a large number of relations with different semantics to arise in asset specification. Consequently, in "Relations" section each relation name is a "hyper-link" to the cyber location where the relations and their meanings are stored in a directory. By traversing the hyperlinks in a template one can explore all the sectors/assets that are related to the asset/sector specified in it. Because of hyperlinks we can use "web navigation" techniques to conduct threat and risk analysis in an efficient way without having to have a full graph model, as done by many researchers [2, 7, 13, 20]. The representation $\{<a_1, u_1>, <a_2, u_2>, ..., <a_k, u_k>\}$ is used to specify the utility factor u_i for the asset a_i, if the asset described in this CASKDT requires a_i. In the section "Nonfunctional…" we include the information necessary to evaluate the risk if this asset is vulnerable to certain threats. So, we have included threat types, attack incidents and how they were handled, availability of this asset in the market place, and the replacement cost. Any knowledge gained in the past is also included, for it might be used in risk analysis. The "Policy" section is useful for both BM and TS groups. The managers of the asset define the policy for asset sharing, determine where and how the asset is to be located (distributed), and define rules for releasing the assets for any application. The structure and semantics of CASKDT are both concise and precise. It is concise because the hyperlinks in the template inherently represent interconnectedness of assets. The description is precise because of the semantics. Each section can include only the information that respects the prescribed semantics. The two primary disadvantages of graph and network models [2, 7, 13, 20] are (1) they do not scale up to large CIs, and (2) they lack precise semantics. Although labeled graphs have been used [19, 21, 25] to assign semantics of interrelatedness the inherent complexity in exploring large graphs make such approaches difficult to adapt for assessing risk in large CIs. Another important virtue of our template representation is that a GUI developed by us can automatically generate the XML version which can be used for communication across system components and cyber communication network. Finally, the most important advantage of our approach is the inclusion of priority and context information in CASKDTs. They help in reducing the amount of information to be gathered dynamically for risk analysis, and in monitoring CI system performance in order to ensure the prevention of attacks in dynamic (such as mobile) situations. No other modeling approach has considered these two aspects; the exception is the recent work [1] which has considered context in the modeling stage. However, no specific method for risk analysis has been reported.

4 Vulnerability and Risk

The two broad categories of Risk Assessment (RA) methodologies *are Structural Analysis (SA), and Behavioral Analysis (BA)*. Structural approaches view the CI as a SoS topology [14] in accounting their interdependencies. A SA approach starts developing an interdependency model [18, 21, 23, 25] for use in vulnerability analysis. The graph models annotate edges with information on demand/supply, utility factor, and constraints on service flow. Behavioral approaches [23] focus on developing mathematical definitions to analyze failure (fault) models, failure analysis (causes for failures), and failure propagation (cascading). Probabilistic methods and linear optimization [13] have been used to assess cascading behavior. Given threat probabilities, these approaches attempt to generate all possible event (threat) combinations and compute the probability of failure under each scenario.

Our approach, explained in Fig. 1, is a modification of the approach suggested by Baker [3]. In our approach both static and dynamic flow of threats are combined from the asset model, which is both a structural and behavior model. As a motivation, we explain the vulnerability assessment procedure for a simple case of "Electric Power System". The full vulnerability assessment procedure of this problem can be found in The National Security Telecommunications Advisory Committee Information Task Force Report [19].

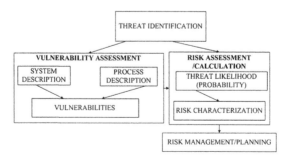

Fig. 1. Risk Assessment Process

The first step is to identify the subsystems for the Power Grid. For the sake of simplicity, we consider the four subsystems "The Control Center, The Substation,

The Communications Infrastructure, and The Distribution Network". For each one of these subsystems the CASKDT asset models are created, and vulnerabilities based upon their services/functionality, are assessed. For example, the primary functionality of the Control Center is to monitor power system operations, save history of its operations, and allow the functioning of the control unit under its domain. Other operations include monitoring transmission systems, power distribution networks, and customer distribution loads. We emphasize that support systems are often easy to attack, because they may not be protected as securely as the main mission-critical subsystems. Therefore, it is essential to include the vulnerable points/events and assign priority to them. Key points of vulnerability are (1) remote ports, (2) links to Internet and Internal Information Management Systems, (3) tools that enable support functions to vendors and customers, (4) flow of power to the central facility, and (5) support system links. In a similar fashion the functionality and vulnerability are assessed for other units. From the list of CASKDTS and the functionality of assets we identify the interrelatedness between the assets in the subsystems by navigating the hyperlinks in the CASKDT. A fault-tree is constructed for each asset in each chain created by hyperlinks navigation. A parallel activity for system development team and experts is to list a set of threats and hazards that affect each one of these assets (physical and cyber). The experts should associate with each threat its "threat level", which will be required in the calculation of risk factor. Based on this set of information, risk analysis can be done using matrix algebra, as explained next.

Navigating through the hyperlinks on "Relations" and "Priority" we will get a set of chains, where a chain is a sequence of hyperlinks. We need to analyze each chain for risk. Let us write $\Sigma = \{\tau_1, \tau_2, ..., \tau_m\}$ to denote a finite set of threats, and $\Lambda = \{\alpha_1, \alpha_2, ..., \alpha_n\}$ to denote the set of assets in a chain. A threat matrix A of size m × n, whose rows are threats and columns are assets can be constructed. The ith row, j^{th} column entry A[i, j] is the severity level of threat τ_i on the asset α_j. Below we explain a few operations on threat matrices that are sufficient to calculate risks arising from interrelatedness of any order.

Subsystem Threat Matrix: Every subsystem s_k has a finite set α_k of assets. Hence $\alpha_k \subset \Lambda$. The threat matrix A_k for the subsystem s_k is the projection of matrix A on the columns defined by α_k. The set Σ_k of threats that are associated with the subsystem s_k is the set of rows in A_k.

Interrelatedness: Subsystems s_k and s_r have *first-order* interrelatedness if their threat matrices have a non-empty intersection, in the sense that either (1) $\alpha_k \cap \alpha_r \neq$ empty, or (2) $\Sigma_k \cap \Sigma_r \neq$ empty.

Cascading: We say s_k and s_r have nth-*order interrelatedness* if they are both in the same chain and there exists n > 1 links between them. The progression of threat from one to the other is called cascading effect. First we discuss the cascading effect for one type of first-order interconnection relation, namely for relation *"requires"*. Next, we explain how to handle cascading effect. Assume that s_k requires s_r, meaning that the subsystem s_k has one or more assets and each one of them requires the services (functionality) of one or more assets from the subsystem s_r. Each "requesting asset" in

s_k is more vulnerable to the threat attack that will infest the "requested asset" in s_r. Hence it is prudent to "raise" the "severity level" of the "requesting asset". This is accomplished as follows: (1) Denote by A_k and A_r the threat matrices corresponding to the subsystems s_k and s_r. (2) If the matrices are "disjoint" (do not have common assets or common threats) then no cascading happens, because "request" is void. If the matrices have common assets then for each common asset α_p we "augment" the matrix A_k as follows. Let the column number in A_k that corresponds to the asset α_p be p_k, and the column number in A_r that corresponds to the asset α_r be p_r. Notice that p_r and p_k are threat vectors, and some of the threats in p_r may or may not be in p_k. In the former case we will replace the vulnerability value by "the sum of the vulnerability values of the threats common to both", and in the latter case we will augment the matrix by adding a new row for the new threat being cascaded from p_r. These two cases are explained next:

Case 1: Let the threat τ_r^i be the ith row of matrix A_r, and let this threat be in the jth row of matrix A_k. So, we must raise the 'severity level' of this threat in matrix A_k by the amount $A_r[i, p_r]$. That is, the new augmented matrix entry is

$$A_k'[j, p_k] = A_k[j, p_k] + A_r[i, p_r]$$

Case 2: Let the threat τ_r^i be the ith row of matrix A_r, and let this be not a threat in A_k.

Then, we should augment matrix A_k, which we do by (1) introducing it in A_k as a new row, and (2) copying the severity level $A_r[i, p_r]$ in column p_k of A_k, while setting the other entries of the new row to 0. To achieve this effect, we let the number of rows and columns in matrix A_k to be r_k and c_k. The augmented matrix entries are

$$A_k'[r_k+1, p_k] = A_r[i, p_r], A_k'[r_k+1, j] = 0, 1 \leq j \leq c_k, j \neq p_k. \tag{1}$$

Example 1. We illustrate threat level calculation when subsystem s_k *requires* (first-order relation) subsystem s_r. The subsystem s_k has three assets and two threats on them. The *subsystem* s_r has two assets and three threats on them. One asset and one threat are common to both subsystems. Two threats that are new in s_r will affect s_k because "s_k *requires* s_r". The threat matrices for the systems are

$$A_k = \begin{bmatrix} & a_1 & a_2 & a_3 \\ t_1 & 0.2 & 0 & 0.6 \\ t_2 & 0.1 & 0.7 & 0.8 \end{bmatrix} \text{ or } B_r = \begin{bmatrix} & b_1 & a_3 \\ t_1' & 0.8 & 0.1 \\ t_1 & 0.7 & 0.9 \\ t_3 & 0.8 & 0.4 \end{bmatrix}$$

Let A_k' denote the cascaded threat matrix. It will have three columns corresponding to the assets a_1, a_2, a_3, because the numbers of assets do not change. Since threat t_1 (first row of A_k) affects the common asset a_3 (in the second row of A_r) the severity level of a_3 becomes 1.5. However, assets a_1 and b_1 are mutually exclusive, meaning that the threats affecting b_1 will not contribute to the "requesting" subsystem. Hence, the threat vector for t_1 in the cascaded system matrix A_k' is (0.2, 0, 1.5). The threat vector t_2 (second row in A_k) is not influenced by the threat matrix A_r. Hence the t_2 row in matrix A_k' is the same as the t_2 row in A_k. Both threats t_1' (first row of A_r) and t_3' (third row of

A_r) affect asset a_3, which is also an asset of system s_k (third column in matrix A_k). They have a cascading effect, which is captured by two new rows introduced in matrix A'_k, one for t'_1 and another for t_3. The threat levels in these two rows are calculated using Eq. (1). The resulting matrix is

$$A'_k = \begin{bmatrix} & a_1 & a_2 & a_3 \\ t_1 & 0.2 & 0 & 1.5 \\ t_2 & 0.1 & 0.7 & 0.8 \\ t'_1 & 0 & 0 & 0.7 \\ t_3 & 0 & 0 & 1.2 \end{bmatrix}$$

Example 2. We explain here threat level calculation when subsystem s_1 requires (nth-order relation) subsystem s_n. Let us denote the chain by $s_1 \rightarrow s_2 \rightarrow \ldots \rightarrow s_{1(n-1)} \rightarrow s_n$. Let their threat matrices be $A_1, A_2, \ldots, A_{(n-1)}, A_n$. We start computing the augmented matrix $A'_{(n-1)}$ for the pair $(A_{1(n-1)}, A_n.)$, by using the steps shown in Example 1. Let us write $A'_{(n-1)} = \text{Aug}\left(A_{(n-1)}, A_n\right)$. Next we calculate $A'_{(n-2)} = \text{Aug}\left(A_{(n-2)}, A'_{(n-1)}\right)$. Proceeding backwards in the chain until we calculate $A'_{(n-1)} = \text{Aug}\left(A_1, A'_2\right)$.

These results are sent by the "Vulnerability Assessment" component to "Risk Calculation Component". Essentially risk implies "uncertainty" and hence risk calculation is based on probability distribution on the set of threats. Expected loss or damage to an asset or a subsystem is calculated by "multiplying the threat severity levels (which are measures of adverse impacts caused by successful threat events) with the probabilities associated with threat/vulnerability". For each category a severity level (say in the range [1, 10]) may be assigned. In order to calculate the "expected risk" of an asset (facility or process), one determines (1) the probability p that the asset will be attacked, (2) the conditional probability q (given the attack probability) that a specific means (such as using a remotely enabled device that blows up) is employed, (3) the probability r of success when that means is used, and (4) damage estimate C (damage to asset). Then, the expected loss E is $p \times q \times r \times C$. Because this number E is only an estimate, often it is "normalized" and expressed qualitatively (Severe, High, Medium, Low). Risk management activity will investigate methods that can reduce risks, such as disabling the adversary before the attack, minimizing the vulnerabilities, and reducing the impact or consequences of an attack. Another way to reduce risk is to distribute the assets geographically to locations where the likelihood of threat is a minimum.

5   Conclusion

The primary merits of CASKDT modeling are its conciseness and precision. It has a rich expressive power to include a wealth of diverse information. It can be expanded and viewed either graphically or mathematically or virtually. As suggested in [25] we can have a multi-agent system consisting of heterogeneous, autonomous, and decentralized agents for template-based CI modeling and risk analysis. Agents can express templates into XML files that can be communicated across sites in a cyber medium. Agents can be taught the semantics of CASKDT and graph algorithms to expand the hyperlinks, generate partial subgraphs for given semantics of interconnectedness, and explore large sparse matrices in order to perform different vulnerability/risk calculations. A Robot is an executable engine in which agents collaborate. Consequently, we can train the agents in a multi-agent system to acquire sector-specific knowledge (CASKDT), facility topologies, terrestrial and aerial views in the surrounding of a facility, and embed it in a robot. Such a robot can be specialized to navigate hazardous terrains and adapt to different attack scenarios. We believe that robot systems approach can be strengthened by automatic learning techniques [17] and "deep learning algorithms [5]".

We also recognize that infrastructure vulnerability and resilience are related to community resilience. Hence it is necessary to include social, organizational, economic, and environment factors in the CI model. However, none of the methodologies practiced today have incorporated the social aspects into the analysis of CI vulnerability. However, there is a great interdependency between humans and CI components in modern times. Bea et al. [4] argue that no system, regardless of how technical or physical it is, is not solely physical and technical. According to their report 80 % of the failure in the Oil and Gas production platform in the North Sea was firmly rooted in human, organizational, and institutional failures. Human factors related to reviews and decision making in CI monitoring and maintenance operations have also contributed to malfunctions in the engineering and control aspects of the system [11]. In a recent monograph, Lazari [16] has discussed the human factor in CI protection, and emphatically states that "the relevance of the human factor in the field of CIP cannot be hidden or underestimated". We reckon that robots, embedding of multi-agent systems, may be given initial models of social, and human aspects related to CI protection, and be trained to combine them with CI model through "deep learning techniques". They might produce unpredictable and surprisingly genuine solutions, after playing against its adversary repeatedly and improving its strategy in every step of their learning cycle.

Acknowledgement. We thank the reviewers for their insightful comments that helped us to improve the paper to the current form.

References

1. Alcaraz, C., Lopez, J.: Wide-area situational awareness for critical infrastructure protection. IEEE Comput. **46**, 30–37 (2013)
2. Arboleda, C.A., Abraham, D.M., Richard, J.P., Lublitz, R.: Vulnerability assessment of healthcare facilities during disaster events. J. Infrastruct. Syst. **15**(3), 149–161 (2009)
3. Baker III., G.H.: A vulnerability assessment methodology for critical infrastructure facilities. In: DHS Symposium, R&D Partnerships in Homeland Security, USA, pp. 1–15 (2005)
4. Bea, R., Mitroff, I., Farber, D., Foster, H., Roberts, K.H.: A new approach to risk: the implications of E3. Risk Manage. **11**, 30–43 (2009)

5. Bengio, Y.: Learning deep architectures for AI. Found. Trends Mach. Learn. **2**(1), 1–27 (2009)
6. Canadian Government Report. Critical Infrastructure. http://www.publicsafety.gc.ca/cnt/ ntnl-scrt/crtcl-nfrstrctr/index-en.aspx
7. Eusgeld, J., Kroger, W., Sansavini, G., Schläpfer, M., Zio, E.: The role of network theory and object-oriented modeling within a framework for the vulnerability analysis of critical infrastructures. Reliab. Eng. Syst. Saf. **94**(5), 954–963 (2009)
8. Filippini, R.: Mastering complexity and risks in modern infrastructures: a paradigm shift. GRF Davos Planet@Risk **3**(2), October 2015
9. Giannopoulos, G., Filippini, R., Schimmer, M.: Risk assessment methodologies for Critical Infrastructure Protection. Part I: A state of the art. In: European Commission Joint Research Centre, Institute for the Protection and Security of the Citizen (2012)
10. The Hacker News. http://thehackernews.com/search/label/car
11. Hellstrom, T.: Critical infrastructure and systemic vulnerability: towards a planning framework. Saf. Sci. **45**(3), 415–430 (2007)
12. Homeland Security. What is Critical Infrastructure? http://www.dhs.gov/what-critical-infrastructure
13. Ibanez, E., Gkritza, K., McCalley, J., Aliprantis, D., Brown, R., Somani, A., Wang, L.: Interdependencies between energy and transportation systems for national long term planning. In: Gopalakrishnan, K., Peeta, S. (eds.) Sustainable & Resilient Critical Infrastructure Symposium. Springer-Verlag, Heidelberg (2010)
14. Jamshidi, M. (ed.): Systems of Systems Engineering. Innovations for the 21st Century. John Wiley and Sons, New York (2008)
15. Kotzanikolaou, P., Theoharidou, M., Gritzalis, D.: Interdependencies between critical infrastructures: analyzing the risk of cascading effects. In: Bologna, S., Hämmerli, B., Gritzalis, D., Wolthusen, S. (eds.) CRITIS 2011. LNCS, vol. 6983, pp. 104–115. Springer, Heidelberg (2013)
16. Lazari, A.: European Critical Infrastructure. Springer Publications. ISBN 978-3-319-07496-2 ISBN 978-3-319-07497-9 (eBook) (2014)
17. Patrascu, A.: Cyber protection of critical infrastructures using supervised learning. In: 20th International Conference on Control Systems and Computer Science, Bucharest, May 2015, pp. 461–468 (2015)
18. Pederson, P., Dudenhoeffer, D., Hartley, S., Permann, M.: Critical Infrastructure Interdependeny Modeling. A Survey of U.S. and International Research, INL, INL/EXT-06-11464 (2006)
19. Peters, J.C., Kumar, A., Zheng, H., Agarwal, S., Peeta, S.: Integrated Framework to Capture the Interdependencies between Transportation and Energey Sectors due to Policy Decisions. USDOT Region V Regional University Transportation Center Final Project, NEXTRANS Project No. 079PY04, Purdue University, West Lafayette, IN, USA (2014)
20. The President's National Security Telecommunications Advisory Committee. Information Assurance Task Force for Power Risk Assessment (1997). http://www.securitymanagement. com/library/iatf.html
21. Quyang, M., Hong, L., Maoa, Z., Yua, M., Qi, F.: A methodological approach to analyze vulnerability of interdependent infrastructures. Simul. Model. Pract. Theory **17**(5), 817–828 (2009)
22. Rice, M., Shenoim, S. (eds.): Critical Infrastructure Protection IX, IFIPAICT 466 (2015)
23. Rosato, V., Issacharoff, L., Tiriticco, F., Meloni, S., De Porcellinis, S., Setola, R.: Modelling interdependent infrastructures using interacting dynamical models. Int. J. Critical Infrastruct. **4**(1–2), 63–79 (2008)

24. SANS Institute: Can Hackers turn Lights Off? The Vulnerability of the US Power Grid to Electronic Attack. Technical Report, SANS Institute (The Information Security Reading Room) (2001)
25. Stapelberg, R.F.: Infrastructure Systems Interdependencies and Risk Information Decision Making (RIDM). Syst. Cybern. Inform. **6**(5), 21–27 (2008)
26. Stergiopoulos, G., Kotzanikolaou, P., Theocharidou, M., Lykou, G., Gritzalis, D.: Time-based critical infrastructure dependency analysis for large-scale and cross-sectorial failures. Int. J. Critical Infrastructures Protection **12**, 46–60 (2016)

Improved Binary Imperialist Competition Algorithm for Feature Selection from Gene Expression Data

Aorigele[1], Shuaiqun Wang[2(✉)], Zheng Tang[1], Shangce Gao[1], and Yuki Todo[3]

[1] Faculty of Engineering, University of Toyama, Toyama, Japan
[2] Information Engineering College,
Shanghai Maritime University, Shanghai, China
wangsq@shmtu.edu.cn
[3] School of Electrical and Computer Engineering,
Kanazawa University, Kanazawa, Japan

Abstract. Gene expression profiles which represent the state of a cell at a molecular level could likely be important in the progress of classification platforms and proficient cancer diagnoses. In this paper, we attempt to apply imperialist competition algorithm (ICA) with parallel computation and faster convergence speed to select the least number of informative genes. However, ICA same as the other evolutionary algorithms is easy to fall into local optimum. In order to avoid the defect, we propose an improved binary ICA (IBICA) with the idea that the local best city (imperialist) in an empire is reset to the zero position when its fitness value does not change after five consecutive iterations. Then IBICA is empirically applied to a suite of well-known benchmark gene expression datasets. Experimental results show that the classification accuracy and the number of selected genes are superior to other previous related works.

Keywords: Gene expression data · Imperialist competitive algorithm · Feature selection · Local optimum

1 Introduction

DNA microarray technology which can measure the expression levels of thousands of genes simultaneously in the field of biological tissues and produce databases of cancer based on gene expression data [1] has great potential on cancer research. Because the conventional diagnosis method for cancer is inaccurate, gene expression data has been widely used to identify cancer biomarkers closely associated with cancer, which could be strongly complementary for the traditional histopathologic evaluation to increase the accuracy of cancer diagnosis and classification [2] and improve understanding the pathogenesis of cancer for the discovery of new therapy. Therefore, It has been gained popularity that the application of gene expression data on cancer classification, diagnosis and treatment.

Due to the high dimensionality of gene expression data compared to the small number of sample, noisy genes and irrelevant genes, the conventional classification

© Springer International Publishing Switzerland 2016
D.-S. Huang et al. (Eds.): ICIC 2016, Part III, LNAI 9773, pp. 67–78, 2016.
DOI: 10.1007/978-3-319-42297-8_7

methods cannot be effectively applied to gene classification due to the poor classification accuracy. With the inherent property of gene data, efficient algorithms are needed to solve this problem in reasonable computational time. Therefore, many supervised machine learning algorithms such as, bayesian networks, neural net-works and support vector machines (SVMs), combined with feature selection techniques, have been used to process the gene data [3]. Gene selection is the process of selecting the smallest subset of informative genes that are most predictive to its relative class using a classification model. The objective of feature selection problems is maximizing the classifier ability and minimizing the gene subsets to classify samples. The optimal feature selection problem from gene data has been shown NP-hard problem. Hence, it is more effective to use meta-heuristics approaches, such as nature inspired evolutionary algorithm, to solve this problem. In recent years, meta-heuristic algorithms based on global search strategy rather than local search strategy have shown their advantages in solving combinatorial optimization problems, and a number of meta-heuristic approaches have been proposed for feature selection, for example, genetic algorithm (GA), particle swarm optimization (PSO), tabu search (TS) and artificial bee colony (ABC).

Meta-heuristic algorithms as a kind of random search technique, it can't guarantee to find the optimal solution every time. Due to a single meta-heuristic algorithm often trapped into an immature solution, the recent trend of research has been shifted towards the several hybrid methods. Kabiret et al. [4] introduced a new hybrid genetic algorithm incorporated a local search to fine-tune the search for feature selection. Shen et al. [5] presented a hybrid PSO and tabu search for feature selection. However, the experiment results were less meaning due to tabu approaches unable to search for an optimal solution or a near-optimal solution. Next, Li et al. proposed a hybrid of PSO and GA [6]. Unfortunately, the experiment results don't obtain high classification accuracy. Alshamlan et al. brought out an idea of ABC to solve feature selection. He first attempted at applying ABC algorithm in analyzing microarray gene expression combined with minimum redundancy maximum relevance (mRMR) [7]. Then, he also hybridized ABC and GA algorithm to select genetic feature for cancer classification and the goal was to integrate the advantages of both algorithm [8]. The result obtained by ABC algorithm was improved to someextent but the small number of genes can't get the high accuracy. Chuang et al. [9] introduced an improved binary PSO which the global best particle is reset to zero position when its fitness values do not change after three consecutive iterations. With a large number of selected genes, the result of the proposed algorithm obtained 100 % classification accuracy in many datasets.

This paper adopts a novel optimization algorithm named imperialist competitive algorithm (ICA) to address the process of feature selection using gene expression data. ICA is a population-based evolutionary algorithm inspired by imperialistic competition. It starts with an initial population and effectively searches the solution space through some specially designed operators to converge to optimal or near-optimal solutions. ICA same as the other evolution algorithm, is easy to trap into local optimal. In order to track this problem, an improved binary ICA (IBICA) that retires the imperialist when trapped in a local optimum is proposed. By resetting local imperialist, it can avoid IBICA getting trapped into a local optimum and superior classification result can be achieved with a reduced number of selected genes. The superiority of

IBICA over some existing approaches developed for selecting feature is validated through extensive experiments using benchmark database taken from literature. Computational results show that IBICA is very effective and efficient in solving feature selection problem with high dimensionality and small sample.

The rest of the paper is organized as follows: Sect. 2 elaborates proposed IBICA included individual representation, empire initialization, colonies assimilation, fitness function evaluation and the framework of IBICA. Section 3 describes the parameter setting and the experiment result based on several benchmark gene datasets including comparative results between IBICA and other variants of PSO. Finally, concluding remarks are presented in Sect. 4.

2 Proposed IBICA

ICA is a population-based stochastic optimization technique, which was proposed by Atashpaz-Gargari and Lucas [10]. ICA as one of the recent meta-heuristic optimization methods, is inspired by socio-political behavior and a multi-agent algorithm in which each agent is a country. The empire is composed of the countries that would be either an imperialist or a colony. Assimilation of colonies by their related imperialist and imperialist competition among the empires form the basis of the ICA. The powerful imperialists are reinforced and the weak ones are weakened, gradually collapsed during these movements. Finally, the algorithm converges to the optimal solution. ICA has been successfully applied in many areas: fuzzy system control, function optimization, artificial neural network training, and other application problems.

ICA was originally introduced as an optimization technique for real-number problem. However, many optimization spaces are discrete in which variables and levels of variables have qualitative distinctions. In this paper, feature selection for gene expression data is a discrete optimization problem. So moving the colonies of an empire toward their imperialist may produce solutions which contain the same genes. So the steps of solution representation, colonies assimilation and fitness evaluation are the main differences between original ICA and our proposed IBICA.

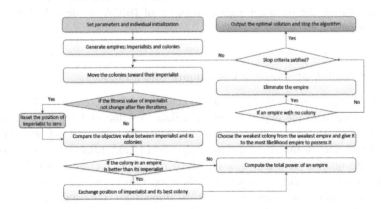

Fig. 1. Flowchart of the proposed algorithm IBICA.

In order to overcome the weakness of premature convergence for ICA, in this paper, our proposed IBICA combining jumping out of local optimal strategy assists the imperialist to explore more potential region. The flowchart of IBICA can be shown in Fig. 1, which are described further as follows:

Step 1: Set parameters of the algorithm and initialize countries with binary representation 0 and 1. Evaluate each country in the population which utilize support vector machine classifier (SVM). The fitness is decided by the percentage classification accuracy of SVM and the number of feature subsets. Then empires are generated depending on their fitness values.

Step 2: Apply a learning mechanism on colonies which is the same as baldwinian learning (BL) mechanism [11]. Find out the different genes between imperialist and one colony, then use the predefined learning probability to decide the number of selected genes to colony. This strategy makes the colonies move toward their imperialist.

Step 3: If the fitness value of imperialist in an empire does not change after five iterations, it can be considered stuck at a local optimum and reset the position of imperialist to zero.

Step 4: Compare the objective value between imperialist and its colonies in the same empire. Exchange the position of imperialist and its colony when the colony is better than its imperialist.

Step 5: Computer the total power of an empire and compete among all the empires, then eliminate the weakest empire when it loses all of its colonies.

Step 6: If the termination condition (the predefined max iterations) is not fulfilled, go back to step 2. Otherwise, output the optimal solution in the current population and stop the algorithm.

It is clear that IBICA integrates two quite different strategies in ICA for feature selection, i.e., the operation on jumping out of local optimal and colonies assimilation. The fitness function, combing the performance of SVM with the number of feature genes, assists IBICA to find the most salient features with less redundancy of information. A reliable selection method for genes relevant for sample classification should be have higher classification accuracy and contain less redundant genes. For more comprehensibility, details about each component of IBICA are described in the following sections.

2.1   Individual Representation and Empire Initialization

In this study, we utilize random approach to generate a string composed of 0 and 1 and its length is equal to the dimension of gene data. A value of 1 represents this gene (bit) should be selected in individual (country) while the value of 0 indicates this gene has not been selected. Initialized country $X_{country} = \{1,1,0,0,1,0,1,0,0,1\}$ is encoded by 0 and 1, the number of 0 and 1 is 10 which represents the dimensions of optimal problem. In order to clearly understand these operations, we take an example for explanation. Assume the gene data with 10 dimensions (10 characters) represents $f_1, f_2, f_3, f_4, f_5, f_6, f_7, f_8, f_9$ and f_{10}. A string is randomly generated and the number of 1 is 5.

Hence, the features f_1, f_2, f_5, f_7 and f_{10} are selected to generate a country which is shown in Fig. 2.

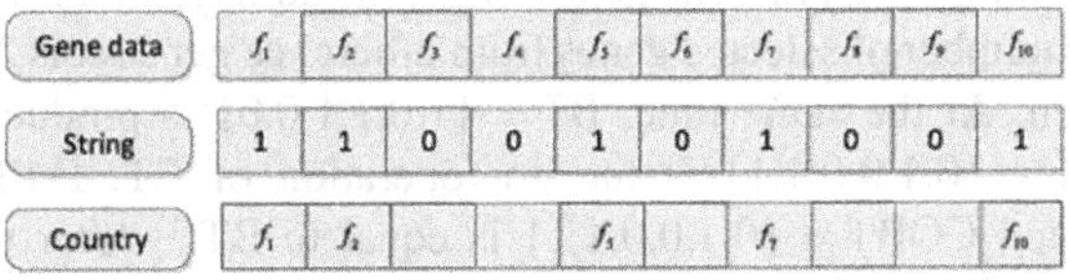

Gene data	f_1	f_2	f_3	f_4	f_5	f_6	f_7	f_8	f_9	f_{10}
String	1	1	0	0	1	0	1	0	0	1
Country	f_1	f_2			f_5		f_7			f_{10}

Fig. 2. An illustrated example with generated subset and individual representation.

2.2 Colonies Assimilation

In IBICA, assimilation is an important operation and could likely be a momentous help in the progress of colonies evolutionary. In this paper, the idea of continuous BL is introduced into the IBICA for colonies assimilation by their imperialist. This strategy can utilize some specific differential information from the imperialist, i.e. by the differential information between imperialist and colony $(X_{IM} - X_{CO})$, indicating that a more effective way to learn from the excellent solution. It can be defined as follows:

$$X_{CO} = X_{CO} + \lfloor \beta \cdot (X_{IM} - X_{CO}) \rfloor$$

the operation of difference states that 1 subtracting 0, the result is 1; 1 subtracting 1, the result is 0; 0 subtracting 1, the result is 0; 0 subtracting 0, the result is 0. The learning rate $\beta \in (0,1)$ is a randomly generated the proportion of selected genes from the differences. $\lfloor \cdot \rfloor$ is the operator that rounds its argument towards the closest integer and $\lfloor \beta \cdot (X_{IM} - X_{CO}) \rfloor$ represents the selected genes. In order to reduce the dimension of the country, our research adopts a randomly generated template depends on the larger dimension of imperialist and colonies. Imperialist (IM) with five characteristics $f_1, f_2, f_5, f_7, f_{10}$ and one of its colonies with six characteristics $f_1, f_3, f_6, f_8, f_9, f_{10}$ in an empire, the dimension of binary template (BT) is 6. The template of colony is generated from not operation of BT, denoted BTF. Because the number of IM feature genes is less than the template BT, the IMBT just takes one part of BT. In order to describe how it

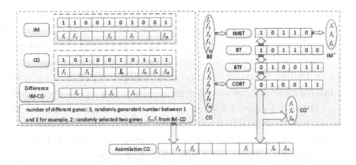

Fig. 3. A colony is assimilated by an imperialist.

works, we will illustrate the following numerical example. Imperialist represents $X_{IM} = \{1,1,0,0,1,0,1,0,0,1\}$, one of its colonies encodes with $X_{CO} = \{1,0,1,0,0,1,0,1,1,1\}$, then the differential information is described as $X_{IM} - X_{CO} = \{0,1,0,0,1,0,1,0,0,0\}$. It is obvious that the number of different genes is 3 (the number of 1). According to the parameter (β), the number of selected genes from difference gene set is 2 and the features of f_2, f_7 are chosen. At the same time, BT = $\{1,0,1,1,0,0\}$ is produced by randomly strategy and BTF = $\{0,1,0,0,1,1\}$ is the not operation of BT. IMBT = $\{1,0,1,1,0\}$ comes from BT and COBT = $\{0,1,0,0,1,1\}$ is equal to BTF. After this process, CO becomes a country with the feature f_3, f_9, f_{10} and the assimilated CO combined different genes between CO and IM, with five features $f_2, f_3, f_7, f_9, f_{10}$ is produced. Therefore, assimilated colony is generated by BL operation which is shown in Fig. 3.

2.3   Fitness Function

The feature selection of gene expression data need to consider the classification accuracy and the number of extraction informative genes. Hence, the fitness function is defined as follows:

$$fitness(X_i) = w_1 \cdot A(X_i) + (w_2 \cdot (n - D(X_i))) / n$$

in which $A(X_i) \in [0, 1]$ is the leave-one-out-cross-validation (LOOCV) classification accuracy that utilizes one country X_i (gene subset) and obtained by SVM model. n is the dimension of optimal problem, in other words, is the total number of genes for each sample and $D(X_i)$ is the number of selected genes in X_i. We use the parameters w_1 and w_2 to measure the importance of classification accuracy and the number of selected genes respectively. The accuracy is more crucial than the number of selected genes and setting of the parameter satisfies constraint condition as follows: $w_1 \in [0, 1]$, $w_2 = 1 - w_1$.

2.4   Improved ICA

Gene expression data with high dimensionality and small sample, so we expect accurate classification results in different gene datasets. In GICA, the colonies are assimilated by their imperialist to search solution space. This operation can increase the speed of convergence and avoid getting trapped into a local optimum by fine-tuning the inertia weight. However, if the imperialist itself is trapped into a local optimum, the search space of corresponding empire will occur in the same area and prevent optimal results of classification. Hence, in this paper, we utilize a method that resets premature imperialist to zero embed in IBICA to avoid the algorithm being trapped in a local optimal and enhance the classification accuracy with a reduced number of selected genes.

Figure 4 (a) shows an empire converged to a region after a certain iteration. Solid dark circle represents the imperialist and hollow white circles represent the colonies in the same empire. If the imperialist does not change after five iterations, it can be regarded as trapped into local optimal. Under this circumstances, this imperialist is reset to zero which means should be reinitialized, shown in Fig. 4(b). Imperialist will skip the local optimum and the superior classification results with small number of genes can be obtained by such operation. Figure 4(c) shows that the colonies will leave the local optimum by assimilated and converge towards the reset imperialist. Figure 4 (d) shows the empire will explore for new region with lower number of genes. By utilizing this idea, IBICA can effectively avoid the local optimum and achieve excellent result including higher classification accuracy and lower number of genes.

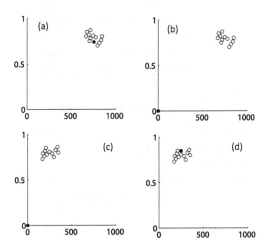

Fig. 4. (a) One imperialist is trapped in a local. (b) Imperialist is reset to zero. (c) Colonies-movement after resetting of their imperialist. (d) Empire congregated towards the updated imperialist value, improving the individual position.

3   Experiment

3.1   Gene Datasets and Parameter Setting

The gene expression datasets are listed in Table 1. The second column and the third column list the name of gene data sets and the sample size in each data set, the fourth column and the fifth column give the dimension and the class size of datasets, respectively. These data sets contained binary-classes and multi-classes data own thousands of genes. They can downloaded from http://www.gems-system.org.

The parameter values for IBICA are shown in Table 2. It is very clear that the parameters of the proposed algorithm is less than binary particle swarm optimization (BPSO). Hence, the influence of parameter setting for IBICA is relatively small and the robustness of algorithm is better.

The size of the population affects the performance of the algorithm and computation efficiency. Large number of countries would require more computational times for completing feature selection while if the number is too small, although the algorithm can take place in a relatively short period of time, the performance of the algorithm is not guaranteed. Therefore, the intermediate vales for the size of population and iteration are chosen 15 and 50, respectively. Due to population composed of imperialists and

Table 1. Description of gene expression datasets

Dataset number	Dataset number	Number of		
		samples	genes	classes
1	Brain_Tumors1	90	5920	5
2	Brain_Tumors2	50	10367	4
3	Leukemia1	72	5327	3
4	Leukemia2	72	11225	3
5	Lung_Cancer	203	12600	5
6	SRBCT	83	2308	4
7	Prostate_Tumor	102	10509	2
8	DLBCL	77	5469	2

Table 2. Parameter settings for IBICA.

Parameters	Values
The number of country	15
The number of imperialist	4
The number of colonies	11
The number of iteration (generation)	50
w_1	0.8
w_2	0.2

colonies, so the number of imperialists also need to determine. If the number of imperialist is 1, the IBICA is transformed single population evolutionary algorithm instead of multi-subpopulation while if the number of imperialist is too large, the number of colonies cannot be guaranteed. The number of imperialist is chosen 4 in our experiment. In Sect. 2.3, the parameter of w_1 and w_2 are introduced and given the range of value. In order to guarantee w_1 larger than w_2, the values of w_1 and w_2 is setting 0.8 and 0.2 in our proposed algorithm with the same parameter setting of EPSO [12].

3.2 Experiment Results

The results obtained by IBICA based on average values and the standard deviations over 10 independent runs for 8 data sets are shown in Tables 3 and 4. It is found that the classification accuracy of IBICA achieves 100 % with less than 10 informative genes for Leukemia1, Leukemia2 and DLBCL while with less than 20 selected genes for SRBCT. The average classification accuracy is more than 92.22 % for all the datasets. In other word, it is strongly demonstrated that IBICA can select informative genes efficiently from high-dimension, binary classes or multi-classes data for classi-fication. For all best classification result, the selected genes is less than 10 while for the average classification result, the informative genes in subset is also less than 10 except for SRBCT data set. Furthermore, the standard deviations is less than 5 for all data sets. From the classification accuracy and the selected informative genes, it is not difficult to find that IBICA is an efficient algorithm for feature selection and produces a near-optimal gene subset from gene expression data.

Table 3. The computational results obtained by our proposed algorithms IBICA for each run on Leukemia1, Leukemia2, SRBCT and DLBCL datasets.

Runs	Leukemia1		Leukemia2		SRBCT		DLBCL	
	Acc (%)	Selected genes	Acc (%)	Selected genes	Acc (%)	Selected genes	Acc (%)	Selected genes
1	100	3	100	9	100	11	100	5
2	100	4	100	13	100	15	100	3
3	100	3	100	7	100	13	100	4
4	100	3	100	7	100	12	100	4
5	100	4	100	8	100	9	100	4
6	100	3	100	7	100	17	100	4
7	100	3	100	5	100	18	100	5
8	100	3	100	7	100	12	100	4
9	100	3	100	8	100	9	100	4
10	100	4	100	6	100	11	100	6
Ave.	100	3.30	100	7.70	100	12.70	100	4.30
Std.	0	0.90	0	2.16	0	3.09	0	0.82

Table 4. The computational results obtained by our proposed algorithms IBICA for each run on Prostate_Tumor, Lung_Cancer, Brain_Tumor1 and Brain_Tumor2 datasets.

Runs	Prostate_Tumor		Lung_Cancer		Brain_Tumor1		Brain_Tumor2	
	Acc (%)	Selected genes	Acc (%)	Selected genes	Acc (%)	Selected genes	Acc (%)	Selected genes
1	98.04	5	96.06	9	93.33	9	94.00	5
2	98.04	6	95.57	14	92.22	15	90.00	7
3	98.04	8	96.06	8	93.33	8	92.00	5
4	97.06	7	95.57	11	91.11	12	92.00	6
5	97.06	8	96.06	6	93.33	9	92.00	4
6	98.04	7	96.06	8	92.22	11	90.00	7
7	97.06	9	95.57	9	91.11	9	94.00	3
8	98.04	10	95.57	7	92.22	7	92.00	4
9	97.06	8	96.06	8	93.33	8	92.00	8
10	98.04	6	95.57	8	93.33	6	94.00	8
Ave.	97.65	7.40	95.82	8.80	92.55	9.40	92.20	5.70
Std.		1.51	0.26	2.25	0.91	2.63	1.48	1.77

In order to verify the effectiveness of the algorithm, we will compare IBICA with other optimization algorithms on several benchmark classification datasets. Table 5 compares experiment results obtained by other approaches from the literature and the proposed method IBICA. Various method included Non-SVM and MC-SVM were used to compare our proposed method. The experiment results listed in Table 5 were taken from Chuang et al. [9] for comparison. Non-SVM contain the K-nearest neighbor method [9, 13, 14], back propagation neural networks and probabilistic neural networks. MC-SVM include one-versus-one and one-versus-rest [15], DAGSVM [16, 17] and the method by Crammer and Singer [18]. The average highest classification accuracy of KNN, OVR and IBICA is 97.14, 93.63 and 97.81, respectively. It is obvious that Chuang [9] proposed IBPSO with the higher accuracy among all algorithms except for our proposed algorithm IBICA.

Table 5. Classification accuracies of gene expression data obtained via different classification methods.

Datasets	Methods			IBICA			
	Non-SVM			Mc-SVM			SVM
	KNN	NN	PNN	OVR	DAG	CS	OVR
Brain_Tumor1	94.44	84.72	79.61	91.67	90.56	90.56	94.44
Brain_Tumor2	94.00	60.33	62.83	77.00	77.83	72.83	94
Leukemia1	100	76.61	85.00	97.50	96.07	97.50	100
Leukemia2	100	91.03	83.21	97.32	95.89	95.89	100
Lung_Cancer	96.55	87.80	85.66	96.05	95.59	96.55	96.06
SRBCT	100	91.03	79.50	100	100	100	100
Prostate_Tumor	92.16	79.18	79.18	92.00	92.00	92.00	98.04
DLBCL	100	89.64	80.89	97.50	97.50	97.50	100
Average	97.14	82.54	79.49	93.63	93.22	92.85	97.81

Table 6. The number of selected genes from data sets between IBICA and IBPSO.

Datasets	IBICA		IBPSO	
	Genes	Percentage of genes	Genes	Percentage of genes
Brain_Tumor1	6	0.001	754	0.13
Brain_Tumor2	3	0.0003	1197	0.12
Leukemia1	3	0.0006	1034	0.19
Leukemia2	5	0.0004	1292	0.12
Lung_Cancer	6	0.0005	1897	0.15
SRBCT	9	0.004	431	0.19
Prostate Tumor	5	0.0005	1294	0.12
DLBCL	3	0.0005	1042	0.19
Average	5	0.00097	1117.6	0.15

Our proposed approach IBICA obtained seven of the highest classification accuracies for the eight benchmark datasets, i.e., for Brain_Tumor, Brain_Tumor2, Leukemia1, Leukemia2, SRBCT, Prostate_Tumor and DLBCL except for Lung_-Cancer. For the datasets of Leukemia1, Leukemia2, SRBCT and DLBCL, the classification accuracy can reach 100 %. However, the average classification accuracy of

IBICA and IBPSO seems the same while the selected genes of IBICA is significantly less than IBPSO listed in Table 6.

The convergence graphs of the best and average classification accuracy obtained by IBICA for SRBCT and DLBCL are shown in Fig. 5. It can be seen that the best classification accuracy is achieved to 100 % less than 10 iterations for SRBCT and between 10 and 20 generations for DLBCL. Therefore, IBICA possesses a faster convergence speed and achieves the optimal solution rapidly.

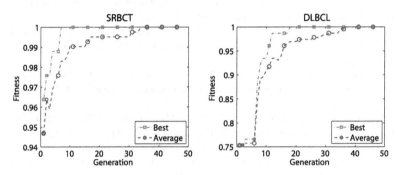

Fig. 5. The convergence graphs of the best and average accuracy classification by IBICA algorithm on SRBCT and DLBCL data sets.

4 Conclusions

In this paper, improved binary imperialist competition algorithm (IBICA) is used to implement feature selection and SVM with one-versus-rest serves as an evaluator of IBICA for gene expression data classification problems. This work attempts to apply ICA for analyzing gene expression data which is crucial for researchers. The classification accuracy obtained by the proposed method IBICA was the highest seven out of the 8 gene expression data test problems. In general, the classification performance of IBICA as good as IBPSO, however, IBICA is superior to IBPSO and other methods in terms of selected genes. In our proposed algorithm, in order to avoid imperialist premature convergence, a jumping out of local optimum strategy is embed in IBICA while a discrete BL mechanism mapping to colonies assimilation operator speeds the algorithm convergence and assists the colonies evolutionary. Experimental results show that our method effectively classify the samples with reduced feature genes. In the future work, imperialist competition algorithm combined other intelligent strategy can used to select informative genes.

References

1. Feilotter, H.: A biologist's guide to analysis of DNA microarray data. Am. J. Hum. Genet. **71**(6), 1483–1484 (2002)
2. Simon, R.: Analysis of DNA microarray expression data. Best Pract. Res. Clin. Haematol. **22**(2), 271–282 (2009)

3. Alshamlan, H., Badr, G., Alohali, Y.: A comparative study of cancer classification methods using microarray gene expression profile. In: Lecture Notes in Electrical Engineering, vol. 285, pp. 389–398 (2014)

4. Kabir, M.M., Shahjahan, M., Murase, K.: A new local search based hybrid genetic algorithm for feature selection. Neurocomputing. **74**(17), 2914–2928 (2011)

5. Shen, Q., Shi, W.M., Kong, W.: Hybrid particle swarm optimization and tabu search approach for selecting genes for tumor classification using gene expression data. Comput. Biol. Chem. **24**, 213–222 (2008)

6. Li, S., Wu, X., Tan, M.: Gene selection using hybrid particle swarm optimization and genetic algorithm. Soft. Comput. **12**(11), 1039–1048 (2008)

7. Alshamlan, H., Badr, G., Alohali, Y.A.: mRMR-ABC: a hybrid gene selection algorithm for cancer classification using microarray gene expression profiling. Biomed Res. Int. **4**, 1–15 (2015)

8. Alshamlan, H.M., Badr, G.H., Alohali, Y.: Genetic Bee Colony (GBC) algorithm: a new gene selection method for microarray cancer classification. Comput. Biol. Chem. **56**, 49–60 (2015)

9. Chuang, L.Y., Chang, H.W., Tu, C.J., Yang, C.H.: Improved binary PSO for feature selection using gene expression data. Comput. Biol. Chem. **32**(1), 29–38 (2008)

10. Atashpaz-Gargari, E., Lucas, C.: Imperialist competitive algorithm: an algorithm for optimization inspired by imperialistic competition. In: IEEE Congress on Evolutionary Computation, CEC 2007, vol. 7, pp. 4661–4667 (2007)

11. Gong, M.G., Jiao, L.C., Zhang, L.N.: Baldwinian learning in clonal selection algorithm for optimization. Inf. Sci. **180**(8), 1218–1236 (2010)

12. Mohamad, M.S., Omatu, S., Deris, S., Yoshioka, M.: An enhancement of binary particle swarm optimization for gene selection in classifying cancer classes. Algorithms Mol. Biol. **8** (4), 1584–1618 (2013)

13. Dasarathy, B.V.: NN concepts and techniques: nearest neighbours (NN) norms. IEEE Computer Society Press, pp. 1–30 (1991)

14. Cover, T., Hart, P.: Nearest neighbor pattern classification. IEEE Trans. Inf. Theor. **13**(1), 21–27 (1967)

15. Orsenigo, C., Vercellis, C.: Multicategory classification via discrete support vector machines. Comput. Manage. Sci. **6**(1), 101–114 (2009)

16. Platt, J.C., Cristianini, N., Shawe-Taylor, J.: Large margin dags for multiclass classification. Adv. Neural Inf. Process. Syst. **12**(3), 547–553 (2010)

17. Hsu, C.W., Lin, C.J.: Errata to "a comparison of methods for multiclass support vector machines". IEEE Trans. Neural Netw. **13**(4), 415–425 (2002)

18. Crammer, K., Singer, Y.: On the learnability and design of output codes for multiclass problems. Mach. Learn. **47**(2–3), 201–233 (2001)

A Novel Multi-objective Bionic Algorithm Based on Plant Root System Growth Mechanism

Lianbo Ma[1(✉)], Xu Li[2], Jia Liu[3], and Yang Gao[4]

[1] College of Software, Northeastern University, Shenyang 110819, China
mlb_vip@hotmail.com
[2] Benedictine University, Lisle, IL, USA
[3] College of Management, Shenzhen University, Shenzhen 518060, China
1191610412@qq.com
[4] Academy of Information Technology, Northeastern University,
Shenyang 110819, China

Abstract. This paper proposes and develops a novel multi-objective optimization scheme called MORSGO based on iterative adaptation of plant root growth behaviors. In MORSGO, the basic local and global search operators are designed deliberately based on auxin-regulated tropism of the natural root system, including branching, regrowing of different types of roots. The fast non-dominated sorting approach is employed to get priority of non-dominated solutions obtained during the search process, and the diversity over archived individuals is maintained by using dynamical crowded distance estimation strategy. Accordingly, Pareto-optimal solutions obtained by MORSGO have merits of better diversity and lower computation cost. The proposed MORSGO is evaluated on a set of bio-objective and tri-objective test functions taken from the ZDT benchmarks in terms of two commonly used metrics IGD and SPREAD, and it is compared with NSGA-II and MOEA/D. Test results verify the superiority and effectiveness of the proposed algorithm.

Keywords: Multi-objective algorithm · Plant root system · Optimization

1 Introduction

Many real-world optimization problems often involve simultaneously optimizing over two or more mutually conflicting objective functions with some uncertain constrains instead of transforming them into a single objective, and accordingly they are generally named as multi-objective optimization problems (MOPs) [1, 2]. These MOPs are more difficult to be handled than the single-objective issues due to the fact that decision makers would find a set of Pareto-optimal solutions (PS) or non-dominated solutions with a trade-off among objectives [3]. Inspired by different backgrounds, an increasing number of multi-objective evolutionary algorithms (EAs) or swarm intelligence (SI) algorithms have been proposed and extended to tackle the MOPs, prominent examples being non-dominated sorting genetic algorithm II (NSGA-II) [2], multi-objective evolutionary algorithm based on decomposition (MOEA/D) [3], strength Pareto evolutionary

© Springer International Publishing Switzerland 2016
D.-S. Huang et al. (Eds.): ICIC 2016, Part III, LNAI 9773, pp. 79–87, 2016.
DOI: 10.1007/978-3-319-42297-8_8

algorithm (SPEA2) [4], and multi-objective particle swarm optimization (MOPSO) [5]. However, how to improve the diversity of population or overcome the local convergence of algorithms is still a challenging issue in MO optimization [6, 7]. In this paper a new multi-objective bionic algorithm is proposed derived from plant root growth behaviors and optimal foraging, namely multi-objective root system growth optimizer (MORSGO), which adopts the branching, regrowing, mortality and tropism operations of the root system. Intuitively, the basic search technique of MORSGO can acquire appropriate balance of local search and global search and dynamical variance of population size [6, 7]. Generally, this elite retention strategy can effectively eliminate the infeasible individuals and keep feasible individuals throughout the search process [2]. Accordingly, Pareto-optimal solutions obtained by MORSGO have better diversity and lower computation cost.

2 Multi-objective Root System Growth Optimizer

2.1 Root System Growth Model (RSGO)

2.1.1 Auxin Information

Assume that the total sum of auxin concentration F_i is defined as 1 in the root system, and each root's F_i value can be calculated as:

Assume that the total sum of auxin concentration F_i is defined as 1 in the root system, and each root's F_i value can be calculated as:

$$f_i = \frac{fit_i - f_{worst}}{f_{best} - f_{worst}} \tag{1}$$

$$N_i = \frac{Nutition_i - Nutition_{worst}}{Nutition_{best} - Nutition_{worst}} \tag{2}$$

$$F_i = \frac{f_i}{\sum\limits_{j=1}^{N} f_i} * \xi + \frac{N_i}{\sum\limits_{j=1}^{N} N_i} (1 - \xi), \ \xi \subset (0, 1) \tag{3}$$

where i is the position of the growing point, ξ is a uniform random quantity, N is the total number of the points, fit (.) is the fitness value of the point and f_{worst} and f_{best} are the maximum and minimum of the current points, respectively. $Nutition_i$ is the current nutrient concentration of individual i and it can be expressed as:

$$Nutition_i^{t+1} = \begin{cases} Nutition_i^t + 1 & if \ f_i^{t+1} < f_i^t \\ Nutition_i^t - 1 & if \ f_i^{t+1} > f_i^t \end{cases} \tag{4}$$

In each growing cycle, all individuals are sorted by their auxin concentrations in descending order [7, 8]. That is, the strong individuals with higher auxin concentrations can be selected as main roots to branch. In our model, half of current sorted roots are selected as main roots:

$$S_m^t = P^t/2 \tag{5}$$

where S_m^t is the number of selected main roots, P^t is the size of current population.

2.1.2   Mainroots Growth Operations

- **Regrowing**

  Step 1. In each cycle, the group of main roots is constructed by selecting half population sorted according to auxin concentration.
  Step 2. Considering effect of hydrotropism, select half of current main roots to search towards the optimal position of individuals, given by:

  $$x_i^{t+1} = x_i^t + R_3(x_{best}^t - x_i^t) \tag{6}$$

  where $i \subseteq [1, S_m^t/2]$, R_3 is random value in the range $(0, 1)$, and x_{best}^t is the best position in the root tip group.
  Step 3. Considering gravitropism, the rest of main roots will grow along their original directions as:

  $$x_i^{t+1} = x_i^t + R_4 l_{max}(\phi_i^t) \; if \; x_i^t > x_i^{t-1} \tag{7}$$

  where $i \subseteq [S_m^t/2, S_m^t]$, l_{max} is the maximum of root elongation length, R_4 is a normally distributed random number with mean 0 and standard deviation 1; $H(\phi_i^t)$ is a D-dimensional growth direction of the main root i; $\phi_i^t = (\phi_{i1}^t, \phi_{i2}^t, \ldots, \phi_{i(D-1)}^t) \in R^{D-1}$ is a D-1-dimensional growth angle, given by:

  $$\phi_i^{t+1} = \phi_i^t + R_5 * \omega_{max}, \; 0 < \omega_{max} < \pi \tag{8}$$

  where $R_5 \in R^{D-1}$ is a uniformly distributed random sequence in the range $(0, 1)$; ω_{max} is the maximum of growing angle, which is limited to π.

- **Branching**

  Step 1. The nutrient concentration (*Nutrition_i*) of mainroot i is compared with *BranchG* $(0 < BranchG < 1)$ to determine whether it performs branching operator:

$$\begin{cases} branching & \text{if } Nutition_i > BranchG \\ nobranching & \text{otherelse} \end{cases} \tag{9}$$

Step 2. Calculated the new branching points as:

$$X_i^{t+1} = X_i^t + R_1 * D_i(\varphi_i), \tag{10}$$

where X_i^{t+1} is the new growing point from X_i^t, R_1, is the elongate-length unit, which is a random varying from 0 to 1, φ_i is the growth angle (φ_{i1}, φ_{i2}... $\varphi_{i(D-1)}$).

The growth angle φ_i is calculated as follows:

$$\varphi^{t+1} = \varphi^t + R_2\alpha_{init} + K * \beta\max/S_{\max}, \tag{11}$$

where R_2 is a random coefficient varying from 0 to 1, α_{init} is original growth angle of the initial mainroot as zero degree, K is randomly parameter selecting the subzone, S_{max} is subzones number, and $\beta_{\max}$ is the maximum growing turning angle. Empirically, $\beta_{\max}$ is limited to π.

2.1.3 Lateral Roots Growth Operation

At the ith iteration, each lateral root tip generates a random head angle and a random elongation length, given by:

$$\phi_i^{t+1} = \phi_i^t + R_5\phi_{\max} \tag{12}$$

$$x_i^{t+1} = x_i^t + R_6 l_{\max} H(\phi_{\max}^{t+1}) \tag{13}$$

where $i \subseteq [0, S_i^t]$, R_5 and R_6 are random values in the range (0, 1), $\phi_{\max}$ is the maximum growing turning angle, and $l_{\max}$ is the maximum of root elongation length.

2.1.4 Dead-Roots Elimination

In the proposed root foraging model, it is assumed that N_i is the current population size, N_i will increase by one if a root tip splits and reduce by one if a root dies determined by auxin distribution, and it will vary in the searching process [9, 10].

2.2 The MORSGO Algorithm

2.2.1 Fast Non-dominated Sorting

In each computational iteration, two significant operations need to be calculated, namely domination count D_x, and the set of solutions S_x dominated by solution x. D_x donates the number of solutions dominated by x from current population P. Detailed procedures of this process is given in *Algorithm 1*.

Algorithm 1. Fast non-dominated sorting
Step 1: For each $x \in P$, initialize
$\quad\quad\quad D_x = 0;$ $\quad\quad\quad$ //Domination counter for solution x
$\quad\quad\quad S_x = \phi;$ $\quad\quad\quad$ //Set of solutions dominated by x
Step 2: Let $q \in P$. for each q,
$\quad\quad\quad$ if $x \prec q$, let $S_x = S_x \cup \{q\};$ // q is added into the solution set of S_x
$\quad\quad\quad$ else if $x \prec q$, let $D_x = D_x + 1;$ //Domination counter for x is accordingly
added
Step 3: If $D_x = 0,$ $\quad\quad\quad$ //x falls within the first front set
$\quad\quad\quad$ let $x_{rank} = 1$ and $F_1 = F_1 \cup \{x\};$ //x is added into the Pareto front set
Step 4: i=1;
Step 5: Let $Q = \phi$, for each $x \in F_i,$ //Q is defined to memorize the solutions of the
next front
$\quad\quad\quad$ for each $q \in S_x,$
$\quad\quad\quad$ Let $D_x = D_x - 1;$
$\quad\quad\quad$ If $D_x = 0,$ let $x_{rank} = i+1,$ $Q = Q \cup \{x\};$
Step 6: Let i=i+1 and $F_i = Q,$
Step 7: If $F_i \neq \phi$, return 5; Else, stop.

2.2.2   The Dynamical Crowded Distance Estimation Method

In order to pick out the redundant PS archive and maintain a certain number of solutions, the crowding-distance estimation method [2] is usually used to calculate the average distance of two points on either side of this point along each of the objectives, and it can be formulated as follows:

$$C_i = \sum_{j=1}^{M} \left(|f_{i+1,j} - f_{i-1,j}| \right) \tag{14}$$

where C_i is the crowding distance of individual i, M is the number of objective functions, $f_{i,j}$ is the j-th objective function value of individual i.

Then, an improved selection method, namely the dynamical crowded distance estimation method is adopted instead of the traditional crowded distance method [2]. Detailed procedures of this method are given in Algorithm 2.

Algorithm 2. The dynamical crowded distance estimation method
Step1: Initialize $C_i = 0;$
Step2: Sort population according to each objective value;
$\quad\quad\quad$ Pre-set the boundary points to an infinite value, and this can ensure the availability of next selection process.
Step3: Calculate the crowding distance of population individuals;
Step4: Determine the minimum individual called *ID* in the population, then re-move it;
Step5: Re-compute the crowding distance of the individual *ID+1* and *ID−1* by:

Finally, based on above mechanisms and strategies, the main procedures of MORSGO are listed in Table 1.

Table 1. Main procedures of MORSGA

MORSGA algorithm
Step. 1: Initialize Pt and set t = 0;
Step. 2: Classify current population into main roots group GR_t and lateral roots group GN_t by Eqs.(1)-(5).
Step. 3: Generate new solutions by Eq.(6)-Eq.(8), and Eq.(17) and then form a combined population $GC_t = GR_t \cup GN_t$.
Step. 4: Calculate the fitness, auxin concentration values of GC_t and sort it by executing Algorithm 1. Then each solution has a domination leve. The size of GC_t is size(GR_t)+ size(GN_t) where size(GN_t) usually becomes bigger than before.
Step. 5: Select exactly size(GR_t) best individuals by Algorithm 2 into new MR_t from current GC_t.
Step. 6: Implement lateral roots growth operators and generate new solutions as LN_t and then form a combined population $LC_t = LR_t \cup LN_t$.
Step. 7: Calculate the fitness, auxin concentration values of LC_t and sort it by implementing Algorithm 1.
Step. 8: Select exactly size(LR_t) best roots by Algorithm 2 as new LR_t from current main root population LC_t.
Step. 9: Remove the deteriorated roots if the corresponding auxin concentration values are 0.
Step. 10: Set t = t + 1.
Step. 11: If the termination conditions are met, stop; otherwise, return to step 2.

3 Benchmark Test

3.1 Test Problems and Performance Measures

Five representative benchmarks, including bio-objective ZDT1, ZDT2, and ZDT6, and tri-objective DTLZ1 and DTLZ6, are selected to evaluate the performance of the proposed algorithm [9]. Detailed formulas of these test instances are referred in [9].

Two performance measures are considered: (1) convergence metric Υ based on IGD-metric [10], (2) spread metric Δ proposed in [2]. Accordingly the detailed information about the two metrics can refer to [2, 10] respectively.

3.2 Experimental Setup

Experiments are conducted with MORSGO, NSGA-II [2], and MOEA/D [3]. The common parameters for all algorithms are set as following: the population size is 100, the maximum iteration number is set to 1000, and the number of independent algorithmic runs is 20. For other parameters in MOEA/D and NSGA-II, they keep the same with their original Refs. [2, 3].

Table 2. Results on 30-D ZDT1, ZDT2 and 10-D ZDT6 by MORSGO, NSGA-II and MOEA/D.

Test problem		ZDT1			ZDT2			ZDT6		
		MO-RSGO	NSGA-II	MOEA/D	MO-RSGO	NSGA-II	MOEA/D	MO-RSGO	NSGA-II	MOEA/D
IGD Metric	Mean	1.24E-3	2.49E-2	4.11E-3	6.65E-2	4.22E-3	8.58E-3	4.71E-3	6.62E-2	1.95E-2
	Median	1.24E-3	2.53E-2	4.2E-03	6.45E-2	3.32E-2	8.87E-3	4.69E-3	2.10E-2	1.95E-2
	Best	9.80E-4	2.64E-3	3.76E-3	5.26E-2	6.93E-3	8.18E-3	4.15E-03	1.42E-2	1.93E-2
	Worst	1.37E-3	9.98E-2	5.4133E-03	7.14E-02	1.14E-01	9.09E-3	5.37E-3	8.21E-2	2.03E-2
	Std	1.35E-4	2.99E-2	4.93E-4	1.15E-02	3.06E-2	2.42E-2	3.70E-4	2.00E-2	6.21E-2
Spread Metric	Mean	6.86E-1	8.20E-1	8.31E-1	7.76E-2	7.89E-1	7.46E-1	7.29E-1	7.49E-01	7.25E-1
	Median	6.80E-1	8.23E-1	8.15E-1	2.70E-2	7.94E-1	7.24E-1	7.36E-1	7.47E-1	7.34E-1
	Best	6.15E-1	7.11E-1	6.99E-1	6.73E-2	7.16E-1	6.36E-01	6.38E-1	6.54E-1	6.25E-1
	Worst	7.72E-1	9.38E-1	1.03	8.73E-2	9.08E-01	9.3262E-01	7.67E-1	8.29E-1	8.27E-1
	Std	5.25E-2	6.72E-2	1.19E-1	1.82E-2	5.81E-1	8.81E-2	3.87E-2	5.06E-2	6.68E-2

Table 3. Results on 7-D DTLZ1 and 22-D DTLZ6 by MORSGO, NSGA-II and MOEA/D.

Test problem		DTLZ1			DTLZ6		
		MORSGO	NSGA-II	MOEA/D	MORSGO	NSGA-II	MOEA/D
IGD Metric	Mean	3.5094	1.40E + 01	2.65E + 01	2.2848E-2	1.28E-1	6.060E-2
	Median	1.5798	1.47E + 01	2.680E + 01	2.2339E-2	1.03E-1	6.171E-2
	Best	8.002E-1	9.298	2.18E + 01	1.8376E-2	5.86E-2	4.012E-2
	Worst	1.187E + 1	2.013E + 1	3.5285E + 1	2.905E-2	3.1055E-01	7.662E-2
	Std	4.3211	3.7561	4.306	3.27E-3	7.45E-2	1.286E-2
Spread Metric	Mean	4.999E-01	5.948E-1	5.131E-01	6.3503E-1	6.78E-1	6.685E-1
	Median	4.816E-01	5.565E-1	5.030E-01	6.301E-1	6.84E-1	6.569E-01
	Best	4.426E-01	5.095E-01	4.385E-01	5.4812E-1	6.30E-1	5.316E-01
	Worst	5.711E-01	7.413E + 02	6.2429E-01	7.0279E-1	7.36E-01	8.032E-01
	Std	5.036E-02	9.017E-2	5.375E-2	4.6848E-2	3.1935E-02	8.91E-2

3.3   Results and Analysis

Tables 2 and 3 give computational results obtained by MORSGO, NSGA-II and MOEA/D in 20 runs on bi-objective and tri-objective ZDTs. It is seen from Table 2 that MORGO is able to find better mean, best and standard deviation values of the IGD metric in all bio-objective test functions except ZDT2. MOEA/D and NSGA-II obtain a little better performance than MORSGO on ZDT2 only in terms of best and median values of IGD-metric. For the Δ-metric, in all cases with MORGO, the standard deviation and mean values are significantly satisfactory, except in NSGA-II with ZDT2 and in MOEA/D with ZDT6. This means that MORGO can get a better spread of non-dominated solutions than other algorithms. From Table 2, it is clearly seen that MORSGO yields similar performance in terms of diversity and convergence to the bio-objective test case. Specially, MORSGO is able to obtain the best spread of solutions on DTLZ1 and DTLZ6. For the IGD metric, on DTLZ1, MORGA obtains the best mean and standard deviation values, but NSGA-II also obtains the best mean values. On DTLZ2, MORGA obtain better performance in terms of mean, best and median values than the others, but MOEA/D obtains the best standard deviation.

4   Conclusions

In this paper, a new plant-inspired algorithm called MORSGO is designed to handle multi-objective optimization problems. In MORSGO, several basic search operators are designed and developed inspired from the plant root growth behaviors. To effectively handle non-dominated solutions, the fast non-dominated sorting approach is employed to get priority of non-dominated solutions obtained during the search process, and the dynamical crowded distance estimation strategy is used to maintain diversity of Pareto optimal solution.

A set of test functions including bio-objective and tri-objective instances have been employed to evaluate the computational performance of the proposed algorithm. Experimental results show that MORSGO has a promising ability of maintaining better population diversity and accordingly obtains better convergence, which indicates that MORSGO has potential ability of effectively tackling real-world problems in the near future.

Acknowledgements. This work is supported by National Natural Science Foundation of China under Grant No. 61503373; Natural Science Foundation of Liaoning Province under Grand 2015020002.

References

1. Yi, X., Zhou, Y.: A dynamic multi-colony artificial bee colony algorithm for multi-objective optimization. Appl. Soft Comput. **35**, 766–785 (2015)
2. Deb, K., Pratap, A., Agarwal, S., Meyarivan, T.: A fast and elitist multiobjective genetic algorithm: NSGA-II. IEEE Trans. Evol. Comput. **6**(2), 182–197 (2002)

3. Zhang, Q., Liu, W., Li, H.: The performance of a new version of MOEA/D on CEC09 unconstrained MOP test instances. In: Proceedings of Congress on Evolutionary Computation (CEC 2009), Norway, pp. 203–208 (2009)
4. Zitzler, E., Laumanns, M., Thiele, L.: SPEA2: improving the strength pareto evolutionary algorithm. In: Proceedings of EUROGEN 2001: Evolutionary Methods Design Optimization Control with Applications to Industrial Problems, Athens, Greece, pp. 95–100 (2002)
5. Coello Coello, C.A., Pulido, G.T., Lechuga, M.S.: Handling multiple objectives with particle swarm optimization. IEEE Trans. Evol. Comput. 8(3), 256–279 (2004)
6. Ma, L., Hu, K., Zhu, Y., Chen, H.: A hybrid artificial bee colony optimizer by combining with life-cycle. Powell's Search Crossover Appl. Math. Comput. 252, 133–154 (2015)
7. Hodge, A., Berta, G., Doussan, C., Merchan, F., Crespi, M.: Plant root growth, architecture and function. Plant Soil 321(1–2), 153–187 (2009)
8. Ma, L., Zhu, Y., Liu, Y., Tian, L.: A novel bionic algorithm inspired by plant root foraging behaviors. Appl. Soft Comput. 37, 95–113 (2015)
9. Deb, K., Thiele, L., Laumanns, M., Zitzler, E.: Scalable multi-objective optimization test problems. In: Proceedings of Congress on Evolutionary Computation, pp. 825–830 (2002)
10. Zhou, A.M., Jin, Y.C., Zhang, Q.F., Sendhoff, B., Tsang, E.: Combining model-based and genetics-based offspring generation for multi-objective optimization using a convergence criterion. In: Proceedings of Congress on Evolutionary Computation, pp. 3234–3241 (2001)

Group Discussion Mechanism
Based Particle Swarm Optimization

L.J. Tan[1], J. Liu[2(✉)], and W.J. Yi[2(✉)]

[1] Department of Business Management,
Shenzhen Institute of Information Technology, Shenzhen 518172, China
[2] College of Management, Shenzhen University, Shenzhen, China
dorcasl@163.com, yiwenjiel006@gmail.com

Abstract. Inspired by the group discussion behavior of students in class, a new group topology is designed and incorporated into original particle swarm optimization (PSO). And thus, a novel modified PSO, called group discussion mechanism based particle swarm optimization (GDPSO), is proposed. Using a group discussion mechanism, GDPSO divides a swarm into several groups for local search, in which some smaller teams with a dynamic change topology are included. Particles with the best fitness value in each group will be selected to learn from each other for global search. To evaluate the performance of GDPSO, four benchmark functions are selected as test functions. In the simulation studies, the performance of GDPSO is compared with some variants of PSOs, including the standard PSO (SPSO), PSO-Ring and PSO-Square. The results confirm the effectiveness of GDPSO in some of the benchmarks.

Keywords: Group discussion · Topology · GDPSO

1 Introduction

Inspired by a swarm behavior of bird flock and fish school, particle swarm optimization (PSO) was originally proposed by Kenney and Eberhart [1, 2]. Since its inception, numerous scholars have been increasingly interested in the work of employing PSO to solve various complicated optimization problems and putting forward a series of methods to improve the performance of PSO in case of trapping in local optimum. The analysis and improvement of PSO can be mainly summarized into three categories: parameters adjustment [3, 4], new population topology design [5–9] and hybrid strategies [10, 11]. In PSO, each particle searches for a better position in accordance with its own experience and the best experience of its neighbors [12]. Accordingly, a variety of researches have been dedicated to modifying the information exchange mechanisms between neighbors (learning exemplars) with various population topological structures. Kennedy proposed a Ring topology, with which each particle is only connected to its immediate neighbors [6]. Mendes presented three other topologies, i.e., four clusters, Pyramid and Square to guarantee every individual fully informed [7]. Jiang proposed a novel age-based PSO with age-group topology [8]. Lim proposed a new variant of PSO with increasing topology connectivity that increases the particle's topology connectivity with time as well as performs the shuffling mechanism [9].

© Springer International Publishing Switzerland 2016
D.-S. Huang et al. (Eds.): ICIC 2016, Part III, LNAI 9773, pp. 88–95, 2016.
DOI: 10.1007/978-3-319-42297-8_9

By mimicking the group discussion behavior of students in class, a new topology is designed. And thus an improved PSO, named GDPSO is proposed. Through some of the benchmarks, the experimental results showed that the proposed GDPSO algorithm adjusts the balance between the local search and global search and it can improve the performance of PSO significantly.

The remainder of this paper can be outlined as follows. In Sect. 2, the substance of standard PSO is presented. Afterwards, we introduce the origins and procedures of the proposed GDPSO in detail in Sect. 3. In Sect. 4, we describe the experimental settings and discuss the results. Finally, conclusions are drawn in the last section.

2 Standard Particle Swarm Optimization

In SPSO, a potential solution of a problem is represented by the position of each particle x_i. The position x_i is updated by a velocity v_i, which controls not only the distance and but also the direction of a particle. The position and velocity are updated for the next iteration according to the following equations:

$$x_{id}^{t+1} = x_{id}^{t} + v_{id}^{t} \tag{1}$$

$$v_{id}^{t+1} = w \cdot v_{id}^{t} + c_1 \cdot r_1 \cdot (p_{id}^{t} - x_{id}^{t}) + c_2 \cdot r_2 \cdot (p_{id}^{t} - x_{id}^{t}) \tag{2}$$

where $x_{id} \in [l_d, u_d]$, d is the dimension of the search space, l_d is the lower bound and u_d is the upper bound of the dth dimension. The t means that the algorithm is going on the tth generation. The inertia weight w, which is linearly reduced during the search time, controls how much the previous velocity influences the new velocity [13]. The w is updated by

$$w = w_{start} - \frac{w_{start} - w_{end}}{t_{max}} \cdot t \tag{3}$$

The c_1 and c_2 are called positive acceleration coefficients, which are taken as 2 normally. The r_1 and r_2 are random values in the range [0, 1]. The best previous position of the ith particle is called personal best particle (P_{best}). The best one of all the P_{best} is called global best particle (G_{best}), denoting the best previous position of the swarm. p_{id} presents the P_{best} while p_{gd} presents the G_{best}.

3 Group Discussion Mechanism Based Particle Swarm Optimization

Group discussion means that all the members in a group take an active part in discussion around a specific theme, during which face to face communication activities, the team spirit, sense of responsibility and mutual benefit are contained. A group member should reflect on their own learning and thinking, and then seriously consider

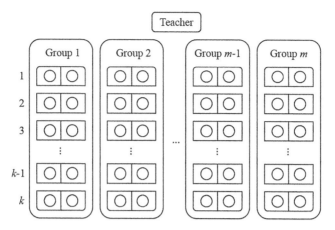

Fig. 1. The structure of a class

the others' view. With every member adhering to the correct opinion and correcting the error, a better idea is more likely to be come up with than ever before.

In general, there is a teacher and a number of students involved in the group discussion in a classroom. The students are divided into several groups randomly by a teacher and each group includes some smaller teams to discuss a specific problem. The structure of a class is shown in Fig. 1. A team may consist of two classmates who are seated in the same desk. During the discussion, the tasks of a teacher are to evaluate everyone's idea and control the discussion time. In order to increase the diversity of the ideas, it's better to enlarge the scope of a team after desk-mates discussing with each other for many times. Hence the teacher let students discuss with someone whose seats are before or after their desks so that there are six members in a new team totally. The team transformation is shown in Fig. 2.

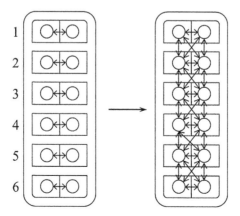

Fig. 2. The transformation of a team

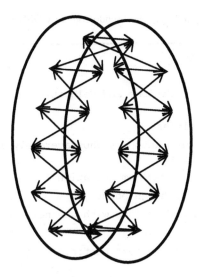

Fig. 3. The topology structure

After the teacher's evaluation, a student with the best idea in each group is ordinarily chosen as a group leader. Each group leader then makes a presentation in the class. Considering the selfishness of human being, we suppose that all group leaders discuss with each other without the participation of other group members.

Incorporating the aforementioned idea into PSO, we designed a new topology and proposed group discussion mechanism based particle swarm optimization (GDPSO). The procedure of GDPSO is described as follow:

Step 1: generate n particles randomly and evaluate each particle according to their fitness function.

Step 2: separate these n particles into m groups. The number of each group is set as the same for the sake of fairness.

Step 3: update the positions and velocities using Eqs. (1) and (2).

Step 4: calculate the fitness value of the updated particles.

Step 5: select a particle as a group leader with the best fitness value from each group.

Step 6: After m group leaders learning from each other, evaluate them again. If the fitness value is not better than before, the previous position should be still remained.

In order to ensure that every particle has the same number of neighbors, we design a topology as shown in Fig. 3. The pseudo-code of GDPSO is given in Table 1.

4   Simulation Experiments and Analysis

To measure the performance of GDPSO, the experiments were conducted to compare three PSOs on four benchmark functions listed in Table 2. These three PSOs are: standard PSO, PSO-Ring [5] and PSO-Square [6]. All these four heuristic algorithms

Table 1. The Pseudo-code of GDBSO

Begin

    For (Each run)

        Initialize the positions and velocities of n particles randomly

        Evaluate n particles

        Divide n particles into m groups averagely so that each group has k (k=n/m) particles.

        While iteration < maximum iteration

          If iteration < change point

          Each particle discusses with one neighbor

          If mod (index, 2)==1

            Discuss with the next particle

          Else

            Discuss with the previous particle

          End If

        Else

          Each particle discusses with the five nearest neighbors

          If mod (index, 2)==1

            Discuss with previous two particles and the next three particles

          Else

            Discuss with the previous three particles and the next two particles

          End If

        End If

          Evaluate n newly particles

          Choose the best particle in each group as the group leader

          m group leaders discuss with each other and then compute the fitness values of m group leaders

          Update Pbest and Gbest

        End While

    End For

End

---

presented in this paper were coded in MATLAB language. And the experiments were implemented on an Intel Core i5 processor with 2.27 GHz CPU speed and 2.93 GB RAM machine, running Windows 7. All experiments were run 20 times.

Table 2. Benchmark functions tested in this paper

	Function	Range	Shape
f_1	Rosenbrock	[–2.048, 2.048]	Unimodal
f_2	Easom2D	[–2,2]	Multimodal
f_3	Himmelblan	[–500,500]	Multimodal
f_4	Michalewicz10	[–50,50]	Multimodal

4.1 Parameter Settings

The set of parameters are listed in Table 3 below.

Table 3. Set of parameters for experiments

Method	n	Dim	Max iteration	c_1	c_2	w_{start}	w_{end}	m	Change point
GDPSO	100	30	1000	2	2	0.9	0.4	5	500
SBSO	100	30	1000	2	2	0.9	0.4	–	–
PSO-Ring	100	30	1000	2	2	0.9	0.4	–	–
PSO-Square	100	30	1000	2	2	0.9	0.4	–	–

4.2 Experimental Results

The results of the experiment are presented in Table 4. In particular, the best results are shown in bold. Figure 4 shows the average best fitness convergent curves. As can be clearly seen from the tables and figures, the performance of GDPSO algorithms is not always better than other PSOs in different functions. From the data of the minimum value, it is obvious that GDPSO can find the best solution in three benchmark functions, i.e., Easom2D, Himmelblan and Michalewicz10 which are all multimodal functions. The standard deviation of GDPSO in Rosenbrock is better than other algorithms. From the curves of these three benchmarks, it is obvious that the improvement we suggest has effect on some of optimization problems.

Table 4. Experimental results

		GDPSO	SPSO	PSO-Ring	PSO-Square
f_1	Mean	5.7189e + 000	5.6727e + 000	5.6020e + 000	**5.3767e + 000**
	Std	9.91	12.68	12.53	12.02
f_2	Mean	**–3.3333e-001**	–2.0000e-001	–2.0000e-001	–2.0000e-001
	Std	0.58	0.45	0.45	0.45
f_3	Mean	**–2.4633e + 001**	–1.5289e + 001	–1.4480e + 001	–1.5098e + 001
	Std	42.67	34.19	32.38	33.76
f_4	Mean	**–3.1351e + 000**	–1.8823e + 000	–2.2789e + 000	–2.5914e + 000
	Std	5.43	4.21	5.10	5.79

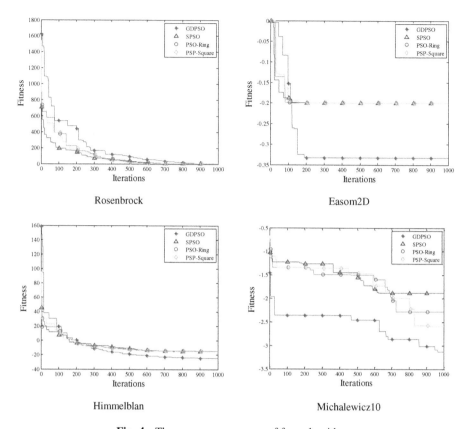

Fig. 4. The convergence curve of four algorithms

5 Conclusion

In this paper, we proposed an improved PSO based on group discussion behaviors in class, which provides a new insight to adjust the balance between the local search and global search. GDPSO divides a swarm into several groups for local search and particles with the best fitness value in each group will be selected to learn from each other for global search. Moreover, GDPSO provides a dynamic topology instead of static topology in SPSO. By testing on four benchmark functions, GDPSO demonstrates better performance than other PSOs in three benchmarks. Although GDPSO doesn't perform best all the time, we believe that it still has a potentiality and capability to solve other different kinds of optimization problems.

However, only applying in four benchmark problems to GDPSO are not enough. Hence more experiments on quantities of benchmark functions must be investigated in the future. Besides, we will focus on utilizing the new topology to other swarm intelligence algorithms as possible. We are also setting about applying the proposed GDPSO algorithm to more applications like portfolio optimization to verify its effectiveness in solving real-world problems.

Acknowledgments. This work is partially supported by The National Natural Science Foundation of China (Grants Nos. 71571120, 71001072, 71271140, 71471158, 71501132, 2016A030310067) and the Natural Science Foundation of Guangdong Province (Grant no. 2016A030310074).

References

1. Eberchart, R.C., Kennedy, J.: Particle swarm optimization. In: IEEE International Conference on Neural Networks, Perth, Australia (1995)
2. Eberchart, R.C., Kennedy, J.: A new optimizer using particle swarm theory. In: Proceedings of the 6th International Symposium on Micromachine and Human Science, Nagoya, Japan, pp. 39–43 (1995)
3. Clerc, M., Kennedy, J.: The particle swarm: explosion, stability, and convergence in multidimensional complex space. IEEE Trans. Evol. Comput. **6**, 58–73 (2002)
4. Clerc, M.: The swarm and the queen: towards a deterministic and adaptive particle swarm optimization. In: Proceedings of the 1999 Congress on Evolutionary Computation, pp. 1927–1930 (1999)
5. Suganthan, P.N.: Particle swarm optimizer with neighborhood operator. In: Proceedings of the IEEE Congress of Evolutionary Computation, pp. 1958–1961 (1999)
6. Kennedy, J.: Small worlds and mega-minds: effects of neighborhood topology on particle swarm performance. In: Proceedings of the 1999 Congress on Evolutionary Computation, Washington, DC, pp. 1931–1938 (1999)
7. Mendes, R., Kennedy, J., Neves, J.: The fully informed particle swarm: simpler, maybe better. IEEE Trans. Evol. Comput. **8**, 204–210 (2004)
8. Jiang, B., Wang, N., Wang, L.: Particle swarm optimization with age-group topology for multimodal functions and data clustering. Commun. Nonlinear Sci. Numer. Simul. **18**, 3134–3145 (2013)
9. Wei, H.L., Isa, N.A.M.: Particle swarm optimization with increasing topology connectivity. Eng. Appl. Artif. Intell. **27**, 80–102 (2014)
10. Angeline, P.J.: Using selection to improve particle swarm optimization. In: Proceedings of IEEE World Congress on Computational Intelligence, Anchorage, Alaska, pp. 84–89 (1998)
11. Li, L.L., Wang, L., Liu, L.H.: An effective hybrid PSOSA strategy for optimization and its application to parameter estimation. Appl. Math. Comput. **179**, 135–146 (2006)
12. Kennedy, J., Mendes, R.: Population structure and particle swarm performance. In: Proceedings of the 2002 Congress on Evolutionary Computation, pp. 1671–1676 (2002)
13. Shi, Y., Eberhart, R.: A modified particle swarm optimizer. In: Proceedings of IEEE World Congress on Computational Intelligence Evolutionary Computation (1998)

Biomedical Informatics
and Image Processing

Srrr-cluster: Using Sparse Reduced-Rank Regression to Optimize iCluster

Shu-Guang Ge[1], Jun-Feng Xia[2,3],
Pi-Jing Wei[1], and Chun-Hou Zheng[1(✉)]

[1] College of Electrical Engineering and Automation,
Anhui University, Hefei 230601, Anhui, China
zhengch99@126.com
[2] Institute of Health Sciences, Anhui University, Hefei 230601, Anhui, China
[3] Center of Information Support and Assurance Technology,
Anhui University, Hefei 230601, Anhui, China

Abstract. Cancer genome projects can provide different types of data on the genetic level, which is significant for cancer research and biological processes in computational methods. Thus, computational methods used to identify cancer subtypes should fully focus on integrating these multidimensional data (e.g., DNA methylation data, mRNA expression data, etc.). Sparse reduced-rank regression (Srrr) method, a state-of-the-art multiple response linear regression method, can easily deal with high dimensional statistical data. In this paper, we introduced Srrr method combining iCluster (Srrr-cluster) to discovery cancer subtypes. Firstly, we used Srrr to estimate the coefficient matrix and then cancer subtypes were clustered by iCluster. Finally, we used our Srrr-cluster method to analyze glioblastoma and breast cancer data. The results show that our Srrr-cluster method is effective for cancer subtype identification.

Keywords: Cancer subtypes · Clustering · iCluster · Sparse reduced-rank regression · High dimensional statistical data

1 Introduction

The genomes of cancer often contain a large number of somatical aberrations information, e.g., DNA copy number aberrations are closely relation to tumor gene by gene amplification or tumor suppressor loss because of genomic instability and deregulation [1, 2]. Other cases, epigenetic aberrations also result in oncogene such as genomic methylation [3]. DNA sequence change will directly affect the mRNA expression levels even other non-coding microRNA, and then change the outcome of the transcriptome, eventually produce individual heterogeneity and lead to distortion of cancer cells. The same cancer may have diverse somatic mutation and transcriptional level, so that the formation of different kinds of subtypes has diverse heterogeneity of biological progresses and phenotypes [4]. For example, glioblastoma (GBM) can be defined as the Classical, Mesenchymal, and Proneual subtypes by aberrations and gene expression of EGFR, NF1 and PDGFRA/IDH1 [5].

© Springer International Publishing Switzerland 2016
D.-S. Huang et al. (Eds.): ICIC 2016, Part III, LNAI 9773, pp. 99–106, 2016.
DOI: 10.1007/978-3-319-42297-8_10

Recently, many cancer genome projects are established and amassed a large number of various types of data. For example, The Cancer Genome Atlas (TCGA) (http://cancergenome.nih.gov/) contains genome, transcriptome and expression information for over 20 cancers from thousands of patients, which produced several types of data, such as methylation data, mRNA expression data, DNA copy number data and so on. Currently, some integrative methods have been proposed which combine different biological data for cancer subtype classification. For example, iCluster is a integrating probability model of multiple data based on Gaussian latent variable model. Which first structures an optimizing penalized log-likelihood function to estimate using Expectation-Maximization algorithm with lasso-type sparse [6, 7], then using K-means to get subtypes. However, bulky datasets also bring about many challenges for subtypes classification. Firstly, the key of data-integration clustering method tend to construct a variance - covariance structure within data types, namely coefficient matrix solving, which is equivalent to a feature selection process. The coefficient matrix is a projection matrix that projects the original data onto an eigengene-eigenarray subspace. Secondly, high-dimensional datasets have a common feature that the number of samples small yet the number of genes is large, so the dimension reduction of coefficient matrix is essential. iCluster used PCA method to estimate the coefficient matrix that defined the first k−1 eigenvectors by a pivoted QR decomposition [8, 9]. However, in the Gaussian latent variable models, PCA has many deficiencies: (i) significant features can't completely be extracted when facing high dimensional statistical data. (ii) eigenvalue of the first principal component is much larger than the eigenvalues of the other main components.

Briefly, we can see that the estimator of coefficient matrix is very important. Considering Gaussian latent variable model, sparse reduced-rank regression (Srrr) is a useful parsimony model when facing a large number of data for multiple response regression [10–14]. Generally, Srrr with the purpose of solving an indicator matrix can be divided into three steps in different algorithm: (i) Working out reduced-rank matrix that can reduce the noise of the model and improve the robustness. (ii) Constructing sparse group lasso, group bridge or group MCP term, which can solve the problem that the sample volume is pretty smaller than the gene volume [10, 11]. (iii) Establishing minimum optimization function to solve the coefficient matrix. Until now, Srrr method has been applied in several research area. E.g., Lin et al. (2013) used it to detect genetic networks associated with brain functional networks in schizophrenia [12]. Chen et al. (2012) proposed a weighted rank-constrained group lasso approach with two heuristic numerical algorithms and studied its large sample asymptotics [13].

In this paper, we used subspace assisted regression with row sparsity (SARRS) algorithm that proposed by Ma et al. (2014) [14], combining with iCluster (Srrr-cluster) to discovery cancer subtypes. Srrr-cluster can be regarded as a data-integration clustering method which first estimating the coefficient matrix of the latent variable model using the Srrr method, and then solving the estimator of the desigen matrix through optimizing a penalized complete-data log-likelihood with sparse term using the Expectation-Maximization (EM) algorithm.

2 Srrr-cluster Methods

2.1 Data Types Integration and a Gaussian Latent Variable Model Representation

Mo et al. summarized the different data types to adapt to different mathematical probability models [15]. For example, mutation status is defined binary variable that is suit for logistic regression model; copy number loss, gain, and normal status are defined multicategory variable that are suit for multilogit regression; NDA copy number data, DNA methylation data, mRNA expression data and so on are defined continuous variable that are suit for Gaussian latent model. In this paper, different types of continuous data are regressed using Srrr model to discovery cancer subtypes. We can fuse the same samples with different types of continuous data into a multiple genomic data. Therefore, we employ an integrating genomic data that harbor different levels of expression and transcriptome information to search subtypes.

Firstly, we establish a Gaussian latent variable model:

$$X = ZW + \varepsilon \tag{1}$$

here $X = \{X_1, \ldots X_P\}$ is the original integration data of dimension $n \times m$, where X_1 can denote DNA methylation data of dimension $n \times m_1$, X_2 can denote DNA copy number data of dimension $n \times m_2$, X_p can denote mRNA expression of dimension $n \times m_p$ and so forth. Z is the design matrix of dimension $n \times l$, W is the coefficient matrix of dimension $l \times m$, ε is the error term and make the additional assumption that $Z \sim N(0, I)$ and $\varepsilon \sim N(0, \psi)$. p is the number of genomic data types, n is the number of samples, m is the number of the genes, l is the number of predictors. Ding et al. (2004) noted that the K-means solution of Z can directly be selected using the first k–1 eigenvectores that span a low-dimensional latent space where the original data are projected onto each of the first K–1principal directions such that the total variance is maximized by PCA. So, Z is the design matrix of dimension $n \times (k - 1)$ that is finally clusters latent tumor subtypes and the initial value of Z is the first k–1 eigenvectors by PCA, where k is the number of clusters [9].

2.2 An Adaptive Srrr Method and Srrr-cluster

Following Eq. (1), we can afford to estimate the solution of the coefficient matrix W using an adaptive Srrr method. The goal is to reduce the rank r of W under the Gaussian latent variable model. Firstly, two error parameters, i.e., a noise level σ expressed as $\sigma = median(\sigma(X))/\sqrt{\min(n,m)}$, where $\sigma(X)$ is the collection of all nonzero singular values of X, and a noise rank level η, expressed as $\eta = \sqrt{2m} + \sqrt{2(\min(n,k))}$, are estimated to work out the reduced-rank r and an orthonormal matrix $V_{(0)}$ that is non-orthogonal to the right singular subspace of W. The estimator of r is computed by:

$$r = \max\{j : \sigma_j(Z(Z'Z)^- Z'X) \geq \sigma\eta\} \tag{2}$$

Where $(ZZ')^-$ is Moore-Penrose pseudo-inverse. So, the Srrr method use the first r-th right vector of $Z(Z'Z)^- Z'X$ to estimate the orthonormal matrix $V_{(0)}$:

$$V_{(0)} = (V_1^{(0)}, \ldots, V_r^{(0)}) \tag{3}$$

Depending on characters of the orthonormal matrix $V_{(0)}$, such as $W = WV_{(0)}V_{(0)}'$, the reduced-rank matrix B can be expressed as:

$$B = WV_{(0)} \tag{4}$$

with dimension $(k - 1) \times r$ which columns being the estimator of rank. What more, VV' is a projection matrix that approximatively maps onto the right singular subspace of W.

For the sake of simplicity, Ma et al. take sparse group lasso in this model, where each row of the B is regarded as a group and all groups are of the same size r [14]. Each row takes sparse process by the ℓ_2 matrix norm as follows:

$$\rho(B; \lambda) = \lambda \sum_{j=1}^{k-1} \|B_{j*}\|_2 \tag{5}$$

where λ is the penalty level.

Following these, Srrr method constructs a right bias-variance tradeoff function with reduced-rank term representing the variance part and sparse lasso term representing the bias part using SARRS algorithm:

$$W = \underset{Z \in \mathfrak{R}^{(k-1) \times n}}{\arg\min} \left\{ \left\|XV_{(0)}V_{(0)}' - ZWV_{(0)}V_{(0)}'\right\|_F^2 \middle/ 2 + \rho(WV_{(0)}V_{(0)}'; \lambda) \right\} \tag{6}$$

we can further reduce the computation cost by first solving:

$$B_{(1)} = \underset{B \in \mathfrak{R}^{(k-1) \times r}}{\arg\min} \left\{ \left\|XV_{(0)} - ZB\right\|_F^2 \middle/ 2 + \rho(B; \lambda) \right\} \tag{7}$$

However, $B_{(1)}$ is not accurate but close to $WV_{(0)}$ because the columns of $V_{(0)}$ is just approximate to the right singular subspace of W.

It is worth noting that the right singular subspace of W is exactly the same as that of ZW. Next step, we can estimate the left singular subspace $U_{(1)} \in \mathbb{R}^{n \times r}$ of $ZB_{(1)}$. Due to (4), $U_{(1)}$ is exactly the left singular subspace of $ZWV_{(0)}$, which in turn equals the left subspace of ZW. Through the same line of logic, $U_{(1)}U_{(1)}'$ is a projection matrix that accurately maps onto the left singular subspace of WZ. Then, we can easily compute the right singular vectors $V_{(1)} \in \mathbb{R}^{m \times r}$ of $U_{(1)}U_{(1)}'X$, which in turn equals the right subspace of ZW. Successfully, a pretty accurate right singular vectors of W is esti-mated. Finally, using $V_{(1)}$ instead of $V_{(0)}$ to solve the equation:

$$B_{(2)} = \arg\min_{B \in \Re^{(k-1) \times r}} \{\|XV_{(1)} - ZB\|_F^2/2 + \rho(B; \lambda)\} \tag{8}$$

Hopefully, we compute the estimated indictor matrix by $W = B_{(2)} V'_{(1)}$.

Given two or more types of data from the same cohort of patients, our Srrr-cluster method first fuse these data into an integrative matrix, and then use the optimized PCA to compute a design matrix for the integrative data. The next step is to use the adaptive Srrr method to calculate the coefficient matrix under the Gaussian latent variable model, which can project sample × gene space of the original data into eigenarray × eigengene subspace. Finally, we use iCluster method to discovery cancer subtypes.

2.3   Evaluation Metric

We use three commonly used metrics to evaluate Srrr-cluster performance by identi-fying subtypes in these cancers. (i) Silhouette score, a measure of cluster homogeneity, which is defined as $s(i) = (b(i) - a(i))/(\max(a(i), b(i)))$, where $a(i)$ is average dis-similarity between i and all the other points of the same subtypes, $b(i)$ is average dissimilarity between i and all the other points of the different subtypes, i is an arbitrary sample. If sihouette value is close to 1, it means that the data are appropriate [16]. (ii) P value in Cox log-rank test, which is used to assess the significance of the different in survival profiles between subtypes [17]. (iii) The proportion of deviance (POD), which is a score of evaluating cluster degree of separation by a diagonal block structure. We set a matrix $A = Z^T Z, A \in \Re^{n \times n}$. Then the elements of A is defined as $a_{ij}/\sqrt{a_{ii}a_{jj}}$ for $i = 1,..., n$ and $j = 1,...,n$, and set negative values to zero, which can order cancers belonging to the same clusters into a adjacent structure. If the diagonal block matrices were prefect, all elements of the diagonal blocks would be non-negative and all ele-ments of the off-diagonal blocks would be zero. So, compared A with the prefect diagonal block structure, we define a deviance measure d, which is the sum of quan-tities that the diagonal blocks' elements of A appear zero and the off-diagonal blocks' elements of A appear non-negative values. POD is defined as d/n^2 so that POD is between 0 and 1. Small values of POD indicate strong cluster separability, and large values of POD indicate of POD indicate poor cluster separability[6].

3   Results

3.1   Subtypes Discovery in Breast Cancer

Using DNA copy number and mRNA expression on the same cDNA microarrys that contain 6691 genes form Pollack et al. [1] from 37 primary breast cancers and four breast cancer cell lines, we compared the Srrr-cluster results with iCluster. As well known, the expression profiles of the four cell line samples (BT474, T47D, MCF7 and

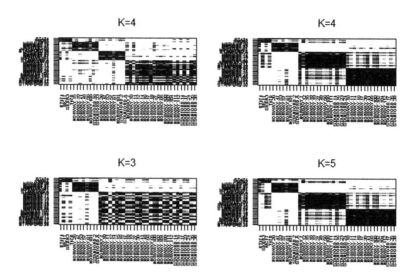

Fig. 1. Diagonal blocks structures obtained using iCluster (k = 4) and Srrr-cluster (k = 4, k = 3 and k = 5) methods.

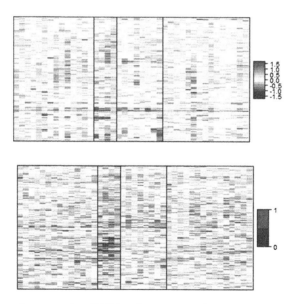

Fig. 2. Heatmaps of copy number (DNA) and gene expression (mRNA) with samples arranged by the integrated cluster assignment under the Srrr-Cluster model. (Color figure online)

SKBR3) should be similar so that they can be considered as a subtype from the rest of the tumor samples. Additional, HER2/ERBB2 is an important prognostic factor for breast cancer near17q12, the clinical features and biological behavior of a special performance, treatment modalities HER2-positive (overexpression or amplification) of breast cancer also have a great difference with other types of breast cancer.

Figures 1 and 2 shows the diagonal blocks structures under the sparse solution $\lambda = (0.2, 0.2)$, r corresponding to Cluster (k = 4) and Srrr-cluster (k = 4, k = 3 and k = 5) respectively. POD values of the clustering solutions are 0.1519533, 0.1254317, 0.2478124 and 0.1259145 respectively. Considering the values of POD, the four clusters obtained by Srrr-cluster method should be the best one. Figure shows the heatmaps of the profiles of DNA copy number data and mRNA expression data when samples were splitted four clusters using Srrr-cluster method. Carefully analysing the four clusters combined heatmaps, we can see that cluster 1 is composed of the four cell lines and cluster 2 is amplification in the DNA and overexpression in the mRNA associated with the HER2/ERBB2.

3.2 Subtypes Discovery in GBM

The GBM dataset contains miRNA (534 genes) and mRNA expression (1740 genes) data from 73 patients with GBM [18]. We used three evaluation metrics to evaluate the result of the Srrr-cluster and iCluster Scluster: (i) The sihouette scores. (ii) The P values. (iii) The POD values. The results of these three metrics are listed in Table 1. According to these metrics, we can see that, using Srrr-cluster method, the within-clusters have stronger coherence and the between-clusters have well separability.

Table 1. Three evaluation metrics to evaluate iCluster and Scluster (3 clustering solution)

Evaluation values	iCluster	Srrr-cluster
Sihouette scores	0.42	0.48
P values	0.31	0.04
POD values	0.20	0.17

4 Discussion

Srrr-cluster method can find more suitable coefficient matrix which can project the original data onto an eigengene-eigenarray subspace when analyzing dataset with small sample size and large variables. In this paper, we proposed to use Srrr-cluster method for cancer subtypes discovery. Compared with iCluster method, our method can identify more stable clusters. However, because Srrr-cluster is established on the basis of iCluster, it has a major limitation that it needs a priori gene selection. In future, we will explore how to solve this problem.

Acknowledgments. This work was supported by National Natural Science Foundation of China (31301101 and 61272339), the Anhui Provincial Natural Science Foundation (1408085QF106), the Specialized Research Fund for the Doctoral Program of Higher Education (20133401120011), and the Technology Foundation for Selected Overseas Chinese Scholars from Department of Human Resources and Social Security of Anhui Province (No. [2014]-243).

References

1. Pollack, J.R., et al.: Microarray analysis reveals a major direct role of DNA copy number alteration in the transcriptional program of human breast tumors. Proc. Natl. Acad. Sci. **99** (20), 12963–12968 (2002)
2. Stratton, M.R., Campbell, P.J., Futreal, P.A.: The cancer genome. Nature **458**(7239), 719–724 (2009)
3. Jones, P.A., Baylin, S.B.: The fundamental role of epigenetic events in cancer. Nat. Rev. Genet. **3**(6), 415–428 (2002)
4. Hoadley, K.A., et al.: Multiplatform analysis of 12 cancer types reveals molecular classification within and across tissues of origin. Cell **158**(4), 929–944 (2014)
5. Verhaak, R.G., et al.: Integrated genomic analysis identifies clinically relevant subtypes of glioblastoma characterized by abnormalities in PDGFRA, IDH1, EGFR, and NF1. Cancer Cell **17**(1), 98–110 (2010)
6. Shen, R., Olshen, A.B., Ladanyi, M.: Integrative clustering of multiple genomic data types using a joint latent variable model with application to breast and lung cancer subtype analysis. Bioinformatics **25**(22), 2906–2912 (2009)
7. Shen, R., et al.: Integrative subtype discovery in glioblastoma using iCluster. PLoS ONE **7** (4), e35236 (2012)
8. Zha, H., et al.: Spectral relaxation for K-means clustering. Neural Inf. Process. Syst. 1057–1064 (2001)
9. Ding, C., He, X.F.: Cluster structure of K-means clustering via principal component analysis. Adv. Knowl. Discov. Data Min. Proc. **3056**, 414–418 (2004)
10. Simon, N., et al.: A sparse-group lasso. J. Comput. Graph. Stat. **22**(2), 231–245 (2013)
11. Huang, J., Breheny P., Ma S.: A selective review of group selection in high-dimensional models. Stat. Sci. **27**(4) (2012)
12. Lin, D.D., et al.: Network-based investigation of genetic modules associated with functional brain networks in schizophrenia. In: 2013 IEEE International Conference on Bioinformatics and Biomedicine (Bibm) (2013)
13. Chen, L., Huang, J.Z.: Sparse reduced-rank regression for simultaneous dimension reduction and variable selection. J. Am. Stat. Assoc. **107**(500), 1533–1545 (2012)
14. Ma, Z., Sun T.: Adaptive sparse reduced-rank regression (2014). arXiv preprint arXiv:1403. 1922
15. Mo, Q., et al.: Pattern discovery and cancer gene identification in integrated cancer genomic data. Proc. Natl. Acad. Sci. USA **110**(11), 4245–4250 (2013)
16. Rousseeuw, P.J.: Silhouettes: a graphical aid to the interpretation and validation of cluster analysis. J. Comput. Appl. Math. **20**, 53–65 (1987)
17. Hosmer Jr, D.W., Lemeshow, S.: Applied Survival Analysis: Regression Modelling of Time to Event Data. European Orthodontic Society (1999)
18. Wang, B., et al.: Similarity network fusion for aggregating data types on a genomic scale. Nat. Methods **11**(3), 333–337 (2014)

An Interactive Segmentation Algorithm for Thyroid Nodules in Ultrasound Images

Waleed M.H. Alrubaidi[1], Bo Peng[1(✉)], Yan Yang[1], and Qin Chen[2]

[1] School of Information Science and Technology,
Southwest Jiaotong University, Chengdu 610031, China
bpeng@swjtu.edu.cn
[2] Department of Ultrasound, Sichuan Provincial People's Hospital,
Chengdu 610072, China

Abstract. Thyroid disease is extremely common and of concern because of the risk of malignancies and hyper-function and they may become malignant if not diagnosed at the right time. Ultrasound is one of the most often used methods for thyroid nodule detection. However, node detection is very difficult in ultrasound images due to their flaming nature and low quality. In this paper, an algorithm for the formalization of the contour of the nodule using the variance reduction statistic is proposed where cut points are determined, then a method of selecting the nearest neighbor points which form the shape of the nodule is generated, later B-spline method is applied to improve the accuracy of the curve shape. The extracted results are been compared with graph_cut and watershed methods for efficiency. Experiments show that the algorithm can improve the accuracy of the appearance of modality and maximum significance of data in the images is also protected.

Keywords: Thyroid nodules · Nodule detection · B_spline curves · VR-Statistics

1 Introduction

The thyroid is a small gland located near the bottom of the neck. It produces hormones that affect heart rate, cholesterol level, body weight, energy level, mental state and a host of other conditions. Epidemiologic studies have shown that palpable thyroid nodules occur in approximately seven percent of the population, but nodules found incidentally on ultrasonography suggest prevalence up to 67 percent [1]. Thyroid nodules are abnormal lumps growing within the thyroid gland which may represent various different conditions including cancer. The risk of developing a palpable thyroid nodule in a lifetime ranges between 5 and 10 % while 50 % of people with solitary nodules detected by experienced physicians have additional nodules detected when further examined by ultrasonography [2]. Ultrasound imaging (US) can be used to detect thyroid nodules that are clinically occult due to their size or shape. However, the interpretation of US images, as performed by the experts, is still subjective. An image analysis scheme for computer aided detection of thyroid nodules would contribute to the objectification of the US interpretation and the reduction of the misdiagnosis rates.

© Springer International Publishing Switzerland 2016
D.-S. Huang et al. (Eds.): ICIC 2016, Part III, LNAI 9773, pp. 107–115, 2016.
DOI: 10.1007/978-3-319-42297-8_11

Ultrasound (US) images contain echo perturbations and speckle noise, which could make the diagnostic task harder. Additionally, image interpretation, as performed by the experts, is subjective. Therefore, a method for computer aided thyroid nodule detection should take into consideration the inherent noise characteristics of the US images and be capable of interpreting these images, based on explicit image features [3]. In order to accomplish computer-aided diagnosis of thyroid nodules on ultrasound images, the nodule's location and its margin must be clearly defined. Some methods based on fully automatic algorithms have been used in Ultrasound images but the lack of efficiency and loss of data might occur. These methods also can produce unacceptable segmentation results for complex scene images. However, manual locating of nodule boundary highly relies on physician's subjective decision. Other methods based on Graph_cut algorithms are also been popular in the last few years [4–7]. Though these sorts of methods might work well to some extent in the form of dividing the "object" and "background" of the image but still when noise diffused edges or occluded objects occur, traditional graph cut algorithm cannot get satisfied segmented object [5].

Poor US image quality and drawbacks caused by the nature of ultrasound limit the performance of various segmentation methods. The final result of the tumor location often depends on region selection. In this study, we proposed a method for thyroid nodule boundary detection based on the combination of radial gradient and Variance-Reduction statistics. We then implement an algorithm to automatically detect the nodule boundary mainly by connecting the most neighbour cut points located in the area of the boundary. Results of the proposed method are then improved with B-Spline method to elaborate the appearance of the curve.

2 Proposed Method

2.1 Analysis

The method implemented here is used to find the maximum and minimum manifest variation of the colors in the original pictures, the region of interest (ROI) is automatically generated based on the selected major and minor axis of the nodule manually inputted by the medical staff. Its radial lines are start growing from the center of the selected ROI to the maximum points that can be generated as shown in Fig. 1 (a–d).

Three colored of cut points on each radial line are searched by the Variance-Reduction statistics so that each cut point resulting in a minimum sum of two group's total sum of square, as in (1).

$$\min_a \left\{ \sum_{i=1}^{a-1} (x_i - \mu_1)^2 + \sum_{i-a+1}^{N} (x_i - \mu_2)^2 \right\} \tag{1}$$

In (1), if a radial line has N number of pixels and their grayscale values are represented in the formula as $\{x_1, x_2, .., x_a, .., x_{N-1}, x_N\}$, where x_i is the grayscale value of the pixel point in order from the center to the boundary of ROI, and x_a is the grayscale value of the cut point that results in the minimum sum of two total sums of squares in which µ1 and µ2 are respectively, the average of the grayscale values of the inner pixels and outer pixels separated by the cut point. Moreover, two additional cut points are found using (1) from the two segments of the radial line separated by the first cut point as shown in Fig. 2.

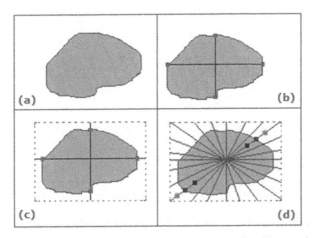

Fig. 1. Diagram of the working of the analysis (a) the main image. (b) input selected by the physician. (c) ROI new matrix based on the selection of major and minor axis. (d) process of cut points starting with the four nodes of the selected axis-es.

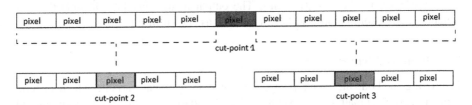

Fig. 2. Three colored cut-points generated by the variance reduction statistic (Color figure online)

2.2   Filtering

The proposed algorithm is applied to the previously generated method taking into account that we need to obtain the most desired cut points located in the area of the boundary. The inferred cut points later shaped the contour. An Input has been generated in the form of lines Matrix where the cut points are represented by three different colors considering Red points as primary.

Firstly, we created a constant value that represented the maximum number of cut points located into different groups called sections. The longest section will start the iteration here and during this process, a thresholding value is maintained in a sense of connecting last point of the last selected section to the next one. After selecting the next connected point we removed the other unwanted points and again start the next phase simultaneously as shown in Fig. 3(c).

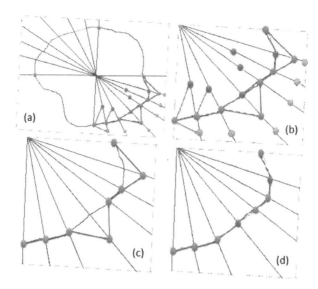

Fig. 3. (a) Creation of section (b) deciding the expected boundary (c) boundary without filter (d) boundary after filtered points.

To avoid some abnormal points that appeared due to the expected noise, the abnormal coordinates been evaluated too to determine the weight of availability and select the most desired closely points. The output matrix will later include the best suitable points coordinators and been extracted as shown in Fig. 3(d).

2.3 B-Spline Method

In filtering method, the result points are stored in a global variable which in term formed the (n * 2) matrix. The exterior of the boundary is appeared to be tortuous and hence the spline stage is generated. Input points are selected by a certain ratio generated and controlled by a changeable constant said to be K.

Depending on how many control points we decide the value of K. Different parameters are occurred using the B-Spline interpolation formula as in (2).

$$x(u) = \sum_{i=0}^{i=n} N_{i,k}(u)x_i \tag{2}$$

For closing the curve, a certain number of control points were added in which spline range started from that number to $(e + 1)$ point. Then all points of that range are appended and the result spline points are drawn as shown in Fig. 4(b).

2.4 Performance Analysis and Evaluation

The performance of the proposed method is evaluated for the delineation of thyroid nodules of 26 ultrasound images. The process for segmenting organs and structures

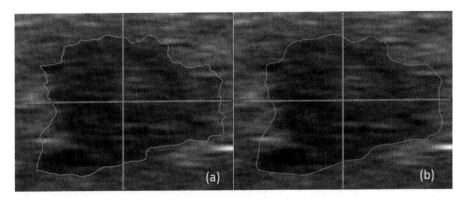

Fig. 4. (a) The output result of filtered method. (b) B-spline resulted in method applied on the extracted result

from medical images is gaining increased importance in the diagnosis of diseases specifically in Thyroid Nodules. And in guiding minimally invasive surgical and therapeutic procedures, the evaluation protocol for the further studies should take into consideration the boundary shape of the contour.

To evaluate the capability of the proposed method, the corresponding performances of our method, graph_cut method [7] and automatic tumor detection based on watershed method [8] were measurable compared. A measure of the accuracy of the thyroid tumor location is measured and the TPF, TNF, FPF, FNF, and Accuracy were also calculated. Each equation is defined as (3–7):

If VS and VT represent the regions enclosed by the segmented boundary and the "true" boundary respectively, we define the true positive volume (TP) as the volume enclosed by both the "true" and algorithm segmented boundaries i.e., VTP = VS ∩ VT, the false positive (FP) volume is VFP = VS − VT, the false negative (FN) volume is VFN = VT − VS, and the true negative (TN) volume is VTN = SCENE − VS − VT briefly explained as in Fig. 5.

(TPF) is the volume fraction in the "true" segmented boundary that is also enclosed by the algorithm segmented boundary; the (FNF) is volume fraction enclosed by the "true" boundary that was missed by the segmentation algorithm, the (FPF) is the volume fraction enclosed by the algorithm segmented boundary that was not enclosed

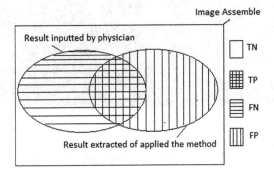

Fig. 5. Evaluation process structure

by the "true" boundary, and the (TNF) is the volume fraction in the background scene that was not enclosed by the "true" boundary and was not enclosed by the algorithm boundaries. After all the accuracy factor is generated. Accordingly the result of this comparison is been studied and shown in Table 1.

Table 1. Comparison of the proposed method and other published methods

Method	TPF (%)	FNF (%)	FPF (%)	TNF (%)	Accuracy (%)
Graph_cut	83.08	16.92	0.19	99.81	99.45
Watershed method	82.96	17.04	0.15	99.85	99.52
Proposed method	82.49	17.51	0.10	99.90	99.63

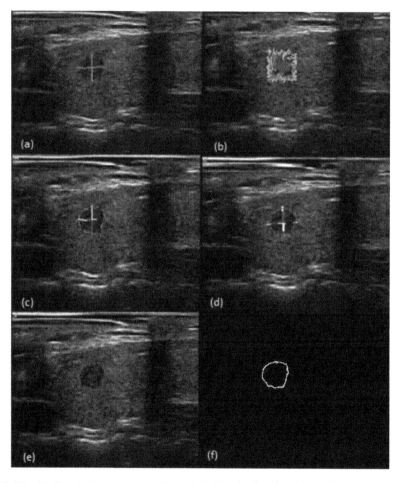

Fig. 6. Result of applying our proposed method (a) selecting the axis on the expected nodule (b) variance reduction method (c) applying the proposed filtering and shape the result (d) applying the b-spline method (e) redraw the final result (f) segment the result for further study.

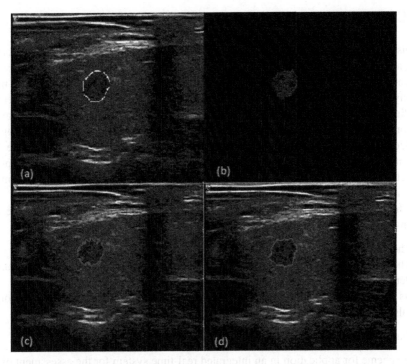

Fig. 7. (a) Result inputted by the doctor. (b) the result of graph_cut method. (c) the result of the automatic watershed method (d) result of our proposed method.

$$TPF = \frac{|A_A \cap A_G|}{|A_G|} \tag{3}$$

$$FNF = \frac{|A_G - A_A|}{|A_G|} \tag{4}$$

$$FPF = \frac{|A_A - A_G|}{|A_G|} \tag{5}$$

$$TNF = \frac{|\overline{A_A \cup A_G}|}{|\overline{A_G}|} \tag{6}$$

$$Accuracy = \frac{TP + TN}{TP + TN + FP + FN} \tag{7}$$

Where A_G represent the area of the ground truth of foreground and its complement is $\overline{A_G}$, A_A represent the area of segmented foreground by the tested segmentation method. Table 1 lists the results of the corresponding formulas of the three measured methods applied on 26 images. We can see that our method achieved the best of the accuracy results (Figs. 6 and 7).

3 Conclusion

In this paper, an algorithm for detecting the contour of thyroid nodule is applied based on parameters that been extracted from the combination of radial gradient and variance-reduction statistic. The proposed method is applied depending on the most closely connected cut points which are located in the area of the boundary. Later the b-spline method is applied so that the final result would be more apparent and to avoid the tortuous resulted from the filtered method. The information of current US image is needed where the image can be changed to a binary image. This takes over the disadvantage of the graph_cut method as some of the data would be lost when applied. Detection accuracy of our method appeared clearer than the watershed automatic method.

The results of the experimental study presented in this paper also led to the following conclusions for the proposed scheme:

- The proposed scheme can improve nodule detection accuracy.
- The proposed scheme can considerably decrease processing time needed for nodule detection.
- Its application clinical practice is feasible and could contribute to the reduction of false medical decisions.

Future work and perspectives include:

- A scheme for application in an integrated real-time system for the assessment of the thyroid gland.
- Experimentation to determine the optimal feature extraction and classification methods for Thyroid Nodule detection.
- Performance of the proposed scheme will be investigated on images acquired from different US imaging system.
- An open problem is still precision, the efficiency of the results of our proposed algorithm compared to the real results inputted by the medical staff is a bit less where the process of finding compatible smoothing methods and reducing the noise of the images would be targeted.

Acknowledgments. This work is supported by the National Science Foundation of China (No. 61572407) and Technology Planning Project of Sichuan Province (No. 2014SZ0207).

References

1. Welker, M.J., Orlov, D.: Thyroid nodules. Am. Fam. Physician **67**(3), 559–566 (2003)
2. Feld, S., et al.: AACE Clinical Practice Guidelines for the Diagnosis and Management of Thyroid Nodules, Endocrine Practice, pp. 78–84, January/February 1996
3. Smutek, D., Sara, R., Sucharda, P., Tjahjadi, T., Svec, M.: Image texture analysis of sonograms in chronic inflammations of the thyroid gland. Ultrasound Med. Biol. **29**(11), 1531–1543 (2003). Wong, R.Y., Hall, E.L.: Sequential hierarchical scene matching. IEEE Trans. Comput. **27**, 359–366 (1978)

4. Boykov, Y.Y., Jolly, M.P.: Interactive graph cuts for optimal boundary and region segmentation of objects in N-D images. In: Eighth IEEE International Conference on Computer Vision, pp. 105–102. IEEE Press, Vancouver (2001)
5. Boykov, Y., Veksler, O., Zabih, R.: Fast approximate energy minimization via graph cuts. Pattern Anal. Mach. Intell. **23**(11), 1222–1239 (2001)
6. Kwatra, V., Schödl, A., Essa, I., Turk, G., Bobick, A.: Graphcut textures: image and video synthesis using graph cuts. ACM Trans. Graphics **22**(3), 277–286 (2003)
7. Tian, H., Peng, B., Li, T., Chen, Q.: A novel graph cut algorithm for weak boundary object segmentation. Found. Intell. Syst. **227**, 263–271 (2014)
8. Zhang, L., Ren, Y., Huang, C., Liu, F.: A novel automatic tumor detection for breast cancer ultrasound images. In: 2011 Eighth International Conference on Fuzzy Systems and Knowledge Discovery (FSKD), vol. 1, pp. 401–404 (2011)
9. Tsantis, S., Dimitropoulos, N., Cavouras, D., Nikiforidis, G.: A hybrid multi-scale model for thyroid nodule boundary detection on ultrasound images. Comput. Method Program Biomed. **84**, 86–98 (2006)
10. Madabhushi, A., Metaxas, D.N.: Combining low-, high-level and empirical domain knowledge for automated segmentation of ultrasonic breast lesions. IEEE Trans. Med. Imaging **22**(2), 155–169 (2003)
11. Blue, J., Chen, A.: Spatial variance spectrum analysis and its applications to unsupervised detection of systematic wafer spatial variations. IEEE Trans. Autom. Sci. Eng. **8**, 56–66 (2010)
12. Abbas, A., Nasri, A., Maekawa, T.: Generating B-spline curves with points, normals, and curvature: a constructive approach. Vis. Comput. **26**, 823–829 (2010)
13. Tsantis, S., Dimitropoulos, N., Cavouras, D., Nikiforidis, G.: A hybrid multi-scale model for thyroid nodule boundary detection on ultrasound images. Comput. Method Programs Biomed. **84**, 86–98 (2006)
14. Chalana, V., Kim, Y.: A methodology for evaluation of boundary detection algorithms on medical images. IEEE Trans. Med. Imaging **16**, 642–652 (1997)

Coordinating Discernibility and Independence Scores of Variables in a 2D Space for Efficient and Accurate Feature Selection

Juanying Xie[1(✉)], Mingzhao Wang[1], Ying Zhou[1], and Jinyan Li[2]

[1] School of Computer Science, Shaanxi Normal University,
Xi'an 710062, People's Republic of China
xiejuany@snnu.edu.cn
[2] Faculty of Engineering and Information Technology,
University of Technology Sydney,
P.O. Box 123, Broadway, NSW 2007, Australia
Jinyan.Li@uts.edu.au

Abstract. Feature selection is to remove redundant and irrelevant features from original ones of exemplars, so that a sparse and representative feature subset can be detected for building a more efficient and accurate classifier. This paper presents a novel definition for the discernibility and independence scores of a feature, and then constructs a two dimensional (2D) space with the feature's independence as y-axis and discernibility as x-axis to rank features' importance. This new method is named FSDI (Feature Selection based on Discernibility and Independence of a feature). The discernibility score of a feature is to measure the distinguishability of the feature to detect instances from different classes. The independence score is to measure the redundancy of a feature. All features are plotted in the 2D space according to their discernibility and independence coordinates. The area of the rectangular corresponding to a feature's discernibility and independence in the 2D space is used as a criterion to rank the importance of the features. Top-k features with much higher importance than the rest ones are selected to form the sparse and representative feature subset for building an efficient and accurate classifier. Experimental results on 5 classical gene expression datasets demonstrate that our proposed FSDI algorithm can select the gene subset efficiently and has the best performance in classification. Our method provides a good solution to the bottleneck issues related to the high time complexity of the existing gene subset selection algorithms.

Keywords: Discernibility · Independence · Feature selection · Gene subset selection

1 Introduction

The fast growing of high-dimensional data sets with lots of redundant and irrelevant features brings great challenges to machine learning and data mining algorithms. Feature selection methods can choose those features which are highly correlated to labels and lowly redundant between them, without sacrificing the classification performance of the

© Springer International Publishing Switzerland 2016
D.-S. Huang et al. (Eds.): ICIC 2016, Part III, LNAI 9773, pp. 116–127, 2016.
DOI: 10.1007/978-3-319-42297-8_12

learning algorithm. Very often, classification models built on the selected feature subset are more accurate and easier to understand, and have a better generalization capacity, higher efficiency, reduced curse of dimensionality, and more intuitive visualization analysis [4].

A feature selection algorithm usually has two parts: feature subset search and feature subset evaluation [14]. According to the dependencies between a feature selection process and the learning algorithm, feature selection approaches can be divided into two categories: the Filters and the Wrappers [1, 11]. The filters are independent to learning algorithms, and their feature selection processes are done via an evaluation criterion which defines the feature importance without considering the learning algorithms [1]. As a consequence filters identify all the relevant features, and these features are all considered as important features to constitute the feature subset. Filters are always fast and with good generalization capability, such as Relief [10], correlation based feature selector (CFS) [6] and maximal relevance-minimal redundancy (mRMR) [15] which are the classical filter feature selection methods. Wrappers rely on the learning algorithms [11] and use the predictive accuracy of the learning algorithms on the validation datasets to test the power of the related feature subset. In general, wrappers select a sparse and representative feature subset for building a more accurate classifier. However, the computational load of wrappers is heavier than that of filters because the classification models need to be trained repeatedly in wrappers. In addition wrappers may lead to over-fitting effects on small datasets. SVM-RFE (SVM recursive feature elimination) [5] and SVM-SFS (SVM sequential forward search) [21] are typical wrappers and they have got good performance on gene expression microarray data analysis. The hybrid methods combine the filters and wrappers together to achieve a better performance. The hybrid approach has become a widely studied area for feature selection [4, 8, 21].

Gene expression data sets having tens of thousands of features but with small numbers of samples contain a high level of redundant and irrelevant gene variables for disease diagnosis purposes [3, 12, 16]. Feature selection is the primary task to analyze this type of data [4]. Time complexity bottleneck is the main issue of the available gene selection algorithms, especially for the wrappers. The cluster analysis can be applied to feature selection by choosing typical features from each cluster to construct the selected gene subset, which can partially solve the time bottleneck problem in gene selection algorithms [17, 20, 21]. However, how to detect the correct clusters is still an open question.

To select a sparse and representative feature subset and to avoid the bottleneck problems in the gene selection process, we propose a new feature selection algorithm named FSDI short for feature selection based on the discernibility and independence of a feature. FSDI defines a score for the feature discernibility and feature independence. It then uses feature independence as the y-axis and feature discernibility as x-axis to construct a two dimensional space. All features are scattered in the two dimensional space with their discernibility and independence scores as coordinates. The features at the top-right corner of the two dimensional space are those ones whose discernibility and independence are relatively large. We adopt the area of the rectangular corresponding to each feature's discernibility and independence to measure the importance of the feature. The rectangular area corresponding to the top-right corner features is always much larger than the remaining ones, so the area for each feature can be used as a weight of the feature. Those features with larger weights are detected by FSDI to construct the sparse and representative feature subset. The classifier built on the feature subset will be more accurate. FSDI takes into account the ability of a feature to identify instances from different classes in its discernibility and the redundancy of a feature in its independence simultaneously, so that it can guarantee the selected genes are the representative ones with high relevancy to the classes and the smallest redundancy as much as possible.

On the high-dimensional gene expression data sets, we first cluster all of the genes to select those typical genes from each cluster to form the preselected gene subset. FSDI is carried out on the preselected gene subsets to get the optimal gene subsets. Experimental results on 5 classical gene expression datasets demonstrate that our FSDI method can detect gene subsets with a high efficiency, and the classifiers built on the selected gene subsets have got a better classification performance for the diagnosis purposes than those classical gene selection algorithms that are available now.

This paper is organized as follows: Sect. 2 describes the proposed FSDI method. Section 3 presents the performance results of our FSDI method on 5 high-dimensional gene expression datasets. Section 4 draws our conclusions.

2   The FSDI Algorithm

The most contribution of our FSDI is that it defines the discernibility and independence scores for each feature and constructs the two dimensional space in the feature's discernibility and independence, so that all features are scattered in the two dimensional space and the features lying at the top-right corner of the two dimensional space will be automatically detected to construct the selected feature subset to build a classifier with higher accuracy.

2.1   Preliminary Feature Selection

K-means algorithm [13] is a fast and classical clustering algorithm. It can be used to cluster big data [9, 21]. In this paper, K-means algorithm is used to cluster all genes of data into clusters of $k' = 30$. Then we use Wilcoxon Signed-rank test to measure the weight of genes and calculate the average weight of each cluster. The genes above the average weight of its cluster are preserved to constitute the preselected gene subset.

According to the principles of clustering, features in same clusters are highly similar to each other and those in different clusters are dissimilar as much as possible. Therefore, we can say that features in same clusters are highly redundant and in different clusters are relatively independent with little redundancy. Therefore the preliminary selection can reduce feature redundancy and retain feature discriminative ability, which will speed up the feature selection process and reduce the requirement of storage space and the curse of dimensionality.

Wilcoxon Signed-rank test is a nonparametric statistical method in statistic hypothesis. It can avoid the influence from parameters, so we adopt it in our research work. It is calculated in Eq. (1).

$$S(f_i) = \sum_{k=1}^{N_0} \sum_{j=1}^{N_1} \chi\left((X_{j,f_i} - X_{k,f_i}) \leq 0\right) \tag{1}$$

where $\chi(\cdot)$ is the discriminant function, and $\chi(\cdot) = 1$ if $(X_{j,f_i} - X_{k,f_i}) \leq 0$, otherwise $\chi(\cdot) = 0$. X_{j,f_i} is the expression value of gene f_i in sample j. The number of samples in two classes of a dataset are respectively denoted by N_0 and N_1. It can be seen from Eq. (1) that the feature f_i has got good ability to detect samples from two classes when its Wilcoxon rank sum statistic value is close to 0 or $N_0 * N_1$ without considering that it has got the same value in all samples. The weight w_i of f_i is calculated by Eq. (2). The higher the value of w_i is the stronger the ability of feature f_i to discriminate samples from different classes, and the greater contribution of it to the classification task.

$$w_i = max[N_0 * N_1 - S(f_i), S(f_i)] \tag{2}$$

2.2   The Main Idea of FSDI Algorithm

Here the discernibility and independence scores of a feature are introduced, then the area of the rectangular is calculated corresponding to a feature's coordinates. FSDI algorithm selects features with much higher values of their rectangular area than the rest ones to constitute the selected feature subset.

Feature Discernibility and Independence and Importance. Suppose $\mathbf{D} = \{X_1; X_2; \cdots; X_m\} \in R^{m \times n}$ to be a train subset with m samples and n features for each. Feature discernibility and independence of feature f_i are defined as follows.

- **Feature discernibility:** we adopt Wilcoxon Signed-rank test to assess the discernibility dis_i of feature f_i and define it as the distinguishable ability of f_i to detect samples from different classes in Eq. (3). We can learn from Eq. (3) that the higher the distinguishable ability of feature f_i, is the higher value of its dis_i, that means the more importance is the feature f_i to classification.

$$dis_i = w_i \tag{3}$$

- **Feature independence:** independence of feature f_i, shown in Eq. (4), is defined as the distance of correlation between features. If a feature has the maximum discernibility value, its independence is defined by Eq. (5).

$$ind_i = \min_{j:dis_j \succ dis_i} \left(\exp\left(-r(f_i, f_j)\right)\right) \tag{4}$$

$$ind_i = \max_j \left(\exp\left(-r(f_i, f_j)\right)\right) \tag{5}$$

where r is the absolute value of Pearson correlation coefficient between features and it is calculated in Eq. (6).

$$r = \frac{\left|(\mathbf{X} - \bar{\mathbf{X}})^T (\mathbf{Y} - \bar{\mathbf{Y}})\right|}{\sqrt{\|\mathbf{X} - \bar{\mathbf{X}}\|^2 \|\mathbf{Y} - \bar{\mathbf{Y}}\|^2}} \tag{6}$$

where $\mathbf{X}, \mathbf{Y}$ indicate two feature vectors, and $\bar{\mathbf{X}}, \bar{\mathbf{Y}}$ are respectively the mean value of feature vector $\mathbf{X}$ and $\mathbf{Y}$.

It can be seen from (4) that the independence ind_i of feature f_i means the relevance of feature f_i with feature f_j whose discernibility is just higher than f_i. If feature f_i has got global maximum discernibility, its independence is the maximum distance of relevance with other features. The formulae (4)–(6) reveal that the feature with stronger correlation with others will obtain the weaker independence, which means the stronger the independence of a feature is, the very lower correlation of the feature with others is. This is coincidence with the real world situation.

It can be seen from (3)–(5) that the stronger distinguishability of feature f_i, is the bigger its discernibility dis_i; and the smaller the redundancy with other features, is the bigger its independence ind_i. Therefore we can construct a two dimensional co-ordinate system with features' independence as y-axis and discernibility as the x-axis, and adopt the area of the rectangle with the points $(0,0)$ and (ind_i, dis_i) as opposite vertices to denote the importance of feature f_i. Features with large value of rectangle area possess informative information for classification whilst with little redundancy. Such kind of features are the ones we are seeking for to constitute the feature subset, which coincide with the original meaning of feature selection. The importance of feature f_i is defined in (7).

Feature Importance: the importance of feature f_i denoted as $Score_i$ is defined as the area of the rectangle surrounded by its discernibility and independence and the axes, which is calculated in Eq. (7).

$$Score_i = dis_i \times ind_i \tag{7}$$

It can be seen from formula (7) that the bigger the discernibility and independence of a feature are, the larger is the value of the importance $Score$ of the feature, and at the same time the greater will the feature contribute to the classification. Calculating the importance of each feature and choosing the top features with much higher importance than the rest features to construct the feature subset, guarantee that the selected feature subset will have a better performance in classification and with very lower redundancy between features.

The FSDI Algorithm. The ideal feature subset is the one with features strong correlated to class labels whilst less redundant between features [15]. In this paper, we propose feature discernibility to measure the distinguishability of a feature between classes and feature independence to value the redundancy of a feature, and construct the two dimensional space in the feature's discernibility and independence. All features are scattered in the two dimensional space. It can be seen from the definitions of a feature's discernibility and independence in formulae (3)–(5) that the stronger the distinguishability of a feature possesses, then the larger of its discernibility is; whilst the smaller the feature redundancy is, then the stronger of its independence is. The features with larger discernibility and independence are always scattered at the top-right corner of the two dimensional space with higher importance. FSDI will automatically detect those features with higher importance than the rest ones to construct the feature subset on which to build the classifier with more accurate. As a consequence that FSDI to some extend solved the problem of how many number of features should be selected in the feature selection algorithms.

Here are the detail steps of our FSDI algorithm.
Input: train subset data D with m samples and n features for each sample, vector Y for class labels, parameter k be the number of features in the selected feature subset.
Output: the selected feature subset S.
Initialize $S=\varnothing$, F be the feature set with all features;
for i = 1 to n do
    calculate dis_i and ind_i respectively by(3) and (4) or (5);
    calculate the importance $Score_i$ of feature f_i by (7);
end for;
features are sorted in descending order in their $Score$;
add top k features with much larger $Score$ than the rest ones to S .

2.3   An Illustrating Example

Here we will test our FSDI in a random synthetic case. The synthetic dataset contains 20 samples from 2 classes, and with 50 features for each sample. We partition the synthetic dataset into train and test subset in bootstrap [7], and run our FSDI on the train subset, then build the SVM classifier on the selected feature subset. We calculate the classification accuracy in formula (8) to balance the overfitting and generalization, where M is the classifier built on the selected gene subset. We use the SVM library [2] to conduct experiments with parameter C for the linear kernel be 20. Here we do not conduct preliminary selection to features for the number of them is only 50. Figure 1(a) displays the 50 features in the two dimensional space in their discernibility and independence with their number in original dataset. Figure 1(b) shows the importance of 50 features in descending order where the y-axis is the feature importance and x-axis is the number of features.

$$Acc = 0.632 \times Acc(M)_{test} + 0.368 \times Acc(M)_{train} \qquad (8)$$

It can be seen from Fig. 1(a) that the discernibility and independence of the 49th feature are very small whist the features of 48th, 39th and 26th have got the much higher discernibility and independence than the rest features. The results displayed in Fig. 1(b) disclose that the 48th, 39th and 26th features have got the highest, the second

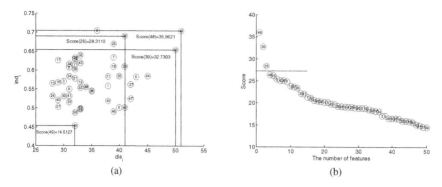

Fig. 1. The descriptions of feature importance, (a) features are scattered in their (dis_i, ind_i) in two dimensional space, (b) feature are scattered in their importance in descending order.

and the third rank of *Scores* whilst the 49th feature the smallest one. These results shown in Fig. 1(a) and (b) agree with each other.

We build the SVM classifier on the feature subset of features 48, 39 and 26. The mean accuracy of the SVM classifier is 100 %. We further analyze the performance of feature subset with top 1 and 2 features, then we respectively get the mean accuracy of 74.72 % and 89.81 %. From the analysis to the case study we can say that the definition of feature importance in this paper is reasonable and it can be used to detect those informative features for classification.

3   Experiments and Analysis

Experiments are conducted on 5 intensively studied gene expression datasets: the Colon, CNS, Leukemia, Carcinoma, and Breast Cancer datasets[1]. Table 1 describes detailed information of these datasets, where Ng and Ns denote the number of features and instances respectively. We first do the preliminary gene selection to the original genes, then run FSDI to detect the optimal gene subset from the preselected gene subset, after that we construct the classifier on the optimal gene subset.

We adopted KNN as a classification tool with K equal to 5, and compared the experimental results of our FSDI with that of the classical methods such as Weight [21], mRMR [15], ARCO [18], SVM-RFE [5] and Relief [10] when the same number of genes are selected (Table 2). Figures 2, 3, 4, 5 and 6 displayed the genes in the two dimensional space, and tagged the genes detected by our FSDI on the 5 gene expression datasets.

From the experimental results of our FSDI on these gene expression datasets shown in Figs. 2, 3, 4, 5 and 6, we can say that the features with high discernibility and independence are scattered at the top-right corner of the two dimensional space with features' discernibility and independence as coordinates. These features have got much

---

[1] http://levis.tongji.edu.cn/gzli/data/mirror-kentridge.html.

Table 1. The description of datasets

Gene datasets	Source	Ng	Ns
Colon	Alon *et al.*	2000	62(40 + 22)
CNS	Notterman *et al.*	7129	90(60 + 30)
Leukemia	Golub *et al.*	7129	72(47 + 25)
Carcinoma	Notterman *et al.*	7458	36(18 + 18)
Breast Cancer	Van't Veer L.J. *et al.*	24481	97(51 + 46)

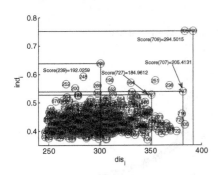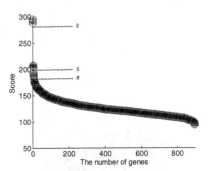

Fig. 2. The results of FSDI on Colon dataset

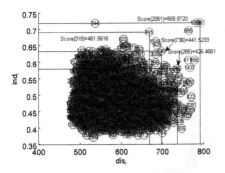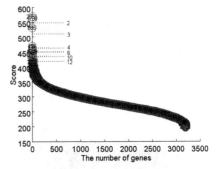

Fig. 3. The results of FSDI on CNS dataset

higher importance than the rest features which are scattered at the bottom-left corner of the two dimensional space. Our FSDI can automatically detect those features which possess much higher importance than the rest ones to construct the selected gene subsets on which to build the classifier with more accurate.

From the figures in Table 2, we can see that our FSDI dominates the other gene selection algorithms in Colon, CNS and Carcinoma datasets, and its performance on Leukemia is similar to ARCO followed by Weight, mRMR, Relief and SVM-RFE, and its performance on Breast Cancer is similar to mRMR, followed by ARCO, Weight,

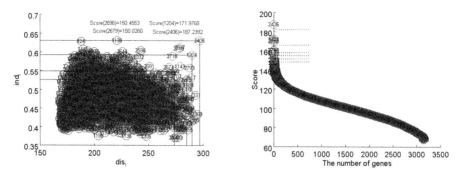

Fig. 4. The results of FSDI on Leukemia dataset

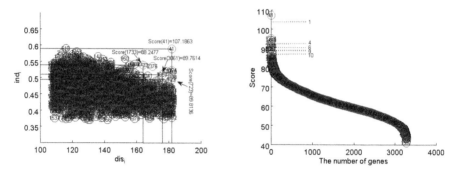

Fig. 5. The results of FSDI on Carcinoma dataset

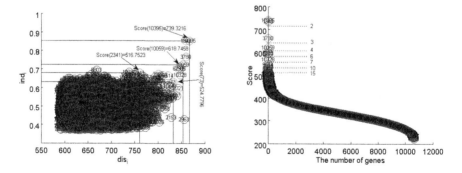

Fig. 6. The results of FSDI on Breast Cancer dataset

Relief and SVM-RFE. These results shown in Table 2 disclose that our FSDI is the best one among the compared gene selection algorithms, and also reveal that the popular gene selection algorithm SVM-RFE is the worst one among the compared gene selection algorithms. Followed our FSDI are the gene selection algorithms mRMR and

Table 2. The classification accuracy comparison of FSDI with other gene selection algorithms

Dataset	FSDI	Weight	mRMR	SVM-RFE	Relief	ARCO	Genes numbers
Colon	**0.8963**	0.5735	0.7743	0.6455	0.8010	0.8094	2
	0.8696	0.7982	0.8345	0.6606	0.8361	0.8445	5
	0.9131	0.7262	0.8361	0.8094	0.8094	0.8712	8
CNS	**0.8575**	0.7315	0.7042	0.7149	0.7423	0.8183	2
	0.9085	0.7120	0.7791	0.8230	0.7660	0.8183	3
	0.9145	0.8173	0.7957	0.8860	0.7838	0.8919	4
	0.9145	0.8173	0.8468	0.8753	0.8183	0.8860	6
	0.9323	0.8270	0.8016	0.8408	0.8242	0.9038	10
	0.9656	0.8451	0.8242	0.8242	0.8468	0.9264	12
Leukemia	**0.9346**	0.7649	0.8230	0.7374	0.8505	**0.9346**	1
	0.9256	0.7576	**0.9442**	0.8513	0.9063	0.9160	4
	0.9256	0.7762	0.9071	0.8230	0.8877	**0.9346**	6
	0.9256	0.8416	0.9071	0.7947	0.9160	0.9160	7
	0.9256	0.9249	0.9256	0.8044	0.8885	**0.9346**	8
	0.9814	**0.9814**	0.9071	0.8416	0.9071	0.9160	11
Carcinoma	**0.9864**	0.7127	0.8459	0.7369	0.9591	0.7893	1
	1	0.9847	**1**	0.9161	0.9864	0.9025	4
	1	**1**	**1**	0.9298	**1**	0.8596	6
	1	**1**	**1**	0.9298	**1**	0.9298	9
	1	0.8947	**1**	0.9298	**1**	0.8596	10
Brest Cancer	0.6922	0.6365	0.6132	0.6188	0.6245	**0.7770**	2
	0.7402	0.5902	**0.8108**	0.6161	0.6951	0.7854	3
	0.7995	0.6848	**0.8305**	0.6866	0.6640	0.7855	4
	0.7967	0.7775	**0.7967**	0.7234	0.6697	0.7544	6
	0.7601	0.7182	**0.7910**	0.6980	0.6697	0.7544	7
	0.8532	0.7200	0.8024	0.6923	0.6837	0.7996	10
	0.8237	0.7683	0.7742	0.7290	0.6753	0.8194	15

ARCO. Relief and Weight algorithms have got the similar performance when used to do gene selection for gene expression data analysis. In addition the size of the selected gene subset shown in Table 2 reveals that selected genes are sparse compared to the original ones.

From the above analysis we can state that our FSDI can find the informative genes to construct the optimal gene subset on which to build the KNN classifier with more accurate than the compared gene selection algorithms, and we can say that our FSDI can detect the gene subset with features relevant to classes and sparse and representative. Therefore we can conclude that our FSDI implement the destination of filters and wrapper simultaneously.

4 Conclusions

We have proposed a new feature selection algorithm named FSDI for informative gene selection, and defined the discernibility and independence scores of a feature to value the distinguishability and redundancy of a feature. The main contribution of this work is the construction of a two dimensional space coordinating a feature's discernibility and independence scores at the x-axis and y-axis. All the relevant features can be found and the sparse and representative feature subset can be formed by collecting those features which are always scattered at the top-right corner of the two dimensional space. Experimental results on 5 widely studied benchmark gene expression datasets demonstrate that our FSDI method has achieved better performance than the typical gene selection algorithms in terms of classification accuracy and efficiency. FSDI combines the merits of both the filter and wrapper feature selection approaches. Our method provides a good solution to the bottleneck issues related to the high time complexity of the existing gene subset selection algorithms.

Acknowledgements. We are much obliged to those who share the gene expression datasets with us. This work is supported in part by the National Natural Science Foundation of China under Grant No. 31372250, is also supported by the Key Science and Technology Program of Shaanxi Province of China under Grant No. 2013K12-03-24, and is at the same time supported by the Fundamental Research Funds for the Central Universities under Grant No. GK201503067, and by the Innovation Funds of Graduate Programs at Shaanxi Normal University under Grant No. 2015CXS028.

References

1. Blum, A.L., Langley, P.: Selection of relevant features and examples in machine learning. Artif. Intell. **97**(1), 245–271 (1997)
2. Chang, C.-C., Lin, C.-J.: LIBSVM: a library for support vector machines. ACM Trans. Intell. Syst. Technol. (TIST) **2**(3), 27 (2011)
3. Ding, C., Peng, H.: Minimum redundancy feature selection from microarray gene expression data. J. Bioinform. Comput. Biol. **3**(2), 185–205 (2005)
4. Guyon, I., Elisseeff, A.: An introduction to variable and feature selection. J. Mach. Learn. Res. **3**, 1157–1182 (2003)
5. Guyon, I., Weston, J., Barnhill, S., Vapnik, V.: Gene selection for cancer classification using support vector machines. Mach. Learn. **46**(1–3), 389–422 (2002)
6. Hall, M.A.: Correlation-based feature selection for machine learning. The University of Waikato (1999)
7. Han, J., Kamber, M., Pei, J.: Data mining: concepts and techniques: concepts and techniques. Elsevier (2011)
8. Hu, Q., Pedrycz, W., Yu, D., Lang, J.: Selecting discrete and continuous features based on neighborhood decision error minimization. IEEE Trans. Syst. Man Cybern. Part B Cybern. **40**(1), 137–150 (2010)
9. Huang, Z.: Extensions to the k-means algorithm for clustering large data sets with categorical values. Data Min. Knowl. Discov. **2**(3), 283–304 (1998)

10. Kira, K., Rendell, L.A.: The feature selection problem: Traditional methods and a new algorithm. Paper presented at the AAAI (1992)
11. Kohavi, R., John, G.H.: Wrappers for feature subset selection. Artif. Intell. **97**(1), 273–324 (1997)
12. Li, Y.-X., Li, J.-G., Ruan, X.-G.: Study of informative gene selection for tissue classification based on tumor gene expression profiles. Chin. J. Comput. Chin. Ed. **29**(2), 324 (2006)
13. MacQueen, J.: Some methods for classification and analysis of multivariate observations. Paper presented at the Proceedings of the Fifth Berkeley Symposium on Mathematical Statistics and Probability (1967)
14. Mao, Y., Zhou, X., Xia, Z., Yi, Z., Sun, Y.: A survey for study of feature selection. Algorithm **20**(2), 211–218 (2007). (in Chinese)
15. Peng, H., Long, F., Ding, C.: Feature selection based on mutual information criteria of max-dependency, max-relevance, and min-redundancy. IEEE Trans. Pattern Anal. Mach. Intell. **27**(8), 1226–1238 (2005)
16. Shah, M., Marchand, M., Corbeil, J.: Feature selection with conjunctions of decision stumps and learning from microarray data. IEEE Trans. Pattern Anal. Mach. Intell. **34**(1), 174–186 (2012)
17. Song, Q., Ni, J., Wang, G.: A fast clustering-based feature subset selection algorithm for high-dimensional data. IEEE Trans. Knowl. Data Eng. **25**(1), 1–14 (2013)
18. Wang, R., Tang, K.: Feature Selection for Maximizing the Area Under the ROC Curve, pp. 400–405 (2009)
19. Xie, J., Gao, H.: Statistical correlation and k-means based distinguishable gene subset selection algorithms. J. Softw. **9**, 013 (2014). (in Chinese)
20. Xie, J., Gao, H.: A stable gene subset selection algorithm for cancers. In: Yin, X., Ho, K., Zeng, D., Aickelin, U., Zhou, R., Wang, H. (eds.) HIS 2015. LNCS, vol. 9085, pp. 111–122. Springer, Heidelberg (2015)
21. Xie, J., Xie, W.: Several feature selection algorithms based on the discernibility of a feature subset and support vector machines. Chin. J. Comput. Chin. Ed. **37**(8), 1704–1718 (2014). (in Chinese)

Palm Image Classification Using Multiple Kernel Sparse Representation Based Dictionary Learning

Pingang Su$^{(\boxtimes)}$ and Tao Liu

Department of Electrical Automation,
College of Electronic Information Engineering,
Suzhou Vocational University, Suzhou 215104, Jiangsu, China
{supg,lt}@jssvc.edu.cn

Abstract. Sparse representation (SR) can effectively represent structure features of images and has been used in image processing field. A new palmprint image classification method by using multiple kernel sparse representation (MKSR) is proposed in this paper. Kernel sparse representation (KSR) behaves good robust and occlusion like as sparse representation (SR) methods. Especially, KSR behaves better classification property than common sparse representation methods and used widely in pattern recognition task. In KSR based classification methods, the selection of a kernel function and its parameters is very important. Usually, the kernel selected is not the most suitable and can not contain complete information. Therefore, MKSR methods are developed currently and used widely in image classification task. Here, multiple kernel functions select the weighted of Gauss kernel and polynomial kernel. In test, all palmprint images are selected from PolyU palmprint database. The palm classification task is implemented by the extreme learning machine (ELM) classifier. Compared with methods of SR and single kernel based SR, experimental results show that our method proposed has better calcification performance.

Keywords: Sparse coding · Dictionary learning · Multiple kernel function · ELM classifier · Sparse representation · Palmprint classification

1 Introduction

Good features that can describe salient aspects in an image are very crucial step in the task of object recognition systems [1, 2]. Moreover, adapting to variations in the visual appearance of images beyond the training set is crucial [1]. So, recognition systems often employ feature extraction methods, which can provide high discrimination between classes. But, no such feature descriptor can provide good discrimination for all classes of images. So, feature extraction methods are still challenging issues to this day. Currently, sparse representation (SR) based feature extraction methods have been the research hot. SR has undergone rapid development both in theory and in algorithms and has been used widely in the signal or image processing field. This is partly due to the fact that signals or images of interest, through high dimensional, can often be coded using a few representative atoms in some dictionary [2]. In recent years, SR based

© Springer International Publishing Switzerland 2016
D.-S. Huang et al. (Eds.): ICIC 2016, Part III, LNAI 9773, pp. 128–138, 2016.
DOI: 10.1007/978-3-319-42297-8_13

classification algorithms have also been explored by more and more researchers [3–5]. Although the SR based classification method is robust to noise and occlusion and can well find a linear representation of a signal or an image, it is not suitable to represent non-linear structures of this signal or this image, which arises in many practical applications. To deal with this problem, the kernel trick is introduced into SR based algorithms so as to map the non-linear data into high dimensional feature space, thus, the data of the same distribution can be easily grouped together and are linearly separable in high dimensional kernel feature space [6–8], in other words, for the same class of test samples and training samples, test samples can be truly linear represented by the training samples. Therefore, in high dimensional feature space, SR algorithms can improve the recognition rate and discrimination performance. And kernel based sparse representation (KSR) algorithms have been proved to produce better classification results than common SR based ones [3]. KSR based classification algorithms usually require a predetermined kernel function such as the Gauss kernel, Sigmoid kernel, radial basis function (RBF) kernel, the polynomial kernel and so on [7, 8]. In classification task, the selection of kernel function and its parameters is an important issue in learning sparse coefficients and dictionaries. Because it is not clear which kernel function is the best in dictionary learning, the combination of several probable kernel functions, namely the multiple kernel learning (MKL) method, is very meaningful in application. In other word, MKL methods allow one to use multiple kernels instead of using a specific kernel function [2]. Finding appropriate feature combinations entails designing good kernel functions among a set of candidate kernels. So, many classification algorithms based on the multiple kernel sparse representation (MKSR) dictionary learning have been developed at present [5], for example, Huang, H.C., et al. used the MKSR method to solve well the cluster problem [1, 2]. In this paper, we use the MKSR based dictionary learning method to learn sparse dictionaries and discuss the classification performance of palmprint images in different classifiers. The Gauss kernel and the polynomial kernel are selected to make up the combination of multiple kernels. In test, the PolyU palmprint database used widely in palmprint recognition research is utilized, and the classification task is implemented by using the extreme learning machine (ELM) classifier. Compared with single kernel based SR methods, experimental results show that palmprint classification using the method of MKSR based dictionary learning behaves the best recognition effect.

2 Sparse Representation Based Classification Method

The method of SR based classification can represent sparsely and linearly test samples by using original training samples, and realizes the classification task according to linear class reconstruction errors [3, 4]. Assumed that the training sample set is denoted by $X = \left[x_1^1, \cdots, x_{N1}^1, \cdots, x_1^C, \cdots, x_{N_C}^C \right] \triangleq [x_1, \cdots, x_N] \in \Re^{d \times N}$, where $X_C \in \Re^{d \times N_C}$ is the matrix of training images from the cth class and $N = \sum_C N_C$, parameter d is the dimension number of each image, C is the number of sample classes and N_C is the number of training samples per class. Then, a new test sample $x_t \in \Re^d$ of unknown class can be linearly represented by the following form:

$$x_t = \sum_{c=1}^{C} \sum_{i=1}^{N_C} s_i^c \, d_i^c . \tag{1}$$

where $s_i^c \in \Re$ is the *ith* coefficient vector of the *cth* class, and d_i^c is the *ith* column vector of the *cth* class. This means that many coefficients without association with class c will be close to zero. Usually, a sparse vector can be restored by using the following optimization equation:

$$s_t = \arg\min_s \|x_t - DS\|_2^2 + \lambda \|S\|_1 . \tag{2}$$

Where $S = \left[s_1^1, \cdots, s_{N_1}^1, \cdots, s_1^C, \cdots, s_{N_C}^C \right]^T \triangleq [s_1, s_2, \cdots, s_N]^T \in \Re^{d \times N}$, D is the dictionary and λ is a parameter. Then, the sparse code s_t is used to determine the class of the test sample x_t by calculating the error $e_c = \left\| x_t - D_c S_t^c \right\|_2$ for each class. Here, s_t^c is the *cth* class coefficient vector corresponding to D_c. Then, the class c^* associated to the test sample x_t can be thought as the one that produces the smallest error:

$$c^* = class \ of \ x_t = \arg\min_c E_c . \tag{3}$$

Where c^* is the class label of x_t. The SR based classification was originally proposed for face biometric, and was then extended for cancelable iris biometric, as well as for automatic target recognition.

3   MKSR Based Dictionary Learning Classification

3.1   Kernel SR Method

In Kernel SR (KSR), the main idea is to map data in the high dimensional feature space and solve Eq. (2) using the kernel trick. Let the mapping function be $\phi(\cdot) : \Re^N \rightarrow \Re^K (N < K)$. Utilizing this function $\phi(\cdot)$, the new high dimensional feature space can be obtained, namely $x \rightarrow \phi(x)$, and for the *cth* class samples $\phi(X_C) = \left[\phi(x_1^C), \phi(x_2^C), \cdots, \phi(x_{N_C}^C) \right]$. Utilizing the mapping relationship, Eq. (3) is substituted and the KSR object function can be obtained:

$$s_t = \arg\min_s \|\phi(x_t) - \phi(D)S\|_2^2 + \lambda \|S\|_1 . \tag{4}$$

where x_t is the test sample and $\phi(D) = \left[\phi(d_1^1), \cdots, \phi(d_{N_1}^1), \cdots, \phi(d_1^C), \cdots, \phi(d_{N_C}^C) \right]$. The first term is the reconstructed error denoted by E_k that is calculated as follows:

$$
\begin{aligned}
E_k(S, D, x_t) &= [\phi(x_t)]^T \phi(x_t) + S^T [\phi(D)]^T \phi(D)S - 2[\phi(x_t)]^T \phi(D)S \\
&= K(x_t, x_t) + S^T K(D, D)S - 2K(x_t, D)S
\end{aligned}
\tag{5}
$$

where $K(D, D) \in \Re^{N \times N}$ is a positive semi-definite kernel Gram matrix whose elements are computed as

$$[K(D, D)]_{i,j} = [\langle \phi(D), \phi(D) \rangle]_{i,j} = \phi(d_i)^T \phi(d_j) = \kappa(d_i, d_j) . \tag{6}$$

and according to Eq. (6), we can obtained the following formulas:

$$\begin{cases} K(x_t, x_t) = \kappa(x_t, x_t) \\ K(x_t, D) \triangleq [\kappa(x_t, d_1), \kappa(x_t, d_2), \cdots, \kappa(x_t, d_N)] \in \Re^{1 \times N} \end{cases} . \tag{7}$$

where $\kappa(\cdot)$ is the kernel function and $K(\cdot)$ is the Gram matrix of kernel function. Some commonly used kernel functions include polynomial kernels with the form of $\kappa(u, v) = (u \cdot v + a)^b$ and Gauss kernels with the form of $\kappa(u, v) = \exp\left(-\beta \|u - v\|_2^2\right)$, where u and v are different two point in kernel space. Because the dimension number in kernel feature space is very high, the Mercer kernel defined as $\kappa(u, v)$ is often used, namely, for $(x_i)_{i=1}^N$, the equations of $\tilde{\kappa}(u, v) = \langle \psi(u), \psi(v) \rangle$.

3.2 MKSR Based Dictionary Learning

3.2.1 The Cost Function

Multiple KSR (MKSR) methods also utilize the non-linear mapping function $\psi(\cdot)$ to map data samples from the original input space to a high dimension or multiple kernel feature space. Let $\tilde{D} = PA$ denote the dictionary, where matrix P is the predefined base dictionary and A is the atomic representation dictionary. Given training sample set $X = [x_1, x_2, \cdots, x_N]$ and dictionary $\tilde{D}$, the corresponding mapping matrixes are respectively $\psi(X)$ and $\psi(\tilde{D})$, where $\psi(X) = [\psi(x_1), \psi(x_2), \cdots, \psi(x_N)]$. Selected $\psi(X)$ as the base dictionary, $\psi(\tilde{D})$ can be represented as $\psi(X)A$. Then the object function of KSR based dictionary learning can be written as

$$\arg \min_{A,S} \|\psi(X) - \psi(X)AS\|_2^2 + \lambda \|S\|_1 . \tag{8}$$

Where the first term in Eq. (8) can be represented as $tr\left[(I - AS)^T K(X, X)(I - AS)\right]$, so, it has nothing to do with the non-linear mapping function $\psi(\cdot)$. When considering the case of MKSR based dictionary learning, the dictionary representation and the training object function are the same as those used in the KSR method. The different is that function $\psi(\cdot)$ corresponds to the multiple kernels' one. Assumed that the set of base kernel functions is denoted by $\kappa_1, \kappa_2, \cdots, \kappa_M$. Then, the linear combinations of the base kernels in MKL framework are described as

$$\tilde{\kappa}(x_i, x_j) = \sum_{i=1}^{M} \omega_m \kappa_m(x_i, x_j) . \tag{9}$$

where ω_m are the weight coefficients of base kernels constrained to $\sum\limits_{m=1}^{M} \omega_m = 1$ and $\omega > 0$. In our MKSR, the multiple kernel functions are chosen as the Gauss kernel function $\kappa_1(x_i, x_j)$ and the polynomial kernel function $\kappa_2(x_i, x_j)$ and the corresponding kernel matrixes are respectively denoted by $K_1(X, X)$ and $K_2(X, X)$, therefore, the multiple kernel matrix $\tilde{K}(X, X)$ can be written as follows

$$\left[\tilde{K}(X, Y)\right]_{ij} = \omega [K_1(X, Y)]_{ij} + (1 - \omega)[K_2(X, X)]_{ij} . \tag{10}$$

and

$$K_m(X, X) = \begin{bmatrix} k_m\left(x_1^1, x_1^1\right) & k_m\left(x_1^1, x_2^1\right) & \cdots & k_m\left(x_1^1, x_{N_c}^c\right) \\ k_m\left(x_2^1, x_1^1\right) & k_m\left(x_2^1, x_2^1\right) & \cdots & k_m\left(x_2^1, x_{N_c}^c\right) \\ \cdots & \vdots & \vdots & \cdots \\ k_m\left(x_{N_c}^c, x_1^1\right) & k_m\left(x_{N_c}^c, x_2^1\right) & \cdots & k_m\left(x_{N_c}^c, x_{N_c}^c\right) \end{bmatrix} \quad (m = 1, 2) . \tag{11}$$

$$K_m(x_i, X) = \left[k_m\left(x_1^1, x_i\right) \quad k_m\left(x_2^1, x_i\right) \quad \cdots \quad k_m\left(x_{N_c}^c, x_i\right) \right]^T . \tag{12}$$

3.2.2   Updating Sparse Coefficients and Dictionary

Based on Eq. (8), the object function of KSR based dictionary learning can also be deduced as follows

$$\arg\min_{s_j} \sum_{j=1}^{N} \left\| \psi(x_j) - \psi(X) A s_j \right\|_2^2 + \gamma \left\| s_j \right\|_1 . \tag{13}$$

Namely, solving the first term in Eq. (14) is equal to solve N equations of $\arg\min\limits_{s_i} \left\| \psi(x_i) - \psi(X) A s_i \right\|_2^2$. Here, based on K-SVD algorithm, we use the kernel orthogonal matching pursuit (KOMP) algorithm to train the sparse coefficients. For the given test sample x and kernel dictionary A, let $\hat{x}_t$ be the tth step estimation, r_t be the residual error of the current tth iteration and I_t be the atomic index set. Then, $\psi(x)$ can be represented by $\psi(X)\hat{x}_t + r_t$. And the first step in KOMP is to project I_t into remaining atoms by the following form

$$r_t^T (\psi(X) a_i) = [\psi(x) - \psi(X)\hat{x}_t]^T (\psi(X) a_i) . \tag{14}$$

where $\tilde{\kappa}(\cdot, \cdot) = \omega k_1(\cdot, \cdot) + (1 - \omega) k_2(\cdot, \cdot)$ and a_i is the ith column of A. According to the kernel function relation, Eq. (14) can be rewritten as

$$r_t^T(\psi(X)\,a_i) = \left[\tilde{K}(x,X) - (\hat{x}_t)^T \tilde{K}(X,X)\right] a_i \quad i \notin I_t \ . \tag{15}$$

Then the *kth* updating of sparse coefficient matrix S^t can be deduced

$$S^t = \left[A_{I_t}^T \tilde{K}(X,X)\,A_{I_t}\right]^{-1} \left[\tilde{K}(x,X)\,A_{I_t}\right]^T \ . \tag{16}$$

According to Eq. (16), $\hat{x}_i$, i.e. the estimation of x_i can be obtained by the form $\hat{x}_i = A_{I_t}\,s_i$. And the residual error $r_t = \psi(x) - \psi(X)\hat{x}_t$ also can be obtained.

Fixed sparse coefficient matrix S, and let a_k denote the *kth* column of A and s_j denote the *jth* row of S corresponding to a_k, then the object function of training dictionary is written as follows

$$J(a_j, s_j) = \left\| \psi(X)\left(I - \sum_{j \neq k} a_j\,s_j\right) - \psi(X)(a_k\,s_k) \right\|_2^2 = \|\psi(X)\,E_k - \psi(X)\,F_k\|_2^2 \ . \tag{17}$$

where E_k is the residual error between real and approximate data when the *kth* atom is removed from the dictionary, and F_k represents the contribution of the *kth* atom to the approximate data. Let $\widehat{E}_k$, $\widehat{F}_k$ and $\widehat{s}_k$ denote respectively vectors reducing zero elements corresponding to E_k, F_k and s_k, then Eq. (17) can be rewritten as the form of $\left\| \psi(X)\,\widehat{E}_k - \psi(X)\,a_k\,\widehat{s}_k \right\|_2^2$. Because the singular value decomposition (SVD) is similar to that of Gram matrix's features, the following formula can be obtained

$$\left[\psi(X)\,\widehat{E}_k\right]^T \left[\psi(X)\,\widehat{E}_k\right] = \left(\widehat{E}_k\right)^T \tilde{K}(X,X)\,\widehat{E}_k = V\Delta V^T \ . \tag{18}$$

Using K-means based SVD (K-SVD) algorithm to decompose Eq. (18), the atoms are updated by $a_k = \sigma_1^{-1}\,\widehat{E}_k\,\eta_1$, where η_1 is the first column of V corresponding to the maximal singular value $\sigma_1^2 = \Delta(1,1)$. Thus, the equation of $\widehat{s}_k = \sigma_1\,\eta_1^T$ can be deduced. When A_i and s_i are trained finally, for each class samples, the residual error can be obtained

$$r_i = \tilde{K}(x,x) + s_i^T A_i^T \tilde{K}(X,X)\,A_i\,s_i - 2\tilde{K}(x,X)\,A_i\,s_i \ . \tag{19}$$

And then the class of unknown samples can be denoted by $I_{inx}(x) = \arg\min_k r_k(x)$.

4 ELM Classifier

So far, ELM learning has been developed to work at a much faster learning speed with the higher generalization performance, especially in the pattern recognition and the regression problem. ELM was explored for the single hidden layer feed forward networks (SLFNs) instead of the classical gradient-based algorithms. When ELM is used as the task of classification, the principle of classification is based on two classification problems as the same as the support vector machines (SVM) classifier. Assumed that V

samples $\{x_i, y_i\}_{i=1}^V$ are given, where $x_i = [x_{i1}, x_{i2}, \cdots, x_{iv}]^T$ and $y_i = [y_{i1}, y_{i2}, \cdots, y_{iu}]^T$, the decision function can be defined as follows:

$$f(x) = \text{sgn}\left[\sum_{i=1}^L \beta_i\, G(\omega_i, b_i\, x)\right] = \text{sgn}[\beta H(x)]. \tag{20}$$

where $H(x)$ is represented as follows:

$$
H(x) = [h_1, h_2, \cdots, h_J] = \begin{bmatrix} h_1(x_1) & \cdots & h_J(x_1) \\ \cdots & \cdots & \cdots \\ h_1(x_V) & \cdots & h_J(x_V) \end{bmatrix}
$$
$$
= \begin{bmatrix} g(\omega_1 \times x_1 + b_1) & \cdots & g(\omega_J \times x_1 + b_J) \\ \cdots & \cdots & \cdots \\ g(\omega_1 \times x_V + b_1) & \cdots & g(\omega_J \times x_V + b_J) \end{bmatrix}_{V \times J}. \tag{21}
$$

where $\omega_i = [\omega_{i1}, \omega_{i2}, \cdots, \omega_{in}]^T$ is the weight vector connecting the ith hidden neuron and the input neurons, and $\beta = [\beta_1, \beta_2, \cdots, \beta_J]_{m \times J}^T$ is the weight vector connecting the ith hidden neuron and the output neurons, and there are J hidden neurons with the activation function $g(x)$, which can be chosen as the Sigmoid function, the hard-limit function, the multiquadric function and so on. The generalization performance of ELM is optimal when the following Equation is optimized:

$$\arg\min\left(\|H\beta - Y\|^2, \|\beta\|\right). \tag{22}$$

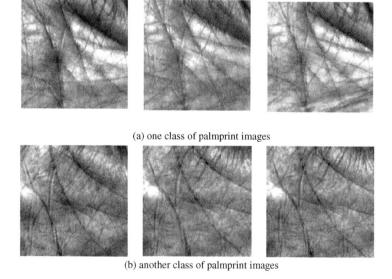

(a) one class of palmprint images

(b) another class of palmprint images

Fig. 1. Two classes of original palmprint images selected randomly from PolyU database.

and it is easy to know that when output samples are mapped to the feature space of ELM from input samples' space, the solution of $H(x)$ is linearly separable and unique. According to the optimization theory, the minimization problem of the ELM's classification hyperplane is defined as

$$\min \quad E = \frac{1}{2}\|\beta\|^2 + \gamma \sum_{i=1}^{V} \xi_i . \tag{23}$$

subject to $y_i\beta \cdot h(x) \geq 1 - \xi_i$, $\xi_i \geq 0$, $i = 1, 2, \cdots, V$.

5 Experimental Results and Analysis

5.1 Testing on PolyU Palmprint Database

In test, the PolyU palmprint database from the Hong Kong Polytechnic University is used to test the classification results using the MKSR based dictionary learning proposed by us in this paper. The PolyU database used here includes 600 palmprint images with the size of 284 × 384 pixels from 100 individuals, namely, each person has six samples collected in two sessions. Several images of two classes were shown in Fig. 1. For each person, the first three images are used as training samples and the other three for testing samples. We extract the central portion of the palm image for palmprint classification. The extracted palmprint image has a size of 128 × 128 pixels. Thus the training set $\mathbf{X}_{train}$ and the testing set $\mathbf{X}_{test}$ have the same size of 300 × (128 × 128) pixels. Here, $\mathbf{X}_{train}$ and $\mathbf{X}_{test}$ are first preprocessed to be centered and have zero-mean. To reduce the computational cost, each palmprint image is scaled to the size of 64 × 64 pixels by using the wavelet method, and then this palmprint image is converted a column vector. Thus, the size of the training set and test set are both changed to be 4096 × 300 pixels. Furthermore, in order to reduce the calculation work, the principal component analysis (PCA) method is used to reduce the dimension. Considering the different number of principal components (PCs) from 20 to 100, then the corresponding PCA features are extracted. And PCA feature basis images of 16 dimensions are shown in Fig. 1. The performance recognition rates of PCA using the distance classifier, ELM classifier and radial basis function neural network (RBFNN) classifier is shown in Table 1. According to Table 1, it is clear to see that the larger the number of PCs is, the better of the recognition rate of ELM classifier is, but for other two classifier, the recognition rate does not always increase with the increasing of the number of PCs. Compared the Recognition rate of each classifier, clearly, the ELM classifier is better fit to the case that the number of PCs is large. Moreover, when the number of PCs is less than 256, the recognition results of three classifiers have a little difference. Simultaneously considering the recognition rate and the computation work, experimental result shows that feature length with 324 PCs can yield the best performance. Thus, instead of processing on images with 324 PCs, the computational load can be reduced significantly in our MKSR (Fig. 2).

After the optimal number of PCs is selected, the MKSR based dictionary can be trained in the given dimension PCA feature spatial. In our MKSR method, the

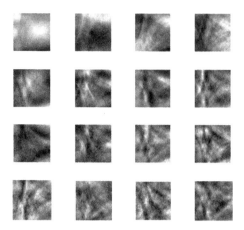

Fig. 2. PCA feature basis images of 16 dimensions.

Table 1. Recognition rate of PCA using three types of classifiers.

Number of PCs	Euclidean distance (%)	RBFNN (%)	ELM (%)
36	75.67	78.67	73.67
49	79.33	83.67	83.00
64	79.67	85.00	84.33
81	78.67	86.00	85.67
121	76.67	86.67	86.67
144	78.67	86.67	86.67
256	80.67	86.67	86.33
324	81.67	87.00	93.67
361	82.00	90.33	93.67
400	82.00	90.33	93.33
512	81.67	90.00	91.00
1024	82.33	90.00	92.00

combination of multiple kernels is made up of Gauss kernel and polynomial kernel. The feature basis images obtained by our MKSR of 36 dimensions are shown in Fig. 3. And the sparse features are very distinct in Fig. 3.

Otherwise, in order to prove our method's advantage, the compared test using other three different SR based algorithms, such as the common K-mean based singular value decomposition (KSVD) model, Gaussian kernel based SR method denoted by GKSR, polynomial kernel based SR method denoted by PKSR, are also done here. When the sparse coefficient vector and sparse dictionary are trained, based on the theory of face recognition used in the face recognition field, the residual error r_k can be calculated. Furthermore, for an unknown class test sample x, using the formula of $I_{inx}(x) = \arg\min_k r_k(x)$, x can be classified well. In the same way, the three classifiers used in PCA based classification are also used in SR based classification task, the

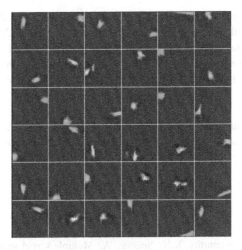

Fig. 3. MKSR feature basis images of 36 dimensions.

Table 2. Classification results using different algorithm and classifiers (324 PCs).

Classifiers Algorithms	ELM (%)	RBFNN (%)	Distance (%)
MKSR	94.23	89.73	87.62
GKSR (Gaussian kernel)	93.54	87.74	86.16
PKSR (Polynomial kernel)	93.32	85.31	84.26
KSVD	91.27	82.87	82.05

recognition results are listed in Table 2. In terms of experimental data from Table 2, it is clear to see that, when fixed the classifier, kernel based SR methods clearly outperform the KSVD method. And MKSR based method is better than single kernel based SR methods of GKSR and PKSR, but the recognition results of GKSR and PKSR have a litter difference. While when fixed an algorithm, it is also distinct to see that the recognition rate of ELM classifier is the best. Therefore, according to test results, it can be concluded that our MKSR method behave the best recognition rate.

6 Conclusions

A new palmprint image classification method using MKSR based dictionary learning is discussed in this paper. In processing high dimension data, MKSRC methods distinctly outperform ones of KSR based classification and SR based classification. In test, the

PolyU database is used, and the combination of Gauss kernel and polynomial kernel are selected as multiple kernels. Using different classifiers of ELM, RBFNN and distance to compared the recognition rates of our MKSR method, GKSR, PKSR and KSVD, experimental results prove that the MKSR proposed in this paper behaves the best recognition rate. Especially, when use the ELM classifier, the recognition result is the best.

Acknowledgement. This work was supported by the grants from National Nature Science Foundation of China (No. 61373098), and the science and technology planning project of Suzhou (No. SZP201310).

References

1. Thiagarajan, J.T., Ramamurthy, K.N., Spanias, A.: Multiple kernel sparse representations for supervised and unsupervised learning. IEEE Trans. Image Process. **23**(7), 2905–2915 (2014)
2. Lin, Y., Liu, T., Fuh, C.: Mutiple kernel learning for dimensionality reduction. IEEE Trans. Pattern Anal. Mach. Intell. **33**(6), 1147–1160 (2011)
3. Rubinstein, R., Bruckstein, A., Elad, M.: Dictionaries for sparse representation modeling. IEEE Proc. **98**(6), 1045–1057 (2010)
4. Mairal, J., Bach, F., Ponce, J.: Task-driven dictionary learning. IEEE Trans. Pattern Anal. Mach. Intell. **34**(4), 791–804 (2012)
5. Nguyen, H., Patel, V., Nasrabad, N., Chellappa, R.: Design of nonlinear kernel dictionaries for object recognition. IEEE Trans. Image Process. **22**(12), 5123–5135 (2013)
6. Cheng, B., Yang, J., Yan, S., Fu, Y., Huang, T.: Learning with L_1-graph for image analysis. IEEE Trans. Image Process. **19**(4), 858–866 (2010)
7. Shrivastava, A., Patel, V., Chellappa, M.: Multiple kernel learning for sparse representation-based classification. IEEE Trans. Image Process. **23**(7), 3013–3024 (2014)
8. Aharon, M., Elad, M., Bruckstein, A.: K-SVD: an algorithm for designing over complete dictionaries for sparse representation. IEEE Trans. Signal Process. **54**(11), 4311–4322 (2006)

One Novel Rate Control Scheme for Region of Interest Coding

Chen Xi[1], Wu Zongze[1,2], Zhang Xie[1(✉)], Xiang Youjun[1],
and Xie Shengli[2]

[1] College of Electronic and Information Engineering, South China University
of Technology, Guangzhou 510641, Guangdong, China
5173204@qq.com, {zzwu,zhangxie,yjxiang}@scut.edu.cn
[2] College of Automation, Guangdong University of Technology,
Guangzhou 510006, Guangdong, China
shlxie@gdut.edu.cn

Abstract. For video communication, Region of Interest (ROI) coding offers an efficient way of improving the quality of videos. In this paper, we propose a novel rate control scheme for Region of Interest coding. We set the area covering people are interested in as the Region of Interest (ROI) to preserve better quality than the background. In addition, we set a coefficient ω to evaluate the significance of ROI, and then use it to calculate the mean absolute distortion (MAD) of the ROI and the background. Finally, the quantization parameter (QP) can be decided by the quadratic model. The difference of QP can be used to distinguish the background and the ROI with the QP which is obtained in advance, so the ROI mask cannot be coded and transmitted. Moreover, through adjusting the coefficient ω to control the degree of the protection of the ROI, our method can protect the ROI effectively and control the difference between the background and the ROI expediently. Experimental results show that the scheme works well in protecting the ROI, the objective quality of the ROI will be improved and the bit-rate can be controlled in a certain range.

Keywords: Rate control · Region of Interest (ROI) · Quantization parameter · Video coding

1 Introduction

With the development of 3C (Communication, Computer and Consumer electronics) technology, more and more people think about the confliction between the storage or bandwidth resources and the demand for information. In fact, the demand cannot be meted forever. Rate control plays an important role for video coding and transmitting, because it helps to achieve the balance between the video quality and the code stream size.

However, the regions appearing in the scenes which people are interested in, called as Region of Interest (ROI), are more important than the background. For the given bit-rate, bits allocated for ROI should be more than that for the background. For example, the area of the shoulder and head of characters in video conference, the area of some small target which carry more important information in the aviation images,

© Springer International Publishing Switzerland 2016
D.-S. Huang et al. (Eds.): ICIC 2016, Part III, LNAI 9773, pp. 139–148, 2016.
DOI: 10.1007/978-3-319-42297-8_14

the area of lesions in the medical images and so on, so that its quality should be better preserved. Therefore, we consider applying the ROI coding to those scenes.

In the recent years, people have made many researches in the ROI coding and gained many beneficial results. The processing methods of the ROI coding include several categories:

(a) The first category method treats the ROI as an independent part, then coding them independently, protecting and transmitting them selectively. For example, Hannuksela et al. [1] treats the ROI as a sub-picture to code independently and puts forward a gradual allocation scheme of bit rate and unequal error protection. This method can achieve the target of protecting the ROI, but it must transmit a mask to mark the ROI which will add the extra cost of bits.

(b) The method of layer ascension is used to improve the quality of the ROI. Shin et al. [2] handles ROI based on fine granular scalability (FGS) coding by the method of layer ascension to improve the compression quality of the ROI's enhancement layer. But this method cannot improve the quality of the ROI (especially the ROI in the image with complex motion) from the angle of compression or be effective to the utilization of the rate resources based on the content of the actual image.

(c) Flexible Macroblock Ordering (FMO) is adopted to improve the error resilience and error protection in the ROI in the H.264/AVC standard [3] and the SVC [4–7] extensions. However, this method must mark the location of the macroblock, which will add the cost of bits.

(d) The last kind method adjusts the quantization parameter (QP) to achieve the protection of the quality of the ROI. The smaller the QP is, the deeper the degree of protection of the region is. So adjusting the QP of the ROI smaller than the QP of the background will improve the quality of the ROI.

And we can use the rate distortion models to adjust the QP. This method can achieve the protection of the ROI. But it must have a mark to distinguish the macroblock if belongs to the ROI which will add the cost of bits and there is no coefficient to decide the degree of the protection of the ROI.

There are also some additional state-of-the art researches in the area of ROI coding [8–10]. In [8], the authors introduce a description-driven content adaptation framework. In [9], the authors provide an approach for extracting and obtaining the desired ROI scalability of a SVC video stream, according to predefined settings, by cropping the ROI from the original image and performing inter layer prediction or using the FMO. In [10], the authors propose a method to improve subjective QoS (Quality of Service) with low transmission overhead by using a fixed size of enhancement layers.

In order to let the method referred to above work better, we suggest a novel Rate-Control Scheme for Region of Interest coding, enabling to control the degree of the protection of the ROI by adjusting the coefficient ω, at the same time, the ROI mask cannot be coded and transmitted because it can be calculated by the known parameters.

In this paper, we present the novel Rate-Control Scheme for Region of Interest coding. The rest paper is organized as follows. Section 2 briefly presents the simple introduction of the rate control; Sect. 3 presents the proposed scheme; the experimental results are discussed in Sect. 4 and the conclusion is given in Sect. 5.

2 Rate Control in JSVM

Rate control is used to create a bit stream to meet with the available bandwidth and be compliant to HRD/VB. Generally, it can be divided into three consecutive levels: group of pictures (GOP) level, picture level and the optional basic unit level.

2.1 GOP Level Rate Control

GOP level rate control calculates the remaining bits for the rest pictures, and initializes the quantization parameter of the first picture (I/P) in the current GOP.

- The remaining bits for the rest pictures

When the ith GOP is to code, the total allocated bits is

$$T_r\left(n_{i,0}\right) = \frac{u\left(n_{i,1}\right)}{F_r} \times N_{gop} - \left(\frac{B_S}{8} - B_c\left(n_{i-1,N_{gop}}\right)\right)$$

Where F_r is the predefined frame rate, N_{gop} is the size of the ith GOP, B_s is the size of the buffer, $u(n_{i,1})$ is the instant bandwidth, $B_c(n_{i-1,N_{gop}})$ is the size of buffer after the $i-1^{th}$ GOP has been coded.

Because the bandwidth doesn't keep still, so the T_r is update by a frame as follow:

$$T_r\left(n_{i,j}\right) = T_r\left(n_{i,j-1}\right) - \frac{u\left(n_{i,j}\right) - u\left(n_{i,j-1}\right)}{F_r} \times \left(N_{gop} - j\right) - A\left(n_{i,j-1}\right)$$

When the bandwidth keeps still, then $u\left(n_{i,j}\right) = u\left(n_{i,j-1}\right)$ so

$$T_r\left(n_{i,j}\right) = T_r\left(n_{i,j-1}\right) - A\left(n_{i,j-1}\right)$$

- The quantization parameter of the first picture (I/P) in the current GOP

Assumed that QP_0 is the QP of the first GOP, the I or first P frame is coded by the QP_0. The QP_0 is decided by the available bandwidth and the size of the GOP.

2.2 Picture Level Rate Control

Picture level rate control includes two stages: Pre-encoding stage and Post-encoding stage. The quantization parameter of each picture is calculated in the Pre-encoding stage.

- Pre-encoding stage

In this stage, the quantization parameters (QPs) of each picture is calculated. First, the QPs of the B frame are calculated.

(1) QP of B frame

The B frame is not used to predict the other picture, so the value of QP of the B frame is bigger than others'. At the same time, in order to keep the quality of the picture smoothly, the different value between the QPs of the adjacent frames cannot exceed 2.

Assuming the number of B frames between two adjacent P frames is L, and the QP of the two P frames are QP_1 and QP_2, then the QP of the B frame can be calculated as follow:

If L = 1, then

$$\overline{QB_1} = \frac{QP_1 + QP_2 + 2}{2}$$

If L > 1, then

$$\overline{QB_l} = QP_1 + \alpha + max\left\{ min\left\{\frac{QP_2 - QP_1}{L - 1}, 2(i - 1)\right\}, -2(i - 1)\right\}$$

Where α is the D-value between the first B frame and QP_1, and set as follow:

$$\alpha = \begin{cases} -3 & QP_2 - QP_1 \le -2L - 3 \\ -2 & QP_2 - QP_1 = -2L - 2 \\ -1 & QP_2 - QP_1 = -2L - 1 \\ 0 & QP_2 - QP_1 = -2L \\ 1 & Qp_2 - QP_1 = -2L + 1 \\ 2 & Otherwise \end{cases}$$

In the end, the QB_i is adjusted by:

$$\overline{QB_l} = min\left\{max\left\{\overline{QB_l}, 1\right\}, 51\right\}$$

(2) QP of P frame

Step 1: Allocating the target bits for the current picture
Step 2: Calculating the QP

Getting the target bits for a picture, the QP can be obtained with pre-defined R-D model. After motion estimation and mode selection, the MAD of I/P pictures is still unable to be determined, so it is predicted from the closet picture coded previously by a linear model:

$$MAD_i(j) = \alpha_1 \times MAD_i(j - 1) + \alpha_2$$

Where α_1 and α_2 are two coefficients with initial value 1 and 0. And then, the quantization parameter corresponding to the target bits is computed as:

$$T_i(j) = X1 \times \frac{MAD_{i(j)}}{QP_{i(j)}} + X2 \times \frac{MAD_{i(j)}}{QP_{i(j)^2}}$$

Where X1 and X2 are two coefficients. Since a drop in peak signal-to-noise ratios (PSNR) among successive pictures will deteriorates the visual quality of the whole sequence, the quantization parameter $QP_i(j)$ is adjusted by:

$$QP_i(j) = min\{QP_i(1) + 2, max\{QP_i(1) - 2, QP_i(j)\}\}$$

In the end, the final quantization parameter is further bounded by 51 and 0.

$$Q_{pc} = min\{51, max\{\overline{Q_{pc}}, 1\}\}$$

3 One Novel Rate Control for ROI

A new rate control for ROI is proposed, which add a coefficient ω to control the degree of the protection of the ROI. The coefficient ω is greater, the degree of the protection of the ROI is deeper. This scheme mainly protect the ROI by adjust the determination of the QP, so we describe the QP of the I, P, B frame separately, i.e. the QP of I frame determination, the QP of P frame determination, the QP of B frame determination. The details of each part are described as follows:

3.1 QP of I Frame

If the QP of the I frame is QP_0 by using the GOP layer rate control, then the QP of the ROI and background in I frame are calculated as follow:

$$QP_R = QP_0 - \log_2 \omega \cdot \frac{S_B}{S} \cdot 6$$

$$QP_B = QP_0 + \log_2 \omega \cdot \frac{S_R}{S} \cdot 6$$

Where QP_R is the QP of the ROI, QP_B is the QP of the background, S_R is the number of pixels of the ROI, S_B is the number of pixels of the background, and S is the sum pixels of the frame.

In order to achieve rate control, we should adjust the QP above.

In the area of ROI, the QP: QP'_R is bounded by:

$$QP'_R = min(QP_0 - 1, QP_R \pm \alpha)$$

In the area of background, the QP: QP'_B is bounded by:

$$QP'_B = \max(QP_o + 1, QP_N \pm \beta)$$

Where α and β are vary linearly with ω, i.e.

$$\alpha = \rho 1 \cdot \omega \quad \beta = \rho 2 \cdot \omega$$

At the experiment, we set $\rho 1 = 0.25$, $\rho 2 = 1$.

In the end, if the adjusted QP'_R or QP'_B is less than 0, we set it to 0. And if the adjusted QP'_R or QP'_B is greater than 51, we use the allocated bits to determine set it to 51 or skip the area to encode.

3.2   QP of P Frame

If the target bits of the current frame is R, and we predict the MAD of the basic unit in the current frame by linear MAD prediction model according to the actual MAD of the same position of the former frame. We get the MAD of the ROI is MAD_R and the MAD of the background is MAD_B, so the sum MAD of the frame is

$$MAD_S = MAD_R + MAD_B$$

In order to make the MAD of the ROI more important in the whole MAD, the MAD of the ROI is weighted,

$$MAD_R 1 = \omega \cdot MAD_R$$

Where ω is a coefficient, and its value express the degree of the protection of the ROI. It can be set 2,4,8...

Then in order to make the sum MAD of the frame is not change, we set the MAD of the background as:

$$MAD_B 1 = MAD_S - MAD_R 1$$

We use the adjusted MAD to allocate the bits to the ROI and the background. We set that the allocated bits of the ROI is RR and the allocated bits of the background is BR, and then RR and BR are quantized as:

$$RR = R \cdot \frac{MAD_R 1}{MAD}$$

$$BR = R - RR$$

The quadratic R-D model is used to calculate the QP the same as the H.246, the quadratic R-D model is

$$R_i = \frac{X1 \times MAD}{QP_i} + \frac{X2 \times MAD}{QP_i^2}$$

Where R_i is the allocated bits, X1 and X2 are defined in H.264. When the other parameters are invariant, MAD is smaller, then QP is smaller. So the MAD can be treated as the allowable distortion. In order to make the quality of the ROI is higher than the quality of the background, the allowable distortion of the ROI should be smaller, so the MAD of the ROI is weighted as follow:

$$MAD_R 2 = \frac{1}{\omega} \cdot MAD_R$$

Then in order to make the sum MAD of the frame is not change, we set the MAD of the background as:

$$MAD_B 2 = MAD_S - MAD_R 2$$

After the adjustment, substituting RR, BR, $MAD_R 2$ and $MAD_B 2$ in the quadratic R-D model, we will get QP_R and QP_B:

$$RR = \frac{X1 \cdot MAD_R 2}{QP_R} + \frac{X2 \cdot MAD_R 2}{QP_R^2}$$

$$BR = \frac{X1 \cdot MAD_B 2}{QP_B} + \frac{X2 \cdot MAD_B 2}{QP_B^2}$$

Then MAD_S and R are substituted in the quadratic R-D model to calculate the original QP_o which is the same as the QP calculated by H.264:

$$R = \frac{X1 \cdot MAD_S}{QP_o} + \frac{X2 \cdot MAD_S}{QP_o^2}$$

In order to achieve rate control, we should adjust the QP_R and QP_B above. In the area of ROI, the QP: QP'_R is bounded by:

$$QP'_R = \min(QP_o - 1, QP_R \pm \alpha)$$

In the area of background, the QP: QP'_B is bounded by:

$$QP'_B = \max(QP_o + 1, QP_N \pm \beta)$$

Where α and β are vary linearly with ω, i.e.

$$\alpha = \rho 1 \cdot \omega \qquad \beta = \rho 2 \cdot \omega$$

At the experiment, we set $\rho 1 = 3.25$, $\rho 2 = 2$.

Finally if the adjusted QP'_R or QP'_B is less than 0, we set it to 0. And if the adjusted QP'_R or QP'_B is greater than 51, we use the allocated bits to determine set it to 51 or skip the area to encode.

3.3   QP of B Frame

The QP in the B frame is related to the QPs of the adjacent I frame or P frame, and the calculation method of the QP in the proposed scheme is similar to the method in H.264. However, in order to calculate the QP of ROI, we choose the QP_R of the adjacent I frame or P frame, and we choose the QP_B of the adjacent I frame or P frame. We also need to adjust the QP of B frame according to the coefficient ω.

4   Experimental Results

The performance of the proposed novel rate-control scheme for region of interest coding has been evaluated. The algorithm works correctly with all sequences recorded with CIF and QCIF resolutions and at different bit-rates. We use the short vide FOREMAN as the test video in JSVM9.6. The video is recorded with CIF and QCIF resolutions, respectively, the GOP is composed of 30 frames and there are 3B frames between I/P frames.

In order to verify whether our scheme can achieve rate control, we set the bit-rate at different rates. In the Table 1, the test sequence is FOREMAN CIF and the bitrate is set at 120 kbps and 200 kbps. In the Table 2, the test sequence is FOREMAN QCIF, and the bitrate is set at 50 kbps and 65 kbps. We set ω to 0, 2, 4 and 8. From the Tables 1 and 2, our scheme can control the coding rate near to the set bitrate in advance.

Table 1. Bit-rate for FOREMAN CIF coding

ω	Bitrate (kbps)	
	120	200
0	123.528	202.552
2	126.856	194.072
4	120.168	195.480
8	126.248	196.976

Table 2. Bit-rate for FOREMAN QCIF coding

ω	Bitrate (kbps)	
	50	65
0	47.704	66.072
2	48.592	66.872
4	50.744	66.696
8	54.920	66.184

The solid line in the Figs. 1 and 2 show the Peak Signal to Noise Ratio (PSNR) curve of our proposed scheme. The quality our ROI coded by our scheme is much better than that of the original JVT-GO12. The blue line is the PSNR when $\omega = 8$; The red line is the PSNR when $\omega = 4$; The cyan line is the PSNR when $\omega = 2$ and the black dotted line is the original result. We can find that the bigger the value of the ω is, the higher the PSNR of the ROI is, which shows that the ROI is protected better. So we can draw the conclusion: the proposed scheme can protect the ROI and can control the degree of the protection of the ROI by adjusting the coefficient ω.

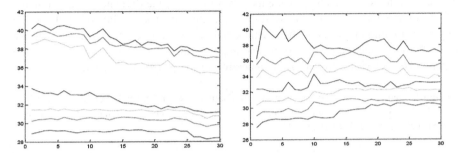

Fig. 1. PSNR of FOREMAN CIF coding at 100 Kbps (Color figure online)

Fig. 2. PSNR of FOREMAN QCIF coding at 65 Kbps (Color figure online)

At the same time, the visual codec videos are shows in Fig. 3 to verify our proposed scheme. Figure 3(a) is the original frame, Fig. 3(b) is the frame when ω = 2, Fig. 3(c) is the frame when ω = 4 and Fig. 3(d) is the frame when ω = 8. We choose a detail (in the red box) in the ROI to compare. From the amplification of the red box, it is obviously that the eye of the men becomes clearer with the increase of the value of the ω. The proposed scheme can make the visual effects of the ROI better.

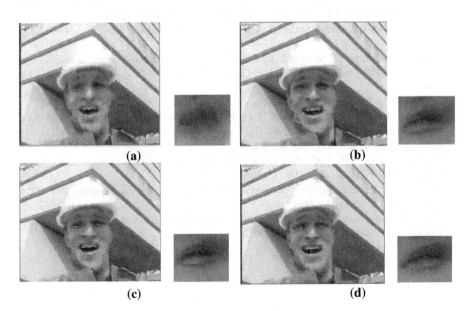

(a)

(b)

(c)

(d)

Fig. 3. Visual effects of the ROI (Color figure online)

5   Conclusion

The proposed scheme for ROI coding can save the code stream with not transmitting the mask of the ROI, thereby provide more bits to protect the ROI and the adjusting the degree of the protection of the ROI by a coefficient. The quality of ROI can be improved to further preserve their details. Future work will be giving an adaptive algorithm to decide the coefficient ω, and applying the scheme to various videos with different characteristics to verify the generality of the proposed algorithm.

Acknowledgement. The work is supported by National Natural Science Foundation of China (61271210) and Major science and technology projects in Guangdong Province (2015B010131014, 2014B010117005).

References

1. Hannuksela, M.M., Wang, Y.K., Gabbouj, M.: Sub-picture: ROI coding and unequal error protection. In: IEEE 2002 International Conference on Image Processing (ICIP 2002), pp. 537–540. IEEE, New York (2002)
2. Shin, C., Seo, K.-D., Kim, J.-K.: Rectangular region based selective enhancement for MPEG-4 fine granular scalability. In: International Packet Video Workshop, Pittsburgh, USA, pp. 101–107, April 2002
3. H.264/AVC, Draft ITU-T Rec. and Final Draft Intl. Std. of Joint Video Spec. (H.264/AVC), Joint Video Team, Doc. JVT-G050, March 2003
4. Wiegand, T., et al.: ISO/IEC 14496-10:200X/Amd.3 Part 10: Advanced Video Coding AMENDMENT 3: Scalable Video Coding Joint Draft ITU-T Rec. H.264/ISO/IEC 14496-10/Amd.3 Scalable video coding, Joint Video Team, Doc. JVT-X201, July 2007
5. Schwarz, H., Marpe, D., Wiegand, T.: Overview of the scalable video coding extension of the H.264/AVC standard. IEEE Trans. Circuits Syst. Video Technol. **17**(9), 1103–1120 (2007)
6. JSVM, JSVM Software Manual, Ver. JSVM 9.18 (CVS tag: JSVM_9_18), June 2009
7. Wiegand, T., Sullivan, G., Reichel, J., Schwarz, H., Wien, M.: Joint draft 8 of SVC amendment, ISO/IEC JTC1/SC29/WG11 and ITU-T SG16 Q.6 9 (JVT-U201), 21st Meeting, Hangzhou, China, October 2006
8. Liu, Y., Li, Z.G., Soh, Y.C.: Rate control of H.264/AVC scalable extension. IEEE Trans. Circuits Syst. Video Technol. **18**(1), 116–121 (2008)
9. Lu, Z., et al.: Perceptual Region-of-Interest (ROI) based scalable video coding, JVT-O056, Busan, KR, 16–22 April 2005
10. Grois, D., Kaminsky, E., Hadar, O.: Dynamically Adjustable and Scalable ROI Video Coding, IEEE BMSB Shanghai, March 2010

Heart Rate Variability Estimation in Electrocardiogram Signals Interferences Based on Photoplethysmography Signals

Aihua Zhang[1(✉)], Qian Wang[1], and Yongxin Chou[2]

[1] College of Electrical and Information Engineering,
Lanzhou University of Technology, Lanzhou 730050, China
zhangaihua@lut.cn
[2] School of Electrical and Automatic Engineering,
Changshu Institute of Technology, Changshu 215500, China

Abstract. In order to improve the accuracy and real-timelines of heart rate variability (HRV) estimation in electrocardiogram (ECG) signals interferences, a novel HRV estimation method based on photoplethysmography (PPG) signals is proposed. The short-time autocorrelation principle is used to detect interferences in ECG signals, then, the improving sliding window iterative Discrete Fourier Transform (DFT) is used to estimate HRV in ECG interferences from the synchronously acquisitioned PPG signals. The international commonly used MIT-BIT Arrhythmia Database/Challenge 2014 Training Set is used to verify the interferences detection algorithm and HRV estimation algorithm which are proposed. At the same time, the proposed algorithms are compared with recently existing representative interferences detection algorithm based on RR intervals and PRV directly replaced HRV algorithm, respectively. The results show that the proposed methods are more accurate and more real-time.

Keywords: Interferences · Heart rate variability · Photoplethysmography signals

1 Introduction

Heart rate variability (HRV) is produced in the periodical change of heart beat intervals, which is one of the important indices for reflecting the sympathetic nerve and vagus nerve activity's balance in the autonomic nervous system. It can be used for many diseases' prediction or diagnoses, such as sudden cardiac death, coronary heart disease, heart failure, hypertension, diabetes, Parkinson's disease and apnea disease, etc. [1]. However, HRV is derived from ECG signals, ECG signals' acquisition needs many electrodes and multifarious wires. At the same time, ECG signals acquisitioned by monitoring equipment often contain interferences caused by many factors, including human movement. In recent years, many scholars have carried out extensive research on interferences detection, such as W. Karlen proposed a method of repeatedly using Gaussian filter and cross-correlation for PPG signals quality evaluation which can realize the interferences detection in PPG signals further [2]. The method needs segment PPG signals earlier. But because of the noise and artifacts, the segmentation

© Springer International Publishing Switzerland 2016
D.-S. Huang et al. (Eds.): ICIC 2016, Part III, LNAI 9773, pp. 149–159, 2016.
DOI: 10.1007/978-3-319-42297-8_15

accuracy is hard to guarantee, which leads the accuracy of algorithm is not high. C. Orphanidou proposed quality evaluation method based on RR intervals [3]. The method is achieved by adaptive template matching theory, but template generation process is very complicated, thus the algorithm is very complex and the real-timeliness becomes poor. Li Qiao proposed machine learning method to classify multichannel ECG signals based on signal quality indices (SQI) [4]. The SQI of each channel signal is extracted first, then using SVM to complete the classification. Although the accuracy of the algorithm is high, as it is aiming at off-line data, the real-timeliness of the algorithm is not so good.

In addition, HRV in ECG signals interferences is difficult to directly be extracted, nowadays, which is mainly estimated by means of the synchronously acquisitioned PPG signals. In 2014, Physionet web site launched a competition entitled "Robust Detection of Heart Beats in Multimodal Data", aiming at estimating heart rate in ECG interferences through multi-channel signal fusion. Because pulse rate variability (PRV) is also derived from small changes in the period of heart beat, contains body's physiological and pathological rich information. Compared with HRV, PRV is extracted from PPG signals, the acquisition of PPG signals doesn't need many electrodes and multifarious wires, so it's easy for mobile portable medical instrument. A large number of studies have shown that PRV and HRV have a clear correlation. Even in the condition of the body at rest, PRV can directly replace HRV to reflect the characteristics of heart beat [5]. Therefore, massive papers in the competition use PRV directly as HRV in ECG signals interferences. But in fact there are tiny differences between HRV and PRV, directly using PRV to estimate HRV in ECG signals interferences will inevitably produce errors, which makes some disease detection methods or system based on HRV to be less reliable or even invalid.

To solve above problems, on the basis of existing research, we proposed ECG signals interferences detection algorithm based on the principle of short-time autocorrelation and the HRV estimation method in ECG signals interferences based on PPG signals, then applied them to 2014 competition data in Physionet to analyze their accuracy and real-timeliness.

2 Materials and Method

2.1 Data

The Challenge 2014 Training Set data (Challenge/2014/Set-p) [6] in the Physionet web site is used as the experimental data. This database provides 100 groups of data from patients with cardiovascular disease and healthy adults, each group includes seven channels synchronously acquisitioned data, including ECG, blood pressure (BP) and EEG, etc., the first of which is ECG signals, the remaining signals can be any of a variety of simultaneously recorded physiologic signals that may be useful for robust beat detection. The sampling frequency is 250 Hz. The database contains signals at most 10 min in length or occasionally shorter. Because BP signals is the other form of PPG, we regarded BP as PPG, and then ECG and PPG signals were chosen as the experimental data.

2.2   Method

The HRV estimation method in ECG signals interferences based on PPG signals which is proposed, mainly includes ECG signals and PPG signals filtering, ECG interferences detection and HRV estimation.

ECG Interferences Detection. According to the characteristics of ECG and PPG signals, the integral coefficient notch filter and low-pass filter are designed in paper [7], which are used to filter out the repressible noises and interferences in ECG and PPG signals first. Then, according to the short-time autocorrelation of ECG signals, smooth degree, and dynamic variation coefficient based on short-time autocorrelation, setting threshold detects interferences. The overall process is shown in Fig. 1.

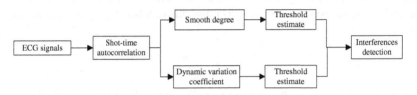

Fig. 1. Overall block diagram of ECG signals interferences detection

ECG Signals Short-Time Autocorrelation. Signal's autocorrelation function can be used in the analysis of the similarity of the same signal at different time or the different signals. Using the sliding window iterative method calculates the autocorrelation function called short-term autocorrelation function. As shown in Fig. 2, the core of the sliding window iteration is: adding a new sampling data and weeding out the old one in one fixed window (window width is N, here, $N = 2*f_s$). That is, when window moves every time, the latest sampling data will be in the window of Nth bit, the original old data will move left with one bit, the first data (the earliest data) will be removed.

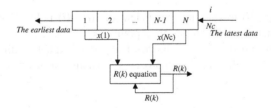

Fig. 2. Sliding window iteration schematic diagram

Setting the window width as N sampling points, ensuring that every window has two full of ECG signals waveforms at least. The sliding window iterative algorithm for calculating the autocorrelation function is defined as:

$$R(k) = \begin{cases} \displaystyle\sum_{n=1}^{N-k} x(n)x(n+k), 1 \leq n \leq N \\ \displaystyle\sum_{n=1}^{N-k} x(n)x(n+k) + x(n-k)x(n) - x(n-N)x(n-N+k), n > N \end{cases} \quad (1)$$

In Eq. (1), k is signal delay points (namely autocorrelation sequence number); N is sampling points for each window sequence (i.e., the length of each window). Then, normalizing the autocorrelation function $R(k)$ to $r(k)$ with the min-max normalization method.

Smooth Degree. Using the ratio of signal extreme value point number and the signal length evaluates smooth degree (SD), for one dimensional signal with a length of m, the more extreme value points (up to $m - 2$) it has, the rougher it will be. Defined smooth degree as:

$$SD = 1 - \frac{N_e}{m - 2} \quad (2)$$

In Eq. (2), Ne is the number of extreme value points. The closer to 1 the SD be, the smoother the signal will be, when $SD = 1$, the signal is a straight line.

Using sliding window method detects the SD of ECG signals short-time autocorrelation function, considering the algorithm's real-timeliness and accuracy requirements, determining the window width is 30 sampling points through a large number of experiments.

Dynamic Variation Coefficient. Defining the dynamic variation coefficient with local mean and variance of ECG signals as:

$$DVC(i) = \frac{Var(i)}{Mean(i)} \quad (3)$$

In Eq. (3), $DVC(i)$ is variation coefficient of $r(i)$, $r(i)$ is ECG signals short-time autocorrelation function of the ith sampling point. $i = 1, 2, 3, \ldots$; $Var(i)$ and $Mean(i)$ are the dynamic mean and dynamic variance of a part of signals with the end of $r(i)$, respectively. Figure 3 is the calculation process of dynamic mean and dynamic variance.

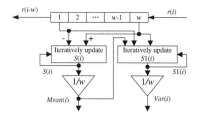

Fig. 3. Calculation process of dynamic mean and dynamic variance

Using sliding window iterative method calculates dynamic mean and variance. The dynamic mean is:

$$Mean(i) = \begin{cases} \dfrac{S(i)}{i} & i \leq w \\ \dfrac{S(i)}{w} & i > w \end{cases} \tag{4}$$

In Eq. (4), w is sliding window width. Being similar to the calculation of smooth degree, setting window width $w = 30$ sample points; $Mean(i)$ represents the mean of the sampling points in the current window; $S(i)$ is the total sample values in the window which can be obtained by iteration method, especially, $S(1) = r(1)$:

$$S(i) = \begin{cases} S(i-1) + r(i); & i \leq w \\ S(i-1) + r(i) - r(i-w) & i > w \end{cases} \tag{5}$$

Similarly, dynamic variance is:

$$Var(i) = \begin{cases} \dfrac{S1(i)}{i} & i \leq w \\ \dfrac{S1(i)}{w} & i > w \end{cases} \tag{6}$$

In Eq. (6),

$$S1(i) = \begin{cases} S1(i-1) + (r(i) - Mean(i))^2 & i \leq w \\ S1(i-1) + (r(i) - Mean(i))^2 - (r(i-w) - Mean(i-w))^2 & i > w \end{cases} \tag{7}$$

In Eq. (7), $S1(1) = 0$. Then, normalizing dynamic variance to the scope of $[0, 1]$ with the method of min-max normalization.

ECG Signal Interferences Detection. The threshold value is set to judge ECG signals interferences with the combination of ECG signals' short-term autocorrelation function value, smooth degree and dynamic variation coefficient. Through calculating the mean of smooth degree thresholds and dynamic variation coefficient thresholds of 100 groups of data in the database, respectively, the final thresholds are obtained. That is, the final smooth degree threshold is 0.85 and dynamic variation coefficient threshold is 0.7. ECG signals interferences are detected by Eq. (8):

$$ECG = \begin{cases} 1, & SD(w) \leq 0.85 \quad and \quad DCV(w) \geq 0.7 \\ 0, & others \end{cases} \tag{8}$$

In Eq. (8), '1' represents interferences, '0' represents the normal; $SD(w)$ is smooth degree of the signal, $DVC(w)$ is dynamic variation coefficient of the signal.

HRV Estimation in ECG Interferences. At present, the estimation of HRV in ECG signals interferences is directly replaced with PRV of synchronously acquisitioned PPG signals, but the prediction accuracy and real-timeliness is not high. So, in order to improve the estimation accuracy and real-timeliness, the improved sliding window iterative DFT algorithm is used to estimate instantaneous PRV real-timely, and then the estimated PRV is regarded as HRV in ECG interferences.

Improved Sliding Window Iterative DFT Algorithm. For periodic signal $\{x(k)\}, k = 0,\dots,M$, M is the length of signal. Assuming that there are N sampling points in the period T, the sampling period $\tau = T/N$, angular frequency $\omega = 2\pi/T$. So the signal's fundamental component $\{x_1(k)\}$ can be obtained by:

$$x_1(k) = A_1 \cos(\omega k \tau) + B_1 \sin(\omega k \tau) \tag{9}$$

$$A_1 = \frac{2}{N} \sum_{i=N_{cur}}^{N_{cur}-N+1} x(i\tau) \cos(\omega i \tau) \tag{10}$$

$$B_1 = \frac{2}{N} \sum_{i=N_{cur}}^{N_{cur}-N+1} x(i\tau) \sin(\omega i \tau) \tag{11}$$

Being based on this theory, an improved method of sliding window iterative DFT had been proposed in paper [8], which can improve the real-timeliness of algorithm further. Figure 4 is the schematic diagram of the improved sliding window iterative DFT.

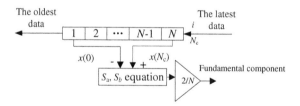

Fig. 4. Improved sliding window iterative DFT schematic diagram

In Fig. 4, S_a and S_b are two iterative process variables of sliding window DFT.

$$S_a(k) = \sum_{i=N-k+1}^{N} x(i) \cos\left(\frac{2\pi}{N} i\right) \tag{12}$$

$$S_b(k) = \sum_{i=N-k+1}^{N} x(i) \sin\left(\frac{2\pi}{N} i\right) \tag{13}$$

In the process of the movement of the window, new data point $x(k)$ is located in the window of Nth bit all the time, that is, $x(k) = x(N_c)$ thus:

$$
\begin{aligned}
x_1(k) &= x_1(N_c) \\
&= A_1 \cos(\omega \tau N_c) + B_1 \sin(\omega \tau N_c) \\
&= A_1 \cos\left(\frac{2\pi}{N} \times N\right) + B_1 \sin\left(\frac{2\pi}{N} \times N\right) \\
&= A_1 = \frac{2}{N} S_a(k)
\end{aligned}
\tag{14}
$$

Compared to the traditional sliding window iterative DFT algorithm, it can significantly reduce the calculation amount and can be effectively used to extract fundamental wave signal real-timely.

The Extraction of PRV. Through applying the improved sliding window iterative DFT algorithm for PPG signals which are synchronously acquisitioned with ECG interferences, the fundamental wave are obtained, which can reflect the periodic change of PPG signals. So, period $\Delta t'$ of fundamental wave signal scan effectively reflect the change of period Δt of PPG signals. Its maximum value is the peak (P$'$ wave), which is easy to detect. So the fundamental wave can replace the PPG signals to realize the detection of PRV signal. The main peak of the fundamental wave signals is denoted as $pp(i), i = 1, 2, \ldots, n - 1, n$ is the amount of P$'$ wave, then, $PRV(i)$ is:

$$
PRV(i) = \frac{60}{pp(i)/f_s}
\tag{15}
$$

In Eq. (15), f_s is sampling frequency. The PRV extracted is regarded as HRV in ECG interferences.

3 Results and Discussion

3.1 ECG Interferences Detection Results

The signals interferences are annotated by Yang Xiaohua first, who is the chief physician in hospital of Lanzhou University of Technology. Then, 100 groups of data in the database are tested, and the algorithm accuracy is evaluated. Counting the number of interferences sections which are detected correctly by proposed algorithm, namely true positives (TP); the number of interferences sections which are detected falsely, namely false positives (FP); as well as the missing interferences sections number, namely false negatives (FN). Evaluating the accuracy (Positive Predictivity, +P) and sensitivity (Se) of the algorithm.

$$+P = \frac{TP}{TP+FP} \tag{16}$$

$$S_e = \frac{TP}{TP+FN} \tag{17}$$

In addition, the proposed method is compared with ECG signals interferences detection algorithm based on repeated gaussian filter and cross-correlation in paper [2] and the algorithm based on RR intervals in paper [3] which are commonly used. The result is shown in Table 1. The '*Time*' in Table 1 represents the running time of a length of 2 min ECG signals. The simulation results of the ECG intercepted from the database are shown in Figs. 5, 6 and 7, respectively.

Table 1. 100 groups of ECG signals interferences detection results statistics

Interferences detection	The total number of interferences	TP	FP	FN	+P/%	Se/%	Time/s
Proposed algorithm	1766	1740	38	26	97.86	98.53	6.73
paper [2]	1766	1600	200	166	88.89	90.60	18.62
paper [3]	1766	1648	172	108	90.51	93.88	12.25

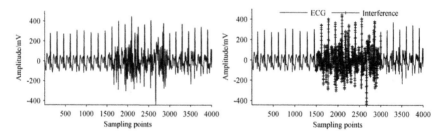

(a) A part interception of ECG signals in database   (b) ECG interferences detection

Fig. 5. ECG and ECG interferences detection result

It can be seen that detection accuracy and sensitivity are both higher than the method proposed in paper [2] and paper [3] from Table 1. In addition, the proposed method takes shorter time. This is due to the algorithm which proposed in paper [2] mainly relies on the accurate segmentation of signals earlier. The algorithm proposed in paper [3] essentially depends on the correct detection of R wave. But they are both hard to segment or detect accurately because of noises and interferences. Also, we can see that the algorithms they proposed detect the normal ECG waves as interferences in Figs. 6 and 7, so the algorithms accuracy is not higher than algorithm proposed. In addition, as the algorithm in paper [2] needs filter with Gaussian filter repeatedly during the period of heart beat segmentation, so the real-timeliness becomes poor. Similarly, the algorithm in paper [3] needs to perform adaptive template matching after R waves

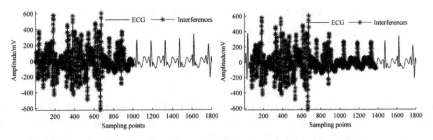

(a) Result of the proposed algorithm   (b) Result of the algorithm in paper [2]

Fig. 6. ECG interferences detection results compared with the algorithm in paper [2]

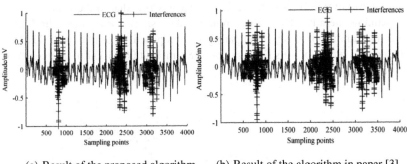

(a) Result of the proposed algorithm   (b) Result of the algorithm in paper [3]

Fig. 7. ECG interferences detection results compared with the algorithm in paper [3]

detection, but the template generation process is complex. So the algorithm complexity is enhanced, and real-timeliness becomes poor. In a word, the algorithm we proposed can detect ECG interferences more accurately and more real-timely.

3.2 HRV Estimation Results in ECG Interferences

Experiments. Selecting the number of 200 clean ECG signals sections from the database, after filtering process, using dynamic difference threshold method [9] extracts R wave, getting RR interval sequences further, then $HRV = 60/(RR(i)/f_s)$(bpm), the sampling frequency is f_s, this HRV is considered as true heart rate variability (THRV); 200 clean sections of PPG signals which are synchronous with ECG signals interferences are intercepted, too. Then, using the proposed method estimates HRV from them, which is regarded as estimated heart rate variability (EHRV).

Experiments Results and Analysis. EHRV is compared with THRV from the aspects of absolute error, relative error, the maximum absolute error, mean relative error and root mean square error, etc. The mean of each parameters is calculated further, respectively, that is, calculating mean maximum absolute error (MMAE); mean relative

Table 2. Error analysis of HRV estimation

Parameters	MMAE	MMRE	MRMSE	MSD	Time/s
Proposed algorithm	6.00	0.02	1.24	0.96	1.010
Directlyreplaced algorithm	16.96	0.04	5.04	4.27	5.159

error (MRE); mean root mean square error (MRMSE) and mean standard deviation (MSD). Similarly, PRV which are extracted with the same method with THRV from PPG signals is compared with THRV; Finally, analyzing the error of proposed algorithm and the traditional PRV directly replaced HRV algorithm. The results are shown in Table 2. The '*Time*' represents the running time of a length of 1 min signal. It can be seen from Table 2, accuracy of the algorithm proposed is higher than PRV directly replaced algorithm. This is because PRV and HRV have tiny difference, so directly replaced algorithm will reduce the accuracy. Meanwhile, due to the characteristics of sliding window iterative DFT algorithm, it takes shorter time than PRV directly replaced algorithm, obviously. Comparing the proposed algorithm with THRV and PRV based on a length of 40 s ECG signals (188 m(100000:110000)) intercepted from database, the result is shown in Fig. 8.

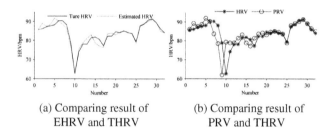

(a) Comparing result of   (b) Comparing result of
    EHRV and THRV   PRV and THRV

Fig. 8. Comparing result of HRV estimation algorithms

4   Conclusion

Aiming at the problem of HRV estimation in ECG interferences, a new method is proposed to estimate HRV in ECG interferences from PPG signals segments which are synchronously acquisitioned with ECG interferences. The proposed method is compared with the PRV directly replaced HRV algorithm. The results show that the method proposed can more accurately and more real-timely estimate HRV in ECG signals interferences. To sum up, the method we proposed can avoid influence of interferences, reduce the false alarm rate of monitor, improve the quality of diagnosis and emergency levels, meanwhile, which has an important application value in areas such as clinical diagnosis, disease monitoring and prevention.

Acknowledgments. This work was supported by the National Natural Science Foundation (grant 81360229) of China, the National Key Laboratory Open Project Foundation (grant 201407347) of Pattern Recognition in China and the Basic Research Innovation Group Project in Gansu Province (grant 1506RJIA031).

References

1. Stein, P.K.: Challenges of heart rate variability research in the ICU. Crit. Care Med. **41**(2), 666–667 (2013)
2. Karlen, W., Kobayashi, K.: Photoplethysmogram signal quality estimation using repeated Gaussian filters and cross-correlation. Physiol. Measur. **33**, 1617–1629 (2012)
3. Orphanidou, C., Bonnici, T., Charlton, P., et al.: Signal-quality indices for the electrocardiogram and Photoplethysmogram: derivation and applications to wireless monitoring. IEEE J. Biomed. Health Inf. **19**(3), 832–838 (2015)
4. Li, Q., Cadathur, R., Clifford, G.D.: A machine learning approach to multi-level ECG signal quality classification. Comput. Methods Programs Biomed. **117**, 435–447 (2014)
5. Yu, E.Z., He, D.N., Su, Y.F., et al.: Feasibility analysis for pulse rate variability to replace heart rate variability of the healthy subjects. In: International Conference on Robotics and Biomimetics (ROBIO), pp. 1065–1070 (2013)
6. Schulte, R., Krug, J., Rose, G.: Identification of a signal for an optimal heart beat detection in multimodal physiological datasets. In: Computing in Cardiology Conference (CinC), pp. 273–276. IEEE (2014)
7. Zhang, A.H., Chou, Y.X.: Dynamic pulse signal acquisition and processing. Med. Apparatus Instrum. J. China **36**(2), 79–84 (2012)
8. Lin, Y.Y., Xu, Y.H., Liu, X.B.: Harmonic current detection method of sliding window iteration DFT. Power Syst. Prot. Control **39**(3), 78–90 (2011)
9. Zhang, A.H., Wang, P., Chou, Y.X.: Pulse signal peak detection algorithm based on dynamic differential threshold. J. Jilin Univ. Eng. Sci. **44**(3), 847–853 (2014)

Predicting Progression of ALS Disease with Random Frog and Support Vector Regression Method

Shu-Lin Wang[1(✉)], Jin Li[1], and Jianwen Fang[2]

[1] College of Computer Science and Electronics Engineering, Hunan University,
Changsha 410082, Hunan, China
smartforesting@gmail.com
[2] Division of Cancer Treatment and Diagnosis, National Cancer Institute,
Rockville, MD 20850, USA

Abstract. Amyotrophic lateral sclerosis (ALS) is a fatal neurodegenerative disease that involves the degeneration and death of the nerve cells in brain and spinal cord that control voluntary muscle movement. This disease can cause patients struggling with a progressive loss of motor function while typically leaving cognitive functions intact. This paper presents a novel predication method that combines a dimension reduction (integrating partial least square into random frog algorithm) with support vector regression to predict the progression of ALS in the next 3–12 months according to the data collected from the patients over the latest three months. The experiment on the actual data from the PRO-ACT database indicates that the proposed method is effective and robust and can predict the clinical outcome by means of the slope of ALS progression, as measured using the ALS functional rating scale (ALSFRS) and the score used for monitoring ALS patients. Especially, the features selected can effectively distinguish the clinical outcome targets. It is of great benefit to aid clinical care, identify new disease predictors and potentially significantly reduce the costs of future ALS clinical trials.

Keywords: Amyotrophic lateral sclerosis · Feature selection · Random frog · SVM

1 Introduction

Amyotrophic lateral sclerosis (ALS) (also known in the US as Lou Gehrig's Disease and as Motor Neuron in the UK) is an idiopathic, fatal neurodegenerative disease of the human motor system [1], and its symptoms include muscle weakness, paralysis and eventually death, usually within 3 to 5 years from disease onset. Approximately one out of 400 people is diagnosed with, and dies of ALS [2]. The modern medicine faces with a major challenge in finding an effective treatment. At present Riluzole is the only approved medication for ALS, and has a limited effect on survival [3].

One substantial obstacle for understanding and developing a treatment for ALS is due to the heterogeneity of the disease. The more heterogeneous the disease, the more difficult it is to predict how a given patient's disease will progress. It is gratifying that

© Springer International Publishing Switzerland 2016
D.-S. Huang et al. (Eds.): ICIC 2016, Part III, LNAI 9773, pp. 160–170, 2016.
DOI: 10.1007/978-3-319-42297-8_16

more accurate way to anticipate disease progression, as measured by a clinical scale (ALS Functional Rating Scale: ALSFRS, or the revised version ALSFRS-R), can lead to great significance in clinical practice and clinical trial management.

Pooled clinical trial data sets have proven invaluable for researchers seeking to unravel complex diseases such as multiple sclerosis, Alzheimer's and others [4]. The data presented were collected from ALS patients in the course of their participation in Phase II and Phase III ALS clinical trials. However, the structure of these data is very complex, so data processing and feature selection are required for analyzing these high dimension data. The selected features are applied not only to construct prediction model but also to aid clinical care and reduce the costs of future ALS clinical trials.

Feature selection algorithms have been studied extensively. For example, information gains, rank sum test, relief-F, random forest [5] *et al.* have been proposed and applied to feature selection. After decades of development in the machine learning and data mining fields, feature selection techniques has shifted from being an illustrative example to becoming a real prerequisite for building model. At present, ensemble feature selection approaches have related research, the evidence that there is often not a single universally optimal feature selection technique, and due to the possible existence of more than one subset of features that discriminates the data equally well. We apply the Random Frog Algorithm coupled with the Partial Least Squares (RFA-PLS) [6] to select features and adopt Support Vector Regression (SVR) [7] to predict the ALSFRS slope to predict the clinical outcome.

2   Methods

2.1   Problem Description

ALS clinical trials accumulated consist of patients from clinical trials available open access on the PRO-ACT database (www.ALSdatabase.org). The goal of analyzing these data is to predict the ALSFRS slope as disease progression. Concretely speaking, our goal is to predict the 3–12 months ALSFRS slope using the clinical trial data measured between 0–3 months. Subsequently, we can transform the original descriptive text of ALS clinical data to the quantitative data that can be represented as a matrix, where denotes the number of patients, and denotes the number of features.

To determine the ALSFRS slope of the patient, the first visit after month three of participation in the clinical trial is assigned as. If there were visits through month 12, the first such visit after month 12 is assigned as. If there was no such visit, the subject was removed from consideration. Then, the ALSFRS slope of the training set can be calculated as

$$y_{slope} = \frac{ALSFRS(m_2) - ALSFRS(m_1)}{m_2 - m_1} \tag{1}$$

thus, we can describe the dataset as the matrix. The -th patient can also be described as a vector, where represent the extracted features, respectively, and the represents the corresponding slope value.

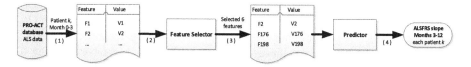

Fig. 1. The flowchart of constructing the prediction model.

2.2 Flowchart of Analysis

The flowchart of our analysis method includes four steps that can be shown in Fig. 1. (1) Data preprocessing; (2) feature selection; (3) prediction; (4) the evaluation of results. The data from a given patient is firstly fed into the feature selection algorithm ("selector" in short). Then the selector selects a subset of features. Next, the prediction program ("predictor" in short) reads selected features in order to predict ALSFRS slope. Finally, our prediction model is evaluated by an independent validation dataset.

2.3 Data Preprocessing

From the raw data associated with a given patient, we must extract a vector of numeric features (shown in Fig. 2) to be used to construct prediction model. The raw data are divided into two types: static data without a temporal element and time series data, so we must apply different feature selection methods to integrate these data.

Accordingly, the raw static data must be digitized. For example, the values "Limb" and "Bulbar" in the "onset_site" field are replaced with the values "1" and "2", respectively. In addition, "0" and "1" represent the values "Male" and "Female" in the "Sex" field, respectively. For the time series data, we extract their statistical features. For example, the fields ALSFRS total score, FVC (forced vital capacity) subject liters, and vital signs data (contains weight, height, respiratory rate, and systolic blood pressure) *et al.* are time series data, and they are summarized by the maximum, minimum, and mean measurement values, the slope of the time series, *etc.*, respectively.

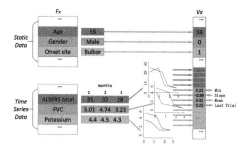

Fig. 2. The diagram of constructing feature vector for each patient. The static data means the fields are constant during the measurement time and the time series data means the changes of physiological function during the clinical trials, represented by the following statistics as each patient features: slope, min, max, mean, last trial, standard deviation, *etc.*

2.4   Feature Selection

After a large set of candidate features from the provided clinical data are extracted and these data are preprocessed for each feature firstly. Then we apply a feature selection technique to select informative features from this set of candidate features.

Random Frog Algorithm coupled with PLS.   Random frog is a kind of method that works in an iterative manner. Briefly, random frog works in three steps mainly: (1) initialized feature subset containing features randomly; (2) select a candidate feature subset containing features from, accept with a certain probability as, then replace with, and loop this step until iterations; (3) compute the selection probability of each feature whose value can be used as a measure of feature importance.

After iterations, feature subsets can be obtained totally. N_j denote the frequency of the jth feature, $j = 1, 2, \ldots, n$, selected from the features. For each feature, its selection probability can be calculated as

$$P_j = \frac{N_j}{N}, j = 1, 2, \ldots, n \tag{2}$$

where is feature subsets after iteration, is the frequency of the feature.

As can be expected, the more optimal a feature is, the more likely to be selected into these feature subsets. That is to say, the measure of feature importance can be used as an index for feature selection.

At the same time, to be able to single out a subset of informative features this can lessen over-fitting and improve the performance of model [8]. We have integrates the PLS modeling together to facilitate the modeling procedure. The library package LibPLS 1.95 downloaded from www.libpls.netis used to our analysis [9].

2.5   ALSFRS Slope Prediction

Our ultimate goal is the prediction of the expected value of the ALSFRS slope during months 3 to 12. The selector chooses a small subset of features to be used to the clinical target predictor. Support vector machines regression is applied to the prediction of ALSFRS slope.

Support Vector Machines Regression.   Usually, support vector machines (SVM) is applied to classification problem, while SVR can be formulated as an optimization problem as follows to predict continue value.

$$min \frac{1}{2} \|f\|^2 + C \sum_{i=1}^{N} \left(\xi_i + \xi_j^* \right)$$

$$subject\,to \begin{cases} y_i - f(x_i) \leqq \varepsilon + \xi_i \\ f(x_i) - y_i \leqq \varepsilon + \xi_j^*, \\ \xi_i, \xi_j^* \leqq 0 \end{cases} \tag{3}$$

where is the regularization parameter that determines the trade-off between the margin and prediction error, and are error items. Only difference between the SVM and the SVR is the loss function or called the [10].

According to a set of training data, where denotes feature vector and represents its corresponding ALSFRS slope. Thus, the expected function of SVR can be formulated as

$$f(x) = \sum_{i=1}^{M} \alpha_i K(x_i, x) + b, \tag{4}$$

where is the kernel function. In our study, we train and build the SVR with the Radial Basis Function (RBF kernel), which can be given by

$$K(x_i, x) = exp\left(-\gamma \|x_i - x\|^2\right) \tag{5}$$

In order to optimize the SVR training, there are some parameters needed to be determined properly such as the regularization parameter and the kernel parameter. For the implementation of SVR algorithm, we used the Online SVR (Francesco Parrella, 2007) software package.

2.6   Performance Evaluation

The root mean square deviation () and Pearson's correlation coefficient (PCC,) are used to evaluate the performance of prediction models. The measures the differences between corresponding slope pair values predicted by a model and the values actually observed. The measurement formula can be denoted as

$$RMSD = \sqrt{\frac{1}{M} \sum_{i=1}^{M} |s_i - p_i|^2}, \tag{6}$$

where is the actual ALSFRS slope and is from the ALSFRS slope prediction.

In addition, Pearson's correlation coefficient that evaluates how well a prediction model is able to reveal ALSFRS trends can be expressed in

$$\rho_{S,P} = \frac{cov(S,P)}{\sigma_S \sigma_P}, \tag{7}$$

where is the covariance of the two variables and, and are the product of their standard deviations. Usually, the smaller the value of is, the better the method performs, while the bigger the value of the PCC is, the better the method performs.

3   Experiments

3.1   Data Collection

The experimental data from ALS clinical trials is from the PRO-ACT database, in which each patient is identified by a PatientID and the patient-specific assessment is

Table 1. Partial data format from the PRO-ACT database.

Patient ID	Data type	Feature name	Feature value	Feature unit	Delta
7824	ALSFRS(R)	ALSFRS Total	30	NA	0
7824	Vital signs	Blood pressure	140	MMHG	14
7824	ALSHX	Onset_Site	Limb	NA	0

identified by a record (each patient has multiple records). Some of assessments are separated into different data types as follows: ALSFRS(R), Laboratory Data, Vital Signs, Demographics, Riluzole use, Adverse events and so on. Table 1 describes part of the data structure. For example, Patient 7824 had, at (day 14 from beginning of measurement), the following vital signs: a blood pressure of 140 MMHG. At (first day of measurements) their ALSFRS total is 30, and the onset site of disease is limb. Overall, from patients of the clinical data we extract features including the ALSFRS scores, personal assessments as well as laboratory measurements, *etc.* All 2187 samples are divided into two groups: training set and validation set. The training set includes 1500 samples randomly selected, and the remaining samples are the validation set.

3.2   Experimental Results

The features selected by the selection model are used to predict the clinical outcome or the ALSFRS slope. In our experiments two kinds of feature selection methods RF and RFA-PLS are adopted to select ALS-related features, we constrain that only six features in each subset of features are selected. For evaluating the relevance between the selected features and the ALSFRS slope, we adopt three regression methods, e.g., RF-regression, PLS-regression and SVR.

Feature Selection Results. RF method has two parameters to determine in this experiment: one is the number of features selected in bootstrap sample called as, and another one is the number of total decision trees in the ensemble called as. The number of trees could affect the used to calculate the percent variance. In the experiment, once the number of trees reaches 1200, will become stable. Therefore, we set (sqrt the number of features) and to construct the predictor model with 10-fold cross-validation (CV). According to the result of each feature, the top-ranked six features with the maximum value are selected and they are Nos. 2, 35, 37, 60, 221, and 222, respectively, where the digital numbers represent the series number corresponding to the features in raw data.

RFA-PLS method has several parameters affecting the performance of RFA. These parameters as well as their settings are given as follows. The number of iterations is set to. The parameter Q represents the number of features contained in the initialized feature vector. Here, for determining the optimal parameter, value is limited to range from 5 to 20 and the selection probability of each feature is used to measure its importance. We design two methods to select informative features. (1) 10-Fold CV is applied to optimize parameter on training set, and the values of and PCC are used to

Table 2. Part of results obtained with different parameter. The features in each line rank by ascending order of importance.

#	Top six features					
5	49	7	176	43	3	222
6	48	7	176	43	3	222
7	158	176	43	7	3	222
8	43	158	176	7	3	222
9	196	7	158	176	3	222
10	196	176	7	158	3	222
11	196	176	7	158	3	222
12	158	198	196	176	3	222
13	196	7	176	158	3	222
14	158	7	196	176	3	222
15	198	176	7	196	3	222
16	198	176	7	196	3	222
17	198	7	157	196	3	222
18	157	176	7	196	3	222
19	157	7	176	196	3	222
20	7	157	196	176	3	222

evaluate the performance of the selected features. The experimental results shown in Table 2 indicate that the optimal number of component is determined to be 7, and its corresponding optimal feature subset selected is Nos. 3, 7, 43, 158, 176, and 222, shown in Fig. 3. (2) For avoiding over-fitting, we just count the occurrence frequency of each feature in all of the selected feature subsets, and then we can get the most frequent six features (Nos. 3, 7, 158, 176, 196, and 222).

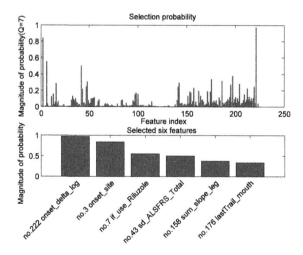

Fig. 3. The upper subplot describes the selection probability of each feature when factor and the lower one is the magnitude of probability of the selected six features via RFA-PLS.

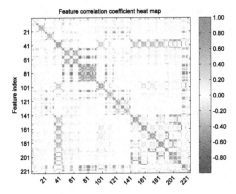

Fig. 4. The correlation coefficient heat map between two arbitrary features.

In raw data exists feature correlation shown in Fig. 4, which can degrade the prediction performance, so the finally selected informative feature set should not contain relevant features. By comparing the feature set selected by RF and RFA-PLS, we find that the feature set selected by RF often contain relevant features, while the feature set selected by RFA-PLS do not contain relevant features. For example, the onset_delta_log (feature No. 222, calculated by) is one of the most important feature for every patients. Surprisingly, the onset_delta (feature No. 2) has a high value in RF but a lower selection probability in RFA-PLS (). It is obvious that the two features have a high correlation. However, in final feature set RF select this feature, while RFA-PLS discard this feature.

Prediction Results. We adopt three regression methods (RF-regression, PLS-regression, and SVR) to predict the ALSFRS slope. For the RF-regression, the parameters are set as follows. (1) The number of trees is set to, owing to against the number of trees no longer fluctuates. (2) The number of candidate predictors at each split node is set to. For PLS-regression, we apply 10-fold CV (each sample consists of only 6 features) to determine the number of components.

SVM has its excellent ability to control error without causing over-fitting to the dataset. In generally, SVM has two practical models: support vector for classification and SVR. Usually, SVR predict continuous value, while SVM predict label value. As for the setting of parameters of SVR, we select radial basis kernel (RBF kernel) function at to build the SVR model. We have tried several parameter sets and determined this combination of parameters has been yield relatively better performance. Figure 5 intuitively illustrates the scatter diagram of the actual slope and predicted one of each validation sample. As can be seen from Fig. 5, the results of the ALSFRS slope predicted by SVR with RF prediction model is very bad, while the best performance results are obtained by SVR with RFA-PLS.

We also adopt and PCC to evaluate the performance of different methods, shown in Table 3. By comparing these results with the evaluation items, it is obvious that our method combining RFA-PLS with SVR performs the best on the validation dataset achieve ideal effect RMSD=0.5243 and ρ=0.4086. The top-ranked features are Nos. 3, 7, 43, 158, 176, and 222, respectively. Their corresponding names areonset_site (location

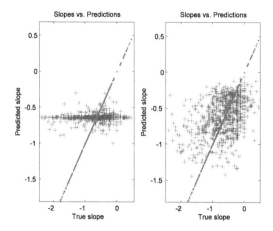

Fig. 5. The scatter diagram of prediction results based on SVR. The left subplot describes the prediction result based on RF selector, the right subplot describes the prediction result based on RFA-PLS selector.

Table 3. The prediction results with different predictors.

Selector	Predictor		
Random forests	RF-regression	0.5315	0.3795
	PLS-regression	0.5364	0.3388
	SVR	0.5674	0.1032
RFA-PLS	RF-regression	0.5253	0.3909
	PLS-regression	0.5312	0.3603
	SVR	**0.5243**	**0.4086**

of the onset of the disease), if_use_Riluzole (whether or not to use Riluzole drugs), sd_ALSFRS_Total (standard deviation on ALSFRS_Total), sum_slope_leg (the sum slope of the function of the leg), lastTrail_mouth (mouth function at the last measurement of the first three months), onset_delta_log (), respectively.

4 Conclusions

This paper aims to identify subgroups of patients with distinct clinical outcomes that can distinguished by the clinical features, which is of great benefit to aid clinical care and identify new disease predictors, thus we presents a novel method of predicting ALS progression including three steps to predict the outcome clinical targets. Firstly, the clinical data collected is preprocessed to construct the feature vector for all patients. Secondly, we design a novel dimension reduction that integrates partial least square into random frog algorithm to select and construct candidate features. Lastly, the support vector regression with the candidate features is applied to predict the slope of ALSFRS to further analyze the ALS progression in the next 3–12 months according to the data collected from the patients over the latest three months. The experimental results indicate that the proposed method can predict the clinical outcomes effectively and robustly. By comparing with the results of random forests method, our method is competitive in two evaluation items including and PCC.

The merits of the proposed method include two aspects. One is that the most important feature can be selected by RFA-PLS method from a group of relevant features while all features in one subset of relevant features are selected by RF method. Another is that the proposed method is time-saving numerical method compared with RF method. The demerit of the proposed method is that it is difficult to determine the optimal parameters combination for SVR model with radial base function kernel. In conclusion, our method can estimate the future disease progression of ALS patients, is helpful to understand the ALS disease mechanisms, and play important role in making decisions regarding the test of the novel therapeutic approaches in clinical trials.

Acknowledgements. This research was supported by the National Natural Science Foundation of China (Grant Nos. 61472467, 60973153 and 61471169) and the Collaboration and Innovation Center for Digital Chinese Medicine of 2011 Project of Colleges and Universities in Hunan Province. What's more, acknowledge both Prize4 Life and Sage Bionetworks-DREAM for supplied the experiment data.

References

1. Kiernan, M.C., Vucic, S., Cheah, B.C., Turner, M.R., Eisen, A., Hardiman, O., Burrell, J.R., Zoing, M.C.: Amyotrophic lateral sclerosis. Lancet **377**(9769), 942–955 (2011)
2. Drigo, D., Verriello, L., Clagnan, E., Eleopra, R., Pizzolato, G., Bratina, A., D'Amico, D., Sartori, A., Mase, G., Simonetto, M., de Lorenzo, L.L., Cecotti, L., Zanier, L., Pisa, F., Barbone, F.: The incidence of amyotrophic lateral sclerosis in Friuli Venezia Giulia, Italy, from 2002 to 2009: a retrospective population-based study. Neuroepidemiology **41**(1), 54–61 (2013)
3. Miller, R.G., Mitchell, J.D., Moore, D.H.: Riluzole for amyotrophic lateral sclerosis (ALS)/motor neuron disease (MND). Cochrane Database Syst. Rev. (3) (2012)
4. Kuffner, R., Zach, N., Norel, R., Hawe, J., Schoenfeld, D., Wang, L.X., Li, G., Fang, L., Mackey, L., Hardiman, O., Cudkowicz, M., Sherman, A., Ertaylan, G., Grosse-Wentrup, M., Hothorn, T., van Ligtenberg, J., Macke, J.H., Meyer, T., Scholkopf, B., Tran, L., Vaughan, R., Stolovitzky, G., Leitner, M.L.: Crowdsourced analysis of clinical trial data to predict amyotrophic lateral sclerosis progression. Nat. Biotechnol. **33**(1), 51–57 (2015)
5. Cutler, A., Cutler, D.R., Stevens, J.R.: Random forests. Mach. Learn. **45**(1), 157–176 (2011)
6. Li, H.D., Xu, Q.S., Liang, Y.Z.: Random frog: an efficient reversible jump Markov Chain Monte Carlo-like approach for variable selection with applications to gene selection and disease classification. Anal. Chim. Acta **740**, 20–26 (2012)
7. Awad, M., Khanna, R.: Support vector regression. Neural Inf. Proc. Lett. Rev. **11**(10), 203–224 (2007)
8. Jiang, J.H., Berry, R.J., Siesler, H.W., Ozaki, Y.: Wavelength interval selection in multicomponent spectral analysis by moving window partial least-squares regression with applications to mid-infrared and hear-infrared spectroscopic data. Anal. Chem. **74**(14), 3555–3565 (2002)

9. Li, H., Xu, Q., Liang, Y.: libPLS: an integrated library for partial least squares regression and discriminant analysis, PeerJ (2014)
10. Mordelet, F., Horton, J., Hartemink, A.J., Engelhardt, B.E., Gordan, R.: Stability selection for regression-based models of transcription factor-DNA binding specificity. Bioinformatics **29**(13), 117–125 (2013)

Chinese Historic Image Threshold Using Adaptive K-means Cluster and Bradley's

Zhi-Kai Huang$^{(\boxtimes)}$, Yong-Li Ma, Li Lu, Fan-Xing Rao,
and Ling-Ying Hou

College of Mechanical and Electrical Engineering,
Nanchang Institute of Technology, Nanchang 330099, Jiangxi, China
huangzhik2001@163.com

Abstract. Resorting to extraction text techniques for Chinese heritage documents becomes an increasing need. Historic documents such as Chinese calligraphy usually were handwritten or scanned in low contrast so that an automatic optical character recognition procedure for document images analysis is difficult to apply. In this paper, we present a historic document image threshold based on a combination of Bradley's algorithm and K-means. An adaptive K-means cluster as a pre-processing methods for document image has been used for automatically grouping the pixels of a document image into different homogeneous regions. In Bradley's methods, every image's pixel is set to black if its brightness is T percent lower than the average brightness of surrounding pixels in the window of the specified size, otherwise it is set to white. Finally, text bounding boxes are generated by concatenating neighboring word clusters with mathematical morphology method. Experimental results show that this algorithm is robust in dealing with non-uniform illuminated, low contrast historic document images in terms of both accuracy and efficiency.

Keywords: Chinese historical image · K-means cluster · Bradley's method

1 Introduction

The digital rubbing is a novel approach to promote and pass on Chinese traditional arts, as well as a new idea to protect stone relics. Historical printed documents, such as old books and rubbings, are being digitized and made available through software interfaces such as web-based libraries, for instance, there is a large collection of Chinese rubbing database keeps in UC Berkeley east Asian library [14, 15]. These documents are challenging for OCR (Optical Character Recognition, OCR) because it use non-standard fonts and suffer from printing noise, artifacts due to aging, varying kerning (space between letters), varying leading (space between lines), frequent line break hyphenation, and other image problems due to the conversion from print-to-microfiche-to-digital [1]. Segmenting heritage documents images into text from background is a crucial pre-processing step for automated reading of historical documents. A Chinese historical rubbing image has been shown in Fig. 1. Many stone texture patches have shown in background of rubbing image because of the nature factor (all characters have been carved out of stone, ink rubbings has reproduced from

© Springer International Publishing Switzerland 2016
D.-S. Huang et al. (Eds.): ICIC 2016, Part III, LNAI 9773, pp. 171–179, 2016.
DOI: 10.1007/978-3-319-42297-8_17

that stone). The histogram has showed in three bands with a unimodal distribution, it is difficult to calculate the specific threshold.

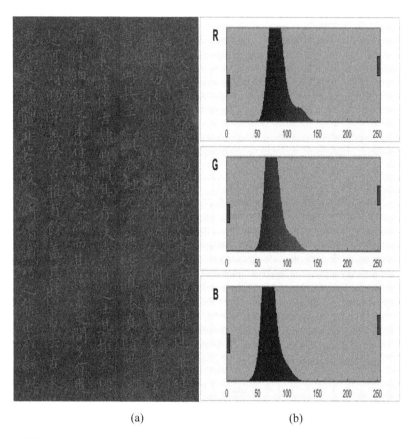

(a)   (b)

Fig. 1. Historical handwritten document image with low contrast and histogram (a) Original image, (b) The Histogram of original image

Document image binarization is a key step from the image processing to decreases the computational load and enables the utilization of the simplified analysis methods compared to 256 levels of grey-scale or color image information, which is also a basic technique in the computer vision. It makes the high-level computer vision tasks possible, so it plays an important role in the technology of the document image processing [2]. A number of promising techniques for document image binarization were implemented at the different literatures [3]. Generally, the methods that deal with document image binarization may be broadly categorized either globally or locally. Whether it is global or local binarization, threshold choice is a sensitive function of the local reflectance map. Specially, for low contrast scanned historic document image, it is difficult to improve on a fixed threshold centered between the extreme observed values, since too low a value will swamp the difference map with spurious changes, while too high a value will suppress significant change [11].The most classical threshold algorithm that is Otsu's method [5], which maximizes the values of class variances to get optimal threshold. Maximum entropy was firstly proposed by Pun [12]. The purpose is to divide the gray-level histogram of image into separate classes and make maximum total entropy of all kinds of classes. An automatic threshold algorithm based on an iterative threshold selection is employed in the study [6–8]. At iteration [10], a new threshold Tn is established using the average of the foreground and background class means. The iterations terminate when the changes |Tn–Tn + 1| become sufficiently small [9]. Sawaki and Hagita demonstrated another specialized binarization method for textured and reverse-video (white-on-black) Japanese headlines [13]. Their method is based on the complementary relationship between characters and their backgrounds as indicated by similarity measure for black and white runs. From Fig. 1(b), the gray-level distribution did not revealed bimodal, it is important to take into consideration the amplitude transfer function of the specific scanner, as well as the spatial and gray-scale characteristics of the image. That is to say, the pixels on an image are highly correlated, i.e. the pixels in the immediate neighborhood possess nearly the same feature data. Therefore, the spatial relationship of neighboring pixels is an important characteristic that can be of great aid in imaging segmentation. Cluster techniques have taken advantage of this spatial information for image segmentation [4].

In this paper, an adaptive foreground and background clustering (FBC) approach to document image binarization, each pixel is assigned to a foreground cluster or a background cluster. The algorithm is based on adaptive K-means algorithm, where the cluster means are updated each time a data point is assigned to a cluster. Since only one background and one foreground is assumed (K = 2), that is to say, only two clusters are considered, which makes the overall implementation easier. Following by that, a median filter has been employed for salt and pepper noise removing. Finally, a Connected components labeling technique is devised to locate possible positions of Chinese character.

2 Proposed Work

Figure 1 shows a scan of a page from a Chinese historical book that the rubbing image gets darker shows a low-contrast image with its histogram. As it can be observed from the luminance histogram, all the values gather in the left of the three bands, so it is impossible to reliably locate a local minimum between histogram valleys.

K-means clustering is one of the popular algorithms in clustering and segmentation. It treats each image pixel (with R,G,B values) as a feature point having a location in space. The basic K-means algorithm then arbitrarily locates, that number of cluster centers in multidimensional measurement space. Each point is then assigned to the cluster whose arbitrary mean vector is closest. The procedure continues until there is no significant change in the location of class mean vectors between successive iterations of the algorithms. Firstly, we use an adaptive K-means cluster based segmentation to improve the performance of threshold image.

2.1   Bradley's Algorithm

The main idea in Bradley's algorithm is that compute the sum of real numbers $f(x, y)$ (for instance, pixel intensity) over a rectangular region of the image. It could be called as integral images. To compute the integral image, we store at each location, $I(x, y)$, the sum of all $f(x, y)$ terms to the left and above the pixel (x, y). This is accomplished in linear time using the following equation for each pixel (taking into account the border cases),

$$I(x, y) = f(x, y) + I(x - 1, y) + I(x, y - 1) - I(x - 1, y - 1) \tag{1}$$

After that, compute the $s \times s$ average using the integral image for each pixel in constant time and then perform the comparison. If the value of the current pixel is t percent less than this average then it is set to black, otherwise it is set to white. The pseudo code for Bradley's algorithm has been showing as following:

Input image In, output binary image out, image width w and image height h.

```
 1: for i = 0 to w do
 2:   sum ← 0
 3:   for j = 0 to h do
 4:     sum ← sum+in[i, j]
 5:     if i = 0 then
 6:       intImg[i, j] ← sum
 7:     else
 8:       intImg[i, j] ← intImg[i−1, j]+sum
 9:     end if
10:   end for
11: end for
12: for i = 0 to w do
13:   for j = 0 to h do
14:     x1 ← i−s/2 {border checking is not shown}
15:     x2 ← i+s/2
16:     y1 ← j−s/2
17:     y2 ← j+s/2
18:     count ← (x2−x1)×(y2−y1)
19:                              sum                                          ←
          intImg[x2,y2]−intImg[x2,y1−1]−intImg[x1−1,y2]+intImg[x1−1,y1−1]
20:     if (in[i, j]×count) ≤ (sum×(100−t)/100) then
21:       out[i, j] ← 0
22:     else
23:       out[i, j] ← 255
24:     end if
25:   end for
26: end for
```

2.2   Overview of the Binarization Technique

Because there are some amount of 'salt & pepper' noise exist in the document image after Bradley's binarization, median filtering is conducted. Finally, a morphology-based technique is devised to locate possible positions of Chinese Character. The detail process is shown in Fig. 2.

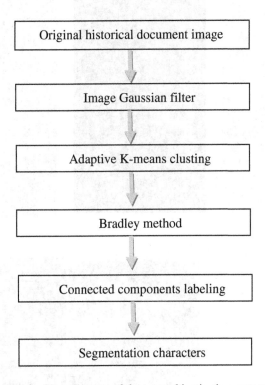

Fig. 2. Flow diagram of document binarization system

3   Experiments Performed on Chinese Rubbing Images

The algorithm is implemented in MATLAB. The algorithm is tested with many scanned Chinese rubbing document images, which contained characters of different fonts and size. The parameters in the Bradley's binarization experiments are set as follows: W = 15, H = 15, and T = 5. The median filtering using MATLAB's medfilt2 function, we have used neighborhoods of size 3-by-3 to remove noising. There chosen two examples of our adaptive threshold result are presented in Figs. 3 and 4. In order to compare our method with other different algorithm, the results of Sauvola's, Ostu's and Isodata's method are shown in same figure, also. Figures 3 and 4 illustrate a text example with a very low contrast. Our method is able to segment most of the text in the image.

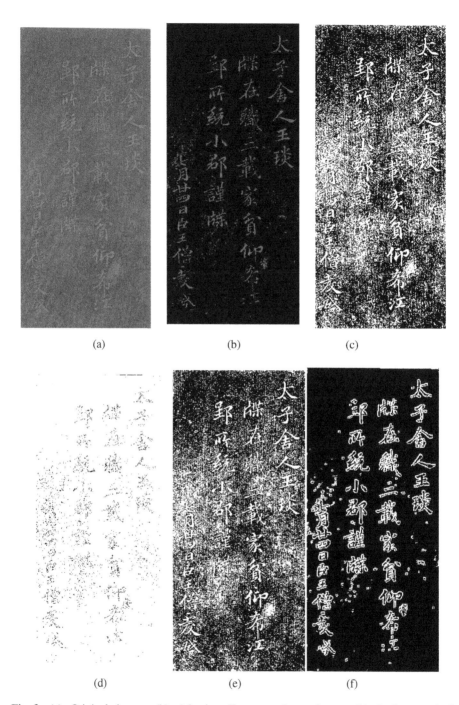

(a)   (b)   (c)

(d)   (e)   (f)

Fig. 3. (a) Original image; (b) Adaptive K-means cluster image; (c) Isodata method; (d) Sauvola's method; (e) Ostu's method; (f) Our method

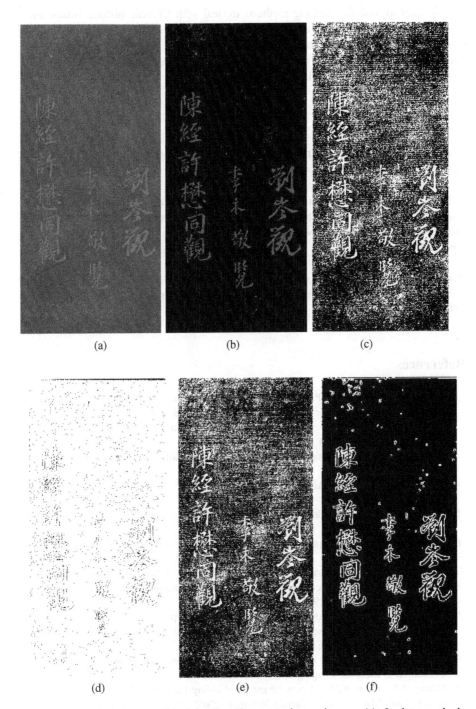

Fig. 4. (a) Original image; (b) Adaptive K-means cluster image; (c) Isodata method; (d) Sauvola's method; (e) Ostu's method; (f) Our method

We tried to look for existing methods to deal with Chinese historic image binarization problem. After many experiments have been implemented, our technique can provide near perfect segmentation despite the very low illumination in the image. Sauvola's technique fails at the characters detection. Ostu's global technique and isodata method keep more noising in image.

4   Conclusion

In this paper, a Chinese rubbing document image binarization algorithm is developed from low contrast Chinese rubbing images. The proposed scheme is developed based on an adaptive K-means combined Bradley operation. We tested our scheme on many different Chinese rubbing, and obtained encouraging results. Because the validation procedure should handle the variety of handwritten characters and the ambiguity in the distinction of characters, extracted them from images is not always easy and is still a topic of future research.

Acknowledgements. This work was supported by the National Natural Science Foundation of China (Grant No. 61472173), the grants from the Educational Commission of Jiangxi province of China, No. GJJ151134.

References

1. Gupta, M.R., Jacobson, N.P., Garcia, E.K.: OCR binarization and image pre-processing for searching historical documents. Pattern Recogn. **40**(2), 389–397 (2007)
2. Sauvola, J., Pietikäinen, M.: Adaptive document image binarization. Pattern Recogn. **33**(2), 225–236 (2000)
3. Yan, H.: Unified formulation of a class of image thresholding techniques. Pattern Recogn. **29**(12), 2025–2032 (1996)
4. Jain, A.K., Dubes, R.C.: Algorithms for Clustering Data. Prentice Hall, Englewood Cliffs (1988)
5. Otsu, N.: A threshold selection using gray level histograms. IEEE Trans. Syst. Man Cybernet. **9**, 62–69 (1979)
6. Bradley, D., Roth, G.: Adaptive thresholding using the integral image. J. Graph. GPU Game Tools **12**(2), 13–21 (2007)
7. Wellner, P.D.: Adaptive thresholding for the DigitalDesk. Xerox, EPC1993-110 (1993)
8. Pappas, T.N.: An adaptive clustering algorithm for image segmentation. IEEE Trans. Signal Process. **40**(4), 901–914 (1992)
9. Chang, C.I., Du, Y., Wang, J., et al.: Survey and comparative analysis of entropy and relative entropy thresholding techniques. In: IEE Proceedings - Vision, Image and Signal Processing, IET, vol. 153(6), pp. 837–850 (2006)
10. Sezgin, M.: Survey over image thresholding techniques and quantitative performance evaluation. J. Electron. Imaging **13**(1), 146–168 (2004)
11. Huang, Z.K., Chau, K.W.: A new image thresholding method based on Gaussian mixture model. Appl. Math. Comput. **205**(2), 899–907 (2008)

12. Hayes, B., Wilson, C.: A maximum entropy model of phonotactics and phonotactic learning. Linguist. Inq. **39**(3), 379–440 (2008)
13. Mori, M., Sawaki, M., Yamato, J.: Robust character recognition using adaptive feature extraction. In: 23rd International Conference on Image and Vision Computing New Zealand, IVCNZ 2008, pp. 1–6. IEEE (2008)
14. http://www.lib.berkeley.edu/EAL/stone/rubbings.html
15. http://vc.lib.harvard.edu/vc/deliver/home?_collection=rubbings

Pavement Transverse Profile Roughness
via Weakly Calibrated Laser Triangulation

Yingkui Du[✉], Panli He, Nan Wang, Xiaowei Han,
and Zhonghu Yuan

Key Lab of Equipment Manufacturing Comprehensive Automation,
School of Information Engineering, Shenyang University,
Shenyang 110044, China
syu_dyk@163.com

Abstract. A weakly calibrated monocular and line structured light based pavement transverse profile measurement method is proposed. Combining the linearization approximation of the laser triangulation, the weakly correlation of the road roughness national standard and the system calibration parameters was deduced. The problem of complicate system calibration in existed methods was transferred into a single factor calibration process. A second-order Gaussian filter was designed to enhance the laser curve and suppress the disturbance of lane markings efficiently utilizing the unique tubular feature of the laser curve intensity distribution. The results of the virtual camera simulation and real image experiments validated the robustness of proposed method and the error can be limited in the 10^{-4} order of magnitudes.

Keywords: Pavement roughness · Laser triangulation · Anisotropic diffusion · Second-order gaussian · Cubic spline

1 Introduction

As the rapid development of Chinese highway construction in recent decades, the efficient and low cost automation roughness measurement techniques were still relatively scarce [1]. The existed pavement roughness detection methods were mainly include ruler measurement, continuous measurement and inertial measurement method [1, 2]. In currently, the vehicle mounted roughness measurement system combining of laser sensor and acceleration sensor has become the primary one, which was a sparse sampling approximate measurement technology. The main problems of this kind of methods were that the measurement of the transverse profile and the roughness were not continuous and the price is very expensive.

The laser triangulation technology belongs to the one of the non-contact measurement methods. For the advantage of continuous profile elevation measurement, it has been more widely applied in practice. In the existing mature systems such as the TRRI and the Way-Link, additional specific spectral band filter was commonly utilized in the measurement module that leads an exorbitant price. Another roughness measurement method that based on low cost common vision sensor and structured laser source has become a hot topic in recent years [3–5]. The key problem of this type of

© Springer International Publishing Switzerland 2016
D.-S. Huang et al. (Eds.): ICIC 2016, Part III, LNAI 9773, pp. 180–191, 2016.
DOI: 10.1007/978-3-319-42297-8_18

measurement method is the non-preprocessing of the information, which causing more uncertainty and relative lower robustness. In the existed works, two basic key problems were still needed to solve. The first one is the complex calibration procedure of traditional 3D dimensional optical measurement method that always brings usability in engineering applications. The second one is the instability of laser curve recognition that caused by the serious impact of sunshine, shadow, landmarks and dynamic disturbance factors. Especially, the highly reflective property of pavement landmark line was still a challenging.

This paper presents a simple pavement transverse profile roughness measurement method by a weakly calibrated laser triangulation system that was composed of low-cost common vision sensor and structure of the light source. The inspiration was that we found there was a weak correlation between the roughness calculation and the calibration parameters as setting a reasonable approximation of linearization of the triangulation model, which provides a possibility to simplify the calibration process. In additional, The characteristics of the laser curve gray distribution was a unique tubular structure that was quite different from the disturbance factors such as markings and shadows and was very similar to the characteristics of the blood vessels gray distribution in medical CT image [6, 7]. The classical second order Gaussian filter in medical image processing was improved and applied in laser curve recognition during pavement roughness measurements to obtain more robustness.

2   Roughness and System Calibration Weak Correlation

2.1   Cross-Sectional Laser Triangulation

The basic principle of optical triangulation is that line laser forms light plane through the lens and forms laser line intersecting with the measured object surface. The CCD camera was fixed rigidly with the laser. The curve center coordinate of the laser line in the image and the corresponding three-dimensional height in the laser line space can be estimated accurately by triangulation displacement principle [8, 9]. The height variation relative to the datum plane of one point of the intersection line formed by pavement and vertical laser light plane was denoted as ΔH. The corresponding movement amount of image pixels in the image plane was Δh.

$$\Delta H = \frac{L\Delta h}{f \sin \theta - \Delta h \cos \theta} \tag{1}$$

The parameter L was the distance of the lens center and the intersection. The parameter θ was the angle of laser structured light and camera optical axis. The focal length f is usually millimeters level in a real system, yet the movement Δh was typically micron level by converting the pixel coordinates into metric coordinates. So the $f \gg \Delta h$ in the Cartesian coordinates and the Eq. (1) can be rewritten by an approximate linearization as:

$$\Delta H = \frac{L}{f \sin \theta} \Delta h \tag{2}$$

Assuming the laser structured light and the camera were fixed rigidly, the parameters L, θ and f were constant parameters and the Eq. (2) can be rewritten as:

$$\Delta H = \lambda \Delta h \tag{3}$$

denoting $\lambda = \frac{L}{f \sin \theta}$, the correlation of height ΔH and movement Δh was linearity (Fig. 1).

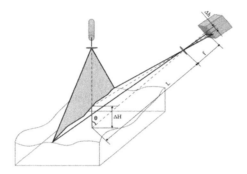

Fig. 1. The principle of laser triangulation

2.2   Chinese Standard of Road Roughness

The deviation indicator σ is the most common standard of pavement roughness evaluation in china. This indicator shows that pavement roughness change was satisfied with normal distribution statistical regularity and was consistent with the international standard indicator called as IRI. the $\bar{y}$ can be obtained by $\bar{y} = \frac{1}{n} \sum_{i=1}^{n} y_i$. The deviation σ can be calculated as

$$\sigma = \sqrt{\frac{1}{n-1} \sum_{i=1}^{n} (\bar{y} - y_i)^2} \tag{4}$$

2.3   Weak Correlation Analysis

The *ith* ΔH of the Eq. (3) was corresponding to the y_i of the Eq. (4), the $\Delta \bar{H}$ was denoted as $\Delta \bar{H} = \frac{1}{n} \sum_{i=1}^{n} \Delta H_i$. The Eq. (4) can be rewritten as

$$\sigma = \sqrt{\frac{1}{n-1}\sum_{i=1}^{n}(\Delta\bar{H} - \Delta H_i)^2} \tag{5}$$

To establish a direct calculation method of Δh and standard deviation indicator σ by the Eqs. (3) and (5), the problem of complex system calibration process must be solved. From the Eq. (3), the following equations can be obtained

$$\Delta\bar{H} = \frac{\lambda}{n}\sum_{i=1}^{n}\Delta h_i \tag{6}$$

$$\Delta H_i = \lambda\Delta h_i \tag{7}$$

Denoting $\Delta\bar{h} = \frac{1}{n}\sum_{i=1}^{n}\Delta h_i$, the Eq. (5) can be rewritten as

$$\sigma = \lambda\sqrt{\frac{1}{n-1}\sum_{i=1}^{n}(\Delta\bar{h} - \Delta h_i)^2} \tag{8}$$

The standard deviation indicator σ has nothing to do with ΔH by Eq. (8). Complex system calibration problem is transformed into coefficient λ calibration problem. We carry out one time standard sample measurement and the λ can be solved by Eq. (3).

3 Laser Curve Recognition

3.1 Anisotropic PM Denoising

The laser curve in mage was generally a bright fine curve. The common Median and Gaussian filtering algorithm blurred image details as image smoothing [10, 11]. The anisotropic diffusion PM filtering algorithm based on partial differential was utilized to smooth image and protects the details [10]. The diffusion is carried out along the direction of perpendicular to the image gradient vector. The ξ was denoted as an unit vector with perpendicular to image gradient ∇I, the diffusion equation was as follow

$$\frac{\partial I}{\partial t} = \frac{\partial^2 I}{\partial\xi^2} \tag{9}$$

According to the definition of directional derivative and denoting the θ as the angle between ξ and horizontal axis of image, the Eq. (9) can be rewritten as

$$\frac{\partial I}{\partial t} = \frac{\partial}{\partial x}\left(\frac{\partial I}{\partial x}\cos\theta + \frac{\partial I}{\partial y}\sin\theta\right)(-\sin\theta) + \frac{\partial}{\partial y}\left(\frac{\partial I}{\partial x}\cos\theta + \frac{\partial I}{\partial y}\sin\theta\right)\cos\theta \tag{10}$$

The finite difference method was employed to solve it approximately as

$$I_{(x,y)}^{n+1} = I_{(x,y)}^n + \Delta t \frac{\partial I}{\partial t} \qquad (11)$$

the $I_{(x,y)}^n$ is pixel gray value of point (x,y) after nth iterations. the Δt is the length of iteration step.

The actual image contrast experiments of PM filtering and Median filtering are shown in Fig. 2. The left image is the original image after noise smoothing and the right is the magnifying show of the red box area.

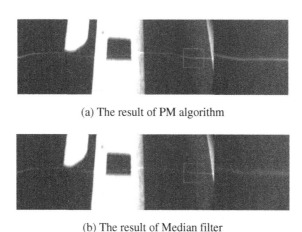

(a) The result of PM algorithm

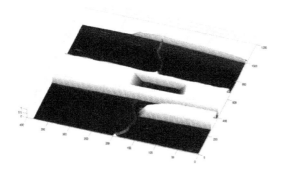

(b) The result of Median filter

Fig. 2. Experimental results of PM algorithm and Median filter (Color figure online)

3.2 Laser Curve Enhancement

The gray distribution of laser curve and landmarks disturbance in the image was shown in Fig. 3. For the serious influence of road surface landmarks disturbance [12], the common edge detection operators cannot well-done as shown in Fig. 3.

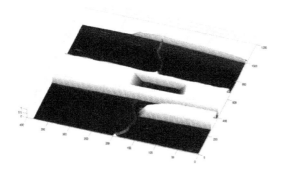

Fig. 3. The tubular structure of the gray distribution of the laser curve

We found that gray distribution of laser curve shows unique tubular structure features with respect to road surface disturbance. The multi-scale Gaussian second derivative enhancement algorithm of reference [6] can enhance effectively typical vessel tubular structure gray distribution in the human ray image, which gives us inspiration.

Considering the real time problem in engineering application, the limited scale range constraints was added to multi-scale Gaussian second derivative filter proposed by reference [6] to enhance laser curve. The laser curve enhancement experimental results of our algorithm and common edge detection operators were shown in Fig. 4.

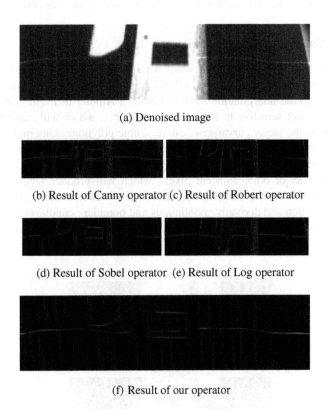

(a) Denoised image

(b) Result of Canny operator (c) Result of Robert operator

(d) Result of Sobel operator (e) Result of Log operator

(f) Result of our operator

Fig. 4. Experimental results of contrasted to the popular edge operators

We can see that the most of disturbances are effectively filter out from the Fig. 4 (f), but obviously not enough and need a further edge disturbance suppression. The edge suppression algorithm proposed by reference [13] was utilized to do secondary enhancement. The effective experimental results were shown in Fig. 5.

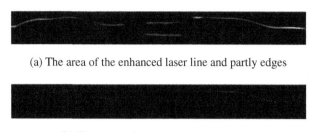

(a) The area of the enhanced laser line and partly edges

(b) The result of edge suppress processing

Fig. 5. Results of the secondary enhancement of the laser curve

3.3   Cubic Spline Curve Interpolation

The classical curve interpolation algorithms have linear interpolation method [14–16], cubic polynomial interpolation method [15, 17], nearest neighbor interpolation method [18] and cubic spline interpolation method [19–21]. Among them, the linear interpolation method is not sensitive to the curvature of curve, which will cause large interpolation error on the larger curvature of curve. Cubic polynomial interpolation method executes slowly, which is not suitable for real-time processing. Nearest neighbor interpolation method use the nearest pixel gray value from the target point, not considering the impact of other adjacent pixels, which will produce jagged and mosaic phenomenon. Cubic spline interpolation use three-moment method to construct the cubic spline function and derivative continuous and boundary conditions as constraints, which is suitable for curve fitting of free curvature. This paper will use cubic spline interpolation to fit laser curve.

(a) The result of linear interpolation

(b) The result of nearest neighbor interpolation

(c) The result of cubic spline interpolation

Fig. 6. Experimental results of curve spline interpolation

The results of the comparative experiment of linear interpolation, nearest neighbor interpolation and cubic spline interpolation method are shown in Fig. 6. Experimental results show that the cubic spline interpolation algorithm has the highest precision when it is fitting large curvature laser curve.

4 Experimental Results

4.1 Weak Correlation and Calibration Parameters Simulation

The purpose of the simulation experiments is: use sinusoidal of maximum value as 10 mm to simulate the three-dimensional space height change of transverse profile discrete points. Equations (4) and (8) of roughness standard deviation indicator calculation results are analyzed to assess the effectiveness and error controllability of roughness calculation methods in this paper. Using three-dimensional coordinates of discrete dense sampling points in highway transverse virtual profile, calculate roughness standard deviation indicator by Eq. (4). The two-dimensional coordinates of corresponding image pixel in image plane coordinates are utilized to estimate the approximate roughness standard deviation indicators according to Eq. (8).

In order to avoid the uncertainty of three-dimensional measurement caused by system calibration errors, a virtual camera and laser structure light roughness measurement system was designed, according to main system calibration parameters and computing model of Eq. (3). System parameters are shown in Table 1. Structure light scanning surface coincides with tangent plane of transverse profile. The three-dimensional coordinates of discrete data points in transverse profile is defined as follows: The vertical direction X coordinate of transverse profile is 0. The maximum coordinate range of transverse profile direction Y is [2000 mm – 3200 mm]. The sampling interval is 10 mm. Virtual camera focal length is 4 mm. The calibration parameters can be obtained according to Eq. (1) in simulation experiment.

Table 1. Parameters of virtual structured laser light roughness measurement system

Parameter	L (mm)	θ (°)	f (mm)
1	1000	5	4
2	1200	10	4
3	1400	15	4
4	1600	20	4

Discrete data points of road surface transverse profile observed by virtual camera in different parameters L constraints is shown in Fig. 7. Corresponding values of different colors were 1000 mm (red), 1200 mm (green), 1400 mm (blue), 1600 mm (yellow). The virtual simulation experimental results of different virtual parameters and comparative will be analyzed by accordance with roughness standard deviation indicator calculation method of the Eqs. (4) and (8) is shown in Table 2.

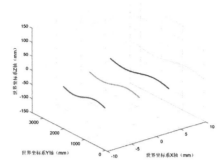

Fig. 7. Pavement transverse profile curve under the constraints of different *L* (Color figure online)

Table 2. Pavement transverse profile curve under the constraints of different *L* and θ

absolute error	$\theta=5°$	$\theta=10°$	$\theta=15°$	$\theta=20°$
l =1000mm	2.0193E6	8.0158E6	1.7807E5	3.1096E5
l =1200mm	2.0184E6	8.0124E6	1.7800E5	3.1083E5
l =1400mm	2.0174E6	8.0100E6	1.7795E5	3.1074E5
l =1600mm	2.0174E6	8.0082E6	1.7791E5	3.1067E5

As we can be seen from Table 2, the linear approximation of the laser triangulation principle introduced into roughness national standard calculation method can achieve the maximum error of 10^{-4}, verifying fully rationality and effectiveness of weak calibration roughness calculations. The proposed method make roughness precision three-dimensional measurement problems converted to pixels shift amount estimation problems in image plane, which greatly simplify calibration cumbersome complex process and system architecture of the existing methods in the premise of achieving the same accuracy.

On the other hand, it can be seen that the angle between camera and laser structure light source should not be too large in the practical application from Table 2. We found that measurement error of this method and the increasing angle amount showed non-linear exponential growth relationship by simple statistical calculation.

(a) Open curve active contour method   (b) Our method

Fig. 8. Experimental results of laser line recognition with disturbance of vehicle shadow

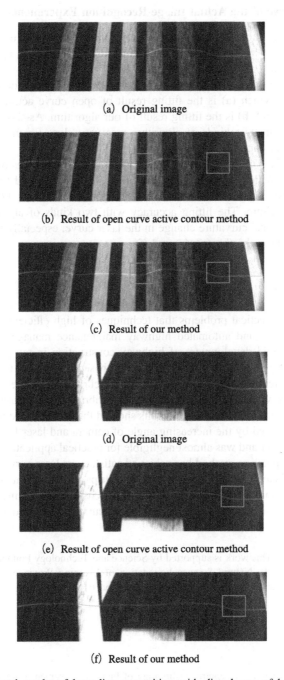

(a) Original image

(b) Result of open curve active contour method

(c) Result of our method

(d) Original image

(e) Result of open curve active contour method

(f) Result of our method

Fig. 9. Experimental results of laser line recognition with disturbance of land marks (Color figure online)

4.2   Laser Curve of the Actual Image Recognition Experiments

We conducted laser curve fitting precision comparative experiments using actual image with vehicle shadow and white runway markings disturbances with open curve active contour evolutionary algorithm in reference [13].

The laser curve fitting precision contrast results with vehicle shadow disturbance is shown in Fig. 8. Which (a) is the fitting result of open curve active contour evolutionary algorithm and (b) is the fitting result of our algorithm. As shown in Fig. 8, the fitting accuracy with two kinds of algorithm was very close for the small curvature change in the laser curve.

The results of contrasting with pavement white runway markings disturbance were shown in Fig. 9. Which (a) and (d) are the original image, (b) and (e) are the fitting result of open curve active contour evolutionary algorithm and (c) and (f) are the fitting result of our algorithm. The fitting accuracy with two kinds of algorithm was very obvious for large local curvature change in the laser curve, especially the area marked by green rectangle in the (b), (e), (c) and (f).

5   Conclusion

In order to solve practical problems that techniques of high efficiency, simple usage, low application cost and automated highway maintenance management are relative scarcity for the rapid development of highway construction in our country. A weak calibrated monocular and line structured light based pavement transverse profile roughness measurement method is proposed to accurately estimate dense pavement profile. Compared with the existing methods, the calibration was greatly simplified and achieving similar accuracy. Although results showed that the measurement accuracy of our method is affected by the increasing angle of camera and laser light source. But it was very weak lower and was almost negligible for practical applications. Although the cubic spline interpolation method has achieved satisfactory laser curve fitting accuracy but it depends heavily on the results of light strips enhancement and background suppression algorithm. The next focus of our work is using cubic spline interpolation to fit laser curve as initial values and utilizing open curve Snake model to improve the robustness.

Acknowledgment. This work is supported by Science and Technology Foundation of Liaoning Province (LT2013024), Natural Science Foundation of Liaoning Province (2015010158-301) and Science and Technology Program of Shenyang City (F16-155-9-00).

References

1. Xu, W.B.: Detection and control technology of highway pavement roughness. Highway **9** (9), 150–151 (2014)
2. Zhang, Z.Y.: Laser Pavement Detection System. Chang'an University (2015)

3. Ma, Y.K., Wang, Z.Y., Yang, G.W., et al.: A system based on structured-light sensors for measurement of pavement evenness. Chin. J. Sens. Actuators **26**(11), 1597–1603 (2013)

4. Xie, J.: Study on Surface Flatness Detection Method Based on Machine Vision. Central South University, Changsha (2014)

5. Kumar, P., Lewis, P., McElhinney, C.P., et al.: An algorithm for automated estimation of road roughness from mobile laser scanning data. Photogram. Rec. **30**(149), 30–45 (2015)

6. Frangi, A.F., Niessen, W.J., Vincken, K.L., Viergever, M.A.: Multiscale vessel enhancement filtering. In: Wells, W.M., Colchester, A.C., Delp, S.L. (eds.) MICCAI 1998. LNCS, vol. 1496, pp. 130–137. Springer, Heidelberg (1998)

7. Yuan, Y., Luo, Y., Chung, A.C.S.: VE-LLI-VO: vessel enhancement using local line integrals and variational optimization. IEEE Trans. Image Process. **20**(7), 1912–1924 (2011)

8. Yan, S., Li, D.G., Yu, Z.L.: Analysis of laser triangulation displacement sensor and parameters optimization design. Electron. Meas. Technol. **35**(10), 21–24 (2012)

9. Li, X.Q., Lu, Z.X., Zhao, L.Y., et al.: The measurement of three-dimensional road roughness based on laser triangulation. Adv. Mater. Res. **346**, 812–816 (2012)

10. Cohen, E., Cohen, L.D., Zeevi, Y.Y.: Texture enhancement using diffusion process with potential. In: 2014 IEEE 28th Convention of Electrical & Electronics Engineers in Israel (IEEEI), pp. 1–5. IEEE (2014)

11. Tian, D., Xue, D., Wang, D.: A fractional-order adaptive regularization primal-dual algorithm for image denoising. Inf. Sci. **296**(1), 147–159 (2015)

12. He, Q., Yan, L.: Algorithm of edge detection based on Log and Canny operator. Comput. Eng. **37**(3), 210–212 (2011)

13. Fan, H.J., et al.: A new open curve detection algorithm for extracting the laser lines on the road (2011)

14. Zhao, Y., Lin, H.-W., Bao, H.-J.: Local progressive interpolation for subdivision surface fitting. J. Comput. Res. Dev. **49**(8), 1699–1707 (2015)

15. Akima, H.: A new method of interpolation and smooth curve fitting based on local procedures. J. ACM (JACM) **17**(4), 589–602 (1970)

16. Athawale, T., Entezari, A.: Uncertainty quantification in linear interpolation for isosurface extraction. IEEE Trans. Vis. Comput. Graph. **19**(12), 2723–2732 (2013)

17. Chkifa, A., Cohen, A., Schwab, C.: High-dimensional adaptive sparse polynomial interpolation and applications to parametric PDEs. Found. Comput. Math. **14**(4), 601–633 (2014)

18. Hu, S.-B., Shao, P.: Improved nearest neighbor interpolators based on confidence region in medical image registration. Biomed. Signal Process. Control **7**(5), 525–536 (2012)

19. Hong, S.-H., Truong, T.-K., Lin, T.-C., et al.: Novel approaches to the parametric cubic-spline interpolation. IEEE Trans. Image Process. **22**(3), 1233–1241 (2013)

20. Angeles, J., López-Cajún, C.S.: Optimization of Cam Mechanisms. Springer Science & Business Media, Dordrecht (2012)

21. Sun, Q.C., Zhou, Y.Z., Ning, C., et al.: A sub-pixel edge detection method based on cubic spline interpolation. Control Eng. China **21**(2), 290–293 (2014)

A Computer Vision and Control Algorithm to Follow a Human Target in a Generic Environment Using a Drone

Vitoantonio Bevilacqua$^{(\boxtimes)}$ and Antonio Di Maio

Department of Information and Electrical Engineering,
Polytechnic of Bari, Bari, Italy
`vitoantonio.bevilacqua@poliba.it`,
`dimaioantonio90@gmail.com`

Abstract. This work proposes an innovative technique to solve the problem of tracking and following a generic human target by a drone in a natural, possibly dark scene. The algorithm does not rely on color information but mainly on shape information, using the HOG classifier, and on local brightness information, using the optical flow algorithm. We tried to keep the algorithm as light as possible, envisioning its future application on embedded or mobile devices. After several tests, performed modeling the system as a set of SISO feedback-controlled systems and calculating the Integral Squared Error as quality indicator, we noticed that the final performance, overall satisfactory, degrades as the background complexity and the presence of disturbance sources, such as sharp edges and moving objects that cross the target, increase .

Keywords: Drones · Computer vision · Control theory · Optical flow · Classifiers · Histogram of oriented gradients

1 Introduction

The interest in using Unmanned Aerial Vehicles (UAVs) to accomplish a series of tasks that can be uncomfortable or risky to be performed by humans has been constantly increasing in the recent years for different reasons. In first place, we can notice the conspicuous increase of the number of commercial drones' sales, due to the dramatic decrease of their price and the raising amount of ways in which they can be used, ranging from games and sports to small utility applications.

This has brought to the scientific community a new interest in investigating the capabilities of UAVs [2, 3, 13, 17–21], with a particular attention to quadcopters in the consumer sector. This is because of their potential wide spread and their applications for solving generic issues of the mass consumer.

Because of these reasons, this work studies and proposes a new algorithm that makes it possible, for a commercial quadcopter with a built-in camera, to follow a walking human target relying exclusively on information extracted from the captured video stream.

© Springer International Publishing Switzerland 2016
D.-S. Huang et al. (Eds.): ICIC 2016, Part III, LNAI 9773, pp. 192–202, 2016.
DOI: 10.1007/978-3-319-42297-8_19

Currently, the great majority of the algorithms that accomplish similar tasks relies on color information or on using an external device to track the position of the target [7–11, 14].

For example, this link https://www.youtube.com/watch?v=JNEZmV8yONQ illustrates the outcome of a similar study, in which the algorithm tracks the target through segmentation of a blob of pixels of similar color that lie over of a piece of clothing of the target. The main drawback of this problem is also noticeable in the video: it stops working when the target is in a dark environment or when it crosses a shadow.

Different approaches use an external device attached to the target's body to track it, but it increases the overall cost and it limits the possibility to easily switch the target of the drone.

In contrast to most of the current related works, this one proposes an innovative technique to accomplish the tracking task, using no external tracking device and only the video stream coming from the drone's built-in camera. Our algorithm does not use color information, but only single-channel images extracted from the video stream frames. In case the target must be followed in total darkness, the drone can be equipped with an IR illuminator and an IR camera.

Therefore, the presented approach eliminates the drawbacks of the algorithms that solve the same problem and it is also computationally lighter, which is good when the algorithm must run on a system embedded in the drone's body.

2   Overview

This work can be divided in the resolution of three distinct sub-problems: the problem of tracking a target in a generic scene, the problem of controlling the drone and the problem of hypotheses testing.

In order to track the movement of the target in the video stream provided from the drone's camera, we need a tool that returns the position of human targets in a picture. The algorithm we have developed uses a Histogram of Oriented Gradients (HOG) classifier, which the scientific literature proved to be more successful for this specific task [6].

Unfortunately, the use of this algorithm on its own is not enough because of its relative slowness and because the video stream can reach up to 30 frames per second. The acquisition framerate is variable and depends on the speed of the algorithm: a new frame is acquired only when all the computation of the previous one has been carried out. This is why launching the classifier for each frame would decrease the frame rate to unacceptable values for controlling the drone. For this reason, the indispensable classification step has been combined with a faster tracking method, using the optical flow algorithm.

Regarding the subproblem of drone control, the intrinsic MIMO system is modeled as a set of 4 SISO systems. This is possible due to the hypothesis of independence of the components of the system state vector, which combined with the assumption of linearity, allow us to use well-known techniques of modeling and control.

For sake of simplicity, we have chosen to use proportional regulators only on 3 degrees of freedom on the total 4 controllable ones. The first and most important control is implemented on the yaw (the rotation of the aircraft around the vertical axis Z): the distance from the vertical axis of the video frame of the center of mass of the target's fiduciary points is multiplied by a multiplicative constant, to be chosen depending on the specifications of the desired response, and provided as input to the drone SDK, which in this case represents the plant. Similarly, the control is performed for the heave motion (along the vertical axis Z).

The surge motion control (along the longitudinal axis X) is performed using a more complex algorithm. The height of the target-enclosing rectangle is linked proportionally to the height of the rectangle that encloses the cloud of fiduciary points, fixed to the target's body. As a result, when the target moves away from the drone, it occurs a contraction of the points cloud for perspective effect. This contraction results in a corresponding reduction of the height of the rectangle enclosing the target and vice versa, when the points cloud expands because the target is getting closer to the drone, there is a height increase of the target box. The difference between the current height of the target box and the height we want it to reach in the video frame will be the input to the proportional controller.

From the geometric point of view, this height is proportional to the distance of the target from the drone and therefore, adjusting the coefficient of the regulator, we modify the distance from the target to which the drone will tend, as well as its reactivity to any change of the setpoint.

The evaluation of hypotheses and algorithm is performed calculating the temporal average of the ISE (Integral Squared Error) as a measure of quality. This evaluator has been chosen because it is impossible to provide canonical inputs to the system due to strong disturbances caused by various environmental and technical reasons. Therefore, the transient part of the error signal does not fade over time. This makes the classic ISE strongly dependent on the duration of the video, while a temporal average makes the evaluator less sensitive to it.

3   The Project Structure

The developed algorithm can be modeled using the following scheme, where the arrows point to the direction in which the data flows (Fig. 1).

The Acquisition thread deals mainly with the acquisition of the frame from the drone's camera and its storage on the shared global memory.

The Send Feedback Thread acquires and sends to the drone, at a fixed frequency, the values of a shared global variable containing the velocities along the three axes x, y, z and the yaw rotating speed. Actually, only 3 of 4 degrees of freedom will be controlled, because the lateral movement (y axis) will not be needed as it is redundant.

Considering the movement of the drone on a 2D horizontal plane, 2 degrees of freedom are enough to cover all of its surface. A degree of freedom we want to keep is along the x axis, because the camera is pointing along the direction of surge motion. Yaw has been chosen as second degree of freedom over the sway motion (y axis) because it changes the orientation of the x axis and therefore the direction in which the camera points. This allows the drone to follow the target always from behind and also when it walks over a curved path, things that would not be possible choosing the y axis

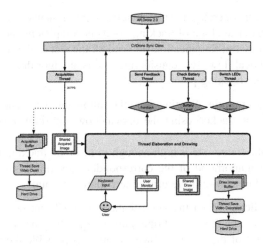

Fig. 1.   General scheme of the algorithm

as second degree of freedom and therefore making the camera point constantly in the same direction.

The Elaboration and Drawing Thread is the biggest and most complex thread of the program, and it requires a particularly thorough description, which will be proposed in the next chapter.

4   The Tracking Algorithm

The elaboration thread encompasses all the algorithms to perform the core operations to accomplish the tracking task, that is HOG Classification, Good Features to Track extraction [4] and Optical Flow computation.

Hereafter, a description of the different fundamental stages through which the thread undergoes will be presented.

HOG Classification mode. When the program starts, the HOG Classifier is launched on the original frame. If the classifier does not find any target, the frame is discarded and the cycle restarts, otherwise a list of rectangles that can potentially contain the target is returned. Among all of the rectangles, the algorithm will choose the one that is closer to the last valid center of mass of the corners cloud, initially set at the frame center. This guarantees the selection of the right target in a scene with multiple ones, because when the target is lost and the classifier is relaunched, the probability to select the same target is maximum according to the principle of spatial locality.

Good features to track. After having selected the target, the Good Features to Track algorithm is launched on the area of the original image that lies under the target rectangle. If no point is returned by the algorithm, the cycle is aborted and restarted. Otherwise, the list of good points to track is stored and the center of mass is calculated.

This algorithm can operate in very cluttered environments with lots of different objects in movement. The cloud of tracking points is completely enclosed in the target box that is returned by the classification step and all the rest of the scene has no impact on the following optical flow step, considering that is an algorithm that works on local neighborhoods of the tracking points.

Optical flow tracking mode. Starting from the following cycle, and until the target is lost, the target will be tracked using the optical flow algorithm instead of the HOG classifier.

Assuming that the frame rate is high enough to keep the tracking points of the new frame in a close neighborhood of the correspondent tracking points of the previous frame, the optical flow algorithm returns a new cloud of points.

Getting rid of outliers. When the good features to track are detected on the background instead that on the foreground (the human target) they behave like outliers. The easiest way to get rid of them is to reduce as much as possible the area in which the corners will be detected [5] (virtually only on a ROI that lies perfectly over the target) and to remove the corners with a suspicious behavior.

We can locate the temporary new target box starting from the position of the old target box and translating it of the average drift vector between the two centers of mass.

All the corners that lie outside the box are discarded because they are likely to be outliers.

New Target Box Location and Dimensions. After the outliers cleaning phase, we need to adapt the dimensions of the new target box according to the actual dimensions' variation of the target. Assuming that the tracking points on the target move jointly with it, and remain bonded to the original detected points, we can say that the variation of the target box dimensions is linearly dependent on the geometrical variations of the cloud of points.

We assume that the aspect ratio of the target box is constant, so we can use just one between width and height of the bounding box of the cloud of points to modify both the target box dimensions.

In the following figure, the big red rectangle is the old target box and the small red one is the bounding box of the old cloud of points. The big blue rectangle is the new target box and the small one is the bounding box of the new cloud of points (Fig. 2).

Our goal is calculating the new value of H and W, relying upon the values of all the other variables H', h', h, W', w' and w.

Our hypothesis is that the ratio between the height of the target box and the height of the cloud bounding box is the same across two frames (the same is valid for the width), so the final formula can be written as following:

$$W = W' \frac{H}{H'} = W' \frac{h}{h'} \tag{1}$$

After calculating the two values H and W, the new target box is evenly enlarged or shrunk along all the borders by the specified quantities and stored.

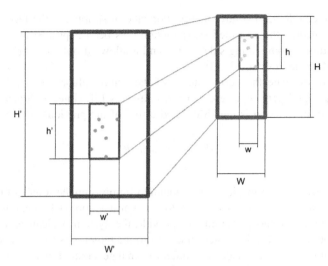

Fig. 2.   The height update of the target box (Color figure online)

5   Modeling and Technical Evaluation

A method to evaluate the algorithm performance in a quantitative, numeric way is desirable. This method might be useful not only to obtain a number that expresses how good the algorithm is performing, but also a practical way to tune the algorithm parameters in order to maximize its performances, according to user's requirements.

These parameters can be considered, similarly to the well-known theories of automatic control, as coefficients of a P (proportional) controller, treating the drone as a plant with unknown transfer function G(s) or, seeing it under the digital control point of view, G(z).

The whole system can be modeled using the following classic schema:

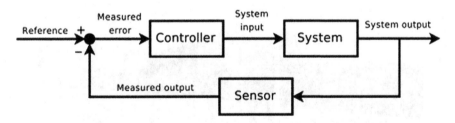

Fig. 3.   The general schema of the modeled MIMO system

Actually, the whole schema can be divided in 4 similar and independent sub-schemas, focusing on each of the 4 feedback dimensions. The independence of the 4 schemas makes it much simpler to deal with the problem, because it makes a complex MIMO system, just a set of 4 easy SISO systems, which can be treated using the well-known methods from control theory. For sake of simplicity, the feedback on the y-axis, that concerns the so-called sway motion, has been kept at 0 by default. This prevents the drone from sliding side to side and it allows it only to rotate, going back and forth and up and down (Fig. 3).

In order to evaluate the performances of the control algorithm, we have used a well-known integral performance estimator, known as ISE (Integral Square Error), in which the epsilon function is the "Measured error" of the previous schema.

$$ISE = \int\limits_{0}^{+\infty} \varepsilon^2(t)dt \tag{2}$$

For time-discrete systems, it will be enough to simply sum the square of the discrete values along the time axis. In contrast to the normal approach regarding the ISE calculation, there was no possibility to provide the system a clean and theoretical canonical input (i.e. a Heaviside step function) so the target tries to keep a steady pace when walking in the two test videos, in order to simulate a ramp function in input in the x axis system.

As well as the classic ISE, calculated on the whole data of the single test flight, an average ISE is also provided by dividing the ISE by the number of frames.

6 Testing

After having modeled the system and chosen a significant quality estimator, according to the selected model of the system, a series of practical tests have been carried out.

A brief summary of two test flights' results is shown hereafter, showing the total flight duration, graphs and a table with the ISE and Average ISE for each controlled axis.

The minimum number of tracking points for the two following test flights has been empirically set to 15. This threshold has been proven to guarantee a good compromise between tracking stability and calling the HOG classifier as less as possible (Figs. 4 and 5).

Fig. 4.   The environment of test case 1 with the algorithm's graphic decorations

Fig. 5. The environment of test case 2 with the algorithm's graphic decorations

6.1 Test Case 1

Flight duration: 1:42

Link to the Video: https://www.youtube.com/watch?v=RzrA1blOfI4 (see Table 1 and Fig. 6)

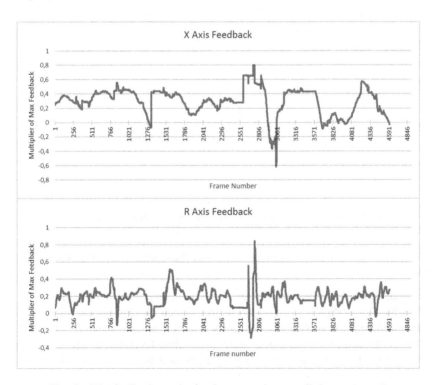

Fig. 6. The feedback graphs for the two main controlled axes (case 1)

Table 1. The performance indexes and for test case 1

AXIS	X	Z	R
ISE	569.660	593.993	213.534
ISE/NFRAMES	0.124	0.129	0.046

6.2 Test Case 2

Flight duration: 4:09

Link to the Video: https://www.youtube.com/watch?v=zEmeSw1BP4I (see Table 2 and Fig. 7)

Table 2. The performance indexes and the proportional regulators' factors for test case 2

AXIS	X	Z	R
ISE	1906.590	556.571	688.656
ISE/NFRAMES	0.179	0.052	0.065

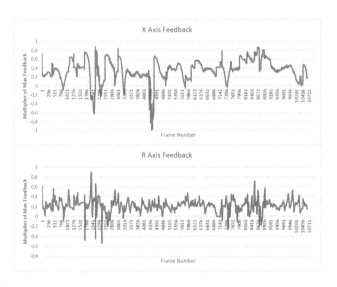

Fig. 7. The feedback graphs for the two main controlled axes (case 2)

In both of the shown test cases we can notice that the graphs have some noticeable features:

- They have some flat regions: this is due to the uneven power of Wi-Fi signal.
- They have several gaps ad leaps: this happens when the target is lost by the tracking algorithm and it must be redetected, or when the tracker is manually restarted.
- They settle, apart from the noise, steady at a certain level: this is due to the type of the system. Providing a ramp signal as input, which is a canonical input of first

order, means having a finite error of $1/|Kb|$ when the system is Type 1 (Type 1 systems have one pole in the origin. $|Kb|$ is the Bode gain).

Considering that the controller has been purposely designed as proportional, without poles in the origin, we can infer that the plant is an intrinsic low-pass filter, in particular a Type 1 system.

In order to contrast the effect of the position error and increase the static precision, it is possible to raise the Bode gain or add a pure integrator behavior to the controller in the control loop. The first strategy is unsuitable because of the bad reaction of the system to disturbances and because its dynamic response can be undesirable for specific user requirements.

Therefore, the best way to achieve disturbances rejection and static error minimization is introducing an integral action aside the proportional one. This increases the controller complexity and therefore its study will be put off to the next stages of the investigation.

7   Conclusion

In this work, a new technique for tracking a human target with a drone has been proposed and tested.

The main advantages of using this technique over the other ones are the few constraints on the environmental and target characteristics, the small computational requirements and the possibility of night tracking [15, 16], because of the independence of the algorithm from color information. Considering the lightness of the devised algorithm, one of the future expansion of the work could be embedding the algorithm in the drone's on-board card to accomplish general track-and-follow tasks. For example, as shown in [2], the system can be used to follow the position of an athlete but without the constraint of a colored shirt.

Furthermore, the target occlusion represents a problem for this as for all the other tracking algorithms. In this case, it has been solved using the variance threshold for the cloud of tracking points but it can be improved using the well-known filters such as Kalman and Particle.

In conclusion, we can say that even though the project still has a wide range of possibilities of expansion, the testing phase has shown quantitatively that the drone performs reasonably well at the task of tracking a human target in a general environment, increasing its performances as the disturbance sources decrease.

References

1. Higuchi, K., Shimada, T., Rekimoto, J.: Flying sports assistant: external visual imagery. In: AH, 12–14 March 2011
2. Nagi, J., Giusti, A., Di Caro, G.A., Gambardella, L.M.: Human control of UAVs using face pose estimates and hand gestures. In: HRI, 03–06 March 2014

3. Kos'myna, N., Tarpin-Bernard, F., Rivet, B.: Bidirectional feedback in motor imagery BCIs: learn to control a drone within 5 minutes. In: CHI, 26 April–1 May 2014
4. Shi, J., Tomasi, C.: Good features to track. In: IEEE Conference on Computer Vision and Pattern Recognition, pp. 593–600 (1994)
5. Harris, C., Stephens, M.: A combined corner and edge detector. In: Proceedings of the 4th Alvey Vision Conference, pp. 147–151 (2001)
6. Dalal, N., Triggs, B.: Histograms of oriented gradients for human detection. In: Computer Vision and Pattern Recognition, vol. 1, pp. 886–893, 25 June 2005
7. Miyoshi, K., Konomura, R., Hori, K.: Above Your Hand: direct and natural interaction with aerial robot. In: ACM, 10–14 August 2014
8. Hansen, J.P., Alapetite, A., Scott MacKenzie, I., Møllenbach, E.: The use of gaze to control drones. In: ETRA, 26–28 March 2014
9. Pittman, C., LaViola Jr., J.J.: Exploring head tracked head mounted displays for first person robot teleoperation. In: IUI, 24–27 February 2014
10. Pfeil, K.P., Koh, S.L., LaViola Jr., J.J.: Exploring 3D gesture metaphors for interaction with unmanned aerial vehicles. In: IUI, 19–22 March 2013
11. Mueller, F., Muirhead, M.: Understanding the design of a flying jogging companion. In: UIST, 05–08 October 2014
12. Mueller, F., Muirhead, M.: Jogging with a Quadcopter. In: CHI 2015 Proceedings of the 33rd Annual ACM Conference on Human Factors in Computing Systems, pp. 2023–2032, 18 August 2015
13. Sefidgari, B.L.: Feed-back method based on image processing for detecting human body via flying robot. Int. J. Artif. Intell. Appl. (IJAIA) 4(6), 35–44 (2013)
14. Munoz, C.A., Sobh, T.M.: Object tracking using autonomous Quad Copter. Robotics, Intelligent Sensing & Control (RISC) Lab., University of Bridgeport, 28 March 2014
15. Liang, N.S., Wan Yusoff, W.A., Dhinesh, R., Sak, J.S.: Low cost night vision system for intruder detection. In: IOP Conference Series: Materials Science and Engineering, vol. 114(1) (2016)
16. Jeong, M.R., Kwak, J.Y., Son, J.E., Ko, B., Nam, J.Y.: Fast pedestrian detection using a night vision system for safety driving. In: 2014 11th International Conference on (IEEE) Computer Graphics, Imaging and Visualization, CGIV, pp. 69–72 (2014)
17. Kimura, M., Shibasaki, R., Shao, X., Nagai, M.: Automatic extraction of moving objects from uav-borne monocular images using multi-view geometric constraints. In: IMAV 2014: International Micro Air Vehicle Conference and Competition 2014, Delft, The Netherlands, 12–15 August 2014
18. Chan, W.S.: Autonomous Quadcopter Flight System with Object Tracking. Department of Electronic Engineering, Undergraduate Final Year Projects, City University of Hong Kong (2015)
19. Mercado, D.A., Castillo, P., Lozano, R.: Quadrotor's trajectory tracking control using monocular vision navigation. In: 2015 International Conference on Unmanned Aircraft Systems (ICUAS), 9–12 June 2015
20. Boudjit, K., Larbes, C.: Detection and implementation autonomous target tracking with a Quadrotor AR.Drone. In: 2015 12th International Conference on Informatics in Control, Automation and Robotics (ICINCO), vol. 02, 21–23 July 2015
21. Harik, E.H.C., Guérin, F., Guinand, F., Brethé, J.F., Pelvillain, H., Zentout, A.: Vision based target tracking using an unmanned aerial vehicle. In: 2015 IEEE International Workshop on Advanced Robotics and its Social Impacts, ARSO 2015, Lyon, France, July 2015

Machine Learning

Multikernel Recursive Least-Squares Temporal Difference Learning

Chunyuan Zhang[1,2(✉)], Qingxin Zhu[1], and Xinzheng Niu[1]

[1] School of Computer Science and Engineering,
University of Electronic Science and Technology of China, Chengdu, China
zcy7566@126.com, {qxzhu,xinzhengniu}@uestc.edu.cn
[2] College of Information Science and Technology,
Hainan University, Haikou, China

Abstract. Traditional least-squares temporal difference (LSTD) algorithms provide an efficient way for policy evaluation, but their performance is greatly influenced by the manual selection of state features and their approximation ability is often limited. To overcome these problems, we propose a multikernel recursive LSTD algorithm in this paper. Different from the previous kernel-based LSTD algorithms, the proposed algorithm uses Bellman operator along with projection operator, and constructs the sparse dictionary online. To avoid caching all history samples and reduce the computational cost, it uses the sliding-window technique. To avoid overfitting and reduce the bias caused by the sliding window, it also considers L_2 regularization. In particular, to improve the approximation ability, it uses the multikernel technique, which may be the first time to be used for value-function prediction. Experimental results on a 50-state chain problem show the good performance of the proposed algorithm in terms of convergence speed and prediction accuracy.

Keywords: Reinforcement learning · Temporal difference · Recursive least squares · Multiple kernels · Online sparsification · L_2 regularization

1 Introduction

Reinforcement learning (RL) is a promising methodology for solving sequential decision problems. By interacting with an initially unknown environment, the agent aims to learn an optimal policy for maximizing its cumulative rewards [1]. In RL, a popular way to compute the optimal policy is through value functions, which are often estimated by temporal difference (TD) algorithms. For small discrete state-space problems, value functions can be stored in a table. But for many real-world problems, which have large or continuous state spaces, tabular TD algorithms will suffer from the "curse of dimensionality". In such cases, a common strategy is to employ linear function approximators, which are better understood and have more convergence guarantees than nonlinear function approximators [2]. One dominant linear TD formulation is the least-squares TD (LSTD) [3–5]. Compared with the standard TD algorithm, LSTD algorithms use samples more efficiently and eliminate all step-size parameters [6, 7]. However, their performance is greatly influenced by the manual

© Springer International Publishing Switzerland 2016
D.-S. Huang et al. (Eds.): ICIC 2016, Part III, LNAI 9773, pp. 205–217, 2016.
DOI: 10.1007/978-3-319-42297-8_20

selection of state features, and their approximation ability is often limited [8]. In this paper, we focus on how to overcome these two drawbacks.

In the last two decades, kernel methods have been extensively studied in supervised and unsupervised learning [9, 10]. One appeal of kernel methods is the ability to handle nonlinear operations on origin data by indirectly computing an underlying nonlinear mapping to a high-dimensional feature space where linear operations can be performed [11]. The other appeal of kernel methods is the ability to automatically construct a compact dictionary by combining with various sparsification techniques. Several works have been done for kernelizing least-squares algorithms. Engel et al. proposed a sparse kernel recursive least-squares (SKRLS) algorithm, where an approximate linear dependency (ALD) criterion was presented for online sparsification [12]. Similarly, Fan et al. proposed another SKRLS algorithm by using a significance criterion for sparsification [13]. Compared with the traditional RLS algorithm, SKRLS algorithms not only have a good nonlinear approximation ability, but also can avoid selecting features manually. Recently, multikernel methods have become more and more popular. By exploiting the advantages of different types of kernels, multikernel methods can further improve the nonlinear approximation ability. Tobar et al. proposed a multikernel least-squares (MKLS) algorithm [14, 15], which has proven to be better than the monokernel least-squares algorithm.

Intuitively, we can bring the benefits of SKRLS and MKLS algorithms to LSTD algorithms. In fact, there have been some works on kernelizing LSTD algorithms in recent years. Xu first proposed a sparse kernel-based LSTD(λ) (SKLSTD(λ)) algorithm by using the ALD criterion [8]. However, this algorithm is only suitable for batch learning, and its computation is very complicated. After that, Xu et al. proposed an incremental SKLSTD(λ) algorithm for policy iteration [16]. Unfortunately, this new algorithm requires a matrix inversion at each time step, which makes its computational cost expensive, and it requires the sparse dictionary to be constructed offline and beforehand, which makes it only approximate the value function correctly in the area of the state space that is covered by the training samples. On this basis, Jakab et al. proposed a sparse kernel RLSTD (SKRLSTD) algorithm by using a proximity graph sparsification method [17], but it still requires the dictionary to be constructed offline. Besides the above problems, these works don't consider regularization either, whereas many real-world problems exhibit high nonlinearity and online kernel learning often has the risk of overfitting [18, 19]. In addition, these works don't consider multikernel methods. To the best of our knowledge, there is no research on function approximation by using multikernel methods in RL.

In this paper, we propose a multikernel recursive least-squares temporal difference (MKRLSTD) algorithm. Compared with the previous monokernel LSTD algorithms, our algorithm has many advantages by combining the following techniques: (i) Our derivation uses Bellman operator along with projection operator, and thus is more simple. (ii) The mulkernel technique further improves the nonlinear approximation ability. (iii) Online sparsification makes our algorithm more suitable for online policy evaluation in a large or continuous state space. (iv) The sliding-window technique avoids caching all history samples and reduces the computational cost. (v) L_2 regularization can avoid overfitting and reduce the bias caused by the sliding window. Finally, we use a chain problem to demonstrate the effectiveness of our algorithm.

2 Background

2.1 Preliminaries

In RL, an underlying sequential decision-making problem is often modeled as a Markov decision process (MDP). An MDP can be defined as $\mathcal{M} = \langle \mathcal{S}, \mathcal{A}, P, r, \gamma, d \rangle$ [6], where $\mathcal{S}$ is a set of states, $\mathcal{A}$ is a set of actions, $P: \mathcal{S} \times \mathcal{A} \times \mathcal{S} \to [0,1]$ is a state transition probability function where $P(s, a, s')$ denotes the probability of transitioning to state s' when taking action a in state s, $r \in \mathbb{R}$ is a reward function, $\gamma \in [0, 1]$ is the discount factor, and d is an initial state distribution. For simplicity of presentation, we assume that $\mathcal{S}$ and $\mathcal{A}$ are finite. Under a fixed policy $\pi: \mathcal{S} \to \mathcal{A}$, an MDP $\mathcal{M}$ will reduce to a Markov reward process (MRP) $\mathcal{R}^\pi = \langle \mathcal{S}, P^\pi, R^\pi, \gamma, d \rangle$, where $P^\pi(s, s') = \sum_{a \in \mathcal{A}} \pi(a|s) P(s, a, s')$ and $R^\pi(s) = \sum_{a \in \mathcal{A}} \pi(a|s) \sum_{s \in \mathcal{S}} \pi(a|s) r(s, a, s')$.

Given an $\mathcal{R}^\pi$, the state-value function can be defined as $V^\pi(s) = \mathbf{E}\left[\sum_{t=1}^{\infty} \gamma^t r_t | s_1 = s, \pi \right]$. It is well known that $V^\pi(s)$ must obey the Bellman equation [1], which can be expressed in vector form

$$V^\pi = R^\pi + \gamma P^\pi V^\pi \tag{1}$$

If P^π and R^π are known, V^π can be solved analytically, i.e.

$$V^\pi = (I - \gamma P^\pi)^{-1} R^\pi \tag{2}$$

where I is the $|\mathcal{S}| \times |\mathcal{S}|$ identity matrix.

However, the environment is initially unknown in RL, namely, P^π and R^π are assumed to be unknown in RL. That means we can't solve V^π by (2) directly. In RL, we often use TD algorithms to tackle this problem. But as mentioned in Sect. 1, many real-world RL problems have a large or continuous state space, which makes $V^\pi(s)$ hard to be expressed explicitly. In order to overcome this problem, we often resort to linear function approximation, i.e. $\hat{V}^\pi(s) = w^T \phi(s)$ and $\hat{V}^\pi = \Phi w$, where $w \in \mathbb{R}^m$ is a parameter vector, $\phi(s) \in \mathbb{R}^m$ is the feature vector of state s, and $\Phi = [\phi(s_1), ..., \phi(s_{|\mathcal{S}|})]^T$ is an $|\mathcal{S}| \times m$ feature matrix. Unfortunately, when approximating V^π in this manner, there is usually no way to satisfy the Bellman equation exactly, because $R^\pi + \gamma P^\pi \Phi w$ may lie outside the span of Φ [6].

2.2 LSTD Algorithm

The LSTD algorithm presents an efficient way to find w such that $\hat{V}^\pi$ approximately satisfies the Bellman equation [6]. By solving $\min_{u \in \mathbb{R}^m} \| \Phi u - (R^\pi + \gamma P^\pi \Phi w) \|_D^2$, we can project $R^\pi + \gamma P^\pi \Phi w$ to the span of Φ and obtain its closest approximation Φu^*. Thus, from (1) and $\hat{V}^\pi = \Phi w$, we can use u^* to replace w for approximating V^π. That means we can find w by solving the following fixed-point equation

$$w = f(w) = \text{argmin}_{u \in \mathbb{R}^m} \| \Phi u - (R^\pi + \gamma P^\pi \Phi w) \|_D^2 \tag{3}$$

where D is a diagonal matrix indicating a distribution over states. Nevertheless, we can't solve (3) exactly, since P^π and R^π are unknown and the full Φ is too large to form anyway in a large or continuous state space. Instead, considering a trajectory $T_t^\pi = \left\{ (s_i, s_i', r_i) | i = 1, \cdots, t \right\}$ following policy π, we use $\hat{\Phi}_t = [\phi(s_1), \cdots, \phi(s_t)]^T$, $\hat{\Phi}_t' = [\phi(s_1'), \cdots, \phi(s_t')]^T$ and $\hat{R}_t = [r_1, \cdots, r_t]^T$ to replace Φ, $P^\pi \Phi$ and R^π, respectively. Then, (3) can be approximately rewritten as

$$w_t = \hat{f}(w_t) = \text{argmin}_{u \in \mathbb{R}^m} \left\| \hat{\Phi}_t u - (\hat{R}_t + \gamma \hat{\Phi}_t' w_t) \right\|^2 \tag{4}$$

Let $\partial \left\| \hat{\Phi}_t u - \hat{R}_t + \gamma \hat{\Phi}_t' w_t \right\|^2 / \partial u = 0$, we get $u^* = (\hat{\Phi}_t^T \hat{\Phi}_t)^{-1} \hat{\Phi}_t^T (\hat{R}_t + \hat{\Phi}_t' w_t)$. Thus, the fixed point $w_t = \hat{f}(w_t) = u^*$ can be solved by

$$w_t = \hat{f}(w_t) = (\hat{\Phi}_t^T (\hat{\Phi}_t - \gamma \hat{\Phi}_t'))^{-1} \hat{\Phi}_t^T \hat{R}_t \tag{5}$$

3   MKRLSTD Learning

3.1   Multifeature LSTD Algorithm

In multikernel methods, the reproducing kernel Hilbert space is a multiple feature space [14, 15]. For facilitating the derivation of our MKRLSTD algorithm, we extend the LSTD algorithm to the case of multiple feature vectors in this subsection.

Consider a feature-vector set $\{ \phi_1(s), \cdots, \phi_L(s) \}$ where $\phi_z(s)$ denotes the z-th feature vector, and define the multifeature vector $\bar{\phi}(s) = \left[\phi_1(s)^T, \cdots, \phi_L(s)^T \right]^T$. Then, V^π and $V^\pi(s)$ can be approximated by $\hat{V}^\pi = \bar{\Phi} \bar{w}$ and $\hat{V}^\pi(s) = \sum_{z=1}^{L} w_z^T \phi_z(s)$, where $\bar{w} = [w_1^T, \cdots, w_L^T]^T$ is the multiparameter vector, $\bar{\Phi} = [\bar{\phi}(s_1), \cdots, \bar{\phi}(s_{|s|})]^T$ is the multifeature matrix, and $w_z \in \mathbb{R}^{m_z}$ is the z-th parameter vector. Correspondingly, (3) needs to be modified as

$$\bar{w} = f(\bar{w}) = \text{argmin}_{\bar{u} \in \mathbb{R}^{\bar{m}}} \| \bar{\Phi} \bar{u} - (R^\pi + \gamma P^\pi \bar{\Phi} \bar{w}) \|_D^2 \tag{6}$$

where $\bar{u} = [u_1^T, \cdots, u_L^T]^T$, $u_z \in \mathbb{R}^{m_z}$ and $\bar{m} = \sum_{z=1}^{L} m_z$. The following derivation is similar to that of the LSTD algorithm in Sect. 2.2. Given a trajectory T_t^π, we use $\tilde{\Phi}_t = \left[\bar{\phi}(s_1), \cdots, \bar{\phi}(s_t) \right]^T$, $\tilde{\Phi}_t' = \left[\bar{\phi}(s_1'), \cdots, \bar{\phi}(s_1') \right]^T$, $\tilde{\Phi}_t' = \left[\bar{\phi}(s_1'), \cdots, \bar{\phi}(s_{|s|}') \right]^T$ and $\hat{R}_t$ to replace $\bar{\Phi}$, $P^\pi \bar{\Phi}$ and R^π. Then, (6) can be approximately rewritten as

$$\bar{w}_t = \tilde{f}(\bar{w}_t) = \mathrm{argmin}_{\bar{u}\in\mathbb{R}^{\bar{m}}} \parallel \tilde{\Phi}_t\bar{u} - (\hat{R}_t + \gamma\tilde{\Phi}'_t\bar{w}_t) \parallel^2 \qquad (7)$$

Let $\partial \left\|\tilde{\Phi}_t\bar{u} - \hat{R}_t + \gamma\tilde{\Phi}'_t\bar{w}_t\right\|^2 \Big/ \partial\bar{u} = 0$, and thus $\bar{w}_t$ can be found by

$$\bar{w}_t = (\tilde{\Phi}_t^{\mathrm{T}}(\tilde{\Phi}_t^{\mathrm{T}} - \gamma\tilde{\Phi}'_t))^{-1}\tilde{\Phi}_t^{\mathrm{T}}\hat{R}_t \qquad (8)$$

3.2   Online Sparse MKRLSTD Algorithm

Now we derive an online sparse MKRLSTD algorithm. The whole derivation includes four parts: L_2 regularization, kernelization, recursion and sparsification.

In the first part, we add an L_2-norm penalty into (7), i.e.

$$\bar{w}_t = \tilde{f}(\bar{w}_t) = \mathrm{argmin}_{\bar{u}\in\mathbb{R}^{\bar{m}}}\left\|\tilde{\Phi}_t\bar{u} - \left(\hat{R}_t + \gamma\tilde{\Phi}'_t\bar{w}_t\right)\right\|^2 + \sum_{z=1}^{L}\eta_z\|u_z\|_2^2 \qquad (9)$$

where $\eta_z \in [0, +\infty)$ is the z-th regularization parameter. For $\forall z \in \{1, \cdots, L\}$, we let $\partial\left(\left\|\tilde{\Phi}_t\bar{u} - (\hat{R}_t + \gamma\tilde{\Phi}'_t\bar{w}_t)\right\|^2 + \sum_{z=1}^{L}\eta_z\|u_z\|_2^2\right)\Big/\partial u_z = 0$ and thus have

$$\tilde{\Phi}_{z,t}^{\mathrm{T}}(\tilde{\Phi}_t\bar{u}^* - \gamma\tilde{\Phi}'_t\bar{w}_t) + \eta_z u_z^* = \tilde{\Phi}_{z,t}^{\mathrm{T}}\hat{R}_t \quad \forall z \qquad (10)$$

where $\tilde{\Phi}_{z,t}^{\mathrm{T}} = \left[\phi_z(s_1), \cdots, \phi_z(s_t)\right]$. Since $\bar{w}_t = \tilde{f}(\bar{w}_t) = \bar{u}^*$, (10) can be rewritten as

$$\tilde{\Phi}_{z,t}^{\mathrm{T}}(\tilde{\Phi}_t - \gamma\tilde{\Phi}'_t)\bar{w}_t + \eta_z w_{z,t} = \tilde{\Phi}_{z,t}^{\mathrm{T}}\hat{R}_t \quad \forall z \qquad (11)$$

In the second part, we try to kernelize (11). Suppose the sparse dictionary $\mathcal{D}_t = \{d_j | d_j \in \mathcal{S}, j = 1, \cdots, n_t\}$. Then, the corresponding multifeature matrix can be defined as $\Psi_t = \left[\bar{\phi}(d_1), \cdots, \bar{\phi}(d_{n_t})\right]^{\mathrm{T}}$. By the Representer Theorem [20], we have $w_{z,t} = \Psi_{z,t}^{\mathrm{T}}\alpha_{z,t}$, where $\Psi_{z,t}^{\mathrm{T}} = \left[\phi_z(d_1), \cdots, \phi_z(d_{n_t})\right]$ and $\alpha_{z,t} = [\alpha_{z,t}^1, \cdots, \alpha_{z,t}^{n_t}]^{\mathrm{T}}$. Then, plugging $w_{z,t} = \Psi_{z,t}^{\mathrm{T}}\alpha_{z,t}$ into (11) and multiplying both sides by $\Psi_{z,t}$ give

$$\Psi_{z,t}\tilde{\Phi}_{z,t}^{\mathrm{T}}\sum_{z=1}^{L}(\tilde{\Phi}_{z,t} - \gamma\tilde{\Phi}'_{z,t})\Psi_{z,t}^{\mathrm{T}}\alpha_{z,t} + \eta_z\Psi_{z,t}\Psi_{z,t}^{\mathrm{T}}\alpha_{z,t} = \Psi_{z,t}\tilde{\Phi}_{z,t}^{\mathrm{T}}\hat{R}_t \quad \forall z \qquad (12)$$

By the Mercer Theorem [20], $k_z(s_i, s_j) = \langle\phi_z(s_i), \phi_z(s_j)\rangle$. Thus, we can define $K_{z,t} = \Psi_{z,t}\Psi_{z,t}^{\mathrm{T}}$, $\tilde{K}_{z,t} = \Psi_{z,t}\tilde{\Phi}_{z,t}^{\mathrm{T}}$ and $\tilde{K}'_{z,t} = \Psi_{z,t}(\tilde{\Phi}'_{z,t})^{\mathrm{T}}$. Then, (12) can be rewritten in multikernel form, i.e.

$$\tilde{K}_{z,t}\sum_{z=1}^{L}(\tilde{K}_{z,t} - \gamma\tilde{K}'_{z,t})^{\mathrm{T}}\alpha_{z,t} + \eta_z K_{z,t}\alpha_{z,t} = \tilde{K}_{z,t}\hat{R}_t \quad \forall z \qquad (13)$$

Let

$$
\boldsymbol{\alpha}_t = \begin{bmatrix} \boldsymbol{\alpha}_{1,t} \\ \vdots \\ \boldsymbol{\alpha}_{L,t} \end{bmatrix}, \tilde{K}_t = \begin{bmatrix} \tilde{K}_{1,t} \\ \vdots \\ \tilde{K}_{L,t} \end{bmatrix}, \tilde{K}_t' = \begin{bmatrix} \tilde{K}_{1,t}' \\ \vdots \\ \tilde{K}_{L,t}' \end{bmatrix}, \Theta_t = \begin{bmatrix} \eta_1 K_{1,t} & \cdots & 0 \\ \vdots & \ddots & \vdots \\ 0 & \cdots & \eta_L K_{L,t} \end{bmatrix} \tag{14}
$$

Equation (13) can be further expressed as

$$
(\tilde{K}_t(\tilde{K}_t - \gamma \tilde{K}_t')^\mathrm{T} + \Theta_t)\boldsymbol{\alpha}_t = \tilde{K}_t \hat{R}_t \tag{15}
$$

However, (15) is inconvenient for the subsequent derivation, since the components of $\boldsymbol{\alpha}_t$, $\tilde{K}_t$, $\tilde{K}_t'$ and Θ_t are sorted by the multikernel index rather than the dictionary index. Fortunately, since (15) is a linear system, we can swap its rows without changing its solution set [21]. Let $\boldsymbol{\beta}_{j,t}^\mathrm{T} = [\alpha_{1,t}^j, \cdots, \alpha_{L,t}^j]$, $\tilde{H}_{j,t}^\mathrm{T} = [\tilde{\boldsymbol{k}}_{1,t}(\boldsymbol{d}_j), \cdots, \tilde{\boldsymbol{k}}_{L,t}(\boldsymbol{d}_j)]$, $(\tilde{H}_{j,t}')^\mathrm{T} = [\tilde{\boldsymbol{k}}_{1,t}'(\boldsymbol{d}_j), \cdots, \tilde{\boldsymbol{k}}_{L,t}'(\boldsymbol{d}_j)]$ and $\mathrm{X}_{j,t} = [\mathrm{X}_{j,t}^1, \cdots, \mathrm{X}_{j,t}^{n_t}]$, where $\tilde{\boldsymbol{k}}_{z,t}(\boldsymbol{d}_j) = [k_z(\boldsymbol{d}_j, \boldsymbol{s}_1), \cdots, k_z(\boldsymbol{d}_j, \boldsymbol{s}_t)]^\mathrm{T}$, $\tilde{\boldsymbol{k}}_{z,t}'(\boldsymbol{d}_j) = [k_z(\boldsymbol{d}_j, \boldsymbol{s}_1'), \cdots, k_z(\boldsymbol{d}_j, \boldsymbol{s}_t')]^\mathrm{T}$, and

$$
\mathrm{X}_{j,t}^v = \begin{bmatrix} \eta_1 k_1(\boldsymbol{d}_j, \boldsymbol{d}_v) & \cdots & 0 \\ \vdots & \ddots & \vdots \\ 0 & \cdots & \eta_L k_L(\boldsymbol{d}_j, \boldsymbol{d}_v) \end{bmatrix} \tag{16}
$$

By swapping rows, we transform $\boldsymbol{\alpha}_t$, $\tilde{K}_t$, $\tilde{K}_t'$ and Θ_t into $\boldsymbol{\beta}_t = [\boldsymbol{\beta}_{1,t}^\mathrm{T}, \cdots, \boldsymbol{\beta}_{n_t,t}^\mathrm{T}]^\mathrm{T}$, $\tilde{H}_t = [\tilde{H}_{1,t}^\mathrm{T}, \cdots, \tilde{H}_{n_t,t}^\mathrm{T}]^\mathrm{T}$, $\tilde{H}_t' = [(\tilde{H}_{1,t}')^\mathrm{T}, \cdots, (\tilde{H}_{n_t,t}')^\mathrm{T}]^\mathrm{T}$ and $\mathrm{X}_t = [\mathrm{X}_{1,t}^\mathrm{T}, \cdots, \mathrm{X}_{n_t,t}^\mathrm{T}]^\mathrm{T}$. Then, (15) can be transformed into

$$
\boldsymbol{\beta}_t = (\tilde{H}_t(\tilde{H}_t - \gamma \tilde{H}_t')^\mathrm{T} + \mathrm{X}_t)^{-1}\tilde{H}_t \hat{R}_t = A_t^{-1}\boldsymbol{b}_t \tag{17}
$$

where $A_t = \tilde{H}_t(\tilde{H}_t - \gamma \tilde{H}_t')^\mathrm{T} + \mathrm{X}_t$ and $\boldsymbol{b}_t = \tilde{H}_t \hat{R}_t$ can be further rewritten as

$$
A_t = \begin{bmatrix} \sum_{i=1}^t \boldsymbol{k}_{t,i}^1 \Delta \boldsymbol{k}_{t,i}^1 + \mathrm{X}_{1,t} & \cdots & \sum_{i=1}^t \boldsymbol{k}_{t,i}^1 \Delta \boldsymbol{k}_{t,i}^{n_t} + \mathrm{X}_{1,t}^{n_t} \\ \vdots & \ddots & \vdots \\ \sum_{i=1}^t \boldsymbol{k}_{t,i}^{n_t} \Delta \boldsymbol{k}_{t,i}^1 + \mathrm{X}_{n_t,t}^1 & \cdots & \sum_{i=1}^t \boldsymbol{k}_{t,i}^{n_t} \Delta \boldsymbol{k}_{t,i}^{n_t} + \mathrm{X}_{n_t,t} \end{bmatrix}, \boldsymbol{b}_t = \begin{bmatrix} \sum_{i=1}^t \boldsymbol{k}_{t,i}^1 r_i \\ \vdots \\ \sum_{i=1}^t \boldsymbol{k}_{t,i}^{n_t} r_i \end{bmatrix} \tag{18}
$$

where $\boldsymbol{k}_{t,i}^j = [k_1(\boldsymbol{d}_j, \boldsymbol{s}_i), \cdots, k_L(\boldsymbol{d}_j, \boldsymbol{s}_i)]^\mathrm{T}$ and $\Delta \boldsymbol{k}_{t,i}^j = [k_1(\boldsymbol{d}_j, \boldsymbol{s}_i) - \gamma k_1(\boldsymbol{d}_j, \boldsymbol{s}_i'), \cdots, k_L(\boldsymbol{d}_j, \boldsymbol{s}_i) - \gamma k_L(\boldsymbol{d}_j, \boldsymbol{s}_i')]$.

In the third part, we derive the recursive formulas of A_t^{-1} and $\boldsymbol{\beta}_t$. Under online sparsification, there are two cases: (1) $\mathcal{D}_t = \mathcal{D}_{t-1}$; (2) $\mathcal{D}_t = \mathcal{D}_{t-1} \cup \{\boldsymbol{s}_t\}$.

For the first case, we have $n_t = n_{t-1}$, $k_{t,i}^j = k_{t-1,i}^j$, $\Delta k_{t,i}^j = \Delta k_{t-1,i}^j$ and $X_t = X_{t-1}$.

Let $k_{t-1,i} = \left[(k_{t-1,i}^1)^T, \cdots, (k_{t-1,i}^{n_{t-1}})^T \right]^T$ and $\Delta k_{t-1,i} = \left[\Delta k_{t-1,i}^1, \cdots, \Delta k_{t-1,i}^{n_{t-1}} \right]$. Then, (18) can be rewritten as

$$A_t = A_{t-1} + k_{t-1,t}\Delta k_{t-1,t}, \quad b_t = b_{t-1} + k_{t-1,t}r_t \tag{19}$$

Using the matrix inversion Lemma [22] for A_t^{-1}, we get

$$A_t^{-1} = A_{t-1}^{-1} - \frac{A_{t-1}^{-1}k_{t-1,t}\Delta k_{t-1,t}A_{t-1}^{-1}}{1 + \Delta k_{t-1,t}A_{t-1}^{-1}k_{t-1,t}} \tag{20}$$

Thus, plugging (20) and $b_t = b_{t-1} + k_{t-1,t}r_t$ into (17), we can obtain

$$\beta_t = \beta_{t-1} - \frac{(r_t - \Delta k_{t-1,t}\beta_{t-1})A_{t-1}^{-1}k_{t-1,t}}{1 + \Delta k_{t-1,t}A_{t-1}^{-1}k_{t-1,t}} \tag{21}$$

For the second case, we have $n_t = n_{t-1} + 1$, $k_{t,i}^j = k_{t-1,i}^j$, $\Delta k_{t,i}^j = \Delta k_{t-1,i}^j$ and $X_{j,t}^y = X_{j,t-1}^y (j \leq n_{t-1})$. Here X_t is expanded as

$$X_t = \begin{bmatrix} X_{t-1} & X_{t-1}^{n_t} \\ (X_{t-1}^{n_t})^T & X_{n_t,t}^{n_t} \end{bmatrix} \tag{22}$$

where $X_{t-1}^{n_t} = [X_{1,t-1}^{n_t}, \cdots, X_{n_{t-1},t-1}^{n_t}]^T$. Then, (18) can be rewritten as

$$A_t = \begin{bmatrix} \tilde{A}_t & B_t \\ C_t & G_t \end{bmatrix}, b_t = \begin{bmatrix} \tilde{b}_t \\ q_t \end{bmatrix} \tag{23}$$

where $\tilde{A}_t$ and $\tilde{b}_t$ are the same as the updated A_t and b_t when the dictionary keeps unchanged, $B_t = \sum_{i=1}^t k_{t-1,i}\Delta k_{t-1,i}^{n_t} + X_{t-1}^{n_t}$, $C_t = \sum_{i=1}^t k_{t-1,i}^{n_t}\Delta k_{t-1,i} + (X_{t-1}^{n_t})^T$, $G_t = \sum_{i=1}^t k_{t,i}^{n_t}\Delta k_{t,i}^{n_t} + X_{n_t,t}^{n_t}$, and $q_t = \sum_{i=1}^t k_{t,i}^{n_t}r_i$. However, the calculation of B_t, C_t, G_t and q_t requires caching all history samples, and their computational cost becomes more and more expensive as t increases. To avoid this situation, here we only use the recent samples to evaluate these terms. This is similar to the sliding-window method introduced in [23] and the truncation method introduced in [18]. Define a sliding window $\mathcal{W}_t = \{(s_u, s_u', r_u) | u = \max(1, t - N + 1), \cdots, t\}$, where N is the window size. Then, B_t, C_t, G_t and q_t can be evaluated as

$$
\begin{cases}
\hat{B}_t = \sum_u k_{t-1,u} \Delta k_{t-1,u}^{n_t} + X_{t-1}^{n_t} \\
\hat{C}_t = \sum_u k_{t-1,u}^{n_t} \Delta k_{t-1,u} + (X_{t-1}^{n_t})^{\mathrm{T}} \\
\hat{G}_t = \sum_u k_{t,u}^{n_t} \Delta k_{t,u}^{n_t} + X_{n_t,t} \\
\hat{q}_t = \sum_u k_{t,u}^{n_t} r_u
\end{cases}
\tag{24}
$$

Similar to those in the first case, A_t^{-1} and $\boldsymbol{\beta}_t$ can be derived as follows

$$
A_t^{-1} = \begin{bmatrix} \tilde{A}_t^{-1} + \tilde{A}_t^{-1}\hat{B}_t M_t^{-1}\hat{C}_t\tilde{A}_t^{-1} & -\tilde{A}_t^{-1}\hat{B}_t M_t^{-1} \\ -M_t^{-1}\hat{C}_t\tilde{A}_t^{-1} & M_t^{-1} \end{bmatrix}
\tag{25}
$$

$$
\boldsymbol{\beta}_t = \begin{bmatrix} \tilde{\boldsymbol{\beta}}_t - \tilde{A}_t^{-1}\hat{B}_t M_t^{-1}(\hat{q}_t - \hat{C}_t\tilde{\boldsymbol{\beta}}_t) \\ M_t^{-1}(\hat{q}_t - \hat{C}_t\tilde{\boldsymbol{\beta}}_t) \end{bmatrix}
\tag{26}
$$

where $M_t^{-1} = (\hat{G}_t - \hat{C}_t\tilde{A}_t^{-1}\hat{B}_t)^{-1}$, and $\tilde{\boldsymbol{\beta}}_t$ is the same as the updated $\boldsymbol{\beta}_t$ when the dictionary keeps unchanged.

In the final part, we discuss the sparsification criterion for the MKRLSTD algorithm. In the kernel machine learning, there have been many sparsification methods such as the novelty criterion (NC) [24] and the ALD criterion [12], but they are mainly designed for monokernel algorithms. Here we only consider extending the ALD criterion to the MKRLSTD algorithm.

Suppose that all types of features are independent each other, and predefine the sparsification thresholds $\varepsilon_1, \cdots, \varepsilon_L$ corresponding to $\phi_1(s), \cdots, \phi_L(s)$. The ALD criterion for MKRLSTD can be expressed as

$$
\begin{cases}
\mathcal{D}_t = \mathcal{D}_{t-1} \cup \{s_t\} & \text{if } \delta_{1,t} > \varepsilon_1 \text{ or } \cdots \text{or } \delta_{L,t} > \varepsilon_L \\
\mathcal{D}_t = \mathcal{D}_{t-1} & \text{otherwise}
\end{cases}
\tag{27}
$$

where $\delta_{z,t}$ is the ALD error of $\phi_z(s_t)$, i.e.

$$
\delta_{z,t} = k_z(s_t, s_t) - k_{z,t-1}^{\mathrm{T}}(s_t)c_{z,t} \quad \forall z
\tag{28}
$$

where $k_{z,t-1}(s_t) = [k_z(s_t, d_1), \cdots, k_z(s_t, d_{n_{t-1}})]^{\mathrm{T}}$ and $c_{z,t} = K_{z,t-1}^{-1}k_{z,t-1}(s_t)$. If $\mathcal{D}_t = \mathcal{D}_{t-1}$, then $K_{z,t}^{-1} = K_{z,t-1}^{-1}$; otherwise, $K_{z,t}^{-1}$ should be expanded as

$$
K_{z,t}^{-1} = \frac{1}{\delta_{z,t}} \begin{bmatrix} \delta_{z,t}K_{z,t-1}^{-1} + c_{z,t}c_{z,t}^{\mathrm{T}} & -c_{z,t} \\ -c_{z,t}^{\mathrm{T}} & 1 \end{bmatrix} \quad \forall z
\tag{29}
$$

Based on the above derivation, the MKRLSTD algorithm with the ALD criterion can be summarized in Table 1. If $\mathcal{D}_t = \mathcal{D}_{t-1}$, the overall computational cost per time step is $O((Ln_t)^2)$; otherwise, it is $O(\max((Ln_t)^2, L^3))$. Note that A_t^{-1} and M_t^{-1} don't become computational bottlenecks, since they are only performed when s_t is added into

Table 1. MKRLSTD algorithm with the ALD criterion

	Cost [1]
Input: $\pi, \gamma, N, \eta_1, \cdots, \eta_L, k_1(\cdot,\cdot), \cdots, k_L(\cdot,\cdot), \varepsilon_1, \cdots, \varepsilon_L$	
for $t = 1, 2, \cdots$	
if $t == 1$ **then**	
Initialize $s_1, \mathcal{D}_1 = \{s_1\}, K_{z,t}^{-1} = 1/k_z(s_1, s_1)$ ($\forall z$)	$O(L)$
Take a_1 given by π, and observe s_1', r_1	
Initialize $\mathcal{W}_1 = \{(s_1, s_1', r_1)\}$	
Compute A_1, b_1 by (18)	$O(L^2)$
Compute $A_1^{-1}, \beta_1 = A_1^{-1} b_1$	$O(L^3)$
else	
Take a_t given by π, and observe s_t', r_t	
Update $\mathcal{W}_t, \mathcal{D}_t = \mathcal{D}_{t-1}, K_{z,t}^{-1} = K_{z,t-1}^{-1}$ ($\forall z$)	
Compute $\delta_{z,t}$ ($\forall z$) by (28)	$O(L(n_{t-1})^2)$
Update β_t, A_t^{-1} by (21), (20)	$O((Ln_{t-1})^2)$
if $\delta_{1,t} > \varepsilon_1$ **or** $\cdots$ **or** $\delta_{L,t} > \varepsilon_L$ **then**	
Update $\mathcal{D}_t = \mathcal{D}_{t-1} \cup \{s_t\}$	
Let $\tilde{A}_t^{-1} = A_t^{-1}, \tilde{\beta}_t = \beta_t$	
Compute $\hat{B}_t, \hat{C}_t, \hat{G}_t, \hat{q}_t$ by (24)	$O(L^2 n_{t-1})$
$M_t^{-1} = (\hat{G}_t - \hat{C}_t \tilde{A}_t^{-1} \hat{B}_t)^{-1}$	$O(\max((Ln_{t-1})^2, L^3))$
Update β_t, A_t^{-1} by (26), (25)	$O(Ln_t^2)$
Update $K_{z,t}^{-1}$ ($\forall z$) by (29)	
end if	
end if	
end for	

[1] We assume that the computational cost of a kernel function is $O(1)$.

$\mathcal{D}_t$ and since L is often small. If $L = 1$, the MKRLSTD algorithm will degenerate into the monokernel RLSTD algorithm. The MKRLSTD algorithm can be modified for solving episodic tasks. When s_t' is an absorbing state, it only requires setting $\gamma = 0$ temporally and setting s_{t+1} as the start state of next episode. In addition, it can also be modified for combining with other online sparsification methods since our derivation doesn't limit the type of sparsification methods in the third part.

4 Experiments

4.1 Experimental Settings

In this section, we use a 50-state chain problem [25] to demonstrate the effectiveness of the MKRLSTD algorithm. For comparison purposes, we also test three other LSTD algorithms (RLSTD [3], SKRLSTD [17] and monokernel RLSTD) on this problem. In this problem, the chain consists of 50 states, which are numbered from 1 to 50. For each state, there are two actions available, i.e., "left" and "right". Each action succeeds with probability 0.9, changing the state in the intended direction, and fails with probability 0.1, changing the state in the opposite direction. Note that the state transition probabilities are available only for solving the true state-value functions V^π, and they are unknown for all the algorithms tested here. The two boundaries of the chain are dead-ends, and the reward is +1 only in states 10 and 41. Due to the symmetry, the optimal policy is to go right in states 1–9 and 26–41 and left in states 10–25 and 42–50. Here we use it as the policy π to be evaluated.

The settings of all the algorithms are summarized as follows: (i) For the MKRLSTD algorithm, we set the sliding-window size $N = 5$, define the multikernel by a radius basis function (RBF) kernel $k_1(x, y) = \exp\left(-\|x - y\|^2 \big/ (2\sigma^2)\right)$ and a triangular kernel $k_2(x, y) = \max(b - \|x - y\|/h, v)$ where $\sigma = 5\sqrt{2}/2$, $b = 1$, $h = 7$ and $v = 0$, and set the sparsification thresholds $\varepsilon_1 = 0.5$ and $\varepsilon_2 = 0.65$. By sweeping $\{0.1, 0.2, \cdots, 1.0\}$, we set the regularization parameters $\eta_1 = 1.0$ and $\eta_2 = 0.6$. (ii) For the RLSTD algorithm, we let a feature vector $\phi(x)$ consist of 19 RBFs plus a constant term 1, resulting in a total of 20 basis functions. The centers of the RBFs are evenly distributed over $[1, 50]$, and the width σ of each RBF is also set to $5\sqrt{2}/2$. In addition, we initialize the variance matrix $C_0 = 0.4I$, where I is the 20×20 identity matrix. (iii) For the SKRLSTD algorithm, we use the same RBF kernel as the MKRLSTD algorithm, and also use the ALD criterion where the sparsification threshold is set to 0.5. (iv) For the monokernel RLSTD algorithm, there are two versions, one with the RBF kernel and the other with the triangular kernel. Here we call the former GKRLSTD, and call the latter TKRLSTD. Both of them have the same settings as the counterparts in the MKRLSTD algorithm, and their sliding-window sizes are also set to 5. (v) For each algorithm, the discount factor γ is set to 0.9, each experiment performs 50 runs, each run includes 100 episodes, and each episode is truncated after 100 time steps. Note that there is an extra run for offline sparsification before each regular run in the SKRLSTD algorithm.

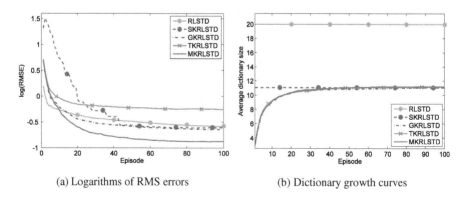

(a) Logarithms of RMS errors   (b) Dictionary growth curves

Fig. 1. Performance comparisons among different LSTD algorithms

4.2   Experimental Results

The main experimental results averaged over 50 runs are illustrated in Fig. 1. Figure 1 (a) shows the logarithms of root mean square (RMS) errors of all tested algorithms. At each episode, the RMS error of each algorithm is calculated by

$$RMSE = \frac{1}{50} \sum\nolimits_{j=1}^{50} \left(\frac{1}{50} \sum\nolimits_{s=1}^{50} \left(\hat{V}_j^\pi(s) - V^\pi(s) \right) \right)^{0.5} \tag{30}$$

where $V^\pi(s)$ is the true state-value function solved by (1), and $\hat{V}_j^\pi(s)$ is the predicted value of the j-th run. Figure 1(b) shows the dictionary growth curves of all tested algorithms. Since the feature vector of the RLSTD algorithm is designed manually, and since the sparse dictionary of the SKRLSTD algorithm is constructed offline and beforehand, their average dictionary sizes don't change with episodes.

From Fig. 1, we can observe: (i) Compared with RLSTD, MKRLSTD can avoid manually selecting state features, and can greatly improve the value-function approximation ability with a smaller dictionary. (ii) Compared with SKRLSTD, MKRLSTD can construct the sparse dictionary online, and can obtain better performance in convergence speed and prediction accuracy. In addition, GKRLSTD is superior to SKRLSTD but TKRLSTD is not, which means monokernel RLSTD is more sensitive to the kernel type than MKRLSTD. (iii) Compared with GKRLSTD and TKRLSTD, MKRLSTD can further improve the convergence speed and prediction accuracy. Although its computational cost is almost L^2 times larger than that of monokernel RLSTD, MKRLSTD uses samples more efficiently by exploiting advantages of different types of kernels. This is very important when samples are expensive.

There is another experimental result not included in Fig. 1, i.e., a big sliding-window size makes the performance of MKRLSTD become worse rather than better. This result will seem ridiculous if we only use (23)–(26) to analyze this problem. Intuitively, a small sliding window will introduce more bias into A_t^{-1} and β_t than a big window. However, here A_t^{-1} is calculated by recursive update and samples are used one by one. Thus, using too many history samples together is not suitable for updating A_t^{-1}. From (9), MKRLSTD has considered L_2 regularization. Therefore, even if the small window introduces some bias into A_t^{-1} and β_t, we can also reduce the influence of this bias by selecting moderate regularization parameters [18].

5   Conclusion

In this paper, we propose a multikernel recursive least-squares temporal difference algorithm, called MKRLSTD. By using Bellman operator along with projection operator, our derivation is more simple. By combining L_2 regularization, recursive least squares, online sparsification, and the sliding-window technique, our algorithm is more suitable for online policy evaluation in a large or continuous state space. In particular, by combining the multikernel technique, our algorithm can obtain better performance in value-function approximation than the monokernel RLSTD algorithm. Finally, experimental results demonstrate the effectiveness of our algorithm.

In future work, we will combine our algorithm with eligibility traces, and extend our algorithm to policy control problems.

Acknowledgments. This work is supported in part by the National Natural Science Foundation of China under Grant Nos. 61300192 and 11261015, the Fundamental Research Funds for the Central Universities under Grant No. ZYGX2014J052, and the Natural Science Foundation of Hainan Province, China, under Grant No. 613153.

References

1. Sutton, R.S., Barto, A.G.: Reinforcement Learning: An Introduction. MIT Press, Cambridge (1998)
2. Tsitsiklis, J.N., Roy, B.V.: An analysis of temporal-difference learning with function approximation. IEEE Trans. Autom. Control **42**(5), 674–690 (1997)
3. Bradtke, S.J., Barto, A.G.: Linear least-squares algorithms for temporal difference learning. Mach. Learn. **22**(1–3), 33–57 (1996)
4. Boyan, J.A.: Technical update: least-squares temporal difference learning. Mach. Learn. **49**(2–3), 233–246 (2002)
5. Xu, X., He, H., Hu, D.: Efficient reinforcement learning using recursive least-squares methods. J. Artif. Intell. Res. **16**, 259–292 (2002)
6. Kolter, J.Z., Ng, A.Y.: Regularization and feature selection in least-squares temporal difference learning. In: Proceedings of the 26th Annual International Conference on Machine Learning, pp. 521–528. ACM Press, New York (2009)
7. Chen, S., Chen, G., Gu, R.: An efficient L_2-norm regularized least-squares temporal difference learning algorithm. Knowl. Based Syst. **45**, 94–99 (2013)
8. Xu, X.: A sparse Kernel-Based Least-Squares temporal difference algorithm for reinforcement learning. In: Jiao, L., Wang, L., Gao, X.-b., Liu, J., Wu, F. (eds.) ICNC 2006. LNCS, vol. 4221, pp. 47–56. Springer, Heidelberg (2006)
9. Shawe-Taylor, J., Cristianini, N.: Kernel Methods for Pattern Analysis. Cambridge University Press, Cambridge (2004)
10. Liu, W., Príncipe, J.C., Haykin, S.: Kernel Adaptive Filtering: A Comprehensive Introduction. Wiley, Hoboken (2010)
11. Bae, J., Chhatbar, P., Francis, J.T., Sanchez, J.C., Príncipe, J.C.: Reinforcement learning via kernel temporal difference. In: 33rd Annual International Conference of the IEEE EMBS, pp. 5662–5665. IEEE Press, New York (2011)
12. Engel, Y., Mannor, S., Meir, R.: The kernel recursive least-squares algorithm. IEEE Trans. Signal Process. **52**(8), 2275–2285 (2004)
13. Fan, H., Song, Q.: A sparse kernel algorithm for online time series data prediction. Expert Syst. Appl. **40**(6), 2174–2181 (2013)
14. Tobar, F.A., Mandic, D.P.: Multikernel least squares estimation. In: 3rd Conference on Sensor Signal Processing for Defence, pp. 1–5. IET, London (2012)
15. Tobar, F.A., Kung, S., Mandic, D.P.: Multikernel least mean square algorithm. IEEE Trans. Neural Netw. Learn. Syst. **25**(2), 265–277 (2014)
16. Xu, X., Hu, D., Lu, X.: Kernel-based least squares policy iteration for reinforcement learning. IEEE Trans. Neural Networks **18**(4), 973–992 (2007)

17. Jakab, H.S., Csató, L.: Novel feature selection and kernel-based value approximation method for reinforcement learning. In: Mladenov, V., Koprinkova-Hristova, P., Palm, G., Villa, A.E., Appollini, B., Kasabov, N. (eds.) ICANN 2013. LNCS, vol. 8131, pp. 170–177. Springer, Heidelberg (2013)
18. Kivinen, J., Smola, A.J., Williamson, R.C.: Online learning with kernels. In: Advances in Neural Information Processing Systems, vol. 14, pp. 785–792. MIT Press, Cambridge (2001)
19. Taylor, G., Parr, R.: Kernelized value function approximation for reinforcement learning. In: Proceedings of the 26th Annual International Conference on Machine Learning, pp. 1017–1024. ACM Press, New York (2009)
20. Schölkopf, B., Smola, A.J.: Learning with Kernels: Support Vector Machines, Regularization, Optimization, and Beyond. MIT Press, Cambridge (2002)
21. Hefferon, J.: Linear Algebra. Orthogonal Publishing L3c, Ann Arbor (2014)
22. Rasmussen, C.E., Williams, C.K.I.: Gaussian Processes for Machine Learning. MIT Press, Cambridge (2006)
23. Vaerenbergh, S.V., Vía, J., Santamaría, I.: A sliding-window kernel RLS algorithm and its application to nonlinear channel identification. In: 2006 IEEE International Conference on Acoustics Speech and Signal Processing, pp. 789–792. IEEE Press, New York (2006)
24. Platt, J.: A resource-allocating network for function interpolation. Neural Comput. 3(2), 213–225 (1991)
25. Lagoudakis, M.G., Parr, R.: Least-squares policy iteration. J. Mach. Learn. Res. 4, 1107–1149 (2003)

Hyperspectral Image Classification with Polynomial Laplacian Embedding

Peng Zhang[1(✉)], Chunbo Fan[1], Haixia He[2], and He Huang[2]

[1] Data Center, National Disaster Reduction Center of China,
Beijing, People's Republic of China
{zhangpeng, fanchunbo}@ndrcc.gov.cn
[2] Department of Satellite Remote Sensing,
National Disaster Reduction Center of China,
Beijing, People's Republic of China
{hehaixia, huanghe}@ndrcc.gov.cn

Abstract. Hyperspectral remote sensing has drawn great interests in earth observation, since contiguous spectrum can provide rich information of ground objects. However, such numerous bands also pose great challenges to efficient processing of hyperspectral image (HSI). Manifold learning, as a promising tool for nonlinear dimensionality reduction, has been widely used in feature extraction of HSI data to find meaningful representations of original spectrum. Nevertheless, lack of explicit and nonlinear mapping is still a limitation. In the paper, we propose a novel manifold learning based method for the classification of ground targets in HSI data, named as Polynomial Laplacian Embedding (PLAE). We first encode spatial information into spectral signal to obtain a fused representation. Then we model the local geometry of high-dimensional HSI data with graph Laplacian, and we introduce a nonlinear and explicit polynomial mapping to find a compact low-dimensional feature space, in which efficient classification of ground targets can be achieved. Experiments conducted on benchmark data sets demonstrate that high classification accuracy can be obtained by using the features extracted by PLAE.

Keywords: Supervised learning · Manifold learning · Feature extraction · Hyperspectral image classification · Graph laplacian

1 Introduction

In the past decades, the advances in spectral sensors have greatly promoted the spatial and spectral resolution of remote sensing images, by which large number of contiguous and narrow bands can be captured within a wide range of spectrum. As a consequence, such rich spectral information enables hyperspectral image (HSI) more capable to classify ground covers [1, 2] in various fields, for example, agriculture [3], mineralogy [4], and military [5], to name just a few.

Compared with single-spectral and multi-spectral images, each pixel in HSI is composed of tens or hundreds of bands recording spectral reflections of ground-covers, which form a high-dimensional vector in the spectral space. Such high dimensionality would exponentially increase computational complexity in classifying targets in HSI

© Springer International Publishing Switzerland 2016
D.-S. Huang et al. (Eds.): ICIC 2016, Part III, LNAI 9773, pp. 218–228, 2016.
DOI: 10.1007/978-3-319-42297-8_21

data. Such issues are also called "curse of dimensionality" or Hughes phenomenon [6]. Moreover, the characteristic spectrum of specific targets concentrates on a very small range of bands, while most bands in HSI are redundant. Such redundancy results in complex data structure, which adds up to the difficulties of designing an efficient classifier.

Recent investigations revealed that high-dimensional and contiguous spectral bands often concentrate on a low-dimensional manifold, which is nonlinearly embedded in the ambient spectral space [9–11]. Motived by such observation, manifold learning (ML) [18, 19], as a promising tool for nonlinear dimensionality reduction, has been successfully applied to finding low-dimensional yet meaningful representations of high-dimensional hyperspectral data, which were used in feature extraction [12, 17], segmentation [13], anomaly detection [14], and classification [15, 16].

Among various manifold learning methods, locality preserving projections (LPP) [21, 22] has drawn great research interests due to its intuitive motivation and simple implementation [7, 8, 23, 24]. Nevertheless, there are two issues with existent approaches. One is the lack of consideration of natural spatial correlations among pixels, and the other is the limitation of linear mapping relationship, which losses valuable nonlinear information encoded in original data.

To address the above issues, in this paper we propose a novel manifold learning based method for the classification of ground targets in HSI data, named as Polynomial Laplacian Embedding (PLAE). We propose an intuitive strategy to simultaneously consider spatial and spectral information, which is simple to implement. We use graph laplacian to model the local geometry of high-dimensional HSI data in a supervised mode. We propose a nonlinear yet explicit mapping to learn a compact embedding, which can optimally preserve the inter-class and intra-class topologies. Then fast feature extraction of unlabeled pixels in the scene can be achieved and the extracted features are useful for classification task. We conduct experiments on benchmark data sets to validate the performance of PLAE, and classification results demonstrate that PLAE has superior performance and achieves high accuracy even with a simple one-nearest-neighbor classifier.

The rest parts of this paper are organized as follows. Section 2 describes the proposed PLAE method in details. Section 3 demonstrates experimental results conducted on benchmark data sets. Section 4 gives some concluding remarks.

2 Feature Extraction with Polynomial Laplacian Embedding

2.1 Encoding Spatial-Spectral Information

A hyperspectral image (HSI) can be viewed as a data cube. At each pixel there is a high-dimensional vector recording the reflectance value within corresponding band. Let $(p(r, c))$ be the pixel at the r-th row and c-th column of a hyperspectral image, which consists of B discrete values of spectra. We take a simple yet intuitive strategy to encode both pixel coordination $c = (r, c)$ and array of bands $b = (b_1, b_1, \cdots, b_B)$. Formally, we use vector

$$x(r,c) = (c,b)^T = (r,c,b_1,b_1,\cdots,b_B)^T = \left(x^1,x^2,\cdots,x^{B+2}\right)^T, \tag{1}$$

as an augmented representation for pixel $p(r,c)$. By doing so, the natural spatial correlations among pixels' coordinates and contiguous spectrum can be simultaneously considered in feature extraction.

2.2 Modeling Local Topology in Supervised Mode

For all labeled pixels, we reformulate $x(r,c)$'s as a linear sequence by ranking along columns and denote the i-th labeled pixel by x_i. Here we use subscripts for indices of pixels (or samples) and superscripts for components of a vector.

For each x_i, we first identify its local neighborhood $\mathcal{N}_i$ in a supervised mode. We define $x_j \in \mathcal{N}_i$ if and only if x_j is among x_i's k nearest neighbors and they are in the same class, in other words, each sample's neighborhood is restricted within the class it belongs to.

Next, we model the local topology of x_i in the ambient space by heat weights. For any x_i and x_j, their pairwise similarity is defined as

$$S_{ij} = \begin{cases} \exp\left(- \parallel x_i - x_j \parallel_2^2 /t\right) & \text{if } x_i \in \mathcal{N}_j \text{ or } x_j \in \mathcal{N}_i \\ 0 & \text{otherwise} \end{cases}. \tag{2}$$

The weights $\{S_{ij}\}$ encode the discrete local Laplacian-Beltrami operator defined on the data manifold, and they character the topology of HSI data by emphasizing nearby neighbors.

2.3 Manifold Learning by Polynomial Laplacian Embedding

The local topology, represented by similarity weights $\{S_{ij}\}$, codes intra-class similarity and inter-class separability. Then the aim of dimensionality reduction is to find a low-dimensional feature space which best preserves such weights. Let y_i and y_j be the learned low-dimensional embedding of x_i and x_j, respectively. We request that each y_i and y_j are still close to each other if x_i and x_j are being so. This is achieved by solving the following optimization problem

$$\begin{aligned} \min \quad & \sum_i \left\| y_i - y_j \right\|_2^2 S_{ij}, \\ \text{s.t.} \quad & \sum_i y_i y_i^T = I_m \end{aligned} \tag{3}$$

where m is the dimension of the feature space and I_m is an identity matrix of order m. Let $Y = [y_1\ y_2\ \cdots\ y_N]$ and Y_i be the i-th row vector of Y. By straightforward algebra deduction, (3) can be transformed into the following optimization problem

$$\begin{aligned} \min \quad & \text{tr}(YLY^T) \\ \text{s.t.} \quad & YY^T = I_m \end{aligned}, \tag{4}$$

where $L = D - S$ and D is a diagonal matrix with $D_{ii} = \sum_j S_{ij}$. By Rayleigh-Reitz theorem, Y_i's $(i = 1, 2, \cdots, m)$ are the eigenvectors of L corresponding to the second to the $(m + 1)$-st smallest eigenvalues.

Traditional manifold learning method [27] learns low-dimensional Y defined in previous subsection implicitly. No explicit mapping relationship from x_i to y_i can be obtained after training. Locality preserving projection (LPP) [21, 22] method introduces a linear mapping to address such issue, where it is assumed that $y_i = U^T x_i$ with U being a linear projection. However, the nonlinear geometric structure is no longer preserved as a compromise.

In our previous work [20], we build up an explicit and nonlinear mapping for locally linear embedding (LLE) [19], which can provide accurate learning results. Now we apply such mapping to learn the aforementioned supervised model of spatial-spectral information and local Laplacian.

Formally, we use a polynomial model to approximate the nonlinear mapping from x_i to y_i. We assume that the k-th component of y_i is a p-th polynomial of x_i, that is,

$$y_i^k = \sum_{j=1}^{B+2} \sum_{l=1}^{p} c_{kjl} (x_i^j)^l , \tag{5}$$

where c_{kjl} is the polynomial coefficient. Let c_k be the $p(B+2)$-dimensional column vector formed by c_{kjl}'s and

$$X_p^{(i)} = \begin{pmatrix} \overbrace{x_i \odot x_i \odot \cdots \odot x_i}^{p} \\ \vdots \\ x_i \end{pmatrix}, \tag{6}$$

where $\odot$ refers to entrywise matrix multiplication. Then (5) can be rewritten as

$$y_i = (c_1^T X_p^{(i)}, c_2^T X_p^{(i)}, \cdots, c_m^T X_p^{(i)})^T. \tag{7}$$

By writing $X_p = [X_p^{(1)} \ X_p^{(2)} \ \cdots \ X_p^{(N)}]$ and substituting (7) into (4), we have the following optimization problem

$$\begin{aligned} \min \quad & \sum_k c_k^T X_p M X_p^T c_k \\ \text{s.t.} \quad & c_i^T X_p X_p^T c_j = \delta_{ij}, \end{aligned} \tag{8}$$

where δ_{ij} equals to one if $i = j$ and zero otherwise. By Rayleigh–Ritz theorem, optimal solutions c_k $(k = 1, 2, \cdots, m)$ are the eigenvectors of the following generalized eigenvalue problem corresponding to the m smallest eigenvalues:

$$X_p M X_p^T c_k = \lambda X_p X_p^T c_k$$

For an unlabeled sample, its low-dimensional feature can be easily computed with Eqs. (6) and (7).

3   Experimental Results

In this section, we test the performance of the proposed PLAE method on three HSI data sets, that are indinepine, ksc, and botswana. These three data sets are commonly used as benchmarks in the literature. The Indine Pines (indinepine) data set is from the AVIRIS (Airborne Visible/Infrared Imaging Spectrometer) built by JPL and flown by NASA/Ames [25]. The scene is over an area 6 miles west of West Lafayette and covers a 2 × 2 mile portion of Northwest Tippecanoe County, Indiana. The Kennedy Space Center (ksc) data set is also from AVIRIS over the Kennedy Space Center (KSC), Florida. The botswana data set is from NASA EO-1 satellite over a 7.7 km strip in the Okavango Delta, Botswana. An overall description of these three data sets is given in Table 1, and graphical illustrations of selected bands as well as ground truth are shown in Fig. 1.

Table 1. Description of the data sets used in experiments.

Data set	Resolution	Number of bands	Labelled pixels	Number of classes
indinepine	145*145	220	10366	16
ksc	1476*256	145	3284	14
botswana	512*614	176	5211	13

Before learning, the components of all pixels in the image are normalized by the following principle

$$x_i^j \leftarrow (x_i^j - \min_i x_i^j)/(\max_i x_i^j - \min_i x_i^j).$$

We apply PCA, LPP, orthogonal LPP (OLPP) [26], and PLAE to classification task on these data. For LPP and OLPP, local neighborhoods are constructed using the supervised approach stated in Sect. 2.2, which is the same to PLAE. Trivial one-nearest-neighbor classifier is implemented in the dimension reduced feature space.

There are three parameters, which need to be tuned, namely, t in the heat weights, ratio r of preserved principal components for PCA, and dimension d of embedding. t is computed using training data as

$$t = \frac{1}{n^2} \sum_{i,j} \|x_i - x_j\|_2$$

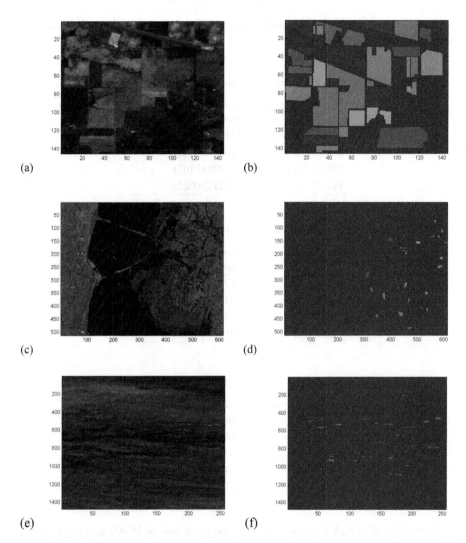

Fig. 1. Gray scale illustration of the imagery and color illustration of ground truth. (a) and (b) indinepine with band 15. (c) and (d) ksc with band 15. (e) and (f) Botswana with band 50.

r and d are tuned within a range. r varies from 0.90 to 0.98 with step length 0.01, and d varies from 3 to 27 with step length 3. For each parameter value, we use a 50-50 percent random split of labeled data to generate training and testing data. Then they are used in the classification task for the aforementioned four methods, and such random test is repeated 20 times. In each test, error rates of these methods, that are, the portions of mis-classified samples in all testing samples, are recorded.

The plots of error rate (mean ± standard deviation) versus parameter for all four methods are demonstrated in Figs. 2, 3, and 4. In each figure, the ticks on the horizontal axis are the ratios for PCA and dimensions for LPP, OLPP, and PLAE. Lowest mean error rates together with corresponding parameter values for each method are summarized in Tables 2, 3, and 4.

Table 2. Best classification results of all four methods on indinepine data set.

Method	Parameter	Lowest mean error rate (standard deviation)
PCA	$r = 0.92$	0.3157(0.0071)
LPP	$d = 12$	0.1037(0.0045)
OLPP	$d = 27$	0.3927(0.0107)
PLAE	$d = 18$	**0.0221**(0.0022)

Table 3. Best classification results of all four methods on ksc data set.

Method	Parameter	Lowest mean error rate (standard deviation)
PCA	$r = 0.98$	0.1764(0.0329)
LPP	$d = 27$	0.0487(0.0048)
OLPP	$d = 27$	0.4685(0.0166)
PLAE	$d = 21$	**0.0141**(0.0017)

Table 4. Best classification results of all four methods on botswana data set.

Method	Parameter	Lowest mean error rate (standard deviation)
PCA	$r = 0.98$	0.1816(0.0440)
LPP	$d = 27$	0.0440(0.0042)
OLPP	$d = 27$	0.3408(0.0263)
PLAE	$d = 18$	**0.0122**(0.0035)

From Figs. 2, 3 and 4, we can see that the error rate of PLAE quickly decreases while d increases and gets to optimal classification rate. The performance of PLAE is quite stable with the dimension of learned embedding after $d = 6$ and consistently outperforms all its counterparts. LPP has similar tendency with PLAE but higher error rates than PLAE. The performance of PCA and OLPP are not satisfactory in these three tests. Their errors rates are much higher than those of LPP and PLAE. Specifically, the orthogonal constraints on LPP do not improve its separability. These observations are also validated in Tables 2, 3, and 4. We can see that PLAE achieves the lowest error rates on all three data sets, which are much lower than the other three methods. This demonstrates that by fusing spatial-spectral information and using nonlinear mapping, the feature space learned by PLAE owns better separating ability for hyperspectral image classification.

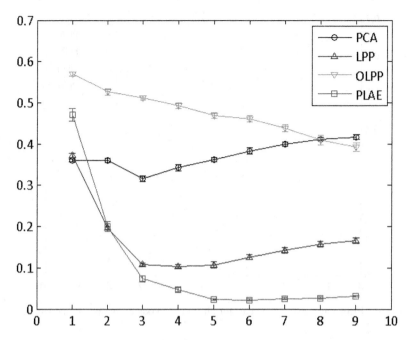

Fig. 2. Error rate versus parameter for all four methods on the indinepine data set.

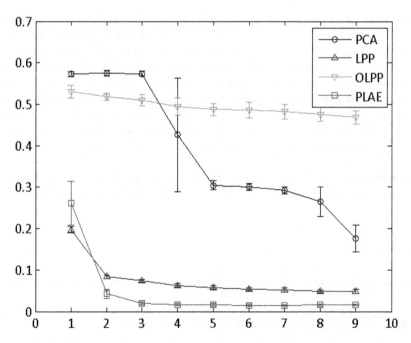

Fig. 3. Error rate versus parameter for all four methods on the ksc data set.

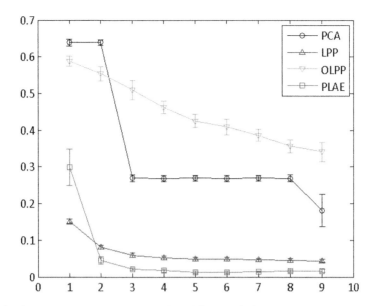

Fig. 4. Error rate versus parameter for all four methods on the botswana data set.

4   Conclusion

In this paper, we propose a supervised feature extraction method for hyperspectral image classification based on manifold learning. We use an intuitive strategy to fuse spatial and spectral information of ground objects, which is simple to implement. We model the geometry underlying HSI data with supervised local Laplacian. Furthermore, we introduce a nonlinear and explicit mapping to efficiently learn a compact low-dimensional feature space in which fast feature extraction and classification for testing samples can be achieved. Experiments conducted on a series of benchmark data sets show that the proposed method has high classification accuracy and superior performance over principle component analysis as well as its linear counterparts.

Acknowledgement. This work was supported by the National Natural Science Foundation (NNSF) of China under Grant nos. 41201552, 41174013, 41401605, and 41301590.

References

1. Zhang, L., Du, B., Zhong, Y.: Hybrid detectors based on selective endmembers. IEEE Trans. Geosci. Remote Sens. **48**(6), 2633–2646 (2010)
2. Du, B., Zhang, L.: Random selection based anomaly detector for hyperspectral imagery. IEEE Trans. Geosci. Remote Sens. **49**(5), 1578–1589 (2011)

3. Datt, B., McVicar, T., Van Niel, T., Jupp, D., Pearlman, J.: Preprocessing EO-1 hyperion hyperspectral data to support the application of agricultural indexes. IEEE Trans. Geosci. Remote Sens. **41**(6), 1246–1259 (2003)

4. Hörig, B., Kühn, F., Oschütz, F., Lehmann, F.: HyMap hyperspectral remote sensing to detect hydrocarbons. Int. J. Remote Sens. **22**(8), 1413–1422 (2001)

5. Eismann, M., Stocker, A., Nasrabadi, N.: Automated hyperspectral cueing for civilian search and rescue. Proc. IEEE **97**(6), 1031–1055 (2009)

6. Hughes, G.: On the mean accuracy of statistical pattern recognizers. IEEE Trans. Inf. Theor. **14**(1), 55–63 (1968)

7. Imani, M., Ghassemian, H.: Feature extraction using weighted training samples. IEEE Geosci. Remote Sens. Lett. **12**(7), 1387–1391 (2015)

8. Li, X., Pan, J., He, Y., Liu, C.: Bilateral filtering inspired locality preserving projections for hyperspectral images. Neurocomputing **164**, 300–306 (2015)

9. Bachmann, C., Ainsworth, T., Fusina, R.: Exploiting manifold geometry in hyperspectral imagery. IEEE Trans. Geosci. Remote Sens. **43**(3), 441–454 (2005)

10. Bachmann, C., Ainsworth, T., Fusina, R.: Improved manifold coordinate representations of large-scale hyperspectral scenes. IEEE Trans. Geosci. Remote Sens. **44**(10), 2786–2803 (2006)

11. Lunga, D., Prasad, S., Crawford, M., Ersoy, O.: Manifold-learning-based feature extraction for classification of hyperspectral data. IEEE Sig. Process. Mag. **1**, 55–66 (2014)

12. He, J., Zhang, L., Wang, Q., Li, Z.: Using diffusion geometric coordinates for hyperspectral imagery representation. IEEE Geosci. Remote Sens. Lett. **6**(4), 767–771 (2009)

13. Mohan, A., Sapiro, G., Bosch, E.: Spatially coherent nonlinear dimensionality reduction and segmentation of hyperspectral images. IEEE Geosci. Remote Sens. Lett. **4**(2), 206–210 (2007)

14. Ma, L., Crawford, M., Tian, J.: Anomaly detection for hyperspectral images based on robust locally linear embedding. J. Infrared Millimeter Terahertz Waves **31**(6), 753–762 (2010)

15. Crawford, M.M., Ma, L., Kim, W.: Exploring nonlinear manifold learning for classification of hyperspectral data. In: Prasad, S., Bruce, L.M., Chanussot, J. (eds.) Optical Remote Sensing. Augmented Vision and Reality, vol. 3, pp. 207–234. Springer, Heidelberg (2011)

16. Li, W., Prasad, S., Fowler, J., Bruce, L.: Locality preserving dimensionality reduction and classification for hyperspectral image analysis. IEEE Trans. Geosci. Remote Sens. **50**(4), 1185–1198 (2012)

17. Jia, X., Kuo, B., Crawford, M.: Feature mining for hyperspectral image classification. Proc. IEEE **101**(3), 676–697 (2013)

18. Tenenbaum, J.B., Silva, V., Langford, J.C.: A global geometric framework for nonline-ar dimensionality reduction. Science **290**(5500), 2319–2323 (2000)

19. Roweis, S.T., Saul, L.K.: Nonlinear dimensionality reduction by locally linear embed-ding. Science **290**(5500), 2323–2326 (2000)

20. Qiao, H., Zhang, P., Wang, D., Zhang, B.: An explicit and nonlinear mapping for manifold learning. IEEE Trans. Cybern. **43**(1), 51–63 (2013)

21. He, X., Niyogi, P.: Locality preserving projections. Adv. Neural Inf. Process. Syst. **16**, 153–160 (2003)

22. He, X., Yan, S., Hu, Y., Niyogi, P., Zhang, H.: Face recognition using Laplacianfaces. IEEE Trans. Pattern Anal. Mach. Intell. **27**(3), 328–340 (2005)

23. Cai, D., He, X., Han, J., Zhang, H.: Orthogonal Laplacianfaces for face recognition. IEEE Trans. Image Process. **15**(11), 3608–3614 (2006)

24. Hou, B., Zhang, X., Ye, Q., Zheng, Y.: A novel method for hyperspectral image classification based on Laplacian eigenmap pixels distribution-flow. IEEE J. Sel. Top. Appl. Earth Obs. Remote Sens. **6**(3), 1602–1618 (2013)
25. Zhou, L., Zhang, X.: Discriminative spatial-spectral manifold embedding for hyperspectral image classification. Remote Sens. Lett. **6**(9), 715–724 (2015)
26. Baumgardner, M., Biehl, L., Landgrebe, D.: 220 Band AVIRIS Hyperspectral Image Data Set: June 12, 1992 Indian Pine Test Site 3 Purdue University Research Repository (2015). doi:10.4231/R7RX991C
27. Belkin, M., Niyogi, P.: Laplacian eigenmaps for dimensionality reduction and data representation. Neural Comput. **15**(6), 1373–1396 (2003)

Improving Deep Learning Accuracy with Noisy Autoencoders Embedded Perturbative Layers

Lin Xia[1,2], Xiaolong Zhang[1,2(✉)], and Bo Li[1,2]

[1] School of Computer Science and Technology,
Wuhan University of Science and Technology,
Wuhan 430065, China
xiaolong.zhang@wust.edu.cn
[2] Intelligent Information Processing and Real-Time Industrial Systems
Hubei Province Key Laboratory, Wuhan 430065, China
iamxialin@hotmail.com, liberol@126.com

Abstract. Autoencoder has been successfully used as an unsupervised learning framework to learn some useful representations in deep learning tasks. Based on it, a wide variety of regularization techniques have been proposed such as early stopping, weight decay and contraction. This paper presents a new training principle for autoencoder based on denoising autoencoder and dropout training method. We extend denoising autoencoder by both partial corruption of the input pattern and adding noise to its hidden units. This kind of noisy autoencoder can be stacked to initialize deep learning architectures. Moreover, we show that in the full noisy network the activations of hidden units are sparser. Furthermore, the method significantly improves learning accuracy when conducting classification experiments on benchmark data sets.

Keywords: Deep learning · Unsupervised pre-training · Autoencoder · Noise · Dropout

1 Introduction

Training deep neural networks with gradient-based optimization is worthy of serious research in deep learning. When the training is starting from random initialization, it gets stuck in undesired local optima easily [1], which prevents the lower layers from learning useful features. This problem can be partially settled by initializing weights in a region near a good local minimum with the greedy layer-wise unsupervised training strategy [2, 3]. Autoencoder is one of these unsupervised learning techniques. Autoencoders can be stacked one beside the other to initialize deep architectures [4].

Autoencoder can improve learning accuracy with regularization, which can be a sparsity regularizer, either a contractive regularizer [5], or a denoising form of regularization [6]. Recent work [7] has shown that regularization can be used to prevent feature co-adaptation by dropout training. The key idea of dropout is to drop units from a neural network by setting a randomly selected subset of activations to zero during training. Dropout can also be interpreted as a way of regularizing a neural network by adding noise to its hidden units. This idea has previously been used in the context of

© Springer International Publishing Switzerland 2016
D.-S. Huang et al. (Eds.): ICIC 2016, Part III, LNAI 9773, pp. 229–238, 2016.
DOI: 10.1007/978-3-319-42297-8_22

denoising autoencoder where noise is added to the inputs of an autoencoder and the target is noise-free. The work of [8] proves that the deletion of hidden units is an effective way to prevent deep architectures from overfitting, which prevents the units from co-adaption too much.

Motivated by noise injection at the input layer in autoencoder, and dropout training strategy, this paper presents a novel autoencoder that takes advantage of denoising autoencoder and dropout regularization by both partial corruption of its input pattern and adding noise to its hidden units. The proposed algorithm effectively learn representations that help training better classifiers. The experimental results demonstrate that the algorithm significantly improved learning accuracy.

2   Various Kinds of Autoencoders

The autoencoder learning algorithm has the ability to automatically learn features from unlabeled data, which was introduced as a technique for dimensionality reduction in the late 80's. It can be used as the building block of deep neural network [9]. There are many sophisticated versions of autoencoder and in many cases they are competitive to or even superior to the best deep learning algorithm.

2.1   Basic Autoencoder

The basic autoencoder (Fig. 1) is a one layer neural network that is trained to reconstruct its inputs. Hence, its hidden layer is forced to learn a structure from the input training examples.

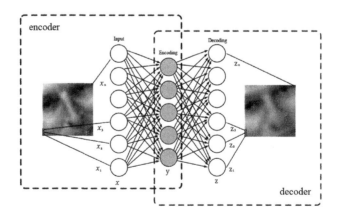

Fig. 1.   Standard autoencoder

Autoencoder is composed of two parts, including an encoder and a decoder. The encoder part can be considered as a deterministic mapping $f_\theta(x) = s(Wx + b)$ that transforms an input vector $x \in [0, 1]^{d_x}$ into a hidden representation $y \in [0, 1]^{d_h}$, where s

is a nonlinear activation function such as a sigmoid function. Its parameter set is described as $\{W, b\}$, where W is a $d_h \times d_x$ weight matrix and b is a bias vector of dimensionality d_h. The decoder function $g_{\theta'}(y) = s(W'y + b')$ maps hidden representation y back to a reconstruction $z \in [0, 1]^{d_x}$ with $\theta' = \{W', b'\}$. In case of tied weight, weight matrix W' will be constrained by $W' = W^T$.

With a set of unlabeled training examples $\{x^{(1)}, x^{(2)}, x^{(3)}, \cdots, x^{(n)}\}$, we compute activations at hidden layer to get a corresponding $y^{(i)}$ for each input $x^{(i)}$, where $x^{(i)} \in R^{d_x}$. We also obtain a reconstruction $z^{(i)}$ at the output layer. The deviation of z from input x can be described as:

$$\mathcal{J}_{AE}\left(\theta, \theta'\right) = \frac{1}{n}\sum_{i=1}^{n} L\left(x^{(i)}, g\left(f\left(x^{(i)}\right)\right)\right) \tag{1}$$

where L is a loss function, usually defined to the traditional squared error $L(x, z) = \|x - z\|^2$. If the activation function is sigmoid and input is interpreted as either bit vectors or vectors of bit probabilities, cross-entropy of the reconstruction [10] can be used as following:

$$L(x, z) = -\sum_{k=1}^{d_x}[x_k \log z_k + (1 - x_k) \log(1 - z_k)] \tag{2}$$

2.2 Denoising Autoencoder

Denoising autoencoder is an extension of the classical autoencoder. It was proposed by Vincent et al. [11]. Its basic idea is to force the hidden layer to discover more robust features and prevent it from simply learning the identity by training the auto-encoder to reconstruct the input data with a corrupted version of it. This yields the following function:

$$\mathcal{J}_{DAE}\left(\theta, \theta'\right) = \frac{1}{n}\sum_{i=1}^{n} \mathbb{E}_{q(x^{(i)}, \tilde{x}^{(i)})} L\left(x^{(i)}, g\left(f\left(\tilde{x}^{(i)}\right)\right)\right) \tag{3}$$

The corrupted input $\tilde{x}$ can be simply gotten by randomly disable some input data or obtained from a corruption process $q(x, \tilde{x})$. For example, if input data is an image, we randomly set some pixels zero and then send it through the autoencoder, to train the encoder and decoder by reconstructing the original image (without noise). We can also consider isotropic Gaussian input noise: $\tilde{x}|x \sim \mathcal{N}(x, \sigma^2 I)$ and control the degree of the corruption by σ.

2.3 Contractive Autoencoder

Contractive autoencoder reconstructs input with an additional constraint on the latent space to improve robustness of the representation. There is a contraction by adding an additional penalization term to the reconstruction cost, which is the Frobenius norm of Jacobian of the hidden layer representation with respect to input values:

$$\left\| J_f(x) \right\|_F^2 = \sum_{ij} \left(\frac{\partial h_j(x)}{\partial x_i} \right)^2 \tag{4}$$

Now the objective function can be described:

$$\mathcal{J}_{DAE}\left(\theta, \theta'\right) = \frac{1}{n} \sum_{i=1}^{n} \left(L\left(x^{(i)}, g\left(f\left(x^{(i)}\right)\right)\right) + \lambda \left\| J_f\left(x^{(i)}\right) \right\|_F^2 \right) \tag{5}$$

3 Deep Learning Framework with Noisy Autoencoder

This paper proposes a deep learning framework with noisy autoencoder, which is extended from a standard denoising autoencoder by embedding a corruption layer between the encoding layer and the decoding layer. One motivation of denoising autoencoder (Fig. 2A) is to increase the robustness of input perturbations. Unlike denoising autoencoder, contractive autoencoder explicitly encourage the robustness of representation. According to the experimental results in [5], it seems that robustness of the extracted features is more important than robustness of the reconstruction. In order to increase the robustness of representation of the hidden units we try to inject different kinds of additional noise (Gaussian noise and dropout noise) at hidden units (Fig. 2B).

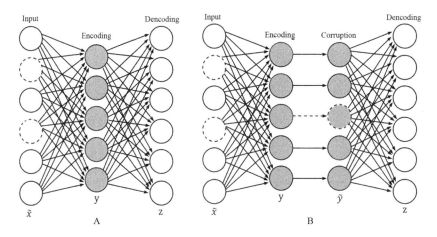

Fig. 2. (A) Standard denoising autoencoder; (B) Noisy autoencoder

The noisy model in the encoder part is the same as that in the denoising autoencoder. However, there exists difference with dropout noise which means temporarily removing some units from the hidden layer, then the input of decoder part becomes the following:

$$\tilde{y} = \frac{\delta * y}{p}, \delta \in R^{d_h} \text{ and } \delta \sim Bernoulli(p) \tag{6}$$

where δ is a vector of Bernoulli selector random variables which are assumed to be independent of each other so the dropping units can be randomly selected. Each unit is retained with a fixed probability $P(\delta_i = 1) = p$, which is independent to other units thus $P(\delta_i = 0) = 1 - p = q$. With probability q = 0.5 of deletion, an autoencoder with n units can be considered as a collection of 2^n possible thinned autoencoders and these networks all share the same weights. In this method dropout noise divides $\delta * y$ by p so output is close to the arithmetic mean of output produced by all the possible sub-autoencoders.

As for Gaussian noise, it can be stated below:

$$\tilde{y} = y + \epsilon, \quad \epsilon \sim N(0, \sigma^2 I) \tag{7}$$

then we use decoder function to maps the corrupted hidden representation back to inputs

$$z = g(\tilde{y}) = s\left(W'\tilde{y} + b'\right) \tag{8}$$

The parameters throughout the deep learning framework then can be updated via stochastic gradient descent strategy by back propagating gradients of the loss function with respect to the parameters.

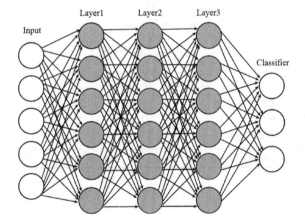

Fig. 3. Stacked noisy autoencoders

A greedy layer-wise training algorithm was proposed to train a deep belief nets one layer at a time. It also can be applied when using an autoencoder instead of the restricted Boltzmann machine as a layer building block of a deep neural network. After unsupervised pre-training of the layers of a stacked noisy autoencoders as a last training stage, it is possible to optimize the parameters of all the layers together with respect to a supervised training criterion in order to minimizing the classification error. The parameter optimization treats all the layers of the stacked noisy autoencoders as a single model (Fig. 3), which fine-tunes all the weights in the deep learning framework at every iteration. The details of how to train the stacked noisy autoencoders (SNAE) are presented as follows:

Inputs: training set $\{x^{(1)}, x^{(2)}, x^{(3)}, \cdots, x^{(n)}\}$ as input data

---

Steps :

For mini-batch data in input-data

    Data=mini-batch data

    1. Using unsupervised Greedy layer-wise pre-training Strategy to get the initial weight matrix W= $\{W^1, W^2, W^3\}$ and the bias vector b= $\{b^1, b^2, b^3\}$. First, Train the lowermost noisy autoencoder.

    2. After training, remove the decoder part of the bottommost noisy autoencoder, construct a new noisy autoencoder by taking the latent representation of the previous noisy autoencoder as input.

    3. Train the new noisy autoencoder. Note the weights and bias of the encoder from the previously trained noisy autoencoders are fixed when training the newly constructed noisy autoencoder.

    4. Repeat step 2 and 3 until enough layers are pre-trained.

    5. After pre-training, do globally supervised fine-tuning. Update the SNAE's parameter

        For i=1 to hidden layer ($l-1$)' numb

           For j=1 to hidden layer l' numb

$$\Delta w_{ij}^{(l)} = -\gamma o^{(l-1)} \delta_j^{(l)}$$

$$\Delta b_i^{(l)} = -\gamma \delta_j^{(l)}$$

        End for j;

      End for i;

    End for mini-batch data

Outputs: The model parameter of SNAE $\{W^1, W^2, W^3\}$ and $\{b^1, b^2, b^3\}$

---

4   Experimental Results

The proposed deep learning method is evaluated with the benchmark image sets. One is MNIST, which is well-known for digit classification. It consists of two datasets, one for training (60,000 images) and one for testing (10,000 images). The training set is divided into two sets consisting of 50,000 images for training and 10,000 for validation. The same benchmark as Minmin Chen et al. [12] is used in our experiment, which includes five harder digit recognition problems derived by adding extra factors of variation to MNIST digits, each with 10000 examples for training, 2000 for validation and 50000 for test. All the models were applied on the training set without using the labels to extract the first layer of features then fine-tuned by a gradient descent on a supervised objective appropriate for classification, using the labels in the training set.

To better evaluate the impact of hidden activation noise, we trained the same network which includes 1000 hidden units with a sigmoidal encoder, a linear decoder and squared error loss. We fix the input noise to be Gaussian noise with the level of 0.1. Figure 4 shows the influence of the level of corruption using a fine-grained grids. It also shows that noisy autoencoder performs better than denoising autoencoder for a rather wide range of noise levels. Noisy autoencoder with dropout noise in hidden layer get the best performance at the level of 0.5. Gaussian hidden noise also improves performance. However, as the noise level increase the performance become worse.

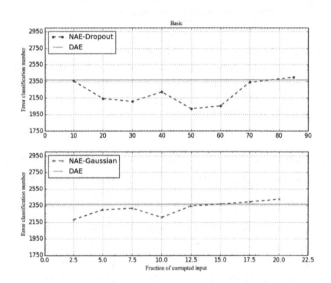

Fig. 4. The effectivness of the level of corruption

To better understand why hidden activation noise can increase the performance we randomly selected 100 images from the test-set and looked at the sparsity of hidden unit activations for the models trained previously. If the output value of a neuron is close to 1 we think of it as being active, or as being inactive if its value close to 0. We would like to constrain the neurons to be inactive by using some regularizers such as KL divergence.

When representations are learnt in a way that encourages sparsity, improved performance is obtained on classification tasks. Comparing to the histograms of activations we can see that the activations of the hidden units become sparse (Fig. 5C) when dropout noise is used without any sparsity inducing regularizer. That is to say, dropout automatically leads to sparse representations.

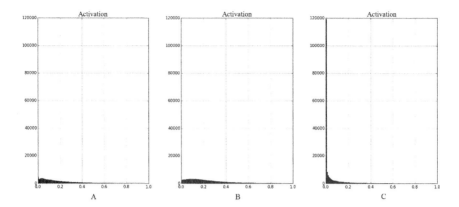

Fig. 5 (A) The activations of denoising autoencoder; (B) The effectiveness of Gaussian noise on activations; (C) The effectiveness of dropout noise on activations

To evaluate the performance we trained 3 hidden layers neural network pretrained with the five different algorithms. Stacked regular autoencoders (SAE-3), stacked restricted Boltzmann machines corresponds to deep belief networks (DBN-3), stacked denoising autoencoders (SDAE-3), stacked contractive autoencoders (SCAE-3) and stacked noisy autoencoders (SNAE-3).

Table 1. Classification error rate on the test set of MNIST for different models

Dataset	SAE-3	DBN-3	SDAE-b-3	SCAE-3	SNAE-3
MNIST	1.4	1.24	1.28	1.24	1.13
basic	3.46	3.11	2.84	2.84	2.56
rot	10.3	10.30	9.53	9.36	9.39
bg-rand	11.28	6.73	10.3	12.25	10.86
bg-img	23.00	16.31	16.68	17.01	16.23
bg-img-rot	51.93	47.39	43.76	47.50	43.35

All autoencoder variants used tied weights that all starting from the same initial random point in weight space. We consider autoencoders with sigmoid activation function for both encoder and decoder, and a cross entropy reconstruction error. They were trained by optimizing their regularized objective function on the training set by stochastic gradient descent. The hyper-parameters for these different algorithms are chosen on the validation set. As for Deep Belief Networks, they were trained by Contrastive Divergence.

From Table 1 we know that Stacked Noisy Autoencoders (10 % dropout hidden noise for both first and second hidden layer, 50 % dropout hidden noise for the third layer) acquires the best learning accuracy among the 5 algorithms on the 6 data sets. On the MNIST dataset we achieve the best test error rate of 1.13 %, perform better than Deep Belief Networks and Stacked Denoising Autoencoders in [4], as well as Stacked

Contractive Autoencoders in [13]. On the other 5 data sets Stacked Noisy Autoencoders appears to achieve performance superior or equivalent to the best other model. This shows clearly that noisy autoencoder pre-training with input noise and hidden activation noise is a better strategy than pre-training with regular Denoise Autoencoder.

5 Conclusion

This papers presents a deep learning algorithm. The learning accuracy can be improved by injecting noise to input layer and hidden layer of autoencoder. We study the influence of different kinds of noise and the level of corruption on the performance of network in details. Our results suggest that dropout noise which is usually used in supervised learning can also be effective used in unsupervised learning. The experiment results demonstrate that the algorithm proposed in this paper performs better than autoencoder and its variations on a range of classification tasks.

Acknowledgements. This work was supported in part by National Natural Science Foundation of China (61273225, 61273303, 61373109), Program for Outstanding Young Science and Technology Innovation Teams in Higher Education Institutions of Hubei Province (No. T201202), as well as National "Twelfth Five-Year" Plan for Science & Technology Support (2012BAC22B01).

References

1. Glorot, X., Bengio, Y.: Understanding the difficulty of training deep feedforward neural networks. In: Proceedings of the International Conference on Artificial Intelligence and Statistics, pp. 249–256 (2010)
2. Erhan, D., Bengio, Y., Courville, A., et al.: Why does unsupervised pre-training help deep learning. J. Mach. Learn. Res. **11**, 625–660 (2010)
3. Bengio, Y.: Learning deep architectures for AI. Found. Trends Mach. Learn. **2**, 1–127 (2009)
4. Vincent, P., Larochelle, H., Lajoie, I., Bengio, Y., Manzagol, P.-A.: Stacked denoising autoencoders: learning useful representations in a deep network with a local denoising criterion. J. Mach. Learn. Res. **11**, 3371–3408 (2010)
5. Rifai, S., Vincent, P., Muller, X., Glorot, X., Bengio,Y.: Contractive auto-encoders: explicit invariance during feature extraction. In: Proceedings of the 28th International Conference on Machine Learning, pp. 833–840 (2011)
6. Chen, M., Xu, Z., Weinberger, Z., Sha, F.: Marginalized denoising autoencoders for domain adaptation. In: Langford, J., Pineau, J. (eds.) Proceedings of the 29th International Conference on Machine Learning, pp. 767–774 (2012)
7. Hinton, G.E., Srivastava, N., Krizhevsky, A., Sutskever, I., Salakhutdinov, R.R.: Improving neural networks by preventing co-adaptation of feature detectors. arXiv:1207.0580 (2012)
8. Srivastava, N., Hinton, G., Krizhevsky, A., Sutskever, I., Salakhutdinov, R.: Dropout: a simple way to prevent neural networks from overfitting. J. Mach. Learn. Res. **15**, 1929–1958 (2014)

9. Bengio, Y., Courville, A., Vincent, P.: Unsupervised feature learning and deep learning: a review and new perspectives. arXiv:1206.5538 (2012)
10. Bengio, Y.: Deep Learning of Representations: Looking Forward. arXiv:1305.0445 (2013)
11. Vincent, P., Larochelle, H., Bengio, Y., Manzagol, P.A.: Extracting and composing robust features with denoising autoencoders. In: Proceedings of the 25th International Conference on Machine Learning, pp. 1096–1103 (2008)
12. Chen, M., Weinberger, K., Sha, F., Bengio, Y.: Marginalized denoising autoencoders for nonlinear representation. In: Proceedings of the 31th International Conference on Machine Learning, pp. 1476–1484 (2014)
13. Zhou, Y., Arpit, D., Nwogu, I., Govindaraju, V.: Is Joint Training Better for Deep Auto-Encoders? arXiv:1405.1380 (2015)

Two Approaches on Accelerating Bayesian Two Action Learning Automata

Hao Ge[1], Haiyu Huang[2], Yulin Li[3], Shenghong Li[1(✉)], and Jianhua Li[1]

[1] Shanghai Jiao Tong University, Shanghai, China
{sjtu_gehao, shli, lijh888}@sjtu.edu.cn
[2] The 28th Research Institute of China Electronic Technology
Group Corporation, Nanjing, China
abel89@126.com
[3] Dalian No. 24 High School, Dalian, China
yulinlin981@hotmail.com

Abstract. Bayesian Learning Automata (BLA) are demonstrated to be as efficient as the state-of-the-art automaton in two action environments, and it has parameter-free property. However, BLA need the explicit computation of a beta inequality, which is time-consuming, to judge its convergence.

In this paper, the running time of BLA is concerned and two approaches are proposed to accelerate the computation of the beta inequality. One takes advantage of recurrence relation of the beta inequality, the other uses a normal distributions to approximate the beta distributions. Numeric simulation are performed to verify the effectiveness and efficiency of those two approaches. The results shows these two approaches reduce the running time substantially.

Keywords: Bayesian Learning Automata · Recurrence relation · Moment matching · Normal approximation

1 Introduction

Learning Automata (LA) are simple self-adaptive decision units that were firstly investigated to mimic the learning behavior of natural organism [1]. The pioneer work can be traced back to 1960s by the Soviet scholar Tsetlin [2, 3]. Since then, LA has been extensively explored and it is still under investigation as well in methodological aspects [4–9] as in concrete applications [10–17]. One intriguing property that popularize the learning automata based approaches in engineering is that LA can learn the stochastic characteristics of the external environment it interacts with, and maximizing the long term reward it obtains through interacting with the environment.

On the development of learning automata, accuracy and convergence rate becomes two major measurement to evaluate the goodness of an algorithm. The former is defined as the probability of a correct convergence and the latter is defined as the average iterations for a learning automaton to get converged.

Most of the reported schemes in the field of LA has two or more tunable parameters, making themselves capable of adapting to different environments. The performance of

© Springer International Publishing Switzerland 2016
D.-S. Huang et al. (Eds.): ICIC 2016, Part III, LNAI 9773, pp. 239–247, 2016.
DOI: 10.1007/978-3-319-42297-8_23

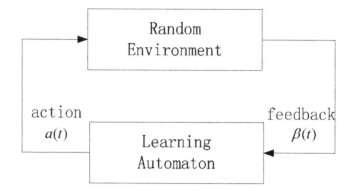

Fig. 1. Feedback connection of automaton and environment.

an automaton is highly dependent on the selection of those parameters. Generally, according to the ε-optimality property of LA, the probability of converging to the optimal action can be arbitrarily close to one. More specifically, as long as the learning resolution is large enough, we can obtain the correct result with an arbitrary accuracy.

Parameter tuning can be a complicated and time consuming procedure, and learning in two action stationary environments is the most fundamental scenario. In order to explore new mechanism to reduce the difficulty in parameter tuning, we took the first step by proposing a parameter-free Bayesian learning automata in two action environment [18]. The proposed BLA achieves the same accuracy and convergence rate as the sate-of-art automaton, besides, it holds the parameter-free property.

The parameter-free property is defined as a property that the performance of a automaton are independent of the selection of parameters, or a set of selected parameters can be universally applied to all environments. This implies the procedure can be omitted without scarifying accuracy and convergence rate. However, BLA also suffers from a heavy computational burden due to the explicit computation of a beta inequality.

In this paper, we propose two approaches to reduce the computational burden. One takes advantage of recurrence relation of the beta inequality, the other uses a normal distributions to approximate the beta distributions. Then we will verify the two approaches via simulations and analyze the simulation results.

The rest of this paper organized as follows. Section 2 introduces the BLA briefly, points out its drawback and advocates two possible ways to overcome it. Extensive simulations are given in Sects. 3 and 4 concludes the paper.

2 Parameter-Free Two Action Learning Automaton

First, we shall introduce the algorithm we proposed in [18] briefly.

Algorithm Bayesian Learning Automaton

1: Initial $\alpha_1 = 1, \beta_1 = 1, \alpha_2 = 1, \beta_2 = 1$
2: **Repeat**
3: Compute the probability $P_1 = Pr(e_1 > e_2)$, where e_1 and e_2 are random variables that follow distribution $Beta(\alpha_1, \beta_1)$ and $Beta(\alpha_2, \beta_2)$ respectively, and $P_2 = 1 - P_1$.
4: Choose an action α_i according to equation (1).
5: Receive a feedback from the environment and update the parameter of Beta distributions: $\alpha_i = \alpha_i + 1$ if a reward is received and $\beta_i = \beta_i + 1$ if a penalty is received.
6: **Until** $max(P_1, P_2) > 0.99$

$$a_i = \begin{cases} argmin_i(\alpha_i + \beta_i) & \text{when } P_1 \neq P_2 \\ \text{randomly chosen} & \text{when } P_1 = P_2 \end{cases} \tag{1}$$

We may notice the exact value of probability $P_1 = Pr(e_1 > e_2)$ is evaluated in step 3 and utilized in step 6 to judge if the convergence criteria is satisfied. The computation of probability $P_1 = Pr(e_1 > e_2)$ is implemented by the following equations.

If we regard $P_1 = Pr(e_1 > e_2)$ as a function of $\alpha_1, \beta_1, \alpha_2, \beta_2$ and denote it as $g(\alpha_1, \beta_1, \alpha_2, \beta_2)$, we have:

$$g(\alpha_1, \beta_1, \alpha_2, \beta_2)$$

$$= \sum_{i=0}^{\alpha_1 - 1} \frac{B(\alpha_2 + i, \beta_1 + \beta_2)}{(\beta_1 + i)B(1 + i, \beta_1)B(\alpha_2, \beta_2)} \tag{2}$$

$$= \sum_{i=0}^{\beta_2 - 1} \frac{B(\beta_1 + i, \alpha_1 + \alpha_2)}{(\alpha_2 + i)B(1 + i, \alpha_2)B(\alpha_1, \beta_1)} \tag{3}$$

$$= 1 - \sum_{i=0}^{\alpha_2 - 1} \frac{B(\alpha_1 + i, \beta_1 + \beta_2)}{(\beta_2 + i)B(1 + i, \beta_2)B(\alpha_1, \beta_1)} \tag{4}$$

$$= 1 - \sum_{i=0}^{\beta_1 - 1} \frac{B(\beta_2 + i, \alpha_1 + \alpha_2)}{(\alpha_1 + i)B(1 + i, \alpha_1)B(\alpha_2, \beta_2)} \tag{5}$$

The above four equivalent equations indicate that the $g(\alpha_1, \beta_1, \alpha_2, \beta_2)$ can be evaluated with $\mathcal{O}(\min(\alpha_1, \alpha_2, \beta_1, \beta_2))$ by any programming language where the log-beta function is well defined. However, it's still can be improved through the following two approaches.

2.1 Recurrence Relationship

The recurrence relations that given in [19] are:

$$g(\alpha_1 + 1, \beta_1, \alpha_2, \beta_2) = g(\alpha_1, \beta_1, \alpha_2, \beta_2) + \frac{h(\alpha_1, \beta_1, \alpha_2, \beta_2)}{\alpha_1} \tag{6}$$

$$g(\alpha_1, \beta_1 + 1, \alpha_2, \beta_2) = g(\alpha_1, \beta_1, \alpha_2, \beta_2) - \frac{h(\alpha_1, \beta_1, \alpha_2, \beta_2)}{\beta_1} \tag{7}$$

$$g(\alpha_1, \beta_1, \alpha_2 + 1, \beta_2) = g(\alpha_1, \beta_1, \alpha_2, \beta_2) - \frac{h(\alpha_1, \beta_1, \alpha_2, \beta_2)}{\alpha_2} \tag{8}$$

$$g(\alpha_1, \beta_1, \alpha_2, \beta_2 + 1) = g(\alpha_1, \beta_1, \alpha_2, \beta_2) + \frac{h(\alpha_1, \beta_1, \alpha_2, \beta_2)}{\beta_2} \tag{9}$$

where $h(\alpha_1, \beta_1, \alpha_2, \beta_2) = \frac{B(\alpha_1 + \alpha_2, \beta_1 + \beta_2)}{B(\alpha_1, \beta_1)B(\alpha_2, \beta_2)}$.

Since one of the four equation can be used to update $g(\alpha_1, \beta_1, \alpha_2, \beta_2)$ after each iteration, and α_1, β_1, α_2, β_2 are all initialized as one according to the algorithm. So we can simply compute any $g(\alpha_1, \beta_1, \alpha_2, \beta_2)$ iteratively starting from $g(1, 1, 1, 1) = 0.5$. A demo of learning process is illustrated in Fig. 2. In Fig. 2, the red-colored number means it changes at the corresponding time instance. Since the terms with green background are known at previous time instance, so only the term with yellow background should be evaluate to get a new value that may be used in the next time instance. By such way, it is obvious that $g(\alpha_1, \beta_1, \alpha_2, \beta_2)$ could be computed with $\mathcal{O}(1)$ by any programming language where the log-beta function is well defined. That is doubtlessly a big improvement and can be expected to reduce the computational burden significantly.

2.2 Normal Approximation

Normal distribution, also known as Gaussian distribution, is a very common continuous probability distribution in probability theory. As advocated in [20], beta distribution may sometimes be approximated by a normal distribution and this approximation becomes exact asymptotically as the beta distribution parameters increase.

The technique used here is called moment matching, which maps a general probability distribution $\mathcal{G}_1$ to another general probability distribution $\mathcal{G}_2$. The two probability distributions share the same moment (mean, variance, etc.).

So the key idea is $\Pr(X_B > Y_B) \approx \Pr(X_N > Y_N)$, where X_B and Y_B are independent beta random variables X_N and Y_N their corresponding normal approximations. We denote the shared mean of X_B and X_N as μ_X, then

$$\mu_X = \frac{\alpha_X}{\alpha_X + \beta_X} \tag{10}$$

And we denote the shared variance of X_B and X_N as σ_X^2, then

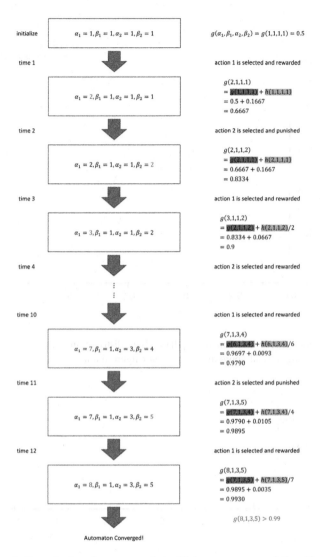

Fig. 2. Demonstration of a learning process that utilizing recurrence relationship (Color figure online)

$$\sigma_X^2 = \frac{\alpha_X \beta_X}{(\alpha_X + \beta_X)^2 (\alpha_X + \beta_X + 1)} \tag{11}$$

Then

$$\Pr(X_N > Y_N) = \Phi\left(\frac{\mu_X - \mu_Y}{(\sigma_X^2 + \sigma_Y^2)^{1/2}}\right) \tag{12}$$

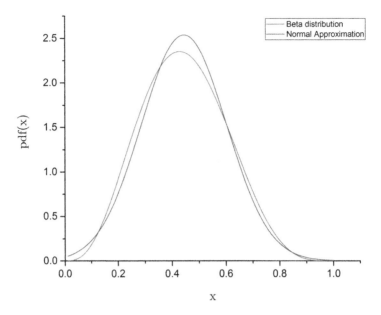

Fig. 3. The probability density function (pdf) of Beta distribution with $\alpha = 4$, $\beta = 5$ and the pdf its corresponding normal approximation. (Color figure online)

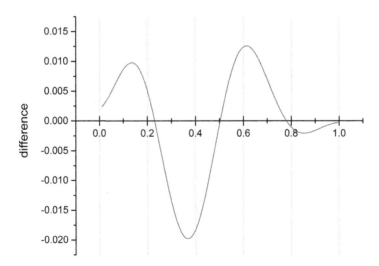

Fig. 4. The difference between the cumulative probability functions of beta distribution with $\alpha = 4$, $\beta = 5$ and its normal approximation

Where Φ denote the cumulative distribution function of the standard normal distribution. As $\Phi(x) = \frac{1}{2}\left[1 + erf\left(\frac{x}{\sqrt{2}}\right)\right]$, so the target formula could be computed by using the relative error function.

It was reported by Cook [20] that if the distribution parameters of X_B and Y_B take on integer values between 1 and 10 inclusive, the maximum absolute error is 0.05069 and the average absolute error over the parameter values is 0.006676. Figure 3 illustrates the probability density functions of a beta distribution and its normal approximation and Fig. 4 shows the differences between the cumulative probability function of the two distributions. It's worth mentioning that the difference will approaches zero asymptotically as the beta distribution parameters increase. That's the foundation of an accurate convergence.

3 Experiments

In this section, simulations are performed to verify the effectiveness of the two approaches. To make a comparison, the environments are the same as environments used in [18], which are:

E1:{0.90, 0.60}.
E2:{0.80, 0.50}.
E3:{0.80, 0.60}.
E4:{0.20, 0.50}.

The iterations and accuracy (number of correctly converged/number of experiments) of each algorithm are summarized in Table 1, where 250000 independent simulations are performed to get averaged iterations and accuracy.

The simulation codes are programmed by C++ with Boost library, which provides the beta random number generator, relative error function, beta function and log-beta function. Then the codes are performed on our workstation, which has Dual-CPU 2.6 GHz Intel Xeon E5-2670 with 4 Gbytes of RAM and a 500 Gbytes hard disk.

The running time, accuracy and averaged iterations of the three different approaches in the four environments are summarized in Tables 1, 2 and 3, respectively.

Table 1. The running time of 250000 independent simulations in E1 to E4

Environment	Original approach	Recurrence relation approach	Normal approximation approach
E_1	151.508 s	0.945 s	0.377 s
E_2	342.406 s	1.123 s	0.399 s
E_3	3004.4 s	2.19 s	0.725 s
E_4	493.156 s	1.113 s	0.395 s

From the tables, we can see Normal Approximation approach needs the least interactions with the environments, but at a cost of a small decline in accuracy. One distinct advantage is that it's about 401 to 4144 times faster than original approach with respect to running time.

Table 2. The accuracy of 250000 independent simulations in E1 to E4

Environment	Original approach	Recurrence relation approach	Normal approximation approach
E_1	99.9 %	99.9 %	99.9 %
E_2	99.9 %	99.9 %	99.8 %
E_3	99.7 %	99.7 %	99.5 %
E_4	99.9 %	99.9 %	99.7 %

Table 3. The averaged iterations to get converged in E1 to E4

Environment	Original approach	Recurrence relation approach	Normal approximation approach
E_1	42	42	40
E_2	49	49	45
E_3	99	99	93
E_4	49	49	45

Recurrence Relation approach keeps the same accuracy and required interactions with original approach, but still be about 160 to 1371 times faster than original approach.

4   Conclusion

In this paper, two approaches for reducing the computation time is proposed and experimentally verified. The simulation results shows both the two approaches are faster than the original approach, one with a slight loss in accuracy but another without any loss. These two approaches are expected to enhance the applicability of BLA in engineering.

However, the problem still remains open, problems such as how can the scheme be extended to be applicable in multi-action environments are supposed to be our future work.

Acknowledgement. This research work is funded by the National Science Foundation of China (61271316), 973 Program of China (2013CB329605), Key Laboratory for Shanghai Integrated Information Security Management Technology Research, and Chinese National Engineering Laboratory for Information Content Analysis Technology.

References

1. Narendra, K.S., Thathachar, M.: Learning automata-a survey. IEEE Trans. Syst. Man Cybern. **4**, 323–334 (1974)
2. Tsetlin, M.L.: On the behavior of finite automata in random media. Avtomatika i Telemekhanika **22**, 1345–1354 (1961)

3. Tsetlin, M.: Automaton Theory and Modeling of Biological Systems. Academic Press, New York (1973)
4. Agache, M., Oommen, B.J.: Generalized pursuit learning schemes: new families of continuous and discretized learning automata. IEEE Trans. Syst. Man Cybern. Part B Cybern. **32**(6), 738–749 (2002)
5. Papadimitriou, G.I., Sklira, M., Pomportsis, A.S.: A new class of ε-optimal learning automata. IEEE Trans. Syst. Man Cybern. Part B Cybern. **34**(1), 246–254 (2004)
6. Zhang, X., Granmo, O.-C., Oommen, B.J.: On incorporating the paradigms of discretization and bayesian estimation to create a new family of pursuit learning automata. Appl. Intell. **39** (4), 782–792 (2013)
7. Zhang, J., Wang, C., Zhou, M.: Last-position elimination-based learning automata. IEEE Trans. Cybern. **44**(12), 2484–2492 (2014)
8. Ge, H., Jiang, W., Li, S., Li, J., Wang, Y., Jing, Y.: A novel estimator based learning automata algorithm. Appl. Intell. **42**(2), 262–275 (2015)
9. Jiang, W., Li, B., Tang, Y., Philip Chen, C.L.: A new prospective for learning automata: a machine learning approach. Neurocomputing (2015)
10. Song, Y., Fang, Y., Zhang, Y.: Stochastic channel selection in cognitive radio networks. In: IEEE Global Telecommunications Conference, GLOBECOM 2007, pp. 4878–4882, November 2007
11. Oommen, B., Hashem, M.: Modeling a student-classroom interaction in a tutorial-like system using learning automata. IEEE Trans. Syst. Man Cybern. Part B Cybern. **40**(1), 29–42 (2010)
12. Horn, G., Oommen, B.: Solving multiconstraint assignment problems using learning automata. IEEE Trans. Syst. Man Cybern. Part B Cybern. **40**(1), 6–18 (2010)
13. Cuevas, E., Wario, F., Zaldivar, D., Perez-Cisneros, M.: Circle detection on images using learning automata. IET Comput. Vision **6**(2), 121–132 (2012)
14. Yazidi, A., Granmo, O.-C., Oommen, B.: Learning automaton based online discovery and tracking of spatiotemporal event patterns. IEEE Trans. Cybern. **43**(3), 1118–1130 (2013)
15. Misra, S., Krishna, P., Kalaiselvan, K., Saritha, V., Obaidat, M.: Learning automata based QoS framework for cloud IaaS. IEEE Trans. Netw. Serv. Manage. **11**(1), 15–24 (2014)
16. Kumar, N., Misra, S., Obaidat, M.: Collaborative learning automata based routing for rescue operations in dense urban regions using vehicular sensor networks. IEEE Syst. J. **9**(3), 1081–1090 (2015)
17. Vahidipour, S., Meybodi, M., Esnaashari, M.: Learning automata based adaptive Petri net and its application to priority assignment in queuing systems with unknown parameters. IEEE Trans. Syst. Man Cybern. Syst. **45**(10), 1373–1384 (2015)
18. Ge, H., Yan, Y., Li, J., Guo, Y., Li, S.: A parameter-free gradient Bayesian two-action learning automaton scheme. In: Proceedings of International Conference on Communications Signal Processing and Systems (2015)
19. Cook, J.D.: Exact calculation of beta inequalities. Technical report, UT MD Anderson Cancer Center Department of Biostatistics, Technical report (2005)
20. Cook, J.D.: Fast approximation of Beta inequalities. Technical report, UT MD Anderson Cancer Center Department of Biostatistics, Technical report (2012)

Adaptive Bi-objective Genetic Programming for Data-Driven System Modeling

Vitoantonio Bevilacqua[1(✉)], Nicola Nuzzolese[1], Ernesto Mininno[2], and Giovanni Iacca[2]

[1] Dipartimento di Ingegneria Elettrica e dell'Informazione, Politecnico di Bari, via Orabona 4, 70125 Bari, Italy
vitoantonio.bevilacqua@poliba.it
[2] Cyber Dyne S.r.l., Via Scipione Crisanzio 119, 70123 Bari, Italy

Abstract. We propose in this paper a modification of one of the modern state-of-the-art genetic programming algorithms used for data-driven modeling, namely the Bi-objective Genetic Programming (BioGP). The original method is based on a concurrent minimization of both the training error and complexity of multiple candidate models encoded as Genetic Programming trees. Also, BioGP is empowered by a predator-prey co-evolutionary model where virtual predators are used to suppress solutions (preys) characterized by a poor trade-off error vs complexity. In this work, we incorporate in the original BioGP an adaptive mechanism that automatically tunes the mutation rate, based on a characterization of the current population (in terms of entropy) and on the information that can be extracted from it. We show through numerical experiments on two different datasets from the energy domain that the proposed method, named BioAGP (where "A" stands for "Adaptive"), performs better than the original BioGP, allowing the search to maintain a good diversity level in the population, without affecting the convergence rate.

Keywords: Multi-objective evolutionary algorithms · Adaptive genetic programming · Machine learning · Home automation · Energy efficiency

1 Introduction

Computational intelligence techniques, such as Evolutionary Computation (EC) and Machine Learning (ML), have proven to be powerful yet general-purpose tools in a broad range of optimization applications [6, 7, 26]. One of the most important fields of application of such tools is energy management, with successful examples in complex problems such as energy management in buildings, renewable energy systems, heating, ventilation and air conditioning control methodologies, and forecasting energy consumption [9, 28]. Of special interest in this domain is the forecasting e.g. of the air temperature in a room [31] as a function of both weather parameters (mainly solar radiation and air temperature) and actuator states or manipulated variables (heating, ventilating, cooling), with the subsequent use of these mid/long-range prediction models for a more efficient temperature control, both in terms of regulation and energy consumption.

© Springer International Publishing Switzerland 2016
D.-S. Huang et al. (Eds.): ICIC 2016, Part III, LNAI 9773, pp. 248–259, 2016.
DOI: 10.1007/978-3-319-42297-8_24

An example of application of these predictive models is efficient building design, where the computation of the heating load and the cooling load is required to determine the specifications of the heating and cooling equipment needed to maintain comfortable indoor air conditions. However, first-principles modeling of indoor air temperature can be a complicated task: it involves a non-linear dynamical system whose inputs are the weather parameters and actuators manipulated variables, and the output is the predicted room temperature. An additional complexity is due to that fact that for each room in a building and for each variable of interest a separate model may be needed, to approximate the complex relationship between the system inputs (in this case the weather parameters and actuators manipulated variables) and the indoor air temperature. To overcome these issues, data-driven modeling techniques, such as Genetic Programming (GP) or Neural Networks (NN), are viable alternatives[1].

In this paper, we propose an adaptive Genetic Programming algorithm built upon a state-of-the-art method from the literature, namely the Bi-objective Genetic Programming (BioGP) [11]. Motivated by the empirical observation that the performance of BioGP highly depends on the chose parameter setting -particularly the mutation rate- we introduce in the original algorithm an adaptive mutation scheme that automatically tunes the mutation rate, based on a characterization of the current population (in terms of entropy) and on the information that can be extracted from it. We then show through numerical experiments on two different real world datasets that the proposed method, named BioAGP (where "A" stands for "Adaptive"), performs better than the original BioGP, allowing the search to maintain a good diversity level in the population, without affecting the convergence rate.

The remainder of this paper is structured as follows. The next section briefly presents the related works on Genetic Programming and details the working principles of BioGP. Section 3 describes the adaptive mutation mechanism proposed, and the motivations behind that. Then, in Sect. 4 we present the numerical results obtained by BioAGP on two different datasets from the energy domain, in comparison with the original BioGP and with alternative data-driven modeling techniques. Finally, in Sect. 5 we give the conclusions of this work.

2   Related Work

In this section, we first recapitulate the main principles of Genetic Programming (Sect. 2.1) then we describe in detail the original algorithm that is at the base of the proposed BioAGP (Sect. 2.2).

2.1   Genetic Programming

Genetic Programming is an evolutionary algorithm originally designed to find computer programs that perform a user-defined task [17]. Similar to other genetic

---

[1] We should observe, however, that the main drawback of data-driven modeling methods is that they depend entirely on experimental data. Therefore, such methods can only be applied after the actual building is built and measurements are available.

techniques used as optimization tools [5, 21], GP conducts a parallel search on a population of candidate solutions in the search space. In GP, however, a solution represents a set of mathematical functions, which can be considered as an approximate model of the system at hand. A GP solution is typically encoded as a *binary tree*, whose nodes can be arbitrary mathematical functions, and leaves can be variables, or constants (as opposed to binary or real-valued genetic algorithms where solutions are encoded as arrays of bits or real values). The elements of the trees are, in general, predetermined, as they are extracted from two user-defined sets initialized before the beginning of the evolutionary process: a *function set*, containing user-defined mathematical functions, and a *terminal set*, containing variables and constants. Notably, one of the main advantages of GP e.g. in comparison to Neural Networks is the possibility of adding custom functions, which makes the algorithm extremely flexible [10].

GP starts off by creating an initial population of trees randomly initialized with elements from the function and terminal sets. Then, the population is evolved over a sequence of generations through mechanisms that mimic genetic recombination (crossover), mutation and selection. When used as a data-driven modeling method, typically a candidate GP tree is fed with a dataset containing n samples $\{(i_1, o_1), (i_2, o_2), \ldots, (i_n, o_n)\}$ (where each sample consists, in general, of m input variables and p output variables[2], i.e. $i_k \in R^m$ and $o_k \in R^p$ $k \in [1; n]$), then the root mean square error (RMSE) of each tree is computed as:

$$RMSE = \sqrt{\frac{1}{n} \sum_{k=1}^{n} (o_k - \hat{o}_k)^2} \qquad (1)$$

where $\hat{o}_k$ is the estimated output calculated by that tree when fed with the k-th sample. The fitness of the tree is then its RMSE, and conventional GP algorithms simply try to find the minimum-RMSE tree [1].

2.2 Bi-objective Genetic Programming

We focus now on the Bi-objective Genetic Programming BioGP) [11], the algorithm at the base of the present work. BioGP differs from conventional GP in that it does not minimize only the RMSE, rather it performs a bi-objective optimization. In particular, the algorithm tries to find a trade-off between the training error ξ (i.e., the RMSE calculated on the training set) and the model complexity ζ (described below) -both to be minimized- by employing the predator-prey scheme proposed in [25]. The algorithm then returns a set of Pareto-optimal solutions characterized by different trade-offs between model complexity and error, considering that more complex models tend to show lower errors (with the risk over-fitting) while simpler models are characterized by higher errors (under-fitting).

---

[2] In the rest of the paper, we will consider problems with $p = 1$ output variable. Nonetheless multiple output variables can be approximated by multiple GP trees, one per variable. An extension of the analysis for $p > 1$ is also possible and will be considered in future studies.

In BioGP, each solution is represented as a weighed sum of sub-trees (each encoding a mathematical expression in Polish notation). More specifically, a linear sum is introduced as parent node, from which r roots emerge (one per sub-tree), where the number of roots r is a parameter of the algorithm. Consequently, the estimated output of a candidate solution fed with a sample i_k can be expressed as follows:

$$\hat{o}_k = f(i_k) = \sum_{j=1}^{r} \omega_j f_j(i_k) + \theta \qquad (2)$$

where $f_j(i_k)$ is the function corresponding to each j-th sub-tree calculated on the sample i_k, ω_j is its associated weight, $j = 1, 2, \ldots, r$, and Θ is a bias value. For each candidate solution, the weights and the bias value are calculated by the linear least square technique, such that the RMSE across all samples is minimized. Then, the solution's error ξ is simply determined by its RMSE. As for the complexity ζ, this is calculated as the weighted sum of the maximum depth of among all the sub-trees in the solution (δ), and the total number of function nodes (v), since both terms contribute to the parameterization in the model. The objective function denoting the complexity ζ is then defined as:

$$\zeta = \lambda\delta + (1 - \lambda)v \qquad (3)$$

where λ is a scalar, set equal to 0.5 as suggested in [30]. The parameter λ can be set to a different value to effectively control the growth of trees, in case the complexity goes beyond an acceptable limit. Furthermore, an additional parameter of the algorithm, d, allows to set the maximum depth of the sub-trees.

In a nutshell, the BioGP algorithm consists of two optimization steps. The first phase is a single-objective minimization of the error ξ, which continues until a prefixed error level is reached, or, alternatively, until a predefined number of generations is completed. Once this first step ends, the bi-objective optimization predator-prey genetic algorithm [25] starts. This mimics the interaction between a population of preys (each one corresponding to a candidate solution) and a population of predators, both moving on a two-dimensional lattice. The demography of preys is controlled by predators, which kill the least fit prey in their neighborhood.

It should be noted that the bi-objective is particularly computationally expensive, since it tends to generate solutions over a wide range of model complexity and error. In this sense, the introduction of the single-objective error reduction phase significantly reduces the computing cost of the algorithm, by guiding the solutions towards an acceptable limit of error that the user would be able to specify and find acceptable, without losing the choice and flexibility of the Pareto solutions.

One final note about BioGP regards its genetic operators. During the single-objective phase, BioGP applies a tournament selection (in the original paper [11] of size 5). For the bi-objective part, selection is performed by means of non-dominated sorting and ranking [8]. As for crossover, BioGP uses a combination of standard and height-fair crossover [23]: in the former, two sub-trees are randomly interchanged between the participating parents, while in the latter case the exchange takes place at a selected depth. Also, BioGP use different types of mutations, namely standard, small and mono parental exchange: in case of a standard mutation, a sub-tree is deleted and then randomly regrown; small mutation implies replacing a terminal set by another (e.g. a numerical value in the terminal set is slightly altered, or a function is replaced with another having the same arity - e.g. "×" by "/"); finally, mono parental exchange involves swapping two sub-trees belonging to the same tree. In BioGP, the probability of activating one of the crossover or mutation operators is fixed.

3   Adaptive Mutation Mechanism

Just like for any other evolutionary algorithm, a crucial aspect in GP algorithms (including BioGP) is the balance between exploration and exploitation [29], i.e., respectively, the ability to visit entirely new regions of the search space, and to refine the search within the neighborhood of previously visited points. In general this is obtained by a careful tuning of the activation probability of the genetic operators, i.e. mutation and crossover. In the following, we refer to the mutation and crossover probability as p_m and p_c, respectively. In general, both p_m and p_c depend on the specific problem at hand, and in turn the overall effectiveness of a GP algorithm significantly depends on their setting. This might impair the use of GP especially for practitioners and users who may be not fully aware of the importance of this tuning of else are not sufficiently knowledgeable about the influence of such parameters.

To overcome the need for application-specific parameter tuning, a current trend in evolutionary algorithms is the use of adaptive mechanisms that are able to automatically tune the activation probability of the genetic operators [15, 22], such that the algorithm can flexibly adjust its behavior to a broad range of problems. The present work falls in this research area, as we focus specifically on the auto-adaptation of the mutation probability in BioGP.

As we saw in the previous section, BioGP uses three mutation operators (standard, small and mono parental exchange): in the following, we will refer to their activation probabilities respectively as p_{stand}, p_{small}, p_{mono}. In the original BioGP algorithm, $p_{stand} = p_{small} = p_{mono} = 1/3 \times 0.1$. Our intuition is that while these predetermined values might be efficient in some cases, the same could be sub-optimal in others. Therefore, we aim here at endowing BioGP with adaptive capabilities, such that the mutation rates can automatically change during evolution, rather than being fixed as in the original algorithm. Allowing for dynamic changes of the mutation rates, we could obtain a better balance between exploration and exploitation as the algorithm would adapt automatically its behavior to different fitness landscapes and search conditions: in fact, higher mutation rates might be needed when the algorithm is stuck into a local optimum and the search should move to new regions of the solution space in the attempt to find the global optimum; vice versa, lower mutation rates are needed to improve exploitation when the algorithm is converging towards the optimum. With these considerations in mind, we introduce an adaptive mutation scheme as follows.

At each generation, the intrinsic information content of the population is calculated taking into account the worst (f_{max}) and best (f_{min}) individual training error ξ among all the trees in the current population, as well as the average error in the population (f_{mean}). Therefore, the normalized deviation between the worst individual and the population mean is calculated as:

$$diversity = \frac{f_{max} - f_{mean}}{f_{max} - f_{min}} \tag{4}$$

which is defined for $f_{max} \neq f_{min}$. If this value is low, the population is homogeneous (i.e. the worst individual is close to the population mean, indicating that all individuals have similar error values), otherwise the population contains individuals with a higher diversity in terms of errors. In the presence of over-fitting, all GP trees tend to show a very small training error: however, keeping some higher-error solution in the population (thus allowing for a higher level of diversity) might prevent an excessive convergence towards similar over-fitting GP trees characterized by small errors. In addition to the diversity index, at each generation the Shannon entropy of the population is measured as:

$$-\sum_k p_k \log(p_k) \tag{5}$$

where the sum is calculated over all the groups in the population having the same value of training error ξ, and p_k is the fraction of the population having the k-th error value. Higher entropy values correspond to a more heterogeneous population, and vice versa. Once the above metrics in Eq. (4) and (5) are calculated, their values are compared with the values from the previous generation. If both the new values are lower than the previous ones (which happens when the entropy of the population is decreasing and the worst individual is getting closer to the population mean), the mutation rates are re-sampled as follows:

$$\begin{cases} p_{stand} = 0.1 + (0.33 - 0.1) \times rand() \\ p_{small} = 0.1 + (0.33 - 0.1) \times rand() \\ p_{mono} = 0.1 + (0.33 - 0.1) \times rand() \end{cases} \tag{6}$$

otherwise they are kept at their current values (starting from the original values used in BioGP). Here $rand()$ indicates a uniform random number sampled in [0, 1]. The rationale behind this adaptive scheme is that an increase of the mutation rates when the diversity in the population is decreasing might be beneficial to counterbalance premature convergence. Otherwise, the current mutation rates are maintained when the population has already a sufficient level of diversity.

4 Numerical Results

To assess the performance of the proposed BioAGP (and determine the performance benefit due to the adaptive scheme), we performed numerical experiments on two different datasets from the energy domain, and compared the performance of BioAGP with that of the original BioGP and with alternative machine learning methods. In the following, first we describe our experimental setup (i.e. the datasets), and then we analyze the numerical results obtained on the two datasets.

4.1   Datasets

The datasets considered in our study are both taken from the UCI repository [18]:

- The **Energy Efficiency** dataset. It is composed of 768 samples and 8 features, namely: compactness, surface area, wall area, roof area, overall height, orientation, glazing area, glazing area distribution. The goal is to predict one real valued response: heating load or cooling load.
- The **SML2010** dataset. It contains 4137 samples and 18 features, namely: date, time, weather forecast temperature, indoor temperature (dinning-room); relative to dinning room and room: carbon dioxide, relative humidity, lighting; sun dusk, wind, sun light in west/east/south facade, sun irradiance, outdoor temperature, outdoor relative humidity. The goal is to predict the indoor temperature.

In the experiments, we split each dataset into three sets (with sizes 60 %, 20 %, 20 %) respectively for training, validation and test.

4.2   Experiments

Several tests were performed on the two real datasets with the main objective to obtain mathematical models with minimum test error and complexity.

- **BioAGP:** the algorithm proposed here, with parameter setting as suggested in [11], apart for the mutation rates that are replaced by adaptive mutation.
- **BioGP:** the original algorithm presented in [11], with the parameter setting suggested in that paper.
- **NEAT:** NeuroEvolution of Augmenting Topologies (NEAT) method pro-posed in [27], with population size set to 500. All other parameters were set as in the original paper.
- **Neural Network (NN):** a NN trained by Resilient Propagation [24]. The NN configuration was chosen applying the methodology described in [2] by means of the optimization software Kimeme [16], resulting in a network with three hidden layers, respectively with 91 (Elliot symmetric activation function), 84 (with ramp activation function) and 68 nodes (with Gaussian activation function). The single output node used a hyperbolic tangent activation function.
- **Multiple Regression (MR):** the ordinary least squares method that estimates the parameters of a multiple linear regression model.

All the algorithms above were implemented in Java code, with multi-thread parallelization at the level of each model evaluation. The proposed BioAGP was implemented by porting the original Matlab code available from [11] and adding the new adaptive mutation scheme. In both BioGP and BioAGP, we used as a function set $\{+, -, /, \times, \wedge, \sqrt{x}, \ln(x)\}$ where "$\wedge$" indicates the power function x^y. Since the computational cost of the GP algorithms is considerable higher than the other techniques (due to the parsing of a very large number of trees generated during the evolutionary process), we decided to run the two GP algorithms for a small number of generations (20) and a large number of predators and preys (respectively 100 and 500), to test their convergence under hard computational constraints and have a fair comparison with the other methods. As for the NEAT and NN, we used the open-source Java library Encog [12], coupled with Kimeme [16] as explained in [2]. Finally, the Multiple Regression algorithm was taken from the Apache Commons Java math library[3]. Each algorithm execution was repeated five times on both datasets, to calculate statistics on training, validation and test error.

First, we analyze the performance of BioAGP for different values of maximum depth and number of roots, see Tables 1, 2, 3. In the tables, we show the mean and std. dev. (over five repetitions) of training, validation and test error[4]. Also, we show in boldface the best tree configurations, i.e. the trees whose test errors are lower than those of the other configurations. We consider values smaller than $\varepsilon = 10^{-16}$ equal to zero. We can see that the optimal configuration of tree depth and number of roots depend on the specific dataset and system to model: indicating with (d,r) a configuration (max depth, number of roots), we observe that the configurations (11,7) and (8,5) allow to obtain the minimum test error on the Energy Efficiency dataset, respectively for heating

Table 1. Energy Efficiency dataset, heating load response

GP tree configuration (depth, roots)	Training error (Mean ± Std. Dev.)	Validation error (Mean ± Std. Dev.)	Test error (Mean ± Std. Dev.)
(6, 4)	0.0719 ± 0.0945	2.7733 ± 0.0648	2.9912 ± 0.1357
(8, 5)	0.0701 ± 0.1746	2.6555 ± 0.2184	3.0557 ± 0.1567
(9, 6)	0.0747 ± 0.1456	2.8550 ± 0.3360	3.1421 ± 0.2664
(10, 6)	0.0682 ± 0.1136	2.6166 ± 0.1221	2.8645 ± 0.1495
(11, 7)	**0.0647 ± 0.1169**	**2.4665 ± 0.1470**	**2.7391 ± 0.2373**
(12, 7)	0.0692 ± 0.2721	2.6330 ± 0.2546	2.9217 ± 0.2147

Table 2. Energy Efficiency dataset, cooling load response

GP tree configuration (depth, roots)	Training error (Mean ± Std. Dev.)	Validation error (Mean ± Std. Dev.)	Test error (Mean ± Std. Dev.)
(6, 4)	0.0841 ± 0.0225	2.9994 ± 0.1808	3.0786 ± 0.1217
(8, 5)	**0.0839 ± 0.0146**	**3.0379 ± 0.1378**	**2.7965 ± 0.0449**
(9, 6)	0.0795 ± 0.0536	2.8104 ± 0.1496	2.8773 ± 0.1315
(10, 6)	0.0808 ± 0.0136	2.8795 ± 0.1349	2.9788 ± 0.0999
(11, 7)	0.0802 ± 0.0565	2.8554 ± 0.0889	2.8998 ± 0.0857
(12, 7)	0.0788 ± 0.0357	2.7607 ± 0.1439	2.9463 ± 0.0748

---

[3] http://commons.apache.org.

[4] We should remark that among all the Pareto-optimal solutions returned by BioAGP, we report here the ones characterized by the lowest training error, following the approach suggested in [11]. However other choices e.g. based on information criteria can also be made.

Table 3. SML2010 dataset, temperature response

GP tree configuration (depth, roots)	Training error (Mean ± Std. Dev.)	Validation error (Mean ± Std. Dev.)	Test error (Mean ± Std. Dev.)
(6, 4)	0.0170 ± 0.0435	0.3609 ± 0.3438	0.2866 ± 0.1217
(8, 5)	0.0164 ± 0.0564	0.3519 ± 0.3519	0.2817 ± 0.0034
(9, 6)	0.0163 ± 0.0239	0.3518 ± 0.3355	0.2793 ± 0.0041
(10, 6)	0.0165 ± 0.0594	0.3543 ± 0.3378	0.2839 ± 0.0026
(11, 7)	0.0165 ± 0.0882	0.3512 ± 0.3347	0.2842 ± 0.0125
(12, 7)	**0.0164 ± 0.0357**	**0.3433 ± 0.3271**	**0.2773 ± 0.0030**

load and cooling load response; on the SML2010 dataset, the configuration (12,7) performs best. Despite the different optimal configurations, we can also observe qualitatively that the variance of the performance across all the tested configurations is quite small.

As a second part of the analysis, we compare the performance of BioAGP with that of the other four methods, see Tables 4, 5, 6. In the tables, the values corresponding to BioAGP were taken from the best configurations shown in the previous tables. The comparative analysis shows that BioAGP performs consistently better than the original BioGP on all the tested datasets and variables of interest. However, when compared against other methods (NN, NEAT and MR), both GP methods produce higher validation and test error, especially on the Energy Efficient dataset. On the other hand, the GP methods are able to fit the data reasonably well even in a short number of generations, producing training errors which are almost on par with the other methods, if not better (compare the training error of BioAGP with that of the other methods in Table 4). Furthermore, the GP algorithms have the advantage of providing an actual mathematical model (instead of a black-box system such as NN and NEAT) that could be used for further analysis. Another interesting aspect is that both GP algorithms seem to perform better for larger datasets (see the validation and test error on the SML2010 dataset in Table 6, which are is much smaller than the errors on the Energy Efficiency dataset shown in Tables 4, 5), producing errors that are almost on par with the other techniques, if not better in some cases (compare e.g. the test error of BioAGP against that of NN in Table 6). This might indicate that both BioGP and BioAGP are able to leverage larger amounts of training data to build mathematical models with higher

Table 4. Performance on Energy Efficiency dataset (heating load response)

Algorithm	Training error (Mean ± Std. Dev.)	Validation error (Mean ± Std. Dev.)	Test error (Mean ± Std. Dev.)
BioAGP	0.0647 ± 0.1169	2.4665 ± 0.1470	2.7391 ± 0.2373
BioGP	0.0739 ± 0.0575	2.8612 ± 0.0644	3.1364 ± 0.0835
NEAT	0.0762 ± 0.0000	0.0747 ± 0.0053	0.0844 ± 0.0073
NN	0.0763 ± 0.0000	0.1667 ± 0.0000	0.1568 ± 0.0000
MR	0.0770 ± 0.0000	0.0801 ± 0.0000	0.0948 ± 0.0000

Table 5. Performance on Energy Efficiency dataset (cooling load response)

Algorithm	Training error (Mean ± Std. Dev.)	Validation error (Mean ± Std. Dev.)	Test error (Mean ± Std. Dev.)
BioAGP	0.0839 ± 0.0146	3.0379 ± 0.1378	2.7965 ± 0.0449
BioGP	0.0867 ± 0.3467	3.1349 ± 0.0320	3.1076 ± 0.0399
NEAT	0.0796 ± 0.0000	0.0821 ± 0.0011	0.0877 ± 0.0183
NN	0.0697 ± 0.0000	0.2208 ± 0.0000	0.2248 ± 0.0000
MR	0.0886 ± 0.0000	0.0931 ± 0.0000	0.0964 ± 0.0000

Table 6. Performance on SML2010 dataset

Algorithm	Training error (Mean ± Std. Dev.)	Validation error (Mean ± Std. Dev.)	Test error (Mean ± Std. Dev.)
BioAGP	0.0164 ± 0.0357	0.3433 ± 0.3271	0.2773 ± 0.003
BioGP	0.0172 ± 0.0009	0.3613 ± 0.0164	0.2939 ± 0.0128
NEAT	0.0993 ± 0.0932	0.1038 ± 0.0023	0.1042 ± 0.0069
NN	0.0694 ± 0.0000	0.2952 ± 0.0000	0.3267 ± 0.0000
MR	0.0148 ± 0.0000	0.0181 ± 0.0000	0.1448 ± 0.0000

generalization capabilities, and explain why on the Energy Efficiency dataset (which has only 768 samples, while the SML2010 dataset has 4137) the performance is lower.

5 Conclusion

In this paper we proposed an adaptive Genetic Programming method, BioAGP. The method is based on the state-of-the-art Bi-objective Genetic Programming algorithm (BioGP), and improves upon it by introducing an adaptive mutation mechanism that adjusts the mutation rates according to the current population diversity level. The proposed algorithm is specifically designed for data-driven modeling, as one of its main advantages is the ability to construct models with different levels of complexity and modeling error.

We tested BioAGP on two datasets related to indoor temperature prediction and energy efficiency in domestic environments. In a first part of the experiments, we showed that the algorithm is fairly robust for various configurations of maximum tree depth and number of roots; nevertheless, it was possible to identify an optimal configuration for each dataset and variable of interest. In the second part of the experimentation, we compared the performance of the best configurations of BioAGP against that one of alternative machine learning algorithms (the original BioGP, NN, NEAT and multiple linear regression). The numerical experiments showed that BioAGP consistently performs better than the original BioGP algorithm. The comparison against the other algorithms highlighted that the performance of BioAGP improves when a larger dataset is available, with comparable errors with respect to the other methods.

Future works will focus on the adaptation of the additional parameters of the algorithm (crossover probabilities, maximum depth and number of roots). Furthermore, we will test the method on modeling problems from other domains, such as bioinformatics [19, 20], sensor systems [3, 13], or robotics [4, 14].

References

1. Behbahani, S., De Silva, C.W.: Mechatronic design evolution using bond graphs and hybrid genetic algorithm with genetic programming. IEEE/ASME Trans. Mechatron. **18**(1), 190–199 (2013)
2. Bevilacqua, V., Cassano, F., Mininno, E., Iacca, G.: Optimizing feed-forward neural network topology by multi-objective evolutionary algorithms: a comparative study on biomedical datasets. In: Rossi, F., Mavelli, F., Stano, P., Caivano, D. (eds.) WIVACE 2015. CCIS, vol. 587, pp. 53–64. Springer, Heidelberg (2016). doi:10.1007/978-3-319-32695-5_5
3. Caraffini, F., Neri, F., Iacca, G., Mol, A.: Parallel memetic structures. Inf. Sci. **227**, 60–82 (2013)
4. Caraffini, F., Neri, F., Passow, B.N., Iacca, G.: Re-sampled inheritance search: high performance despite the simplicity. Soft. Comput. **17**(12), 2235–2256 (2013)
5. Coello, C.A.C.: Multi-objective evolutionary algorithms in real-world applications: some recent results and current challenges. In: Advances in Evolutionary and Deterministic Methods for Design, Optimization and Control in Engineering and Sciences, pp. 3–18. Springer (2015)
6. Dasgupta, D., Michalewicz, Z.: Evolutionary Algorithms in Engineering Applications. Springer Science & Business Media (2013)
7. Deb, K.: Multi-objective Optimization Using Evolutionary Algorithms, vol. 16. Wiley, Chichester (2001)
8. Deb, K., Pratap, A., Agarwal, S., Meyarivan, T.: A fast and elitist multiobjective genetic algorithm: NSGA-II. IEEE Trans. Evol. Comput. **6**(2), 182–197 (2002)
9. Ferreira, P., Ruano, A., Silva, S., Conceio, E.: Neural networks based predictive control for thermal comfort and energy savings in public buildings. Energy Build. **55**, 238–251 (2012)
10. Garg, A., Tai, K.: Comparison of regression analysis, artificial neural network and genetic programming in handling the multicollinearity problem. In: International Conference on Modelling, Identification & Control (ICMIC), pp. 353–358. IEEE (2012)
11. Giri, B.K., Hakanen, J., Miettinen, K., Chakraborti, N.: Genetic programming through bi-objective genetic algorithms with a study of a simulated moving bed process involving multiple objectives. Appl. Soft Comput. **13**(5), 2613–2623 (2013)
12. Heaton, J.: Programming Neural Networks with Encog 2 in Java (2010)
13. Iacca, G.: Distributed optimization in wireless sensor networks: an island-model framework. Soft. Comput. **17**(12), 2257–2277 (2013)
14. Iacca, G., Caraffini, F., Neri, F.: Memory-saving memetic computing for path-following mobile robots. Appl. Soft Comput. **13**(4), 2003–2016 (2013)
15. Iacca, G., Caraffini, F., Neri, F.: Multi-strategy coevolving aging particle optimization. Int. J. Neural Syst. **24**(01), 1450008 (2014)
16. Iacca, G., Mininno, E.: Introducing Kimeme, a novel platform for multi-disciplinary multi-objective optimization. In: Rossi, F., Mavelli, F., Stano, P., Caivano, D. (eds.) WIVACE 2015. CCIS, vol. 587, pp. 40–52. Springer, Heidelberg (2016). doi:10.1007/978-3-319-32695-5_4

17. Koza, J.R.: Genetic Programming: On the Programming of Computers by Means of Natural Selection. MIT Press, Cambridge (1992)
18. Lichman, M.: UCI Machine Learning Repository (2013). http://archive.ics.uci.edu/ml
19. Menolascina, F., Tommasi, S., Paradiso, A., Cortellino, M., Bevilacqua, V., Mastronardi, G.: Novel data mining techniques in aCGH based breast cancer subtypes profiling: the biological perspective. In: IEEE Symposium on Computational Intelligence and Bioinformatics and Computational Biology (CIBCB), pp. 9–16, April 2007
20. Menolascina, F., Bellomo, D., Maiwald, T., Bevilacqua, V., Ciminelli, C., Paradiso, A., Tommasi, S.: Developing optimal input design strategies in cancer systems biology with applications to microfluidic device engineering. BMC Bioinform. **10**(12), 1 (2009)
21. Onwubolu, G.C., Babu, B.: New optimization techniques in engineering, **141** (2013). Springer
22. Parmee, I.C.: Evolutionary and Adaptive Computing in Engineering Design. Springer Science & Business Media, London (2012)
23. Rennard, J.P.: Handbook of research on nature-inspired computing for economics and management. IGI Global (2006)
24. Riedmiller, M., Braun, H.: RPROP-a fast adaptive learning algorithm. In: Proceedings of ISCIS VII, Universitat (1992)
25. Costa e Silva, M.A., Coelho, L.d.S., Lebensztajn, L.: Multiobjective biogeography-based optimization based on predator-prey approach. IEEE Trans. Magn. **48**(2), 951–954 (2012)
26. Stadler, W.: Multicriteria Optimization in Engineering and in the Sciences, vol. 37. Springer Science & Business Media, New York (2013)
27. Stanley, K.O., Miikkulainen, R.: Evolving neural networks through augmenting topologies. Evol. Comput. **10**(2), 99–127 (2002)
28. Tsanas, A., Xifara, A.: Accurate quantitative estimation of energy performance of residential buildings using statistical machine learning tools. Energy Build. **49**, 560–567 (2012)
29. Crepinsek, M., Liu, S.H., Mernik, M.: Exploration and exploitation in evolutionary algorithms: a survey. ACM Comput. Surv. **45**(3), 35:1–35:33 (2013)
30. Vrieze, S.I.: Model selection and psychological theory: a discussion of the differences between the Akaike Information Criterion (AIC) and the Bayesian information criterion (BIC). Psychol. Methods **17**(2), 228 (2012)
31. Zamora-Martnez, F., Romeu, P., Botella-Rocamora, P., Pardo, J.: On-line learning of indoor temperature forecasting models towards energy efficiency. Energy Build. **83**, 162–172 (2014)

Gaussian Iteration:
A Novel Way to Collaborative Filtering

Xiaochun Li[(⊠)], Fangqi Li, Ying Guo, and Jinchao Huang

Department of Electronic Engineering, Shanghai Jiao Tong University,
800 Dongchuan Road, Min Hang, Shanghai 200240, People's Republic of China
{xiaochun_lee, solour_lfq}@sjtu.edu.cn

Abstract. Based on the missing not at random assumption and central limit theorem, this paper presents a novel way to accelerate the iteration speed in the collaborative filtering models called Gaussian iteration. In the proposed model, adding the Gaussian distribution to the estimation error makes the falling direction more credible, which significantly reduces the running time with the ideal accuracy. For evaluation, we compare the performance of the proposed model with three existing collaborative filtering models on two kinds of Movielens datasets. The results indicate that the novel method outperforms the existing models and it is easy to implement and faster. Moreover, the proposed model is scalable to the analogous objective function in other models.

Keywords: Collaborative filtering · Gaussian iteration · Recommender system

1 Introduction

With the rapid development of computers and computer networks, we are facing the revolution of big data [1]. People get stuck by the mass data when they surf the Internet. Customers cannot find out their favorite items effectively and the qualified commodity can't be known by more person. In the recent decades, thousands of scholars and researchers have focused on solving the problem. Generally speaking, recommender systems have been one of effective technologies to address such challenges [2].

Recommender systems can be divided into two different strategies, the content based approach [3] and the collaborative filtering (CF) [4]. The former needs a profile for each user or item to describe its feature. Researchers analyze the content and figure out the matching products or users. However, the shortcoming about the content based approach is that it requires gathering further information that might not be easily collected. The latter, CF, is the subject that we focus on, which analyses the relationship between users and products through users' historical record. On the whole, neighborhood models and latent factor models are the major approach in CF. Neighborhood models mean to locate the most parallel users set for the target user. In order to identify the neighborhood, researchers can calculate the similarity based on user or item. The user-oriented approach and item-oriented approach have a lot of practices, see [5, 6] for more details. The neighborhood models have many ways to calculate the similarity, e.g. locate the group/community [7, 8], and modify the correlation coefficient [9, 10]. Latent factor models comprise an alternative approach to CF with the more holistic goal to

© Springer International Publishing Switzerland 2016
D.-S. Huang et al. (Eds.): ICIC 2016, Part III, LNAI 9773, pp. 260–270, 2016.
DOI: 10.1007/978-3-319-42297-8_25

uncover latent features that explain observed ratings; examples include pLSA [11], neural networks [12], and Latent Dirichlet Allocation [13]. The main models in latent factor models were induced by Singular Value Decomposition (SVD) [14] on the user-item ratings matrix. A typical model of latent factor models could be decomposed into user-factors matrix and item-factors matrix. We get the prediction by taking an inner product with a user-factors vector and an item-factors vector. More information can be imported into the model, such as the user/item bias, explicit feedback data [14].

Neighborhood models and latent factor models can be integrated into one integrated model. Paper [15] reveals the outperformance. The integrated model introduces more parameters into the model, therefore, represents the observations in more details. The integrated models, or ensemble learning have been a hot topic in the academia.

In this paper, we concentrate upon the collaborative filtering approach, including neighborhood models, latent factor models and the integrated models [15].

The rest of this paper is organized as follows. Section 2 reviews the related work including global neighborhood models, SVD++ and an integrated model. Section 3 introduces the proposed Gaussian iteration model. Experiments are demonstrated in Sect. 4 to verify the effectiveness of the proposed model. Last section concludes this paper.

2 Previous Work

In order to give the readers clear understanding of our proposed model, we make a brief introduction about the main collaborative filtering. At first, we reserve special indexing letters for distinguishing users from items: for users u, v and for items i, j. A rating r_{ui} indicates the preference by user u of item i, where high values mean stronger preference. For example, values can be integers ranging from 1 (star) indicating no interest to 5 (star) indicating a strong interest. We distinguish predicted ratings from known ones, by using the notation $\hat{r}_{ui}$ for the predicted value of r_{ui}. The (u, i) pairs for which r_{ui} is known are stored in the set $\kappa = \{(u, i) | r_{ui} \text{ is known}\}$. Usually the vast majority of ratings are unknown.

2.1 Neighborhood Models

The most common approach to CF is based on neighborhood models. The used models are limited of the local neighborhood, which cause to the deviation about the prediction. In this part, we introduce the global neighborhood model [16], which uses the all users in the data set. First, let us make a definition of the baseline estimates:

$$b_{ui} = \mu + b_u + b_i \tag{1}$$

The parameters b_u and b_i indicate the observed biases of user u and item i, respectively, from average. Here, μ denotes the overall average rating. Such a simple model would not represent the data set entirely. In this case, Koren [16] came up with the global neighborhood model, and he considered the effect of the overall user in the data set. An improved model can be expressed with the formula as the following:

$$\hat{r}_{ui} = \mu + b_u + b_i + |R(u)|^{-\frac{1}{2}} \sum_{j \in R(u)} (r_{uj} - b_{uj})w_{ij} + |N(u)|^{-\frac{1}{2}} \sum_{j \in N(u)} c_{ij} \qquad (2)$$

Here, $\hat{r}_{ui}$ means the prediction of user u to the item i, and $R(u)$ is the set that users give their ratings to the target item i. On the meantime, $|R(u)|$ is the size of the set, in other words, is the number of the users that have given their preferences to the item i. Compared with the baseline estimates, the $(r_{uj} - b_{uj})$ can be regarded as the bias of user j. Then, w_{ij} reveals the correlation coefficient of the bias. And c_{ij} is the implicit feedback, which provides an alternative way to learn user preferences. In order to weaken the dichotomy between heavy raters and those rare raters, we use $|N(u)|^{-\frac{1}{2}}$ to moderate the behavior. Here, $N(u)$ is the set that other users show their preferences to the target item i, and $|N(u)|$ is the size of the set. In our experiments, mostly, c_{ij} would indicate the rated users' preferences for the target item. In other words, $|R(u)|$ is equal to $|N(u)|$.

2.2 Latent Factor Models

A popular approach to latent factor models is induced by a SVD-like lower rank decomposition of the ratings matrix. Each user u is associated with a user-factors vector $p_u \in \mathbb{R}^f$, and each item i with an item-factors vector $q_i \in \mathbb{R}^f$. Prediction is done by the rule:

$$\hat{r}_{ui} = b_{ui} + p_u^T q_i \qquad (3)$$

Koren [16] suggested the SVD++ model, which takes the implicit feedback into consideration. With the modification of the implicit feedback, we could get more accurate prediction about the missing ratings. The model can be described as the following formula:

$$\hat{r}_{ui} = b_{ui} + q_i^T \left(p_u + |N(u)|^{-\frac{1}{2}} \sum_{j \in N(u)} y_j \right) \qquad (4)$$

Formula (4) adds the implicit feedback y_j into the influences of the user-factors vector p_u, where y_j denotes the preference of the user who had given his rating to the target item.

2.3 An Integrated Models

As mentioned in [15], there is no perfect model. Instead, the best results came from combining predictions of models that complemented each other. Koren [16] came up with the integrated model that combined neighborhood models and latent factor models, as following:

$$\widehat{r}_{ui} = \mu + b_u + b_i + q_i^T \left(p_u + |N(u)|^{-\frac{1}{2}} \sum_{j \in N(u)} y_j \right)$$
$$+ |R(u)|^{-\frac{1}{2}} \sum_{j \in R(u)} \left(r_{uj} - b_{uj} \right) w_{ij} + |N(u)|^{-\frac{1}{2}} \sum_{j \in N(u)} c_{ij} \tag{5}$$

In general, the parameters in the models mentioned above can be learnt by solving the regularized least squares problem with stochastic gradient descent. Taking model (5) as an example, the objective function can be described as formula (6). Here, the objective function can be divided into estimation errors $\left(r_{ui} - \widehat{r}_{ui} \right)^2$ and regularization term $\lambda(b_u^2 + \cdots + c_{ij}^2)$. It is obvious that the learning process takes a long time and many iterations [16]. Further, the better accuracy can be learned from the observations.

$$\min_{b_*,q_*,p_*,y_*,w_*,c_*} \sum_{(u,i) \in \kappa} \left(r_{ui} - \widehat{r}_{ui} \right)^2 + \lambda \left(b_u^2 + b_i^2 + \|q_i\|^2 + \|p_u\|^2 + \|y_j\|^2 + w_{ij}^2 + c_{ij}^2 \right) \tag{6}$$

3 Proposed Model

[17–19] proposed the missing not at random assumption, which means the missing/unknown ratings don't follow the same distribution about the observed ratings. The collaborative filtering approach split the observed data into training set and test set, using the known information to predict the unknown data. Inspired by the missing not at random assumption, we propose a novel approach to complement the bias between the observed data and the missing data. It is necessary to add some kinds of noise into the model, making the parameters more generalization. In the meanwhile, we would like to reduce the iterations and running time of approach, because of the massive data. Hence, we come up with the Gaussian iteration collaborative filtering, an approach to revise the loss function. Next section would give unambiguous presentation on the proposed method.

3.1 Revised Objective Function

As mentioned in Sect. 2, the objective function is combined with two parts: the estimation errors and the regularization term. The regularization term can prevent the model from overfitting. Considering the biases between observed data and unknown data, our proposed approach adds the Gaussian distribution to the estimation errors. The probability density of the normal distribution is:

$$f(x \mid \mu, \sigma^2) = \frac{1}{\sigma\sqrt{2\pi}} e^{\frac{(x-\mu)^2}{2\sigma^2}} \tag{7}$$

Here, μ is the mean or expectation of the distribution. The parameter σ is its standard deviation with its variance then σ^2. Let us denote the prediction error, $r_{ui} - \widehat{r}_{ui}$, by e_{ui}. In our model, the estimation errors in the objective function become:

$$\left(1 + f(\widehat{r}_{ui} \mid \mu, \sigma^2)\right) \cdot (r_{ui} - \widehat{r}_{ui})^2 \tag{8}$$

The mean μ and the standard deviation σ^2 can be tuned by the cross validation. According to the heuristic algorithm, we can start with the mean and the standard deviation of the ratings in the training set. In this case, the objective function of the global neighborhood model (2) would change into the following formula:

$$\min_{b_*, w_*, c_*} \sum_{(u,i) \in \kappa} \left(1 + f(\widehat{r}_{ui} \mid \mu, \sigma^2)\right) \cdot \left(\begin{matrix} r_{ui} - \mu - b_u - b_i - |R(u)|^{-\frac{1}{2}} \sum_{j \in R(u)} (r_{uj} - b_{uj}) w_{ij} + \\ |N(u)|^{-\frac{1}{2}} \sum_{j \in N(u)} c_{ij} \end{matrix}\right)^2$$
$$+ \lambda_1 (b_u^2 + b_i^2) + \lambda_2 \left(\sum_{j \in R(u)} w_{ij}^2 + \sum_{j \in N(u)} c_{ij}^2\right) \tag{9}$$

In the meanwhile, the objective function of the SVD++ (4) and the integrated model (5) have been revised as (9) and (10), respectively.

$$\min_{b_*, q_*, p_*} \sum_{(u,i) \in \kappa} \left(1 + f(\widehat{r}_{ui} \mid \mu, \sigma^2)\right) \cdot \left(r_{ui} - \mu - b_u - b_i - q_i^T \left(p_u + |N(u)|^{-\frac{1}{2}} \sum_{j \in N(u)} y_j\right)\right)^2$$
$$+ \lambda_3 (b_u^2 + b_i^2) + \lambda_4 \left(\|q_i\|^2 + \|p_u\|^2 + \|y_j\|^2\right) \tag{10}$$

$$\min_{b_*, q_*, p_*, y_*, w_*, c_*} \sum_{(u,i) \in \kappa} \left(1 + f(\widehat{r}_{ui} \mid \mu, \sigma^2)\right) \cdot \left(r_{ui} - \widehat{r}_{ui}\right)^2 + \lambda_5 (b_u^2 + b_i^2)$$
$$+ \lambda_6 \left(\|q_i\|^2 + \|p_u\|^2 + \|y_j\|^2\right) + \lambda_7 \left(w_{ij}^2 + c_{ij}^2\right) \tag{11}$$

3.2   Gaussian Iteration

In the revised objective function, the new parameters in the Gaussian distribution $\widehat{r}_{ui}, \mu, \sigma^2$ are regarded as constants in each iteration. Therefore, the gradients of the collaborative filtering approaches would be described as shown in Table 1.

We still use the stochastic gradient descent to update the parameters, randomly select the user's rating to train the model.

Table 1. The gradients of the collaborative filtering approaches

---

The Gaussian iteration on Global Neighborhood Model:

$\bullet b_u \leftarrow b_u + \gamma_1 \bullet \left(\left(1 + f(\hat{r}_{ui} \mid \mu, \sigma^2)\right) \bullet e_{ui} - \lambda_1 \bullet b_u \right)$

$\bullet b_i \leftarrow b_i + \gamma_1 \bullet \left(\left(1 + f(\hat{r}_{ui} \mid \mu, \sigma^2)\right) \bullet e_{ui} - \lambda_1 \bullet b_i \right)$

$\bullet \forall j \in R(u):$

$w_{ij} \leftarrow w_{ij} + \gamma_2 \bullet (\mid R(u) \mid^{-\frac{1}{2}} \bullet \left(1 + f(\hat{r}_{ui} \mid \mu, \sigma^2)\right) \bullet e_{ui} \bullet (r_{uj} - b_{uj}) - \lambda_2 \bullet w_{ij})$

$\bullet \forall j \in N(u):$

$c_{ij} \leftarrow c_{ij} + \gamma_2 \bullet (\mid N(u) \mid^{-\frac{1}{2}} \bullet \left(1 + f(\hat{r}_{ui} \mid \mu, \sigma^2)\right) \bullet e_{ui} - \lambda_2 \bullet c_{ij})$

The Gaussian iteration on Latent Factor Model:

$\bullet b_u \leftarrow b_u + \gamma_3 \bullet \left(\left(1 + f(\hat{r}_{ui} \mid \mu, \sigma^2)\right) \bullet e_{ui} - \lambda_3 \bullet b_u \right)$

$\bullet b_i \leftarrow b_i + \gamma_3 \bullet \left(\left(1 + f(\hat{r}_{ui} \mid \mu, \sigma^2)\right) \bullet e_{ui} - \lambda_3 \bullet b_i \right)$

$\bullet q_i \leftarrow q_i + \gamma_4 \bullet \left(\left(1 + f(\hat{r}_{ui} \mid \mu, \sigma^2)\right) \bullet e_{ui} \bullet \left(p_u + \mid N(u) \mid^{-\frac{1}{2}} \sum_{j \in N(u)} y_j \right) - \lambda_4 \bullet q_i \right)$

$\bullet p_u \leftarrow p_u + \gamma_4 \bullet \left(\left(1 + f(\hat{r}_{ui} \mid \mu, \sigma^2)\right) \bullet e_{ui} \bullet q_i - \lambda_4 \bullet p_u \right)$

$\bullet \forall j \in N(u):$

$y_j \leftarrow y_j + \gamma_4 \bullet \left(\left(1 + f(\hat{r}_{ui} \mid \mu, \sigma^2)\right) \bullet e_{ui} \bullet q_i \bullet \mid N(u) \mid^{-\frac{1}{2}} - \lambda_4 \bullet y_j \right)$

The Gaussian iteration on the integrated model:

$\bullet b_u \leftarrow b_u + \gamma_5 \bullet \left(1 + f(\hat{r}_{ui} \mid \mu, \sigma^2)\right) \bullet e_{ui} - \lambda_5 \bullet b_u)$

$\bullet b_i \leftarrow b_i + \gamma_5 \bullet \left(\left(1 + f(\hat{r}_{ui} \mid \mu, \sigma^2)\right) \bullet e_{ui} - \lambda_5 \bullet b_i \right)$

$\bullet q_i \leftarrow q_i + \gamma_6 \bullet \left(\left(1 + f(\hat{r}_{ui} \mid \mu, \sigma^2)\right) \bullet e_{ui} \bullet \left(p_u + \mid N(u) \mid^{-\frac{1}{2}} \sum_{j \in N(u)} y_j \right) - \lambda_6 \bullet q_i \right)$

$\bullet p_u \leftarrow p_u + \gamma_6 \bullet \left(\left(1 + f(\hat{r}_{ui} \mid \mu, \sigma^2)\right) \bullet e_{ui} \bullet q_i - \lambda_6 \bullet p_u \right)$

$\bullet \forall j \in N(u):$

$y_j \leftarrow y_j + \gamma_6 \bullet \left(\left(1 + f(\hat{r}_{ui} \mid \mu, \sigma^2)\right) \bullet e_{ui} \bullet \mid N(u) \mid^{-\frac{1}{2}} \bullet q_i - \lambda_6 \bullet y_j \right)$

$\bullet \forall j \in R(u):$

$w_{ij} \leftarrow w_{ij} + \gamma_7 \bullet \left(\mid R(u) \mid^{-\frac{1}{2}} \bullet \left(1 + f(\hat{r}_{ui} \mid \mu, \sigma^2)\right) \bullet e_{ui} \bullet (r_{uj} - b_{uj}) - \lambda_7 \bullet w_{ij} \right)$

$\bullet \forall j \in N(u):$

$c_{ij} \leftarrow c_{ij} + \gamma_7 \bullet \left(\mid N(u) \mid^{-\frac{1}{2}} \bullet \left(1 + f(\hat{r}_{ui} \mid \mu, \sigma^2)\right) \bullet e_{ui} - \lambda_7 \bullet c_{ij} \right)$

---

4 Experiments

In this section, we have conducted the experiments on three main collaborative filtering models to examine the performance of our proposed algorithm. Section 4.1 gives the description of the experiment datas, Sect. 4.2 introduces the adjustment process of the parameters in each model and Sect. 4.3 shows the experiment results.

4.1 Data Description

In order to verify the proposed approach, experiments were conducted on the three models mentioned before, which can make a valid conclusion. The datasets we used for the experiments are Movielens 1 M and Movielens 10 M [20]. In the experiments, we use the user ID, item ID and the ratings. Table 2 shows the characteristics of the datasets. The ratings in 10 M dataset and 1 M dataset are made on a 5-star scale, with half-star increments and one-star increments, respectively. Analyzing the 10 M data, it is easy to find out that the number of the movies is not consequent, which wastes the space of the storage. For saving the memory of computer, we pretreat the dataset by mapping the movie number into the successive one ranged 1 to 10677. The number of rated users had the same treatment ranged 1 to 69878. The datasets are split randomly into training set and test set, 80 % ratings on each user for training set and 20 % ratings for test set.

Table 2. Datasets information

Dataset	Number of users	Number of movies	Number of ratings	Rating range
1 M	6040	3952	1000209	[1, 5]
10 M	71567	10681	10000054	[1, 5]

In this paper, we use Root Mean Square Error (RMSE) to compare the performance of the models. The RMSE is defined as (12). Here, T_κ is the set of known ratings of the test set and $n(T_\kappa)$ is the number of ratings in (T_κ). For the dataset on three models, we ran the algorithm 5 times. In each experiment, the training set and test set were randomly selected. The final result of the dataset in the model is the average RMSE on the test set. In each model, we stop the updating of the parameters once the RMSE of the test set start increasing.

$$\sqrt{\frac{\sum_{(u,i)\in T_\kappa} \left(r_{ui} - \widehat{r}_{ui}\right)^2}{n(T_\kappa)}} \qquad (12)$$

4.2 Parameters Tuning

As mentioned above, there are some kinds of hyper parameter in the models. In general, there are learning rate, regularization coefficient, the mean and the standard deviation, which can be tuned by cross validation. The adjustment of the standard deviation σ is in reference to heuristic algorithm. For example, the mean of one Movielens 1 M training set is 3.58177, and the standard deviation is 1.125. Keeping the mean of the training set in the model, σ_1 is started from the standard deviation of the training set. Through the experiment, we select $\sigma_1 = 1.8$ as the optional choice. Table 3 shows the settings of experiment in the validation models. In both of latent factor models and the integrated models, the dimension of the user-factors vector $p_u \in \mathbb{R}^f$ and item-factor vector $q_i \in \mathbb{R}^f$ were set to 50.

Table 3. The hyper parameters for each dataset

	Neighborhood model	Latent factor model	Integrated model
1 M	$\gamma_1 = \gamma_2 = 0.005$, $\lambda_1 = \lambda_2 = 0.002$, $\sigma_1 = 1.8$	$\gamma_3 = 0.009, \gamma_4 = 0.008$, $\lambda_3 = 0.005, \lambda_4 = 0.02$, $\sigma_1 = 1.8$	$\gamma_5 = 0.007, \gamma_6 = 0.007$, $\gamma_7 = 0.005, \lambda_5 = 0.015$, $\lambda_6 = 0.015, \lambda_7 = 0.05$, $\sigma_1 = 1.8$
10 M	$\gamma_1 = \gamma_2 = 0.005$, $\lambda_1 = \lambda_2 = 0.002$, $\sigma_1 = 1.3$	$\gamma_3 = 0.0045, \gamma_4 = 0.0045$, $\lambda_3 = 0.015, \lambda_4 = 0.015$, $\sigma_1 = 1.3$	$\gamma_5 = 0.007, \gamma_6 = 0.007$, $\gamma_7 = 0.005, \lambda_5 = 0.015$, $\lambda_6 = 0.05, \lambda_7 = 0.002$, $\sigma_1 = 1.3$

4.3 Experiment Results

Figures 1, 2 and 3 show the experiment results of the Gaussian iteration on the Movielens datasets on three CF models, the lower RMSE the better, the less time the better and the fewer iteration number the better. All the experiments were done on a single processor 3.2 GHz Intel(R) Core(TM) PC. As shown in Figs. 1, 2 and 3, we can see that the RMSE on the Gaussian models is slightly worse than the original ones, but the running time and iteration number have significantly reduced. The decrease rate on the running time is more than 20 %. Considering that the amount of information generated from the Internet is tremendous, it is necessary to cut down the running time.

Next, we make an explanation of the experiment results. First, how the proposed models can significantly reduce the running time? The central limit theorem (CLT) states that, given certain conditions, the arithmetic mean of a sufficiently large number of iterates of independent random variables, each with a well-defined expected value and well-defined variance, will be approximately normally distributed, regardless of the underlying distribution [21]. In this case, we can conclude that most of the user will give the same ratings on the certain item. In the process of the stochastic gradient descent, once the prediction $\hat{r}_{ui}$ is near to the mean of the training set, we have the confidence that the descent direction is trustable and the declined step can be increased. As we all know, the Gaussian distribution is bell-shaped, symmetrical about the mean and the larger the variance the steeper the curve. Hence, adding the factor $\left(1+f(\widehat{r}_{ui}|\mu,\sigma^2)\right)$ to the prediction error is the good way to accelerate the speed of the iteration. In each iteration, the nearer the prediction to the mean, the larger the falling gradient, vice versa. Secondly, why adding the new factor would not make the prediction become worse? As mentioned in Sect. 3, our proposed models are based on the missing not at random assumption, which indicates the deviation between the missing data and the observed data. The observation is just a small part of the whole massive data. It is feasible to normalize the objective function with Gaussian distribution. In other words, the proposed approach has the ability to balance the noise among data, making the models more robust.

Figures 1, 2 and 3 show the performance of the proposed model in the collaborative filtering models with the Movielens datasets. Here, we denote the global neighborhood

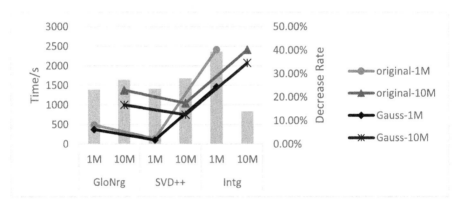

Fig. 1. Comparison on running time on 1 M dataset and 10 M dataset. (Color figure online)

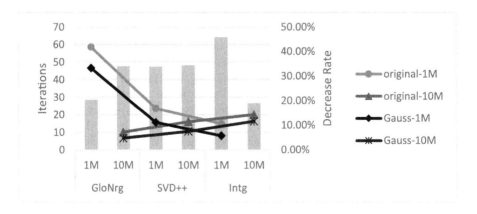

Fig. 2. Comparison on iterations on 1 M dataset and 10 M dataset. (Color figure online)

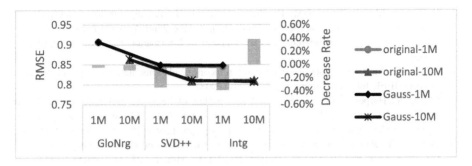

Fig. 3. Comparison on RMSE on 1 M dataset and 10 M dataset. (Color figure online)

model 2 as GloNrg, the SVD++ model (4) as SVD++ and the integrated model (5) as Intg. The line with circular and with diamond represent the original collaborative filtering approaches on the 1 M dataset and 10 M dataset, respectively. The line with diamond and with star represent our proposed model on the 1 M dataset and 10 M dataset, respectively.

5   Conclusions

In this work, we proposed a new, simple and efficient way to accelerate the iteration speed and optimize the performance on the collaborative filtering approach called Gaussian iteration. In this method, a Gaussian factor was added to the prediction error in the objective function, based on the missing not at random assumption and the central limit theorem. The practical experiment results on three kinds of models with the Movielens datasets showed that our proposed approach was better than the original method in performance and convergence speed.

Further work is to apply the Gaussian iteration to other models with the similar objective function, not only in the recommender system. Because the missing not at random assumption and the central limit theorem are based on the attribution of the data. In the experiment, parameter tuning is a thorny problem. Other future work will be to come up with an automatic tuning method.

References

1. James, M., Michael, C., Brad, B., et al: Big data: the next frontier for innovation, competition, and productivity (2011)
2. Ricci, F., Rokach, L., Shapira, B.: Introduction to Recommender Systems Handbook. Springer, US (2011)
3. Melville, P., Mooney, R.J., Nagarajan, R.: Content-boosted collaborative filtering for improved recommendations. In: AAAI/IAAI, pp.187–192 (2002)
4. Goldberg, D., Nichols, D., Oki, B.M., Terry, D.: Using collaborative filtering to weave an information tapestry. Commun. ACM **35**(12), 61–70 (1992)

5. Herlocker, J.L., Konstan, J.A., Borchers, A., Riedl, J.: An algorithmic framework for performing collaborative filtering. In: Proceedings of the 22nd Annual International ACM SIGIR Conference on Research and Development in Information Retrieval, pp. 230–237 (1999)
6. Linden, G., Smith, B., York, J.: Amazon.com recommendations: item-to-item collaborative filtering. IEEE Internet Comput. **7**(1), 76–80 (2003)
7. Chang, S., Harper, F.M., Terveen, L.: Using groups of items for preference elicitation in recommender systems. In: Proceedings of the 18th ACM Conference on Computer Supported Cooperative Work & Social Computing, pp. 1258–1269, New York (2015)
8. Pujahari, A., Padmanabhan, V.: A new grouping method based on social choice strategies for group recommender system. Comput. Intell. Data Min. **1**, 325–332 (2015)
9. Chen, M.H., Teng, C.H., Chang, P.C.: Applying artificial immune systems to collaborative filtering for movie recommendation. Adv. Eng. Inform. **29**(4), 830–839 (2015)
10. Patra, B.K., Launonen, R., Ollikainen, V., et al.: A new similarity measure using bhattacharyya coefficient for collaborative filtering in sparse data. Knowl.-Based Syst. **82**, 163–177 (2015)
11. Hofmann, T.: Latent semantic models for collaborative filtering. ACM Trans. Inf. Syst. **22**(1), 89–115 (2004)
12. Salakhutdinov, R., Mnih, A., Hinton, G.: Restricted Boltzmann machines for collaborative filtering. In: Proceedings of the 24th International Conference on Machine Learning, pp. 791–798. ACM (2007)
13. Blei, D.M., Ng, A.Y., Jordan, M.I.: Latent dirichlet allocation. J. Mach. Learn. Res. **3**, 993–1022 (2003)
14. Koren, Y., Bell, R., Volinsky, C.: Matrix factorization techniques for recommender systems. Computer **42**(8), 30–37 (2009)
15. Bell, R.M., Koren, Y.: Lessons from the Netflix prize challenge. ACM SIGKDD Explor. Newsl. **9**(2), 75–79 (2007)
16. Koren, Y.: Factorization meets the neighborhood: a multifaceted collaborative filtering model. In: Proceedings of the 14th ACM SIGKDD International Conference on Knowledge Discovery and Data Mining, pp. 426–434. ACM (2008)
17. Steck, H.: Training and testing of recommender systems on data missing not at random. In: Proceedings of the 16th ACM SIGKDD International Conference on Knowledge Discovery and Data Mining, pp. 713–722. ACM (2010)
18. McAuley, J., Leskovec, J.: Hidden factors and hidden topics: understanding rating dimensions with review text. In: Proceedings of the 7th ACM Conference on Recommender Systems, pp. 165–172. ACM (2013)
19. Devooght, D., Kourtellis, N., Mantrach, A.: Dynamic matrix factorization with priors on unknown values. In: Proceedings of the 21th ACM SIGKDD Conference on Knowledge Discovery and Data Mining, pp. 189–198. ACM (2015)
20. Movielens Dataset. http://grouplens.org/datasets/movielens/1m/
21. Rice, J.A.: Mathematical Statistics and Data Analysis, 3rd edn, pp. 181–188. Duxbury Press, Belmont (2007)

A Novel Image Segmentation Approach Based on Truncated Infinite Student's t-mixture Model

Lu Li, Wentao Fan, JiXiang Du$^{(\boxtimes)}$, and Jing Wang

Department of Computer Science and Technology,
Huaqiao University, Xiamen, China
{1400214003, fwt, jxdu, wroaring}@hqu.edu.cn

Abstract. Mixture models have been used as efficient techniques in the application of image segmentation. In order to segment images automatically without knowing the number of true image components, the framework of Dirichlet process mixture model (DPMM, also known as the infinite mixture model) has been introduced into conventional mixture models. In this paper, we propose a novel approach for image segmentation by considering the truncated Dirichlet Process of Student's t-mixture model (tDPSMM). We also develop a novel Expectation Maximization (EM) algorithm for parameter estimation in our model. The proposed model is tested on the application of images segmentation with both brain MR images and natural images. According to the experimental results, our method can segment images effectively and automatically by comparing it with other state-of-the-art image segmentation methods based on mixture models.

Keywords: Image segmentation · Truncated Dirichlet process · Student's t-mixture model · Novel Expectation Maximization algorithm

1 Introduction

Image segmentation is an analyzing function in most image processing. Varieties of methods have been adopted to perform image segmentation [1], such as clustering, which is a widely used unsupervised machine learning method. In image segmentation, clustering means dividing pixels of an image into several groups and assigning labels to each group [2, 3]. In this paper, we focus on developing a novel image segmentation approach based on a clustering technique known as mixture modeling. Our motivation is that, each component in a mixture model can be considered as a region of an image, which is an intuitive assumption for image segmentation.

Among many existing mixture models, Gaussian mixture model (GMM) [4] is one of the most well-known models and has been used a lot in image segmentation [5–8]. However, the segmentation results of GMM are extremely sensitive to noise in images [5–8]. In order to solve this problem, image segmentation method using Student's t-mixture model (SMM) [9] has been proposed in [10, 11]. This approach assumes that pixels of an image can be divided into certain regions, where each region is considered

© Springer International Publishing Switzerland 2016
D.-S. Huang et al. (Eds.): ICIC 2016, Part III, LNAI 9773, pp. 271–281, 2016.
DOI: 10.1007/978-3-319-42297-8_26

following a Student's t-distribution, and thus each region corresponds to a component of the SMM. Since Student's t-distribution has an additional parameter called degree of freedom which contains heavier tail than Gaussian distribution, SMM is more robust to noise than GMM for image segmentation. Expectation Maximization (EM) algorithm is available to solve SMM and the model has been used effectively to perform image segmentation, such as in [10, 11]. However, there exists a significant limitation in either GMM or SMM approach for image segmentation. That is, the number of segmentation components normally cannot be selected automatically. Since the user-specified number of segments may cause limitation and inconvenience in practice, Dirichlet process mixture models (DPMM) [12] are then adopted as an effective solution to this problem. DPMM can be depicted as a mixture model with an infinite number of components. In our case, rather than the finite SMM, we perform image segmentation using an infinite Student's t-mixture model (iSMM) in which the Dirichlet process is used as the prior for components. The DPMM has become popular in machine learning recently [12–14]. But common learning techniques to learn the DPMM are always complicated both in theory and computing, such as variational Bayes (VB) [15] or Markov chain Monte Carlo (MCMC) [12]. Thus, more efficient methods for learning DPMM are desired.

Based on the considerations above, we propose a novel image segmentation approach based on truncated Dirichlet Process Student's t-mixture model (tDPSMM). In our tDPSMM, the number of mixing components is limited to a reasonably large number by adopting a truncation approach [16, 17]. And we propose a novel EM algorithm based on the idea proposed in [18] to learn the model parameters by derivation of the objective function. The method can segment images effectively and automatically by comparing it with some other state-of-the-art image segmentation methods.

The rest of the paper is organized as follows. In Sect. 2, an introduction of SMM is presented. Our tDPSMM and the proposed novel EM algorithm are described in detail in Sect. 3. Performances of our model are evaluated in Sect. 4. Conclusions are made in Sect. 5.

2   Finite Student's t-mixture Model

Let $\vec{x}_1, \cdots, \vec{x}_N \in X$ denote observations or pixels in an image. The values of pixels show intensity or color of an image. The jth pixel $\vec{x}_j$ is a vector of d dimension. SMM assumes that these pixels are independent and identically distributed according to a mixture of K Student's t-distributions. The log-likelihood function of SMM is given by

$$L(\Psi) = \sum_{j=1}^{N} \sum_{i=1}^{K} \log \pi_i t(\vec{x}_j \mid \Theta_i) \tag{1}$$

where $\Psi = \{\pi_{1\ldots K}, \Theta_{1\ldots K}\}$, and Student's t-distribution $t(\vec{x}_j \mid \Theta_i)$ with mean $\vec{\mu}_i$, covariance $\sum_i$, degree of freedom v_i is the probability density function of the ith component, which is given by

$$t(\vec{x}_j|\Theta_i) = \frac{\Gamma(v_i+d)|\Sigma_i|^{-\frac{1}{2}}}{(\pi_i v_i)^{\frac{1}{2}}\Gamma(\frac{v_i}{2})\{\frac{\delta(\vec{x}_j,\vec{\mu}_i;\Sigma_i)}{v_i}\}^{\frac{v_i+d}{2}}} \tag{2}$$

where $\Theta_i = \{\vec{\mu}_i, \Sigma_i, v_i\}$, $\delta(\vec{x}_j, \vec{\mu}_i; \Sigma_i) = (\vec{x}_j - \vec{\mu}_i)^T \Sigma_i^{-1}(\vec{x}_j - \vec{\mu}_i)$ and $\Gamma(.)$ is Gamma function.

And mixing weights $\pi_{1...K}$ satisfy

$$\pi_i \geq 0 \text{ and } \sum_{i=1}^{K} \pi_i = 1 \tag{3}$$

According to the complete data condition, maximizing the log-likelihood function in (1) will lead to an increase in the value of the objective function $J(\Psi)$ [8]:

$$J(\Psi) = \sum_{j=1}^{N} \sum_{i=1}^{K} z_{ij} \log \pi_i t(\vec{x}_j; \vec{\mu}_i, \Sigma_i, v_i) \tag{4}$$

To maximize the function in (4), Student's t-distribution is considered as infinite mixture of scaled Gaussian distribution [9] because there is no closed-form solution for the Student's t-mixture model. So, we have

$$J(\Psi) = \sum_{j=1}^{N} \sum_{i=1}^{K} z_{ij} \log \pi_i N(\vec{x}_j|\vec{\mu}_i, \Sigma_i/\vec{u}_j) \Gamma(\vec{u}_j|v_i/2, v_i/2) \tag{5}$$

Details about SMM may refer to [9].

3 Truncated Dirichlet Process Student's t-mixture Model and a Novel EM Algorithm

The number of components for SMM in Sect. 2 is assumed to be fixed and known. However, in practice it needs to be estimated from data. Thus, Dirichlet process has been introduced for finite SMM in order to form an infinite mixture model. Now, for the Dirichlet Process of Student's t-mixture model (DPSMM), the objective function $J(\Psi)$ becomes

$$J(\Psi) = \sum_{j=1}^{N} \sum_{i=1}^{\infty} z_{ij} \log \pi_i N(\vec{x}_j|\vec{\mu}_i, \Sigma_i/\vec{u}_j) \Gamma(\vec{u}_j|v_i/2, v_i/2) \tag{6}$$

where infinite mixing weights $\pi_{1...\infty}$ are defined as below

$$\pi_i = \begin{cases} w_i \prod_{m=1}^{i-1}(1-w_m) & i > 1 \\ w_1 & i = 1 \end{cases} \tag{7}$$

The equation in (7) corresponds to stick-breaking prior representation of the Dirichlet process. $w_{1...\infty}$ are random variables from a beta distribution, $w_i \sim \text{beta}(1,\alpha)$. The higher hyper parameter α is, the higher number of components in DPSMM is.

For DPMM, it is shown in [16, 17] that a truncated Dirichlet prior always has virtually no effect if the number of components is reasonably large. So, we consider a truncated Dirichlet prior with a reasonably large K and set $w_k = 1$. Figure 1 shows the graphical model representation of tDPSMM. Within this graph, hollow circles in this graph represent stochastic variables; shadow circle denotes observations; rounded rectangle denotes hyper parameter; frames indicate repetitions with the number of repetitions in the corners and arrows describe conditional dependencies between variables.

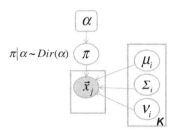

Fig. 1. Probabilistic graphical model of tDPSMM

Based on equation in (7), the objective function of tDPSMM is

$$J(\Psi) = \sum_{j=1}^{N} \sum_{i=1}^{K} z_{ij} \log N(\vec{x}_j | \vec{\mu}_i, \Sigma_i / \vec{u}_j) \Gamma(\vec{u}_j | v_i/2, v_i/2)$$
$$+ \sum_{j=1}^{N} \sum_{i=1}^{K} z_{ij} \log w_i + \sum_{m=1}^{K-1} \log(1 - w_m) \left(\sum_{j=1}^{N} \sum_{i=m+1}^{K} z_{ij} + \alpha - 1 \right) \tag{8}$$

where $\Psi = \{\vec{\mu}_{1\cdots K}, \Sigma_{1\cdots K}, w_{1\cdots K}, v_{1\cdots K}\}$, and the conditional expectation values z_{ij} of hidden variables are computed as follows

$$z_{ij} = \frac{\pi_i t(\vec{x}_j; \vec{\mu}_i, \Sigma_i, v_i)}{\sum_{m=1}^{K} \pi_m t(\vec{x}_j; \vec{\mu}_m, \Sigma_m, v_m)} \tag{9}$$

and u_{ij} can be computed by

$$u_{ij} = \frac{v_i + d}{v_i + (\vec{x}_j - \vec{\mu}_i)^T (\Sigma_i)^{-1} (\vec{x}_j - \vec{\mu}_i)} \tag{10}$$

Instead of using MCMC or VB for DPMM learning, we propose a novel EM algorithm for tDPSMM learning. By maximizing the objective function (8) of parameters in Ψ, we obtain

$$\vec{\mu}_i = \frac{\sum_{j=1}^{N} z_{ij} u_{ij} \vec{x}_j}{\sum_{j=1}^{N} z_{ij} u_{ij}} \tag{11}$$

$$\Sigma_i = \frac{\sum_{j=1}^{N} z_{ij} u_{ij} (\vec{x}_j - \vec{\mu}_i)(\vec{x}_j - \vec{\mu}_i)^T}{\sum_{j=1}^{N} z_{ij} u_{ij}} \tag{12}$$

$$w_i = \frac{\sum_{j=1}^{N} z_{ij}}{\sum_{j=1}^{N} \sum_{m=i+1}^{K} z_{mj} + \alpha - 1 + \sum_{j=1}^{N} z_{ij}} \quad i = 1:K-1 \tag{13}$$

v_i is not close-formed and can be updated using equation below in iterative EM algorithm:

$$\log(\frac{v_i}{2}) - \psi(\frac{v_i}{2}) + 1 - \log(\frac{v_i^{old} + d}{2}) + \frac{\sum_{j=1}^{N} z_{ij} (\log u_{ij} - u_{ij})}{\sum_{j=1}^{N} z_{ij}} + \psi(\frac{v_i^{old} + d}{2}) = 0 \tag{14}$$

where v_i^{old} represents old value of v_i and is $\psi(.)$ is Digamma function.

Hence, the main steps of the proposed novel EM algorithm for tDPSMM are listed as follows

Step (1)   Initialize the parameters $\pi_{1...K}, \vec{\mu}_{1...K}, \sum_{1...K}, v_{1...K} = 1$

Step (2)   E step: Update z_{ij} and u_{ij} using (9) and (10)

Step (3)   M step: Update $\vec{\mu}_i, \sum_i, w_i$ and v_i using (11), (12), (13) and (14). Compute π_i using (7)

Step (4)   Evaluate the log-likelihood function in (8) and check the convergence of it. If not converge, go to step 2

Once the algorithm converges, parameters of tDPSMM are determined. Then pixels $\vec{x}_1, \cdots, \vec{x}_N$ in an image can be divided according to largest posterior probability, which is illustrated in detail below.

In tDPSMM, we build a Student's t-mixture model for the distribution of pixels in an image. Each component (Student's t-distribution) in the mixture model can be considered as a region of an image. The number of significant components (i.e. whose estimated mixing weights $\pi_i > 0.01$) and the parameters of these components are estimated by the novel EM algorithm. For each pixel, the selection of Student's t-distribution with the highest posterior probability produces segmentation. According to Bayes rule, the posterior probability for a pixel $\vec{x}_j$ to be classified into the ith component is

$$p(\text{component } i \,|\, \vec{x}_j) = \frac{p(\vec{x}_j \,|\, \text{component } i)p(\text{component } i)}{p(\vec{x}_j)} \tag{15}$$

where $i = 1 : K, j = 1 : N$ and $p(\text{component } i \,|\, \vec{x}_j)$ is equivalent to z_{ij} in (9).

Hence, the jth pixel is divided into the ith component. The value of i depends on equation below.

$$\text{component label } i = \underset{i \in \{1,...,K\}}{\arg \max} \, p(\text{component } i \,|\, \vec{x}_j) = \underset{i \in \{1,...,K\}}{\max} z_{ij} \tag{16}$$

4 Experimental Results

In this section, the performance of the proposed algorithm is compared with those of Dirichlet Process Gaussian mixture model (DPGMM) implemented by MCMC [12] and EM algorithm [18], fuzzy c-means based hidden Markov random field (HMRF-FCM) [19] and Student's-t mixture model with spatial constraints (SMM-SC) [20] on brain MR images in Sect. 4.1. In order to evaluate the results of different algorithms, misclassification ratio (MCR) [20] is employed. MCR is defined as

$$MCR = \frac{\text{number of misclassified pixels}}{\text{total number of pixels}} \times 100 \qquad (17)$$

The value of MCR varies from 0 to 100 and lower MCR indicates better segmentation. For brain MR images, total number of pixels in (17) means brain tissue pixels. In Sect. 4.1, Dice metric [20, 21] is also used to quantify the overlap between segmented results with ground truth. We define that TP refers to the number of true positives, FP to false positives, FN to false negatives and TN to true negatives. So, Dice metric is written as

$$Dice\ metric = \frac{2TP}{2TP + FP + FN} \times 100 \qquad (18)$$

Where TP + FP + FN + TN = number of brain tissue pixels in a MR image. The value of Dice metric ranges from 0 to 100, where higher value indicates better agreement. In order to further test the performace of our method, natural images are tested in Sect. 4.2. We fit our tDPSMM with α=2 based on the experimental results.

4.1 Brain MR Images

In the first experiment of this section, a real brain MR image set (IBSR05, 256 × 256 × 128) obtained from the Center for Morphometric Analysis at Massachusetts General Hospital [22] is used to test effectiveness of different algorithms. Figure 2(a) shows an image (index = 47) in IBSR05. The objective is to segment the image into three parts: cerebral spinal fluid (CSF), gray matter (GM) and white matter (WM). A low contrast between GM and WM increases the difficulty of segmentation. The ground truth of Fig. 2(a) with three classes is shown in Fig. 2(b). We first compare our tDPSMM with two state-of-the-art segmentation methods also based on infinite mixture models: DPGMM with MCMC [12] and EM algorithm [18]. Figure 2(c)–(e) shows the results. All the three methods segment the image into three components without knowing the number of true image components in advance. However, the result in Fig. 2(c) is sensitive to noise. Many pixels in WM and GM region have been misclassified in both Fig. 2(c) and (d), which is also shown in Table 1 that Dice2 and Dice3 obtained by DPGMM with MCMC and EM algorithm are poor. Figure 2(e) shows that our method produces better segmentation with lower MCR.

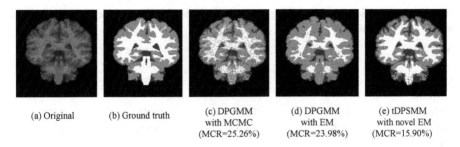

| (a) Original | (b) Ground truth | (c) DPGMM with MCMC (MCR=25.26%) | (d) DPGMM with EM (MCR=23.98%) | (e) tDPSMM with novel EM (MCR=15.90%) |

Fig. 2. Segmentation results of a brain MR image (IBSR05, index = 47)

In the above three iterative infinite mixture models, the time complexity of each iteration are all $O(K \times N)$, where K is the number of components in mixture models and N is the number of pixels in an image. However, the proposed tDPSMM converges faster than the other two models as shown in Fig. 3.

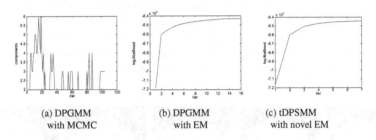

| (a) DPGMM with MCMC | (b) DPGMM with EM | (c) tDPSMM with novel EM |

Fig. 3. Convergences of segmentation (IBSR05, index = 47)

Table 1. Comparision for different segmentation methods on IBSR05 (index = 47): Dice metric (Dice$_j$ %), MCR (%)

Methods	Dice$_1$(CSF)	Dice$_2$(GM)	Dice$_3$(WM)	MCR
DPGMM with MCMC	71.63	76.19	73.37	25.26
HMRF-FCM	17.28	75.37	91.21	24.04
DPGMM with EM	82.63	78.50	67.84	23.98
SMM-SC	23.42	81.95	91.87	18.11
tDPSMM with novel EM	70.45	82.74	90.92	15.90

We use the same image as in [19, 20], so we also compare our tDPSMM with HMRF-FCM [19] and SMM-SC [20] in Table 1. Dice1 (CSF) of these two finite mixture models is poor, which means that CSF region is segmented poorly through HMRF-FCM and SMM-SC. Within methods in the table, the proposed tDPSMM produces best segmentation with high Dice metric and lowest MCR.

The average Dice metric and MCR on IBSR05 obtained by different segmentation methods are shown in Table 2, which also convince that our program performs well.

Table 2. Comparision for different segmentation methods on IBSR05 (index = $1 \sim 128$): Dice metric (Dice$_j$ %), MCR (%)

Methods	Average Dice$_1$(CSF)	Average Dice$_2$(GM)	Average Dice$_3$(WM)	Average MCR
DPGMM with MCMC	62.43	80.41	71.93	24.15
HMRF-FCM	17.88	80.91	85.91	21.72
DPGMM with EM	69.81	84.90	71.85	18.93
SMM-SC	24.72	84.86	84.08	17.81
tDPSMM with novel EM	73.60	85.90	85.27	16.24

In the second experiment, another dataset (IBSR01) is used. Image slices at index 30 and 50 are shown in the first column of Figs. 4 and 5. As seen from the last three columns in Figs. 4 and 5, DPGMM with EM algorithm reduces the effect of noise compared with DPGMM with MCMC. However, many pixels are still misclassified. The proposed tDPSMM in the last column has a lower MCR. The average Dice metric and MCR for images in IBSR01 by different algorithms are listed in Table 3. It can be seen again that the proposed method performs effectively.

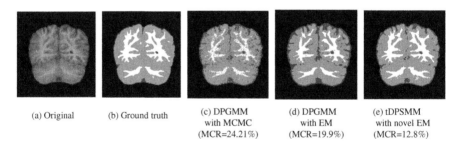

(a) Original   (b) Ground truth   (c) DPGMM with MCMC (MCR=24.21%)   (d) DPGMM with EM (MCR=19.9%)   (e) tDPSMM with novel EM (MCR=12.8%)

Fig. 4. Segmentation results of a brain MR image (IBSR01, index = 30)

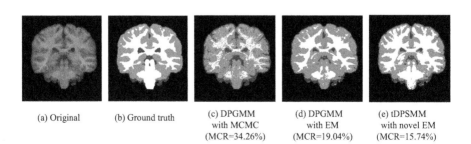

(a) Original   (b) Ground truth   (c) DPGMM with MCMC (MCR=34.26%)   (d) DPGMM with EM (MCR=19.04%)   (e) tDPSMM with novel EM (MCR=15.74%)

Fig. 5. Segmentation results of a real brain image (IBSR01, index = 50)

Table 3. Comparison for different segmentation methods on IBSR01 (index = 1 ∼ 128): Dice metric (Dicej %), MCR (%)

Methods	Average Dice$_1$(CSF)	Average Dice$_2$(GM)	Average Dice$_3$(WM)	Average MCR
DPGMM with MCMC	44.27	80.49	69.70	24.36
HMRF-FCM	16.56	82.15	86.01	21.37
DPGMM with EM	45.48	85.03	85.33	18.35
SMM-SC	24.42	86.49	84.79	16.90
tDPSMM with novel EM	47.19	89.09	84.93	12.48

In the third experiment, all brain MR images (from IBSR01 to IBSR18) from the Center for Morphometric Analysis at Massachusetts General Hospital are used. The average misclassification ratio of DPGMM with MCMC, HMRF-FCM, DPGMM with EM algorithm, SMM-SC and the proposed tDPSMM with novel EM are 24.62, 21.18, 19.70, 17.39, and 13.86 respectively. Our proposed method achieves the best accuracy.

4.2 Natural Images

Natural images from the Berkeley's image dataset [23] are used to compare proposed algorithm with SMM [10], DPGMM with MCMC [12] and EM algorithm [18]. The initialization is the same. One difficulty is the true number of components is unknown.

A gray image, as shown in Fig. 6(a), is used. The objective is to segment the image into three components: tree and elephants, grasses and sky. In SMM, K is set to 3 in advance. A visual inspection of results indicates that the number of components in Fig. 6(c) is estimated wrongly. The number of components is estimated correctly in Fig. 6(d) and (e). However, the effect of noise on grasses is smalleer in Fig. 6(e).

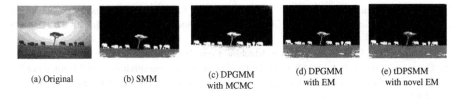

| (a) Original | (b) SMM | (c) DPGMM with MCMC | (d) DPGMM with EM | (e) tDPSMM with novel EM |

Fig. 6. Segmentation examples (image 2530362)

In the next experiment, image in Fig. 7(a) is used. The objective is to segment the image into the insect and grasses. Figure 7(b)–(e) shows the results. In Fig. 7(b), we set K = 2. Some details about the insect are lost. The latter three methods all estimate the number of components successfully. However, the effect of noise on Fig. 7(e) is smaller. So, tDPSMM performs better without knowing the number of components.

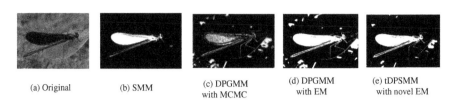

| (a) Original | (b) SMM | (c) DPGMM with MCMC | (d) DPGMM with EM | (e) tDPSMM with novel EM |

Fig. 7. Segmentation examples (image 350702)

5 Conclusion

In this work, we introduced truncated Dirichlet process into finite SMM to achieve automatic image segmentation. Besides, we proposed a novel EM algorithm for learning tDPSMM and showed that it can be effectively used in image segmentation. The proposed method is more effective as compared to other algorithms in our experiments. One limitation is the initialization may affect segmentation. So, extension of the current work may devote to reducing the dependence of the initialization.

Acknowledgements. This work was supported by the Grant of the National Science Foundation of China (No.61175121, 61502183), the Grant of the National Science Foundation of Fujian Province (No.2013J06014), the Promotion Program for Young and Middle-aged Teacher in Science and Technology Research of Huaqiao University (No.ZQN-YX108), the Scientific Research Funds of Huaqiao University (No.600005-Z15Y0016), the National Natural Science Foundation of China (61502183), the Scientific Research Funds of Huaqiao University (600005-Z15Y0016) and Subsidized Project for Cultivating Postgraduates' Innovative Ability in Scientific Research of Huaqiao University (Project No. 1400214003,1400214009).

References

1. Khan, W.: Image segmentation techniques: a survey. J. Image Graph. **1**(4), 166–170 (2013)
2. Xu, R., Wunsch, D.: Survey of Clustering Algorithms. IEEE Trans. Neural Netw. **16**(3), 645–678 (2005)
3. Cai, W., Chen, S., Zhang, D.: Fast and robust fuzzy c-means clustering algorithms incorporating local information for image segmentation. Pattern Recogn. **40**(3), 825–838 (2007)
4. Bishop, C.M.: Pattern Recognition. Machine Learning (2006)
5. Greenspan, H., Ruf, A., Goldberger, J.: Constrained gaussian mixture model framework for automatic segmentation of MR brain images. IEEE Trans. Med. Imaging **25**(9), 1233–1245 (2006)
6. Diplaros, A., Vlassis, N., Gevers, T.: A spatially constrained generative model and an EM algorithm for image segmentation. IEEE Trans. Neural Netw. **18**(3), 798–808 (2007)
7. Nguyen, T.M., Wu, Q.J.: Gaussian-mixture-model-based spatial neighborhood relationships for pixel labeling problem. IEEE Trans. Syst. Man Cybern. Part B Cybern. **42**(1), 193–202 (2012)
8. Nguyen, T.M., Wu, Q.: Fast and robust spatially constrained gaussian mixture model for image segmentation. IEEE Trans. Circ. Syst. Video Techn. **23**(4), 621–635 (2013)

9. Peel, D.G., McLachlan, J.: Robust mixture modelling using the t distribution. Stat. Comput. **10**(4), 339–348 (2000)
10. Sfikas, G., Nikou, C., Galatsanos, N.: Robust image segmentation with mixtures of student's t-distributions. In: IEEE International Conference on Image Processing. IEEE (2007)
11. Phophalia, A., Mitra, S.K.: Bayesian estimation of spatially variant finite mixture model for brain MR image segmentation. Asian J. Comput. Sci. Inf. Technol. **2**(5), 94–98 (2013)
12. Da Silva, A.R.F.: A Dirichlet process mixture model for brain MRI tissue classification. Med. Image Anal. **11**(2), 169–182 (2007)
13. Jbabdi, S., Woolrich, M.W., Behrens, T.E.J.: Multiple-subjects connectivity-based parcellation using hierarchical Dirichlet process mixture models. NeuroImage **44**(2), 373–384 (2009)
14. Wang, J.-C., et al.: Hierarchical Dirichlet process mixture model for music emotion recognition. IEEE Trans. Affect. Comput. **6**(3), 261–271 (2015)
15. Wei, X., Li, C.: The infinite student's t-mixture for robust modeling. Sig. Process. **92**(1), 224–234 (2012)
16. Blei, D.M., Jordan, I.: Variational inference for Dirichlet process mixtures. Bayesian Analy. **1**(1), 121–143 (2006)
17. Ishwaran, H., James, L.F.: Gibbs sampling methods for stick-breaking priors. J. Am. Stat. Assoc. (2011)
18. Kimura, T., et al.: Expectation-maximization algorithms for inference in Dirichlet processes mixture. Pattern Anal. Appl. **16**(1), 55–67 (2013)
19. Chatzis, S.P., Varvarigou, T.A.: A fuzzy clustering approach toward hidden markov random field models for enhanced spatially constrained image segmentation. IEEE Trans. Fuzzy Syst. **16**(5), 1351–1361 (2008)
20. Nguyen, T.M., Wu, Q.J.: Robust student's-t mixture model with spatial constraints and its application in medical image segmentation. IEEE Trans. Med. Imaging **31**(1), 103–116 (2012)
21. Ashburner, J., Friston, K.J.: Unified segmentation. Neuroimage **26**(3), 839–851 (2005)
22. Data Exchange. Center for Morphometric Analysis. Massachusetts General Hospital
23. Martin, D., et al.: A database of human segmented natural images and its application to evaluating segmentation algorithms and measuring ecological statistics. In: Proceedings of the Eighth IEEE International Conference on computer vision, ICCV 2001. IEEE (2001)

Stock Price Prediction Through the Mixture of Gaussian Processes via the Precise Hard-cut EM Algorithm

Shuanglong Liu and Jinwen Ma[✉]

Department of Information Science, School of Mathematical
Sciences and LMAM, Peking University, Beijing 100871, China
jwma@math.pku.edu.cn

Abstract. In this paper, the mixture of Gaussian processes (MGP) is applied to model and predict the time series of stock prices. Methodically, the precise hard-cut expectation maximization (EM) algorithm for MGPs is utilized to learn the parameters of the MGP model from stock prices data. It is demonstrated by the experiments that the MGP model with the precise hard-cut EM algorithm can be successfully applied to the prediction of stock prices, and outperforms the typical regression models and algorithms.

Keywords: Mixture of Gaussian processes · EM algorithm · Parameter learning · Stock price · Times series prediction

1 Introduction

The stock market has the characteristics of high return and high risk [1], which has always been concerned on the analysis and forecast of stock prices. Actually, the complexity of the internal structure in stock price system and the diversity of the external factors (the national policy, the bank rate, price index, the performance of quoted companies and the psychological factors of the investors) determine the complexity of the stock market, uncertainty and difficulty of stock price forecasting task [2]. Because the stock price is collected according to the order of time, it actually forms a complex nonlinear time series [3]. Some traditional stock market analysis methods, such as stock price graph analysis (k line graph [4]), cannot profoundly reveal the stock intrinsic relationship, so that the prediction results are not so ideal on stock price. Stock price prediction methodologies fall into three broad categories which are fundamental analysis, technical analysis (charting) and technological methods.

From the view of mathematics, the key to effective stock price prediction is to discover the intrinsic mapping or function, and to fit and approximate the mapping or the function. As it has been quickly developed, the mixture of Gaussian processes (MGP) model [5] is a powerful tool for solving this problem. But most of the MGP models are very complex and involve a large number of parameters and hyper-parameters, which makes the application of the MGP models very difficult [6]. Thus, we adopt the MGP model which proposed in [7] with excluding unnecessary priors and carefully selecting the model structure and gating function. This MGP model

© Springer International Publishing Switzerland 2016
D.-S. Huang et al. (Eds.): ICIC 2016, Part III, LNAI 9773, pp. 282–293, 2016.
DOI: 10.1007/978-3-319-42297-8_27

remains the main structure, features and advantages of the original MGP model. Moreover, it can be effectively applied to the modeling and prediction of nonlinear time series via the precise hard-cut EM algorithm. In fact, the precise hard-cut EM algorithm is more efficient than the soft EM algorithm since we could get the hyper-parameters of each GP independently in the M-step. It was demonstrated by the experimental results that this precise hard-cut EM algorithm for the MGP model really gives more precise prediction than some typical regression models and algorithms.

Along this direction, we apply the MGP model to the short-term stock price forecasting via the precise hard-cut EM algorithm. The experimental results show that this MGP based method can find potential rules from historical datasets, and their forecasting results are more stable and accurate.

The rest of this paper is organized as follows. In Sect. 2, we give a brief review of the MGP model and introduce the precise hard-cut EM algorithm. Section 3 presents the framework of stock price forecasting and the experimental results of the MGP based method as well as the comparisons of the regression models and algorithms. Finally, we give a brief conclusion in Sect. 4.

2   The Precise Hard-cut EM Algorithm for MGPs

2.1   The MGP Model

We consider the MGP model as described in [7]. In fact, it can be viewed as a special mixture model where each component is a GP. The whole set of indicators $Z = [z_1, z_2, \ldots, z_N]^T$, inputs X and outputs Y are sequentially generated and the MGP model is mathematically defined as follows:

$$p(z_t = c) = \pi_c, t = 1, 2, \ldots, N, \tag{1}$$

$$p(x_t | z_t = c, \theta_c) = N(x_t | \mu_c, S_c), t = 1, 2, \ldots, N, \tag{2}$$

$$p(y | X, \theta) = \prod_{c=1}^{C} N(y_c | 0, K(X_c, X_c | \theta_c) + \sigma_c^2 I_{N_c}) \tag{3}$$

where $K(x_i, x_j) = g^2 \exp\{-\frac{1}{2}(x_i - x_j)^T B(x_i - x_j)\}$, $B = \mathrm{diag}\{b_1^2, b_2^2, \ldots, b_d^2\}$, and Eq. (2) adopts Gaussian inputs in most generative MGP models [8–10]. $\theta_c = \{\pi_c, \mu_c, S_c, g_c, b_{c,1}, b_{c,2}, \ldots, b_{c,d}, \sigma_c\}$ are the parameters in the c-th GP component and $\theta = \{\theta_c\}_{c=1}^{C}$ denotes all the parameters in the mixture model.

The generative structure is prominent and clear for the MGP model, and the model avoids the complicated parameters setting. In various GP components, Gaussian means μ_c are different so that each component concentrates on the different region and this mixture model can fit multimodal dataset.

2.2   The Precise Hard-cut EM Algorithm

To avoid the computational complexity of Q function, it is reasonable to use the hard-cut version of the EM algorithm and we then can efficiently learn the parameters for the MGP model. In fact, the precise hard-cut EM algorithm [7] is a good choice and we summarize its procedures as follows:

Algorithm 1. The Precise Hard-cut EM Algorithm

1: Initialization of indicators:
   Cluster $\{(x_t, y_t)\}_{t=1}^N$ into C classes by the k-means clustering, and set z_t as the indicator of the t-th sample to the cluster.
2: **repeat**
3:    M-step:
      Calculate π_c, μ_c and S_c in the way of the Gaussian mixture model:

$$\pi_c = \frac{1}{N} \sum_{t=1}^N I(z_t = c) \tag{4}$$

$$\mu_c = \frac{\sum_{t=1}^N I(z_t=c)x_t}{\sum_{t=1}^N I(z_t=c)} \tag{5}$$

$$S_c = \frac{\sum_{t=1}^N I(z_t=c)(x_t-\mu_c)(x_t-\mu_c)^T}{\sum_{t=1}^N I(z_t=c)} \tag{6}$$

   and obtain the GP parameters by maximizing the likelihood.
4:    E-step:
      Classify each sample into the corresponding component according to the MAP criterion:

$$z_t = \text{argmax}_c p(z_t = c|x_t, y_t) = \text{argmax}_c \pi_c N(x_t|\mu_c, S_c)N(y_t|0, g_c^2 + \sigma_c^2) \tag{7}$$

5: **until**   Either the component remains the same in the previous iteration, or the iteration number reaches certain threshold.
6: Output the estimated parameters of MGP.

After the convergence of the precise hard-cut EM algorithm, we have obtained the estimates of all the parameters for the MGP. For a test input x^*, we can classify it into the z-th component of the MGP by the MAP criterion as follows:

$$z = \text{argmax}_c p(z^* = c|x^*) = \text{argmax}_c \pi_c N(x^*|\mu_c, S_c) \tag{8}$$

Based on such a classification, we can predict the output of the test input via the corresponding GP using

$$\hat{y}^* = K(x^*, X)\left[K(X,X) + \sigma^2 I\right]^{-1} y \tag{9}$$

In the next section, the precise hard-cut EM algorithm for the MGP model will be used for the stock closing price prediction, and the obtained results will be compared with the classical regression models and algorithms.

3 Stock Price Prediction

3.1 The General Prediction Model

The time series can be denoted as $\{s(t)\}_{t=1}^{\infty}$. For time series prediction task under certain conditions, Taken's Theorem [11] ensures that for some embedding dimension $d \in N^{+}$ and almost all time delay $\tau \in N^{+}$, there is a smooth function $f : R^{d} \rightarrow R$ so that $s(t) = f[s(t - d\tau), \ldots, s(t - 2\tau), s(t - \tau)]$. Thus, a natural choice of the training dataset can be $\{x_t, y_t\}_{t=1}^{N}$ where $x_t = [s(t - d\tau), \ldots, s(t - 2\tau), s(t - \tau)]$ and $y_t = s(t)$, and the test dataset $\{x_t^{*}, y_t^{*}\}_{t=1}^{L}$ can be set in the same way. In this way, time series prediction task can be transformed into the regression problem which aims at estimating and approximating the unknown function f.

We utilize Shanghai Composite Index (stock code: 000001) and Donghua energy (stock code: 002221) stock closing prices datasets from 2011 to 2013 which are downloaded from the Dazhihui software, and generate training datasets and test datasets which are respectively shown in the blue curve and red curve in Fig. 1.

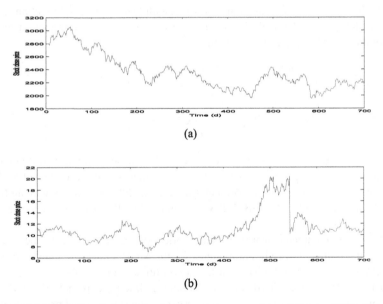

(a)

(b)

Fig. 1. Shanghai and Donghua stock closing price curves from 2011 to 2013, blue curve represents 600 training data and red curve represents 100 test data. (a). Shanghai stock closing price curve. (b). Donghua stock closing price curve. (Color figure online)

For $d = 1, 2, 3, 4$ and $\tau = 1, 2, 3, 4$, firstly we generate 700 samples, and every sample is a $d + 1$ dimensions vector. The first d data are the input sample of our model and the last data is the output. Secondly, we normalize all the training and test outputs by $y \rightarrow (y - m)/\sigma$, where m and σ denote the mean and the standard variance of the training outputs, respectively. Again, 700 samples are divided two parts, including 600 training samples and 100 test samples.

3.2   Prediction Results and Comparisons

We implement the precise hard-cut EM algorithm for MGPs (referred to as PreHard-cut) on the training dataset, and verify its performance on the test dataset. Actually, we implement it on each of the 16 normalized training datasets, get the trained MGP model and make the prediction. We finally de-normalize the prediction by $\hat{y} \rightarrow \hat{y}\sigma + m$. In order to compare its prediction performance, we run the MGP model with the other EM algorithms and some typical regression models and algorithms as follows:

(1)   The LOOCV hard-cut EM algorithm (referred to as LOOCV) proposed in [12] for MGPs, which approximates the posteriors and the Q function via the leave-one-out cross validation mechanism;

(2)   The variational hard-cut EM algorithm (referred to as VarHard-cut) proposed in [13] for MGPs, which approximates the posteriors via the variational inference;

(3)   The Radial Basis Function neural network with Gaussian kernel function (referred to as RBF), the classical regression algorithm which makes prediction by linear combinations of radial basis functions.

The prediction accuracy is evaluated by the root mean squared error (RMSE) on each experiment, which is mathematically defined as follow

$$\text{RMSE} = \sqrt{\frac{1}{L}\sum\nolimits_{t=1}^{L}(\hat{y}_t - y_t)^2}, \tag{10}$$

where y_t and $\hat{y}_t$ denote the output true value of the t-th test sample and its predictive value, respectively. Meanwhile, we compare the efficiency of these algorithms by the total time consumed for both the parameter learning and the prediction, with an Intel(R) Core(TM) i5 CPU and 16.00 GB of RAM running Matlab R2014a source codes for all the experiments.

Before the parameter learning, some prior parameters have to be specified, including the number C of GP components for the MGP model, the number of pseudo inputs (PI) for the variational hard-cut EM algorithm and the number of neurons in the hidden layer (HL) for the RBF model. Without additional explanation, some typical values of these parameters are tested and these ones are selected and presented with the least prediction RMSEs.

The RMSEs as well as the best values of the predetermined parameters for each algorithm on each dataset are listed in Table 1. We find that in terms of prediction accuracy, the precise hard-cut EM algorithm rank the first in the dataset with

$d = 3, \tau = 1$, which demonstrates the advantage in Shanghai and Donghua stock closing price prediction. And the predictive results are better than the results using the generalized RBF neural network in paper [14, 15]. The variational hard-cut EM algorithm for MGP model is comparable with the precise hard-cut algorithm on accuracy. But the last one is more stable and uniformly optimal with $d = 3, \tau = 1$ on Shanghai and Donghua stock price prediction. The LOOCV hard-cut EM algorithm for MGP model and the RBF model are not qualified for stock price prediction. Besides, Table 1 also shows a general decrease trend on prediction RMSEs with the embedding dimension d, since a large d means more information in the inputs.

Table 1. The RMSEs for Shanghai and Donghua stock closing price prediction.

d	τ	PreHard-cut		LOOCV		VarHard-cut		RBF	
		Shanghai	Donghua	Shanghai	Donghua	Shanghai	Donghua	Shanghai	Donghua
1	1	21.1933	0.2224	**21.2151**	0.2218	21.2106	**0.2194**	**21.5791**	**0.2204**
1	2	31.8931	0.3174	32.5150	0.3258	31.5769	0.3134	33.4016	0.3188
1	3	38.8945	0.3713	40.4267	0.3864	39.1431	0.3654	41.3143	0.3784
1	4	43.9104	0.4144	45.9821	0.4207	43.8947	0.4030	47.9299	0.4159
2	1	21.2464	0.2192	21.4326	0.2209	21.2879	0.2195	21.8939	0.2216
2	2	32.0199	0.3133	32.8077	0.3428	31.8553	0.3146	33.5479	0.3232
2	3	39.3107	0.3737	41.6617	0.3918	38.0136	0.3556	47.1278	0.3621
2	4	43.7612	0.3940	45.8229	0.4418	43.7895	0.4016	53.1479	0.3975
3	1	**21.0782**	**0.2183**	21.4797	**0.2203**	21.0879	0.2206	22.9824	0.2272
3	2	32.0317	0.3192	33.1330	0.3547	31.5318	0.3159	36.3263	0.3126
3	3	39.1218	0.3448	42.0259	0.4089	38.5492	0.3545	43.5631	0.3579
3	4	44.9140	0.3901	52.6917	0.4585	43.7351	0.3726	57.8089	0.4096
4	1	21.1804	0.2225	21.8933	0.2212	21.1461	0.2219	22.9345	0.2341
4	2	33.3603	0.3125	33.3099	0.4030	32.1945	0.3104	38.2823	0.3131
4	3	39.5789	0.3695	43.0564	0.4681	39.1205	0.3586	54.2466	0.3705
4	4	45.9405	0.4067	58.1039	0.5645	45.3620	0.3960	72.0162	0.4536

Moreover, the proposed technique has good scalability, but for stock price prediction 600 days stock closing price data are enough on the grounds that time span is up to two years!

Figures 2 and 3 show the best forecasting results with the parameters $d = 3$ and $\tau = 1$, which intuitively show the validity of the predictions. In Shanghai and Donghua test samples, the real and predicted values of the next 100 days are anastomotic and the best prediction RMSEs are 21.0782 and 0.2183 represented in bold in Table 1 respectively. The true values of the test samples are in good agreement with the predicted values, and the corresponding prediction errors are in actual allowable range which are mainly in ± 0.2 and ± 0.4 respectively as shown in Figs. 2 and 3.

The total time consumptions are shown in Table 2. We see that the precise hard-cut EM algorithm takes slightly longer. Nevertheless, no algorithms take longer than 6 min such that the remaining time is adequate for engineers to adjust the output power.

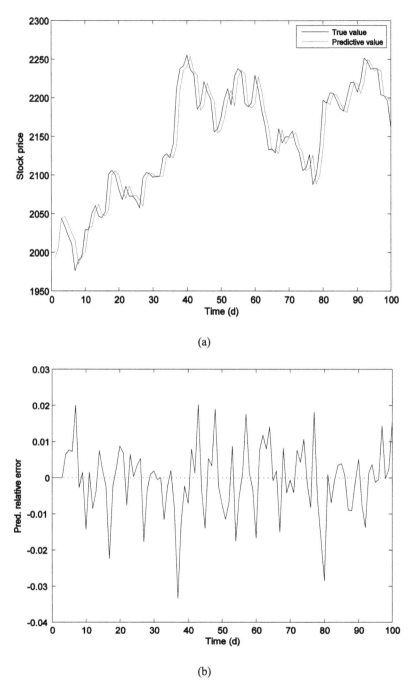

(a)

(b)

Fig. 2. (a). The prediction results of Shanghai stock closing price data; (b). The corresponding errors of Shanghai stock closing price data. (Color figure online)

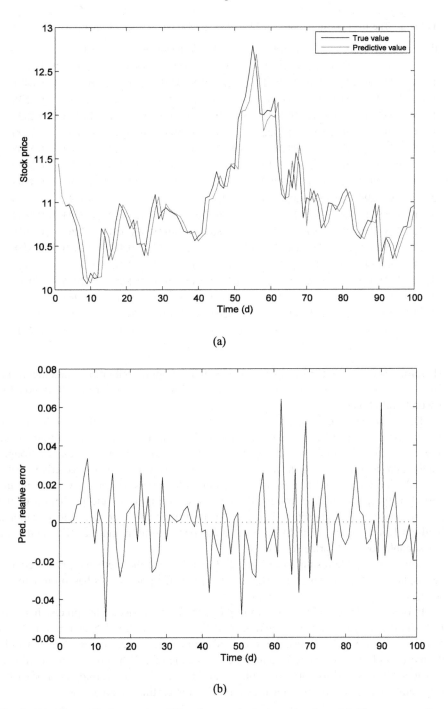

Fig. 3. (a). The prediction results of Donghua stock closing price data; (b). The corresponding errors of Donghua stock closing price data. (Color figure online)

Table 2. The time consumptions for Shanghai and Donghua stock closing price prediction.

d	τ	PreHard-cut		LOOCV		VarHard-cut		RBF	
		Shanghai	Donghua	Shanghai	Donghua	Shanghai	Donghua	Shanghai	Donghua
1	1	59.5766	120.2869	77.7581	103.5980	84.8291	72.0887	6.6121	0.8491
1	2	77.6747	110.2947	66.0087	86.9751	60.2808	86.6801	1.9530	0.5538
1	3	89.4715	105.6588	56.4041	90.7678	71.2045	59.5331	1.3176	1.2327
1	4	110.5455	101.7635	54.8922	87.8987	82.5639	65.4218	0.3662	0.7490
2	1	72.1279	224.2375	88.2539	104.4790	97.8030	105.5201	0.7642	0.5208
2	2	93.3649	140.5822	65.2119	78.5423	122.9084	114.4137	0.7450	0.5049
2	3	112.4934	126.7071	54.1397	89.6782	101.0682	147.6291	1.2788	0.4943
2	4	152.2296	131.7796	55.6939	75.1938	101.6412	134.8176	0.3793	0.5290
3	1	176.9017	211.0905	70.8960	101.1628	129.4165	120.8225	0.7593	0.7829
3	2	115.6750	198.6486	59.2711	76.5386	127.8695	130.1543	1.2261	0.8615
3	3	115.2516	198.8352	64.5232	79.8847	126.0775	130.8770	0.7630	0.6358
3	4	130.5847	180.8102	53.0291	73.4977	117.0582	138.7150	0.3607	0.7125
4	1	247.9665	266.5358	75.6300	86.7739	153.3917	160.6503	1.2926	1.0299
4	2	146.0758	215.2569	57.9361	78.0230	155.2156	169.2637	0.7964	0.7965
4	3	81.0702	255.1925	53.7123	74.4203	155.8308	180.2085	1.2557	0.7503
4	4	129.1127	225.2805	51.6860	76.9758	149.6366	159.0454	1.8375	0.7804

Therefore, accuracy is the key factor in selecting the appropriate model and algorithm for stock price forecasting, so the precise hard-cut EM algorithm for the MGP model is a wonderful choice.

The best predictive curve for each algorithm is shown in Fig. 4. It can be found that the precise hard-cut EM algorithm and the variational hard-cut EM algorithm fit the true stock price extremely well except when the stock price reaches a peak or a trough, where there is a dramatic turn of the stock price. However, the two predictive curves are still within small and acceptable range around the true stock price even during the period of the peak and the trough. Besides, at some moments, the prediction of the precise hard-cut EM algorithm is closer to the true stock price than the variational hard-cut EM algorithm. The LOOCV hard-cut EM algorithm and the RBF model are not suitable for stock price forecasting.

Some remarkable results from Figs. 2, 3 and 4 is that the predicted prices seem to be displaced some constant time. Because the predicted price $s(t)$ is based on before d stock price: $s(t - d\tau), \ldots, s(t - 2\tau), s(t - \tau)$.

In order to further explore how to improve the performance of the precise hard-cut EM algorithm, we plot the prediction RMSEs for $d = 1, 2, 3, 4, 5$ and $\tau = 1, 2, 3, 4$ respectively in Fig. 5. It can be observed from Fig. 5 that the RMSE generally decreases with the increasing of d and the decreasing of τ. When $d \geq 3$, the RMSE is considerably low and its variation with d and τ is very tiny. Therefore, an appropriate large embedding dimension d ensures a precise forecasting in stock price.

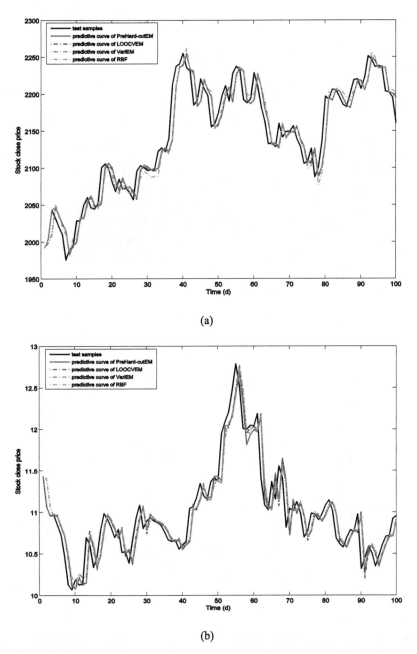

(a)

(b)

Fig. 4. (a). Comparisons of each algorithm for the predictive curves of Shanghai stock closing price in 100d test data. (b). Comparisons of each algorithm for the predictive curves of Donghua stock closing price in 100d test data. (Color figure online)

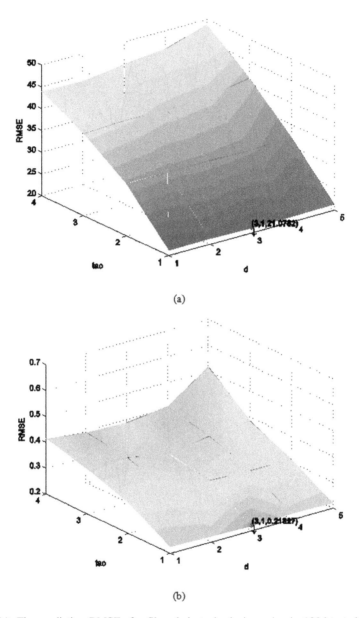

(a)

(b)

Fig. 5. (a). The predictive RMSEs for Shanghai stock closing price in 100d test data in the precise hard-cut EM algorithm with various values of d and τ. (b). The predictive RMSEs for Donghua stock closing price in 100d test data in the precise hard-cut EM algorithm with various values of d and τ.

4 Conclusion

We have successfully applied the MGP model via the precise hard-cut EM algorithm to modeling and predicting the time series of stock prices. The experiment results demonstrate that this MGP based method via the precise hard-cut EM algorithm turns out to be valid, feasible and highly competitive on prediction accuracy with acceptable time consumption, and outperforms some typical regression models and algorithms.

Acknowledgement. This work was supported by the Natural Science Foundation of China for Grant 61171138.

References

1. Sun, W., Guo, J., Xia, B.: Discussion about stock prediction theory based on RBF neural network. Heilongjiang Sci. Technol. Inf. (22), 130 (2010)
2. Fu, C., Fu, M., Que, J.: Prediction of stock price base on radial basic function neural networks. Technol. Dev. Enterp. 23(4), 14–15, 38 (2004)
3. Liu, H., Bai, Y.: Analysis of AR model and neural network for forecasting stock price. Math. Pract. Theory 41(4), 14–19 (2011)
4. Zhu, Y.: Research on Stock Prediction Methods. Xi'an: Master Thesis of Northwestern Polytechnical University (2006)
5. Tresp, V.: Mixtures of Gaussian processes. In: Advances in Neural Information Processing Systems, vol. 13, pp. 654–660 (2000)
6. Meeds, E., Osindero, S.: An alternative infinite mixture of Gaussian process experts. In: Advances in Neural Information Processing Systems, 18, pp. 883–890 (2006)
7. Chen, Z., Ma, J., Zhou, Y.: A precise hard-cut EM algorithm for mixtures of Gaussian processes. In: Huang, D.-S., Jo, K.-H., Wang, L. (eds.) ICIC 2014. LNCS, vol. 8589, pp. 68–75. Springer, Heidelberg (2014)
8. Joseph, J., Doshi-Velez, F., Huang, A., Roy, N.: A Bayesian nonparametric approach to modeling motion patterns. Auton. Robots 31(4), 383–400 (2011)
9. Sun, S.: Infinite mixtures of multivariate Gaussian processes. In: International Conference on IEEE Machine Learning and Cybernetics (ICMLC), vol. 3, pp. 1011–1016 (2013)
10. Tayal, A., Poupart, P., Li, Y.: Hierarchical double dirichlet process mixture of Gaussian processes. In: AAAI, pp. 1126–1133 (2012)
11. Zhou, Y., Zhang, T., Sun, J.: Multi-scale gaussian processes: a novel model for chaotic time series prediction. Chin. Phys. Lett. 24(1), 42–45 (2007)
12. Yang, Y., Ma, J.: An efficient EM approach to parameter learning of the mixture of Gaussian processes. In: Liu, D., Zhang, H., Polycarpou, M., Alippi, C., He, H. (eds.) ISNN 2011, Part II. LNCS, vol. 6676, pp. 165–174. Springer, Heidelberg (2011)
13. Nguyen, T., Bonilla, E.: Fast allocation of Gaussian process experts. In: Proceedings of the 31st International Conference on Machine Learning, pp. 145–153 (2014)
14. Liu, S., Ma, J.: The application of diagonal generalized RBF neural network to stock price prediction. Highlights Sciencepaper Online 7(13), 1296–1306 (2014)
15. Zheng, W., Ma, J.: Diagonal log-normal generalized RBF neural network for stock price prediction. In: Zeng, Z., Li, Y., King, I. (eds.) ISNN 2014. LNCS, vol. 8866, pp. 576–583. Springer, Heidelberg (2014)

A DAEM Algorithm for Mixtures of Gaussian Process Functional Regressions

Di Wu and Jinwen Ma[⊠]

Department of Information Science, School of Mathematical
Sciences and LMAM, Peking University, Beijing 100871, China
jwma@math.pku.edu.cn

Abstract. The mixture of Gaussian process functional regressions (mix-GPFR)
is a powerful tool for curve clustering and prediction. Unfortunately, there
generally exist a large number of local maximums for the Q-function of the
conventional EM algorithm so that the conventional EM algorithm is often
trapped in the local maximum. In order to overcome this problem, we propose a
deterministic annealing EM (DAEM) algorithm for mix-GPFR in this paper.
The experimental results on the simulated and electrical load datasets demonstrate that the DAEM algorithm outperforms the conventional EM algorithm on
parameter estimation, curve clustering and prediction.

Keywords: Gaussian process · Mixture of Gaussian process · EM algorithm ·
Curve clustering · Deterministic annealing

1 Introduction

Gaussian process (GP) [1, 2] is a powerful tool in the fields of signal and information
processing, machine learning and data mining. But the mean function of the GP model
is generally assumed to be zero or a linear function of input variables. Moreover, a
single GP cannot deal with a multimodality dataset. In fact, curve clustering is a typical
multimodality dataset problem. Specifically, each curve can be regarded as one
"sample", referred to as a sample curve or functional datum. The aim of curve clustering is to separate these sample curves into different clusters or classes which can be
modeled by certain Gaussian processes. So, the actual model of curve clustering is a
mixture of Gaussian processes. The mean functions of the Gaussian processes are very
important for curve clustering, but they are generally assumed to be zeros or linear
functions for easy computation. In literature, there are only a few methods for learning
nonlinear mean functions and the Gaussian process functional regression (GPFR)
model [3] provides a feasible way. The mean function of the GPFR is assumed to be a
linear combination of b-spline basis functions [4]. For solving the curve clustering
problem, Shi previously utilized the mixture of GPs (mix-GP) model [5–8] where the
sample curves belong to one cluster are subject to a general GP. In order to improve the
performance of curve clustering, the GPFR models were introduced and the mixture of
GPFRs (mix-GPFR) model [9] was finally utilized.

Although the maximum likelihood estimate (MLE) of the GPFR model can be
calculated by the gradient method, the computation of the MLE for the mix-GP model

© Springer International Publishing Switzerland 2016
D.-S. Huang et al. (Eds.): ICIC 2016, Part III, LNAI 9773, pp. 294–303, 2016.
DOI: 10.1007/978-3-319-42297-8_28

is rather difficult. To overcome this problem, the MCMC approach was applied for the mix-GP with zero mean functions. But about 20000 samples were needed and thus it might take one day time on a small dataset. In addition, the MCMC approach is quite difficult to be applied for the mix-GPFR model since the sampling of the parameters in the mean function is not so easy. Alternatively, the conventional EM algorithm [10–13] was adopted for the mix-GPFR model. Although the conventional EM algorithm has some advantages such as low cost per iteration and ease of programming, it is a local search method and cannot get rid of the local maximum problem.

In this paper, we propose a deterministic annealing EM (DAEM) algorithm for the mix-GPFR model to overcome the local maximum problem. The idea of the DAEM algorithm is to transform the Q-function of the conventional EM algorithm into the U-function which can be flexible in a deterministic annealing way. In the early iterations, the U-function is smoother, i.e., has less local maximum, than the Q-function so that the maximum of the U-function is more global than that of the Q-function. During the following iterations, the U-function gradually tends to the Q-function and the DAEM algorithm has more probable to arrive at the global maximum point. We conduct the experiments on both simulated and electrical load datasets. The experimental results demonstrate that the DAEM algorithm for the mix-GPFR model outperforms the conventional EM algorithm on parameter estimation as well as curve clustering.

The remainder of this paper is organized as follows. The GPFR and mix-GPFR models are introduced in Sect. 2. In Sect. 3, we propose the DAEM algorithm for the mix-GPFR model. The experimental results and comparisons are summarized in Sect. 4. Finally, we give a brief conclusion in Sect. 5.

2 The GPFR and Mix-GPFR Models

In this section, we introduce the GPFR model as well as the mix-GPFR model.

2.1 The GPFR Model

The GP is a common and important stochastic process in which any group of states (as random variables) are subject to a Gaussian distribution. $y(x) \in \mathbb{R}$ (the real number field) is a stochastic process, where $x \in \mathbb{R}$. With any natural number N and any vector $\mathbf{x} = (x_1, \ldots, x_N)^T$, the definition of the Gaussian process can be given as follows. If $\mathbf{y} = [y_1, \ldots, y_N]^T$, where $y_n = y(x_n)$, is subject to an N-dimensional Gaussian distribution $N(\boldsymbol{\mu}, \mathbf{C})$, then $y(x)$ is said to follow a Gaussian process, where $\boldsymbol{\mu} = [\mu(x_1), \ldots, \mu(x_N)]^T$ and $\mathbf{C} = [C(x_n, x_{n'})]_{N \times N}$ represents an $N \times N$ kernel matrix in which $C(x_n, x_{n'})$ is a kernel function. The GP model is written as

$$y(x) \sim GP[\mu(x), C(x, x')].$$

Here, we utilize the kernel function

$$C(x_n, x_{n'}) = (\theta_1)^2 \exp\left[-\frac{1}{2}(\theta_2)^2 (x_n - x_{n'})^2\right] + (\theta_3)^2 \delta_{nn'},$$

where $\delta_{nn'}$ is the Kronecker delta function and $\boldsymbol{\theta} = (\theta_1, \theta_2, \theta_3)$.

However, the mean function of GP is generally assumed to be zero, or a linear function or quite simple nonlinear function. To learn the mean function of the GP model better, Shi proposed the model of GPFR [3]. In this model, the mean function is approximated by a linear combination of b-spline basis functions and we illustrate a set of b-basis functions in Fig. 2(a). Denote a set of b-spline basis functions, $\varphi = [\varphi_1(x), \ldots, \varphi_D(x)]^T$. Then the mean function is approximate by

$$\mu(x) = \mathbf{b}^T \varphi = \sum_{j=1}^{D} b_j \varphi_j(x),$$

where $\mathbf{b} = (b_1, \ldots, b_D)^T$ is a D-dimensional coefficient vector. Thus, the GPFR can be described by

$$y(x) \sim GPFR(x; \mathbf{b}, \boldsymbol{\theta}).$$

2.2 The Mix-GPFR Model

There is heterogeneity among the sample curves sometimes and this kind of dataset cannot be learned by a single GPFR. To overcome this problem, Shi [9] proposed the mix-GPFR model. The M curves generated by mix-GPFR could be separated into K components or classes and the curves belong to each component is subject to a same GPFR model. The mix-GPFR is a powerful model for curve clustering and the detail of mix-GPFR model is given as follows.

We introduce an indicator variable z_{mk}, where $m = 1, \ldots, M$ and $k = 1, \ldots, K$. If the m-th batch belongs to the k-th component, $z_{mk} = 1$; otherwise, $z_{mk} = 0$. All the indicator variables are assumed to share the same prior and the prior is given by

$$P(z_{mk} = 1) = \pi_k,$$

where $\sum_{k=1}^{K} \pi_k = 1$. After these curves are separated into the components by the indicator variables, the output of the k-th component $y(x)$ is subject to a GPFR model.

$$y(x) \sim GPFR(x; \mathbf{b}_k, \boldsymbol{\theta}_k).$$

The total log likelihood of mix-GPFR model is

$$L(\Theta) = \sum_{m=1}^{M} \sum_{k=1}^{K} z_{mk}[\log \pi_k + \log p(\mathbf{y}_m | \mathbf{x}_m, \mathbf{b}_k, \boldsymbol{\theta}_k)],$$

where $\Theta = \{\pi_k, \mathbf{b}_k, \boldsymbol{\theta}_k\}_{k=1}^{K}$.

3   The DAEM Algorithm

The conventional EM algorithm is widely used in machine learning, but it has the local maximum problem. Thus, if the initialization of parameters is not good enough, the performance of the EM algorithm may be very poor. However, the initialization of mix-GPFR is very difficult. So we construct a DAEM algorithm to solve the local maximum problem of the conventional EM algorithm. As these M curves are independent, the Q-function of the conventional EM algorithm for mix-GPFR can be given as follows.

$$Q(\Theta) = \sum_{m=1}^{M} \sum_{k=1}^{k} \tilde{\alpha}_{mk}[\log \pi_k + \log p(\mathbf{y}_m|\mathbf{x}_m, \mathbf{b}_k, \boldsymbol{\theta}_k)],$$

where

$$\tilde{\alpha}_{mk} = \hat{\pi}_k p(\mathbf{y}_m|\mathbf{x}_m, \hat{\mathbf{b}}_k, \hat{\boldsymbol{\theta}}_k) / \sum_{j=1}^{K} \hat{\pi}_j p\left(\mathbf{y}_m|\mathbf{x}_m, \hat{\mathbf{b}}_j, \hat{\boldsymbol{\theta}}_j\right).$$

We introduce an annealing parameter β to the Q-function and then construct the U-function of the DAEM algorithm for mix-GPFR as follows.

$$U(\Theta, \beta) = \sum_{m=1}^{M} \sum_{k=1}^{K} \alpha_{mk}[\log \pi_k + \log p(\mathbf{y}_m|\mathbf{x}_m, \mathbf{b}_k, \boldsymbol{\theta}_k)],$$

where

$$\alpha_{mk} = \left[\hat{\pi}_k p\left(\mathbf{y}_m|\mathbf{x}_m, \hat{\mathbf{b}}_k, \hat{\boldsymbol{\theta}}_k\right)\right]^{\beta} / \sum_{j=1}^{K} \left[\hat{\pi}_j p\left(\mathbf{y}_m|\mathbf{x}_m, \hat{\mathbf{b}}_j, \hat{\boldsymbol{\theta}}_j\right)\right]^{\beta}.$$

Obviously, $U(\Theta, 1)$ is equal to $Q(\Theta)$ so the conventional EM algorithm could be regarded as a special case of the DAEM algorithm with $\beta = 1$. With the U-function, we can show the details of the DAEM algorithm in five steps.

Step 1.   Initialize α_{mk} by a simple curve clustering method (such as the k-means algorithm) and set the initial value $\beta = \beta_{min}$, where $\beta_{min} < 1$

Step 2.   M-step: calculate Θ by maximizing $U(\Theta, \beta)$

Step 3.   E-step: update α_{mk}

Step 4. $\beta = \min(\beta \times const, 1)$

Step 5.   When $\beta = 1$ and the increase of $U(\Theta, 1)$ is small enough, stop; otherwise, return to Step 2.

For the DAEM algorithm, the initialization is not so important and infact most of curve clustering methods can be used. The only distinction between the DAEM algorithm and the conventional EM algorithm is just β. The experiments demonstrate that $\beta_{min} = 0.2$ is small enough to avoid the local maximum problem. But we sometimes use a bigger β_{min} because some useful initialization can be utilized in certain practical applications. Another advantage of the DAEM algorithm is that the time consumer of the DAEM algorithm is about two times of the conventional EM

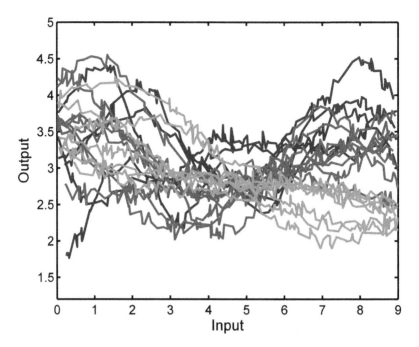

Fig. 1. The curves are the training samples of the simulated dataset for the mix-GPFR model and there are only 20 curves of 3 components in 3 colors. (Color figure online)

algorithm. The M-step of the DAEM algorithm is quite the same as the M-step of the conventional EM algorithm and the only difference is the new parameter β. In [9], there are two types of prediction, but the second type is not very useful. So we just consider the first type prediction for mix-GPFR and the details of the M-step and prediction method can be referred to [9]. In addition, the theory of the DAEM algorithm was described in [11, 12].

Table 1. The parameter estimation of the DAEM algorithm on the simulated dataset for the mix-GPFR: we show the true value (TV), estimated value (EV) and relative error (RE) of the parameters

		π_k	θ_{k1}	θ_{k2}	θ_{k3}
k = 1	TV	0.3333	0.6325	1.0000	0.0632
	EV	0.3256	0.6449	0.9982	0.0631
	RE	2.3 %	2.0 %	0.2 %	0.2 %
k = 2	TV	0.3333	0.4472	0.7071	0.0632
	EV	0.3274	0.4409	0.6999	0.0627
	RE	1.8 %	1.4 %	1.0 %	0.8 %
k = 3	TV	0.3333	0.3162	0.4472	0.0632
	EV	0.3470	0.3124	0.4523	0.0633
	RE	4.1 %	1.2 %	1.1 %	0.2 %

4   Experimental Results

In this section, we demonstrate the experimental results of the DAEM algorithm for the mix-GPFR model on both the simulated and electrical load datasets, being compared with the conventional EM algorithm and related approaches. Note that the mix-GP just means the mixture of the GPs with zero mean functions.

4.1   On the Simulated Dataset

We conduct various experiments on datasets generated by different mix-GPFR models and the DAEM algorithm always performs very well. Typically, we show the results on a simulated dataset generated by the mix-GPFR model with 3 components. The mean functions of GPFR models are $\mu_1(x) = 0.5 \sin\left[0.125(x-4)^2\right] + 3$, $\mu_2(x) = -3(2\pi)^{-0.5} \exp\left[-0.125(x-4)^2\right] + 3.7$ and $\mu_3(x) = 0.5 \arctan(0.5x - 2) + 3$, respectively. The parameters π_k and θ_k of components are shown in Table 1. In Fig. 1,

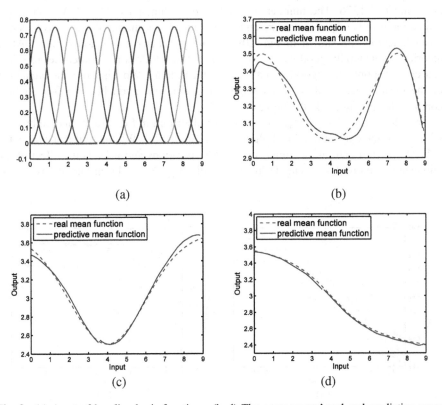

Fig. 2. (a) A set of b-spline basis functions; (b–d) The components' real and predictive mean functions of the simulated dataset for the mix-GPFR trained by the DAEM algorithm, respectively

20 sample curves of training dataset are already illustrated and the curves are obviously difficult to be clustered. Actually, we generate 300 training curves and each of them consists of 50 points. On the other hand, we generate 600 test curves and each of them have 40 known points and 110 test points.

Table 2. The mean RMSEs of the models and algorithms on the simulated and electrical load datasets for the mix-GPFR.

Model	Algorithm	Simulated		Electrical load	
		K	RMSE	K	RMSE
Mix-GPFR	DAEM	2	0.0735	4	0.6394
Mix-GPFR	Conventional EM	2	0.0741	3	0.6647
Mix-GP	Conventional EM	2	0.0793	4	0.9741
GPFR	MLE	–	0.0772	–	0.8608
GP	Gradient method	–	0.0838	–	1.0394

After trial and error, $\beta_{min} = 0.2$ is used at last and the const in Step 4 of the DAEM algorithm is 1.1576. The number of b-spline basis functions $D = 22$. We show the estimation results of parameters π_k and θ_k in Table 1 and the predicted as well as real mean functions are shown in Fig. 2(b–d). The parameter estimation of the DAEM algorithm is generally good except π_k on this dataset. It does not matter, because it is caused by not only the algorithm but also the stochasticity. What is more, the prediction of the mean function of the 1st component is not very good because $\theta_{2,1}$, which controls the amplitude, is the biggest.

We show the values of α_{mk} of the DAEM algorithm in Fig. 4(a–b) and compare it with $\tilde{\alpha}_{mk}$ of the conventional EM algorithm in Fig. 4(c–d). We find out that α_{mk} of the DAEM algorithm are more similar in the early iterations and it leads the effect of the initialization more little. The U-function with $\beta_{min} = 0.2$ during iterations of the

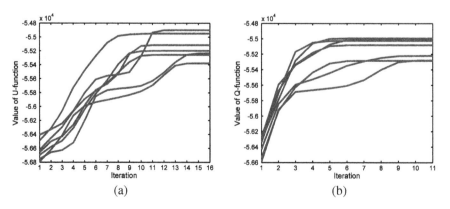

(a)   (b)

Fig. 3. (a) The value of U-function of the DAEM algorithm with $\beta_{min} = 0.2$ during the iterations; (b) The value of Q-function of the conventional EM algorithm during the iterations

DAEM algorithm and the Q-function during iterations of the conventional EM algorithm are illustrated in Fig. 3(a–b), respectively. The two EM algorithms are effective but the DAEM algorithm is better because the value of U-function is a bit bigger after the convergence.

The classification accuracy rate (CAR) of the DAEM algorithm on the curves of the test dataset is 98.17 %, which is bigger than the CAR of the conventional EM algorithm, 97.17 %. The root mean square errors (RMSEs) of the DAEM and conventional EM algorithms for the mix-GPFR are shown in Table 2. In addition, we show the RMSEs of other three models, which are GPFR, mix-GP and GP models. The RMSEs of the mix-GPFR are smaller and the RMSE of the mix-GPFR trained by the DAEM algorithm is the smallest. The GPFR and mix-GP models are not the best but better than the GP.

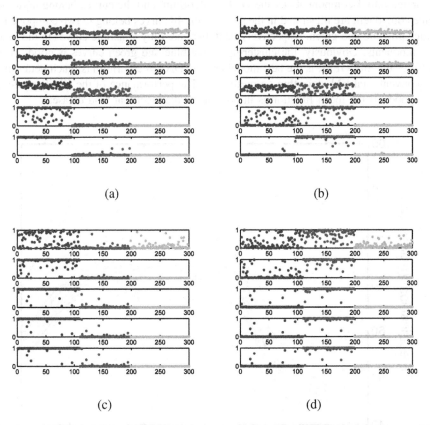

(a)　　　　　　　　　　　　(b)

(c)　　　　　　　　　　　　(d)

Fig. 4. α_{mk} or $\tilde{\alpha}_{mk}$ of batches belonging to the 1st, 2nd and 3rd components are illustrated by blue, red and green points, respectively. (a–d) α_{m1} and α_{m2} of 300 training batches with the iterations, which are the 1st, 4th, 7th, 10th and 13th iterations, of the DAEM algorithm with $\beta_{min} = 0.1$ for mix-GPFR, respectively; (c–d) $\tilde{\alpha}_{m1}$ and $\tilde{\alpha}_{m1}$ of training batches with iterations, which are the 1st, 3th, 5th, 7th and 9th iterations, of the conventional EM algorithm for mix-GPFR, respectively (Color figure online)

4.2   On the Electrical Load Dataset

Electrical load prediction plays a vital role in optimal unit commitment, start up and shut down of thermal plants, control of reserve and exchanging electric power in interconnected systems [14]. The electrical load dataset is from the Northwest China Grid Company. There are 100 sample curves in electrical load dataset and 96 points of each curve are observations of one day. We separate this dataset into 2 groups and each group has 50 sample curves. One group of sample curves is the training dataset and the other is the test dataset. We also separate the points of each curve in test dataset into two groups with 48 points in each. We use one group for training and the other for testing.

We make model selection by the cross validation method on the training dataset with various numbers of components K. In Fig. 5, the 50 training sample curves are separated into 4 components by the DAEM algorithm and the curves belonging to the same component are illustrated in the same color. 48 curves belong to two components, which are blue and green, respectively, and there are only 2 sample curves in the other 2 components. Obviously, the clustering of the blue and green curves is good. From Table 2, the mix-GPFR is the best model and the DAEM algorithm is much better than the conventional EM algorithm on prediction. The performance of the GPFR model is also good, so the mean function is important for the electrical load dataset.

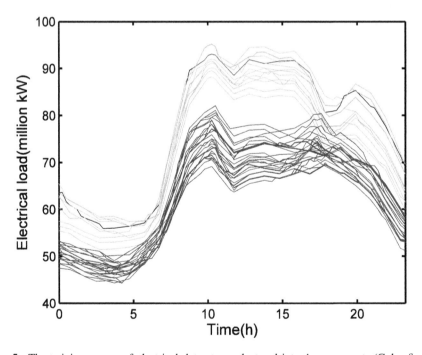

Fig. 5. The training curves of electrical dataset are clustered into 4 components (Color figure online)

5   Conclusions

We have established the DAEM algorithm for the mix-GPFR model to solve the problem of local maximum associated with the conventional EM algorithm. As the key difference between the DAEM algorithm and the conventional EM algorithm, a flexible variable β is introduced in the DAEM algorithm to make the parameter learning process in a deterministic way. On simulated dataset, $\beta_{min} = 0.2$ may be small enough and the DAEM algorithm performs well on parameters estimation. In addition, the DAEM algorithm is better than the conventional EM algorithm on curve clustering and prediction. Moreover, the experimental results on a real-world dataset, i.e., the electrical load dataset, demonstrate that the DAEM algorithm is also good on curve clustering and better than the conventional EM algorithm on prediction.

Acknowledgement. This work is supported by the National Science Foundation of China under Grant 61171138.

References

1. Rasmussen, C.E.: Evaluation of Gaussian Processes and Other Methods for Non-linear Regression. Ph.D. dissertation, Department of Computer Science, University of Toronto (1996)
2. Rasmussen, C.E., Williams, C.K.I.: Regression Gaussian Process for Machine Learning, ch. 2. MIT Press, Cambridge (2006)
3. Shi, J.Q., et al.: Gaussian process functional regression modeling for batch data. Biometrics **63**, 714–723 (2007)
4. Boor, D.E.: On calculating with B-splines. J. Approximation Theory **6**, 50–62 (1972)
5. Shi, J.Q., Murray-Smith, R., Titterington, D.M.: Bayesian regression and classification using mixtures of Gaussian process. Int. J. Adapt. Control Signal Process. **17**, 149–161 (2003)
6. Shi, J.Q., Murray-Smith, R., Titterington, D.M.: Hierarchical Gaussian process mixtures for regression. Stat. Comput. **15**, 31–41 (2005)
7. Kamnik, R., et al.: Nonlinear modeling of FES-supported standing-up in paraplegia for selection of feedback sensors. IEEE Trans. Neural Syst. Rehabil. Eng. **13**(1), 40–52 (2005)
8. Qiang, Zhe, Ma, Jinwen: Automatic model selection of the mixtures of Gaussian processes for regression. In: Hu, X., Xia, Y., Zhang, Y., Zhao, D. (eds.) ISNN 2015. LNCS, vol. 9377, pp. 335–344. Springer, Heidelberg (2015). doi:10.1007/978-3-319-25393-0_37
9. Shi, J.Q., Wang, B.: Curve prediction and clustering with mixtures of Gaussian process functional regression models. Stat. Comput. **18**, 267–283 (2008)
10. Dempster, A.P., Laird, N.M., Rubin, D.B.: Maximum likelihood from incomplete data via the EM algorithm. J. R. Statist. Soc. (B) **39**, 1–38 (1977)
11. Ueda, N., Nakano, R.: Mixture density estimation via EM algorithm with deterministic annealing. In: Proceedings of the IEEE Neural Networks for Signal Processing, pp. 66–77 (1994)
12. Ueda, N., Nakano, R.: Deterministic annealing EM algorithm. Neural Netw. **11**, 271–282 (1998)
13. Ma, J., Xu, L., Jordan, M.I.: Asymptotic convergence rate of the EM algorithm for Gaussian mixtures. Neural Comput. **12**(12), 2881–2907 (2000)
14. Mohandes, M.: Support vector machines for short-term electrical load forecasting. Int. J. Energy Res. **26**, 335–345 (2002)

Simulation Analysis of the Availability of Cloud Computing Data Center

Fangfang Geng[(⊠)] and Chang-ai Chen

Henan University of Traditional Chinese Medicine, Zhengzhou, Henan, China
gfhactcm@126.com

Abstract. This paper analyzes the basic structure of Cloud Computing Data Center. Then, the reliability block diagram and the Monte Carlo simulation method are used to analyze the availability of Cloud Computing Data center. Meanwhile, suggestions on improving the availability of Cloud Computing Data Center are proposed. This method can effectively complete the simulation analysis of the availability and operation and maintenance cost of the Cloud Computing Data Center. It also can provide the theoretical basis for the operation and maintenance of the Cloud Computing Data Center.

Keywords: Cloud computer data center · Availability · Monte Carlo

1 Introduction

With the rapid development of Internet and computer, the market of our country begins to understand the new ICT application mode of cloud computing. The cloud computing will be applied to data center. Data center is the core of information technology and is responsible for the key business of all enterprises and institutions [1]. Its availability is one of the most concerned issues of operation and maintenance personnel. But, the cloud computing data center is the integration of the various parts of the organic interconnection and integration. Through the function integration, database integration, network integration and software integration, the efficiency and management ability are improved. Therefore, establishing the corresponding evaluation index of the cloud computing data center and exploring its availability can provide a basis for the further protection of the business.

2 Analysis of Cloud Computing Data Center

The typical Cloud Computing Data Center is composed of virtualization architecture, network communication system, monitoring system, analyzing system, power supply system and service system. The basic structure is shown in Fig. 1.

Virtualization architecture is the core system of Cloud Computing Data Center. It is mainly composed of servers, virtualization management software, virtual operating system, data storage devices, switches and other components. Its function is to complete the building of virtualization and to provide basic operation platform for business operation and data storage. At the same time, it also provides a management platform for

© Springer International Publishing Switzerland 2016
D.-S. Huang et al. (Eds.): ICIC 2016, Part III, LNAI 9773, pp. 304–312, 2016.
DOI: 10.1007/978-3-319-42297-8_29

Cloud Computing Data Center. Network communication system consists of access switch, core switch, optical fiber communication line, education and scientific research network, telecommunication network, mobile network, China Unicom Network and so on. It is mainly used for the exchange of data center business data and Internet data, while providing service support for campus network users and Internet users. The monitoring system consists of a server, a switch, a monitoring software, a large screen and so on. It is mainly to complete the data center business displayed and submitted of the warning and fault information, so that the operation and maintenance personnel to find and deal with the problem in time. The analyzing system mainly consists of a server, a switch, an analysis software and so on. It is mainly responsible for the analysis of Web business. The power supply system is mainly composed of the municipal electric power, UPS power supply, UPS power supply unit and so on. It is mainly responsible for the power supply of data center servers, storage devices, network devices and so on. Service system is mainly composed of the virtual host, operating system software, service software and so on. It is mainly to meet the needs of users of the data center.

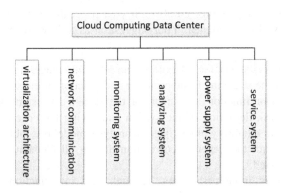

Fig. 1. Cloud computing data center structure

3 The Availability Simulation of Cloud Computing Data Center Based on Monte Carlo Simulation

3.1 Basic Principle

Monte Carlo simulation method is also named stochastic simulation method. It is a mathematical simulation method using probability model as approximate calculation. According to the random sampling method and the functional block diagram of the system, the reliability of the system is predicted [2]. Monte Carlo computer simulation by constructing random probability model and the random variables based on uniform distribution which meet the needs of the corresponding stochastic model of the random number, the design selects suitable sampling method sampling on the random variables, according to system model simulation calculation, statistical analysis simulation test results and other steps to solve engineering problems.

The system reliability block diagram model is a mathematical description of the system fault characteristic. It consists of the box, the logical relation and the connection and the connection of the device in the system. It is the basic input model of system availability [3]. However, when the structure of the system is complex, the equipment is large and the equipment life is arbitrary, the conventional analytical method is difficult to solve the problem of usability analysis. Based on the system reliability block diagram analysis model, the application of Monte Carlo simulation can be very good to complete the system availability analysis.

As the Cloud Computing Data Center has the characteristics of complex structure and multi equipment, Monte Carlo simulation can be applied to the analysis of its availability. In the process of simulation, combined with the characteristics of the failure modes of various equipments, the choice of suitable simulation sampling method, the simulation speed and accuracy can be improved.

3.2 Basic Process

There is no need to include page numbers. If your paper title is too long to serve as a running head, it will be shortened. Your suggestion as to how to shorten it would be most welcome.

1. System function analysis, fault definition and reliability diagram modeling. Through the analysis of the structure, basic principle and function of the system, the function block diagram or function flow chart of the system are established. Then, with Defining the task definition of the system and giving the criterion of the system fault, the system reliability block diagram analysis model is established. In the form of block diagram, the dependence relation among all elements in the system is visually expressed.
2. Determination of input parameters. Through field investigation, data query and other ways to get the various devices in the system reliability, maintainability and other related input parameters.
3. Monte Carlo simulation analysis and calculation. Preferably computer digital simulation software, combined with all devices in the system reliability block diagram model and system failure patterns, used for the simulation of sampling method, system reliability, maintainability parameters and other simulation parameters input, simulation and the statistical analysis was completed.

4 Simulation and Analysis of the Availability of Cloud Computing Data Center in a University

4.1 Structure of Cloud Computing Data Center

Combined with the cloud computing data center structure shown in Fig. 1, Table 1 shows the composition of a cloud computing data center in a university. The data center is composed of five parts which are virtualization architecture, network communication system, monitoring and analyzing system, power supply system and service system. Meanwhile, it contains 103 sets of equipment. The data center should be able to complete the management of the virtual machine, storage device management, Internet based business access, business analysis, business monitoring and so on. Any of these functions could not be implemented as a system failure.

Table 1. The Structure of cloud computing data center in a university.

Subsystem name	Device name	Number	Subsystem name	Device name	Number
Virtualization architecture	virtual host	4	monitoring and analyzing system	server	3
	storage device	2		large screen	1
	switch	2		monitoring software	1
	Virtualization OS	4		analytical software	1
	virtualization management software	1	power supply system	Commercial Power	1
Network communication system	access switch	2		UPS power supply	1
	core switch	2		UPS power supply unit	5
	Optical fiber communication line	1	service system	virtual machine	20
	CERNET	1		Linux OS	16
	CT	1		Windows OS	4
	CU	1		DBMS	3
	CMCC	1		Application software	25

4.2   Reliability Diagram Modeling

According to the structure and function of cloud computing data center, and the fault analysis, the reliability block diagram analysis model of cloud computing data center and each subsystem is established by using mathematical simulation platform software. The reliability diagram models as shown in Figs. 2, 3, 4, 5, 6 and 7.

1. Establish the total reliability diagram of Cloud Computing Data Center.
   The Cloud Computing Data Center is composed of five subsystems in series, as shown in Fig. 2.

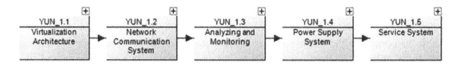

Fig. 2. Reliability diagram model of cloud computing data center

2. Establish the reliability diagram of each subsystem.

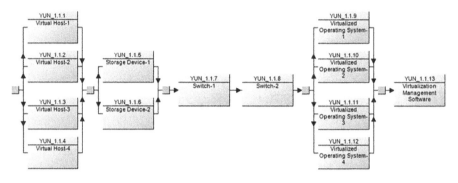

Fig. 3. The reliability diagram model of virtual architecture subsystem

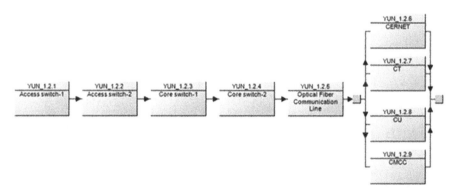

Fig. 4. The reliability diagram model of network communication subsystem

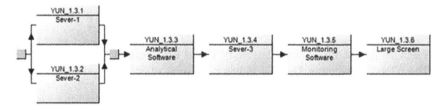

Fig. 5. The reliability diagram model of monitoring and analyzing subsystem

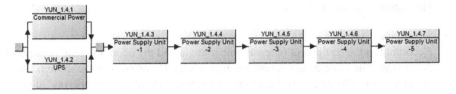

Fig. 6. The reliability diagram model of power supply subsystem

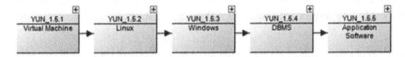

Fig. 7. The reliability diagram model of service subsystem

3. Determination of input parameters

   Combined with the specific equipment selection of the system, and then through the research equipment manufacturers and users of the field of application and maintenance of data, the availability data parameters of the 19 equipment are obtained (including reliability parameters and maintainability data parameters). Reliability data parameters mainly include equipment reliability, failure distribution function, MTBF. Parameters of maintainability data including repair time, repair maintenance personnel and equipment requirements, plan preventive maintenance time, regularly scheduled maintenance personnel and equipment requirements, and various spare parts quantity, cost, delay time and other parameters. The availability data for some devices are shown in Table 2.

Table 2. The availability data for some devices.

Device name	MTBF/h	Corrective maintenance					Spare parts
		Repair time/h	Mechanics	Maintain electric	Technical tool box	Electric tool box	Unit price (yuan)
Virtual host	70080	1	1	–	1	–	12000
Access switch	399193	1	1	–	1	–	7500
Core switch	296088	4	1	–	1	–	–
Optical fiber communication line	87600	6	1	–	1	–	–
Storage device	35040	2	1	–	1	–	5000
Virtual machine	52560	3	1	–	–	–	–
DBMS	43800	4	1	–	–	–	–
Application software	39420	2	1	–	–	–	–
Linux OS	175000	6	1	–	–	–	–
UPS power supply unit	53200	2	–	1	–	1	500

4. Simulation and result analysis

- Simulation

  According to the equipment renewal time interval and the operation requirements of the system, the system simulation running life is 15 years, and the system simulation times is 100 times. Assume that the mobilization of personnel costs 20 yuan, the delay time of 1 h, the fee is 100 yuan per hour, tool mobilization costs 10 yuan fee, 10 yuan/hour, maintenance interval is 720 h, parts arrival delay time of 0.5 h. The availability and maintenance cost of the system life cycle are simulated and analyzed. The results are shown in Tables 3 and 4.

Table 3. Simulation results of the availability of cloud computing data center in a university.

System name	Total shutdown time/(h)	Number of total shutdown	MTBO/ (h)	MTTR/ (h)	Average availability (%)
Cloud computing data center	923.8	239	733.1	3.865	99.47
Virtualization architecture	15.57	5.48	31970	2.841	99.99
Network communication system	22.21	4.2	41710	5.288	99.99
Monitoring and analyzing system	57.86	18.25	11320	3.17	99.97
Power supply system	51.66	17.22	10170	3	99.97
Service system	776.8	194.1	902.7	4.003	99.56

Table 4. Simulation results of operation and maintenance cost of cloud computing data center in a university.

Category	Staff	Equipment	Spare parts procurement	Spare parts storage	Spare parts maintenance	Total
Cost (yuan)	84800	27500	320300	65700	12200	530700

- Availability simulation results analysis

  In 15 years, Cloud computing data center total downtime 239 times, 923.8 h, MTBO of 733.1 h (about 30.55 days, 1 months). The fault rate of the service subsystem is the highest, and the MTBO is 902.7 h. This is mainly due to the large number of components of the subsystem and More operating times. In virtualization architecture, network communication system, monitoring and analysis, power supply system, main node such as server, network, power supply are taken redundant backup and city electricity and communication line failure rate is relatively low, so the four subsystems of fault rate is low.

The MTTR of the total system is 3.865 h. Among them, the MTTR of the network communication subsystem is the longest, 5.288 h. It is mainly due to the interruption of optical fiber communication line. It needs to find the point of failure, and fiber splicing, so that costing a long time. Due to the virtual host and storage are equipped with spare parts and therefore the MTTR of virtualization architecture subsystem is the shortest. 2.841 h.

- Analysis of simulation results of operation and maintenance cost
  In 15 years, Cloud computing data center operation and maintenance of the total cost is 530700 yuan. Among them, the staff costs 84800 yuan, accounting for 15.98 % of the total cost, tools and equipment to use the cost of 27500 yuan, accounting for 5.18 %, The procurement of spare parts, storage and maintenance costs cost of 386000 yuan, accounting for 72.73 %. Therefore, during the operation of the system, system equipment cost, spare parts procurement cost, storage cost and the replacement cost are the most important part of system operation and maintenance.

5 Suggestions on Improving the Availability of Cloud Computing Data Center

In order to improve the availability of the software, data center operating system and application software are carried out according to the month.

Hardware device with high reliability and long service life will be selected such as servers, storage, etc. Take necessary redundant hot backup measures for important equipment to achieve automatic switching function.

The necessary spare equipment and a sufficient number of spare parts can effectively reduce the maintenance time of the system or equipment failure. They can also reduce the safety impact of the fault period.

6 Conclusion

The function structure of the Cloud Computing Data Center is analyzed in this paper. Combining with reliability block diagram analysis, the Monte Carlo simulation method is used to analyze the availability of the Cloud Computing Data Center. Finally, the simulation analysis of typical system is completed by this method. The improvement measures are put forward. This method can also simulate and analyze various parameters of the system during the later period of the system, such as failure rate, MEBO, availability, labor costs, equipment costs, spare parts costs and so on. This provides a theoretical basis for the operation and maintenance, thus further improving the overall availability of the Cloud Computing Data Center.

References

1. Gao, Y.N.: Research on the Planning and Evaluation of Cloud Computing Data Center Base Environment Based on ANP, pp. 20–30. Tianjin University, Tianjin (2013)
2. Chen, W.Y., Wang, G.H., Lv, J.G.: System Reliability Theory. China Building Industry Press, Beijing (2010)
3. Zeng, S.K., Zhao, T.D., Zhang, J.G., Kang, R., Shi, J.Y.: System Reliability Design and Analysis. Beijing University of Aeronautics and Astronautics Press, Beijing (2001)

Data Scheduling for Asynchronous Mini-batch Training on the Data of Sparsity

Xiao Fang[1], Dongbo Dai[1(⊠)], and Huiran Zhang[1,2]

[1] School of Computer Engineering and Science, Shanghai University,
Shanghai 200444, China
dbdai@shu.edu.cn
[2] Materials Genome Institute, Shanghai University, Shanghai 200444, China

Abstract. A fast rate of convergence means that a given fixed amount of data can achieve a high level of prediction performance. In a fully asynchronous system, it is considered that the performance of the learning model could be reduced, because of the collision in parallel gradient computations. In this work, we proposed an algorithm called GreedyAC to reduce the collision. We present the detailed design procedure of GreedyAC, and indicate that using GreedyAC can reduce the regret bound of asynchronous adaptive gradient (AsyncAdaGrad) with mini-batch training when the data is sparse. The experimental results show that the proposed algorithm improves the performance significantly as the number of workers (work nodes) increases, and we found that the results are roughly consistent with our theoretical results.

Keywords: Distributed learning · Mini-batch training · Coordinate selecting · Follow-The-Regularized-Leader Proximal

1 Introduction

With the wide availability of large datasets that exceed the handling capacity of single machines, distributed optimization method is becoming a prerequisite for solving large scale machine learning problems. Stochastic Gradient Descent (SGD) has proved to be well suited to data-intensive machine learning tasks [1–3]. Many variations of SGD have been suggested. The inherent sequential nature of such approaches over SGD becomes a problematic limitation in parallel and distributed settings as the predictor must be updated after each training point is processed, providing very little opportunity for parallelization. However, when the data access is sparse, meaning that individual SGD steps only modify a small part of the decision variable, memory overwrites are rare and that they introduce barely any error into the computation when they do occur [3]. This type of data sparsity is prevalent in statistical optimizations problems and can be leveraged to develop these algorithms in parallel and distributed systems [3–5].

In order to parallelize SGD, employing mini-batch training is needed to reduce the communication cost, which uses several examples at each iteration. The question is that mini-batch versions of both SGD and coordinate descent (CD) [6, 7] suffer from their convergence rate degrading towards the rate of batch gradient descent as the size of the mini-batch gets larger [8, 9]. This follows because mini-batch updates are made based

© Springer International Publishing Switzerland 2016
D.-S. Huang et al. (Eds.): ICIC 2016, Part III, LNAI 9773, pp. 313–325, 2016.
DOI: 10.1007/978-3-319-42297-8_30

on the outdated previous parameter vector, in contrast to methods that allow immediate local updates.

For learning at a massive scale, online logistic regression has many advantages, with applications like click-through-rate prediction for web advertising and estimating the probability that an email message is spam. Given a data set of observations, representing n observations $\{x_t\}$, where $t = 1, \ldots, n$ together with corresponding target $\{y_t \in \{+1, -1\}\}$. x_t is the tth such example, and is a $d \times 1$ vector. As noted earlier, both n and d may be on the order of billions to trillions of examples and coordinates, respectively. But any particular training example has only a small fraction of the features. The goal is to learn a model consisting of parameters w to predict the value of y for a new value of x. Before observing each x_t we predict a point w_t. After making the prediction w_t, we observe x_t and suffer the loss $f_t(w_t)$, where $f_t(w_t)$ is a predefined loss function (e.g., we have $f_t(w_t) = \log(1 + \exp(-y_t w_t \cdot x_t))$ in online logistic regression). On a sequence of $t = 1, \ldots, T$, each round incurs a loss $f_t(w_t)$. We measure the difference between the cumulative loss of our predictions and the cumulative loss of a comparator w^* using the notation of *Regret*, defined as an objective function

$$Regret(w^*) \equiv \sum\nolimits_{t=1}^{T} [f_t(w_t) - f_t(w^*)] \text{ where } w^* = arg\ min_{w \in \mathbb{R}^d} \sum\nolimits_{t=1}^{T} f_t(w). \quad (1)$$

In machine learning, the learning algorithm typically minimizes this objective function to obtain the optimal model.

$$minimize_{w^* \in \mathbb{R}^d} \qquad Regret(w^*) \equiv \sum\nolimits_{t=1}^{T} [f_t(w_t) - f_t(w^*)].$$

In general, learning starts from an initial model; it iteratively improves this model by processing the training data to approach the solution.

Considering a fully asynchronous model, some researchers have shown that standard stochastic algorithms essentially "work" even when updates are applied asynchronously by many threads [3, 10]. McMahan et al. [11] confirms that it does effect for moderate amounts of parallelism, but their experiments show that the performance can degrade significantly for large amounts of parallelism. They presented an algorithm called AdaptiveRevision to solve the problem (1) by choosing learning rates and efficiently revising past learning-rate choices, but AdaptiveRevision incurs a cost as applying adaptive revision of some previous steps would slow down the convergence rate. Duchi et al. [5] has given the upper bound on the problem (1) with sparse condition. They developed two algorithms: asynchronous dual averaging (AsyncDA) and AsyncAdaGrad, which allow asynchronous parallel solution of the problem (1).

Our approach differs from previous work. In order to improve parallelism, we proposed a simple strategy called GreedyAC described in Fig. 1(b): suppose that we have k workers (work nodes) to train a predictive model, then cache $c(c > k)$ mini-batches $(0, 1, \ldots, c - 1)$ in memory; when having received $c \times s$ (s denotes the mini-batch size) training examples, we start to load these training examples into relevant mini-batches instead of filling a mini-batch with consecutive inputs. This paper makes two main contributions as following:

- We proposed an algorithm called GreedyAC to reduce the collision efficiently as the number of the workers increases. Experimental results have proved that using GreedyAC can have better performance.
- Upon AsyncAdaGrad and some assumptions, we provided an analysis to get the upper bound with mini-batch training. The analysis we present can be easily applied to different variation of SGD with minor modification.

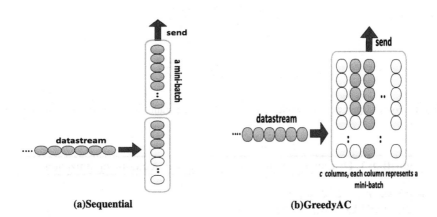

Fig. 1. (a) shows the traditional method that fills a mini-batch sequentially: when a mini-batch has been filled, it will be sent to a request worker; (b) shows the GreedyAC method: when c cloums have been filled, it will be sent to a request worker from left to right.

2   System Overview

In this section, we will introduce the architecture to train the learning model. We build our distributed computing system as a set of k workers to train the above model, each of which is an independent processor. Each worker receives a data stream from an outside source called dispatcher (see Fig. 2).

We will make use of the distributed mini-batch algorithm described in Algorithm 1, a method of converting any serial gradient-based online prediction algorithm into a parallel or distributed algorithm, introduced in [12]. AsyncAdaGrad with mini-batch training can be considered as a concrete implementation of the distributed mini-batch algorithm. The distributed mini-batch algorithm that runs in parallel on each worker processes the short input stream in batches $t = 1, 2, ...T...$, where each batch contains s consecutive inputs. Considering a synchronous model, during each batch t, all of the workers have a same predictor w_t. Once each worker has calculated s stochastic gradients $(g_{t_j})_{j=1}^s$ and accumulates these gradients into a batch-gradient b_t, where b_t denotes the sum of the gradients in tth batch, they start to calculate the sum of these batch-gradients instead of average updates between all workers so as to yield much faster convergence both in practice and in theory [13]. Here, in a realistic asynchronous system, we do not have to process inputs while waiting for the sum operation to complete [12].

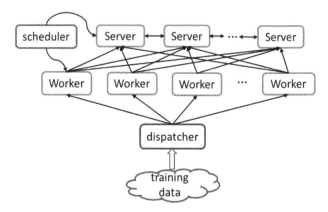

Fig. 2. Workers process data while servers synchronize parameters and perform global updates.

Algorithm 1. Distributed mini-batch algorithm for stochastic optimization

---

Worker $r = 1, 2, \ldots, k$:
  Initialize $w_0^r = 0$;
  for $t = 1, 2, \ldots, \lfloor \frac{n}{s} \rfloor$ do in parallel
    load a part of training examples $(y_{t_j}, x_{t_j})_{j=1}^s$ into
    a mini-batch;
    get w_t^r from Server;
    calculate $b_t^r \leftarrow \sum_{j=1}^s \nabla f_t(x_{t_j}, y_{t_j})$;
    send b_t^r to Server;
  end for
Server:
  aggregate $b_t \leftarrow \sum_{r=1}^k b_t^r$;
  set $(w_{t+1}, \eta_{t+1}) = \phi(w_t, \eta_t, b_t)$;

In Algorithm 1, on round t, each prediction is made by an unspecified update rule: $(w_{t+1}, \eta_{t+1}) = \phi(w_t, \eta_t, b_t)$, where η_t is a learning rate (e.g., $\eta_t = \frac{1}{\sqrt{t}}$) and the batch-gradient

$$b_t = \sum_{j=1}^s g_{t_j} = \sum_{j=1}^s \nabla f_t(x_{t_j}, y_{t_j}) = \sum_{j=1}^s \frac{y_{t_j} x_{t_j}}{1 + exp(y_{t_j} w_t \cdot x_{t_j})}.$$

As concrete examples, SGD performs the update:

$$w_{t+1} = w_t - \eta_t b_t,$$

and AdaGrad algorithm instead uses the update:

$$w_{t+1} = arg\ min_w (w \cdot \sum_{\tau=1}^t b_\tau + \frac{1}{2\eta_t} w \cdot S^{\frac{1}{2}} w),$$

where the matrix S is the diagonal sum of squares of gradient entries (see details in [5]).

In fact, we used an update called Follow-The-Regularized-Leader Proximal (FTRL-Proximal) [14, 15] in our experiments:

$$w_{t+1} = arg\ min_w \left(w \cdot \sum_{\tau=1}^t b_\tau + \frac{1}{2} \sum_{\tau=1}^t \delta_\tau \|w - w_\tau\|_2^2 + \lambda_1 \|w\|_1 \right),$$

where $\delta_1 = \frac{1}{\eta_1}, \delta_t = \frac{1}{\eta_t} - \frac{1}{\eta_{t-1}}$ for $t > 1$ and $\lambda_1 > 0$ is a coefficient that induces sparsity.

3   GreedyAC Algorithm

We have briefly introduced GreedyAC method in Sect. 1, and the key issue is how to dynamically dispatch these $c \times s$ examples into c mini-batches to reduce collision. As each example x_t consists of a set of coordinates S_t, we just load some training examples with a chosen coordinate $i \in S_t$ into a common mini-batch, and try to ensure that the chosen coordinate i should not be occurred in the previous few mini-batches and the next few mini-batches.

In $c \times s$ examples, we select all examples with an optimal coordinate i^* (the number of such examples can't exceed s), and load these examples into the middle mini-batch of c mini-batches. This way guarantees that the update of coordinate i^* would not incur a collision with adjacent $c/2$ mini-batches. Often a large c can be used to improve the performance of the proposed algorithm, but it would incur more memory and computational cost.

In fact, we select more than one coordinate. For example, if having selected an optimal coordinate i_1^*, we can select a second optimal coordinate i_2^* to load associated training examples into a mini-batch near the middle. At the same time, we should ensure that i_2^* is not associated with i_1^*, which means that i_1^* and i_2^* can't be occurred in a same example.

Additionally, when calculating the combined gradient b_t^r, w_t^r should be updated iteratively after one example has been trained.

Denote the importance of coordinate i as $Score(i)$. The analysis of evaluating the importance of coordinate i is given in Sect. 4.

Pseudo-code for the algorithm is given in Algorithm 2.

4   The Analysis of Measuring Coordinate Importance

Denote by $regret_i^T (not\ choose\ i)$ the regret bound incurred if coordinate i isn't chosen at iteration T, and similarly let $regret_i^T (choose\ i)$ be the regret bound incurred if coordinate i is chosen at iteration T. Then we can write the following problem as our objective needed to be solved:

$$Score(i) = regret_i^T(not\ choose\ i) - regret_i^T(choose\ i).$$

Before we can address this problem, we have to first take a look at a few concepts introduced in the following.

Algorithm 2. GreedyAC algorithm for asynchronous mini-batch training

---

```
read c × s examples;
let i be a coordinate and Nᵢ be the number of times that
coordinate i occurs in c × s examples;
initialize threahold = 2, c = 4k;
let U be the set of unchosen coordinates, let C be the
set of chosen coordinates;
for all i do
    if Nᵢ ≤ s and Nᵢ ≥ threahold do
        calculate Score(i);
        add i into U;
end for
for U isn`t ∅ do
    choose i* = arg max∀ᵢ∈U(Score(i));
    if i* is not associated with ∀i ∈ C do:
        load these Nᵢ* examples into a same mini-batch
        near the middle;
        add i* into C;
        delete i* from U;
    else
        delete i* from U;
end for
the rest of examples will be assigned to arbitrary po-
sitions.
```

4.1 Bounding Per-coordinate Regret Without Mini-batch Training

Duchi et al. [5] has given a precise characterization of the (natural) set of sparse optimization problems and shows the regret bound of AsyncAdaGrad. We consider that their approach is to use the same update rule for each coordinate i $(i \in \{1, \ldots, d\})$ as in serials case and each coordinate can be updated independently on the other coordinates, then we can bound (1) by proving a per-coordinate bound (2) for linear functions and then summing across coordinates so as to greatly simplify our analysis.

$$Regret(w^*) \equiv \sum\nolimits_{i=1}^{d} Regret_i(w_i^*) \text{ where } Regret_i(w_i^*) \equiv \sum\nolimits_{t=1}^{T} (f_t(w_{t,i}) - f_t(w_t^*)).$$
(2)

Similar approach can be seen in McMahan et al. [11].

Then, we consider a 1-dimension problem where the upper bound on the delay of any processor $m = k - 1$, the diameter of the feasible set $R_\infty = sup_{w \in \mathbb{R}^d} \|w - w^*\|_\infty$. Let p_i be the marginal probability of appearance of coordinate i in a training example. On the condition of $\delta^2 \geq \sum_{\tau=t}^{t+m-1} g_{\tau,i}^2$ for all t with high probability, where δ is an initial constant diagonal term and $g_{t,i}$ is the ith coordinate of the gradient g_t, Theorem 5 of Duchi et al. [5] becomes

$$\mathbb{E}[Regret_i(w_i^*)] \leq \frac{1}{\eta} R_\infty^2 \mathbb{E}\left[\left(\sum_{\tau=t}^{t+m-1} g_{\tau,i}^2 + \sum_{t=1}^{T} g_{t,i}^2\right)^{\frac{1}{2}}\right] + \eta \mathbb{E}\left[\left(\sum_{t=1}^{T} g_{t,i}^2\right)^{\frac{1}{2}}\right](1 + p_i m). \quad (3)$$

In general, $p_i = P(x_{t,i} \neq 0)$ cannot be known in advance in an online setting, where $P(\cdot)$ is a symbol of probability mass function.

By Jensen's inequality, we have that

$$\mathbb{E}\left[\left(\sum_{\tau=t}^{t+m-1} g_{\tau,i}^2 + \sum_{t=1}^{T} g_{t,i}^2\right)^{\frac{1}{2}}\right] \leq \left(\mathbb{E}\sum_{\tau=t}^{t+m-1} g_{\tau,i}^2 + \mathbb{E}\sum_{t=1}^{T} g_{t,i}^2\right)^{\frac{1}{2}}.$$

Assuming that there exist constants $H_i > 0$ such that for $\mathbb{E}\left[g_{t,i}^2 | x_{t,i} \neq 0\right] \leq H_i^2$, and note that $\mathbb{E}\left[g_{t,i}^2\right] \leq p_i H_i^2 + (1 - p_i) \cdot 0 = p_i H_i^2$, we have

$$\mathbb{E}[Regret_i(w_i^*)] \leq \frac{H_i}{\eta} R_\infty^2 \sqrt{mp_i + Tp_i} + \eta H_i \sqrt{Tp_i}(1 + p_i m).$$

4.2 Bounding Per-coordinate Regret with Mini-batch Training

Now we consider a mini-batch with size s uniformly at random from among all subsets of T training examples. In practice, data is often partitioned between k workers, and at each iteration s points are sampled from each machine. We assume s divides T, so $T' = T/s$ is the number of batches. Thus, with mini-batch training, we can run T' steps of online gradient descent on the combined gradients b_t.

As the non-zero rounds are chosen randomly with probability p_i, $p_{i,s} = P(\xi_i \neq 0) = 1 - (1 - p_i)^s$, where ξ_i is the number of times that coordinate i occurs in a given mini-batch, and $p_{i,s}$ denotes the marginal probability of appearance of coordinate i in a mini-batch with size s. Due to the fact that the weights of the model during training would oscillate over the best when the model nears optimal, we can assume that $g_{t,i} \in \{-H_i, 0, H_i\}$ and $P(g_{t,i} = H_i | x_{t,i} \neq 0) = P(g_{t,i} = -H_i | x_{t,i} \neq 0) = \frac{1}{2}$. Then

$$\mathbb{E}\left[b_{t,i}^2 | \xi_i = l\right] = \mathbb{E}\left[\left(\sum_l g_{t,i}\right)^2\right]$$

$$= \sum_{j=0}^{l} C_l^j \frac{1}{2^l} (j - (l - j))^2 H_i^2 = l H_i^2 (l = 0, 1, \ldots, s),$$

and thus we have

$$\mathbb{E}\left[b_{t,i}^2\right] = \sum_{l=0}^{s} P(\xi_i = l)\mathbb{E}\left[b_{t,i}^2 | \xi_i = l\right] = \sum_{l=0}^{s} C_s^l p_i^l (1-p_i)^{s-l} l H_i^2 = H_i^2 s p_i.$$

Often the derivation of $\mathbb{E}\left[b_{t,i}^2\right]$ is not suitable for the beginning stage of model training, hence, we use a constant ε to correct the bias. Therefore, combined with the above technique applied on mini-batch training, inequality (3) becomes

$$\mathbb{E}\left[Regret_i(w_i^*)\right] \leq \frac{1}{\eta}R_\infty^2 \mathbb{E}\left[\left(\sum_{\tau=t}^{t+m-1} b_{\tau,i}^2 + \sum_{t=1}^{T'} b_{t,i}^2\right)^{\frac{1}{2}}\right] + \eta\mathbb{E}\left[\left(\sum_{t=1}^{T'} b_{t,i}^2\right)^{\frac{1}{2}}\right](1 + p_{i,s}m)$$

$$\leq \frac{H_i}{\eta}R_\infty^2 \sqrt{msp_i + Tp_i + \varepsilon} + H_i\eta\sqrt{Tp_i + \varepsilon}(1 + p_{i,s}m) = A_i,$$

$$(4)$$

where A_i is a notation of the upper bound.

4.3   Per-coordinate Importance

Back to the original problem mentioned at the beginning of this section. Considering such a simple situation that a fixed coordinate i is chosen to a relevant worker on each round, and a data example without coordinate i will be assigned to one of k workers randomly, thus we have

$$p_i' = \frac{Tp_i}{Tp_i + T(1 - p_i)/k} = kp_i/(kp_i + 1 - p_i),$$

where p_i' denotes the marginal probability of appearance of coordinate i in a data example of a relevant worker. Meanwhile, when we run T' steps, the number of steps relevant to coordinate i can be calculated as: $T'sp_i/p_i' = T'(kp_i + 1 - p_i)/k$. This implies that $p_{i,s}' = \left(1 - (1 - p_i')^s\right)(kp_i + 1 - p_i)/k$, where $p_{i,s}'$ denotes the marginal probability of appearance of coordinate i in a mini-batch with size s when taking this strategy. Inequality (4) becomes

$$\mathbb{E}\left[Regret_i(w_i^*)\right] \leq \frac{H_i}{\eta'}R_\infty^2 \sqrt{msp_i + Tp_i + \varepsilon'} + H_i\eta'\sqrt{Tp_i + \varepsilon'}\left(1 + p_{i,s}'m\right) = B_i, \quad (5)$$

where B_i is a notation of the upper bound, η' is a step size, and ε' is a bias term.

In practice, each coordinate $i \in S_t$ has the opportunity to be selected to determine which worker the training example should be assigned to. Moreover, the sparsity of data indicates that the changes of $\varepsilon \to \varepsilon'$ and $\eta \to \eta'$ are small. On the other hand, the influence of such changes could be diminished with the increase of iterations, so we can approximately have $\varepsilon' = \varepsilon$ and $\eta' = \eta$.

Let M_i^t be the number of times that coordinate i has been chosen on round t, and let N_i^t be the number of times that coordinate i occurs on round t. It is obvious to see that the regret upper bound of coordinate i approaches to B_i as $M_i^t \to N_i^t$ and approaches to A_i as $M_i^t \to 0$. Therefore, an approximation to calculate $regret_i^T(not\ choose\ i)$ and $regret_i^T(choose\ i)$ on round $t = T$ can be defined as:

$$regret_i^T(not\ choose\ i) = \left(1 - \frac{M_i^{T-1}}{N_i^{T-1}+1}\right)A_i + \frac{M_i^{T-1}}{N_i^{T-1}+1}B_i,$$

$$regret_i^T(choose\ i) = \left(1 - \frac{M_i^{T-1}+1}{N_i^{T-1}+1}\right)A_i + \frac{M_i^{T-1}+1}{N_i^{T-1}+1}B_i,$$

and it is easy to show that

$$Score(i) = \frac{A_i - B_i}{N_i^T} = \frac{H_i \eta \sqrt{Tp_i} + \varepsilon(p_{i,s}m - p'_{i,s}m)}{N_i^T}. \tag{6}$$

Note that $p'_{i,s} < p_{i,s}$, we can draw a conclusion that theoretically this method can reduce the regret upper bound.

The analysis given above is based on AsyncAdaGrad. Since using FTRL-Proximal in an asynchronous system can achieve a similar regret bound to AsyncAdaGrad, the conclusion is also effective when adopting FTRL-proximal algorithm to update the weight of our predictive model. As a matter of fact, FTRL-Proximal with per-coordinate learning rate can achieve the best prediction accuracy [15, 16], so the learning rate of coordinate i at iteration t in our experiments is set to

$$\eta_{t,i} = \frac{\alpha}{\beta + \sqrt{\sum_{\tau=1}^{t/s} b_{\tau,i}^2}},$$

where α, β are two tunable parameters. By noting that β just ensures that early learning rates are not too high ($\beta = 1$ usually good enough), we can ignore the term β and set $= \eta_{t,i} \approx \alpha / (\sum_{\tau=1}^{t/s} b_{\tau,i}^2)^{\frac{1}{2}}$. Then, on round $t = T$, we obtain

$$\eta \mathbb{E}\left[\left(\sum_{t=1}^{T'} b_{t,i}^2\right)^{\frac{1}{2}}\right] \approx \mathbb{E}\left[\alpha\left(\sum_{t=1}^{T'} b_{t,i}^2\right)^{\frac{1}{2}} \Big/ \left(\sum_{t=1}^{T'} b_{t,i}^2\right)^{\frac{1}{2}}\right] = \alpha.$$

Considering that $\eta\sqrt{Tp_i} + \varepsilon$ is a special case of $\eta \mathbb{E}\left[\left(\sum_{t=1}^{T'} b_{t,i}^2\right)^{\frac{1}{2}}\right]$, then

$$\frac{A_i - B_i}{N_i^T} \approx \frac{m}{N_i^T}\alpha\left(p_{i,s} - p'_{i,s}\right) = \frac{1 - (1 - p_i)^s - \frac{kp_i - p_i + 1}{k}(1 - (1 - \frac{kp_i}{kp_i - p_i + 1})^s)}{Tp_i}(k - 1)\alpha. \tag{7}$$

Considering that T, α, k are same for each coordinate, the above discussion leads us to a solution to measure coordinate importance,

$$Score(i) = \frac{1 - (1 - p_i)^s - \frac{kp_i - p_i + 1}{k}\left(1 - \left(1 - \frac{kp_i}{kp_i - p_i + 1}\right)^s\right)}{p_i}.$$

In $c \times s$ examples, we can obtain approximately $p_i = \frac{N_i}{cs}$, where N_i is the number of times that coordinate i occurs in these $c \times s$ examples.

5   Experiments

To confirm our theoretical results and insights, we experimented on two binary classification datasets of varying scales and high-dimensional sparsity, which are listed on Table 1. The URL dataset aims to detect malicious URLs. CTR is a private dataset containing an ad click prediction. We implemented our algorithm in Parameter Server [17, 18]. We trained logistic regression models using FTRL-Proximal update rule with the training data sets and evaluated with average log-loss (the average logistic loss) and the test error rate averaged over 3 runs over the testing dataset (see Figs. 3 and 4). We performed the experiments on a single machine running Ubuntu Linux. The setting of parameters s and α are based on the outcomes of multirun simulation.

Table 1. A collection of real datasets

Name	URL	CTR
#training examples	$1.8 * 10^6$	$2.238 * 10^5$
#testing examples	$0.6 * 10^6$	$6.355 * 10^4$
#features	$3.2 * 10^6$	$1.314 * 10^7$
#class	2	2
label ratio +1: −1	1:2	1:1

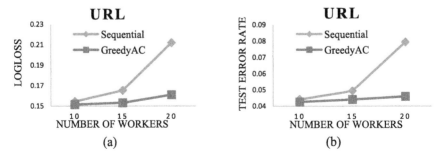

Fig. 3. Experiment results of comparing GreedyAC with sequential method on URL dataset for various $k = 10, 15, 20$.

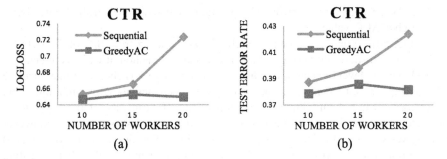

Fig. 4. Experiment results of comparing GreedyAC with sequential method on CTR dataset with various $k = 10, 15, 20$.

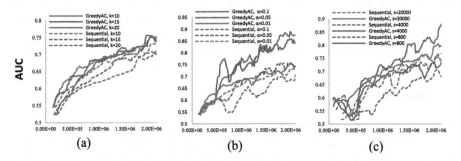

Fig. 5. Accuracy measured by AUC of comparing GreedyAC with sequential method for various number of workers k, various learning rate parameter α, various mini-batch size s. (Color figure online)

In Fig. 3, we compared the performance of GreedyAC and sequential method after doing 10 pass through the URL training dataset with various number of workers k (fixing $s = 8000$ and $\alpha = 0.02$). In Fig. 4, we ran 2 pass through the CTR training dataset. Our experiments confirm that GreedyAC can get a smaller logloss and test error rate for each value of k ($k = 10, 15, 20$). Comparing with sequential method, these two figures demonstrate the better ability of GreedyAC to scale with an increasing number of workers k.

Besides considering the impact of the number of workers k, we also investigated the parameter α that tunes the step size, and the mini-batch size s. We used progressive validation (sometimes called online loss) on the CTR dataset rather than cross-validation or evaluation on a held out dataset and measured with AUC (the area under the ROC curve) in Fig. 5. In all subplots of this figure, the x-axis is computational cost measured by the number of training examples and the y-axis is AUC value, then we found that GreedyAC typically achieves an effective additional speedup over sequential method.

In Fig. 5(a), fixing $s = 10000$ and $\alpha = 0.01$, we concluded our experiments by AUC for various number of worker ($k = 10, 15, 20$). The figure shows that the convergence rate decreases with the increase of k for both GreedyAC and sequential method. When comparing GreedyAC with traditional sequential method, GreedyAC has a higher AUC value.

In Fig. 5(b), fixing $k = 10$ and $s = 10000$, we varied α ($\alpha = 0.1, 0.05, 0.01$) for each method. It shows that using a small α in sequential method, it will take much cost to reach a given level of accuracy. When a relatively large α is chosen, it may increase the convergence rate at first but the training loss oscillates above the minimum and never goes down to the minimum. The optimal value of α can vary a fair bit depending on the coordinates. Sequential method with $\alpha = 0.1$ has worst performance, however GreedyAC with $\alpha = 0.1$ does the opposite result. From Fig. 5(a) and (b), we observed that the difference of AUC between GreedyAC and sequential method before getting the optimal increases with the increase of α and k as Eq. (7) indicates.

Figure 5(c) shows the contribution of GreedyAC with different mini-batch size s ($s = 800, 4000, 20000$) when fixing $k = 10$ and $\alpha = 0.02$. Inequality (4) and (5) show that the regret bound will get larger as the mini-batch size increases. However, our experiments show that a moderate mini-batch size ($s = 4000$) can get better performance than a small size ($s = 800$), which can be expressed by a fact that a large mini-batch size can improve the validity of combined gradients.

6 Conclusion

We have demonstrated that GreedyAC can significantly speed up the convergence rate as the number of workers becomes large, but, it also incurs a preprocessing cost. The key algorithmic technique is based on the fact that the updates for different coordinates are dependent on each other and to aggregate some training examples which have a common chosen coordinate into a mini-batch. Compared with a general analysis, our analysis of measuring coordinate importance may produce less resource consumption.

Acknowledgement. This work is supported by the Shanghai Municipal Science and Technology Commission (Grant No. 14DZ2261200, 15DZ2260300).

References

1. Zhang, T.: Solving large scale linear prediction problems using stochastic gradient descent algorithms. In: ICML, pp. 919–926 (2004)
2. Bottou, L., Bousquet, O.: The tradeoffs of large scale learning. In: NIPS, pp. 161–168 (2008)
3. Niu, F., Recht, B., Re, C., Wright, S.J.: Hogwild: a lock-free approach to parallelizing stochastic gradient descent. In: NIPS, pp. 693–701 (2011). http://www.dblp.org/db/conf/nips/nips2011.html#RechtRWN11
4. Takác, M., Bijral, A.S., Richtárik, P., Srebro, N.: Mini-batch primal and dual methods for SVMs. In: ICML, pp. 1022–1030 (2013). http://www.dblp.org/db/conf/icml/icml2013.html#TakacBRS13
5. Duchi, J., Jordan, M.I., McMahan, B.: Estimation, optimization, and parallelism when data is sparse. In: NIPS, pp. 2832–2840 (2013)

6. Shalev-Shwartz, S., Zhang, T.: Accelerated proximal stochastic dual coordinate ascent for regularized loss minimization. In: ICML, pp. 64–72 (2014). http://www.dblp.org/db/conf/icml/icml2014.html#Shalev-Shwartz014

7. Richtárik, P., Takác, M.: Distributed Coordinate Descent Method for Learning with Big Data. CoRR abs/1310.2059 (2013). http://www.dblp.org/db/journals/corr/corr1310.html#RichtarikT13

8. Byrd, R.H., Chin, G.M., Nocedal, J., et al.: Sample size selection in optimization methods for machine learning. Math. Program. **134**(1), 127–155 (2012)

9. Li, M., Zhang, T., Chen, Y., Smola, A.J.: Efficient mini-batch training for stochastic optimization. In: KDD, pp. 661–670 (2014). http://www.dblp.org/db/conf/kdd/kdd2014.html#LiZCS14

10. Duchi, J.C., Agarwal, A., Wainwright, M.J.: Dual averaging for distributed optimization: convergence analysis and network scaling. IEEE Trans. Automat. Control **57**(3), 592–606 (2012)

11. McMahan, H.B., Streeter, M.J.: Delay-tolerant algorithms for asynchronous distributed online learning. In: NIPS, pp. 2915–2923 (2014). http://www.dblp.org/db/conf/nips/nips2014.html#McMahanS14

12. Dekel, O., Gilad-Bachrach, R., Shamir, O., Xiao, L.: Optimal distributed online prediction using mini-batches. JMLR **13**, 165–202 (2012). http://www.dblp.org/db/journals/jmlr/jmlr13.html#DekelGSX12

13. Ma, C., Smith, V., Jaggi, M., Jordan, M.I., Richtárik, P., Takác, M.: Adding vs. averaging in distributed primal-dual optimization. In: ICML, pp. 1973–1982 (2015)

14. McMahan, H.B.: Follow-the-regularized-leader and mirror descent: equivalence theorems and L1 regularization. In: AISTATS, pp. 525–533 (2011). http://www.dblp.org/db/journals/jmlr/jmlrp15.html#McMahan11a

15. McMahan, H.B., Holt, G., Sculley, D., Young, M., Ebner, D., Grady, J., Nie, L., Phillips, T., Davydov, E., Golovin, D., Chikkerur, S., Liu, D., Wattenberg, M., Hrafnkelsson, A.M., Boulos, T., Kubica, J.: Ad click prediction: a view from the trenches. In: KDD, pp. 1222–1230 (2013). http://www.dblp.org/db/conf/kdd/kdd2013.html#McMahanHSYEGNPDGCLWHBK13

16. Streeter, M., McMahan, H.B.: Less Regret via Online Conditioning. Computer Science – Learning (2010)

17. Li, M., Andersen, D.G., Smola, A.J., Yu, K.: Communication efficient distributed machine learning with the parameter server. In: NIPS, pp. 19–27 (2014). http://www.dblp.org/db/conf/nips/nips2014.html#LiASY14

18. Li, M., Andersen, D.G., Park, J.W., Smola, A.J., Ahmed, A., Josifovski, V., Long, J., Shekita, E.J., Su, B.-Y.: Scaling distributed machine learning with the parameter server. In: OSDI, pp. 583–598 (2014). http://www.dblp.org/db/conf/osdi/osdi2014.html#LiAPSAJLSS14

Knowledge Discovery and Natural Language Processing

The Improved Clustering Algorithm
for Mining User's Preferred Browsing Paths

Xiaojing Li$^{(\boxtimes)}$ and Yanzhen Cheng

Jiyuan Vocational and Technical College,
Jiyuan 459000, Henan, China
Lxj55883@163.com

Abstract. The current mining algorithm only consider the user's access frequency, neglecting the interest of users in their visiting path. Compared to the current algorithms for mining user browsing preferred path, clustering algorithm combines the Jacques ratio coefficient and the longest public path coefficient multiplication. This proposed method can estimate the user similarity of page interest and website access structures matrix more accurately for the element value based on the "three tuple" model. Adopting an improved mining algorithm for preference and interest calculation, the bad impact of mining is removed due to pages idle and links. The experimental results showed that the algorithm had higher efficiency and accuracy in web log mining of big data.

Keywords: Clustering algorithm · Data mining · Web logs · Preferred browsing paths

1 Introduction

Web user browsing preference path mining algorithm is used to analyze of web log records and find the user access rules. This algorithm has been successfully applied to personalized web recommendation, system improvement and business intelligence and so on. At present, the most commonly used algorithms in the acquisition of browsing patterns are the maximum frequent sequence method, the reference length method and the tree topology structure method [1]. But these algorithms in fact belong to an improved association rule algorithm and there exist two issues. Firstly, it simply assumed that the frequency of user's browsing is a representation of the user's interest. Secondly, web log data gradually showed distributed, heterogeneous, dynamic and massive properties with the development of network [2], thus the traditional centralized data mining algorithms can't meet the needs of web log mining process with massive data.

In order to solve the problems mentioned above, this paper combines the clustering algorithm and web user browsing pattern mining algorithm, improves the existing algorithm, together with putting the method of multiplying the Jacobi coefficient and longest common path coefficient into consideration to reflect the similarity between users more accurately. This method uses a three tuple to represent the degree of the page interest, considering the user's access time, the size of the page and the number of visits and so on to construct a data matrix which take the URL address of the reference page as the row, the web page URL address as the column and the degree of access

© Springer International Publishing Switzerland 2016
D.-S. Huang et al. (Eds.): ICIC 2016, Part III, LNAI 9773, pp. 329–337, 2016.
DOI: 10.1007/978-3-319-42297-8_31

interest as the element value, and calculate the preference and interest of the matrix with the improved mining algorithm based on this [3]. When to choose to browse the next page, users can get more accurate preference path since the comprehensive consideration of the number of visits, access time and page size.

2 Improved Clustering Algorithm

2.1 The Basic Definition of Clustering Algorithm

Assume n users access path set $U = \{C_1, C_2, \ldots, C_n\}$, one of the access path for $C_i = \{V_1, V_2, \ldots, V_i\}$, where V_i represents a node to be visited.

Definition 1: The number of modes of user access is equal to the length of the path $|c|$.

Definition 2: The Jacobi coefficient

$$s'_{ij} = \frac{|c_i \cap c_j|}{|c_i \cup c_j|} \tag{1}$$

For example, there are two web user access path C1 = {V1, V2, V3}, C2 = {V2, V1, V3}, and the calculated results are both 1 using the Jacobi coefficient, but the two paths were obviously not the same. It is because the transaction data which the Jacobi coefficient described does not have precedence relation, but the web user access path is a sequential event. So, it can't be simply described the similarity of access paths by the Jacobi coefficients.

Definition 3: The similarity coefficient of access path refer to s''_{ij}.

Assume $comm(c_i, c_j)$ represents the longest length of common path, $\max(|c_i|, |c_j|)$ represents the longest length of path from c_i to c_j, and the similarity coefficient for user access is:

$$s''_{ij} = \frac{|comm(c_i, c_j)|}{\max(|c_i|, |c_j|)} \tag{2}$$

If there are three web access paths where $C_1 = \{V_1, V_2, V_3\}$, $C_2 = \{V_2, V_3, V_4\}$, and $C_3 = \{V_3, V_2, V_4\}$, the longest length of common path from C_1 to C_2 is V_2 or V_3, the length is 2, and the similarity factor is 0.5. Also the longest length of common path from C_1 to C_3 is V_2 or V_3, the length is 1, and the similarity factor is 0.25. The nodes of the paths C_2 and C_3 are exactly the same with the same order, but the similarity coefficient is only 0.33 using the calculation method and it is lower than the similarity coefficient from C_1 to C_2. This is obviously unreasonable. Thus we propose to improve it as follows:

Definition 4: The similarity coefficient of path from C_i to C_j refer to S_{ij}

$$S_{ij} = \left(S'_{ij}\right)^\alpha \left(S''_{ij}\right)^\beta, 0 \leq \alpha, \beta \leq 1 \tag{3}$$

Where α, β refer to the adjusted metric coefficients. The effect of the Jacobi coefficient increases as the increase of the value of α, and the effect of similarity coefficient increases as the increase of value of β, and the effect of sequential increase accordingly.

This similarity coefficient considers the advantages of both the Jacobi coefficient and similarity coefficient. We get the data matrix S of similarity coefficient of user access by using the coefficient to calculate the similarity coefficient of all the paths:

$$S = \begin{bmatrix} S_{11} & S_{12} & \cdots & S_{1n} \\ & S_{22} & \cdots & S_{2n} \\ & & \cdots & \vdots \\ & & & S_{nn} \end{bmatrix}$$

2.2 Improved Clustering Algorithm

In the process of combining the access paths, the maximum similarity coefficient plays a decisive role. The most similarity coefficients less than the threshold value are not effective for clustering [4]. In order to solve the "dimension disaster" problem of the traditional clustering algorithm in the high-dimensional web log data clustering, filtering the smaller similarity coefficient could greatly reduce the data size.

Algorithm 1. output the similar user access path clustering
Enter: The web log file, and set a threshold;
Export: Similar path clustering $C = \{C_i\}$.
Description of algorithm:

```
C = {φ} ; // initialization
While (not to the end of the file)
{
 Read records from a data table;
 While (not to the end of the file)
 {
   Read records from a data table;
   Calculate the similarity coefficient of access path
   S = (S')ᵅ(S'')ᵝ ;
   If (S>θ) //compare of S and threshold θ
   {Keep the current number of path;
   }
 }
 Get temporary clustering Cᵢ ;
 If(Cᵢ is not a subset of set C)
 {
  Put Cᵢ into set C;
 }
}
```

Calculate the intersection of each category of membership, and eliminate duplicates according to the degree of membership.

Output to get set C.

3 Improved Preferred Path Mining Algorithm

3.1 Browsing Frequency Preference

If the user has m ways to leave a web page, the user is more interested in the choice with a higher selected number [5].

Definition 1: Assume S_i represents the frequency of user selected into the next page through i option. Based on the traditional definition of confidence and formula 1, regardless of the site structure in the confidence limits on the traditional case, assume the threshold of i option as:

$$C_K = \frac{S_K}{\left(\sum_{i=1}^{n} S_i\right)} \tag{4}$$

Definition 2: On a website, assume all the URL as set U, all the subset as W. If $w \subset W$, $\forall x \in w$ (x represents the page browsing sequence composed by $\forall u \in U$, where the i represents the i browse page), the m numbers before the page browsing sequence are the same, but the m + 1 exists in n different pages, and it represents there are n different browse ways from m to m + 1. So, we assume the j(j = 1, 2, ..., n) reference of browse way as:

$$P_j = \frac{S_j}{\left(\sum_{i=1}^{n} S_i\right)/n} \tag{5}$$

Thus, when n > 1, the possibility of i approaches to surfing the internet is considered in the preference coefficient of P in n options. Therefore, it could reflect the user's interest degree more accurately compared to the traditional confidence.

3.2 Browsing Interest Preference

The algorithm of formula 5 only consider the frequency of user browsing, and it is not comprehensive [6, 7]. As the interest of users is related to the size, time and frequency of use access, large page results in a long time, and the long browsing time represents the high interest of browsing. At the same time, the interest degree of user browsing also depends on the number of users access.

Definition 3: Set the interest degree of user browsing as:

$I(URL.time, URL.size, URL.num)$, if we assume the access number of page $URL_i \rightarrow URL_j$ as num, the access time of URL_j as $URL_{i \rightarrow j}.time$, and the size of page URL_j as $URL_j.size$. Then the interest degree of a user is:

$$I_{ij} = Int\left(\sum_{i=1}^{num} URL_{i \rightarrow j}.time\right) / URL_j.size) + 1 \qquad (6)$$

3.3   Improved Preferred Path Mining Algorithm

Assume there are n URL pages on a website, then construct a matrix of URL-URL, where the row element is URL-Reference, the column element is URL, and the element value is the interest degree of a user from a reference page links to access page. Set an additional null value in row and column. If the user directly enter the web address or access page links from other sites, null appeared in a row. If the user visits the page on the website or exit from the site links to external sites, null appeared in the column [8, 9]. In addition, all reference web page are not themselves, so the diagonal elements of the matrix is 0 (Table 1).

$$
\begin{array}{c}
\quad\quad Null\ \ URL_1\ URL_2\ \cdots\ URL_n \\
\begin{array}{c}
Null \\
URL_1 \\
URL_2 \\
\vdots \\
URL_n
\end{array}
\left[
\begin{array}{cccccc}
0 & I_{01} & I_{02} & \cdots & I_{0n} \\
I_{10} & 0 & I_{12} & \cdots & I_{1n} \\
I_{20} & I_{21} & 0 & \cdots & I_{2n} \\
\vdots & \vdots & \vdots & \vdots & \vdots \\
I_{n0} & I_{n1} & I_{n2} & \cdots & 0
\end{array}
\right]
\end{array}
$$

Table 1. The process of mining user preferred path in example 1

	Null	A	B	C	D	E	F	G	H	The sum
Null	0	63	15	2	5	0	0	0	5	90
A	6	0	36	35	0	0	0	0	0	77
B	43	0	0	0	6	40	0	0	8	97
C	3	0	0	0	0	10	30	0	0	43
D	16	0	0	0	0	8	0	35	20	79
E	3	0	0	0	0	0	31	0	0	34
F	5	0	0	23	0	0	0	0	0	28
G	4	45	0	0	0	0	0	0	0	49
H	6	0	0	0	0	0	0	0	0	6

Calculation 2: The improved clustering algorithm for mining Web preference path
Input: Assume Web browsing matrix M[n+1][n+1], Sup represents the threshold of browsing support, Pre represents the threshold of browsing preference.
Output: Web preferred browsing path set as NPS.

```
i=0;
while(i<n+1)
{
m=non-zero number of colum;
coun=0;//coun represents the sum of interest expressed
in the row
j=0;
while(j<n+1)
 { if(Sij>0)
   coun+=Supij;
   if((Sij>=Sup)&&( Sij/(coun/m)>=Pre)
   item2= item2+{i,j};
   j++;
 }
i++;
}// merge the same preferred path
x=2;//x represents a set of matrix data item
Flag=0; // detection if X concentration is a merge op-
eration
while(Flag==1)
{
 i=1;
 while(i< path numbers in itemx-1)
 {
  P1= the i preference sub path in itemx-1;
  comb=0;// judge whether to do the merge operation
  j=i+1;
  while(j<= path numbers in itemx-1)
  {
   P2= the j preference sub path in itemx-1;
   if((the (x-2) of before P2)==(the (x-2) after P1))
   {
    Merge the sub path of P1 and P2 to X item of set
    itemx;
    comb=1;
   }
   j++;
  }
```

```
if(comb==0)
Write the preference sub path P1 to set NPS;
Flag=Flag ∪ comb;
i++;
}
x++;
}
```

4   Analysis of Experimental Results

Assume $|URL|$ represents the number of web page, by using the algorithm one can draw that the time complexity of sub path of browsing preference is $O(2(|URL|+1)^2)$. The time to merge with the same paths is $O((|URL|-2)(|URL|+1)^2)$. Thus the total time is $O((|URL|+1)^3)$.

In the process of experiment, 25930 records and 35 pages of web log were experiment objects. The preferred path mining algorithm proposed in this paper and the MFP algorithm in path mining is used to control the threshold setting. In the scenario where the two kinds of mining method were used to explore the same number of preference sub path and frequency browse sub path, the respective accuracy of the

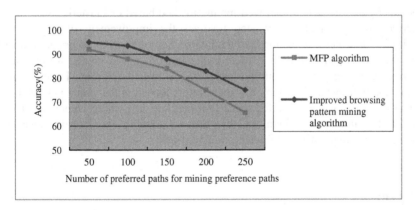

Fig. 1. The accuracy of the algorithm (Color figure online)

proposed algorithm is greater than the known preference path of site access (Fig. 1).

As it can be seen, the improved algorithm mentioned in this paper was more accurate than the MFP algorithm. At the same time, the accuracy of algorithm reduced with the increase of the mining path. It is because that the threshold of mining interest reduced with the increase of the number of paths, which in turn lead to the decrease of credibility of the preferred path. In order to detect the mining time performance of the two methods, we divided the experimental subject into 5,000 records, 15,000 records, 20,000 records and 25,000 records in the experiment. And Fig. 2 showed the comparison of the execution time. We could find that the improved user access pattern mining algorithm had less execution time increase amplitude and better expansibility

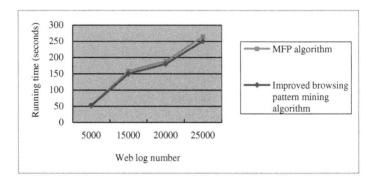

Fig. 2. Time performance comparison of the algorithm (Color figure online)

than the traditional MFP algorithm.

5   Conclusion

This paper puts forward an improved mining method based on clustering web log user preferred browsing paths. First step involves the improvement of the clustering algorithm, duplicates elimination, and the intersection of items to accurately reflect the web user access path similarity. Then, on the basis of a trial model, we explore the preferred browsing paths of multiple pages of a similar user group. Finally, through comparison with other algorithm, the algorithm proposed in this paper has advantages both in accuracy and time performance. Furthermore, it is more comprehensive and accurate data mining algorithm, and has better scalability based on the analysis of different user groups of web browsing preferred path.

References

1. Guan, Y.J., Wang, Y., He, D.N.: A data stream mining approach based on function iterative operation. J. Guangxi Univ. Nationalities **18**(1), 45–49 (2012)
2. Pierrakos, D., Paliouras, G., Papatheodorou, C., Spyropoulos, C.D.: Web usage mining as a tool for personalization: a survey. User Model. User-Adap. Inter. **13**(4), 311–372 (2003). Kluwer Academic Publishers
3. Myra, S., Lukaa, F.: A Data Miner Analyzing the Navigational Behaviour of Web Users, 28 July 2001. http://www.wiwi.hu-berlin.de/ ~ myra/w_acai99.ps.gz

4. Peng, H.Y., Chai, X.G., Chen, X.J.: The clustering algorithm of level iterated theory. J. Tangshan Coll. **24**(3), 86–91 (2011)
5. Caramel, E., Crawford, S., Chen, H.: Browsing in hypertext: a cognitive study. IEEE Trans. Syst. Man Cybern. **22**(5), 865–883 (1992)
6. Salwani, A.: An exponential Monte-Carlo algorithm for feature selection problems. Comput. Ind. Eng. **67**(1), 160–167 (2014)
7. Agrawal, R., Srikant, R.: Mining sequential patterns. In: Proceedings of 11th International Conference Data Engineering, vol. 5, pp. 3–14 (1995)
8. Miao, Y., Song, B.: Research on mining typical anonymous users browsing paths based on web logs. J. Comput. Appl. **29**(10), 2774–2777 (2009)
9. David, D.: Analysis of feature selection stability on high dimension and small sample data. Comput. Stat. Data Anal. **71**, 681–693 (2014)

A Novel Graph Partitioning Criterion Based Short Text Clustering Method

XiaoHong Li[✉], TingNian He, HongYan Ran, and XiaoYong Lu

College of Computer Science and Engineering,
Northwest Normal University, Lanzhou 730070, China
xiaohongli@nwnu.edu.cn

Abstract. A novel clustering method based on spectral clustering theory and spectral cut standard is proposed via analyzing the characteristics of short text and the defects of the existing clustering algorithms. First of all, a weighted undirected graph is created according to spectral clustering theory, similarity between node and node is calculated on graph, and a symmetrical documents similarity matrix is constructed, which provides all information for the clustering algorithm. Inspired by Greedy strategy, we utilize prim to develop PrimMAE algorithm for the purpose of partitioning graph into two parts, in which *RMcut* is termination condition of partitioning process, and then it is fed into CASC algorithm to cut the documents set iteratively. Ultimately, high quality clustering results demonstrate the effectiveness of the new clustering algorithm.

Keywords: Short text · Similarity matrix · *RMcut* criterion · Prim algorithm · Clustering algorithm

1 Introduction

With the increasing growth of the internet, social media based on it such as news title, micro-blog, instant message and so on has become increasingly popular, which produces a large number of information every day, and the key here is that most of the information is appeared in the form of short text. Different from traditional text, short text is usually of the following characteristics [1]: (1) Growing dynamically. That is to say, they have a rapid growth rate. (2) Text is short. Usually there are fewer words in short texts, so they are lack of sufficient statistical information for effective similarity measure. (3) The number of short texts is huge.

The existing text clustering methods cannot meet the need of theoretical analysis for the emerging wide range of short texts. Traditional document clustering methods cannot accomplish the task effectively, we need to research and design more suitable methods. Therefore how to improve the efficiency of clustering the mass of short text through information fusion has become the researching focus [2].

In recent years, clustering algorithms for short texts are constantly raised. Tang et al. [3] extended short text representation via a multi-language representation method by means of machine translation. Wang et al. [4] proposed an extension of VSM (vector space model) for communication short texts, called WR-KMeans. Yin and Wang [5] calculated the semantic distance between two word strings as balance of the

© Springer International Publishing Switzerland 2016
D.-S. Huang et al. (Eds.): ICIC 2016, Part III, LNAI 9773, pp. 338–348, 2016.
DOI: 10.1007/978-3-319-42297-8_32

extent of word sequence alignment and the meaning matching between word strings. Peng et al. [6] explored the text characteristic of high dimensions and sparse space, proposed a novel text clustering algorithm based on semantic inner space model. Methods mentioned above are mainly focused on how to carry out feature representation and similarity calculation of short texts. However, the clustering performance has been improved a little. Xing et al. [7] proposed a novel K-means clustering based on the immune programming algorithm, while the method can only deal with small-scale datasets due to huger search space, and cannot meet the need of practical application on large-scale data. Wang et al. [8] adopted a sliding window for short texts clustering combined hierarchical clustering and clusters merging method based on information gain. Chen et al. [9] proposed an improved CBC algorithm, which adaptively determined the center of each cluster; these methods could not preserve the accurate semantic similarities.

To solve the above problem above, in this paper, we propose a novel algorithm of improving the clustering accuracy and time efficiency taking *RMcut* [10] as cut standard to cluster short text. An overall architecture of the proposed method is illustrated in Fig. 1. Given a short text collection, the goal of this work is to cluster these texts via CASC based on graph model. In order to train CASC, we, inspired by Greedy strategy, utilize prim which is the minimum spanning tree algorithm of an undirected weighted graph for the purpose of developing PrimMAE algorithm, and then it is fed into CASC to cut the document set iteratively, at last, clusters are obtained. The main contributions of this paper are summarized as follows:

(1) To finish one unsupervised learning task, inspired by Greedy strategy, we utilize the minimum spanning tree algorithm based on an undirected weighted graph to partition graph into two parts.
(2) To the best of our knowledge, this is the first attempt to explore the effectiveness of short text clustering combining Prim algorithm concept and spectral cut via iterative process.
(3) We conduct experiments on two corpora, and compare with other kindred algorithms. The results demonstrate that the method proposed achieves excellent performance in terms of both purity and F-measure.

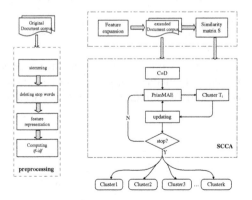

Fig. 1. Clustering algorithm flow

The rest part of the paper is organized as follows. Preliminary knowledge is presented in Sect. 2. Section 3 provides a detailed description of Clustering strategy and algorithm. Section 4 covers the experimental results and analysis. Finally, concluding remarks and discussions about future work are presented in Sect. 5.

2 Preliminary Knowledge

In this section, we briefly introduce some related knowledge in three aspects: text model, similarity matrix, and cut criterion.

2.1 Text Model

The most popular text representation method in present is VSM, but it ignores the associations of text features as well as the context and grammar structure. Since the terms are rare in short text, it is weak for describing specific information. Therefore, we reconstruct the short text model using the method proposed by Qiu et al. [11].

Assuming that all of short text is represented as a text collection, N denotes the number of documents, the vector of a piece of short text snippet can be represented as following:

$$V_{d_i} = \{w_{i1}, w_{i2}, \ldots w_{ij}, \ldots w_{iM}\} \tag{1}$$

Where in w_{ij} is weight of term t_j, in particular, we employ the *tf-idf* term weighting model to calculate wij. M is the number of terms.

$$w_{ij} = tfidf(t_j) = tf_{ij} \times log\frac{N}{n_j} \tag{2}$$

2.2 Document Similarity Matrix

In our method, the dot product is used to compute the similarity between two short text snippets. So the similarity between text d_i and d_j is defined as follows:

$$sim(d_i, d_j) = V_{d_i} \cdot V_{d_j} = \sum_{k=1}^{M} w_{ik} \times w_{jk} \tag{3}$$

There are many other similarity criteria, the reason why we select the dot product is that it can make the RMcut computation simple as will be exhibited in Sect. 2.3.2.

Based on the above conditions, similarity matrix S is constructed as follows:

$$S = \begin{bmatrix} s_{11} & \cdots & s_{1N} \\ \vdots & \vdots & \vdots \\ s_{i1} & s_{ij} & s_{iN} \\ \vdots & \vdots & \vdots \\ s_{1N} & \cdots & s_{NN} \end{bmatrix} \tag{4}$$

In this formula, $s_{ij} = sim(d_i, d_j)$, and $s_{ij} \geqslant 0$, S is symmetrical matrix, which contains the similarity information required by the clustering algorithm in this paper. The greater s_{ij} is, the more similar the content of two text are, thus the two short text is more likely to get into the same cluster [12].

2.3   Cut Criterion for Spectral Clustering

Spectral clustering has many applications in machine learning, data analysis, computer vision and speech processing. The spectral clustering algorithm is derived from the theory of spectral graph partitioning [13–15], which considered the data clustering problem as a multi-way partitioning for undirected weighted graph. The optimal partitioning criterion based on graph theory is to make similarity maximum inside subgraph and minimum between two subgraphs.

Whether the partitioning criterion is good or not, it directly affects the performance of clustering. Common criteria are *Normalized cut, Minimum Cut, Min-max Cut*, Ratio Cut etc.

We introduce two common partitioning criterions in this section.

2.3.1   Min-Max Cut

In the spectrum graph theory, Shi and Malik [16] proposed a two-way objective function which could partition original graph into subgraph C_1 and C_2. Applications of the min-max clustering principle minimize similarity between clusters and maximize similarity within a cluster. The similarity between C_1 and C_2 is the cut size:

$$cut(C_1, C_2) = sim(C_1, C_2) \tag{5}$$

Where in $sim(C_1, C_2) = \sum_{d_i \in C_1, d_j \in C_2} s_{ij}$, $sim(C_1) = \sum_{d_i, d_j \in C_1} s_{ij}$.

$sim(C_1)$ is inter similarity within a partition (subgraph C_1), and its value is the sum of all edge weights within C_1, Thus the min-max clustering principle requires we minimize $cut(C_1, C_2)$ while maximizing $sim(C_1)$ and $sim(C_2)$ at the same time. All these requirements can be simultaneously meet by the following objective function:

$$Mcut(C_1, C_2) = \frac{cut(C_1, C_2)}{sim(C_1)} + \frac{cut(C_1, C_2)}{sim(C_2)} \tag{6}$$

The above new function is min-max cut. *Mcut* is a much desirable objective function for data clustering. However, the *Mcut* criterion may lead to a set of clusters with very biased size; for example, some cluster may be one hundred times larger than others.

2.3.2   Ratio Min-Max Cut

The RMcut criterion is a clustering quality criterion, which measures the quality of clusters according to the following clustering principles—minimizing the inter-cluster similarity while maximizing the intra-cluster similarity. If the collection C is clustered into K clusters $\{C_1, C_2, ..., C_K\}$, their RMcut can be calculated as following:

$$RMcut(C_1, C_2, \ldots C_k) = \sum_{m=1}^{K} \frac{cut(C_m, C - C_m)}{|C_m| \sum_{d_i, d_j \in C_m} \text{sim}(d_i, d_j)} \tag{7}$$

Where in $\sum_{d_i, d_j \in C_m} sim(d_i, d_j)$ can be viewed as an inter-cluster similarity, $cut(C_m, C-C_m)$ is the edge-cut between the vertexes in C_m and the vertexes in $C-C_m$, especially, if $K = 2$, then computation of $RMcut$ is shown in Eq. (8)

$$RMcut(C_1, C_2) = cut(C_1, C_2) \sum_{m=1}^{2} \frac{1}{|C_m| \sum_{d_i, d_j \in C_m} \text{sim}(d_i, d_j)} \tag{8}$$

The similarity between a pair of short text is usually not accurate owing to the limited co-occurring terms between them. The $RMcut$ criterion can weaken impact of the sparseness of short text vectors by summing up all the pair-wise similarities within a cluster.

3   Clustering Strategy and Algorithm

We can describe an undirected weighted graph as G = (V, E), where V = $\{v_1, v_2 \ldots v_N\}$ is a set of all nodes corresponding to dataset D, and the vertex v_i represents short document d_i, E is the set of edges connecting pairs of nodes (v_i, v_j) if v_i is similar to v_j. So the symmetrical matrix S quantified the similarity between points vi and v_j. Then the clustering problem consists in divided V into K disjoints subsets V_i, i = 1,…,K. That is to say, V = $\{V_1, V_2, \ldots V_K\}$, $V_i \cap V_j = \Phi$ and i ≠ j.

3.1   Partitioning Graph Algorithm

In graph theory, Prim's algorithm is a greedy algorithm that finds a minimum spanning tree for a weighted undirected connected graph. The algorithm operates by building this tree one vertex at a time, from an arbitrary starting vertex, at each step adding the cheapest possible connection from the tree to another vertex. In our paper, the original graph is not necessarily connected, so we utilize the idea of prim algorithm only.

According to reference [10], we can find an object cluster by minimizing the value of $RMcut$. The central concept of our strategy, identified as MAEPrim (Maximum Adjacent Edge via Prim), is that we can select at random a vertex v_i in candidate set U as initial value of T, that is to say, T = $\{v_i\}$. Then an iteration procedure can be used to search for vertex v_j (most similar to all vertexes in T) from U–T. If the value of $RMcut$ is greater than ever, then iterative process is stopped. The pseudo-code of partitioning graph algorithm is given:

```
Program.1: MAEPrim(S', U)
Input: similarity matrix S', collection of vertexes U
Output: a cluster T
1. initial T:     T={ vᵢ };     // vᵢ is any vertex
2. minRMcut = ∞;
3. while (1):
       3.1 queried S, search for the maximum weight and   its
corresponding edge (vᵢ, vⱼ),   where vᵢ∈T, vⱼ∈U;
       3.2 according to formula 7 compute RMcut:
               tempRMcut = RMcut(T∪{ vⱼ }, U - { vⱼ });
       3.3 if (tempRMcut > minRMcut)     break;
       3.4 T = T∪{ vⱼ },   U = U - { vⱼ };
       3.5 minRMcut = tempRMcut;
4. // end while
5. return T.
```

3.2   Clustering Algorithm Based on Spectral Cut

An iterative new method identified as CASC (clustering algorithm based on spectral cut) for short text has been proposed in this paper, the key concept of it is to partition graph constructed by all short documents into two parts: one is final target cluster; the other is candidate vertexes set further partitioned again. This process can just terminate as the number of clusters equal to threshold K or when the candidate vertexes set is empty. Moreover, there are two subgraphs in the course of partitioning graph in the whole process, it make computation of complicated *RMcut* simple, that is to say, the number of cut is 2, the value of *RMcut* is computed by formula (8) in Sect. 2.3.2, so our method has a lower computational complexity. In addition, Ω [16] is equivalent to two-dimensional array in the following algorithm.

```
Program2: SCCA (S, V, K)
Input: similarity matrix S, vertexes set V, the number of
clusters K
Output: clusters array Ω[], the number of clusters
1. initial : for k = 1 to K
   Ω[k] = Φ;
2. CNumber = 1, C = V;
3. while (CNumber < K and C!=Φ)
   3.1.   Ω[CNumber] = MAEPrim (S, C) //clustering
   3.2.     CNumber ++;
   3.3.     C = C- Ω[CNumber]; // update C
   3.4.     update S:
      if vᵢ∈Ω[CNumber], then let sᵢ,ⱼ=0, j=1,2,…N;
4. if  (C !=Φ)     Ω[CNumber]= C;
       else    --CNumber
5. return Ω[], CNumber
```

4 Experiment and Analysis

We conduct a series of experiments on two different kinds of short text snippet datasets to evaluate the proposed method, one dataset is Chinese news data come from Sohu website, and the other is English data from DBLP. The datasets, evaluation criterion, experimental setup and results are described in detail, respectively, in the following subsections.

4.1 Datasets

(1) Sohu dataset: it is a collection of 53624 piece of news posted from July to August 2014. We get 14 classes that contain 4057, 3987, 3820, 3421, 3332, 4250, 3240, 4172, 4069, 3856, 3548, 3741, 4025 and 4106 pieces of news, respectively. Eventually, only use the title of each piece of news for clustering. And then, we segment titles by means of ICTCLAS and remove the stop words by checking a stop word list containing 1028 words. The preprocessing steps also include extracting features and enriching features. At last, we keep two thirds pieces of news for each class as training samples and leave the remaining 17874 pieces of news in total as test samples.

(2) DBLP dataset (http://dblp.uni-trier.de/db/conf/mldm): A computer bibliography lists more than two million of tracked articles on computer science in Dec. 2012. Tracked articles are in most journals and conference proceedings. Ten sub-categories (titled from A to J) that cover most fields in data mining obtained the DBLP dataset are used to evaluate the clustering performance; we further extract title records and stem the dataset with Porter stemmer. The resultant words in each snippet are used as the terms for clustering.

4.2 Evaluation Measures

To evaluate the effectiveness of the system performance, F-measure [17] and purity [18] are adopted in this study, they have been widely used to evaluate the performance of unsupervised text clustering. The purity evaluates the coherence of a cluster, that is, the degree to which a cluster contains documents from a single category. F-measure is computed as the harmonic mean of precision and recall.

Assuming the number of the documents is N, let $C = (c_1, \ldots c_s, \ldots c_j)$ be j labeled classes of the documents, and $\Omega = (\omega_1, \ldots \omega_r, \ldots \omega_k)$ be the set of clusters that produced by clustering algorithm.

Suppose documents in each cluster should take the dominant class in the cluster. Purity is the accuracy of this assignment measured by counting the number of correctly assigned documents and divides by the total number of test documents. Formally:

$$\mathrm{purity}(\Omega, C) = \frac{1}{N} \sum_{r=1}^{k} \max_{1 \le s \le j} |\omega_r \cap c_s| \tag{10}$$

Note that when all the documents in each cluster are with the same class, purity is highest with value of 1. Conversely, it is close to 0 for bad clustering.

F-measure is used as the performance measure that is defined as the harmonic precision (P) and recall (R) as:

$$F = \frac{\sum_1^j (|c_s| \times max\{F(c_s, \omega_r)\})}{\sum_1^j |c_s|} \tag{11}$$

$$F(c_s, \omega_r) = \frac{2P(c_s, \omega_r) \times R(c_s, \omega_r)}{P(c_s, \omega_r) + R(c_s, \omega_r)} \tag{12}$$

Where $|c_s|$ denotes the number of documents which are labeled into category c_s.

4.3 Results and Analysis

In order to investigate the quality of the proposed CASC algorithm visually, experiments are conducted on Sohu news dataset and DBLP datasets respectively. To analyzing the partition criterion sensitivity and the efficiency of CASC, we performed two experiments:

(1) The influence of different partition criterion on clustering algorithm;
(2) Comparison of clustering algorithm, the experiment simply analysis the efficiency of CASC by comparing it with other clustering algorithms.

4.3.1 The Effect of Partition Criterion

We have compared our new methods with three other graph partitioning methods: *Cut*, *Mcut* and *Rcut*. For the accuracy of the statistics, we conduct these two cluster experiments on both datasets.

As shown in Tables 1 and 2, firstly, on average, CASC algorithm achieves the highest F-measure and purity value on two dataset when setting partition criterion as *RMcut*. For instance, the results of our experiment have been improved 3.3 % and 4.1 % respectively compared with the minimum in terms of purity and F-measure on DBLP. By observing the clustering results, we can find that, on the one hand, *RMcut* is easier to find the core term for dividing clusters accurately, on the other hand, *RMcut* can balance bias which produced during computing intra and inter cluster similarity. Secondly, the quality of feature words is a key factor, so we should choose words that are very typical and high discriminative as features. Lastly, DBLP is an English corpus, and Sohu news is Chinese corpus, the above experimental performance on the former outperforms the latter, because syntax and semantic differences exist between the Chinese and English.

Table 1. Results on Sohu news

Measures	Cut	Mcut	Rcut	RMcut
Purity	0.651	0.661	0.667	0.681
F-measure	0.731	0.746	0.751	0.783

Table 2. Results on DBLP

Measures	Cut	Mcut	Rcut	RMcut
Purity	0.656	0.684	0.686	0.689
F-measure	0.738	0.747	0.753	0.779

4.3.2   Comparison of Clustering Algorithm

In this part, we compare the performance of CASC with K-means [19], CTVSM [20], and BA [21]. K-means is a popular similarity-based clustering model. CTVSM and BA are all state-of-the-art model-based short text clustering approach. The former is based on word co-occurrence technique, and the latter is based on an immune network.

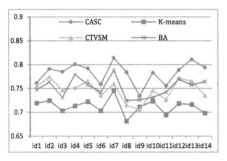 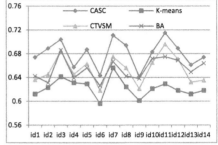

Fig. 2. F-measure of four algorithms on Sohu news   **Fig. 3.** Purity of four algorithms on Sohu news

It can be seen from Figs. 2, 3, 4 and 5 that purity value and F-measure of CASC algorithms are both larger than the other clustering algorithm on two datasets, especially in Figs. 4 and 5, with different F-measure from 0.724 to 0.813, purity from 0.643 to 0.706 respectively on the dataset DBLP, the number of documents reduced increasingly in the process of running CASC. So we speculate that the result of CASC

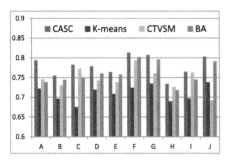 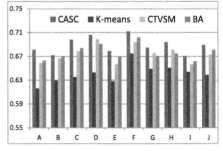

Fig. 4. F-measure of four algorithms on DBLP   **Fig. 5.** Purity of four algorithms on DBLP

clustering algorithm is good because there is little noise data. As is seen, K-means is poor at short text clustering, though its performance is very desirable on traditional clustering. CTVSM raises problem due to counting the term co-occurrence frequency, since the short text are too short to obtain a few co-occurrence word pairs. And we find that BA is not achieved the peak performance too. So our algorithm is slightly higher than K-means, CTVSM and BA in purity and F-measure.

5 Conclusion

In this paper, we propose algorithm based on spectral cut for short text clustering. We also develop the PrimMAE algorithm as a sub procedure of CASC which can help us understand how and why CASC works as well as the meaning of its parameters. We find that CASC can converge well with a good balance whether or not the number of clusters is given accurately; this algorithm also has high time performance. Thorough experimental study shows CASC can achieve significantly better performance than the other kondred methods.

There are several directions for us to explore in the future research. First, we plan to investigate how to incorporate the better weighting strategy for words and similarity metrics as discussed in Sect. 2.2. Second, as discussed in Sect. 3, CASC can achieve good performance and lower time complexity when we take *RMcut* as partition criterion, we plan to explore whether there is a more appropriate rule to partition clusters. Third, we also plan to make a research on how to improve the quality of features and interpretability of clustering results by considering semantic analysis, combined with the characteristics decomposition of Laplacian matrix.

Acknowledgments. This work was supported in part by National Natural Science Foundation of China under Grant No. 61272088, the Natural Science Foundation for Young Scientists of Gansu Province, China (Grant No. 1308TJY085, 145RJYA259), Youth Teacher Scientific Capability Promoting Project of Northwest Normal University (No. NWNU-LKQN-13-23).

References

1. He, H., Chen, B., Xu, W.: Short text feature extraction and clustering for web topic mining. In: Proceedings of IEEE 3rd International Conference on Semantics Knowledge and Grid (SKG 2007), pp. 382–385 (2007)
2. Sun, Q., Wang, Q., Qiao, H.: The algorithm of short message hot topic detection based on feature. Inf. Technol. J. **8**(2), 236–240 (2009)
3. Tang, J., Wang, X., Gao, H., et al.: Enriching short text representation in microblog for clustering. Front. Comput. Sci. **6**(1), 88–101 (2012)
4. Wang, L., Jia, Y., Han, W.: Instant message clustering based on extended vector space model. In: Kang, L., Liu, Y., Zeng, S. (eds.) ISICA 2007. LNCS, vol. 4683, pp. 435–443. Springer, Heidelberg (2007)
5. Yin, J., Wang, J.: A dirichlet multinomial mixture model-based approach for short text clustering. In: SIGKDD, pp. 233–242. ACM (2014)

6. Peng, J., Yang, D.Q., Tang, S.W.: A novel text clustering algorithm based on inner product space model of semantic. Chin. J. Comput. **30**(8), 1354–1363 (2007)

7. Xing, X.S., Pan, J., Jiao, L.C.: A novel K-means clustering based on the immune programming algorithm. Chin. J. Comput. **26**(5), 605–610 (2003)

8. Wang, Y., Wu, L.H., Shao, H.Y.: Clusters merging method for short texts clustering. Open J. Soc. Sci. **2**, 186–192 (2014)

9. Chen, J.C., Hu, G.W., Yang, Z.H., et al.: Text clustering based on global center-determination. Comput. Eng. Appl. **47**, 147–150 (2011)

10. Ni, X., Quan, X., Lu, Z., et al.: Short text clustering by finding core terms. Knowl. Inf. Syst. **27**(3), 345–365 (2011)

11. Qiu, Y., Wang, L., Shao, L.: User interest modeling approach based on short text of micro-blog. Comput. Eng. **40**(2), 275–279 (2014)

12. Man, Y.: Feature extension for short text categorization using frequent term sets. In: Proceedings of 2nd International Conference on Information Technology and Quantitative Management, ITQM 2014. Procedia Computer Science, vol. 31, pp. 663– 670 (2014)

13. Bach, F.R., Jordan, M.I.: Learning spectral clustering. Adv. Neural Inf. Process. Syst. **7**(2), 2006 (2004)

14. Ng, A.Y., Jordan, M.I., Weiss, Y.: On spectral clustering: analysis and an algorithm. Adv. Neural Inf. Process. Syst. **2**(14), 849–856 (2002)

15. Li, J., Tian, Y., Huang, T., et al.: Multi-polarity text segmentation using graph theory. In: International Conference on Information Processing (ICIP), San Diego, American, pp. 3008–3011. IEEE (2008)

16. Shi, J., Malik, J.: Normalized cuts and image segmentation. Pattern Anal. Mach. Intell. **22**(8), 888–905 (2000)

17. Cai, D., He, X., Han, J.: Document clustering using locality preserving indexing. Knowl. Data Eng. **17**(12), 1624–1637 (2005)

18. Zhao, Y., Karypis, G.: Criterion functions for document clustering: experiments and analysis. Mach. Learn. **55**(3), 311–331 (2004)

19. Hartigan, J.A., Wong, M.A.: Algorithm as 136: a K-means clustering algorithm. Appl. Stat. **28**(1), 100–108 (1979)

20. Chang, P., Feng, N., Ma, H.: Document clustering algorithm based on word co-occurrence. Comput. Eng. **38**(2), 213–214, 220 (2012)

21. He, T., Cao, X.-B., Tan, H.: An immune based algorithm for Chinese network short text clustering. Acta Autom. Sin. **35**(7), 896–902 (2009)

A Novel Clustering Algorithm for Large-Scale Graph Processing

Zhaoyang Qu[1], Wei Ding[1], Nan Qu[2], Jia Yan[3], and Ling Wang[1(✉)]

[1] Department of Computer Technology, School of Information Engineering,
Northeast Dianli University, Jilin, China
qzywww@nedu.edu.cn, 649993290@qq.com,
smile2867ling@163.com
[2] The Jiangsu Province electric power overhauls company,
Suzhou, Jiangsu, China
157729173@qq.com
[3] State Grid Jilin Electric Power Co. Ltd., Changchun, Jilin, China
407401293@qq.com

Abstract. The most important issue of big data processing is the relevance of analytical data; thought of this paper is to analyze the data as a graph optimal partitioning problem. Computing all circuit graphics firstly, calculated frequent map and redrawing of the system structure according to the results, the core problem is the time complexity of the algorithm. To solve this problem, researching DEMIX algorithm in non-strongly connected graph and study on relationship between frequent node and adjacency matrix which is strongly connected branches. Gives the corresponding examples, and analyzes the algorithm complexity. On the time complexity of the proposed method DEMIX is retrieving effect faster, more accurate search results.

Keywords: Large data · DEMIX algorithm · Graph

1 Introduction

With the popularization of Internet and Big Data, used graph theory to deal with the growing trend of data structure becomes currently research. Mining frequent subgraphs is an important operation on graphs [1]. Most existing work assumes a database of some small graphs, but modern applications, such as smart grid, sensor networks and energy Internet [3], etc. Intellectual grid is a combination of integrated information, communicated technology, transmission and distributed infrastructure, which is quite obviousness that large graphs constitute millions of vertices and more than tens of millions of edges, as shown in Fig. 1. On the whole, we desired for establishing a well-built graphical structure or design an algorithm for retrieve and update data more efficiently.

In a big data diagram, the method of research on a large graph has become a hot issue, it's axiomatic that more difficult to analyze graph data or to optimize the structure for large data fields, due to the structure of a data complexity and size. User always search result cannot guarantee the most relevant information; the low accuracy rate

© Springer International Publishing Switzerland 2016
D.-S. Huang et al. (Eds.): ICIC 2016, Part III, LNAI 9773, pp. 349–358, 2016.
DOI: 10.1007/978-3-319-42297-8_33

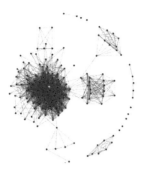

Fig. 1. Large data graph

reduces the user's experience. For large data graph, sub-graph matching is a very important skill. However, only by matching pairs figure does not solve the problem of improving efficiency of update and shortening high-quality search time pursuit.

The emphasis is aiming to make a better consequent of the problem at current circumstances in paper; it turns out working on solving the smart grid energy internet etc., particularly graph G of big-data and this sub-graph, they are possibility constituted of multi-million of points and edges, which are applied to sub-indexes that indexed possible relationship from among all peaks. Notwithstanding, in earlier work, correlated characteristic figured out problem of update searching. On the other hand, we attempted at figured out a model of a hierarchical structure that extracted valuable data, which disassemble complex relationship of big-data diagram and integrated it. Therefore, we devoted ourselves to lessening cost time of architecture of system, which efficiently researching large data set by designed efficient algorithm and structures of index.

2   Related Work

Seeking for large data diagram of smart grid, the key to building a search relevant sub-graph, which, in this paper, speedily searching and accurately matching of graph. There are some work-related studies demonstration that searched by time complexity is more efficiently than the whole graph.

American Electric Reliability Technology Association (CERTS) funded the development of DER-CAM can be micro-grid in the lowest cost for energy or CO_2 emissions as low as optimization target plan single or multiple targets can be determined inside the micro-grid distributed energy excellent combination of capacity and the corresponding operational plans [4]. Hybrid Optimization Based on Genetic Algorithm Design Software (HOGA) Department of Electrical Engineering University of Zaragoza in Spain developed by a single goal or objective can be optimized [5]. Japan's Kyushu Electric Power company uses Hadoop cloud computing platform for the mass consumption of electricity user system data for rapid analysis, and the development of various types of distributed batch processing applications in the platform on the basis of improved data processing speed and efficiency [6, 7]. Thereby improving the search graph pruning effect; MQC algorithms for real-time data to a relatively short period property which help to build the vertex set and edge set. Along with structure points and edges of the graph contained in time series order, the algorithm can visually describe events in real-time to change circumstances in the process [8]; effectively enhance the accuracy and usefulness of the real-time graph. However, in the majority of the data set, the job or task usually has a potential link, thus it is very important in the process of updating and searching when we go through the traditional method to traverse the entire data setting which consumes huge. Although the above method to search and match data is a good approach [9], our updating and searching from the perspective of an approaching relevance to the entire system is a more like a hierarchical diagram conserving system resources, which will figure out the specific relevant naphthalene extraction and build frequent node set, thus greatly improve the efficiency of the search and update.

In the current popular large graph map update process, we are more concerned about how to design a real-time updates algorithm, in other words, process the dynamic updates of data structures. In the real time case, a relatively small amount of data, a time or event nodes balanced distribution system updates the progress effectively will save the consumption of the system, which facilitates the attention of the most popular, the hottest, most valuable topics or events [10–15]. However, due to the increasing amount of data, from the perspective of the future development of the amount of data we are looking at now, perhaps, what we now call the Big Data is only "small data". Therefore we need to update large data structure diagram quickly which we could use in study and explore various methods, such as design a naphthalene structure searching algorithm [16–20]. In the above example, we point out: There is an active node, through which we pass messages very quickly, this indicates that if we go to an event updates via the node groups which are also concerned will keep that situation, Extracting these nodes are often very necessary. At the same time, and it exists in the relationship between the group closed loop, on behalf of certain events focus groups among them, so extracting the key part of these comparative studies of valuable information really matters.

This paper principally focused on studying other method of processing diagrams, discussing their limitations, studying the structure of subgraph, analyzing the correlation relationship which including relational model that building the graph. Finally, we proposed the core methods in this paper.

3 Strongly Connected Graph Structure

In this chapter, the problem is defined; well the primary problem can be solved the investigation of data graph. We present the views about how to build strongly connected structure in our algorithm, which need be satisfied to optimize searching structure and to update purposes. Strongly connected components are actual connection which studied in this paper. Particular, to avoid shortcomings in the search and update, we need to define a few attributes.

Definition 1 (Frequent Node). *Give a large graph SG, if node d have many properties, we should make sure that there is a value introducing graph system structure design, therefore, for the large graph, there is intermediate node d in the layer memory SG, i.e. d∈SG, d is greater than the threshold value of the point.*

Definition 2 (Loop and Trajectory). *Given G, a trajectory X is a sequence ((x1, x2), (x2, x3),..., (xk, x1)) such that there exists a path x1 → x2, → ,..., → xk → x1 on G, we recorded and marked down for this track. Besides, if an x1–xn path is a nonempty graph X.s = (Vi, Ei), and Vi = {x1, x2...xn} and Ei = {(x1, x2),..., (xn − 1, xn), (xn, x1)}, so X.s is a sub-graph of G and the path are not identical.*

Definition 3 (Vertex Frequency). *Given G, vertex set V(i) ∈ G,we distinguish by calculating the thresholds Θ to determine which vertices may be added to the data set SG, considering the follow-up study. In Fig. 2, for example, dotted line is an infrequent vertex while solid line is strong ties, so the strong ties vertexes belong to naphthalene structure graph.*

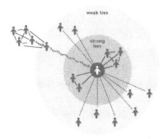

Fig. 2. Strong ties and weak ties

A formal definition should be proposed to facilitate the understanding, here are the main notation used in this paper and the rest of the column in Table 1 we used to explain implication.

According to the above definition clearly expressed, it is the main concept of this paper. The following part will describe how to create a strongly connected graph, as well as implementation of the algorithm.

Table 1. Notations

Notation	Description
G	*Large graph*
SG	*Strongly connected graph*
V(*)/E(*)	*Vertex/edge set of **
X	*Index graph*
X.s	*Meet the conditions index graph*
d(*)	*Frequency of vertex set of **
α	*Frequency of X.s*
β	*Weakly connected graph*
Θ	*Analyzing critical value*

Graphics extract strongly connected structure is the core of DEMIX algorithm. From the performance point of view, frequent node needs to be extracted, because they have a separate high-value associated with the data, which are stored separately and found a similar pattern by looking for contacts and information invisible loop when large data exists. Therefore, this paper provides a construct to calculate the first major design DEMIX graph each point must be determined according to the degree of connectivity frequently find the figure all the loops, generated between the cross find all strongly connected components in graph. If it is lower than the threshold by nodes frequency, not extracted it. If opposite, marking nodes and node records related trajectories.

4   DEMIX Clustering Algorithm

In this section, diagram cluster algorithm DEMIX will be introduced. DEMIX with trim and matching algorithm starts, dealing mainly with issues strongly connected components. DEMIX There are three main steps. First DEMIX Search, and then build a hierarchical sub-graph of the last indexing structure: a sub-query in the graph G.

DEMIX clustering algorithm is a graph theory, it belongs to the category of blind search, which aims to expand the institutional, aims to identify the results of its examination of all the nodes in diagram. The goal which we want to achieve is to handle large data graphs. Each loop represents a group of followers of the event; the results of the first part of the data processing will be the next steps for the polymerization.

4.1   Constructing Stratified Sub-graph

In order to better and faster search and update the entire graph structure, the aim is to construct a sub-graph (SG). The main task is based on the strong graph search system architecture, in order to meet the data to build a secondary threshold criteria and the adjacent node graph closely integrated. Layered system concept mentioned represents an event or joined together collections of other relationships.

In fact, it is necessary to extract the index filter layer data for large data graph nodes frequently set the threshold Θ. Algorithm settings α conditions such frequent node hash table stored in the table; meanwhile, we want to establish a good mapping nodes, and by a weak connection diagram β relationships with other node set in large data graph, we ignore the relationship between a specific point to point represents, on the contrary, the form of the entire extracted storage node list. Because of the relevance of the data extremely strong, so this represents a good structure of this paper studies the value of the smart grid, as long as we judge frequently loop nodes, higher demands on the efficiency of the system can be realized.

4.2   Algorithm

The main indicators of the study are based on the structural strength of the connector assembly, the main indicators index layer, and the first layer is supplemented by the index. Objective to construct the index structure to make it easier and faster access to the data graph, the index is mainly based on the frequent node α stored in a hash table, because the loop is formed by a plurality of nodes, each node has its own letters. Naphthalene map data set mapping structure α and β have also maintained the original relationship between β and the first layer nodes. Frequent node data set α is uniformly distributed, this expression is the key nodes, where α is a key indicator of the first layer and the contact layer data structure, when the search index layer structure cannot get the right results, the search to express the α the first layer, the results continue to match the potential to meet the needs of users. For clarity, the main steps used in this paper are presented in Table 2.

Table 2. Main steps for the DEMIX

1.	Input :Ω |V (G)| |V (G)| matrix, SG;
2.	for each visit[v] 1 to n;
3.	do map[v] > = u;
4.	v Q returned queue array by calling DEMIX (G, map);
5.	G^T TRANSPOSE-GRAPG(G);
6.	w returned queue array by calling DEMIX (G^T, map);
7.	for each u ∈ SG do
8.	adj^T[u] NIL
9.	for each u ∈ SG do
10.	for each v ∈ adj[u] do
11.	INSERT(adj^T[v], u)
12.	repeat steps;
13.	return SG;

5   Experimental

In this section, we analyzed the performance of our method comparing SAPPER by experiment [6]. SAPPER is a very important method for sub-graph matching in graphs of large size due to its excellent timeliness.

In this part, we selected twitter real data-set experiments and detailed data are shown in Table 3 (http://snap.stanford.edu/data/egonets-Twitter.html), data set includes more than 81,000 nodes and 1,768,000 edges, and the experiments used Java programming language based on Vulcan 64G memory graphics processor.

First, since the data set involved is a Stanford University tidied large graph data, experiments aim mainly to compare through the following sections: algorithm query time and the accuracy of the data. As shown in Fig. 3, the comparison index size at the different number of index nodes, Because of advantages in the naphthalene graphical

Table 3. Dataset

Parameter	Default value
Number of peaks in G	81306
Number of edges in G	1768149
Nodes in largest WCC	81306 (1.000)
Edges in largest WCC	1768149 (1.000)
Nodes in largest SCC	68413 (0.841)
Edges in largest SCC	1685163 (0.953)
Average clustering coefficient	0.5653
Number of triangles	13082506
Diameter (longest shortest path)	7
90-percentile effective diameter	4.5

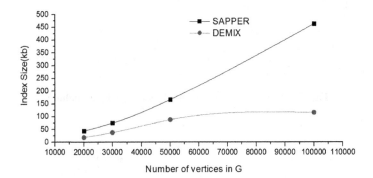

Fig. 3. Index size (Color figure online)

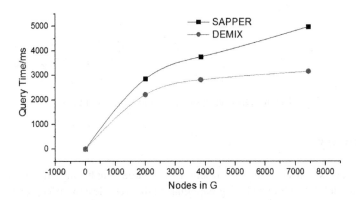

Fig. 4. Query time by different nodes (Color figure online)

structure, DEMIX algorithm has better performance than SAPPER obviously after 5000 nodes as seen from the figure; in Figs. 4 and 5, Comparing the query time by different number of node or edges, and in smaller data sets, this method showed no advantage, but when the number of data increased, our algorithm is much better than SAPPER, because many edges have been cut through DEMIX, reduced dataset infrequently expenses when searching. Figure 6 is the algorithm accuracy in different size of the data sets.

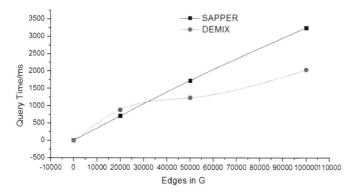

Fig. 5. Query time by different edges (Color figure online)

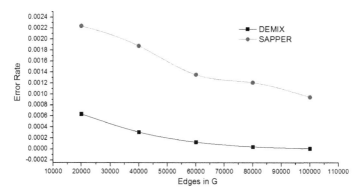

Fig. 6. Data set error rate (Color figure online)

6   Summary

In this paper, we discuss the construction of a large graph diagram in the DEMIX structure model through introducing naphthalene diagram design aiming to achieve fast search and effective algorithm updates. Experimental results show that the method of the research for graph computing that exists in the cluster event concerned and fixed interest groups such as messages delivering faster than conventional sub-graph matching method has great research value. In the following study, we will continue to research the potential of biphenyl existing in a smart grid, and the construction of multilevel, high sensitivity graph structure.

Acknowledgements. This work is supported by the development of National Natural Science Foundation Project (No. 51277023), by the Jilin Province plans to emphasis transformation projects (No. 20140307008GX), and by the Education Department Foundation of Jilin Province (No. 201698).

References

1. Elseidy, M., Abdelhamid, E., Skiadopoulos, S.: GraMi: frequent subgraph and pattern mining in a single large graph. Proc. VLDB Endow. **7**(7), 517–528 (2014)
2. Zhou, Y., Cheng, H., Yu, J.X.: Graph clustering based on structural/attribute similarities. Proc. VLDB Endow. **2**(1), 718–729 (2010)
3. Zhao, F., Tung, A.K.H.: Large scale cohesive subgraphs discovery for social network visual analysis. Proc. VLDB Endow. **6**(2), 85–96 (2012)
4. Martinez, C.: Intelligent real-time tools and visualizations for wide-area electrical grid reliability management, pp. 1–4 (2008)
5. Zhang, Y., Zhang, H., Lu, C.: Study on parameter optimization design of drum brake based on hybrid cellular multiobjective genetic algorithm. Math. Probl. Eng. **2012**(1), 1–18 (2012)
6. Miyake, Y., Tanaka, K., Okubo, H.: Seaweed consumption and prevalence of depressive symptoms during pregnancy in Japan: baseline data from the Kyushu Okinawa maternal and child health study. BMC Pregnancy Childbirth **14**(5), 572–578 (2014)
7. Williams, M., Wallis, S., Komatsu, T.: Dragons, brimstone and the geology of a volcanic arc on the island of the last Samurai, Kyushu, Japan. Geol. Today **32**(1), 21–26 (2016)
8. Li, Y., Liu, Z., Zhu, H.: Enterprise search in the big data era: recent developments and open challenges. Proc. VLDB Endow. **7**(13), 1717–1718 (2014)
9. Agarwal, M.K., Ramamritham, K., Bhide, M.: Real time discovery of dense clusters in highly dynamic graphs: identifying real world events in highly dynamic environments. Proc. VLDB Endow. **5**(10), 980–991 (2012)
10. Gupta, P., Satuluri, V., Grewal, A., et al.: Real-time twitter recommendation: online motif detection in large dynamic graphs. Proc. VLDB Endow. **7**(13), 1379–1380 (2014)
11. Pavan, A., Tangwongsan, K., Tirthapura, S., et al.: Counting and sampling triangles from a graph stream. Proc. VLDB Endow. **6**(14), 1870–1881 (2013)
12. Budak, C., Georgiou, T., Agrawal, D., et al.: Geoscope: online detection of geo-correlated information trends in social networks. Proc. VLDB Endow. **7**(4), 229–240 (2013)
13. Wu, X., Zhu, X., Wu, G.Q., et al.: Data mining with big data. IEEE Trans. Knowl. Data Eng. **26**(1), 97–107 (2014)
14. Cohen, J., Dolan, B., Dunlap, M., et al.: MAD skills: new analysis practices for big data. Proceedings VLDB Endow. **2**(2), 1481–1492 (2009)
15. Agarwal, M.K., Ramamritham, K., Bhide, M.: Real time discovery of dense clusters in highly dynamic graphs: identifying real world events in highly dynamic environments. Proc. VLDB Endow. **5**(10), 980–991 (2012)
16. Kim, Y., Moon, J., Lee, H.-J., Bae, C.-S.: Knowledge Digest Engine for Personal Bigdata Analysis. In: Park, J.H(., Jin, Q., Yeo, MS.-s., Hu, B. (eds.) Human Centric Technology and Service in Smart Space. LNEE, vol. 182, pp. 261–267. Springer, Heidelberg (2012)

17. Hoff, P.D., Raftery, A.E., Handcock, M.S.: Latent space approaches to social network analysis. J. Am. Stat. Assoc. **97**(97), 1090–1098 (2002)
18. Valente, T.W.: Social network thresholds in the diffusion of innovations. Soc. Netw. **18**(1), 69–89 (1996)
19. Wang, W., Yang, J.: Mining sequential patterns from large data sets. Adv. Database Syst. **28** (7), 3–14 (2013)
20. Bullmore, E., Sporns, O.: Complex brain networks: graph theoretical analysis of structural and functional systems. Nat. Rev. Neurosci. **10**(3), 186–198 (2009)

An Alarm Correlation Algorithm Based on Similarity Distance and Deep Network

Boxu Zhao$^{(\boxtimes)}$ and Guiming Luo

School of Software, Tsinghua University, Beijing, China
boxu.zhao@foxmail.com, gluo@tsinghua.edu.cn

Abstract. Currently, a few alarm correlation algorithms are based on a framework involving frequency and support-confidence. These algorithms often fail to address text data in alarm records and cannot handle high-dimensional data. This paper proposes an algorithm based on the similarity distance and deep networks. The algorithm first translates text data in alarms to real number vectors; second, it reconstructs the input, obtains the alarm features through a deep network system and performs dimension reduction; and finally, it presents the alarm distribution visually and helps the administrator determine the new fault. Experimental results demonstrate that it cannot only mine the correlation among alarms but also determine the new fault quickly by comparing the graphs of the alarm distribution.

Keywords: Alarm correlation · Fault discovery · Word to vector · Similarity distance · Auto-encoder · Deep network

1 Introduction

Alarm correlation is one of coral studies in fault location field, especially in fault location of large complex set of devices, since alarms, resulted in by equipment failures, have strong correlation in different devices or different components of device. As a key technology in telecommunication networks management, it obtains the root cause in a large number of alarms by mining alarm information. A few alarm correlation algorithms have been proposed include Rule-Based Correlation [1], Coding Approach [2], Model-Based Reasoning [3], Case-Based Reasoning [4], Fuzzy Logic [5], Bayesian Networks [6], Neural Networks [7] and so forth. However, those algorithms to mining information often ignore unstructured fields or text data containing important information, also cannot handle the high-dimensional alarm data and fail to provide enough information to help the networker locate and clean the fault.

Some other researchers focused on the data mining methods to analyze alarm sequence and mine alarm correlation rules. Correlation rule mining are also often used to compress the alarms and investigate the root reasons, where frequent episodes of alarms are paid close attention and support-confidence are calculated to obtain correlation rules. Mannila et al. [8] presented the efficient algorithms (WINEPI) for detection of frequent episodes from a given amount of episodes. The algorithms are applied in telecommunication alarm management in TASA [9]. Gardner and Harle [10] studied the generalization of a great deal of network performance information gathered everyday and

© Springer International Publishing Switzerland 2016
D.-S. Huang et al. (Eds.): ICIC 2016, Part III, LNAI 9773, pp. 359–368, 2016.
DOI: 10.1007/978-3-319-42297-8_34

made a tool for network fault discovery by using alarm data in SDH network system. Cuppens and Miege [11] constructed a cooperative module to manage, cluster, merge and correlate alarms for intrusion detection system, significantly reducing the amount of alarms to deal with. Shin and Ryu [12] applied the concept of data mining to alarm correlation, which is helpful not only to supervise the terrible users and hosts but also to find out potential alarm sequences. Xu and Guo [13] presented Alarm Association Algorithms based on Spectral Graph Theory (AAASG), which can reduce the search data set and quickly find fault by the changes in the point structure.

Analyzing more than 5500 sequential alarm records in Management Information System of a Chinese telecom service provider between January 2012 and May 2013, we discover that there are 41 network elements generating 8 different type of alarms. In text data field like "alarm summary", there are 1401 different values including great amount of information which need efficient data conversion technology to dimensionality reduction. Therefore, the algorithms should be capable of translating the text expression data to digital expression. In alarm correlation, its the key point to discover and position the associated alarms, so clustering algorithms are often made use of so that alarms are divided into different categories and similar property and features of them are in one category. When considering each alarm record as a vector, the "similar" ones are of relatively close distance in metric space. However, number of categories is hard to decide and can varies from one dataset to another, so it is necessary to transform the data in high dimensional space to lower dimensional space and show them visually to help manager realize the correlation relationship quickly and accurately and then locate the fault.

In the context of the paper, we present a new method based on similarity distance and deep network to find correlation of different alarms and compress redundant alarm in telecom network. Initially, text data are separated to words and transformed to high-dimensional vector. Later, we use deep learning method to do unsupervised learning and obtain another optimal expression represent alarm features. Finally, alarm data after processing is visually shown through alarm distribution graph, then administrator can efficiently find out root fault by the changes of graph.

2   Related Work

2.1   Mathematization of Text Data

When considering that machine deals with natural language, mathematical language is necessary and real vector is usually used. A simple method One-hot Representation expresses word by a vector consisting of '0' and '1' as long as the dictionary. While there is only one '1' in the vector, which represents the position of word in dictionary, leading to curse of dimensionality. Furthermore, One-hot fails to take order and relation between words into consideration. In 1986, Hinton [14] put forward 'Distributed Representation' concept, that is, every word is mapped to real number vector and distance between vectors represents semantic similarity. When finishing training a language model, word vector is obtained. One method producing word vector is Artificial Neural Network, firstly presented by Xu and Rudnicky [15], Institute of Deep learning, Baidu. Subsequently, a series of related research work has been done including Bengio et al. [16] and Mikolov's group [17].

2.2 Alarm Distance

After translating word to vector, alarm data can be considered as a high-dimensional Euclidean space $\Omega = \{a_1, a_2, \ldots, a_N\}, a_i \in R^D$. Each alarm is a point in the space of RD. In [13], correlation alarms (association alarms) are defined as follows:

If $\exists A_1, A_2, \ldots, A_i$ are correlation alarms, we say that

$$min\left(\frac{freq(A_1 \bigwedge A_2 \bigwedge \ldots \bigwedge A_i)}{freq(A_1)}, \frac{freq(A_1 \bigwedge A_2 \bigwedge \ldots \bigwedge A_i)}{freq(A_2)}, \ldots, \frac{freq(A_1 \bigwedge A_2 \bigwedge \ldots \bigwedge A_i)}{freq(A_i)}\right) \geq f_{min} \quad (1)$$

Where $freq(A_i)$ is the frequency of alarm A_i occurred in different associated windows. However, this definition ignores two points: (1) A_i is defined as alarm type of a few alarms, or can be seen as an attribute of alarm data, but not the alarm individual. In this paper, we define the alarm individual as a_i, $(a_1, a_2, \ldots, a_N)$ is the sequence of alarms in order of occurrence time. (2) The associated window and alarm frequency only reflect 'time' attribute of alarm data, while there are other attributes like "IP address", "alarm reason", "alarm level" without consideration. Alarm can be seen as a tuple $(t, X_1, X_2, \ldots, X_M)$, t represents occurrence time, $X_1, X_2, \ldots, X_M$ represent other alarm attributes or features, X_{ji} represents the jth attribute of alarm a_i. The alarm similarity distance between a_i and a_j can be defined as:

$$D(a_i, a_j) = \left[\alpha_0 (t_i - t_j)^2 + \sum\nolimits_{k=1}^{M} (X_{ki} - X_{kj})^2\right]^{\frac{1}{2}} \quad (2)$$

$\alpha_0, \alpha_1, \ldots, \alpha_M$ is weight coefficient. When $D(a_i, a_j) > d$, we consider a_i and a_j are correlate.

2.3 Deeping Learning Method

Usually, an alarm record consists of many dimensions, especially after word to vector, a text data dimension is translated into many digital dimensions. That means data may include too much 'unimportant' information which fails to reflect intrinsical features, then it is necessary to acquire 'good' features in some way and try best to ensure the difference between input and output as small as possible at the same time. In Deep Learning, we stack multiple layers and make the output of this layer as the input of the next layer to express the input hierarchically. Through adjusting the parameters in the system, we make the difference between the input and final output as small as possible. In this paper, we use Auto-encoder model capturing the most important factor of input data to find out the main ingredient representing the original information.

3 Method

The algorithm works as follows: (1) Translate text data in alarm record data to real number vector. (2) Combine the result of last step and Non-text data, reproduce the input signals as far as possible and obtain the most important factor representing the original alarm data lossless. (3) Show the result visually. Figure 1 shows algorithm steps.

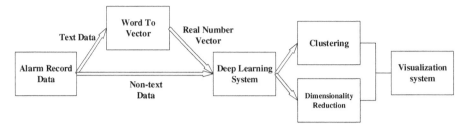

Fig. 1. Algorithm steps

3.1   Text Data to Digital Data

Separate Sentence to Words. Word is the smallest meaningful linguistic component. In Chinese, a sentence consists of several words, a word consists of one or more 'Chinese character' which is basic element of Chinese. Blank space is the natural delimiter in English, but there is no clear delimiter between Chinese words. So Chinese term analysis and parsing is the key of information processing. At present, three main methods have been used for Chinese word segmentation, which include character matching method, statistical method and understanding method [18]. Actual segmentation system usually makes use of character matching method as first step and further improves the accuracy of segmentation through other language information. In character matching, a word is recognized if it is found in a dictionary. Dictionary, text scanning sequence and matching principle are three essential factors. Text scanning order includes forward scanning, reverse scanning and bilateral scanning. Matching principle includes maximum matching, minimum matching, word by word matching and optimal matching. Below is a typical maximum matching algorithm of forward scanning.

Algorithm 1. A typical maximum matching algorithm of forward scanning

---

```
 1: procedure MMAOFS(S, Ws, Ml)
 2:     S represents sentence to be separated, Ws represent
        output words, Ml represents max length of a word.
 3:     Ws = "";
 4:     while S is not null do
 5:         repeat
 6:             if Str is not in dictionary then
 7:                 Remove the character on the Right of Str;
 8:             else
 9:                 Ws = Ws + Str+"\"; S = S + Str;
10:                 Break;
11:             end if
12:         until Str is a single character
13:         Ws = Ws + Str+"\"; S = S + Str;
14:     end while
15:     output Ws;
16: end procedure
```

Translate Words to Vector. After segmentation, next step is translating words into vector so that we can simplify the text processing to vector operations and calculate the similarity in vector space to represent the similarity of text semantics. Word2vec [17] has caused many researchers' attention since it was released free by Google. It uses a three-layer neural network to model the language and get the representation of word in vector space at the same time. Continuous Bag-of-Words Model (CBOW) and Skip-gram Model [19] is used in the language model. The former is a prediction of the current term through known context; on the contrary, the latter is to predict the context through current word. For CBOW and Skip-gram, word2vec gives two framework based on Hierarchical Softmax and Negative Sampling. More details can be seen in [16, 19, 20].

3.2 Autoencoder and Deep Network

An autoencoder is usually a feed-forward neural network aimed to learn a compressed, distributed dataset representation (encoding). The output is trained as a "Representation" of the input, and the input and target data are the same, that is, autoencoder tries to learn a $h(x) \approx x$ function. However, some limits, such as number of hidden layers and hidden neurons, should be added to obtain a meaningful structure. Autoencoder has two advantages: (1) From the output of hidden layers, autoencoder can get some of the compressed representation. For example, if input data has 8 dimensions and one of hidden layers has 4 nodes, 8 dimensional output closed to input should be reconstructed from the 4 dimensional data. (2) Hidden layers' data retains the correlation of the input and makes it easy to observe. Luckily, the features of autoencoder satisfy the need for alarm correlation very well. The compressed data can help show the correlation of alarms visually.

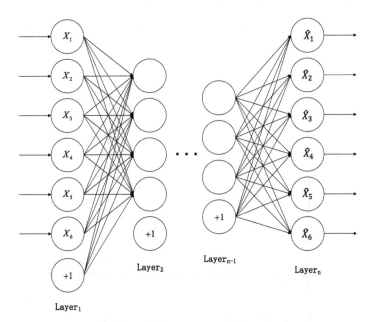

Fig. 2. Deep network of autoencoders

As autoencoder can get the reconstruction of the input data, the hidden layer can be seen as another expression of the original input data. We remove the third layer, and then use the same method to create a three-layer network and make the input of new-constructed network the output of hidden layer. The hidden layer of the second autoencoder is another expression of the input. Following this approach, we create deep neural network composed of a few autoencoders, as shown in Fig. 2.

3.3   Visual System

Alarm Window.   Every alarm $(t, X_1, X_2, \ldots, X_M)$ has a time stamp t, $StartTime < t < EndTime$, representing alarm occurrence time. Divide the time from $StartTime$ to $EndTime$ into several segments, each one has the length w. We set 2w the maximum time interval of related events, so that the first alarm and the last alarm are related when we analyses alarms of adjacent time windows. (See Fig. 3) If there are N windows between $StartTime$ to $EndTime$, they can be expressed as $w_1, w_2, \ldots, w_M$.

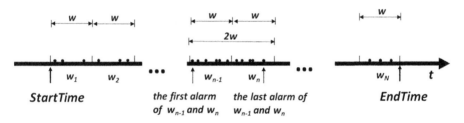

Fig. 3.   Alarm window

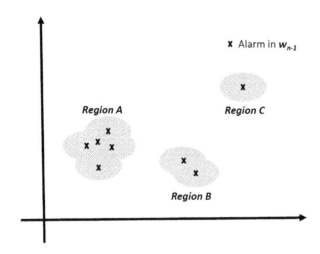

Fig. 4.   Three related regions in alarm window w_{n-1}

Related Region. To show alarm records visually, high dimensional alarms in an alarm window are converted into two dimension through deep learning system. Every alarm data point forms a region nearby less than correlation distance d. Some gathering data points form a larger region as they have common area, some scattered points form single region. There are three Related Region below in Fig. 4.

Alarm Attention. When we have alarm distribution graph and related regions during a continuous period, it is easy to classify the alarm data and find new faults. To measure how close we pay attention to each alarm, we introduce the concept Alarm attention. If $a_i \in A_{w_n}$, Alarm attention of a_i can be defined as:

$$Alarm\ attention(a_i) = min\big(D\big(a_i, a_j\big)\big),\ a_j \in A_{w_{n-1}} \tag{3}$$

That means the farther between a_i and related regions in previous time window, the more attention should be paid.

In Fig. 5 we can see No.324, No.325, No.326, No.330 alarm have larger alarm attention than No.327, No.328, No.329, No.331. As we know, alarm distribution has obvious change when a new fault occurs. Therefore, alarms not in related regions have larger attention, that is, the administrator of telecom system should give priority to these alarms and find the fault leading to these alarms.

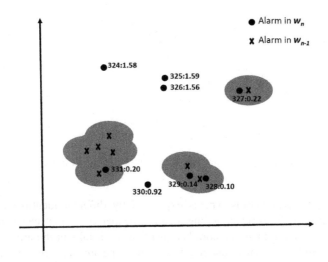

Fig. 5. Alarm in w_n and their alarm attention

4   Experiment

The algorithm is implemented on the dataset in management information system of telecom service providers. Every alarm in the dataset is numbered according to their occurrence order. More than 5500 alarms are produced in 478 days. In the experiment, we set $w = 1\ day$, then there are 478 windows from January 1st, 2012 to April 23rd, 2013.

The key to this algorithm is to compare the alarm distribution between neighbor alarm windows, observe alarm trend and find out system faults leading to these alarms. Alarm identification (IP address of network element, alarm type, alarm level, alarm summary) and the occurrence are selected as alarm attributes. To specifically analyzing two adjacent alarm window we show the alarm point structure of w_{424} and w_{425}. The alarm id and alarm attention is expressed by XXX:XXX in the Fig. 7. The id represents Time sequence of alarm, such that occurs earlier alarm 4855 than alarm 4982.

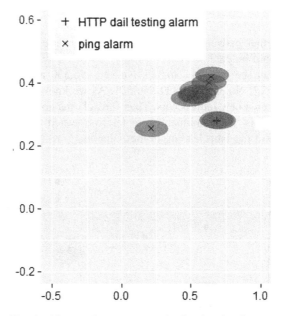

Fig. 6. Alarm point structure and related region in w_{424}

In Fig. 6, different alarms type is expressed by different identifiers. Gray area represents related regions. Overlapping area has deeper color, that means an alarm in next window has more than one correlate alarm when it falls in the area.

Compared to Fig. 6, alarms encircled by dashed line are alarms having larger alarm attention in Fig. 7, that is, the distribution has obviously changed. The change of alarm structure is due to a fault in the system, because the distribution of alarm will be changed when a new fault takes places. When administrator follows alarms of large alarm attention closely, it will be clear how to deal with the new fault.

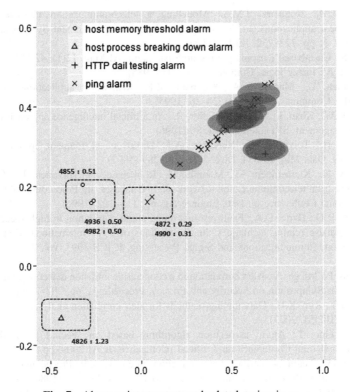

Fig. 7. Alarm point structure and related region in w_{425}

5 Conclusion

This article presents an algorithm based on alarm similarity distance, which is different from the skeleton of frequency and support-confidence. The algorithm focus on each alarm individual, not a type of alarms. Experimental results demonstrate the algorithms has the following advantage: (1) it can deal with the text data of alarm and translate it to vector space, which has not been used before; (2) it can reconstruct the alarm distribution lossless, then get the correlation between alarm accurately according to graph of alarms; (3) it can help administer find out the new fault quickly in system by changes of adjacent windows and clean it without too much effort.

Acknowledgment. This work is supported by the Funds NSFC61572279 and the Science Foundation of Chinese Ministry of Education—China Mobile 2012.

References

1. Cronk, R.N., Callahan, P.H., Bernstein, L.: Rule-based expert systems for network management and operations: an introduction. Network **2**(5), 7–21 (1988)
2. Kliger, S., Yemini, S., Yemini, Y., Ohsie, D., Stolfo, S.J.: A coding approach to event correlation. Integr. Netw. Manage. **95**, 266–277 (1995)

3. Meira, D.M., Nogueira, J.M.S.: Modelling a telecommunication network for fault management applications. In: Network Operations and Management Symposium, NOMS 1998, vol. 3, pp. 723–732. IEEE (1998)
4. Slade, S.: Case-based reasoning: a research paradigm. AI Mag. 12(1), 42 (1991)
5. Zadeh, L.A.: Fuzzy logic. Computer 4, 83–93 (1988)
6. Heckerman, D., Mamdani, A., Wellman, M.P.: Real-world applications of Bayesian networks. Commun. ACM 38(3), 24–26 (1995)
7. Gurer, D.W., Khan, I., Ogier, R., Keffer, R.: An artificial intelligence approach to network fault management. SRI International, 86 (1996)
8. Mannila, H., Toivonen, H., Verkamo, A.I.: Discovery of frequent episodes in event sequences. Data Min. Knowl. Disc. 1(3), 259–289 (1997)
9. Hätönen, K., Klemettinen, M., Mannila, H., Ronkainen, P., Toivonen, H.: Knowledge discovery from telecommunication network alarm databases. In: Proceedings of the Twelfth International Conference on Data Engineering, pp. 115–122 (1996)
10. Gardner, R.D., Harle, D.A.: Fault resolution and alarm correlation in high-speed networks using database mining techniques. In: Proceedings of 1997 International Conference on Information, Communications and Signal Processing, ICICS 1997, vol. 3, pp. 1423–1427 (1997)
11. Cuppens, F., Miege, A.: Alert correlation in a cooperative intrusion detection framework. In: 2002 IEEE Symposium on Security and Privacy. Proceedings, pp. 202–215 (2002)
12. Shin, M.S., Ryu, K.H.: Data mining methods for alert correlation analysis. Int. J. Comput. Inf. Sci. (IJCIS) (2003)
13. Xu, Q., Guo, J.: Alarm association algorithms based on spectral graph theory. In: International Joint Conference on Artificial Intelligence, JCAI 2009, pp. 320–323 (2009)
14. Hinton, G.E.: Learning distributed representations of concepts. In: Proceedings of the Eighth Annual Conference of the Cognitive Science Society, vol. 1, p. 12 (1986)
15. Xu, W., Rudnicky, A.I.: Can artificial neural networks learn language models? (2000)
16. Bengio, Y., Schwenk, H., Senécal, J.S., Morin, F., Gauvain, J.L.: Neural probabilistic language models. In: Holmes, D.E., Jain, L.C. (eds.) Innovations in Machine Learning, pp. 137–186. Springer, Heidelberg (2006)
17. Mikolov, T., Le, Q.V., Sutskever, I.: Exploiting similarities among languages for machine translation. arXiv preprint arXiv:1309.4168 (2013)
18. Huang, C., Zhao, H.: Chinese word segmentation: a decade review. J. Chin. Inf. Process. 21 (3), 8–20 (2007)
19. Guthrie, D., Allison, B., Liu, W., Guthrie, L., Wilks, Y.: A closer look at skip-gram modelling. In: Proceedings of the 5th International Conference on Language Resources and Evaluation (LREC-2006), pp. 1–4 (2006)
20. Collobert, R., Weston, J., Bottou, L., Karlen, M., Kavukcuoglu, K., Kuksa, P.: Natural language processing (almost) from scratch. J. Mach. Learn. Res. 12, 2493–2537 (2011)

Synonym-Based Reordering Model
for Statistical Machine Translation

Zhenxin Yang[1,2(✉)], Miao Li[1], Lei Chen[1], and Kai Sun[2]

[1] Institute of Intelligent Machines,
Chinese Academy of Sciences, Hefei 230031, China
mli@iim.ac.cn, alan.cl@163.com
[2] University of Science and Technology of China, Hefei 230026, China
{xinzyang, sasunkai}@mail.ustc.edu.cn

Abstract. Reordering model is the crucial component in statistical machine translation (SMT), since it plays an important role in the generation of fluent translation results. However, the data sparseness is the key factor that greatly affects the performance of reordering model in SMT. In this paper, we exploit synonymous information to alleviate the data sparseness and take Chinese-Mongolian SMT as example. First, a synonym-based reordering model with Chinese synonym is proposed for Chinese-Mongolian SMT. Then, we flexibly integrate synonym-based reordering model into baseline SMT as additional feature functions. Besides, we present source-side reordering as the pre-processing module to verify the extensibility of our synonym-based reordering model. Experiments on the Chinese-Mongolian dataset show that our synonym-based reordering model achieves significant improvement over baseline SMT system.

Keywords: Synonym · Reordering · Statistical machine translation · Feature function

1 Introduction

Statistical machine translation (SMT), which translates one language into another language automatically by statistical training corpus, breaks the barrier of different languages [1]. Since the fluency of translation results is one of the key evaluation metric in SMT, modeling word order between source and target sentences has been a research focus since the emerging of statistical machine translation [2].

The past decade has witnessed the rapid development of phrase reordering models. Among them, lexicalized reordering models [3–6] and syntax-based reordering models [7–10] have been widely used in practical phrase-based SMT systems.

Lexicalized reordering models have been widely researched in the phrase-based machine translation systems. Generally speaking, they take advantage of lexical information to predict the orientation of current phrase pair by using word alignment sentences. They often distinguish three orientation types of current phrase pair with respect to context: *monotone*, *swap*, and *discontinuous*. However, lexicalized reordering models suffer the data sparseness problem when translating with low-resource languages, such as Mongolian.

© Springer International Publishing Switzerland 2016
D.-S. Huang et al. (Eds.): ICIC 2016, Part III, LNAI 9773, pp. 369–378, 2016.
DOI: 10.1007/978-3-319-42297-8_35

Source-side reordering models [7–9] as the pre-processing modules are the effective syntax-based reordering models. These approaches first parse the source language sentences to create parse trees. Then, syntactic reordering rules, either hand-written or extracted automatically, are applied to these parse trees to adjust the source word order to match the target word order. This method is usually done in a pre-processing step, and then followed by a standard phrase-based SMT system to finish the translation. However, syntax-based reordering models will be affected by the accuracy of the parser easily.

The data sparseness and order difference bring great challenge to phrase reordering in Chinese-Mongolian SMT. Generally speaking, Mongolian is the low-resource minority language in China, so the training data used in SMT system is scarce. Besides, Mongolian is the Subject-Object-Verb (SOV) structure while Chinese is the Subject-Verb-Object (SVO) structure. Figure 1 illustrates word order difference between Chinese and Mongolian.

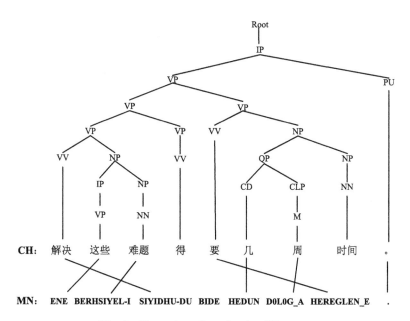

Fig. 1. Illustration of word order difference.

To address the challenge, a novel and effective synonym-based reordering model is proposed to alleviate the data sparseness by making full use the Chinese synonymous information. We integrate synonym-based reordering model into baseline SMT flexibly by additional feature functions. Finally, to further verify the effectiveness and extensibility of our method, some comparable experiments are conducted to demonstrate the effectiveness of our method.

The remainder of this article is organized as follows. Section 2 describes the background of baseline SMT and Sect. 3 provides synonym-based reordering model and its integration method. Extensive experiments are presented in Sect. 4. Section 5 describes the related work. Finally, we conclude in Sect. 6.

2 Background

In this section, we provide some background on baseline statistical machine translation system [11], which has emerged as the dominant paradigm in Chinese-Mongolian SMT research.

Given an input sentence $f = f_1\ f_2\ ...\ f_m$ which is to be translated into a target sentence $e = e_1\ e_2\ ...\ e_n$, the baseline SMT searches the most probable translation e^* according to the following decision rule:

$$e^* = \arg\max_e p(e|f) = \arg\max_e \{\sum_{j=1}^{J} \lambda_j h_j(f, e)\} \tag{1}$$

where $h_j(f,e)$ are J arbitrary feature functions over sentence pairs and λ_j are corresponding feature weights. The feature functions are soft constraints to encourage the decoder to choose correct translations during the decoding step. The feature weights can be learned by Minimum Error Rate Training [12] (MERT) with a separate development data.

Given a set of source sentences F_1^s with corresponding reference translations R_1^s, the goal of MERT is to find a parameter set which minimizes an automated evaluation criterion under a log-linear model:

$$\widehat{\lambda}_1^M = \arg\min_{\lambda_1^M} \left\{ \sum_{s=1}^{S} Err\left(R_s, \widehat{E}(F_s; \lambda_1^M)\right) \right\} \tag{2}$$

$$\widehat{E}(F_s; \lambda_1^M) = \arg\max_E \left\{ \sum_{s=1}^{S} \lambda_m h_m(E, F_s) \right\} \tag{3}$$

The translation results are evaluated automatically by BLEU metric [13], which has been demonstrated by showing that it correlates with human judgment. BLEU compares n-grams of the candidate with the n-grams of the reference translation and count the number of matches. Given the precision p_n of n-grams of size up to N (usually $N = 4$), the length of the test set in words c and the length of the reference translation in words r, the BLEU is computed as follows:

$$BLEU = \min(1, e^{1-r/c}) * \exp \sum_{n=1}^{N} \log p_n \tag{4}$$

3 Reordering Model with Synonym

In this section, we will introduce lexicalized reordering model used in baseline SMT and its drawback at first and then describe our synonym-based reordering model and its integration method.

3.1 Lexicalized Reordering Model

Lexicalized reordering model [4] calculates the reordering orientation of current phrase pairs based on previous word and next word during the phrase extraction. The model takes advantage of lexical information to predict the orientation of current phrase pair by using word alignment sentences. More formally, given a source sentence S, a target sentence T and its word alignment set A, we use a_s^t to denote the alignment from source position s to target position t, S_i^j to denote the phrase between position i and j in S and T_m^n to denote the phrase between position m and n in T. A source phrase S_i^j and a target phrase T_m^n form a phrase pair, lexicalized reordering model aims to predict orientations of a given phrase pair. The most widely used orientations in practice are *monotone* (M), *swap* (S), and *discontinuous* (D), the phrase pair can be classified to one of three orientations with respect to the previous word:

- *Monotone*: if $a_{i-1}^{m-1} \in A \cap a_{j+1}^{m-1} \notin A$
- *Swap*: if $a_{j+1}^{m-1} \in A \cap a_{i-1}^{m-1} \notin A$
- *Discontinuous*: if $(a_{i-1}^{m-1} \in A \cap a_{j+1}^{m-1} \in A) \cup (a_{i-1}^{m-1} \notin A \cap a_{j+1}^{m-1} \notin A)$

The orientations with respect to the next word are analogous. Hence, six lexicalized reordering feature functions are exploited in the baseline SMT. Figure 2 provides an example of the orientations with previous word.

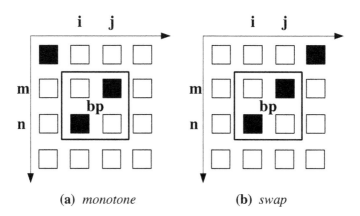

(a) *monotone*   (b) *swap*

Fig. 2. Reordering orientation with previous word.

The probability of a phrase pair bp in the orientation o is calculated as follows:

$$p(o|bp) = \frac{Count(o, bp) + \alpha}{\sum_{o'} (Count(o', bp) + \alpha)} \tag{5}$$

where α is the smoothing value, and we set α to 0.5 in practice.

We can conclude that maximum likelihood estimation can hardly train the lexicalized reordering model accurately because of the data sparseness in Chinese-Mongolian SMT.

3.2   Synonym-Based Reordering Model

In this subsection, a novel and effective synonym-based method will be presented to alleviate the data sparseness in phrase reordering. Inspired by the work [14], which used source language information to handle unknown words in SMT, we exploit monolingual information rather than bilingual information to enrich the reordering instances. Much reordering instances will lead the better probabilities estimation.

Intuitively, words with same meaning have similar reordering rules. Therefore, we use the Chinese synonym to calculate reordering probabilities, which assume that a word and its synonyms can share their reordering instances extracted from training data. We utilize the synonym dictionary named HIT IR-Lab Tongyici Cilin[1] (Extended) to train a synonym-based reordering model.

Synonym-based approach is modeled only condition on source language and synonym dictionary. If A and B are synonymous phrases, the count of B is also used as additional information when calculating the reordering probability of A. The training procedure is similar with the model conditioned on both source and target languages. The probability of a source phrase sp in the orientation o is calculated as follows:

$$p(o|sp) = \frac{Count(o, sp) + \alpha}{\sum_{o'} (Count(o', sp) + \alpha)} \qquad (6)$$

where α is the smoothing value, and we set α to 0.5 like previous equation.

Note that we only have synonym rather than synonymous phrase, so our dictionary is helpless for a phrase with more than one word. However, large percentage of translation units used during decoding are word-by-word or phrases including two words even in a phrase-based system [15]. Therefore, we only consider synonymous information for phrases including one or two words, and ignore the situation of long, uncommonly used phrases. We use a simple method to expand our synonym to synonymous phrase which has only two words.

The key idea of our synonym-based reordering model is that synonymous phrases share the reordering instances, so the sufficient data is used for probabilities calculation. For a phrase including two words, if one of the words in phrase has a synonym, we just use its synonym to replace this word to construct a synonymous phrase. For example, if we have a phrase "飞快 地(feikuai de)", and the word "飞快(feikuai)" has a synonym "迅速(xunsu)" in the dictionary, then the previous phrase has a synonymous phrase "迅速 地(xunsu de)". Both two phrases are found in our training data. The data sparseness can be alleviated since both two phrases can share the reordering instances.

3.3   Integration into Baseline SMT

The model assigns three distinct parameters $(\lambda_m, \lambda_s, \lambda_d)$ for the three feature functions with respect to previous word:

---

[1] http://ir.hit.edu.cn/.

$$f_m = \sum_{i=1}^{n} \log p(o_i = M | \cdots) \tag{7}$$

$$f_s = \sum_{i=1}^{n} \log p(o_i = S | \cdots) \tag{8}$$

$$f_d = \sum_{i=1}^{n} \log p(o_i = D | \cdots) \tag{9}$$

Three feature functions with respect to the next word are analogous. Hence, the above six additional feature functions are added into the log-linear model of the baseline SMT system. Essentially, they are soft constraints to encourage the decoder to choose translations with the correct word order.

4   Experiment

4.1   Data and Setup

The training set which contains 67288 Chinese-Mongolian parallel sentences is obtained from the 5th China Workshop on Machine Translation (CWMT). The test set contains 400 sentences, where each source sentence has four reference sentences translated by native linguistics experts, since the correct translation results are not unique. Besides, the development set is the same with the test set.

We employed GIZA++[2] and the grow-diag-final-and [11] balance strategy to generate the final symmetric word alignments. A 3-gram language model with modified Kneser-Ney smoothing [16] is built by SRILM toolkit. The final translation quality is evaluated in terms of BLEU metric. The feature weights are learned by using Minimum Error Rate Training [12] (MERT) algorithm. We use toolkit ICTCLAS[3] for Chinese word segmentation. The open-source toolkit Moses is used for each translation task. Besides, maximum phrase length is set to 7 when extracting bilingual phrase pairs. We run each experiment 3 times and get the average BLEU score [13] as the experimental result.

4.2   Evaluation for Synonym-Based Approach

We compare the synonym-based reordering model with the baseline. Besides, in order to compare our proposed model with previous work, we re-implement two approaches, which are denoted as system C and system D respectively. Table 1 shows the results. System A is baseline SMT, which is widely used in Chinese-Mongolian SMT research. System B denotes that we integrate our proposed synonym-based reordering model into baseline SMT by six additional feature functions. System C [6] is a improvement of lexicalized reordering model by weighted alignment matrices. System D [9] is a source-side reordering model which is also a research focus in SMT.

---

[2] http://code.google.com/p/giza-pp/.
[3] http://ictclas.nlpir.org/.

Table 1. Experimental results of different translation methods.

System	BLEU(%)
A	24.05
B	**25.27**
C	24.73
D	25.01

From Table 1, it can be noted that our synonym-based reordering model can substantially improve the translation quality compared with the baseline SMT. Besides, our proposed method is better than some typical related work, demonstrating the synonym-based reordering model can alleviate the data sparseness in phrase reordering.

4.3 Source-Side Reordering Model as Pre-processing Module

To further evaluate the effectiveness and extensibility of the proposed synonym-based reordering model, we incorporate a source-side reordering model [9] as a pre-processing module into system B. A phrase structure tree with syntactic information is acquired by stanford parser [17], then we use the manual rules to reorder the source language. The reordering rules described in Table 2, where VP denotes verb phrase, PP denotes prepositional phrase, NP denotes noun phrase, VV denotes verb.

Table 2. Source-side reordering rule.

No.	Original rule	Reordering rule
(1)	VP—> VV PP	VP—> PP VV
(2)	VP—> VV NP	VP—> NP VV

The subtrees are constructed to match with the original linguistic rules. The corresponding reordering rules on these subtrees are exploited to swap their left branches and the right branches. Then, the reordering of the source sentence is achieved.

Figure 3 is a source-side reordering of Fig. 1. The final BLEU scores are given in Table 3.

Table 3. Translation results of combination of two models.

System	BLEU(%)
A	24.05
E	**25.53**

From Table 3, it can be noted that the combination of synonym-based reordering model and source-side reordering model can further improve the translation quality with a maximum improvement of 1.48 BLEU score points increment over baseline. Hence, our synonym-based approach is easily incorporate with other module.

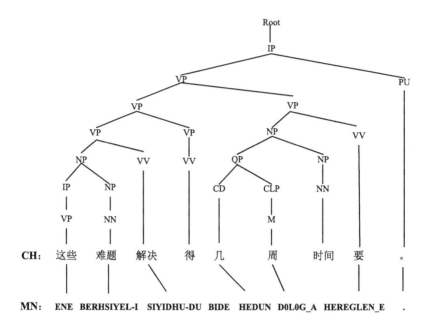

Fig. 3. The reordered phrase structure tree.

5 Related Work

Phrase reordering for statistical machine translation has attracted much attention in recent years. Within the phrase-based SMT framework, such research efforts can be roughly divided into two groups: (1) lexicalized reordering models, which model reordering relationships between adjacent phrases; (2) syntax-based reordering models, which apply some syntactic reordering rules on the phrase structure subtree to select right word order.

In lexicalized reordering work, the classifier will make decision on next phrase's relative position with the context. The classifier can be trained with maximum likelihood like Moses lexicalized reordering [4] and hierarchical lexicalized reordering model [5] or be trained under maximum entropy framework [18, 19]. Some improvement methods on lexicalized reordering, such as graph-based lexicalized reordering model [20], weighted alignment matrices reordering [6], are proposed to further improve the performance of reordering model. Similar work [21] based on maximum entropy exploited Mongolian morphological information as features to train a robust maximum entropy based reordering model. However, all these methods mentioned above are suffered a serious data sparseness in Chinese-Mongolian SMT.

Another direction is to employ syntactic information for word order selection. These methods need a statistical parser to produce the grammatical structure of sentences. Among them, source-side reordering is one of the research focus. In source-side approaches, the source sentence is reordered by heuristics, so that the word order of source and target sentences is similar. Liang et al. [9] proposed a rule-based reordering, which applied some syntactic reordering rules on the phrase structure subtree to reorder source language. Similar work [10] used dependence information to extracted reordering rules automatically. Yang et al. [22] utilized a ranking model based on word order precedence in the target language to reposition nodes in the syntactic parse tree of a source sentence. Visweswariah et al. [23] presented an automatic method to learn rules using only a source side parse tree and automatically generated alignments. However, syntax-based models are easily affected by the accuracy of the statistical parser.

Inspired by the work [14], which used source language information to handle unknown words in statistical machine translation, we make full use of monolingual information to alleviate the data sparseness in reordering model. To the best of our knowledge, our work is the first attempt to exploit the source synonym information to alleviate the data sparseness in reordering model, and we have successfully incorporated the proposed model into an SMT system with significant improvement of BLEU metrics.

6 Conclusion

In this paper, we presented a novel synonym-based reordering model to improve translation quality. We first utilize Chinese synonymous information to alleviate the data sparseness by enrich reordering instances. Then a novel and effective synonym-based reordering model is integrated into current baseline SMT by additional feature functions. The new additional features embedded in the log-linear model can encourage the decoder to produce more fluent translation results. Besides, we employ source-side reordering as a pre-processing module to verify the extensibility of our approach. Our experimental results demonstrate that our approach can significantly improve the translation quality.

Our contributions can be summarized as follows:

(1) To our knowledge, our work is the first attempt to utilize synonym dictionary to alleviate the data sparseness in reordering model while the dictionary is always used for translation model on SMT research.
(2) We investigate the flexible and efficient integration method, which integrates synonym-based reordering model as additional features into current SMT system.
(3) Our model is a general method, besides Chinese-Mongolian SMT, the method can also be adopted for other language pairs.

Acknowledgement. This work is supported by the National Natural Science Foundation of China under No. 61572462, No. 61502445, the Informationization Special Projects of Chinese Academy of Science under No. XXH12504-1-10.

References

1. Tu, M., Zhou, Y., Zong, C.: Exploring diverse features for statistical machine translation model pruning. IEEE/ACM Trans. Audio Speech Lang Process. **23**(11), 1847–1857 (2015)
2. Farzi, S., Faili, H., Khadivi, S.: A syntactically informed reordering model for statistical machine translation. J. Exp. Theor. Artif. Intell. **27**(4), 449–469 (2015)

3. Tillmann, C.: A unigram orientation model for statistical machine translation. In: HLT-NAACL 2004: Short Papers, pp. 101–104. Association for Computational Linguistics (2004)
4. Koehn, P., Hoang, H., Birch, A., et al: Moses: open source toolkit for statistical machine translation. In: ACL, pp. 177–180. Association for Computational Linguistics (2007)
5. Galley, M., Manning, C.D.: A simple and effective hierarchical phrase reordering model. In: EMNLP, pp. 848–856. Association for Computational Linguistics (2008)
6. Ling, W., Luis, T., Graa, J., Coheur, L., Trancoso, I.: Reordering modeling using weighted alignment matrices. In: ACL-HLT, pp. 450–454. Association for Computational Linguistics (2011)
7. Yeon-Soo, L.E.E.: Utilizing global syntactic tree features for phrase reordering. IEICE Trans. Inf. Syst. **97**(6), 1694–1698 (2014)
8. Chen, L., Li, M., He, M., Liu, H.: Dependency parsing on source language with reordering information in SMT. In: IALP, pp. 133–136 (2012)
9. Liang, F., Chen, L., Li, M.: Nasun-urtu: a rule-based source-side reordering on phrase structure subtrees. In: IALP, pp. 173–176 (2011)
10. Cai, J., Utiyama, M., Sumita, E., et al.: Dependency-based pre-ordering for Chinese-English machine translation. In: ACL, pp. 155–160. Association for Computational Linguistics (2014)
11. Koehn, P., Och, F.J., Marcu, D.: Statistical phrase-based translation. In: NAACL-HLT, pp. 48–54. Association for Computational Linguistics (2003)
12. Och, F.J.: Minimum error rate training in statistical machine translation. In: ACL, pp. 160–167. Association for Computational Linguistics (2003)
13. Papineni, K., Roukos, S., Ward, T., Zhu, W.J.: BLEU: a method for automatic evaluation of machine translation. In: ACL, pp. 311–318. Association for Computational Linguistics (2002)
14. Zhang, J., Zhai, F., Zong, C.: A substitution-translation-restoration framework for handling unknown words in statistical machine translation. J. Comput. Sci. Technol. **28**(5), 907–918 (2013)
15. Hoang, H., Koehn, P.: Improving mid-range reordering using templates of factors. In: EMNLP, pp. 372–379. Association for Computational Linguistics (2009)
16. Chen, S.F., Goodman, J.: An empirical study of smoothing techniques for language modeling. In: ACL, pp. 310–318. Association for Computational Linguistics (1996)
17. Levy, R., Manning, C.: Is it harder to parse Chinese, or the Chinese treebank? In: ACL, pp. 439–446. Association for Computational Linguistics (2003)
18. Xiong, D., Liu, Q., Lin, S.: Maximum entropy based phrase reordering model for statistical machine translation. In: ACL, pp. 521–528. Association for Computational Linguistics (2006)
19. He, Z., Meng, Y., Yu, H.: Maximum entropy based phrase reordering for hierarchical phrase-based translation. In: EMNLP, pp. 555–563. Association for Computational Linguistics (2010)
20. Ling, W., Graça, J., de Matos, D.M., Trancoso, I., Black, A.W.: Discriminative phrase-based lexicalized reordering models using weighted reordering graphs. In: IJCNLP, pp. 47–55. Association for Computational Linguistics (2011)
21. Yang, Z., Li, M., Zhu, Z., et al.: A maximum entropy based reordering model for Mongolian-Chinese SMT with morphological information. In: IALP, pp.175–178 (2014)
22. Yang, N., Li, M., Zhang, D., Yu, N.: A ranking-based approach to word reordering for statistical machine translation. In: ACL, pp. 912–920. Association for Computational Linguistics (2012)
23. Visweswariah, K., Navratil, J., Sorensen, J., et al.: Syntax based reordering with automatically derived rules for improved statistical machine translation. In: ICCL, pp. 1119–1127 (2010)

Tree Similarity Measurement for Classifying Questions by Syntactic Structures

Zhiwei Lin$^{(\boxtimes)}$, Hui Wang, and Sally McClean

Faculty of Computing and Engineering, Ulster University, Coleraine, UK
{z.lin, h.wang, si.mcclean}@ulster.ac.uk

Abstract. Question classification plays a key role in question answering systems as the classification result will be useful for effectively locating correct answers. This paper addresses the problem of question classification by syntactic structure. To this end, questions are converted into parsed trees and each corresponding parsed tree is represented as a multi-dimensional sequence (MDS). Under this transformation from questions to MDSs, a new similarity measurement for comparing questions with MDS representations is presented. The new measurement, based on the all common subsequences, is proved to be a kernel, and can be computed in quadratic time. Experiments with kNN and SVM classifiers show that the proposed method is competitive in terms of classification accuracy and efficiency.

Keywords: Question classification · Tree kernel · Tree similarity · Tree edit distance

1 Introduction

In Information retrieval (IR), question answering (QA) systems aim to find answers from a collection of documents, by matching a question with documents and then extracting information from the relevant documents. Key word based systems fail to find sensible answers because they ignore the inherent information from questions, such as syntactic structure information. For example, a keyword-based QA system cannot differentiate the following two questions:

- What is the measurement of the model?
- What is the model of the measurement?

The keyword based approach does not use any structure information from the questions. However, if these two questions are converted into tree structures with natural language processing parser (Fig. 1 shows the parsed trees for these two questions.), the difference between these two questions becomes obvious[1].

Syntactic information has been found very useful in QA systems and tree similarity measurement is key to this approach to utilizing syntactic information [3, 6–11, 13–15, 17, 19]. In this paper, we focus on question classification in QA systems, and present a new method for measuring similarity of questions based on their syntactic structures.

---

[1] In the trees, SBARQ, WHNP, and *etc al* are tags defined in Penn Treebank II.

© Springer International Publishing Switzerland 2016
D.-S. Huang et al. (Eds.): ICIC 2016, Part III, LNAI 9773, pp. 379–390, 2016.
DOI: 10.1007/978-3-319-42297-8_36

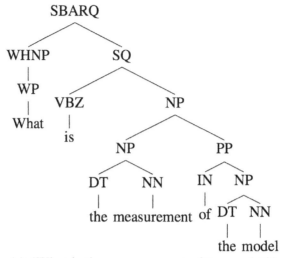

(a) 'What is the measurement of the model?'

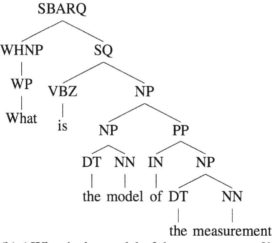

(b) ' What is the model of the measurement?'

Fig. 1. An example of two parsed trees

Figure 2 shows a typical architecture of question answering systems [4]. It is clear that question classification is the fundamental component of an efficient and effective QA system. Question classification is to determine semantic role (i.e., class) for a question so that the potential answers can be found according to the constraints imposed by the semantic roles. For example, to answer the question of 'What city will host the 2012 Olympics?', this question is firstly classified into class of location or class of location:

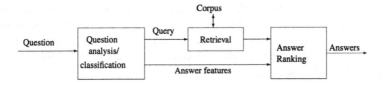

Fig. 2. A general architecture of QA system

city in the fine-grained level. This class is then used to locate relevant candidate answers and the most irrelevant answers will be filtered out.

To improve question classification accuracy, syntactic structure representations for questions, where each question is converted into a tree by using natural language parser, have been found useful [7, 12, 13, 15, 19]. However, most of these studies evaluated their algorithms at a coarse-grained level (e.g., location), and rarely at the fine-grained level (e.g., location:city). For example, for the question 'What city will host the 2012 Olympics?', the location is a coarse-grained class but it has to be further specified into a sub-class, e.g., city, as location alone may indicate different objects, such as city, country or mountain. To find the most sensible answers for QA systems, it is therefore necessary to use the fine-grained class information. Therefore, it is important to classify questions into fine-grained classes. This paper proposes a new algorithm for classifying questions into fine-grained classes. To this end, a novel tree similarity measurement for question classification (QACS for short) is proposed by using a sequence combinatorics approach, which has quadratic time complexity. This measurement is proved to be a valid kernel, which means that it can be used in kernel based classifiers, e.g., support vector machines. Experimental results show that our method is competitive in terms of classification accuracy and efficiency.

2   A Review of Question Classification by Syntactic Structure

This section presents some key concepts about trees and sequences, followed by a review of question classification by syntactic structure.

2.1   Preliminaries

A rooted, labeled and ordered tree T is a tuple T = (V,E), where V and E are the set of nodes and the set of edges of tree T. The root of tree T is denoted by root(T). In this paper, we assume trees are rooted, labeled, and left-right ordered unless otherwise stated. The size of T, denoted by $|T|$, is the number of nodes in T, i.e., $|T| = |V|$. Let v be a node of T, we write $|v|$ for the number of children of v, and a leaf v is a node without any child, i.e., $|v| = 0$. Let v be the i-th leaf in tree T in left-right order, we use path(T, i) or path(T, v) to denote the acyclic path starting from root(T) to leaf v. The depth of v, depth(v), is the number of nodes in the path between root(T) and leaf v. Let $v_1, v_2, \ldots, v_n$ be the children of any node v, the pre-order traversal of T(v) first visits v and then the subtrees routed at v_k (for all $1 \leq k \leq n$).

A *sequence* is a special tree where each node in the tree does not have any siblings. An n-long sequence s is an ordered set $\{s_1, s_2, \ldots, s_n\}$. The length of sequence s is denoted by $|s|$. The empty sequence, which has a length of zero, is denoted by λ, i.e., $|\lambda| = 0$. A k-long sequence y is a subsequence of s, denoted by $y \propto s$ if there exist $k + 1$ sequences $x^1, \ldots, x^{k+1}$ such that

$$s = x^1 y_1 \ldots x^k y_k x^{k+1} \tag{1}$$

2.2 Question Classification by Syntactic Structure

In this section we review the state of art of question classification by syntactic structures and some tree similarity algorithms.

Question classification uses machine learning methods to classify a question into one of the pre-defined classes [3, 12, 14, 15, 19]. These classes impose semantic constraints on candidate answers. There are 6 coarse grained classes and 50 fine grained subclasses in question classifications (more details can be found in [7]). Question classification by syntactic structures converts a question into a tree with natural language parser [3, 8, 9, 12–15, 19]. A constituent tree, containing some predefined tags, is more informative than a dependency tree for the same sentence. However, the study in [12] shows that the dependency structure may lead to better accuracy than constituent tree. Additionally, for a given questions with n words, question classification by dependency structure is more efficient than that by constituent tree because a constituent tree has $O(n \log n)$ nodes whereas a dependency tree has only $O(n)$ nodes.

Tree similarity measurement is key to classify questions in syntactic structure and tree kernel is widely used in question classification and wider applications in language processing [2, 19]. In [2], a pre-terminal tree kernel (PTTK) was proposed and it is of quadratic time complexity. The pre-terminal condition was relaxed and a new tree kernel (called QTK for short) was proposed for question classification in [19]. Different from tree kernel, tree edit distance is a distance function and it is the minimum number of edit operations to transform one tree to another. The common edit operations are: (1) insert a node into a tree; (2) delete a node from a tree; (3) change label of a node [20]. The optimal TED (OTED for short), allowing edit operations on any nodes, can provide a full picture of similarity between two trees and a dynamic programming algorithm by Zhang and Shasha in [20] is commonly used to compute edit distance between two ordered trees. A recent study with TED is the approximate TED algorithm (pq_Gram for short), which is of O(n log n) time complexity [1]. Both OTED and pq_Gram are not kernel functions since their feature spaces do not comply with Hilbert space [16].

3 Subsequence Combinatorics Approach to Tree Similarity

This section introduces how questions can be classified by measuring similarity between them in tree representations. This includes 4 steps: (1) Parse questions into trees with Stanford parser; (2) Convert parsed trees of questions into MDSs; (3) Use QACS algorithm to measure similarity between MDSs; and (4) Integrate QACS measurements for question classification with kNN or SVM classifier.

3.1 Encoding a Question into a Multi-dimensional Sequence

The first step to classify questions is to encode each question into an MDS through its parsed tree. A question is parsed into a tree by Stanford parser. The parsed tree could be a constituent tree or a dependency tree. After that, we encode a parsed tree (either in constituent or dependency structure) into an MDS so that the similarity between questions can be efficiently measured in terms of MDS representation.

We further encode a parsed tree (either in constituent or dependency structure) into an MDS so that the similarity between questions can be efficiently measured in terms of MDS representations. Suppose a parsed tree T has l leaves and let path(T, i) where $1 \leq i \leq l$ be an acyclic path starting from root(T) to the i-th leaf. Let M be a multi-dimensional sequence of T, where $|M| = l$, then M_i ($1 \leq i \leq l$) in M is path(T, i).

3.2 Subsequence Combinatorics for MDSs

This section introduces how to measure the similarity between questions by MDS representations, following the previous section of how to encode each question into MDS through its parsed tree.

Subsequence combinatorics have been extensively studied [5], and one of the popular approaches is to count *all common subsequences* (ACS) in [18]. ACS is a valid kernel function [6] and Lemma 1 shows a dynamic programming algorithm for counting ACSs [5].

Lemma 1 (Theorem 1 in [5]). Consider two sequences s and t, $m = |s|$ and $n = |t|$. For $x \in \Sigma$, if there exists k ($1 \leq k \leq i$), such that $x = s_k$, let $\ell_s(i, x) = \max\{k \mid s_k = x \}$; otherwise, let $\ell_s(i, x) = 0$. Then, for $1 \leq i \leq m$, $1 \leq j \leq n$,

$$
\phi[i,j] = \begin{cases}
\phi[i-1,j], & if\, \ell_t(j, s_i) = 0; \\
\phi[i-1,j] + \phi[i-1, \ell_t(j, s_i) - 1], & if\, \ell_t(j, s(i)) > 0, \ell_s(i-1, s_i) = 0; \\
\phi[i-1,j] + \phi[i-1, \ell_t(j, s_i) - 1] \\
\quad - \phi[\ell_s(i-1, s_i) - 1, \ell_t(j, s_i) - 1], & otherwise.
\end{cases}
\tag{2}
$$

where, $\phi[i,j]$ is an $(m + 1) \times (n + 1)$ matrix, and $\phi[i,0] = 1$, $\phi[0,j] = 1$ and $\phi[0,0] = 1$. Then, $\theta(s,t) = \phi[m,n]$.

The study in [18] presents a simplified version for counting ACSs when there are no repeating symbols in two sequences. Equation 3 shows the algorithm proposed in [18].

$$
\phi[i,j] = \begin{cases}
\phi[i-1, j-1] \times 2, & if\, s_i = t_j; \\
\phi[i-1,j] + \phi[i, j-1] - \phi[i-1, j-1] & otherwise.
\end{cases}
\tag{3}
$$

4 An Extension of Counting All Common Subsequences

Both Eq. 2 and 3 need hard matching for each element when comparing sequences. However, this does not make sense to many applications with numerical elements in sequences or multi-dimensional sequences, since hard matching is not feasible.

To compare numerical sequences with the ACS approach, we need to extend $\theta(s,t)$ to allow soft matching. One straightforward way to extend $\theta(s,t)$ is to estimate similarity degree between s_i and t_j. To this end, a similarity probability $prob(s_i = t_j)$ is defined, referring to the similarity degree between s_i and t_j, where $0 \le prob(s_i = t_j) \le 1$. On the contrary, the dissimilarity probability $prob(s_i \ne t_j)$ is

$$prob(s_i \ne t_j) = 1 - prob(s_i = t_j)$$

Then the similarity between $x = (x_1, \ldots, x_m)$ and $y = (y_1, \ldots, y_n)$, denoted by $\hat{\theta}(x,y)$, with an estimation of $prob(x_i = y_j)$, can be calculated as

$$
\hat{\phi}[i,j] = (\hat{\phi}[i-1,j] + \hat{\phi}[i,j-1] - \hat{\phi}[i-1,j-1]) \times prob(s_i \ne t_j) \\
+ \hat{\phi}[i-1,j-1] \times 2 \times prob(s_i = t_j)
\tag{4}
$$

$$
\hat{\phi}[i,j] = (\hat{\phi}[i-1,j] + \hat{\phi}[i,j-1]) \times \bigl(1 - prob(s_i = t_j)\bigr) \\
+ \hat{\phi}[i-1,j-1] \times \bigl(3 \times prob(s_i = t_j) - 1\bigr)
\tag{5}
$$

and $\hat{\theta}(x,y) = \hat{\phi}[m,n]$.

Therefore, we have

- When x_i and y_j are categorical or ordinal, Eq. 3 is a special case of Eq. 5 if we let $prob(x_i, y_j) = 1$ when $x_i = y_j$, and $prob(x_i, y_j) = 0$, otherwise.
- For numerical case, as an example, we can estimate $prob(x_i, y_j)$ by kernel functions, e.g., Gaussian kernel function:

$$prob(x_i = y_j) = \exp(-\gamma |x_i - y_j|). \tag{6}$$

We further extend this for measuring similarity between multi-dimensional sequences. Consider two MDSs $S = (S_1, \cdots, S_m)$ and $T = (T_1, \cdots, T_n)$, and let $\hat{\theta}(S,T)$ be a similarity function of S and T, we propose to measure $\hat{\theta}(S,T)$ with an estimation function $prob(S_i = T_j)$ by

$$prob(S_i = T_j) = \frac{\theta(S_i, T_j)}{\sqrt{\theta(S_i, S_i) \times \theta(T_j, T_j)}}. \tag{7}$$

However, to apply $\hat{\theta}(S,T)$ into SVM, we need to prove that it is a valid kernel so that an optimal solution can be guaranteed. The following theorem shows that by posing a constraint on estimating $prob(x_i = y_j)$, the function $\hat{\theta}(x,y)$ is a valid kernel.

Theorem 1. $\hat{\theta}(x,y)$ *is a valid kernel if* $prob(x_i, y_j)$ *is estimated by a valid kernel function.*

Proof. The proof of this theorem is trivial. Since $\hat{\theta}(x,y) = \hat{\phi}[m,n]$ and $\hat{\phi}[i,0] = 1$ and $\hat{\phi}[0,j] = 1$ (for $0 \le i \le m, 0 \le j \le n$), the operators (either by addition or multiplication for kernel functions) in Eq. 5 to combine $\hat{\phi}[i-1,j], \hat{\phi}[i,j-1], \hat{\phi}[i-1,j-1]$ and $prob(x_i = y_j)$ for $\hat{\phi}[i,j]$, result in an valid kernel if $prob(x_i = y_j)$ is estimated by a valid kernel function.

4.1   An $O(n^2)$ Algorithm to Compare Trees in MDSs

Equation 2 and its extension in Eq. 5 are of quadratic complexity. If MDSs have m spatial dimensions and n sequential dimensions, then the time complexity of calculating similarity between MDSs by combining Eqs. 5 and 7, is $O(m^2n^2)$. This implies high computational cost for tree similarity measurement. Therefore an efficient algorithm is needed to compare MDSs. In this paper, we find that cost reduction can be achieved through reducing redundant computation.

In QACS approach, nodes in a tree S are aligned in pre-order traversal sequence, and $\hat{i}$ is used to denote node's index in the sequence. Let $p(\hat{i})$ be the index of parent of node $\hat{i}$ (if $\hat{i}$ is a root node, i.e., $\hat{i} = 1$, let $p(\hat{i}) = 0$), and $\sigma(\hat{i})$ be the label of $\hat{i}$, for a given label $x \in \sum$, we define an $|S| + 1$ vector $\hat{\ell}_S$ for tree S:

$$\hat{\ell}_S[\hat{i}, x] = \begin{cases} 0, & \text{if } \hat{i} = 0 \\ \hat{i}, & \text{if } \sigma(\hat{i}) = x; \\ \hat{\ell}_S[p(\hat{i}), x], & \text{otherwise.} \end{cases}$$

Consider trees S and T, and their pre-order index $\hat{i}$ and $\hat{j}$. For $1 \le \hat{i} \le |S|$, given a label $\sigma(\hat{i})$, for brevity, let $\ell_f = \hat{\ell}_f[p(\hat{i}), \sigma(\hat{i})]$, and similarly, for $1 \le \hat{j} \le |T|$, let $\ell_T = \hat{\ell}_T[\hat{j}, \sigma(\hat{i})]$. Let ϕ be an $(|S| + 1) \times (|T| + 1)$, we define:

$$\phi[\hat{i}, \hat{j}] = \begin{cases} \phi[p(\hat{i}), \hat{j}], & \text{if } \ell_T = 0; \\ \phi[p(\hat{i}), \hat{j}] + \phi[p(\hat{i}), p(\ell_T)], & \text{if } \ell_T > 0, \ell_S = 0; \quad (8) \\ \phi[p(\hat{i}), \hat{j}] + \phi[p(\hat{i}), p(\ell_T)] - \phi[p(\ell_S), p(\ell_T)], & \text{if } \ell_T > 0, \ell_S > 0. \end{cases}$$

where $\phi[\hat{i}, 0] = 1$, $\phi[0, \hat{j}] = 1$, and $\phi[0, 0] = 1$.

Based on the definition of Eq. 8, we present the pseudo-code (in Algorithm 3.1) to compute QACS, which is of quadratic time complexity. In Algorithm 3.1, ϕ_S on Line 2 and ϕ_T on Line 3 are initiated and used to normalize $\phi [\hat{i}, \hat{j}]$. Together with the loops from Line 5, the time complexity of QACS is $O(n^2)$.

Input: Two trees S and $\mathcal{T}$
Output: Similarity between S and $\mathcal{T}$ of QACS
// Initiation for $1 \le \hat{i} \le |S|$ and $1 \le \hat{j} \le |\mathcal{T}|$ from Line 1 to 3
1   Initialize $\ell_S[p(\hat{i}), \sigma(\hat{i})]$ and $\ell_{\mathcal{T}}[\hat{j}, \sigma(\hat{i})]$;
2   **for** $\hat{i} \leftarrow 1$ **to** $|S|$ **do** $\phi_S[\hat{i}] = \phi[\hat{i}, \hat{i}]$;;
3   **for** $\hat{j} \leftarrow 1$ **to** $|\mathcal{T}|$ **do** $\phi_{\mathcal{T}}[\hat{j}] = \phi[\hat{j}, \hat{j}]$;;
4   $k = l = 1$;
5   **for** $\hat{i} \leftarrow 1$ **to** $|S|$ **do**
6   **for** $\hat{j} \leftarrow 1$ **to** $|\mathcal{T}|$ **do**
7   Use Equation 8 to calculate $\phi[\hat{i}, \hat{j}]$;
8   **if** $\hat{i}, \hat{j}$ *are leaves* **then**
9            $p = \frac{\phi[\hat{i}, \hat{j}]}{\sqrt{\phi_S[\hat{i}] \times \phi_{\mathcal{T}}[\hat{j}]}}$;
10   Calculate $\widehat{\phi}[k, l]$ with p and Equation 5;
11   l++;
12   **end**
13   **end**
14   **if** $\hat{i}$ *is leaf* **then** k++;;
15 **end**
16 **return** $\widehat{\phi}[k - 1, l - 1]$;

Algorithm 3.1: Pseudocode for calculating QACS

5   Evaluation

In this section, we compare QACS with the optimal tree edit distance (OTED) [20], pq_Gram [1] and tree kernels (PTTK [2] and QTK [19]) in terms of question classification accuracy and runtime.

We use kNN and SVM as classifiers for this evaluation due to the fact that both kNN and SVM are able to accept texts, sequences or trees as input, compared to those parametric algorithms, such as the Bayesian approach, logistic regression or neural network algorithms. Both OTED and pq_Gram are distance functions, but not kernel functions as their feature spaces do not comply with Hilbert space [16]. Therefore we only use QACS and tree kernels in SVM classifier [1, 20]. Tree kernels are usually evaluated in SVM classifier, rarely in other classifiers. To have a balanced comparison, we also use kNN classifier to evaluate these similarity measurements. For kNN, the parameter k is set to 1, 3, 5, 7, 9, 11, 13 and 15.

The dataset used in this evaluation is downloaded from UIUC Cognitive Computation Group[2], which contains 5952 questions. These questions are categorized into 6 coarse grained classes and 50 fine grained subclasses. This paper considers the problems of classifying questions into both coarse and fine grained classes.

We conducted experiments to evaluate the measurements with 10-fold CV strategy by a comprehensive experiments with both constituent and dependency tree structures.

---

[2] Available at http://l2r.cs.uiuc.edu/ ~ cogcomp/Data/QA/QC/.

The coarse-grained classification results are presented in Tables 1 and 2. From the results, we find that SVM has significantly higher accuracy but it has higher time than kNN classifier due to training. In kNN classifier, OTED has a very competitive performance but its runtime is much higher compared to the others. Our method QACS has highest accuracy except that in kNN in terms of dependency structure. The fine-grained classification results are presented in Tables 3 and 4. From the results, we find that with constituent tree structure, kNN classifier can provide a higher classification accuracy than SVM, for example, the accuracy of QACS in SVM is 57.6 % while its accuracy in kNN is 60.4 %.

Table 1. Coarse-grained classification accuracy (%) and runtime (seconds) with dependency tree structure.

	SVM/time	kNN								kNN time
		k=1	3	5	7	9	11	13	15	
QACS	74.5/3122	62.1	59	56.3	53.2	51.3	50	48.2	47.3	211
PTTK [2]	74.1/1079	61.6	60.4	57.8	57.1	55.9	56	55.8	55.6	98
QTK [19]	69.8/2059	53.4	50.9	49.2	48.3	47.3	47.4	47	46.9	185
OTED	N/A	67.9	68.9	68.4	68.3	67.5	66.9	66.6	66.6	1275
pq-Gram	N/A	54.3	46.6	41.4	37.7	35.1	32.8	31.0	29.4	312

Table 2. Coarse-grained classification accuracy (%) and runtime (seconds) with constituent tree structure.

	SVM/time	kNN								kNN time
		k=1	3	5	7	9	11	13	15	
QACS	75.2/18076	71.5	71.0	69.8	69.5	69.3	68.3	67.4	67.1	1140
PTTK [2]	74.9/9498	67.4	66.9	66.1	65.1	63.8	63.4	62.6	61.8	615
QTK [19]	74/23683	66.8	65.9	65.1	64.3	62.9	62.1	61.4	61.1	1480
OTED	N/A	65.3	66	66.6	66.7	66.3	66.3	65.9	65.8	14427
pq-Gram	N/A	68.5	68.9	68.4	68.2	68.0	67.1	66.7	66.3	1280

Table 3. Fine-grained classification accuracy (%) and runtime (seconds) with dependency tree structure.

	SVM/time	kNN								kNN time
		k=1	3	5	7	9	11	13	15	
QACS	75.2/18076	71.5	71.0	69.8	69.5	69.3	68.3	67.4	67.1	1140
PTTK [2]	74.9/9498	67.4	66.9	66.1	65.1	63.8	63.4	62.6	61.8	615
QTK [19]	74/23683	66.8	65.9	65.1	64.3	62.9	62.1	61.4	61.1	1480
OTED	N/A	65.3	66	66.6	66.7	66.3	66.3	65.9	65.8	14427
pq-Gram	N/A	68.5	68.9	68.4	68.2	68.0	67.1	66.7	66.3	1280

Table 4. Fine-grained classification accuracy (%) and runtime (seconds) with constituent tree structure.

	SVM/time	kNN								kNN time
		k=1	3	5	7	9	11	13	15	
QACS	58.6/5682	55.4	53.4	51.6	49.1	47.0	45.2	43.9	42.9	200
PTTK [2]	59.5/1769	50.5	50.3	49.3	47.8	47.2	46.4	46.5	46.4	96
QTK [19]	54.9/3146	41.2	39.7	39.0	38.2	37.3	37.3	37.0	36.5	182
OTED	N/A	56.7	58.7	59.7	59.5	59.1	58.3	57.6	57.0	1269
pq-Gram	N/A	50.5	45.8	41.1	36.9	34.6	32.5	30.7	28.9	299

5.1 Discussion

The experimental results show that our method QACS is competitive in syntactic question classification. There are still some interesting issues behind these results.

First, because constituent trees contain natural language tags from parser, the constituent tree approach has significantly higher accuracy, in both coarse-grained and fine-grained classification. However, a dependency tree, without those tags, implicitly containing some grammatical information, can also produce competitive performance. In terms of efficiency, for a question with n words, the time and space complexities of comparing its constituent tree are O(n log n) but the complexities for dependency tree are O(n). This explains why the runtime for constituent trees are always higher than that for dependency trees (See Tables 1, 2, 3 and 4).

It is well known that SVM outperforms kNN in most cases. In our experiments, for the same similarity measurement, different classifiers could produce significantly different accuracy and SVM again is better than kNN. However, due to training, SVM needs longer run time than kNN.

Finally, we find that OTED has very good performance when it is used in kNN classifier as it can provide a full picture of how similar between two trees are. The main problem with OTED is that it is a time-consuming method and cannot be used in large datasets.

6   Conclusion

This paper addresses question classification by syntactic structure. In this paper, each question is parsed into a tree and the corresponding parsed tree is represented as a multi-dimensional sequence. With MDS representation, we present a new similarity measurement for comparing questions in MDS representation, by extending the all common subsequences algorithm. The new algorithm is of $O(n^2)$ time complexity and is proved to be a valid kernel. The algorithm is integrated into kNN and SVM classifiers for question classification. Evaluation results show that our method is competitive in terms of classification accuracy and efficiency. Future work includes using this framework for analyzing the answers in QA system.

Acknowledgments. The authors would like to thank anonymous reviewers for their helpful comments to this paper by pointing out relevant literature and a number of annoying flaws in the submission. This paper is partially sponsored by EU DESIREE project (http://www.desiree-project.eu/).

References

1. Augsten, N., Bhlen, M., Gamper, J.: Approximate matching of hierarchical data using pq-grams. In: Proceedings of the 31st International Conference on Very Large Data Bases, VLDB 2005, pp. 301–312. VLDB Endowment (2005)
2. Collins, M., Duffy, N.: Convolution kernels for natural language. In: Advances in Neural Information Processing Systems, vol. 14, pp. 625–632. MIT Press (2001)
3. Croce, D., Basili, R., Moschitti, A.: Semantic tree kernels for statistical natural language learning. In: Basili, R., Bosco, C., Delmonte, R., Moschitti, A., Simi, M. (eds.) Harmonization and Development of Resources and Tools for Italian Natural Language Processing within the PARLI Project, pp. 93–113. Springer International Publishing, Cham (2015)
4. Croft, B., Metzler, D., Strohman, T.: Search Engines: Information Retrieval in Practice, 1st edn. Addison-Wesley Publishing Company, Reading (2009)
5. Elzinga, C., Rahmann, S., Wang, H.: Algorithms for subsequence combinatorics. Theor. Comput. Sci. **409**(3), 394–404 (2008)
6. Feng, G., Xiong, K., Tang, Y., Cui, A., Bai, J., Li, H., Yang, Q., Li, M.: Question classification by approximating semantics. In: Proceedings of the 24th International Conference on World Wide Web, pp. 407–417, Companion. ACM, New York (2015)
7. Li, X., Roth, D.: Learning question classifiers. In: Proceedings of the 19th International Conference on Computational Linguistics, pp. 1–7. Association for Computational Linguistics, Morristown, NJ, USA (2002)
8. Lin, Z., Wang, H., McClean, S.: Measuring tree similarity for natural language processing based information retrieval. In: Hopfe, C.J., Rezgui, Y., Métais, E., Preece, A., Li, H. (eds.) NLDB 2010. LNCS, vol. 6177, pp. 13–23. Springer, Heidelberg (2010)
9. Lin, Z., Wang, H., McClean, S.: A multidimensional sequence approach to measuring tree similarity. IEEE Trans. Knowl. Data Eng. **24**(2), 197–208 (2012)
10. Mittendorfer, M., Winiwarter, W.: Exploiting syntactic analysis of queries for information retrieval. J. Data Knowl. Eng. **42**(3), 315–325 (2002)
11. Moschitti, A.: Efficient convolution kernels for dependency and constituent syntactic trees. In: Fürnkranz, J., Scheffer, T., Spiliopoulou, M. (eds.) ECML 2006. LNCS (LNAI), vol. 4212, pp. 318–329. Springer, Heidelberg (2006)
12. Moschitti, A.: Making tree kernels practical for natural language learning. In: Proceedings of the Eleventh International Conference on European Association for Computational Linguistics, Trento, Italy (2006)
13. Moschitti, A., Quarteroni, S., Basili, R., Manandhar, S.: Exploiting syntactic and shallow semantic kernels for question answer classification. In: Proceeding of the Association for Computational Linguistics, pp. 776–783 (2007)
14. Pan, Y., Tang, Y., Lin, L., Luo, Y.: Question classification with semantic tree kernel. In: Proceedings of the 31st Annual International ACM SIGIR Conference on Research and Development in Information Retrieval, SIGIR 2008, pp. 837–838. ACM, New York (2008)

15. Punyakanok, V., Roth, D., Yih, W.-T.: Mapping dependencies trees: an application to question answering. In: Proceedings of the 8th International Symposium on Artificial Intelligence and Mathematics (2004)
16. Shawe-Taylor, J., Cristianini, N.: Kernel Methods for Pattern Analysis. Cambridge University Press, Cambridge (2004)
17. Strzalkowski, T. (ed.): Natural language Information Retrieval. Kluwer, New York (1999)
18. Wang, H.: All common subsequences. In: Proceedings of the 20th International Joint Conference on Artifical Intelligence, pp. 635–640, Hyderabad, India (2007)
19. Zhang, D., Lee, W.S.: Question classification using support vector machines. In: Proceedings of the 26th Annual International ACM SIGIR Conference on Research and Development in Informaion Retrieval, SIGIR 2003, pp. 26–32. ACM, New York (2003)
20. Zhang, K., Shasha, D.: Simple fast algorithms for the editing distance between trees and related problems. SIAM J. Comput. **18**(6), 1245–1262 (1989)

A Novel Approach of Identifying User Intents in Microblog

ChenXing Li$^{(\boxtimes)}$, YaJun Du, Jia Liu, Hao Zheng, and SiDa Wang

School of Computer and Software Engineering,
Xihua University, Chengdu 610039, China
272463637@qq.com

Abstract. Social micro-blogging platforms facilitate the emergence of citizen's needs and desires which reflect a variety of intents ranging from daily life (e.g., food and drink) to leisure life (e.g., travel and physical exercise). Identifying user intents in microblog and distinguishing different types of intents are significant. In this paper, we propose a novel approach to classify user intents into three categories, namely Travel, Food&Drink and Physical Exercise. Our method exploits Wikipedia concepts as the intent representation space, thus, each intent category is represented as a set of Wikipedia concepts. The user intents can be identified through mapping the microblogs into the Wikipedia representation space. Moreover, we develop a Collaborative User Model, which exploits the user's social connections to obtain a comprehensive account of user intents. The quantitative evaluations are conducted in comparison with state-of-the-art baselines, and the experimental results show that our method outperforms baselines in each intent category.

Keywords: User intents · Microblog · Wikipedia · Collaborative user modeling

1 Introduction

With the explosive development of social network, micro-blogging services have become a popular platform for people to express their needs and desires. For example, the microblog "I want to take part in physical exercise to lose weight" explicitly indicates that the user has the intent to do some physical exercises. If the intent is identified accurately, information providers can push related advertisement to the user and recommend relevant microblogs as well as users with similar intents to the user.

Intent stands for a purposeful action. We perform intent behaviors every day from querying a search engine to buying a smartphone. A large number of studies focus on identifying the query intent in Web [1, 2], which can be classified to navigational, informational and transactional intent, however the research of identifying user intents in microblog domain is much different. Microblogs often contain sentences which often

Project supported by the National Nature Science Foundation of China (No. 61271413,61472329, 61532009). Innovation Fund of Postgraduate, Xihua University.

D.-S. Huang et al. (Eds.): ICIC 2016, Part III, LNAI 9773, pp. 391–400, 2016.
DOI: 10.1007/978-3-319-42297-8_37

explicitly express user intents. Moreover, microblogs often contain more information than queries, e.g., friendship and context [9]. We exploit a classification form for identifying specific intent classes of microblogs.

We meet with two key challenges in the intent classification of microblogs. One is that the microblogs cover diverse intents, then how to define the space of intent representation that can precisely identify the intent of the content is critical. Moreover, we need to clarify the semantic boundary of the intent domain so that the intent classifier can accurately detect whether the microblog falls under the specific domain. Another is that the social interaction is an important issue in intent classification and it requires a careful selection of relevant social features.

For the first challenge, we utilize the Wikipedia concepts as the space of intent representation. All of the relevant concepts for specific intent could construct the Wikipedia Link Graph where each concept will be assigned an intent score by using random walk algorithm. The semantic boundary of the intent domain will be clearly identified by the intent scores of the Wikipedia concepts. Moreover, we exploit explicit semantic analysis (ESA) [5] for the words that are not contained by the Wikipedia concepts. For the second challenge, we construct the Collaborative User Model leveraging the interactions (e.g., mentions and re-microblogs) to give a comprehensive account of user intents. At last, we demonstrate the effectiveness of our method in three applications, namely Travel, Food&Drink and Physical Exercise. The numerous evaluations in comparison with state-of-the-art baselines show that our method outperforms other methods in each intent domain. The rest of the paper is organized as follows. We review related work in Sect. 2, Sect. 3 presents our method, Sect. 4 describes our experiments and the conclusion and future work are in Sect. 5.

2 Related Work

2.1 Query Intent Identification

Researchers have designed approaches to mine intent in queries using data from user search logs, including clicks, click sequence graphs and query terms. The query intent can be broadly categorized to navigational, informational and transactional intent [1–4]. However query is different from microblogs. Query is too short to obtain effective characteristics for intent identification. In addition, the intents of queries are typically implicit. For example, a keyword query like "laptop" does not explicitly express user intents. Conversely, microblogs often contain sentences that explicitly express the user intent. For example, the microblog "I want to buy a laptop" explicitly expresses the user intent. Our focus in this paper is to identify the intent of microblogs that explicitly express user intents, which is different from query intent classification.

2.2 Online Intention Identification

Prior research on online intention classification mainly focused on commercial and crisis domains. Dai et al. [6], who first proposed the Online Commercial Intention (OCI), presented the framework of building machine learning models to learn OCI based on Web page content and search queries. Chen et al. [7] aimed at identifying intents expressed in posts of forums. Hollerit et al. [8] firstly defined the commercial intent(CI) and employed traditional classification models like Naïve Bayes with n-gram and part-of-speech tags as features to classify tweets. The most related is the work [9], Wang et al. proposed a semi-supervised learning approach to classify intent tweets into six categories. However they did not take the interactions between users into consideration like the work we do in this paper.

3 Our Method

In this paper, we firstly construct the Wikipedia Link Graph ultilizing Wikipedia concepts to represent the intent space, then the Random Walk are performed on the graph to obtain an intent score (i.e. intent probability) for each concept. For the words that are not covered by the Wikipedia concepts, we exploit Explicit Semantic Analysis (ESA) [5] to obtain the most related Wikipedia concepts for the words. At last, the Collaborative User Model exploiting the interactions between users is constructed to further identify user intents in microblog.

3.1 Wikipedia Link Graph Construction

For each intent, we firstly choose several most representative words for specific intents and then search them in the Wikipedia. Considering the structure of the Wikipedia, for each word, we will get an article describing the word in detail and containing a large number of the words which are associated with the specific word. Through browsing these concepts it contains, the category it belongs as well as its sibling concepts, we can easily get enough concepts to cover the specific intent domain. For example, if we want to acquire the concepts about the "Physical Exercise", the first step would be searching the query "Physical Exercise" in the Wikipedia. Through browsing the corresponding articles, the concepts such as "aerobic exercise", "bodyweight exercise", "walking" would be collected. Moreover, we can obtain more sibling concepts such as "Swimming" and "Yoga".

With the collection of the concepts we could build the Wikipedia Link Graph G = (V, E) where V represents the collected concepts while E presents the link relations within concepts. Only when two concepts include each another then there is one edge between them. Conversely, there is no edge between them. Based on the graph G, we can construct a weight matrix $\mathbf{W}$ where the element w_{ij} equals the link count connecting vertices between v_i and v_j in the matrix. The weight matrix $\mathbf{W}$ is symmetric due to $w_{ij} = w_{ji}$.

3.2 Random Walk on the Wikipedia Link Graph

Taking Figs. 1 and 2 as examples to represent random walk algorithm. The four labeled concepts "Yoga", "Category: Physical exercise", "Aerobic exercise", and "Bodyweight exercise" are seeds for the Physical Exercise intent. For the reason that Wikipedia concepts, which link to each other through article or category links, often share similar topics, we assume that their immediate neighbors also have the same kind of intent to some extent [5]. Starting from a seed concept i, it moves to it's neigborhood node j with probability P_{ij} after the first step. The walk continues until it converges to a stable state, and all the concepts in the graph have probabilities that they belong to Physical Exercise intent. In this example, "Running", "Walking", "Plank", and "Muscle" have a higher Physical Exercise intent probability after first step.

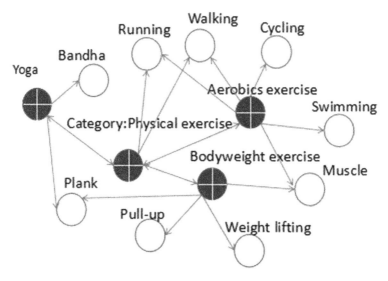

Fig. 1. A link graph where blue nodes represent seed concepts labeled +. (Color figure online)

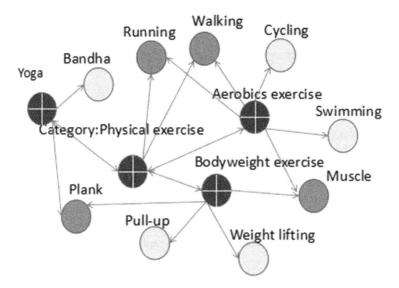

Fig. 2. Label information is propagated from the seed concepts to unlabeled concepts (the darkness of blue denotes the intent probability). (Color figure online)

We define transition probabilities $P_{t+1|t}(v_k|v_i)$ from the vertex v_i at time t to $v_k(v_i, v_k \in V)$ at time $t+1$ by normalizing the score out of node v_i, and it can be represented as:

$$P_{t+1|t}(v_k|v_i) = w_{ik} / \sum_j w_{ij} . \tag{1}$$

where j ranges over all vertices connecting to v_i. The score w_{ik} is symmetric, but the transition probabilities $P_{t+1|t}(w_k|w_i)$ generally are not because of the normalization varies across nodes. We rewrite the one-step transition probabilities in a matrix form as $\mathbf{P} = [P_{t+1|t}(w_k|w_i)]_{ik}$. The matrix $\mathbf{P}$ is row stochastic so that rows sum to 1.

Based on the concepts of weight matrix $\mathbf{W}$, we can select a small set of seed concepts as positive examples and it can be denoted as $S(S \subseteq V)$. The concepts in S are labeled as +1 while the rest are assigned zero. After that, we initiate an intent label vector $v_0 = (p(v_i))_{i=1}^m$ (where m is the total number of vertexes) with values:

$$p(v_i) = \begin{cases} v_i / \sum_j s_j, & \text{if } v_i \in S, \\ 0, & \text{otherwise.} \end{cases} \tag{2}$$

where $s_j \in S$, $p(v_i)$ is the probability that a random walk starts from v_i. The vector v_0 is updated as:

$$v_0 \leftarrow \alpha \mathbf{P}^T v_0 + (1 - \alpha)v_0, \quad \text{where } \alpha \in [0, 1). \tag{3}$$

After t iterations, the transition probability from vertex v_i to vertex v_k, denoted by $P_{t|0}(w_k|w_i)$, is equal to $P_{t|0}(w_k|w_i) = [\mathbf{P}^t]_{ik}$. The random walk sums the probabilities of all paths of length t between two vertices and gives a measure of the amount of the paths from one vertex to another. If there are many paths, the transition probability will be higher. Since the matrix $\mathbf{P}$ is a stochastic matrix, the largest eigenvalue of $\mathbf{P}$ is 1 and all other eigenvalues are in [0,1). Consequently, the vector v_0 will converge to v_*.

The value in the entry of the vector v_* is the probability that the vertex v_i is associated with a specified intent. Each Wikipedia concept is assigned a probability reflecting the degree of intent. We can treat the multiple intents classification as a set of binary intent classification. Consequently, this algorithm can be directly applied for each intent separately.

3.3   Explicit Semantic Analysis (ESA)

Since there are a large number of new words included in microblogs which are not covered by Wikipedia concepts, we address this problem by exploiting explicit semantic analysis (ESA) [5], which will provide a semantic interpreter that maps the fragment of text into some related concepts from Wikipedia ordered by their relevance to the input text fragment. Each Wikipedia concept is represented as a vector of words that occur in the corresponding article. Entries of these vectors are weighted by TFIDF scheme. Then we build an inverted index, which maps each word into a list of concepts in which it appears. Given a text fragment, we first represent it as a vector using TFIDF scheme. The semantic interpreter iterates over the text words, retrieves corresponding entries from the inverted index, and merges them into a weighted vector of concepts ordered by their relevance to the given text.

For words that are not contained in Wikipedia, we firstly use search engine such as Google to augment the representation of the word with top 10 search result snippets which include the titles and description parts. Then, we merge each snippets into sentences. Consequently, we can get a list of relevant Wikipedia concepts for each sentence by exploiting ESA. After the max-min normalization of rank scores of retrieved concepts, we get a desired number of highest-scoring concepts. Thus, we can predict the intent of the word according to the sum of the intent probabilities for the highest-scoring concepts.

For a given microblog, if the words in the microblog are covered by the Wikipedia concepts, we can get the probability that the microblog belongs to the defined intent using the intent probabilities of the mapped Wikipedia concepts. For the words that are not covered by Wikipedia concepts, we exploit ESA to map the words to the most related Wikipedia concepts and make the intent judgment based on the intent probabilities of the mapped Wikipedia concepts.

Let $\mathbf{K}$ be the set of intent categories (three intent categories in this paper). Each microblog will be associated with a vector of $\|\mathbf{K}\|$ elements and each element d_i^k in the vector represents the confidence score (i.e., intent probability) of the microblog belonging to category k estimated by our proposed method. Then the category with the highest intent score for each microblog is chosen as the inferred category, i.e., $\hat{k} = \arg\max_{k \in K} d_i^k$.

3.4 Collaborative User Model (CM)

One of the most important features of microblogs is its social network structure, which enables interactions between users. User u could mention his/her friend f with the symbolic @ sign and repost the microblogs posted by friend f. All the mentions and re-microblogs could be assigned an intent category k according to the intent probabilities of mapped Wikipedia concepts. The Collaborative User Model weights each friend f of user u on intent k by exploiting the Intent-interaction between them.

Intent-interaction: We first retrieve all microblogs which include the mentions of f by u, and re-microblogs of f's microblogs by u and then assign each microblog to an intent k. The intent-interaction weight between user u and the friend f is normalized by $w(u,f,k) = \log_{10}(1 + c(u,f,k))$ where $c(u,f,k)$ is the count of interactions assigned to intent k.

The Collaborative User Model can be denoted as Eq. 4, where F_u is the set of u's friends.

$$\theta_{u,k}^{CM} = \sum_{f \in F_u} w_{u,f,k} .$$

(4)

3.5 Intent Predictor

For a given microblog, it will be associated with a vector of $||K||$ elements and the intent category k is determined by $\hat{k} = \arg\max_{k \in K} d_i^k$. With the Collaborative User Model, we can re-evaluate the d_i^k by simply add the $\theta_{u,k}^{CM}$ to it. If d_i^k is greater than a predefined threshold ι, then the given microblog has the intent k and vice versa.

4 Experiments

4.1 Intent Applications

In this work, we apply our algorithm on three applications, namely Travel, Food&-Drink and Physical Exercise, our approach is general enough to be applied to other applications as well.

- Travel is a complex social activity that involves with various services including agency services, transportation services, accommodation services, and other hospitality industry services. Therefore, we mark a microblog with a Travel intent if it is directly or indirectly related to the services mentioned above.
- The microblog that contains words related to food or drink is considered to have the Food&Drink intent.
- Physical Exercise includes various exercises such as Aerobics exercise, Bodyweight exercise, Walking, Yoga an so on. We mark a microblog with a Physical Exercise intent if it is directly or indirectly related to the exercises mentioned above.

4.2 Data Collection

The data in this paper include Wikipedia data and microblog data. For the Wikipedia data, we identify over million distinct Wikipedia concepts for the three intents. There are 96,000 categories with an average of ten subcategories and 18 articles in each category. The seed concepts and the proportion of the three intents are shown in Table 1.

In order to collect enough intent microblogs, we give the definitions of *Intent-indicator* and *Intent-keyword* [9]. *Intent-indicator* refers a verb or infinitive phrase that expresses intent on a general level. For example, "wanna", "wanna to" are intent-indicators. *Intent-keyword* is a noun, verb, multi-word verb or compound noun (consisting of several nouns) contained in a verb or noun phrase which immediately follows an intent-indicator, e.g., in microblog "I wanna buy a car", "buy" and "car" which are contained in the phrase "buy a car" are intent-keywords. Given the idea that a microblog is more likely to be an intent microblog if it contains Intent-indicator and Intent-keyword. We adopt the bootstrapping based method to retrieve intent microblogs.

Specifically, given a seed set of intent-indicators, (1) we extract the intent-keywords that frequently co-occur with intent-indicators, and (2) we use the extracted intent-keywords to extract more intent-indicators if their co-occurrence frequency is above a certain threshold. We repeat these steps until we cannot extract more intent-indicators and intent-keywords. Finally (3) microblogs which contain these extracted intent-indicators and intent-keywords are kept in our test collection for manual annotation. Finally, 43,584 potential intent microblogs are obtained, two annotators are employed to annotate the microblogs according to three categories. We get 19,714 intent microblogs and 23,870 non-intent microblogs with the same label by two annotators. The Cohens Kappa coefficient between the two annotators is 0.6713. We summarize the statistics of this dataset in Table 2.

Table 1. Number of Wikipedia data

Intent type	(%)	Seed concepts
Travel	21 %	Travel; Hotel; Tourism; Airline tickets; Scenery; Travel Guide
Food&Drink	42 %	Food; Drink; Restaurant; Noodle; Coffee; Juice
Fitness exercise	37 %	Physical exercise; Aerobic; Bodyweight; Running; Swimming; Yoga

Table 2. Number of labeled dataset

Intent type	#(%)	Seed concepts
Travel	6379 (49 % positive)	I wanna travel to Paris. My dream is to travel around the world
Food&Drink	7153 (52 % positive)	I need some coffee. So hungry, it would be nice to have delicious noodles
Fitness exercise	6182 (42 % positive)	I find great pleasure in aerobic activity. I should go to the gym

4.3   Algorithms Compared and Results

- **Hollerit's Method [8]:** It uses a Bayes Complement Naïve Bayes classifier, a classification model which attempts to address the shortcomings of the Naïve Bayes classifier with word and part-of-speech n-grams as attributes. The part-of-speech tagger we use is the Stanford POS Tagger.
- **Maximum entropy(MaxEnt) Method:** It exploits a hybrid feature representation including Bag-of-Words, POS tagging, named entity, and dependency trees.
- **Wang's Method [9]:** It formulates the problem of inferring intent categories from a small number of labeled tweets as an optimization problem. It constructs an intent graph with intent tweet nodes and intent-keyword nodes. With the labeled data, it leverages the association between nodes and propagates the evidence of intent categories via the intent graph.
- **Hassan's Method [10]:** It constructs a word relatedness graph with seed words which have known its polarities and words without knowing its polarities, then the random walk is performed. The confidence of a word being positive/negative is

determined by the percentage of time at which the walk ends at a seed word and the average time the walk takes to hit a seed word. We use the microblogs for the words, and the seed words are the microblogs with known intent categories.

- **Ours:** Our method utilizes Wikipedia concepts and Collaborative User Model.

The metrics we used are F1, the macro-average and the micro-average. From the Table 3 we can see that our method outperformes other methods in every metric for three intent categories. The MaxEnt method is better than Hollerit's method for the linguistically oriented features such as POS tagging, named entity and dependency trees enhance the recognition of the intent categories. Wang's method and Hassan's method are both graph-based algorithms. Wang's method improves the macro-average and micro-average compared with the Hassan's method. The difference is that the results of the Hassan's method changes with the iterations the random walk takes, while Wang's method produces the unique answer to the optimization problem, which is more stable and efficient. Our method is better than Wang's method which does not take the interactions into consideration like our method.

Table 3. The F1 scores on all categories.

Category	Travel	Food&Drink	Fitness exercise	Marco-F1	Micro-F1
Hollerit's	44.74 %	42.14 %	43.15 %	41.91 %	46.13 %
MaxEnt	45.01 %	43.72 %	44.23 %	42.42 %	46.22 %
Hassan's	46.24 %	45.62 %	45.31 %	45.21 %	47.56 %
Wang's	47.12 %	45.82 %	45.81 %	45.53 %	47.82 %
Ours	**50.12 %**	**50.22 %**	**50.31 %**	**57.53 %**	**51.42 %**

To demonstrate the importance of the CM in our method, we compare our method with the method without using CM. Table 4 shows the comparison of precision, recall and F1 of two algorithms. It is shown that our method significantly outperformes the method without using CM in all three intent applications based on the F1 measure, which means that CM is important for intent identification.

Table 4. Comparison of the performance of our method and the method without using CM on the three intent identication task.

Intent type	Our method			Method without using CM		
	Precision	Recall	F1	Precision	Recall	F1
Travel	67.46 %	39.87 %	**50.12 %**	44.95 %	37.99 %	41.18 %
Food&Drink	68.23 %	39.73 %	**50.22 %**	45.35 %	35.60 %	39.89 %
Fitness exercise	65.12 %	40.99 %	**50.31 %**	43.47 %	37.84 %	40.46 %

5 Conclusions

In this paper, we aim to solve the problem of identifying user intents in the microblog. We construct the Wikipedia Link Graph, each Wikipedia concept in the graph is assigned an intent probability. We exploit ESA to acquire most related Wikipedia concepts for the words that are not covered by Wikipedia concepts and consequently get their intent probabilities. Then the intent of the microblog is determined by combing the intent probabilities of the mapped Wikipedia concepts and Collaborative User Model. The experimental results demonstrate that our algorithm achieves much better classification accuracy than baselines. Our work differs from previous works since we aim to use a human knowledge base to identify the user intent in microblog, and we do not collect large quantities of examples to train an intent classifier. This approach allows us to minimize the human effort required to investigate the features of a specified domain.

References

1. Broder, A.: A taxonomy of web search. In: ACM Sigir forum, pp. 3–10. ACM (2002)
2. Rose, D.E., Levinson, D.: Understanding user goals in web search. In: Proceedings of the 13th International Conference on World Wide Web, pp. 13–19. ACM (2004)
3. Shen, D., Pan, R., Sun, J.T., et al.: Query enrichment for web-query classification. ACM Trans. Inf. Syst. (TOIS) 24(3), 320–352 (2006)
4. Ashkan, A., Clarke, C.L.A.: Impact of query intent and search context on clickthrough behavior in sponsored search. Knowl. Inf. Syst. 34(2), 425–452 (2013)
5. Hu, J., Wang, G., Lochovsky, F., et al.: Understanding user's query intent with wikipedia. In: Proceedings of the 18th International Conference on World Wide Web, pp. 471–480. ACM (2009)
6. Dai, H.K., Zhao, L., Nie, Z., et al.: Detecting online commercial intention (OCI). In: Proceedings of the 15th International Conference on World Wide Web, pp. 829–837. ACM (2006)
7. Chen, Z., Liu, B., Hsu, M., et al.: Identifying intention posts in discussion forums. In: HLT-NAACL, pp. 1041–1050 (2013)
8. Hollerit, B., Kröll, M., Strohmaier, M.: Towards linking buyers and sellers: detecting commercial intent on twitter. In: Proceedings of the 22nd International Conference on World Wide Web Companion. International World Wide Web Conferences Steering Committee, pp. 629–632 (2013)
9. Wang, J., Cong, G., Zhao, X.W., et al.: Mining user intents in twitter: a semi-supervised approach to inferring intent categories for tweets. In: Twenty-Ninth AAAI Conference on Artificial Intelligence, pp. 213–222 (2015)
10. Hassan, A., Radev, D.: Identifying text polarity using random walks. In: Proceedings of the 48th Annual Meeting of the Association for Computational Linguistics, pp. 395–403. Association for Computational Linguistics (2013)

A Novel Entity Relation Extraction Approach Based on Micro-Blog

Hao Zheng[(✉)], YaJun Du, SiDa Wang, ChenXing Li,
and JianBo Yang

School of Computer and Software Engineering,
Xihua University, Chengdu 610039, China
1132305430@qq.com

Abstract. Entity relation extraction is a key task in information extraction. The purpose is to find out the semantic relation between entities in the text. An improved tree kernel-based method for relation extraction described in this paper adds the predicate verb information associated with entity, prunes the original parse tree, and removes some redundant structure on the basis of the Path-enclosed Tree. The experiment shows that the proposed method delivered better performance than existing methods.

Keywords: Relation extraction · Tree kernel-based method · Convolution tree kernel

1 Introduction

With the rapid development of science and technology, especially internet technology, the rapidly increasing network information in the real world is far beyond the ability of human reading. The difficulty people face is how to filter the useless information and how to draw out specific information that people need. Information extraction is originally presented at the Message Understanding Conferences (MUC, 1987–1998) by the Defense Advanced Research Projects Agency (DARPA), and seen as an important branch in the field of Natural Language Processing (NLP). After the MUC suspended, the Automatic Content Extraction program (ACE) [1] supported by National Institute of Standards and Technology (NIST) aims at further promoting the development of information extraction. The application scope of relation extraction is becoming more and more wide. Relation extraction has played an important role in the development of information extraction, question answering system, machine translation and so on, and its purpose is to find out semantic relations between entities from the tagged text.

In this paper, we use the tree kernel-based method [2–6] for relation extraction, which calculates the similarity by counting the number of the same sub-structures between two relation instances. As the tree kernel-based method can make full use of

Project supported by the National Nature Science Foundation of China (No. 61271413, 61472329, 61532009), Innovation Fund of Postgraduate, Xihua University.

D.-S. Huang et al. (Eds.): ICIC 2016, Part III, LNAI 9773, pp. 401–410, 2016.
DOI: 10.1007/978-3-319-42297-8_38

the structural information which can't be expressed by the feature vector-based methods, more and more researchers begin to study and use it in recent years. In this paper, the improved tree kernel-based method we propose reduces the redundant information and expands the original tree structure, so that it can contain more semantic information. Experiment shows that the proposed method can significantly improve the performance of the relation extraction.

This paper is organized as follows. In Sect. 2, we review previous work on relation extraction. In Sect. 3, we present our tree kernel-based methods to relation extraction. In Sect. 4, we analyze the results of our experiments. Finally, we present conclusions in Sect. 5.

2   Previous Works

Previous approaches to relation extraction can be classified into three groups: rule-based methods, feature vector-based methods, and tree kernel-based methods.

The rule-based method [7] for relation extraction has a higher accuracy rate in specific domain, but it needs to build a large-scale knowledge base by experts in specific fields. There are obvious deficiencies in terms of portability.

The feature vector-based method [8–10] needs to construct a suitable feature vector, and then uses a variety of machine learning algorithms such as support vector machines (SVM) and Winnow as the learning classifier. In relation extraction, the typical feature vector-based method includes maximum entropy models (MaxEnt) [8] and support vector machines (SVM) [9, 10]. The feature vector-based method focus on how to obtain the effective lexical, syntactic and semantic features. For example, Zhao et al. [9] integrate a variety of words, syntax parsing tree and dependency tree features. The complexity and variability of the expression of semantic relations between entities make machine learning methods which are mainly based on feature vector obtain relatively simple information. Machine learning methods are unable to make full use of the deep syntactic analysis results, stay in the processing stage of frequency and symbol, and rely on a large number of training data, so they limit the final effect of relation extraction.

Different from the feature vector-based methods, the tree kernel-based methods don't need to construct the feature vector space, which do not rely on large-scale corpus. Classifying the relation using the similarity between two syntactic structure trees makes the tree kernel-based methods explore the implicit high dimensional feature space. Zelenko et al. [2] firstly applied the tree kernel-based methods to the field of relation extraction, and they designed a dynamic programming algorithm to extract relations. Their experiment achieved satisfactory results on news corpus. Culotta et al. [3] continue Zelenko's study, and their experiment extracted 5 kinds of relations on ACE2003 corpus, and the F-value achieved 45.8. Zhang et al. [4] proposed a convolution tree kernel which was consisted of rich syntax rules and semantic relations. Their experiment on the Automatic Content Extraction/Relation Detection and Characterization (ACE RDC) corpora shows that their method outperforms other state-of-the-art methods. The kernel function method directly takes the structure tree as the processing object, and calculates the similarity between them.

Above researchers have made a variety of attempts in relation extraction using the tree kernel-based methods. The experimental result shows that the tree kernel-based methods have a good performance in relation extraction. Researchers mainly adjusts the structure of the syntax tree, but they ignore the predicate verb information associated with the entity in the text and redundant structure information in the syntactic tree. In this paper, we use a set of pruning strategy in adjusting the structure of the syntax tree, and we add some semantic information which effectively enrich the structured information. The proposed method enhances the performance of relation extraction system.

3   Tree Kernel for Relation Extraction

In this section we describe the convolution kernel function used in this paper. We also describe the pruning strategy of the syntax parsing tree proposed in this paper. The pruning strategy shows how to remove the redundant structure and how to add semantic information into the syntax parsing tree.

3.1   The Convolution Tree Kernel

A convolution kernel [4, 11] aims to capture structured information in terms of sub-structures. As a specialized convolution kernel, the convolution tree kernel suggested by Collins and Duffy [12] counts the number of common sub-trees (sub-structures) as the syntactic structured similarity between two parse trees. In their vector representation of a parse tree, a parse tree T is represented by a vector of integer counts of each sub-tree type (regardless of its ancestors):

$$\Phi(T) = (\#subtree_1(T), \ldots, \#subtree_i(T), \ldots, \#subtree_n(T)) \tag{1}$$

where $\#subtree_i(T)$ is the occurrence number of the ith sub-tree type ($subtree_i$) in T. Since the number of different sub-trees is exponential with the parse tree size, it is computationally infeasible to directly use the feature vector $\Phi(T)$. To solve this computational issue, Collins and Duffy proposed the following parse tree kernel $K(T_1, T_2)$ to calculate the dot product between the above high-dimensional vectors implicitly.

$$K(T_1, T_2) = \sum_{n_1 \in N_1} \sum_{n_2 \in N_2} \Delta(n_1, n_2) \tag{2}$$

Where N_j is the set of nodes in tree T_j, and $\Delta(n_1, n_2)$ evaluates the common sub-trees rooted at n_1 and n_2, Here, each node n encodes the identity of a sub-tree rooted at n and, if there are two nodes in the tree with the same label, the summation will go over both of them. Therefore, $\Delta(n_1, n_2)$ can be computed recursively as follows:

1. If the context-free productions (Context-Free Grammar[CFG] rules) at n_1 and n_2 are different, $\Delta(n_1, n_2) = 0$; otherwise go to 2.
2. If both n_1 and n_2 are pre-terminals (POS tags), $\Delta(n_1, n_2) = 1 \times \lambda$; otherwise go to 3.
3. Calculate $\Delta(n_1, n_2)$ recursively as:

$$\Delta(n_1, n_2) = \lambda \prod_{k=1}^{\#ch(n_1)} (1 + \Delta(ch(n_1, k), ch(n_2, k))),$$ (3)

where $\#ch(n)$ is the number of children of node n; $ch(n, k)$ is the kth child of node n; and $\lambda(0 < \lambda < 1)$ is the decay factor in order to make the kernel value less variable with respect to different sub-tree sizes. The context-free production at node n is the representation of the node n to its full children. For example, the context-free production at node "IP" is "IP $\longrightarrow$ NP PU NP VP PU" in Fig. 1.

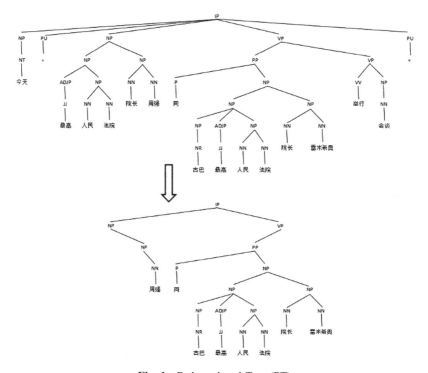

Fig. 1. Path-enclosed Tree (PT)

3.2   Relation Instance

A semantic relation instance between two entities can often be represented as a syntactic parse tree. Different structures of tree have great influence on the calculation results of the similarity, which will affect the effectiveness of relation extraction. Zhang et al. [4, 13] systematically explored five different semantic relation tree structures, including Minimum Complete Tree (MCT), Path-enclosed Tree (PT), Context-Sensitive Path Tree (CPT), Flattened Path-enclosed Tree (FPT) and Flattened CPT (FCPT). They found that the PT performed best. PT is the smallest common sub-tree including the two entities. In other words, the sub-tree is enclosed by the shortest path linking the two entities in the parse tree. Figure 1 shows the structure of the Path-enclosed Tree.

However, since the path between entities is far or the structure of syntactic parse tree is relatively complex, there is still a lot of redundant information in the Path-enclosed Tree. The method of the Path-enclosed Tree cuts out some of the contextual semantic information in the handling process, which reduces the effect of relation extraction. In order to avoid redundant structure and improve the performance of the system, this paper proposed the following methods.

Zhang et al. [4] found that the PT performed best. However, this is contrary to our understanding of semantic relationship in some sentences. For example, in the sentence "扎克伯格和普莉希拉结婚了。 (Zuckerberg and Priscilla are married.)", "结婚 (married)" is critical when determining the relationship between "扎克伯格 (Zuckerberg)" and "普莉希拉 (Priscilla)", as shown in Fig. 2, and the information contained in the PT "扎克伯格和普莉希拉 (Zuckerberg and Priscilla)" is not enough to determine their semantic relationship.

The example shows that the predicate information is necessary to determine the relationship between two entities. We then adapt the PT as the default semantic relation tree and expand the tree structure. In our algorithm, the expansion is made firstly by moving up until a predicate-headed phrase is found and then moving down along the predicated-headed path to the predicate terminal node. Figure 2 shows an example for the expansion algorithm.

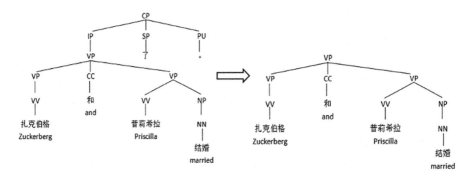

Fig. 2. An example of the rich semantic relation tree structure for the "predicate-linked" category.

After contextual expansion to the widely-used PT structure, the semantic relation tree structure may contain some unnecessary information, such as modifier structure and parallel structure. Based on the observation that it is necessary to remove unnecessary components from the semantic relation tree structure. Modifier redundancy refers to the presence of a modifier such as articles, adjectives, prepositions and other structures between the two entities. These modifiers in the generated PT tree structure is the noise of the classifier. So we will remove these modifiers. Modifier redundancy strategy refers to the modifiers in the PT tree need to be removed. For example, in the sentence "主演者梅丽尔·斯特里普扮演了英国的玛格丽特·撒切尔。 (Actress Meryl Streep plays Britain's Margaret Thatch.)", "英国的 (Britain's)" is an adjective and could be removed as shown in Fig. 3. Likewise, in the sentence "普京在圣彼得堡会见吉尔吉斯总统阿

坦巴耶夫。 (Putin met with the Kyrgyz President Atambayev in St. petersburg.)", "在圣彼得堡 (in St. petersburg)" is a prepositional phrase and could be removed for it did not affect the relationship between "普京 (Putin)" and "阿坦巴耶夫 (Atambayev)".

Parallel redundancy means that the sentence structure has parallel redundancy. For example, in the sentence fragment "德仁和雅子的女儿爱子 (Dukin and Masako's daughter Aiko)", we can determine the relationship between "雅子 (Masako)" and "爱子 (Aiko)". Because "德仁 (Dukin)", "雅子 (Masako)" is the parallel structure, so we can consider there is a relationship between "德仁(Dukin)" and "爱子 (Aiko)". In fact, classifier can't recognize the relation between "德仁 (Dukin)" and "爱子 (Aiko)", and we call this case as parallel redundancy. In other words, the removal of interference of parallel relationship can be a very good solution to this situation as well as improving the accuracy rate of classification effectively. For example, when we identify the relationship between "德仁 (Dukin)" and "爱子 (Aiko)", we can delete ",和雅子 (and Masako)", so that the classifier can correctly identify the relationships between them as shown in Fig. 4. Parallel redundancy strategy refers to the parallel structure in the PT tree need to be removed.

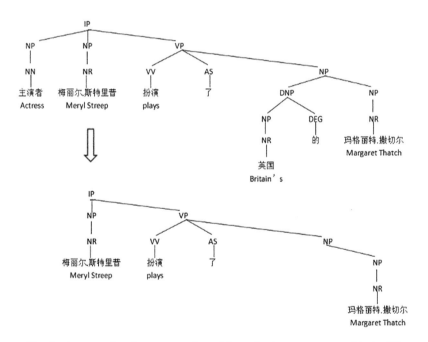

Fig. 3. An example of parse tree of modifier redundancy category with the PT

4   Experiments

This paper uses the real data crawled from the Tencent micro-blog. Tencent micro-blog, the most widely used micro-blog platform in China, allows users to write the post which is no more than 140 Chinese characters. The length of posts brings challenges to relation extraction. Besides, it is not easy to cope with Internet slang for it does not follow language rules.

4.1   Experimental Setting

In this section, we conduct the experiment of relation extraction. We obtain 8446 posts including 21128 sentences about five topics from the Tencent micro-blog, and use LTP (Language Technology Platform Cloud [14]) to pre-process these sentences. We keep the sentences that include at least two entities, and finally get 7090 posts including 15522 sentences and 19686 entities. In the 15522 sentences, we manually marked words associated with relation names in every two entities in each sentence. Finally, we get 14316 positive relations (i.e., there is a relation between two entities.) and 6774 negative relations (i.e., there is no relation between two entities) through the manual annotation. Table 1 shows the detail of data set. Then, Stanford parser [15] is used to parse the data set and get the parse tree of each relation instance. We then performed a 10-fold cross validation on the corpus (15522 sentences). The similarity of the two relation instances is calculated with the method we propose. In our experimentation, SVM [16] is selected as our classifier. In particular, λ in our tree kernel is fine-tuned to 0.4, in accordance with the work of Zhang et al. [4]. In our experiments, we divide the entity relation into two kinds: positive relation and negative relation.

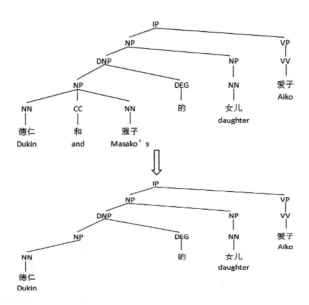

Fig. 4. An example of parse tree of parallel redundancy category with the PT

Table 1. The data set

The topic name	Positive relation	Negative relation	All relation
Education	2610	1134	3744
Sports	2412	1286	3698
Commonweal	3094	1750	4844
Society	3604	1610	5214
Government	2596	994	3590
All	14316	6774	21090

4.2   Experimental Results and Discussion

We compared the performance of the different relationship instance parse tree structures with the untreated tree structure in Table 2. It shows that our method achieves the best performance with 79.6 %/66.9 %/72.7 in precision/recall/F1-measure, respectively. It also indicates that:

- Expanding the predicate verb information associated with entity moderately improves the F1-measure by 5.3 units due to the increase in both precision and recall. This suggests that the predicate verb in the context is very useful, and it should be incorporated.
- Removing the modifier structure and parallel structure of the relation instance slightly improves the F1-measure by 2.6 units and 1.5 units, respectively. This means that proper handling of the modifier structure and parallel structure of the relation instance is beneficial to the simplification of the tree structure.

Table 2. Comparison of performance in different pruning strategies

Semantic relation tree structure	Precision(%)	Recall(%)	F1-measure
PT	67.9	60.7	64.1
PT + the "predicate-linked" category	76.6	63.4	69.4
PT + modifier redundancy category	72.4	61.9	66.7
PT + parallel redundancy category	71.2	60.9	65.6
PT + three proposed categories	79.6	66.9	72.7

Finally, Table 3 compare our method with other methods on our data set. Table 3 shows that our tree kernel-based method greatly outperforms the previous tree kernel-based methods. This is largely due to the rich semantic relation tree structure, which incorporates necessary syntactic and semantic information. The table also show that our tree kernel-based method outperforms the state-of-the-art dependency trigram kernel method. This proves the great potential for semantic relation extraction inherent in the tree structure, even though the total time for learning a tree kernel-based system is usually longer than that required to learn a dependency trigram kernel example.

Table 3. Comparison of relation extraction performance

System	Precision	Recall	F1-measure
Our system	79.6	66.9	72.7
Zhang [4]	67.9	60.7	64.1
Choi [5]	78.8	62.6	69.8

5 Conclusion

In this paper, we use the convolution tree kernel function method to extract the relation between entities. The paper proposes a new method to generate relational instances. The new method cuts redundant information and adds specific semantic information to the parse tree. The experimental results show that our tree kernel-based method outperforms the others. In the experiment we find the predicate verb of the sentence has a great influence on the relation extraction. Due to some reasons, the effect of the experimental is not obvious under the condition of removing the parallel structure, which can be provide as further research direction of other scholars. What pruning strategies of the syntax parsing tree are useful for parallel redundancy? The tree kernel-based method can make full use of flat and structured features, which is of value to study. Hence, the research scope of this paper will focus on this aspect. It is a challenge for us to find a better way to further improve the performance of relation extraction.

Acknowledgements. This research is supported by Projects 61271413, 61472329 and 61532009 under the National Natural Science Foundation of China and Project Innovation Fund of Postgraduate under Xihua University and Project "Xihua Cup" college students innovation and entrepreneurship under Xihua University.

References

1. ACE (1999–2008): Automatic Content Extraction (2008). http://www.ldc.upenn.edu/Projects/ACE/
2. Zelenko, D., Aone, C., Richardella, A.: Kernel methods for relation extraction. J. Mach. Learn. Res. **3**, 1083–1106 (2003)
3. Culotta, A., Sorensen, J.: Dependency tree kernels for relation extraction. In: Proceedings of the Annual Meeting of the Association for Computational Linguistics (ACL 2004), 21–26 July 2004, Barcelona, Spain, pp. 423–429 (2004)
4. Zhang, M., Zhang, J., Su, J., Zhou, G.D.: A composite kernel to extract relations between entities with both flat and structured features. In: Proceedings of the International Conference on Computational Linguistics and the Annual Meeting of the Association for Computational Linguistics (COLINGACL 2006), 17–21 July 2006, Sydney, Australia, pp. 825–832 (2006)
5. Choi, M., Kim, H.: Social relation extraction from texts using a support-vector-machine-based dependency trigram kernel. In: Information Processing and Management, pp. 303–311 (2013)

6. Zhou, G.D., Qian, L.H., Fan, J.X.: Tree kernel-based semantic relation extraction with rich syntactic and semantic information. Inform. Sci. **18**, 1313–1325 (2010)
7. Miller, S., Fox, H., Ramshaw, L.: A novel use of statistical parsing to extract information from text. ANLP 2000, pp. 226–233 (2000)
8. Kambhatla, N.: Combining lexical, syntactic and semantic features with maximum entropy models for extracting relations. In: Proceedings of the Annual Meeting of the Association for Computational Linguistics (ACL 2004) (Poster), 21–26 July 2004, Barcelona, Spain, pp. 178–181 (2004)
9. Zhao, S.B., Grishman, R.: Extracting relations with integrated information using kernel methods. In: ACL, pp. 419–426 (2005)
10. Zhou, G.D., Su, J., Zhang, J., Zhang, M.: Exploring various knowledge in relation extraction. In: ACL, pp. 427–434 (2005)
11. Haussler, D.: Convolution kernels on discrete structures. Technical Report UCS-CRL-99-10, University of California, Santa Cruz (1999)
12. Collins, M., Duffy, N.: Convolution kernels for natural language. In: NIPS-2001, Cambridge, MA, pp, 625–632 (2001)
13. Zhang, M., Zhou, G.D., Aw, A.T.: Exploring syntactic structured features over parse trees for relation extraction using kernel methods. Inform. Proc. Manage. **44**, 687–701 (2008)
14. Language Technology Platform Cloud (2012). http://www.ltp-cloud.com/
15. The Stanford Parser. http://nlp.stanford.edu/software/lex-parser.shtml
16. Joachims, T.: Text Categorization with support vector machine: learning with many relevant features. In: Proceedings of European Conference on Machine Learning (ECML 1998), 21–24 April 1998, Chemnitz, Germany, pp. 137–142(1998)

Nature Inspired Computing
and Optimization

The Robotic Impedance Controller Multi-objective Optimization Design Based on Pareto Optimality

Erchao Li[⊠]

College of Electrical and Information Engineering,
Lanzhou University of Technology, Lanzhou 730050, China
lecstarr@163.com

Abstract. The robotic impedance control is currently one of the main control methods, its main characteristic is that it can make manipulators move to the appointed position quickly and accurately. Due to the high complexity of the robot system, to adjust the impedance controller parameter is always difficult. The impedance controller multi-objective optimization design method is proposed, taking dynamic performances as the optimization objectives, a multi-objective optimization algorithm based on Pareto optimality is applied to the optimal design, obtain Pareto optimal solutions, and get some initial impedance controller adjustment rules, the satisfactory solution is selected in Pareto-optimal solutions according to the requirements of the present system. Simulation results indicate the effectiveness of the proposed algorithm.

Keywords: Impendence control · Multi-objective optimization · Pareto optimal solution

1 Introduction

When the industrial robots implement the grinding, polishing and other tasks, the end need to do the ideal trajectory along the surface of the workpiece, but also to apply force on the workpiece surface, which requires that the position and force are controlled simultaneously, and the position/force hybrid control is a common method [1, 2]. The hybrid position/force control use the orthogonal characteristics of position movement and contact force, respectively control the position and the force, control the contact force along the direction of the robot position movement. Raibert and Craig proposed the hybrid position/force control for the first time in 1981, and achieved good results. Since then, many scholars have improved and perfected the control scheme [3–5]. However, the robot is multi variables system which is nonlinear and strong coupling, in the movement process, due to the presence of friction, load variations and other uncertain factors, meanwhile, it is also a time variable system, it still needs the robot position/force control to do further research to improve the ability to compensate the uncertainty.

The hybrid position/force control theory is clear, but it is difficult to be put into practice, and it will be unstable when the robot hand is contacted.

© Springer International Publishing Switzerland 2016
D.-S. Huang et al. (Eds.): ICIC 2016, Part III, LNAI 9773, pp. 413–423, 2016.
DOI: 10.1007/978-3-319-42297-8_39

The impedance control is widely used in the robotic force control. Not like hybrid position/force control as a direct position and force control, the impedance control is through the stiffness adjustment of the robot end effector to make the force and position to meet the ideal dynamic relationship. Force control and position control is in a frame, used the same strategy to achieve, so it has less task programming workload, and impedance control is to achieve based on simple position control method, so has strong robustness for some uncertainty and disturbance factors. There are two methods to achieve the impedance control: (1) torque impedance control, if the torque information is assumed, then the impedance function can be realized indirectly in the control loop. (2) position impedance control, the input torque information is not available, so the impedance function is directly implemented in the position control outer loop. Thus, the position inner loop is controlled by the position adjustment of the outer loop impedance function. This method is easy to implement in the traditional robotic position control system. However, the impedance control is indirectly controlled by the reference position adjustment, the control accuracy depends on the accurate environment knowledge. In practical applications, due to the various factors influence, the environment is often very imprecise or an inability to know, which leads to the large force error and can't be applied where requires high force control accuracy, which is a major flaw for the impedance control. In order to change this defect, experts and scholars carried out a lot of research. Lasky and Hsia proposed a reference force-tracking impedance control system, consisting of a conventional impedance controller in the inner-loop and a trajectory modifying controller in the outer-loop for force-tracking, the design of the outer-loop is presented and the stability of the two-loop control system is analyzed [6]. Mehrzad Namvar deals with force and motion control of robotic manipulators interacting with a general class of environments whose both geometry and stiffness are not exactly known, using only measurements of robot joint motion variables and the contact force, the proposed adaptive controller generates a torque control signal ensuring asymptotic tracking of desired force and motion trajectories [7]. Seul Jung proposed neural force control scheme is capable of making the robot track a specified desired force as well as of compensating for uncertainties in environment location and stiffness, and in robot dynamics, separate training signals for free-space motion and contact-space motion control are developed to train the neural compensator online [8]. Zhenhua Jiang proposed a novel method of active tracking unknown surface based on force control, uses differential geometry as its main analysis tool, get the geometric properties of surface near the contact point through a six-dimensional force sensor and then give the prediction of trajectory and the compensation of force simultaneously on the basis of the previous force and geometry information [9]. Negar Kharmandar proposed impedance controller is used in a task including an interrupted trajectory tracking to check the ability of the Position-Based Impedance Control in tracking and regulating of the position and the force, fuzzy controller is proposed in the presence of unknown environment dynamic [10].

According to the different environmental conditions, through the reasonable impedance parameters adjustment to adapt to different tasks is the primary goal, but so far, for the reasonable impedance parameters adjustment rule is always the difficult problem to grasp for the operator. In view of the disadvantages of the existing impedance parameter adjustment methods, a multi-objective optimization algorithm based on Pareto optimality is applied to the robotic impedance controller design, obtain Pareto optimal solutions, and get some initial impedance controller adjustment rules, the satisfactory solution is selected in Pareto-optimal solutions according to the requirements of the present system.

2 The Proposed Approach

The diagram of the typical position control system is shown in Fig. 1. The close-loop system can be modeled by the transform function.

$$X(s) = P(s)X_c(s) \tag{1}$$

Where, X is the actual trajectory, X_c is command trajectory.

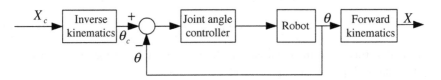

Fig. 1. The position control system of robot

The impedance control builds a relationship between the end-effector and the environment. The impedance control has form of

$$M_d(\ddot{X} - \ddot{X}_d) + B_d(\dot{X} - \dot{X}_d) + K_d(X - X_d) = F_d - F_e \tag{2}$$

Where F_d is the desired force; F_e is the external force from the environment; M_d, B_d and K_d is the symmetric positive definite matrices of desired inertia, damping and stiffness; X_d is the desired trajectory; X is the real trajectory. Based on Eq. (1), the impedance control can be presented in frequency domain as (Fig. 2)

$$(M_d s^2 + B_d s + K_d)(X - X_d) = F_d - F_e \tag{3}$$

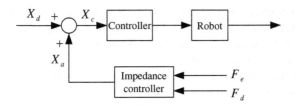

Fig. 2. The structure of position-based impedance control

2.1   Robot Impedance Control

Assuming the adjusted trajectory X_a is

$$X_a = (X - X_d) \tag{4}$$

Substituting Eq. (4) into Eq. (3) yields

$$X_a = \frac{F_d - F_e}{M_d s^2 + B_d s + K_d} \tag{5}$$

To generate the command trajectory X_c as

$$X_c = X_a + X_d \tag{6}$$

The external force is assumed by

$$K_e(X - X_e) = F_e \tag{7}$$

K_e is the stiffness of the environment; X_e is the position of the environment. Substituting Eq. (7) into Eq. (2) renders

$$(M_d s^2 + B_d s + K_d + K_e)X = (M_d s^2 + B_d s + K_d)X_d + K_e X_e + F_d \tag{8}$$

The steady-state position and force tracking accuracy of the proposed impedance controller is studied by Eq. (8). The steady-state position is obtained from Eq. (8)

$$X_{ss} = \lim_{s \to 0} s \frac{1}{1 + K_e/K_d} \left[X_d(s) + \frac{K_e}{K_d} X_e(s) + \frac{1}{K_d} F_d(s) \right] \tag{9}$$

Position-tracking is good if $K_d \gg K_e$, since the $\frac{1}{K_d} F_d(s)$ is smaller than the admissible position error. On the other hand, the external force at steady-state can be presented by

$$F_{e,ss} = \lim_{s \to 0} s \frac{1}{1 + K_d/K_e} \left[F_d(s) + K_d(X_d(s) - X_e(s)) \right] \tag{10}$$

Where force tracking is good if $K_e \gg K_d$, since $K_d(X_d(s) - X_e(s))$ is smaller than the admissible force error. It is obvious that the impedance parameter K_d influences the system steady-state response. In short, $K_d \gg K_e$ is primarily "position control". Conversely, $K_e \gg K_d$ is "force control".

Substituting Eq. (8) into Eq. (2) and assuming $F_d = 0$ and $X_e = 0$ leads to a characteristic equation

$$\Delta(s) = M_d s^2 + B_d s + (K_d + K_e) \tag{11}$$

Where the poles of the system is

$$s_{1,2} = \frac{-B_d \pm \sqrt{B_d^2 - 4M_d(K_d + K_e)}}{2M_d} \tag{12}$$

The characteristic equations of the standard second order system is

$$s^2 + 2\zeta\omega_n s + \omega_n^2 = 0 \tag{13}$$

Where, ζ is the damping ratio and ω_n is the natural undamped frequency. Comparing Eqs. (12) and (13) yields

$$\omega_n = \sqrt{\frac{K_d + K_e}{M_d}} \tag{14}$$

ζ is

$$\zeta = \frac{B_d}{2\sqrt{M_d(K_d + K_e)}} \tag{15}$$

There are three cases for impedance control design goals as follows.

$$\zeta = 1 \Leftrightarrow B_d^2 = 4M_d(K_d + K_e) \tag{16}$$

$$\zeta > 1 \Leftrightarrow B_d^2 > 4M_d(K_d + K_e) \tag{17}$$

$$\zeta < 1 \Leftrightarrow B_d^2 < 4M_d(K_d + K_e) \tag{18}$$

It shows in Eq. (12) that if M_d is large, then the poles are close to the image axis. If M_d is small, then the poles are far away from the image axis. On the other word, the stability of the system is related to the impedance parameter M_d. In this study, small M_d is selected for the stability of the system. Generally speaking, the impedance parameters M_d, B_d and K_d can be chosen to various manipulation task objectives. In practical applications, $F_d \neq 0$, $X_e \neq 0$, the Eqs. (16) – (18) is not completely suitable for the impedance parameter adjustment, but the conclusion can provide reference for the impedance control parameters adjustment.

2.2   Impedance Controller Multi-objective Optimization Design

Impedance control multi-objective optimization design is according to the given performance index, use M_d, B_d, K_d parameter adjustment, when K_e is changed, get the Pareto optimal solution set. The structure of impedance control is shown in Fig. 3 [11].

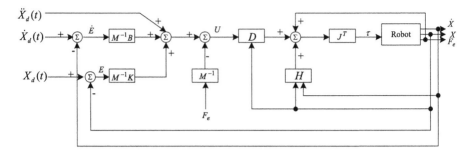

Fig. 3. The structure of impedance control

In this paper, the multi-objective optimization is to achieve system fast, stable, force track output of the overshoot σ, adjust time t_s (error with 2 %), steady-state error $F_{e,ss}$ minimum as the optimization goal based on Pareto optimal NSGA-II algorithm [12], find the combination of M_d, B_d, K_d, establish impedance control multi-objective optimization model:

$$\min \quad f(M_d, B_d, K_d) = (\sigma, t_s, F_{e,ss}) \qquad (19)$$

The individual of M_d, B_d, K_d is real number coding, parameter search space used the literature [13] results as the center, to the left and right sides expand to form. This method makes full use of the existing literature parameters to narrow the actual parameters search space, improve the searching speed and precision. In the evolution operation, tournament selection method, the arithmetic crossover operator and the polynomial mutation operator are used [14], the scale of the tournament selection is 2, the crossover rate is 0.9, the variation rate is 0.1.

3 Experiments and Results

In the literature [13], $M_d = I$, $B_d = \begin{bmatrix} 50 \\ & 50 \end{bmatrix}$, $K_d = \begin{bmatrix} 625 \\ & 625 \end{bmatrix}$, K_e value is 2000, 4000, 40000 in turn, corresponding to different working environment. In the paper, $M_d = I \pm 0.5I$, $B_d = \begin{bmatrix} 50 \\ & 50 \end{bmatrix} \pm 0.5 \begin{bmatrix} 50 \\ & 50 \end{bmatrix}$, $K_d = \begin{bmatrix} 625 \\ & 625 \end{bmatrix} \pm 0.5 \begin{bmatrix} 625 \\ & 625 \end{bmatrix}$, K_e is 2000, 4000, 40000 in turn, $F_d = 50N$, simulation results is shown in Figs. 4, 5 and 6, which coordinate (a) respectively steady-state error, the regulating time and overshoot, the abscissa in (b) is the solution set number, ordinate from bottom to top respectively the steady state error and regulating time and overshoot, the green solid line for steady-state error of the change, the blue dashed for changes in regulation time, the red dashed for changes in overshoot.

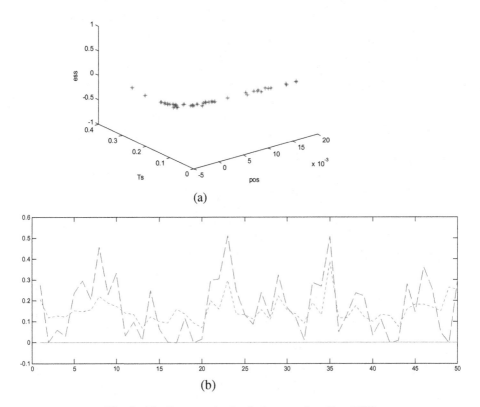

(a)

(b)

Fig. 4. The Pareto optimal solution set when $K_e = 2000$

From the above chart shows, this algorithm only needs to run once, you can search a diversity of Pareto optimal solution set, which can meet the requirements of rapidity and stability of the system, and the solution set in the target function are different, decision makers can select one or some satisfactory solution as the final solution according to specific requirements, once the environment stiffness change, decision makers without rerunning the algorithm, can directly from these solutions, pick out another solution to meet the requirements. And if the simulation results in paper [13] were compared in this paper, if take a relatively satisfactory solution, the simulation of overshoot and the regulation time and other indicators is improved significantly than in the literature [13], will not repeat them here.

The typical Pareto optimal solution set when $K_e = 2000$ is shown in Table 1.

The typical Pareto optimal solution set when $K_e = 4000$ is shown in Table 2.

The typical Pareto optimal solution set when $K_e = 40000$ is shown in Table 3.

From Figs. 4, 5 and 6 can be seen, if the search interval of M_d, B_d, K_d is proper, the solutions meet the system rapidity and stability can be singled out. But K_e value is 40000, from the Pareto optimal solution set obviously can be seen at this time in this paper, the search interval is not appropriate, overshoot is too big, there is no satisfactory solution.

420   E. Li

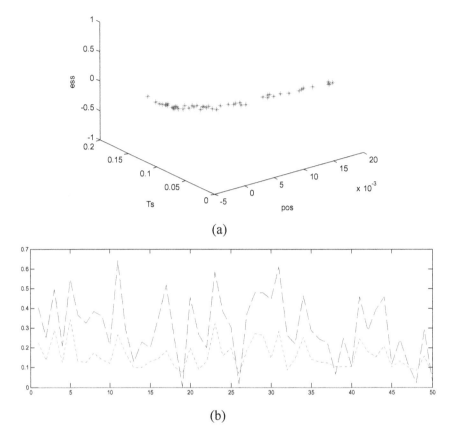

(a)

(b)

Fig. 5. The Pareto optimal solution set when K_e = 4000

Observe the Pareto optimal solution set, impedance control parameters adjustment rules can be summed up preliminary: M_d decreases, overshoot decreases, adjusting time becomes short; B_d becomes large, the overshoot is reduced, adjusting time becomes short; K_d becomes large, overshoot becomes large, adjust time is longer. K_e becomes large, overshoot becomes large, adjust time is longer.

Mentioned in the literature [15], if M_d, B_d, K_d is changed according to the changes of K_e, impedance control effect is better, but the parameters change method proposed requires repeated trial and error, do not follow the law. If use the Pareto optimal solution set in this paper, then these parameters can be convenient choosed. For example, when K_e = 2000, the set of parameters M_d = 1.3353, B_d = 69.7245, K_d = 447.8248 inside the Pareto optimal solution set is selected, retain a decimal point, can be reduce M_d = 1.3, B_d = 69.7, K_d = 447.8. when K_e = 4000, the set of parameters M_d = 0.9529, B_d = 63.3290, K_d = 519.3773 inside the Pareto optimal solution set is selected, retain a decimal point, can be reduce M_d = 1.0, B_d = 63.3, K_d = 519.4. And when t = 5 s, K_e is jumped from 2000 to 4000, at the same time, M_d, B_d, K_d is changed correspondly. The simulation results is shown in Fig. 7, from the figure can

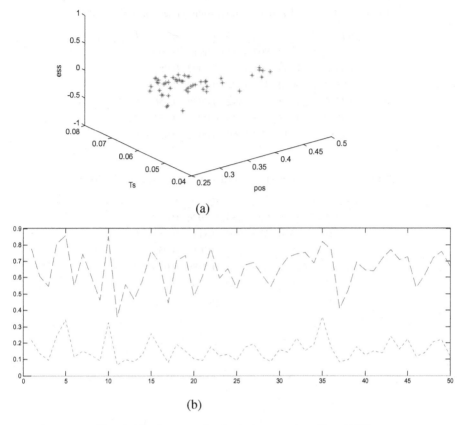

(a)

(b)

Fig. 6. The Pareto optimal solution set when K_e = 40000

Table 1. The typical Pareto optimal solution set when K_e = 2000

M_d	B_d	K_d	σ	t_s	$F_{e,ss}$
1.3353	69.7245	447.8248	0.0046	0.0610	0
1.4125	70.4756	459.6920	0.0008	0.0703	0
0.6153	57.2627	719.5927	0.0084	0.0638	0
0.9292	52.0548	800.9265	0.0014	0.0697	0
0.9248	51.8633	787.7555	0.0003	0.0778	0
0.9194	52.1427	790.0847	0.0061	0.0678	0

seen tracking effect is better. Of course, if this method used in the impedance parameters fuzzy adaptive controller [16], the fuzzy membership function are decided by the Pareto optimal solution set, after experimental verification, the effect is very good, but this is not this paper's purpose.

Table 2. The typical Pareto optimal solution set when $K_e = 4000$

M_d	B_d	K_d	σ	t_s	$F_{e,ss}$
1.0863	69.5740	807.4974	0.0012	0.0604	0
0.6854	33.1859	681.4427	0.0002	0.0620	0
0.5214	63.6090	478.4307	0.0002	0.0651	0
0.7760	50.4594	681.5392	0.0035	0.0562	0
0.9662	62.4343	510.7845	0.0092	0.0511	0
0.7782	50.4747	687.1069	0.0092	0.0536	0

Table 3. The typical Pareto optimal solution set when $K_e = 40000$

M_d	B_d	K_d	σ	t_s	$F_{e,ss}$
0.7401	30.3179	921.9035	0.2926	0.0627	0
0.9228	35.7356	379.4190	0.3145	0.0557	0
1.4058	44.1515	568.1166	0.3612	0.0520	0
1.0407	53.2876	630.7300	0.2780	0.0639	0
0.7336	30.6922	937.5000	0.3207	0.0544	0
0.5162	30.3087	433.3003	0.3112	0.0573	0

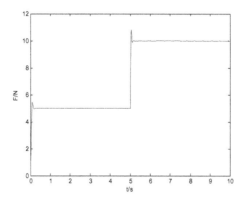

Fig. 7. Force tracking

4 Conclusion

The impedance controller multi-objective optimization design method is proposed, taking dynamic performances as the optimization objectives, a multi-objective optimization algorithm based on Pareto optimality is applied to the optimal design, obtain Pareto optimal solutions, and get some initial impedance controller adjustment rules, the satisfactory solution is selected in Pareto-optimal solutions according to the requirements of the present system. Simulation results indicate the effectiveness of the proposed algorithm.

Acknowledgments. This research was supported by grants from The National Natural Science Fund (No. 61403175), Gansu Province Fundamental Research Funds (No. 2014), Gansu Province Basic Research Innovation Group Project (No. 1506RJIA031), Lanzhou University of Technology Hongliu Project Funds (No. Q201210).

References

1. Gao, H., Lu, S., Tian, G., Tan, J.: Vision-integrated physiotherapy service robot using cooperating two arms. Int. J. Smart Sens. Intell. Syst. **3**(3), 1024–1043 (2014)
2. Wang, W.-C., Lee, C.-H.: Fuzzy neural network-based adaptive impedance force control design of robot manipulator under unknown environment. In: IEEE International Conference on Fuzzy Systems, pp. 1442–1448 (2014)
3. Borghesan, G., De Schutter, J.: Constraint-based specification of hybrid position-impedance-force tasks. In: IEEE International Conference on Robotics & Automation, pp. 2290–2296 (2014)
4. Liu, H., Tang, Y., Zhu, Q., Xie, G.: Present research situations and future prospects on biomimetic robot fish. Int. J. Smart Sens. Intell. Syst. **7**(2), 458–480 (2014)
5. Kim, S., Kim, J.-P., Ryu, J.: Adaptive energy-bounding approach for robustly stable interaction control of impedance-controlled industrial robot with uncertain environments. IEEE/ASME Trans. Mechatron. **19**(4), 1195–1205 (2014)
6. Plius, M.P., Yılmaz, M., Seven, U.: Fuzzy Controller scheduling for robotic manipulator force control. In: The 12th IEEE International Workshop on Advanced Motion Control, pp. 1–8 (2012)
7. Namvar, M., Aghili, F.: Adaptive force control of robots in presence of uncertainty in environment. In: American Control Conference, pp. 3253–3258 (2006)
8. Luo, R.C., Perng, Y.-W., Shih, B.-H., Tsai, Y.-H.: Cartesian position and force control with adaptive impedance/compliance capabilities for a humanoid robot arm. In: 2013 IEEE International Conference on Robotics and Automation, pp. 496–501 (2013)
9. Jiang, Z., Xu, Y., Xie, J.: Active tracking unknown surface based on force control for robot. In: International Conference on Digital Manufacturing & Automation, pp. 143–145 (2011)
10. Kharmandar, N., Khayyat, A.A.: Force impedance control of a robot manipulator using a neuro-fuzzy controller. In: International Conference on Mechatronic Science, Electric Engineering and Computer, pp. 559–563 (2011)
11. Huan, W.S.: The study of uncertain robot force/position intelligent control and trajectory tracking experiment. In: Qinhuangdao: Doctoral Dissertation of Yanshan University, pp. 69–115 (2005)
12. Deb, K., Pratap, A., Agarwal, S., et al.: A fast and elitist multiobjective genetic algorithm: NSGA-II. IEEE Trans. Evol. Comput. **6**(2), 182–197 (2002)
13. Zhen, Y.: An research on the adjustment of parameters of robot based on the impedance control. J. Zaozhuang Univ. **25**(2), 89–93 (2008)
14. Erden, M.S., Billard, A.: End-point impedance measurements at human hand during interactive manual welding with robot. In: 2014 IEEE International Conference on Robotics & Automation (ICRA), pp. 126–133 (2014)
15. Albu-Schäffer, A., Ott, C., Hirzinger, G.: A unified passivity-based control framework for position, torque and impedance control of flexible joint robots. Int. J. Rob. Res. **26**(1), 23–39 (2007)
16. Erchao, L., Zhanming, L., Wei, L.: Robotic fuzzy adaptive impedance control based on neural network visual servoing. Trans. China Electrotechnical Soc. **26**(4), 40–43 (2011)

High-Frequency Trading Strategy
Based on Deep Neural Networks

Andrés Arévalo[1(✉)], Jaime Niño[1], German Hernández[1],
and Javier Sandoval[2]

[1] Universidad Nacional de Colombia, Bogotá, Colombia
{ararevalom,jhninop,gjhernandezp}@unal.edu.co
[2] Algocodex Research Institute, Universidad Externado, Bogotá, Colombia
javier.sandoval@uexternado.edu.co

Abstract. This paper presents a high-frequency strategy based on Deep Neural Networks (Dnns). The DNN was trained on current time (hour and minute), and n-lagged one-minute pseudo-returns, price standard deviations and trend indicators in order to forecast the next one-minute average price. The DNN predictions are used to build a high-frequency trading strategy that buys (sells) when the next predicted average price is above (below) the last closing price. The data used for training and testing are the AAPL tick-by-tick transactions from September to November of 2008. The best-found DNN has a 66 % of directional accuracy. This strategy yields an 81 % successful trades during testing period.

Keywords: Computational finance · High-frequency trading · Deep neural networks

1 Introduction

Financial Markets modelling has caught a lot of attention during the recent years due to the growth of financial markets and the large number of investors around the world in pursuit of profits. However, modelling and predicting prices of Financial Assets is not an easy work, due to the complexity and chaotic dynamics of the markets, and the many non-decidable, non-stationary stochastic variables involved [3, 9]. Many researchers from different areas have studied historical patterns of financial time series and they have proposed various models to predict the next value of time series with a limited precision and accuracy [8].

Since late 1980s, neural networks are a popular theme in data analysis. Artificial Neural Networks (ANNs) are inspired by brain structure; they are composed of many neurons that have connections with each other. Each neuron is a processor unit that performs a weighted aggregation of multiple input signals, and propagates a new output signal depending on its internal configuration. ANNs have the ability to extract essential features and learn complex information patterns in high dimensional spaces. Those features have proven useful for forecasting financial time series. Although neural network models have existed for long time and they have been used in many disciplines, only since early 1990s they are used in the field of finance [7]; The first known application for forecasting financial time series was described in [16].

© Springer International Publishing Switzerland 2016
D.-S. Huang et al. (Eds.): ICIC 2016, Part III, LNAI 9773, pp. 424–436, 2016.
DOI: 10.1007/978-3-319-42297-8_40

A Deep Neural Network (DNN) is an ANN with multiple hidden layers between the input and the output layer, such that data inputs are transformed from low-level to high-level features. The input layer is characterized by having many inputs. At each hidden layer, the data are encoded in features of less dimensions by non-linear transformations; then, the next layers refine the learned patterns in high-level features of less dimensions, and so on until it is capable of learning complex patterns, which are of interest in this work. This type of neural networks can learn high-level abstractions from large quantities raw data through intermediate processing and refinement that occurs in each hidden layer [2, 14].

Traditionally, neural networks are trained with the back-propagation algorithm, which consists in initializing the weights matrices of the model with random values. Then the error between network output and desired output is evaluated. In order to identify the neurons that contributed to the error, the error is propagated backwards from the output layer to all neurons in the hidden layers that contributed to it. This process is repeated layer by layer, until all neurons in the network have received an error signal describing their relative contribution to the total error. Later, the weights are updated in order to try to reduce the error. Then the error is calculated again and this process is repeated until a tolerable error or maximal number of iterations is reached [13].

A serious problem of back-propagation algorithm is that the error is diluted exponentially as it passes through hidden layers on their way to the network beginning. In a DNN that has many hidden layers, only the last layers are trained, while the first ones have barely changed. Although DNNs exist long time ago, they were useless because the challenge of training networks with many layers had remained unsolved. This challenge was solved in 2006 by [5], who successfully included paradigms of Deep Learning in Computer Science.

In recent years Deep Learning (DL) has emerged as a very robust machine learning technique, improving limitations of ANN. Models based on DL have begun to arouse the interest of the general public, because they are able to learn useful representations from raw data and they have shown high performance in complex data, such as text, images and even video. However, applications of DNNs in computational finance are limited [15, 18, 19].

The paper is organized as follows: Sect. 2 presents some important definitions of key concepts in this work. Section 3 describes the dataset used for the experiment. Section 4 presents the DNN modelling for forecasting the next one-minute average price of Apple, Inc. within financial crisis of 2008, when a high volatility behaviour was evidenced. Section 5 describes the proposed trading strategy algorithm. Section 6 presents the strategy performance. Moreover, Sect. 7 presents some conclusions and recommendations for future research.

2   Definitions

Bellow some important definitions are presented:

Definition 1. *Log-return.* It is a term commonly used in finance. Let p_t be the current trade or close price and p_{t-1} the previous trade or close price.

$$R = \ln\left(\frac{p_t}{p_{t-1}}\right) = \ln(p_t) - \ln(p_{t-1}) \tag{1}$$

From a log-return R, the original price p_t can be reconstructed easily:

$$p_t = p_{t-1}e^R \tag{2}$$

Definition 2. *Pseudo-log-return.* It is defined as a logarithmic difference (log of quotient) of between average prices on consecutive minutes. On the other hand, the typical log-return is a logarithmic difference of between closing prices on consecutive minutes.

Definition 3. *Trend Indicator.* It is a new statistical indicator created for this work. All trades within each minute are taken, then a linear model ($y = ax + b$) is fitted. The Trend indicator is equal to the parameter a. A small value, close to zero, means that in the next minute, the price is going to remain stable. A positive value means that the price is going to rise. A negative value means that the price is going to fall. Change is proportional to distance value compared to zero; if distance is too high, the price will rise or fall sharply.

3   Dataset Description

From the TAQ database of the NYSE [6], all trade prices for Apple ordinary stock (ticker: AAPL) were downloaded from the September 2nd to November 7th of the year 2008. Figure 1 shows price behaviour in the selected period.

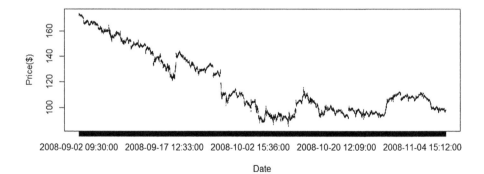

2008-09-02 09:30:00   2008-09-17 12:33:00   2008-10-02 15:36:00   2008-10-20 12:09:00   2008-11-04 15:12:00

Date

Fig. 1. Apple stock price.

The selected period covers stock crash due to the financial crisis of 2008. During this crash, the AAPL price suffered a dramatic fall from 172 to 98 dollars. This period was chosen intentionally to demonstrate the performance of proposed strategy under high volatility conditions. During a financial crisis, market behaviour is strongly impacted by external factors to the system, such as news, rumours, anxiety of traders, among others. If a DNN can identify and learn patterns under these difficult conditions, it can yield equal or even better with other time series without financial crisis.

As it is shown on Fig. 2, the distribution of Tick-by-Tick log-returns is some symmetric with mean $-3.802453e^{-8}$, zero in practical terms. The dataset is composed by 14,839,394 observations, has a maximum value on 0.09429531 and a minimum one on -0.09433554.

Reviewed literature suggests that any stock log-returns follows approximately a normal distribution with mean zero [4, 10, 17]. For this reason, the best variables that describe the behaviour of the market within a minute are mean price and standard deviation of prices.

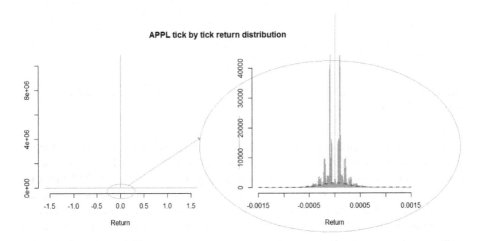

Fig. 2. Distribution of Tick-by-Tick log-returns.

First, the data consistency was verified. All dates were in working days (not holidays and not weekends). All times were during normal hours of trading operation (between 9:30:00 am and 3:59:59 pm EST). All prices and volumes were positive. Therefore, it was not necessary to delete records.

All data are summarized with a one-minute detailed level. Three time series were constructed from trading prices: **Average Price**, **Standard Deviation of Prices** and **Trend Indicator**. Each series has 19110 records (49 trading days × 390 min per day).

4 Deep Neural Network Modelling

4.1 Features Selection

In total four inputs-groups were chosen: Current Time, last n pseudo-log-returns, last n standard deviations of prices and last n trend indicators, where n is the window size. The current time group is composed of two inputs: Current hour and current minute. The others groups are composed of n inputs for each one. In total the number of DNN inputs I is $3n + 2$. The following paragraphs describe each input group:

1. **Current Time:**

The literature reviewed did not include time as an input. However, the hour and minute as integer values were chosen as two additional inputs, due to financial markets are affected by regimes that occurs repeatedly in certain minutes or hours during the trading day. This behaviour may be explained by the fact that both human and automatized (machines) traders have strategies that are running in synchronized periods.

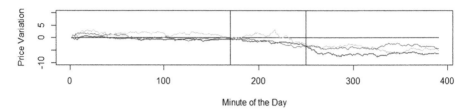

Fig. 3. The price variations that occurred at the first (blue line), third (red line) and sixth (green line) day. (Color figure online)

To illustrate this affirmation, Fig. 3 shows the price variations that occurred at the first, third and sixth day. Approximately, in the minute 170, the stock was traded at the same opening price of corresponding day. Approximately, in the minute 250, the stock price fell 3 dollars relative to the opening price in these days. As these patterns, many more are repeated at certain time of day. In order to identify and to differentiate better these patterns, the current hour and current minute of day were added as additional inputs of DNN. These variables have 7 and 60 possible values ranging from 0 to 6 and from 0 to 59 respectively.

2. **Last n pseudo-log-returns:**

It is common to see works of neural networks used to forecast time series whose inputs are composed principally by the last untransformed observations. This is fine for several types of time series, but it is not advisable in financial time series forecasting. In any dataset and particularly in the one used in this work, if the nominal prices are used, it will be useless because a neural network will train with special conditions (prices fluctuates between 120 and 170 dollars) and then it will be tested against different market conditions (prices fluctuates between 90 and 120 dollars).

In other words, the neural network learns to identify many static patterns that will be not appearing at all. For example, a pattern like when the price is over 150 dollars, raises to 150.25 dollars and falls to 149.75 dollars, then it will change to 150.50 dollars, could be found, but this pattern never will occur because in the closest future the prices fluctuates between 90 and 120 dollars. However, if prices are transformed into differences or logarithmic returns, not only the data variance is stabilized, but also the time series acquire temporal independence. For example at the beginning of the selected period, a pattern, like when the price rises 25 cents and it falls 50 cents, then it will raise 75 cents, could be found and this pattern is more likely to occur in the future. Therefore, the last n one-minute pseudo-log-returns are inputs of DNN.

3. **Last n standard deviations of prices:**

The last n one-minute standard deviations of prices are DNN inputs.

4. **Last n trend indicators:**

The last n one-minute trend indicators are DNN inputs.

4.2   Output Selection of the Deep Neural Network

The DNN forecasts the next one-minute pseudo-log-return. As it is shown on Fig. 4, the average price **(black line)** is the variable that best describes market behaviour. The highest or lowest prices **(blue lines)** usually are found within a confidence range of average price, therefore the next highest and lowest prices can be estimated from a predicted average price. The closing price **(red line)** can be any value close to the average price; it sometimes coincides with the highest or lowest price. Unlike the average price, the highest, lowest and closing ones are exposed largely to noise or external market dynamics, for example, some traders listen a false rumour about bad news that will cause a sudden fall in the price, in order to reduce losses. As a result, they decide to sell at a lower price than the one traded before. This operation could be done at a certain second and it could affect numerically the highest, lowest or closing prices on the minute.

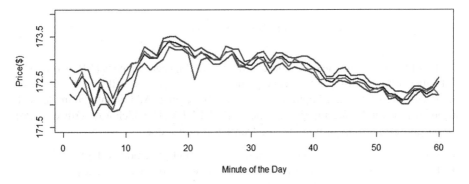

Fig. 4. First 60 one-minute Apple Stock Prices at September 2nd, 2008. Blue: High and Low. Red: Close. Black: Average. (Color figure online)

Since the objective of this work is to learn the dynamics of the market to take advantage of it eventually, the average price forecasts could be more profitable than the closing price forecast. With a good average price forecast, it is known that the stock is going to trade to that predicted value at any moment within the next minute. A real automated trading strategy should wait until the right time (for example, stock price reaches to price forecast) to open or to close their positions.

4.3   Deep Neural Network Architecture

The architecture was selected arbitrarily. It has one input layer, five hidden layers and one output layer. The number of neurons in each layer depends on the number of inputs I. Each layer has I, I, $\lfloor 4I/5 \rfloor$, $\lfloor 3I/5 \rfloor$, $\lfloor 2I/5 \rfloor$, $\lfloor I/5 \rfloor$ and 1 neurons respectively. All neurons use a *tanh* activation function except the output neuron that uses a *linear* activation function.

4.4   Deep Neural Network Training

The final dataset is made up of $19109 - n$ records. Each record contains $3n + 3$ numerical values ($3n + 2$ inputs and 1 output). It should be noted that to construct each record, only information from the past is required. Therefore, there is not look-ahead bias and this DNN could be used for a real trading strategy.

As shown on Fig. 1, the dataset has two regimes: one bearish regime (first 50 % samples) and a no-trending one (last 50 % samples). The final dataset was divided into two parts: In-sample data (first 85 % samples in bearish regime and first 85 % samples in no-trending regime) and out-sample data (last 15 % samples in bearish regime and last 15 % samples in no-trending regime). The Fig. 5 shows the splitting.

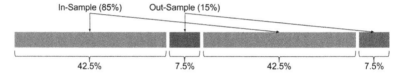

Fig. 5. Data splitting.

For this work, H_2O, an open-source software for big-data analysis [12], was used. It implements algorithms at scale, such as deep learning [1], as well as featuring automatic versions of advanced optimization for DNN training. Additionally, it implements an adaptive learning rate algorithm, called ADADELTA [1], which is described in [20]. It was chosen in order to improve the learning process, due:

• It is a per-dimension adaptive learning rate method for gradient descent.
• It is not necessary to search parameters for gradient descent manually.
• It is robust to large gradients and noise.

4.5   Deep Neural Network Assessment

In order to assess the DNN performance, two statistics were chosen. Let E as the expected series and F as the series forecast:

- **Mean Squared Error:** $MSE = \frac{1}{n}\sum_{t=1}^{n}(E_t - F_t)^2$
- **Directional Accuracy:** Percent of predicted directions that matches with the ideal differences time series. This measure is unaffected by outliers or variables scales. $DA = \frac{100}{n}\sum_{t=1}^{n}(E_t \cdot F_t > 0)$.

5   Proposed Strategy

The DNN predictions are used by the following high-frequency trading strategy: For each trading minute, it always buys (sells) a stock when the next predicted average price is above (below) the last closing price. When the price yields the predicted average price, it sells (buys) the stock in order to ensure the profit. If the price never yields the expected price, it sells (buys) the stock with the closing price of the minute, in order to close the position and potentially stop losing positions. Figure 6 shows the strategy flowchart. Below the algorithm is formally presented in pseudo-code:

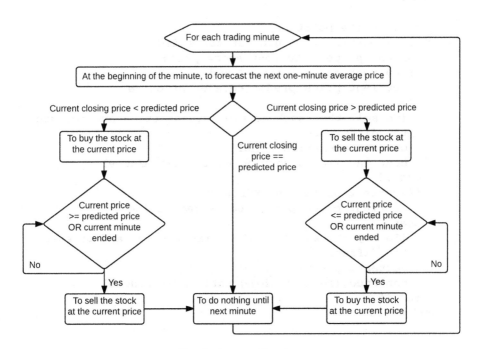

Fig. 6. Strategy flowchart.

```
for each trading minute
  {At the beginning of the minute, to forecast the next
  one-minute average price}
  I ← Current features vector.
  predicted.pseudo.return ← DNN.forecast (I)
  predicted.average.price ← last.average.price
    e^predicted.pseudo.return


  {To open a position}
  if predicted.average.price>previous.closing.price
  then
    {To buy a stock at the current price}
    current.operation ← BUY
  else if predicted.average.price<previous.closing.price
  then
    {To sell a stock at the current price}
    current.operation ← SELL
  else
    continue {To do nothing until next minute}
  end if


  {To close the position}
  while the current minute has not ended
    current.price ← Current stock price
    if current.operation==BUY and
       current.price≥predicted.average.price
    then
      {To sell the stock at the current price and to earn
      the difference}
      current.operation ← null
      continue {To do nothing until next minute}
    end if
    if current.operation==SELL$ and
       current.price≤predicted.average.price
    then
      {To buy the stock at the current price and to earn
      the difference}
      current.operation ← null
      continue {To do nothing until next minute}
    end if
  end while
```

```
{To stop the losses}
if The operation was not closed
then
  if current.operation==BUY
  then
    {To sell the stock at the current price}
  else if current.operation==SELL
  then
    {To buy the stock at the current price}
  end if
  current.operation ← null
  end if
end for
```

6 Experiment

The DNN was trained only with the in-sample data during 50 epochs. The chosen ADADELTA parameters were $\rho = 0.9999$ and $\in = 10^{-10}$. On the other hand, the DNN was tested only with the out-sample date. Table 1 illustrates DNN performance using different windows sizes and the same network architecture. Ten different networks were trained for each parameter. The best results are obtained with small window sizes such as three, four and five.

Table 1. DNN performance.

Window Size	DNN Architecture	Maximum		Minimum		Mean		σ	
		MSE	DA (%)	MSE	DA (%)	MSE	DA (%)	MSE	DA (%)
2	8 8:6:4:3:1 1	0.07832	65.71328	0.06768	61.63236	0.07042	64.47506	0.00294	1.30354
3	11 11:8:6:4:2 1	0.07678	66.15492	0.06823	63.67759	0.07125	65.17794	0.00233	0.71620
4	14 14:11:8:5:2 1	0.09158	65.71328	0.07197	63.30659	0.07576	65.07847	0.00579	0.70564
5	17 17:13:10:6:3 1	0.10132	66.05024	0.07569	64.30565	0.08561	64.83949	0.00729	0.52326
6	20 20:16:12:7:3 1	0.10512	65.74816	0.07574	62.99267	0.08514	64.91105	0.00874	0.83872
7	23 23:18:13:9:4 1	0.10383	65.63154	0.08251	63.22400	0.08929	64.43475	0.00634	0.73761
8	26 26:20:15:10:5 1	0.09873	65.60865	0.07813	63.41123	0.08754	64.40879	0.00648	0.68195
9	29 29:23:17:11:5 1	0.09020	65.49197	0.07628	63.78227	0.08437	64.59874	0.00475	0.45499
10	32 32:25:19:12:6 1	0.10250	65.50401	0.07400	63.09731	0.08476	64.55877	0.00829	0.76088
11	35 35:28:21:13:6 1	0.10565	65.24773	0.07702	62.90997	0.08537	64.25680	0.00892	0.83953
12	38 38:30:22:15:7 1	0.09440	65.32961	0.07746	63.69026	0.08698	64.46110	0.00562	0.62078
13	41 41:32:24:16:8 1	0.09442	64.61967	0.07437	61.89811	0.08491	63.79972	0.00618	0.96908
14	44 44:35:26:17:8 1	0.09833	65.32961	0.07849	62.39972	0.08781	64.25531	0.00635	0.94794
15	47 47:37:28:18:9 1	0.09871	64.86392	0.08201	61.86322	0.08916	63.64270	0.00531	1.13385

Overall, the networks achieved between 63 % and 66 % directional accuracy. Depending on training results, DNN performance may be better, but all networks converge with very similar and homogeneous results. The DNN is able to predict these sudden rises or falls in price. This information may be useful for any trading strategy.

Figures 7, 8 and 9 show the strategy performance during a trading simulation over the testing data. The simulation did not consider transaction costs and it was performed with the best-found DNN (66.15492 % of DA). Buying and selling only one stock, the strategy accumulated 72.3036 dollars at the end of the period. It made 2333 successful trades and 520 unsuccessful ones, approximately 81.77 % successful trades.

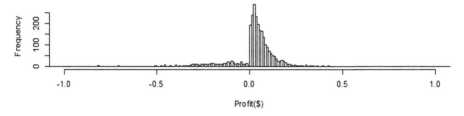

Fig. 7. Profit histogram of the trading strategy

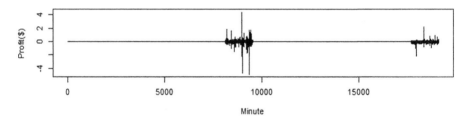

Fig. 8. Profit of the trading strategy

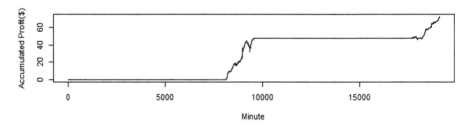

Fig. 9. Cumulated profit of the trading strategy

During the training data period (from 0 % to 42.5 % and from 50 % to 92.5 % in the time series), the strategy did not perform trades, and then it did not yields profits and losses on those minutes. For this reason, Figs. 8 and 9 have a horizontal line during these periods.

7 Conclusions

Although the strategy turns out to be interesting and yields a good performance, it must be refined in order to implement in a real environment, for example, it could analyse whether it closes its position in the next minute or it keeps it open in order to decrease transaction costs.

Traders collectively repeat the behaviour of the traders that preceded them [11]. Those patterns can be learned by a DNN. The proposed strategy replicates the concept of predicting prices for short periods. Furthermore, adding time as a DNN input allows it to differentiate atypical events and repetitive patterns in market dynamics. Moreover, small data windows sizes are able to explain future prices in a simpler way.

Overall, the DNNs can learn the market dynamic with a reasonable precision and accuracy. Within the deep learning arena, the DNN is the simplest model, as a result, a possible research opportunity could be to evaluate the performance of the strategy using other DL model such as Deep Recurrent Neural Networks, Deep Belief Networks, Convolutional Deep Belief Networks, Deep Coding Networks, among others.

References

1. Arora, A., et al.: Deep Learning with H2O (2015)
2. Bengio, Y.: Learning deep architectures for AI. Found. Trends® Mach. Learn. **2**(1), 1–127 (2009)
3. De Gooijer, J.G., Hyndman, R.J.: 25 years of time series forecasting. Int. J. Forecast. **22**(3), 443–473 (2006)
4. Härdle, W., et al.: Applied Quantitative Finance: Theory and Computational Tools. Springer, Heidelberg (2013)
5. Hinton, G.E., et al.: A fast learning algorithm for deep belief nets. Neural Comput. **18**(7), 1527–1554 (2006)
6. Intercontinental Exchange Inc.: TAQ NYSE Trades (2016). http://www.nyxdata.com/data-products/nyse-trades-eod
7. Kaastra, I., Boyd, M.: Designing a neural network for forecasting financial and economic time series. Neurocomputing **10**(3), 215–236 (1996)
8. Li, X., et al.: Enhancing quantitative intra-day stock return prediction by integrating both market news and stock prices information. Neurocomputing **142**, 228–238 (2014)
9. Marszałek, A., Burczyński, T.: Modeling and forecasting financial time series with ordered fuzzy candlesticks. Inf. Sci. (Ny) **273**, 144–155 (2014)
10. Mills, T., Markellos, R.: The econometric modelling of financial time series (2008)
11. Murphy, J.J.: Technical Analysis of the Financial Markets: A Comprehensive Guide to Trading Methods and Applications. Penguin, New York (1999)
12. Nusca, A., et al.: Arno Candel, physicist and hacker, 0xdata. Meet Fortune's 2014 Big Data All-Stars (2014)
13. Riedmiller, M., Braun, H.: A direct adaptive method for faster backpropagation learning: the RPROP algorithm. In: IEEE International Conference on Neural Networks, pp. 586–591. IEEE (1993)
14. Schmidhuber, J.: Deep learning in neural networks: An overview. Neural Netw. **61**, 85–117 (2014)

15. Takeuchi, L., Lee, Y.: Applying Deep Learning to Enhance Momentum Trading Strategies in Stocks. cs229.stanford.edu
16. Trippi, R.R., Turban, E.: Neural Networks in Finance and Investing: Using Artificial Intelligence to Improve Real World Performance. Probus Publishing Company, Chicago (1992)
17. Tsay, R.S.: Analysis of Financial Time Series. Wiley, New York (2005)
18. Yeh, S., et al.: Corporate Default Prediction via Deep Learning (2014)
19. Pham, D.-N., Park, S.-B. (eds.): PRICAI 2014. LNCS, vol. 8862. Springer, Heidelberg (2014)
20. Zeiler, M.D.: ADADELTA: An Adaptive Learning Rate Method, 6 (2012)

Adaptive Extended Computed Torque Control of 3 DOF Planar Parallel Manipulators Using Neural Network and Error Compensator

Quang Dan Le[1], Hee-Jun Kang[2(✉)], and Tien Dung Le[3]

[1] Graduate School of Electrical Engineering,
University of Ulsan, Ulsan, South Korea
ledanmt@gmail.com
[2] School of Electrical Engineering, University of Ulsan, Ulsan, South Korea
hjkang@ulsan.ac.kr
[3] The University of Danang - University of Science
and Technology, Danang, Vietnam
ltdung@dut.udn.vn

Abstract. In this paper, an adaptive extended computed torque controller is proposed for trajectory tracking of 3 degree-of-freedom planar parallel manipulators. The dynamic model, including the modeling errors and uncertainties, is established in the joint space of 3 degree-of-freedom planar parallel manipulators. Based on the dynamic model, an adaptive extended computed torque control scheme is proposed in which a feed-forward neural network is combined with error compensators for adaptive compensating the unknown modeling errors and uncertainties of the parallel manipulators. The weights of the neural network are based on sliding functions and self-tuned online during the tracking control of system without any offline training phase. Using the combination of Sim Mechanics and Solid works, the comparative simulations are conducted for verifying the efficiency of the proposed control scheme.

Keywords: Parallel manipulator · Online self-tuning · Adaptive control · Extended computed torque control · Neural network

1 Introduction

There are two types of robot manipulators: serial robots and parallel robots. Serial robots cannot operate with good accuracy at high-speed due to their serial chain structure. Parallel robots, unlike serial robots, have closed-chain mechanisms consisting of a number of serial kinematic chains formed by rigid links and joints. Parallel manipulators have the following potential advantages over serial manipulators: better accuracy, higher stiffness and payload capability, higher velocity, lower moving inertia, and so on. However, parallel manipulators have some disadvantages because of its complexity. For example, due to closed-chain structure, parallel manipulators have complex kinematic and dynamic model. In addition, there are many singularities inside the workspace of the robot which is usually small. For the above characteristics, the control of parallel manipulators needs more advanced technologies than that of serial manipulators.

© Springer International Publishing Switzerland 2016
D.-S. Huang et al. (Eds.): ICIC 2016, Part III, LNAI 9773, pp. 437–448, 2016.
DOI: 10.1007/978-3-319-42297-8_41

In the last years, many studies have been undertaken in order to realize the potential performances in motion control of parallel manipulators. The proposed approaches can be grouped in two control strategies for parallel manipulators [1–3]: kinematic-based control strategies and dynamic model-based control strategies. In the dynamic model-based control strategies, the computed torque controller has been proposed [4–8] for parallel robots. However, the computed torque control method requires dynamic model which must be as accurate as possible. Due to the complex kinematic model of parallel manipulators, the control loop only considers active joints while the passive joints angular are estimated. Therefore, the accuracy of the positioning of the end-effector relies on the accuracy of the kinematic model and the control system needs a lot of computations. In order to give a solution to this issue, the use of extra sensors in the passive or non-actuated joints is proposed as a good solution. In [9, 10], an extended computed torque controllers were designed for 3 DOF parallel manipulators in which the extra sensor data can be used. This approach was also adopted in other papers [11–13] in order to reduce the estimation error of the pose of the end-effector of parallel manipulators. However, most of these control algorithms did not consider the compensating of modeling error and uncertainties of the parallel robot.

In this paper, a new adaptive extended computed torque controller is proposed and applied to the 3 DOF planar parallel manipulators. The proposed controller is based on the combination of nominal dynamic model of the parallel manipulator, a feed-forward neural network and error compensators. The feed-forward neural network and error compensators work together for compensating modeling error and uncertainties. The self-tuning laws of the feed-forward are based on sliding functions and self-tuned online during the trajectory tracking control of the manipulator without any offline training phase.

The rest of the paper is organized as follows. In Sect. 2, the dynamic model of 3 DOF parallel manipulators is presented. The proposed adaptive computed torque controller together with the self-tuning laws are presented in Sect. 3. The comparative simulations of 3 DOF planar parallel manipulator are conducted to verify the validity of the proposed controller as given in Sect. 4. Finally, a conclusion is reached in Sect. 5.

2   Dynamic Model of 3 DOF Planar Parallel Manipulators

The geometric description of 3 DOF planar parallel manipulators is depicted in Fig. 1. The manipulator includes three active joints, six passive joints and an end-effector. The link lengths of the parallel manipulator are denoted as: $l_1 = A_iB_i$, $l_2 = B_iC_i$, $i = 1, 2, 3$. The end-effector is an equilateral triangle with $l_3 = C_iP$ which is radius of the circle circumscribing the three vertices.

We define the vectors of passive joints and passive joints as:

- $\theta_a = [\theta_{a1}, \theta_{a2}, \theta_{a3}]^T$: Vector of actuated joints A_1, A_2 and A_3.
- $\theta_p = [\theta_{P1}, \theta_{P2}, \theta_{P3}]^T$: Vector of passive joints B_1, B_2 and B_3.
- $P = [x_P, y_P, \phi_P]^T$: Vector of the end-effector co-ordinates.

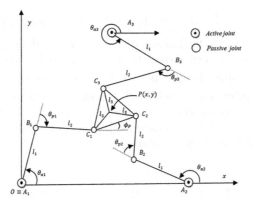

Fig. 1. Thegeometric description of the 3 DOF planarparallel manipulator

From the geometric model shown in the Fig. 1, a vector loop equation can be written for each limb as:

$$OP = OA_i + A_iB_i + C_iP \tag{1}$$

where $i = 1, 2, 3$.

Expanding the Eq. (1), we get:

$$x_P = x_{Ai} + l_1 \cos(\theta_{ai}) + l_2 \cos(\theta_{ai} + \theta_{pi}) + l_3 \cos(\psi_i + \phi_P) \tag{2}$$

$$y_P = y_{Ai} + l_1 \sin(\theta_{ai}) + l_2 \sin(\theta_{ai} + \theta_{pi}) + l_3 \sin(\psi_i + \phi_P) \tag{3}$$

where ψ_i ($i = 1, 2, 3$) are the passive angulars of the end-effector. The values of ψ_i are inside the vector $\psi = [\pi/6, 5\pi/6, \pi/2]^\mathrm{T}$.

Differentiating the Eqs. (2) and (3), we get:

$$\dot{\theta}_a = J_a \dot{X} \tag{4}$$

$$\dot{\theta}_p = J_p \dot{X} \tag{5}$$

$$J_x \dot{X} = J_{ap} \dot{\theta} \tag{6}$$

$$J = \frac{\partial \theta_p}{\partial \theta_a} = \left[\frac{\partial \theta_a}{\partial \theta_x}\right]^{-1} \frac{\partial \theta_p}{\partial \theta_x} \tag{7}$$

$$J_c = J_x^+ J_{ap} \tag{8}$$

$$\dot{\theta} = W \dot{\theta}_a \tag{9}$$

in which $\dot{X} = \left[\dot{x}_p, \dot{y}_p, \dot{\phi}_p\right]^T$; $\dot{\theta} = \left[\dot{\theta}_{ai}, \dot{\theta}_{pi}\right]^T$; $J \in \Re^{3\times3}$ is the Jacobian matrix of the active joints with respect to the passive joints; $J_c \in \Re^{3\times6}$ is the Jacobian matrix of the end-effector co-ordinates with respect to the passive and active joints; J_x^+ is the Moore-Penrose Generalized inverse; $W = [I, J]^T$ is the Jacobian matrix of all the joints with respect to the active joints; and $I \in \Re^{3\times3}$ is identity matrix.

From the virtually cut model shown in Fig. 2, we derive the Lagrangian equation of the open-chain system and then compute the joint torques needed to generate a given motion which satisfies the loop constraints.

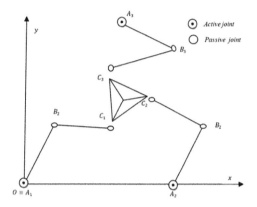

Fig. 2. The equivalent open-chain system obtained by virtually cuts

Dynamic model of each planar 2 DOF serial manipulator and end-effector given by Largange's as the following:

$$\frac{d}{dt}\left(\frac{\partial L_i}{\partial \dot{\theta}_i}\right) - \frac{\partial L_i}{\partial \theta_i} = \tau_{\theta i} - F_i \tag{10}$$

$$\frac{d}{dt}\left(\frac{\partial L_i}{\partial \dot{X}}\right) - \frac{\partial L_i}{\partial X} = \tau_X - F_{xi}. \tag{11}$$

Where L_i is the Lagrangian function for each serial manipulator ($i = 1, 2, 3$); $\theta_i = \left[\theta_{ai}, \theta_{pi}\right]^T$ is the joint vector; $\tau_{\theta i} = \left[\tau_{ai}, \tau_{pi}\right]^T$ is the joint torque vector; $F_i = \left[f_{ai}, f_{pi}\right]^T$ is the vector of the active and passive joint friction torques of ith serial chain; L_i is the Lagrangian function for the end-effector; $X = [x_P, y_P, \phi_P]^T$ is the position of end-effector in the Cartesian co-ordination; $\tau_X = \left[\tau_{xP}, \tau_{yP}, \tau_{\phi P}\right]^T$ is the end-effector torque vector; $F_{xi} = \left[f_{c1}, f_{c2}, f_{c3}\right]^T$ is the friction torques of the end-effector.

As the parallel manipulator operates on a horizontal planar plane, the Lagrangian function only contains the kinetic energy of the mechanism:

$$L_i = \frac{1}{2}\left[m_{i1}(\dot{x}_{i1}^2 + \dot{y}_{i1}^2) + m_{i2}(\dot{x}_{i2}^2 + \dot{y}_{i2}^2) + I_{i1}\dot{\theta}_{ai}^2 + I_{i2}(\dot{\theta}_{ai} + \dot{\theta}_{pi})^2\right] \quad (12)$$

in which m_{i1}, m_{i2} are the masses of links of the serial chain; I_{i1} and I_{i2} are the inertia tensors of links of the serial chain.

From the geometric description shown in the Fig. 1, we have:

$$\dot{x}_{i1} = -l_1 \sin\theta_{ai}\dot{\theta}_{ai}$$

$$\dot{y}_{i1} = l_1 \cos\theta_{ai}\dot{\theta}_{ai}$$

$$\dot{x}_{i2} = -l_1 \sin\theta_{ai}\dot{\theta}_{ai} - l_2 \sin\theta_{pi}(\dot{\theta}_{ai} + \dot{\theta}_{pi})$$

$$\dot{y}_{i2} = l_1 \cos\theta_{ai}\dot{\theta}_{ai} + l_2 \cos\theta_{pi}(\dot{\theta}_{ai} + \dot{\theta}_{pi}).$$

By substituting the above equations into (12), the Lagrangian function becomes:

$$L_i = \frac{1}{2}\alpha_i\dot{\theta}_{ai}^2 + \frac{1}{2}\beta_i(\dot{\theta}_{ai} + \dot{\theta}_{pi})^2 + \gamma_i \cos\theta_{pi}(\dot{\theta}_{ai}^2 + \dot{\theta}_{ai}\dot{\theta}_{pi}) \quad (13)$$

in which $\alpha_i = m_{i1}l_1^2 + m_{i2}l_2^2 + I_{i1}$, $\beta_i = m_{i2}l_2^2 + I_{i2}$, $\gamma_i = m_{i2}l_1l_2$, $i = 1, 2, 3$.

Next, substituting (13) into Lagrange's equation (10), we obtain the dynamic equations of each serial chain of the open-chain system as the following:

$$M_i(\theta_i)\ddot{\theta}_i + C_i(\theta_i,\dot{\theta}_i)\dot{\theta}_i + F_i = \tau_i \quad (14)$$

in which M_i is inertia matrix; C_i is the Coriolis and centrifugal force matrix.

Combining the dynamic models of three open serial chains together, the dynamic model of the equivalent tree-structure mechanism is obtained as:

$$M_t(\theta)\ddot{\theta} + C_t(\theta,\dot{\theta})\dot{\theta} + F_i = \tau_t \quad (15)$$

in which $\theta = \left[\theta_a, \theta_p\right]^T \in \Re^{6 \times 1}$ is the vector of joint angles; $\tau_t = \left[\tau_a, \tau_p\right]^T \in \Re^{6 \times 1}$ is the joint torques vector.

The inertial matrix $M_t \in \Re^{6 \times 6}$ and the Coriolis and centrifugal force matrix $C_t \in \Re^{6 \times 6}$ in (15) are expressed by:

$$M_t = \begin{bmatrix} \delta_1 & 0 & 0 & \varepsilon_1 & 0 & 0 \\ 0 & \delta_2 & 0 & 0 & \varepsilon_2 & 0 \\ 0 & 0 & \delta_3 & 0 & 0 & \varepsilon_3 \\ \gamma_1 c_{p1} & 0 & 0 & \beta_1 & 0 & 0 \\ 0 & \gamma_2 c_{p2} & 0 & 0 & \beta_2 & 0 \\ 0 & 0 & \gamma_3 c_{p3} & 0 & 0 & \beta_3 \end{bmatrix}, C_t = \begin{bmatrix} \vartheta_1 & 0 & 0 & \mu_1 & 0 & 0 \\ 0 & \vartheta_2 & 0 & 0 & \mu_2 & 0 \\ 0 & 0 & \vartheta_3 & 0 & 0 & \mu_3 \\ \rho_1 & 0 & 0 & 0 & 0 & 0 \\ 0 & \rho_2 & 0 & 0 & 0 & 0 \\ 0 & 0 & \rho_3 & 0 & 0 & 0 \end{bmatrix}$$

where $\delta_i = \alpha_i + \beta_i + 2\gamma_i c_{pi}$, $\varepsilon_i = \beta_i + \gamma_i c_{pi}$, $\vartheta_i = -2\gamma_i s_{pi}\dot{\theta}_{pi}$, $\mu_i = -\gamma_i s_{pi}\dot{\theta}_{pi}$, $\rho_i = \gamma_i s_{pi}\dot{\theta}_{ai}$, $c_{pi} = \cos\theta_{pi}$, $s_{pi} = \sin\theta_{pi}$ ($i = 1, 2, 3$).

Lagrangian function for end-effector only contains the kinetic energy of the mechanism:

$$L_p = \frac{1}{2}\left[m_p(\dot{x}_P^2 + \dot{y}_P^2) + I_P\dot{\phi}_P^2\right]. \tag{16}$$

Where m_P is the mass of the end-effector; I_p is the inertia tensor of the end-effector. By substituting the Lagrange function (16) into Lagrange's equation (11), we obtain the dynamic equations of the end-effector:

$$M_P(X)\ddot{X} + F_{xi} = \tau_x. \tag{17}$$

Where M_P is the inertial matrix of the end-effector; $\tau_x \in \Re^{3x1}$ is the torque of the end-effector.

$$M_P = \begin{bmatrix} m_P & 0 & 0 \\ 0 & m_P & 0 \\ 0 & 0 & I_P \end{bmatrix}.$$

By differentiating both sides of (6) we obtain:

$$\ddot{X} = \dot{J}_c\dot{\theta} + J_c\ddot{\theta}. \tag{18}$$

Next, substituting (18) into (17) we have:

$$M_PJ_c\ddot{\theta} + M_PJ_c\dot{\theta} + F_{xi} = \tau_x. \tag{19}$$

Combining the dynamic models (15) and dynamic model of end-effector (19), and using natural orthogonal complement we obtain the dynamic model of 3-DOF planar parallel manipulators as the following:

$$M\ddot{\theta} + C\dot{\theta} + F_i = \tau_a \tag{20}$$

where $M = W^T(M_t + J_c^T M_p J_c)$, $C = W^T(C_t + J_c^T M_p \dot{J}_c)$.

3 Proposed Adaptive Extended Computed Torque Controller

The motion of parallel manipulators is naturally defined in terms of the motion of its end-effector in Cartesian space. Defining the desired trajectory in the task space $X_d = [x_{dp}, y_{dp}, \varphi_{dp}]^T$, then by using the inverse kinematic model we can find the desired trajectory $\theta_d = [\theta_{da}, \theta_{dp}]^T$ in the joint space.

Define the tracking error:

$$e = \theta_d - \theta \tag{21}$$

in which $\theta = [\theta_a, \theta_p]^T$ is the actual joint angular vector; θ_a is the actual active joint vector; θ_p is the actual passive joint vector. All of these actual joint angular values are measured by position sensors.

The traditional extended computed torque controller which was proposed in previous works [9, 10] is expressed by the following:

$$\tau_a = M\left(\ddot{\theta}_d + K_p e + K_v \dot{e}\right) + C\dot{\theta} \tag{22}$$

where K_p, K_v are the positive symmetric position and velocity gain; the matrices M and C are computed in the Eq. 22. This controller does not consider the compensation of modelling error and uncertainties of the robot.

The proposed adaptive controller in this paper is expressed as follows:

$$\tau_a = M\left(\ddot{\theta}_d + K_p e + K_v \dot{e}\right) + C\dot{\theta} + f_{NN} + \Gamma \int sdt \tag{23}$$

in which f_{NN} is the output of a feed-forward neural network for compensating the modeling error and uncertainties of the robot. Since the output of the f_{NN} is not able to approximate $\Delta\tau$ accurately, the error compensator $\Gamma \int sdt$ is used to attenuate the approximation errors. The block diagram of the proposed controller (23) is illustrated in Fig. 3.

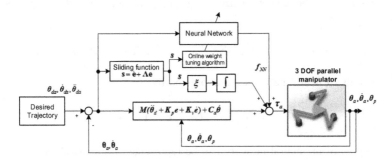

Fig. 3. Block diagram of the proposed adaptive extended computed torque controller.

In the Eq. (23), the sliding function s is defined as:

$$s = \dot{e} + \Lambda e = \dot{\theta}_a - (\theta_{da} - \Lambda e) = \dot{\theta}_a - \dot{\theta}_{ar} \tag{24}$$

Where $\Lambda = diag(\lambda_1, \lambda_2)$. λ_1 and λ_2 are positive constants which determine the motion feature in the sliding surface; and $\dot{\theta}_{ar} = \dot{\theta}_{da} - \Lambda e$ is defined as the reference velocity vector.

The architecture of the neural network used in this study includes the input layer, the hidden layer, and the output layer. The neural network has 3 outputs corresponding to 3 active joints of the considered parallel manipulators.

The Input Layer: The input vector of the NN is denoted by:

$$x = [x_1, x_2, \ldots, x_6]^T = [e_1, \dot{e}_1, e_2, \dot{e}_2, e_3, \dot{e}_3]^T \tag{25}$$

where $e_i = \theta_{ai} - \theta_{dai}$, $\dot{e}_i = \dot{\theta}_{ai} - \dot{\theta}_{dai}$, $i = 1, 2, 3$.

The Hidden Layer: By denoting the number of neurons in hidden layer as N_h, the weight matrix connecting the input and hidden layers is expressed by:

$$V = (v_1, v_2, \ldots, v_{Nh}) \in \Re^{6 \times N_h}, \ v_i = (v_{i1}, v_{i2}, \ldots, v_{iN_i})^T \in \Re^{N_i}, \ i = \overline{1, N_h} \tag{26}$$

The inputs and outputs of the hidden layer are respectively presented as:

$$net_i = \sum_{j=1}^{N_i} v_{ij} x_j \tag{27}$$

$$G_i = g(net_i), \quad i = \overline{1, N_h}. \tag{28}$$

The transfer function in the hidden layer is used as sigmoid function:

$$g(z) = \frac{1}{1 + \exp(-z)}. \tag{29}$$

The Output Layer: The weight matrix connecting the hidden and output layers is expressed by:

$$W = (w_1, w_2, \ldots, w_{N_h})^T \in \Re^{N_h \times 3}, w_i = (w_{i1}, w_{i2}) \in \Re^2, \ i = \overline{1, N_h} \tag{30}$$

The outputs of neural network are expressed by:

$$y_k = \sum_{i=1}^{N_h} w_{ik} G_i, \quad k = 1, 2, 3. \tag{31}$$

The outputs of the neural network can be represented in a vector form:

$$y = W^T G(x, V) \in \Re^{3 \times 1} \tag{32}$$

where $G = [G_1, G_2, \ldots, G_{N_h}]^T \in R^{N_h \times 1}$.

The online tuning laws for the neural network are designed based on the modified of back propagation algorithm [14] as the following:

$$\dot{\hat{W}} = -\eta \hat{G}s^T \tag{33}$$

$$\dot{\hat{V}} = -\mu x \left(G \hat{W} s \right)^T \tag{34}$$

where η and μ are the positive learning rates.

4 Simulation and Results

To illustrate the proposed adaptive extended computed torque controller in this paper, a simulation example is performed for a 3 DOF planar parallel manipulator. Simulation studies were conducted on Matlab-Simulink and the mechanical model of the parallel manipulator was built on SolidWorks and exported to Simmechanics toolbox. By this way, the mechanical model is almost the same with the real model. The comparisons between the performance of the conventional extended computed torque controller and the proposed adaptive extended computed torque controller were performed.

Table 1. Parameters for the parallel manipulator ($i = 1, 2, 3$)

Links	l_i (m)	L_{ci} (m)	m_i (kg)	I_i (kg.m^2)
Active link	$l_1 = 0.4$	0.2	5.118	9139×10^{-6}
Passive link	$l_2 = 0.6$	0.3	7.39	26763×10^{-6}
End-effector	$l_3 = 0.2$	–	1.84	3170×10^{-6}

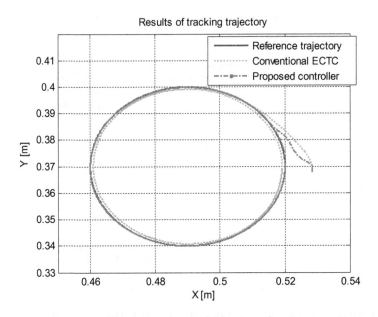

Fig. 4. Results of tracking control

The link parameters of the 3 DOF planar parallel manipulator used in simulations were presented in Table 1, where l_i is length of link i^{th}, l_{ci} is distance from the joint to the center of mass for i^{th} link, m_i is mass of link i^{th}, and I_i is inertia of link i^{th} ($i = 1, 2, 3$). The gains of the conventional extended computed torque controller and the proposed adaptive controller were chosen as $K_p = 150 \times I_{3\times3}$, $K_v = 25 \times I_{3\times3}$ where $I_{3\times3}$ is the identity matrix with dimension equal to 3×3.

In the proposed adaptive extended computed torque controller (23), the number of hidden neurons is $N_h = 10$, and the learning rates are chosen as: $\mu = 0.0018$, $\eta = 0.0013$. All values of the initial weights of the feed-forward neural network are set to 10^{-4}. The design parameters of the error compensator are set as: $\Gamma = diag\{12, 10, 8\}$.

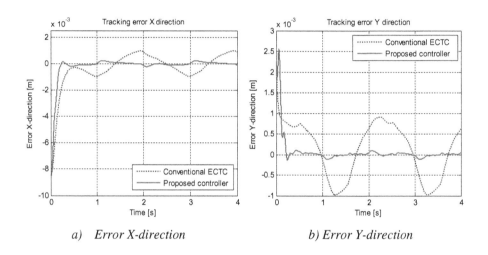

a)   *Error X-direction*   b) *Error Y-direction*

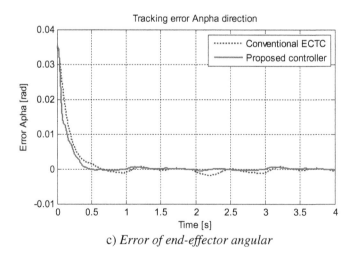

c) *Error of end-effector angular*

Fig. 5. Comparison of tracking errors: (a) X-direction, (b) Y-direction, (c) End-effector angular.

Figure 4 shows the comparison of the performance of trajectory tracking between the conventional extended computed torque controller and the proposed adaptive extended computed torque controller using neural network and error compensator. Because the initial position of the end effector of mechanical model does not lie on the reference circle, the actual trajectory of robot in simulation has a glitch shown at the beginning of motion.

The results of tracking errors of the end-effector on the X-direction, Y-direction and the end-effector angular are shown in Fig. 5. The dash lines are the results of conventional extended computed torque controller case, and the solid lines are the results of the proposed controller case. It can be seen that the proposed controller bring about smaller tracking errors. These results prove the efficiency of the control performance of the proposed adaptive extended computed torque controller in comparison to the conventional extended computed torque controller.

It could be concluded from the above-mentioned simulation results that the proposed adaptive extended computed torque controller using neural network and error compensators is of high efficiency for the control of a 3 DOF planar parallel manipulator.

5   Conclusion

In this paper, an adaptive extended computed torque controller using neural network and error compensator is proposed in order to improve the control performance of a 3DOF planar parallel manipulator. The proposed controller is based on the combination of nominal dynamic model of the parallel manipulator, a feed-forward neural network and error compensators. The feed-forward neural network and error compensators work together for compensating modeling error and uncertainties. The self-tuning laws of the feed-forward are based on sliding functions and self-tuned online during the trajectory tracking control of the manipulator without any offline training phase. Through the control simulations of the 3 DOF planar parallel manipulator tracking a circular trajectory, it can be seen that the proposed adaptive extended computed torque controller leads to get better performances in comparison with conventional extended computed torque controller.

Acknowledgement. Following are results of a study on the "Leaders Industry-university Cooperation" Project, supported by the Ministry of Education (MOE).

References

1. Shang, W., Cong, S.: Nonlinear computed torque control for a high-speed planar parallel manipulator. Mechatronics **19**, 987–992 (2009)
2. Weiwei, S., Shuang, C., Yaoxin, Z., Yanyang, L.: Active joint synchronization control for a 2-DOF redundantly actuated parallel manipulator. IEEE Trans. Control Syst. Technol. **17**, 416–423 (2009)
3. Shang, W., Cong, S.: Nonlinear adaptive task space control for a 2-DOF redundantly actuated parallel manipulator. Nonlinear Dyn. **59**, 61–72 (2010)

4. Hui, C., Yiu-Kuen, Y., Zexiang, L.: Dynamics and control of redundantly actuated parallel manipulators. IEEE/ASME Trans. Mechatron. **8**, 483–491 (2003)
5. Davliakos, I., Papadopoulos, E.: Model-based control of a 6-DOF electrohydraulic Stewart-Gough platform. Mech. Mach. Theory **43**, 1385–1400 (2008)
6. Yu, H.: Modeling and control of hybrid machine systems — a five-bar mechanism case. Int. J. Autom. Comput. **3**, 235–243 (2006)
7. Le, T.D., Kang, H.-J., Suh, Y.-S., Ro, Y.-S.: An online self-gain tuning method using neural networks for nonlinear PD computed torque controller of a 2-DOF parallel manipulator. Neurocomputing **116**, 53–61 (2013)
8. Llama, M.A., Kelly, R., Santibanez, V.: Stable computed-torque control of robot manipulators via fuzzy self-tuning. IEEE Trans. Syst. Man Cybern. Part B Cybern. **30**, 143–150 (2000)
9. Zubizarreta, A., Marcos, M., Cabanes, I., Pinto, C., Portillo, E.: Redundant sensor based control of the 3RRR parallel robot. Mech. Mach. Theory **54**, 1–17 (2012)
10. Zubizarreta, A., Cabanes, I., Marcos, M., Pinto, C., Portillo, E.: Extended CTC control for parallel robots. In: 2010 IEEE Conference on Emerging Technologies and Factory Automation (ETFA), pp. 1–8 (2010)
11. Bengoa, P., Zubizarreta, A., Cabanes, I., Mancisidor, A., Portillo, E.: A stable model-based control scheme for parallel robots using additional sensors. In: 2015 IEEE/RSJ International Conference on Intelligent Robots and Systems (IROS), pp. 3170–3175 (2015)
12. Baron, L., Angles, J.: The kinematic decoupling of parallel manipulators using joint-sensor data. IEEE Trans. Rob. Autom. **16**, 644–651 (2000)
13. Zubizarreta, A., Marcos, M., Cabanes, I., Pinto, C.: A procedure to evaluate Extended Computed Torque Control configurations in the Stewart-Gough platform. Rob. Auton. Syst. **59**, 770–781 (2011)
14. Le, T.D., Kang, H.-J., Suh, Y.-S.: Chattering-free neuro-sliding mode control of 2-DOF planar parallel manipulators. Int. J. Adv. Rob. Syst. **10**(22), 1–15 (2013)

Exploiting Twitter Moods to Boost Financial Trend Prediction Based on Deep Network Models

Yifu Huang[1], Kai Huang[1], Yang Wang[2], Hao Zhang[1], Jihong Guan[3],
and Shuigeng Zhou[1(✉)]

[1] School of Computer Science, and Shanghai Key Lab of Intelligent Information
Processing, Fudan University, Shanghai 200433, China
{huangyifu,kaihuang14,15110240009,
sgzhou}@fudan.edu.cn
[2] School of Software, Jiangxi Normal University, Nanchang 330022, China
yang1995t@163.com
[3] Department of Computer Science and Technology,
Tongji University, Shanghai 201804, China
jhguan@tongji.edu.cn

Abstract. Financial trend prediction is an interesting but also challenging research topic. In this paper, we exploit Twitter moods to boost next-day financial trend prediction performance based on deep network models. First, we summarize six-dimensional society moods from Twitter posts based on the profile of mood states Bipolar lexicon expanded by WordNet. Then, we combine Twitter moods and financial index by Deep Network models, and propose two methods. On the one hand, we utilize a Deep Neural Network of good fitting capability to evaluate and select predictive Twitter moods; On the other hand, we use a Convolutional Neural Network to explore temporal patterns of financial data and Twitter moods through convolution and pooling operations. Extensive experiments over real datasets are carried out to validate the performance of our methods. The results show that Twitter mood can improve prediction performance under the deep network models, and the Convolutional Neural Network based method performs best on most cases.

Keywords: Financial trend prediction · Twitter mood · Deep neural network · Convolutional neural network

1 Introduction

Financial trend prediction, aiming at determining the future value of financial index based on computational methods, is an interesting but also challenging research topic. The challenges of financial trend prediction lie in exploring the major factors behind trend changing and building powerful prediction models, to achieve satisfactory prediction accuracy. In the early years of financial trend prediction, researchers focused on the methods based on Technical Analysis (TA) [21]. According to TA, the future of financial index is established on the past. So we can build models by analyzing the

© Springer International Publishing Switzerland 2016
D.-S. Huang et al. (Eds.): ICIC 2016, Part III, LNAI 9773, pp. 449–460, 2016.
DOI: 10.1007/978-3-319-42297-8_42

financial data of the past, and then make prediction for the future. On the other hand, Efficient Markets Hypothesis (EMH) [9] claims that the stock price is the result of all available information. According to EMH, the price is determined by the recent information rather than the past data.

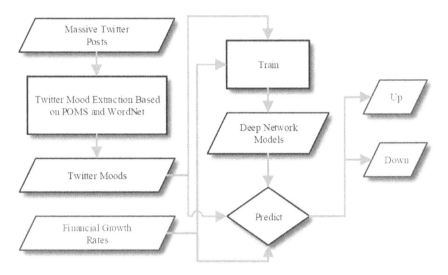

Fig. 1. The workflow of our approach.

Nowadays, Behavior Economics (BE) [6] becomes hot in the area of financial trend prediction. In the perspective of BE, financial trend is influenced by society moods. Because the mood state of an individual investor may influence individual economical behavior, and the mood state of the entire society may aggregately influence the market behavior. BE-based prediction methods had not made significant progress in the past years. Only recently, social media such as Twitter, Facebook and Pinterest are widely used by the entire society. For example, it was estimated that the number of active Twitter users is more than 300 million at the beginning of the 2015 [3]. People share their thoughts, feelings and even particulars of their daily lives in text messages through social platforms. And these huge amounts of text data are the perfect source of society moods.

In this paper, we propose to improve financial trend prediction performance with Twitter moods based on deep network models. To the best of our knowledge, this is the first attempt to exploit Twitter moods to boost financial trend prediction based on deep network models. More specifically, we first extract six-dimensional society moods from Twitter posts based on the profile of mood states (POMS) [18] Bipolar lexicon expanded by WordNet [19]. These Twitter moods are composed-anxious, agreeable-hostile, elated-depressed, confident-unsure, energetic-tired and clearheaded-confused. Then, we combine Twitter moods and financial index by deep network models, and develop two methods. One is based on Deep Neural Network (DNN), which is utilized to evaluate and select predictive Twitter moods by exploiting DNN's good fitting capability.

The other is based on Convolutional Neural Network (CNN), where we take the advantage of CNN to explore temporal patterns hidden in financial data and Twitter moods through convolution and pooling operations. Figure 1 illustrates the workflow of our approach.

We conduct extensive experiments over real datasets. The results show that Twitter moods can indeed improve prediction performance under deep network models, and the CNN-based method performs best on most cases.

The rest of this paper is organized as follows: Sect. 2 reviews the related works, and pinpoints the differences between our work and the most relevant ones. Section 3 elaborates the details of our DNN-based and CNN-based prediction methods. Section 4 presents the experimental results. Finally, Sect. 5 concludes the paper, and highlights some future directions.

2 Related Work

Financial trend prediction has been an active research topic in academia. Various prediction methods have been proposed under different frameworks, including Technical Analysis (TA), Efficient Markets Hypothesis (EMH) and Behavior Economics (BE). Here, we review the related works from two different dimensions: prediction models and additional data used in prediction. Table 1 shows the works using different prediction models, and Table 2 lists the works using different complementary data (in addition to historical financial data).

From Table 1, we can see that many related works focus on Support Vector Machine (SVM), Hidden Markov Model (HMM) and Neural Network (NN). Because these models are complex enough to learn the patterns in the historical data. From Table 2, we can see that several news-driven methods have been proposed, in which social media based methods appeared in recent years. There is only one paper that uses

Table 1. The related works arranged in prediction model dimension.

Prediction model	Related works
Vector Auto Regression	[26, 27]
Matrix Factorization	[20]
Support Vector Machine	[14, 16, 17, 24, 30]
Hidden Markov Model	[12, 22, 23, 25, 32]
Neural Network	[4, 5, 7, 8, 10, 29]

Table 2. The related works arranged in additional data dimension.

Additional data	Related works
None	[10, 17, 22, 23, 25, 32]
Trading Records	[29]
News	[7, 8, 14, 16, 20, 24, 30]
Social Media	[4, 5, 12, 26, 27]

trading records. As trading records are not publicly available data, such work is very difficult to follow. In what follows, we discuss in details the works most relevant to our methods in this paper.

Our previous work [12] combined Twitter moods and selective HMM to predict the next-day trend. We first extracted and evaluated Twitter mood, then extended the selective HMM so that Twitter moods and historical financial data can be combined to train the model. Finally, we make prediction based on the trained model. Different from our previous work, in this paper we focus on utilizing DNN and CNN to weight Twitter moods and financial growth rate automatically to further improve prediction performance.

Ding et al. [7] exploited news data to predict the trend of S&P500. They used Open Information Extraction to extract structured entity-relationships from news, called events. And then they combined news events and historical S&P500 data by a DNN to make prediction. Recently, they proposed prediction models based on Deep Learning [8]. They first used Neural Tensor Networks to make news events denser, then built prediction models based on CNN to further improve prediction accuracy of S&P500. The difference between our methods and Ding et al.'s is that we use Twitter moods as additional data, instead of news events.

3   Methods

Here, we present two methods to combine Twitter moods and historical financial data by DNN and CNN, respectively.

3.1   The DNN-Based Method

Multilayer Perceptron (MLP) containing more than three layers with nonlinear activation is called Deep Neural Network (DNN) [11]. A DNN can be represented as a directed acyclic graph with multiple layers. The nodes of two neighboring layers are fully connected and forward activated. Nonlinear activation functions are used in these nodes, so DNN can discriminate data that cannot be classified linearly. DNN is a supervised learning model, which is trained with labeled training samples (X, Y). We can formalize a DNN as a quintuple $\lambda = \{L, N, W, B, \sigma\}$, where L denotes the number of layers, N denotes the number of nodes in each layer, W denotes the parameters of connections, B denotes the bias terms, and σ denotes the activation functions. The training of DNN is called backward propagation, and the prediction of DNN is called forward propagation. Our DNN-based method is illustrated in Fig. 2.

We use growth rates of financial index and growth rates of Twitter moods in the last week as the input of DNN. Concretely, financial index data consists of open price, close price, highest price, lowest price and trading volume. Twitter mood data consists of composed-anxious, agreeable-hostile, elated-depressed, confident-unsure, energetic-tired and clearheaded-confused. We retrieve Twitter moods as in our previous work [12]. The number of nodes in the input layer is *The number of day* $\times$ (*the number of financial index data + the number of Twitter mood data*) $= 7 \times (5 + 6) = 77$.

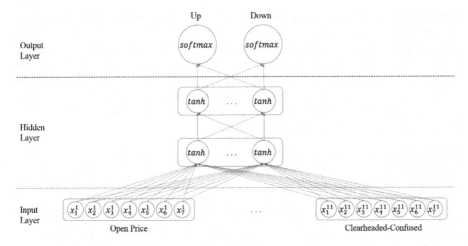

Fig. 2. The DNN-based method.

The changing trend of next-day close price is the output of DNN. Concretely, we consider two kinds of price changing trend: **UP** and **DOWN**. So the node number of the output layer is 2.

For nodes in the hidden layers, we use nonlinear *tanh* function as activation function. Such an activation function provides inputs of 0 average for the next layer, and achieves good convergence in practice. Because nodes from neighboring layers are fully connected, there are a large number of parameters in DNN. On the one hand, these parameters are the foundation of the fitting power of DNN; On the other hand, they may also cause overfitting. We follow the method of [28] to use Dropout to prevent overfitting. The main idea of Dropout is like this: we randomly remove nodes and their connections in the training process of DNN to prevent nodes co-adaptation.

For nodes in the output layer, we use nonlinear *softmax* function as the activation function. It is a generalization of logistic regression, and has shown good fitting capability in practice.

Finally, we use Mean Square Error as the cost function, and use Stochastic Gradient Descent as the optimization method in the training process.

3.2   The CNN-Based Method

DNN is not easy to train with high dimensional data, because it has a large number of parameters. Convolutional Neural Networks (CNN) [13] is proposed to solve this problem. In CNN, only some of input nodes are relevant to the hidden nodes, so it associates a small contiguous region of the input data with the nodes in hidden layers. The idea of this local connected network is inspired by the visual system, that is, neurons in visual cortex have localized receptive fields.

Comparing to DNN, CNN has two specific processes:

Convolution. There is statistic invariance in the input data, and the statistic invariance of a part of data is the same as that of another part. In other words, the features we learn from one part of data are applicable to another part. Convolution is to extract high-level features from a small contiguous region. More specifically, we sample some $a \times b$ size regions X_{small} from the $r \times c$ size input data X_{large}. We train a spare autoencoder by using X_{small} to get k convolution features. Then, we use the trained sparse autoencoder to extract convolution features from all $a \times b$ regions in X_{large}. Finally, we get $k \times (r - a + 1) \times (c - b + 1)$ convolution features.

Pooling. Convolution transforms a large input data to multiple small convolution features. But the total number of convolution features are too large to train efficiently. Considering the statistic invariance of input data, the features of a part of data is the same as that of another part. So we can use pooling to extract aggregated features and these features are invariant even the input data get panned. Formally, we first define pooling region $m \times n$. Then, we spilt the convolution features into multiple $m \times n$ disjoint regions. Finally, we compute the max or average aggregated features in these regions to get the pooling features.

Convolution and pooling both take advantage of statistic invariance in the input data. They can not only reduce the dimensionality of the original input data, but also extract higher-level features. After convolution and pooling, we flatten the features first and then concatenate them with DNN. The training and prediction of CNN are similar to that of DNN. Our CNN-based method is illustrated in Fig. 3.

The input data of CNN are similar to the input data we use for DNN in the last section. It contains 5 financial index features and 6 Twitter mood features within 7 days. DNN treats these 77 features independently, so the temporal relevance among features is neglected.

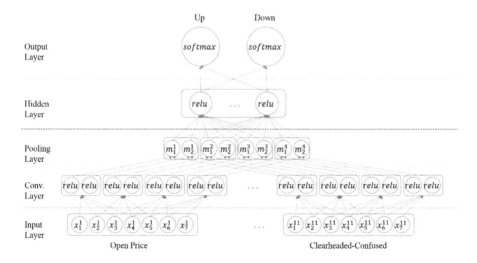

Fig. 3. The CNN-based method.

So we perform convolution operation on the time dimension of each feature. In other words, for every contiguous region of each feature, we extract local features via sliding windows. Furthermore, to obtain the most useful local features, we add a max pooling layer on the top of the convolution layer, which enables the network to get the most useful local features from the output of the convolution layer.

After convolution and pooling, we flatten the high level features as one dimensional features, and then we concatenate them into a forward neural network that contains a hidden layer. Similarly, the output layer of the network generates the changing trend of next-day close price, i.e., **UP** or **DOWN**. So, the node number of the output layer is 2.

We use the *relu* activation function in the nodes of the convolution layer and the pooling layer, and use the *softmax* activation function in the nodes of the final output layer. We use Categorical Cross Entropy as the cost function, and use *adadelta* to optimize the training process.

4   Experimental Evaluation

We present the experimental evaluation results in this section. We first introduce the experimental datasets and environment, then check the effects of some parameters on the performance of our methods. Finally, we compare the performance of our methods with that of some existing methods, and discuss the experimental results.

4.1   Experimental Setup

We use two Twitter datasets for performance evaluation. The first dataset is obtained from [31]. It contains 467 million Twitter posts that were published on Twitter in a seven-month period from Jun. 2009 to Dec. 2009. We denote it Twitter2009. The second dataset is from [15], which contains 50 million tweets that cover a 20 month period from Jan. 2010 to Aug. 2011. We denote it Twitter2011. The splitting of datasets for training and predicting is presented in Table 3. The financial data used are S&P500 Index, NYSE Composite Index, DJIA Index and NASDAQ Composite Index from Yahoo! Finance. We use all the four indexes to train the DNN-based method and the CNN-based method. Our goal is to predict the changing trend of the next-day close S&P500 Index and NYSE Composite Index. The proposed methods are implemented by using Keras [1] library that supports fast prototyping of deep learning models based on Theano [2].

Table 3. Twitter dataset splitting scheme.

Dataset	Train	Predict
Twitter2009	2009/06/12-2009/11/05	2009/11/06-2009/12/21
	100 days	30 days
Twitter2011	2010/01/04-2011/03/16	2011/03/17-2011/07/26
	300 days	90 days

In our experiments, we compare five configurations of our methods:

1. *DNN*: the DNN-based method using only historical financial data.
2. *DNN*$_{agr}$: the DNN-based method using close price and agreeable-hostile Twitter mood in the last three days.
3. *DNN*$_{all}$: the DNN-based method using historical financial data and all Twitter moods.
4. *CNN*: the CNN-based method using only historical financial data.
5. *CNN*$_{all}$: the CNN-based method using historical financial data and all Twitter moods.

4.2   Parameter Effect

Effect of the number of hidden layers of *DNN*$_{all}$. We set the number of hidden layers from 1 to 3 to run *DNN*$_{all}$, the results are given in Table 4. From Table 4, we can see that *DNN*$_{all}$ achieves the best prediction accuracy when the number of hidden layers is 3. For example, with Twitter2011 and **NYSE**, the error rates of *DNN*$_{all}$ are 42.667 %, 42.167 %, 40.667 % when the number of the hidden layers is 1, 2 and 3, respectively. This is because that more hidden layers can uncover more complex relationships between input and output.

Table 4. Performance of *DNN*$_{all}$ with different numbers of hidden layers.

#Hidden layers	Error rate (%)			
	Twitter2009		Twitter2011	
	S&P500	**NYSE**	**S&P500**	**NYSE**
1 layer	45.900	47.233	46.000	42.667
2 layers	44.667	46.000	44.933	42.167
3 layers	**44.444**	**41.111**	**44.667**	**40.667**

Effect of the number of convolution layers of *CNN*$_{all}$. We set the number of convolution layers from 1 to 3 to run *CNN*$_{all}$, and the results are presented in Table 5. From Table 5, we can see that for Twitter2009 and Twitter2011, *CNN*$_{all}$ achieves the best when the number of convolution layers is 1 and 2, respectively. For example, with Twitter2009 and **NYSE**, the error rates of *CNN*$_{all}$ are 43.933 %, 46.000 %, 46.667 % when convolution layers are 1, 2 and 3. With Twitter2011 and **NYSE**, the error rates of *CNN*$_{all}$ are 41.000 %, 39.733 %, 42.222 % when the number of convolution layers is 1, 2 and 3, respectively. It seems that an appropriate number of convolution layers should be set to achieve the best performance, and a large number of convolution layers may cause overfitting, which results in worse performance.

Table 5. Performance of CNN_{all} with different numbers of convolution layers.

#Convolution layers	Error rate (%)			
	Twitter2009		Twitter2011	
	S&P500	**NYSE**	**S&P500**	**NYSE**
1 layer	**43.933**	**43.933**	46.667	41.000
2 layers	45.333	46.000	**43.433**	**39.733**
3 layers	48.667	46.667	47.333	42.222

4.3 Prediction Performance Comparison

We compare our methods (with five configurations) with two existing methods [12]:

1. HMM_{agr}: the HMM-based method using close price and agreeable-hostile Twitter mood in the last three days.
2. SVM_{agr}: the SVM-based method using close price and agreeable-hostile Twitter mood in the last three days.

Here, we do not compare our methods with that proposed by Ding et al. [7], the reason is that they used news events, instead of moods of social media, as additional data. Table 6 presents the prediction performance results of these methods.

From Table 6, we can see that the CNN-based method CNN_{all}, which combines all Twitter moods and historical financial data, achieves the best results on more cases. Concretely, comparing to the other methods, CNN_{all} performs better on 2 of the total 4 cases. The reason is two-fold: on the one hand, deep network models have strong fitting power; on the other hand, the convolution layer and the pooling layer of CNN extract higher level features.

By comparing DNN_{all} and DNN, CNN_{all} and CNN, we can see that DNN_{all} and CNN_{all} perform better. For example, with Twitter2011 and **NYSE**, the error rates of DNN_{all} and DNN are 40.667 % and 47.778 %, and the error rates of CNN_{all} and CNN are 39.733 % and 41.111 %. This shows that Twitter moods can indeed improve financial trend prediction under deep network models. Furthermore, by comparing DNN_{all} and DNN_{agr}, we can see that the performance of DNN_{all} is better. For example, with Twitter2011 and **NYSE**, the error rates of DNN_{all} and DNN_{agr} are 40.667 % and 47.233 %, respectively. This is possibly because the combinatory prediction power of multiple Twitter moods, and DNN can learn appropriate weights for each kind of Twitter moods. So it is better to use multiple Twitter moods than to select just one kind of Twitter moods via Granger Causality Analysis as in our previous work [12].

We also notice that HMM_{agr} and SVM_{agr} perform better than our methods on **S&P500** with Twitter2009, the major reason is that the size of our training data prevents us training large enough neural networks to exert the potential of deep neural networks.

Table 6. Performance comparison of our methods.

Model	Error Rate (%)			
	Twitter2009		Twitter2011	
	S&P500	**NYSE**	**S&P500**	**NYSE**
HMM_{agr}	**36.667**	46.667	44.444	54.444
SVM_{agr}	40.000	46.667	50.000	44.444
DNN	47.778	48.889	46.667	47.778
DNN_{agr}	48.900	49.767	47.133	47.233
DNN_{all}	44.444	**41.111**	44.667	40.667
CNN	45.556	46.667	50.000	41.111
CNN_{all}	43.933	43.933	**43.433**	**39.733**

5   Conclusion

In this paper, we propose two methods to boost financial trend prediction performance with Twitter moods based on deep network models. Twitter moods are extracted from Twitter posts based on a POMS Bipolar lexicon expanded by WordNet. Both DNN and CNN are used to combine Twitter moods and financial index. We make use of DNN's strong fitting power to evaluate and select predictive Twitter moods, and take advantage of CNN to explore the temporal patterns among financial data and Twitter moods through convolution and pooling operations. Experimental results show that Twitter mood is helpful in improving prediction performance under deep network models, and the CNN-based method achieves the best performance on most cases.

For future work, on the one hand, we plan to consider both social media and financial news; on the other hand, we will try other DNN models, such as recursive neural networks.

Acknowledgement. This work was partially supported by the Key Projects of Fundamental Research Program of Shanghai Municipal Commission of Science and Technology under grant No. 14JC1400300. Jihong Guan was partially supported by the Program of Shanghai Subject Chief Scientist.

References

1. Keras: Deep learning library for theano and tensorflow. http://keras.io/
2. Theano is a python library that allows you to define, optimize, and evaluate mathematical expressions involving multi-dimensional arrays efficiently. http://www.deeplearning.net/software/theano/
3. Twitter: number of monthly active users 2010–2015. http://www.statista.com/statistics/282087/number-of-monthly-active-twitter-users/
4. Bollen, J., Mao, H.: Twitter mood as a stock market predictor. Computer **44**(10), 91–94 (2011)

5. Bollen, J., Mao, H., Zeng, X.: Twitter mood predicts the stock market. J. Comput. Sci. **2**(1), 1–8 (2011)
6. Camerer, C.F., Loewenstein, G., Rabin, M.: Advances in Behavioral Economics. Princeton University Press, New Jersey (2011)
7. Ding, X., Zhang, Y., Liu, T., Duan, J.: Using structured events to predict stock price movement: an empirical investigation. In: Proceedings of the 2014 Conference on Empirical Methods in Natural Language Processing, EMNLP 2014, Doha, Qatar, A meeting of SIGDAT, A Special Interest Group of the ACL, 25–29 October 2014, pp. 1415–1425 (2014)
8. Ding, X., Zhang, Y., Liu, T., Duan, J.: Deep learning for event-driven stock prediction. In: Proceedings of the 24th International Conference on Artificial Intelligence, pp. 2327–2333. AAAI Press (2015)
9. Fama, E.F.: The behavior of stock-market prices. J. Bus. **38**(1), 34–105 (1965)
10. Giacomel, F., Pereira, A.C., Galante, R.: Improving financial time series prediction through output classification by a neural network ensemble. In: Chen, Q., Hameurlain, A., Toumani, F., Wagner, R., Decker, H. (eds.) DEXA 2015. LNCS, vol. 9262, pp. 331–338. Springer, Heidelberg (2015)
11. Hinton, G.E., Osindero, S., Teh, Y.W.: A fast learning algorithm for deep belief nets. Neural Comput. **18**(7), 1527–1554 (2006)
12. Huang, Y., Zhou, S., Huang, K., Guan, J.: Boosting financial trend prediction with twitter mood based on selective hidden Markov models. In: Renz, M., Shahabi, C., Zhou, X., Chemma, M.A. (eds.) DASFAA 2015. LNCS, vol. 9050, pp. 435–451. Springer, Heidelberg (2015)
13. LeCun, Y., Bottou, L., Bengio, Y., Haffner, P.: Gradient-based learning applied to document recognition. Proc. IEEE **86**(11), 2278–2324 (1998)
14. Li, Q., Jiang, L., Li, P., Chen, H.: Tensor-based learning for predicting stock movements. In: Twenty-Ninth AAAI Conference on Artificial Intelligence (2015)
15. Li, R., Wang, S., Deng, H., Wang, R., Chang, K.C.C.: Towards social user profiling: unified and discriminative influence model for inferring home locations. In: Proceedings of the 18th ACM SIGKDD International Conference on Knowledge Discovery and Data Mining, pp. 1023–1031 (2012)
16. Li, X., Wang, C., Dong, J., Wang, F., Deng, X., Zhu, S.: Improving stock market prediction by integrating both market news and stock prices. In: Hameurlain, A., Liddle, S.W., Schewe, K.-D., Zhou, X. (eds.) DEXA 2011, Part II. LNCS, vol. 6861, pp. 279–293. Springer, Heidelberg (2011)
17. Lin, Y., Guo, H., Hu, J.: An SVM-based approach for stock market trend prediction. In: The 2013 International Joint Conference on Neural Networks, pp. 1–7 (2013)
18. Mcnair, D., Lorr, M., Droppleman, C.: Profile of mood states. Educational & Industrial Testing Service, San Diego (1971)
19. Miller, G.A.: Wordnet: a lexical database for english. Commun. ACM **38**(11), 39–41 (1995)
20. Ming, F., Wong, F., Liu, Z., Chiang, M.: Stock market prediction from WSJ: text mining via sparse matrix factorization. In: 2014 IEEE International Conference on Data Mining (ICDM), pp. 430–439. IEEE (2014)
21. Murphy, J.J.: Technical analysis of the financial markets: A comprehensive guide to trading methods and applications. New York Institute of Finance, New York (1999)
22. Pidan, D.: Selective Prediction with Hidden Markov Models. Master's thesis, Technion (2013)
23. Pidan, D., El-Yaniv, R.: Selective prediction of financial trends with hidden Markov models. In: Advances in Neural Information Processing Systems, pp. 855–863 (2011)
24. Schumaker, R.P., Chen, H.: A discrete stock price prediction engine based on financial news. Computer **43**(1), 51–56 (2010)

25. Shi, S., Weigend, A.S.: Taking time seriously: Hidden markov experts applied to financial engineering. In: Proceedings of the IEEE/IAFE 1997 Computational Intelligence for Financial Engineering (CIFEr), pp. 244–252. IEEE (1997)
26. Si, J., Mukherjee, A., Liu, B., Li, Q., Li, H., Deng, X.: Exploiting topic based twitter sentiment for stock prediction. In: Proceedings of the 51st Annual Meeting of the Association for Computational Linguistics (Volume 2: Short Papers), pp. 24–29 (2013)
27. Si, J., Mukherjee, A., Liu, B., Pan, S.J., Li, Q., Li, H.: Exploiting social relations and sentiment for stock prediction. In: Proceedings of the 2014 Conference on Empirical Methods in Natural Language Processing, EMNLP 2014, Doha, Qatar, A meeting of SIGDAT, A Special Interest Group of the ACL, 25–29 October 2014, pp. 1139–1145 (2014)
28. Srivastava, N., Hinton, G., Krizhevsky, A., Sutskever, I., Salakhutdinov, R.: Dropout: A simple way to prevent neural networks from overfitting. J. Mach. Learn. Res. **15**(1), 1929–1958 (2014)
29. Sun, X.Q., Shen, H.W., Cheng, X.Q.: Trading network predicts stock price. Sci. Rep. **4** (3711), 1–6 (2014)
30. Xie, B., Passonneau, R.J., Wu, L., Creamer, G.G.: Semantic frames to predict stock price movement. In: Proceedings of the 51st Annual Meeting of the Association for Computational Linguistics, pp. 873–883 (2013)
31. Yang, J., Leskovec, J.: Patterns of temporal variation in online media. In: Proceedings of the Fourth ACM International Conference on Web Search and Data Mining, pp. 177–186 (2011)
32. Zhang, Y.: Prediction of financial time series with Hidden Markov Models. Master's thesis, Simon Fraser University (2004)

Terrain Recognition for Smart Wheelchair

Shamim Al Mamun$^{(\boxtimes)}$, Ryota Suzuki, Antony Lam,
Yoshinori Kobayashi, and Yoshinori Kuno

Graduate School of Science and Engineering, Saitama University, Saitama, Japan
{shamim,suzuryo,antonylam,
yosinori,kuno}@cv.ics.saitama-u.ac.jp

Abstract. Research interest in robotic wheelchairs is driven in part by their potential for improving the independence and quality-of-life of persons with disabilities and the elderly. However the large majority of research to date has focused on indoor operations. In this paper, we aim to develop a smart wheelchair robot system that moves independently in outdoor terrain smoothly. To achive this, we propose a robotic wheelchair system that is able to classify the type of outdoor terrain according to their roughness for the comfort of the user and also control the wheelchair movements to avoid drop-off and watery areas on the road. An artificial neural network based classifier is constructed to classify the patterns and features extracted from the Laser Range Sensor (LRS) intensity and distance data. The overall classification accuracy is 97.24 % using extracted features from the intensity and distance data. These classification results can in turn be used to control the motor of the smart wheelchair.

Keywords: Surface · Feature extraction · Classification · ANFIS · LRS

1 Introduction and Motivation

In recent years, millions of people around the world have exhibited mobility impairments due to aging problems. Several robotic wheelchairs possessing user-friendly interfaces and/or autonomous functions for reaching a goal have been proposed to meet the needs of an aging society [1, 4]. There are two kinds of wheelchairs that are used for the elderly or disabled people. Manual wheelchairs are widely used around the world by those who are able to control the wheelchair by hand. But it needs extra labor to control the direction of the wheel. Powered wheelchairs are more convenient to run but need a high level of attention to control the motor using a joy stick or remote control. Furthermore, such conventional control interfaces are sometimes forbidden by doctors for elderly because of safety concerns. Thus the development of powered wheelchairs that can overcome doctors' safety concerns would be of great benefit to many. In fact, [3] found that around 3 million people could benefit from such wheelchairs. Moreover, elderly wheelchair users are often accompanied by caregivers or companions. Thus, in designing wheelchair technology it is important to consider how we can reduce the caregiver's workload and support communication between user and caregiver [5]. Smart wheelchairs are a potential solution to these issues, which

© Springer International Publishing Switzerland 2016
D.-S. Huang et al. (Eds.): ICIC 2016, Part III, LNAI 9773, pp. 461–470, 2016.
DOI: 10.1007/978-3-319-42297-8_43

could also have the option of granting many elderly wheelchair users the independence of controlling their wheelchairs on their own.

Smart wheelchairs can be either driven by the user or computer system that can follow the caregiver's movements to drive itself [6, 7]. Such a wheelchair follows the caregiver autonomously, by detecting the registered person in the system. But it is more important to detect the surface condition as well for comfortable navigation in outdoor environments. We have found that the wheelchair's terrain surfaces may be classified into four different categories. These are rough surfaces, watery places on the road, indoor plain surfaces, and dirt tracks. These roads may have different conditions which can be dangerous for the wheelchair user due to the risk of sudden accidents when driving over them. Suppose a wheelchair follows the caregiver but s/he misses hazards such as bumpy/pit areas. If the wheelchair were to not actively look for such hazards, it could fall and lead to injury of the occupant. Moreover, the motor speed of the wheelchair for different surface area should not be same. It must be carefully handled by the wheelchair itself. For this reason, detecting the type of surface is a necessity. Another noticeable thing is that, for smooth movement, a wheelchair needs some surface information like the degree of roughness, which helps the system to control the speed of the motor. Although several improvements have been achieved in navigation for indoor settings, only a few proposed solutions are able to achieve safety and smooth motor control by looking at the surface road conditions in outdoor environments. The paper [8] obtained surface information about irregularities using a 2D laser for decreasing the burden of human workers on the road. Moreover, recent projects include an MIT intelligent wheelchair [11] which is able to navigate in indoor environment using voice-commands. The authors of [12, 13] developed a wheelchair which is able to assist the user for some house hold work. In [14], a map based localization approach was taken for outdoor navigation and the same work was extended to classify streets and sidewalks using grass information. Paper [15] studied accessibility in urban terrains and they found common environmental barriers for moving wheelchairs to mainly consist of steep ramps, rough terrains, and sidewalk irregularities. It is worth mentioning that all autonomous wheelchairs employ distance laser sensors and depth cameras but intensity data is also available in the laser range sensor devices. This suggests that we can use all the available information provided by a laser sensor for effective indoor and outdoor navigational assistance to users. Specially, our proposed system detects surface roughness using depth distance and intensity data. The intensity data is first used for classification of the path type such as flat, uneven, dirt track and watery places on the road and then the wheelchair makes a decision to move-on, slow-down, stop or avoid accordingly. In this paper we used an ANFIS (Adaptive Neuro-Fuzzy Interference System) model for classification of surfaces and 3D elevation map (from distance data) generation for measuring the degree of roughness for our proposed wheelchair.

2 Surface Classification Methodology

The general block diagram of the whole procedure is shown in Fig. 1. We first, choose four types of common surfaces where the wheelchair to navigate. For surface rough-ness detection, we give extra attention to distinguish between indoor and outdoor

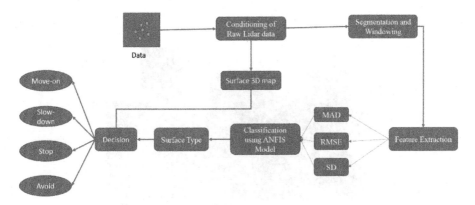

Fig. 1. General block diagram of the whole procedure

environments. In particular, we use the mean value of indoor roughness as a threshold to separate indoor and outdoor surfaces. We further categorize outdoor environments into three types, (1) dirt tracks, (2) asphalt roads, and (3) watery places on the road, which are shown in Fig. 6. For data observation, the wheelchair uses one ULM-30LX LRS in an orientation such that it can directly measure the distance to the surface as well as get the intensity value of any particular point in view. In our proposed system, the surface roughness is calculated using intensity data and corresponding angle of observation. The arrangement of the data collection tools in the wheelchair is illustrated in Figs. 2 and 3.

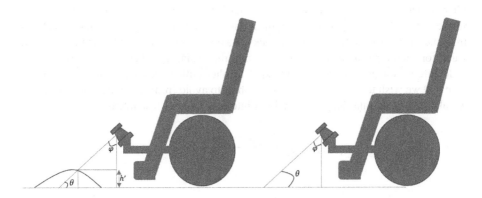

Fig. 2. Arrangement of laser for surface roughness data calculation

The pathways selected in our dataset consist of plain surfaces, uneven asphalt walkway road surfaces, muddy areas, and also observed water on the road. The data was taken from the front region of the wheelchair as shown in the shaded triangle in Fig. 3. The great considerable thing is that the midpoint of laser range sensor devicehas always a reflection point on the road's surfaces. So we consider the data around the

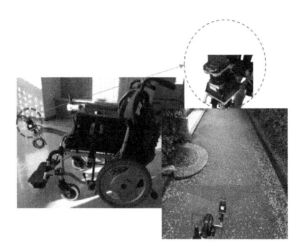

Fig. 3. Proposed system setup on wheelchair for data collection

midpoint segment for surface classification. The following subsections detail our surface detection approach.

2.1   Conditioning of Raw LRS Data

All acquired data per scan sequences including distance and intensity pair $\{inten_i, dist_i\}$ were passed through the chi-square (χ^2) test using Eq. 1. This test is essentially used to filter outlier data from the sensor so that we can more accurately classify the type of the surface. We first collect some prototypical data of flat indoor surfaces, uneven surfaces from the roads, watery areas and dirt tracks in a control environment to get the expected value for each type of surfaces. Thus we find the mean, μ_k, where, $k = \{surfacetype\}$ as feature vector that can be used in Eq. 1 to filter outlier sensor data for each surface type that we wish to detect. In addition, in order to avoid from any undesirable artifacts, we also filter out data that has very high intensity or very low intensity.

$$\chi_k^2 = \sum \frac{(Observeddata - \mu_k)^2}{\mu_k} \qquad (1)$$

2.2   Segmentation and Windowing

Using the entire LRS intensity data as input for pattern recognition and classification is not practical because the huge amount of data has high computational cost. To obtain the surface type the total scanned area is divided into small segments or patches. We define a "surface patch" S_k as a small region on the surface. Patch data block is 10×5 (data points by scanned time) which consists of intensity data and distance data points. The process of segmentation is repeated for checking whether the wheelchair is on a new surface or not. These patches are also used for generating a 3D surface map for measuring the roughness of the surface.

2.3 Surface 3D Map Generation

After preprocessing, we also create a patch wise 3D elevation map of the current surface to measure the roughness of the surface and to find the possible running path of the wheelchair. For this, the surface roughness is calculated from the mean of the average height of the surface compared to the prototypical surface. The height of each pixel is not same in the uneven surface hence we calculate the height of the patch, plot a 3D map, and get the average roughness of the road according to Eq. 2. Sample 3D maps from the first patch of different roads can be seen in Fig. 4

$$h' = (dist_{i+1} - dist_i) \times \sin(\theta) \tag{2}$$

where, i is the scan sequence.

Roughness, R_Z is calculated by Japanese industrial standard based on the five highest peaks and lowest valleys over the entire sampling length, which is formulated as Eq. 3.

$$R_{ZJIS} = \frac{1}{5}\sum_{i=0}^{5}\left(h'_{pi} - h'_{vi}\right) \tag{3}$$

The surface roughness parameter results for different surfaces have been calculated using Eq. 3. The lowest roughness value can be seen in indoor places with a value of

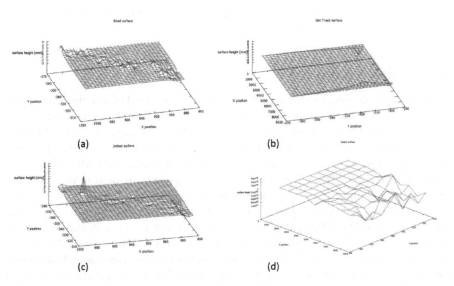

Fig. 4. 3Delevation map fromsurface patches: (a) Asphalt road, (b) Dirt track, (c) Indoor, and (d) Watery places.

27.8 and for road and dirt track and watery places we see 39.0, 85.5 and 43.2 respectively. Though this is not the most principaled way to find out the roughness of the surface, we find it provide enough information for effective use of smart wheelchairs moving through rough terrain.

2.4 Feature Extraction

Feature extraction methods use segmented data as feature vectors. The feature must contain enough information of the data and must be simple enough for fast training and classification. The feature extraction method we chose is based on the best findings in [9] in which the authors showed that the most effective feature extraction for our task uses the Mean Absolute Deviation (MAD), Standard Deviation (SD), and Root Mean Square Error (RMSE) shown in Table 1. In addition, we use other features such as mean, standard deviation, the minimum value and maximum value of each surface for classification.

Table 1. Feature values of different typs of surface data

	Watery		Road		Indoor		Dirt track	
	Intensity	Distance	Intensity	Distance	Intensity	Distance	Intensity	Distance
SD	90.3	8.5	53	6.7	25.8	3.2	230.5	12.1
MAD	70.4	6.7	41.2	5.5	19.3	2.5	183.6	10
RMS	2525.5	948.9	3072.6	948.3	3447.3	931.6	4553.6	911.9

2.5 Classification Model

There are many techniques introduced and proposed in the field of machine learning for object classification such as the Multi-Layer Perceptron (MLP), Fuzzy C-mean (FCM), Adaptive Neuro-fuzzy Interference System (ANFIS), and Support Vector Machine (SVM). Any of these various learning methods could be used but we found ANFIS to be both fast and accurate. Real time decision making is especially vital for a smart wheelchair. For the ANFIS, Singular Value Decomposition is used to determine the number of fuzzy rules. The membership parameters of the Takagi–Sugeno-type ANFIS are adjusted by using neural network. Figure 5 shows the ANFIS structure with three input features, MAD, RMS, and SD denoted by I_i, where $i = [1, 2, 3]$ and one output, denoted by Oi [16, 17]. Bell-shaped membership functions (MF) are used due to their advantage of being smooth, concise, and nonzero at all points. The required parameters for the input-output pairs are calculated using the hybrid back propagation and least mean squares (LMS) algorithms. The membership functions are derived from four

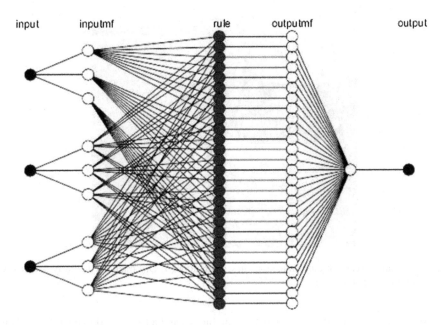

Fig. 5. Structure of adaptive neuro-fuzzy inference system using neuroFuzzyDesigner in Matlab with 3 inputs and one output.

fuzzy rules and are given appropriate names (low, medium, and high). The output is calculated by using Eq. 4.

$$O = \sum_{i=1}^{L} \left[\frac{\prod_{j=1}^{p} MF_i^j (I_i)(Z_i)}{\sum_{i=1}^{L} \left(\prod_{j=1}^{p} MF_i^j (I_i) \right)} \right] \tag{4}$$

where p is the number of linguistic variables and L is the number of rules in the rule base. The output of each rule, denoted by Zi, is given by Eq. 5:

$$Z_i = \alpha_{i1} I_1 + \alpha_{i2} I_2 + \ldots + \alpha_{in} I_n \tag{5}$$

Where α_{ij} is the consequent parameter of input j. The linguistic variable of the inputs is: MF_i^1: Low, MF_i^2: Medium and MF_i^3: High. The multiplexed output MF of the ANFIS is then fed to the speed controller program for controlling the wheelchair's motors.

3   Experiment and Results

In our experiment, we divided our terrain into four classes. All terrains were from the inside and outside of our research building. We chose indoor corridor, the outdoor asphalt walkway road connected to outside of the building, muddy places from a playground and watery spots after rain shows in Fig. 6.

Fig. 6. Example scene of data collection sites: (a) Indoor (b) Dirt track (c) walkway (d) watery places on the road after rain

We segmented our data into two separate blocks and a total of 2000 data were taken at a time. For one block of data, 1600 were used for training the ANFIS classifier and 400 completely different terrain data were used for testing. As the LRS gave a large amount of intensity and distance data in 200 ms of time interval, we took only the middle portion data as described earlier in the segmentation and windowing section. Individual test procedures were performed to interpret the contribution of each LRS data. The tests were validated using several parameters which constituted the overall classification accuracy and were measured as the closeness between the actual class and predicted class using the following equation [6]:

$$Accuracy = (TP + TN)/(TP + FP + FN + TN) \qquad (6)$$

where TP and FP are the number of true and false classifications of a particular surface respectively. TN and FN are the number of true and false classifications of a surface identified by other surface and the correct surface, respectively. Table 2 shows the confusion matrix of our classifier on the different classes of surfaces. It is satisfactory that the identification of each surface is 96 % on average except for watery places, which was perfectly classified. The misclassification rate is within the tolerance for the wheelchair to run through the surface. We note that [14] also achieved 97 % accuracy in classifying the difference between streets and sidewalks however they did not consider the multiclass case.In our case, we achieve similar accuracies on multiple classes and even 100 % on the watery case.

The roughness of the surface is calculated using Eqs. 2 and 3. Using this roughness information the wheelchair can choose its terrain to move on. In our test cases, the surface class is identified correctly and the wheelchair is able to make appropriate decisions for navigation.

Table 2. Confusion matrix for the pattern classification method (4 classes for each test surface)

	Watery	Road	Indoor	Dirt track
Watery	100 %	0	0	0
Road	4 %	96 %	0	0
Indoor	3 %	1 %	96 %	0
Dirt track	1 %	1 %	1 %	96 %

4 Conclusions and Future Work

This paper presents a method for terrain surface classification and roughness calculation using a single LRS sensor's intensity and distance data. The ANFIS classifier gave an overall accuracy 97.24 %. The combined results of classification and roughness are fed into the wheelchair motor controller for smooth running in urban settings. Moreover, the computed 3D surface elevation map could be used for making the decision of whether to move-on, slow down the speed or to avoid watery and muddy places in the road.

We are developing a wheelchair that can move on uneven surface by mechanically adjusted the positions of its wheel. We will implement the proposed method on this wheelchair so that it can autonomously detect surface conditions and move safely and smoothly.

Acknowledgments. This work was supported by Saitama Prefecture Leading-edge Industry Design Project and JSPS KAKENHI Grant Number 26240038.

References

1. Kuno, Y., Shimada, N., Shirai, Y.: Look where you're going: a robotic wheelchair based on the integration of human and environmental observations. IEEE Robot. Autom. **10**(1), 26–34 (2003)
2. Min, J., Lee, K., Lim, S., Kwon, D.: Human friendly interfaces of wheelchair robotic system for handicapped persons. In: Proceedings of the International Conference on Intelligent Robots and Systems (IROS), vol. 2, pp. 1505–1510 (2011)
3. Satoh, Y., Sakaue, K.: An omnidirectional stereo vision-based smart wheelchair. J. Video Process. **2007**, 87646 (2007)
4. Iwase, T., Zhang, R., Kuno, Y.: Robotic wheelchair moving with the caregiver. In: Proceedings of the SICE-ICASE International Joint Conference 2006, pp. 238–243 (2006)
5. Suzuki, R., Yamada, T., Arai, M., Sato, Y., Kobayashi, Y., Kuno, Y.: Multiple robotic wheelchair system considering group communication. In: Bebis, G., et al. (eds.) ISVC 2014, Part I. LNCS, vol. 8887, pp. 805–814. Springer, Heidelberg (2014)
6. Lv, J., et al.: Indoor slope and edge detection by using two-dimensional EKF-SLAM with orthogonal assumption. Int. J. Adv. Robot. Syst. **12**, 1–16 (2015)
7. Yamada, T., Ito, T., Ohya, A.: Detection of road surface damage using mobile robot equipped with 2D laser scanner. In: 2013 IEEE/SICE International Symposium on System Integration (SII). IEEE (2013)

8. Hamedi, M., et al.: Comparison of different time-domain feature extraction methods on facial gestures' EMGs. In: Electromagnetics Research Symposium Proceedings, PIER, KL (2012)
9. Rand, W.M.: Objective criteria for the evaluation of clustering methods. J. Am. Stat. Assoc. **66**(336), 846–850 (1971)
10. Hemachandra, S., Kollar, T., Roy, N., Teller, S.: Following and interpreting narrated guided tours. In: Proceedings of the International Conference on Robotics and Automation (ICRA), Shanghai, China, May 2011
11. Bostelman, R., Albus, J.: Sensor experiments to facilitate robot use in assistive environments. In: Proceedings of the International Conference on Pervasive Technologies Related to Assistive Environments, Athens, Greece, July 2008
12. Xu, J., Grindle, G., Salatin, B., Ding, D., Cooper, R.A.: Manipulability evaluation of the personal mobility and manipulation appliance (permma). In: International Symposium on Quality of Life Technology, Las Vegas, United States, June 2010
13. Montella, C., Pollock, M., Schwesinger, D., Spletzer, J.: Stochastic classification of urban terrain for smart wheelchair navigation. In: Proceedings of the IROS Workshop on Progress, Challenges and Future Perspectives in Navigation and Manipulation Assistance for Robotic Wheelchairs (2012)
14. Meyers, A.R., Anderson, J.J., Miller, D.R., Schipp, K., Hoenig, H.: Barriers, facilitators, and access for wheelchair users: sbstantive and methodologic lessons from a pilot study of environmental effects. Soc. Sci. Med. **55**(8), 1435–1446 (2002)
15. Kaiser, M.S., et al.: A neuro-fuzzy control system based on feature extraction of surface electromyogram signal for solar-powered wheelchair. J. Cogn. Comput. **8**(1), 1–9 (2016)
16. Mamun, S.A., Kaiser, M.S., Ahmed, M.R., Islam, M.S., Islam, M.I.: Performance analysis of optical wireless communication system employing neuro-fuzzy based spot-diffusing techniques. J. Commun. Netw. **5**(3), 260–265 (2013)

Research on Synergistic Selection Decision-Making of Manufacturing Enterprises and the Third Party Logistics in the Cluster Environment

Jin Tang[1], Chundong Guo[2], and Shiwen Zhao[2(✉)]

[1] School of Management and Economics, Beijing Institute of Technology,
Beijing 100081, China
[2] College of Economics and Management, Hebei University of Science
and Technology, Shijiazhuang 050000, China
guochundong007@163.com, zhaoshiwen138332@126.com

Abstract. Reasonable interest distribution and risk-sharing problem are known as the biggest obstacles for manufacturing enterprises and the third party logistics enterprises to choose whether to collaborate and evaluate the operation effect of the existing cooperative system. Firstly, based on the deeply exploration of the interest and risk relationship in the cluster environment, decision model of synergy profit distribution and risk sharing based on the Shapley value has been put forward; Secondly, this paper explores the 2-d curve fluctuations while the logistics cost is taken as the fixed variable and the value of interest distribution and risk quantification as the relative variable, and suggests the breakthrough point and the range of synergistic selection to help enterprises in decision-making; Finally, the case study has verified the fluctuation rule.

Keywords: Synergy · Shapley value · Interest distribution · Risk management

1 Introduction

Setting in the economic globalization, there are two seemingly contradict economic groups, one of which is third party logistics in favor of the geographic cluster and the other is in favor of the geographic expansion, having been found that it exist a combination way for them to promote the development in cluster environment. First of all, with the progress of traffic network and information technology, the cluster constantly take advantage of the similar economic individual to promote industrial clusters into specific economic integration, which has created new conditions and support for the development of regional economy. Meanwhile, the characteristics of the cluster environment determine the relationship between the competition and cooperation of the production enterprises which serve as cells to make up the specific industrial cluster. On the one hand, the industrial cluster has promoted the market demand and made the social division of labor. This tendency requires the production enterprise pay more attention to its core business and continuously improve the core competitiveness. On the other hand, the competitions from the counterparts also force the enterprise to

© Springer International Publishing Switzerland 2016
D.-S. Huang et al. (Eds.): ICIC 2016, Part III, LNAI 9773, pp. 471–481, 2016.
DOI: 10.1007/978-3-319-42297-8_44

change the business process in order to reduce the cost, improve the service level and shorten the delivery cycle time. Consequently, the cluster production enterprises which aim to reduce logistics costs and improve logistics efficiency need to transfer or outsource their non-core logistics business to the professional third party logistics enterprises. Secondly, the third party logistics has advantages of high efficiency integration of logistics resources, which make them be able to fully expand the geographical boundaries of the clusters, so that the scope of the market can be explored effectively. What's more, with the further improvement of the market competition of the enterprises, the third party logistics will play more indispensable and professional role in the enterprises. By the cooperation with the cluster enterprises, the third party logistics not only meet the need of their own development, but also make a win-win situation. The realization of the collaborative model of manufacturing enterprises and third party logistics under the environment of the cluster is like chemical reaction of the enterprises in favor of the geographic cluster and the third party logistics in favor of the geographic expansion.

In the cluster environment, although the practice of the synergistic model between the production enterprises and the third party logistics constantly blazes new trails, the theory is laggard compared with the practice. Abroad Murali researched and realized the professional third party logistics provides a new type of network competitiveness for small and medium-sized enterprises clusters. He has researched deeply and verified this point of view based on a logistics model of a representative Danish furniture industry clusters [1]. Moreover, foreign scholars focus more on the macroscopic synergistic relationships between the third party logistics and clustered enterprises, such as Robert L analyzed the impact on the manufacturing of changes of the third party logistics and the corresponding satisfaction of manufacturing to the process of changes in the third party logistic [2]. Ginanoccaro on the basis of analyzing the logistics collaboration inherent mechanism thinks that logistics collaboration is a new continuation of the development of supply chain. He compared logistics collaboration with modern logistics and traditional logistics, and put forward the basic principles of modern logistics collaboration [3]. Mark Gohl and Argus Aus analyzed deeply the profession treatment of the third party logistics, and gave objective advice on the construction of production companies and third-party logistics information collaboration platform [4].

Domestic scholars' researches about the synergistic model of production enterprises and the third party logistics are most confined to qualitative assessment analysis. By analyzing the necessity of synergy and the existing problems, they put forward the corresponding countermeasures. In quantitative research, they mainly apply the gray association model, the dynamic equation of order parameter, DEA and combining with related data to analyze and evaluate. For example, combining with industrial cluster competitive advantage, Chouyong Chen makes a deep research of the role of the third party logistics to cluster enterprises. He uses the Logistic model to analyze the relationship between cluster enterprises and the third party logistics, and holds the point of view that the third party logistics is the most important industry within the radiation scope of industrial cluster and plays an irreplaceable role in strengthening industrial cluster. By analyzing the dynamic equation for the order parameter of coordination theory [5], Lei Xie thinks that when the two are in synergy state the logistics system works in its best and so does the overall efficiency. At the same time he illustrates that the coordinated development of cluster enterprises and the third party logistics is an inevitable trend, further demonstrates the possibility of the synergy between cluster companies and the third party logistics [6]. Xiao Zou makes a deep research of logistics links in the cluster enterprise logistics businesses and provides theoretical guidance for the development of third-party logistics collaborative model in the cluster environment [7].

Although scholars have carried out explorations and researches for the cooperative relations and strategies of manufacturing enterprises and the third party logistics in the cluster environment, but they lack design and practice of the theory system for the cooperation mode. The relationship of the cluster enterprises and the third party logistics is manufacturing and service industries in essence. In the cluster environment, manufacturers and third-party logistics need to establish long-term and stable cooperative mode to strengthen the cluster companies and third party logistics' advantages in the market competition and protect the common interest of them. However, the distribution of interests in cooperation mode exhibits complex game relationship. How to reasonably analyze the inner laws of the value of interest distribution, risk quantized value and logistics cost, becomes the key of cooperation mode research. Therefore, based on Shapley value model, as for the distribution of interests this paper fully considers the advantage of logistics cost and contributions of participating enterprises, and analyze the curve of interest distribution and risk quantized value, put forward decision curve suggestions of cooperation selection decision curve based on logistics cost.

2   Collaborative Benefit Allocation and Risk Allocation Decision Model Based on Shapley Value

2.1   The Problem Is Put Forward

As a typical dynamic system cooperation mode, aimed at quick response to market, in the process of cooperative construction, operation and integration, participating enterprises create synergistic overall interests by improving product quality and service level, reducing the production cost and so on. Therefore, cooperation interest referred in this paper means cluster enterprises and the third party logistics create and access long-term benefits and effectiveness through cooperation mode. If many independent participation enterprises actively involved in logistics coordination, the sum of the benefits arising from synergies must be greater than the sum of the interests of all participating companies independently produced.

The cooperation mode of cluster enterprises and the third party logistics takes market demand as a center. This Organizational characteristic determines the essence of involved enterprises choosing cooperation is pursuing economic benefits growth. However, the risks always exist in the cooperation mode with the interest. For the involved enterprises in cooperation mode, interest and risk are inseparable.

The cooperative process and complex operation links and relationships of involved enterprises, increases the possibility of the risk of a coordinated to produce. Firstly, as the supporting and protective industrial for economical society development, manufacturing and the third party logistics' special basis status determines the complexity and particularity of cluster companies and third-party logistics collaborative mode, and often have serious risk consequences. Secondly, because of cooperative companies' input and contribution changing with time, manufacturing enterprises and the third party logistics are difficult to grasp each others' runtime behavior. They can only get the final result of synergies and need the original contract forms to adapt, that's why it tends to have a greater process risk. What's more, influenced by external industrial policy and internal adaption and other factors, the inherent contract status and form of cooperation change easily, causing dynamic changes of state variables of cooperation risk and a variety of complex behavioral outcomes. In the cluster environment, manufacturing enterprises and the third party logistics is a typical dynamic system, needs combination of cooperative operation mode, on the basis of objective risk identification and assessment to design risk quantized value models and strategies to control risk.

2.2 Based on Shapley Value of Cooperative Profit Distribution and Risk Decision Model

On the basis of Shapley value mode put forward by Shapley L.S. to solve the problem of n persons cooperate, this paper applies the rule that overrun value compared the cost before cooperation and the cost after cooperation is the logistics synergy value, uses logistics costs as variables into Shapley Value Method, and proposes interest distribution method for n involved enterprises in cluster environment.

The Shapley value is determined by the characteristic function V, Expressed as $\Phi(v) = (\varphi_1(v), \varphi_2(v), \ldots, \varphi_n(v))$, represents the benefit distribution strategy of logistics cooperative mode. The formula for $\varphi_i(v)$ is:

$$\varphi_i(v) = \sum \omega(|s|)[v(s) - v(s\backslash i)], (i = 1, 2, \ldots, n) \tag{1}$$

Among them, $\omega(|s|) = [n - |s|!(|s| - 1)!]/n!$, n is a collection of all subsets of N containing i, $|s|$ is a number of companies involved in S, $\omega(|s|)$ can be considered as a weighting factor. Shapley value is defined as a probabilistic interpretation. If the participating enterprises participate in the collaborative model based on the random order, the probability of all kinds of order is $1/n!$. The participating enterprises and the $|s| - 1$ people In front of it consists coordination s. The contribution of i to the cooperative mode is $V(S) - V(S\text{-}I)$. The player in $S\text{-}i$ and $N\text{-}S$ sequence have $(n - |s|)!(|s| - 1)!$ species, the probability of each sequence is $\frac{(n-|s|)!(|s|-1)!}{n!}$. According to this interpretation, the contribution expectation of the player i is just the Shapley value. Therefore, this method overcomes the shortcoming of distributing interest according to the investment ratio of involved enterprises only. But in the process of business operations, complex operation links and relationship of involved enterprises and cooperation process make the earnings fraught with uncertainty, causing involved enterprises face the risk of loss. However, Shapley value method only assume that in logistics cooperation mode involved enterprises take the same proportion of loss risk, ignoring the risk of loss that each enterprise taking. In logistics cooperation mode, the cost of logistics and the logistics resource owned and other prerequisites for each enterprises are different, which cause the loss risk is different for each enterprise. So we should increase the benefit ratio for the enterprises who take high risk. Therefore, risk factors need to be included in the interests of the distribution mechanism for correcting Shapley value model.

Supposing the logistics cooperation mode is formed by n involved enterprises, the risk coefficient of involved enterprise i is expressed as R_i, the logistics cost is I_i, the loss risk quantized value is $\varphi_i(R)$, then $\varphi_i(R) = R_i I_i$. Namely when suffering mentioned risk, logistics union can't deliver the size of risk to every enterprise smoothly. The risk coefficient R_i can be determined by average success rate and business strength in similar logistics business. The core idea of reasonable logistics risk quantized value is risk-sharing, that's to say the loss risk quantized value should be closely related to the risk the involved enterprise takes. The risk coefficient R_i reflects the size of the risk that a enterprise take in logistics cooperation. And the ascending and descending of risk coefficient are consistent with risk loss quantized value, it Increases with the increase of the risk coefficient R_i.

The total risk of the enterprises participating in the cooperation (TRL) is the following:

$$TRL = \sum_{i=1}^{n} R_i I_i, \ (i = 1, 2, \cdots, n) \tag{2}$$

Thus it can be concluded that the proportion of the total risk loss in the participating enterprises is:

$$R_i' = \left[(I_i R_i) \Big/ \sum_{i=1}^{n} R_i I_i \right], \ (i = 1, 2, \cdots, n) \tag{3}$$

Among them, $R_i \in (0,1)$, R' $\in (0,1)$. The difference between the proportion of the risk loss in the actual logistics cooperative mode and the risk loss in the ideal case is:

$$\Delta R_i = R_i' - \frac{1}{n} \tag{4}$$

Among them, $\sum_{i=1}^{n} R_i' = 1$, $\sum_{i=1}^{n} \Delta R_i = 0$.

The difference of risk loss as the relative variables, average profit distribution who regard of the risk as a fixed variable. The risk loss correction of the enterprise i is:

$$\Delta \varphi(R) = \bar{\varphi}_i(v) \Delta R_i \tag{5}$$

Among them, $\bar{\varphi}_i(v) = \Phi(v)/n$, By formula (5), the formula (6) can be converted to:

$$\Delta \varphi_i(R) = \bar{\varphi}_i(v) \Delta R_i = \bar{\varphi}_i(v)(R_i' - \frac{1}{n}) \tag{6}$$

At the same time, the value of the actual risk of I equals the actual loss value, that is:

$$\varphi_i'(v) = \varphi_i(v) + \Delta\varphi_i(R) = \varphi_i(R) \tag{7}$$

Obviously, the sum of all enterprises still is the total revenue of the logistics synergy obtained.

2.3   Collaborative Decision Analysis Based on Interest Distribution and Risk Allocation

Benefit Distribution Curve. Based on the Shapley value benefit allocation model, the Shapley value is determined by the characteristic function V. In the practical logistics synergy mode, the number of enterprises n is a variable factor, which has a direct influence on the change of $|s|$. When $n \geq 2$, $\omega(|s|)$ increased gradually along with the increasing number of subsets. At the same time, the change of $[v(s) - v(s \setminus i)]$ has a direct influence on the benefit allocation value $\varphi_i(v)$. In the synergistic model, because the logistics cost of enterprise i is proportional increase to the income, benefit allocation value $\varphi_i(v)$ is gradually increasing with the increase of $[v(s) - v(s \setminus i)]$. Therefore, benefit allocation curve can be expressed as a logistics cost investment as the horizontal axis (x), allocation of benefit value for the longitudinal axis (y) monotonically increasing curve is linear and we can get from it that the allocation value of benefit increases with the increase of the logistics cost of I_i. As shown in Fig. 1 black lines.

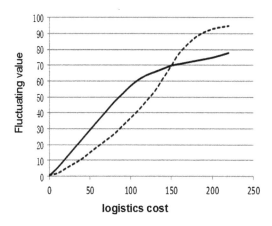

Fig. 1. The fluctuation of interest distribution and risk loss

Risk Quantification Curve. In the actual logistics system, participating enterprises face with different levels of risk in cluster environment, so the risk coefficient Ri must exist. There is no possibility that risk coefficient is zero for any involved enterprises. So the risk loss value $\varphi_i(R)$ must exist and $\varphi_i(R) > 0$. We can also calculate that R_i' is monotone increasing at $(0,1)$. And the participating enterprise i average value of the enterprise's risk loss in the total risk loss$1/n$ move at $(0,1)$, as shown in Fig. 1.

Analysis of Collaborative Selection of Centerprise and Third Party Logistics. In cluster environment, the cooperation of manufacturing enterprises and the third party logistics is a dynamic interactive process. In this dynamic interactive process, there must be benefit and risk combined together, making their relationship more closely. Combined with analysis of benefit and risk in this paper, we know that in the process of cooperation, participating companies' benefit distribution value and risk loss value will fluctuate with the change of logistics cost investment. When participating enterprises' logistics cost in a certain stage, its profit distribution value relative to the size of the risk loss value will change. The corresponding benefit of logistics cost input is greater than the risk of loss value, participating companies will choose synergies; on the contrary, do not select collaboration.

According to the opinion above, combined with analyzing the curve of benefit distribution value and the curve of risk quantized value, we analyze the 2-d curve fluctuations while the logistics cost is taken as the fixed variable and the value of interest distribution and risk quantification as the relative variable, to intuitively and effectively reflect the relationships of the participation enterprises' corresponding benefit of logistics cost and the risk loss value. In order to provide convenient conditions for participating companies to choose cooperation or not in cluster environment, at the same time, providing new ideas for the further researches of the cooperation mode of manufacturing enterprises and the third party logistics.

By formula (6) and (7) it can be known: Only when $R_i' = \frac{1}{n}$, $R_i' - \frac{1}{n} = (R_i I_i) / \sum_{i=1}^{n} (R_i I_i) - \frac{1}{n} = 0$ The enterprise logistics investment cost is: $I_i = \sum_{i=1}^{n} (R_i I_i) / n R_i$, $\varphi_i(R) = \varphi_i(v) + \Delta\varphi_i(R) = \varphi_i(v) + \bar{\varphi}_i(v)(R_i' - \frac{1}{n}) = \varphi_i(v) + 0 = \varphi_i(v)$, The value of the distribution of interests is equal to the value of risk loss;

When $R_i' < \frac{1}{n}$, $R_i' - \frac{1}{n} = (R_i I_i) / \sum_{i=1}^{n} (R_i I_i) - \frac{1}{n} < 0$, The enterprise logistics investment cost is: $\varphi_i(R) = \varphi_i(v) + \Delta\varphi_i(R) = \varphi_i(v) + \bar{\varphi}_i(v)(R_i' - \frac{1}{n}) < \varphi_i(v)$, The value of interest distribution is bigger than the risk loss value;

When $R_i' > \frac{1}{n}$, $R_i' - \frac{1}{n} = (R_i I_i) / \sum_{i=1}^{n} (R_i I_i) - \frac{1}{n} > 0$, The enterprise logistics investment cost is: $I_i > \sum_{i=1}^{n} (R_i I_i) / n R_i \varphi_i(R) = \varphi_i(v) + \Delta\varphi_i(R) = \varphi_i(v) + \bar{\varphi}_i(v)(R_i' - \frac{1}{n}) > \varphi_i(v)$, the distribution value of interest is less than the risk loss value;

When $R_i' = \frac{1}{n}$, the logistics cost of the participating enterprise is called the critical point of cooperation. When participating enterprises' i logistics cost into $I_i = 0$, the corresponding value of the profit distribution $\varphi_i(v) = 0$, the risk quantification value of infinite close to 0, So $I_i = 0$ became another synergistic selection critical point.

Accordingly participating enterprises' i Synergy mode selection critical point: $I_i = 0$ and $I_i = \sum_{i=1}^{n}(R_iI_i) \bigg/ nR_i = TRL/nR_i$. In the same time, because of $1 > R_i > 0$, the minimum input cost for participating enterprises i is: $I_i = TRL/nR_i$, and that is mean value of the alliance total risk loss. The relationship between the distribution of benefits and the risk of loss in the enterprise is shown in Fig. 1. Enterprises can make a decision according to the actual situation.

3 Case Study

With the help of Shijiazhuang equipment manufacturing base industrial management committee and manufacturing enterprises, this paper selects 5 manufacturing enterprises. Shijiazhuang Longda Tools Co. Ltd., Hebei Luan Sen Electromechanical Equipment Co. Ltd., Hebei Electric Motor co., LTD are defined as the supply enterprises, denoted by G; definition of Hebei Endurance Compressor Ltd. and Shijiazhuang Xingda Automobile Limited enterprise for the productive enterprises, denoted by M.

Table 1. $\{G_1, M_1, M_2, L\}$ in or not in the collaborative state of the logistics cost table.

Synergic relationship	pre-S	post-S
G_1	58.5	58.5
M_1	220.1	220.1
M_2	96.9	96.9
$\{G_1, M_1\}$	278.6	233.0
$\{G_1, M_2\}$	155.4	142.2
$\{G_1, L\}$	56.1	47.8
$\{M_1, M_2\}$	317.1	285.8
$\{M_1, L\}$	260.1	224.7
$\{M_2, L\}$	127.3	101.4
$\{M_1, M_2, L\}$	375.9	319.7
$\{G_1, M_1, M_2\}$	375.5	309.7
$\{G_1, M_1, L\}$	301.4	238.2
$\{G_1, M_2, L\}$	199.1	138.3
$\{G_1, M_1, M_2, L\}$	396.4	283.5

The 5 manufacturing enterprises and logistics enterprises constitute a synergistic combination S1 = $\{G_1, G_2, G_3, M_1, M_2, L\}$. Among them, including three kinds of synergistic state: $\{G_1, M_1, M_2, L\}$, $\{G_2, M_1, M_2, L\}$ and $\{G_3, M_1, M_2, L\}$. Logistics costs include the logistics cost of enterprises under the condition of non-synergy and synergy. Among them: In the condition of non-synergy, enterprise logistics cost is the sum of participating enterprise's logistics cost, denoted by pre-S; In the condition of synergy, the logistics cost according to ABC (activity-based-costing) get the participating enterprise's logistics cost, denoted by post-S, as shown in Table 1 (Unit: ten thousand yuan). Under the synergistic state, the allocation of benefit to participating

Table 2. Participation in the distribution of each corporate interest.

Participating enterprises	Cooperative interests $\varphi_i(v)$			
Supply enterprise G_1	27.4	–	–	27.4
Supply enterprise G_2	–	23.3	–	23.3
Supply enterprise G_3	–	–	53.5	53.5
Supply enterprise M_1	35.2	10.1	11.5	56.8
Supply enterprise M_2	25.5	7.5	7.9	40.9
third party logistics L	24.8	8.7	10.3	43.8
total	112.9	49.6	83.2	245.7

Table 3. Risk quantification results.

Participating enterprises	Risk coefficient R_i	Logistics cost I_i	Risk loss value $R_i \times I_i$	Fixed risk quantitative loss $\varphi_i(R)$
G_1	0.556	58.5	32.5	24.7
G_2	0.203	57.5	11.7	18.0
G_3	0.613	173.5	106.4	60.1
M_1	0.479	220.1	105.4	63.3
M_2	0.278	96.9	26.9	37.5
L	0.426	95.7	40.8	42.1
total	2.329	684.2	323.7	245.7

enterprises based on Shapley value model. The benefit allocation value are shown in Table 2 (Unit: ten thousand yuan).

From the above three kinds of synergistic states can get the results of the benefit allocation of the participating enterprises as shown in Table 2. The conclusions and the total benefit of preliminary analysis fully consistent, Shapley value method based on proved to logistics costs as the starting point for the allocation of benefits is reasonable.

In addition, each manufacturing enterprise logistics synergy investment and risk factors in the Shijiazhuang equipment manufacturing base as shown in Table 3. The total risk value is 245.7 ten thousand yuan, the result is in agreement with the previous 245.7 ten thousand yuan of benefit allocation value.

The logistics cost as horizontal coordinates, the benefit allocation value and risk value for vertical coordinates. At $n = 6$ state, we establish a coordinate of the benefit allocation value, as shown in Fig. 2, it's consistent with the foregoing conclusions: The cluster enterprises and enterprises of the third-party logistics synergy mode, Whose benefit allocation curve show an increasing trend with increasing investment in logistics costs. In Fig. 2, benefit allocation curve and risk quantification curve can intersect, means that there is a participating enterprise (i) selection critical point when

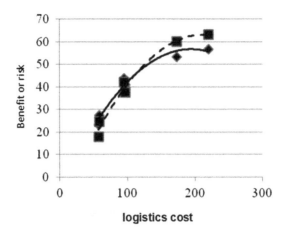

Fig. 2. G1, G2, G3, M1, M2, L, the distribution of benefits and risk

the productive enterprises in cluster environment collaborates with the third party logistics enterprise, it is: $I_i = \sum_{i=1}^{n} (R_i I_i) \Big/ nR_i = TRL/nR_i$.

For example, Shijiazhuang Longda tool limited enterprise G_1 logistics cost is $I_{G1} = 58.5$ in logistics collaboration, the risk coefficient is $R_{G1} = 0.556$. From Table 3, we can see: $TRL = 323.7$, and we can be deduce: $TRL/(nR_i) = 323.7/(6 \times 0.556) = 97.03$, and $I_{G1} \in (0, TRL/(6R_i))$, after verified we get: $R'_{G_1} - \frac{1}{n} = 0.1005 - 0.1667 = -0.0662 < 0$, and also because the difference between the benefit allocation value and risk value is $\varphi_i(v) - \varphi_i(R) = 27.4 - 24.7 = 2.7$. Therefore, Shijiazhuang Longda Tools Co. Ltd. Should selects synergy.

In the same way we can get that Hebei Luansen Electromechanical Equipment Co. Ltd G_2 should selection synergy; Hebei Electric Motor co., LTD G_3 should not select synergy; Hebei endurance Compressor Co., Ltd. M_1 should not select synergy; Shijiazhuang Xingda Automobile Co. Ltd. M_2 should select synergy; In addition, logistics enterprise L should select synergy.

4   Conclusions

A reasonable benefit allocation and risk-sharing mechanism is the key to ensure the synergy of cluster enterprises and the third party logistics operations efficiently. How to reasonably and objectively analyze the inner laws of the benefit allocation value, risk quantized value and logistics cost, becomes the key of making synergistic decision. Therefore, there are some conclusions: firstly, regarding the market requirement as the center is the synergy mode's organizational characteristic. This characteristic determines the essence of participating enterprises' synergy is to pursue economic growth, and risk always exists with benefit in the synergy mode. Secondly, based on the Shapley value mode, as for the benefit allocation, this paper fully considers the advantage of logistics cost and contributions of participating enterprises, and puts forward benefit allocation and risk-sharing decision mode based on Shapley value model; Thirdly, the mentioned synergistic decision model provides a new way for synergy mode research of cluster enterprises and third party logistics. Finally, this paper verifies the benefit allocation curve and risk quantification curve variation rules with the actual data of Shijiazhuang equipment manufacturing base, and verifies the exits of critical point of selection of enterprise synergy and synergistic selection scope, gives suggestions on how participating in enterprises collaboration choose synergy.

References

1. Murali, S.: Study on producer logistics service and it's outsourcing from manufacturing firms: a perspective of industrial cluster. Phys. Distrib. Logistics Manage. (2010)
2. Clerk, R.: Defined Benefit Pension Plan Distribution Decisions by Public Sector Employees. OR Spectrum (2014)
3. Ginanoccaro, I., Pontrandolfo, P.: Supply chain coordination by revenue sharing contracts. Int. J. Prod. Econ. **89**, 131–139 (2004)
4. Goh, M., Ang, A.: Some Logistics Realities in Indochina. MCB UP Ltd (2011)
5. Chen, C.: The Mechanism of the Third Party Logistics and the Coordinated Development of Industry Cluster and Pattern. Science Press (2009)
6. Xie, L., MA, S.: Supply logistics two-dimensional synergy theory and its realization ways empirical research. Manag. Sci. (2012)
7. Zou, X., Zhang, S.: The coordinated development of the logistics industry and manufacturing research were reviewed. Syst. Eng. (2012)

On Cross-Layer Based Handoff
Scheme in Heterogeneous Mobile Networks

Byunghun Song[1], Junho Shin[1], Hana Jang[2], Yongkil Lee[3],
Jongpil Jeong[4], and Jun-Dong Cho[4(✉)]

[1] IoT Convergence Research Center, Korea Electronics Technology
Institute (KETI), Seongnam 463-816, Korea
bhsong@gmail.com, harri94@gmail.com
[2] Graduate School of Information and Communications,
Sungkyunkwan University, Sungkyunkwan-Ro,
Jongno-Gu, Seoul 110-745, Korea
hnsh77@naver.com
[3] Graduate School of Entrepreneurship, Sungkyunkwan University,
2066 Seobu-ro, Jangan-gu, Suwon, Kyunggi-do 440-746, Korea
Reyong22@naver.com
[4] College of Information and Communication Engineering,
Sungkyunkwan University, 2066 Seobu-ro,
Jangan-gu, Suwon, Kyunggi-do, Korea
{jpjeong, jdcho}@skku.edu

Abstract. The necessities of the MNs are generally included in the cellular
modules available in LTE/3G and Wi-Fi for high-speed Internet. Further, many
people with Wi-Fi and LTE/3G use a cross-layer-based handoff as they move
from one place to another. MN mobility management has been handled ade-
quately; however, using the network-based mobility management described in
this paper, carriers can manage and maintain a lower-cost network.

Keywords: PMIPv6 · MIPv6 · Mobility management · Cross-layer · Vertical
handoff

1 Introduction

If various wireless-access techniques co-exist, two types of handovers are possible [1]:
horizontal and vertical. A horizontal handover refers to a handover within the same
network. A vertical handover moves from one network type to another, requiring a
change in the data link layer (L2). In addition to horizontal and vertical handovers,
handovers can be divided into two types, depending on which entity is responsible for
sensing a terminal about to leave one access network and enter another; namely,
host-based and network-based. A host-based handover means that the user's MN
senses the change of network and carries out the handover. A network-based handover
means that the MN does not recognize that it is leaving the network; the network
recognizes it and carries out the handover. Most handovers are host-based horizontal
handovers which are carried out by the MN. Network-based mobility management

© Springer International Publishing Switzerland 2016
D.-S. Huang et al. (Eds.): ICIC 2016, Part III, LNAI 9773, pp. 482–493, 2016.
DOI: 10.1007/978-3-319-42297-8_45

techniques are generally cheaper than host-based mobility management techniques, when calculating the total cost [2].

After carrying out a vertical handover, the MN must bind itself to a new IP address in order to maintain the data transmission with the original protocol. The Internet Engineering Task Force (IETF)established the mobile internet protocol, version 6 (MIPv6) [3] in order to solve the problems associated with updating an IP address when an MN moves from one network to another. Researchers have discussed the problem of the quick handover in a wireless local area network (WLAN) [4]. In order to solve this problem for vertical handovers, the cross-layer-based received signal strength (RSS) prediction method can carry out a vertical handover in a heterogeneous wireless network; i.e., it selects the network with the strongest signal as the new network.

This study analyzes the handover performance of various mobility management techniques to determine which has the lowest cost. Moreover, this study uses a performance analysis environment which is similar to a real environment in which the LTE/3G data standard and Wi-Fi are widely used.

The rest of this paper is organized as follows. In Sect. 2, we discuss previous works. In Sect. 3, the cross-layer prediction method based on RSS (received signal strength) is proposed. In Sect. 4, using a network model, we evaluate the performance of our proposed information. In Sect. 5, we conclude our work.

2 Related Work

A variety of MN protocols have been introduced to support mobile services. In particular, the network layer mobility protocol has been developed by the IETF.

Figure 1 shows the operating structure of MIPv6. Following the improved performance of PMIPv6 [5], fast-handovers for MIPv6 (FMIPv6) [6] and hierarchical MIPv6 (HMIPv6) [7] were developed. Performance analyses of the IPv6 mobility management protocol helped expand the different MIPv6s [8, 9]. For example, performance analyses and research on MIPv6 and PMIPv6 were used for HMIPv6. HMIPv6 can handle the characteristics and performance indicators for each mobility management protocol.

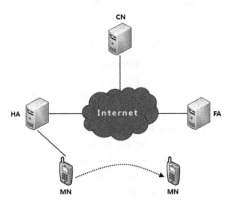

Fig. 1. Operating structure of MIPv6 (clockwise from left: home agent, correspondent node, foreign agent, mobile nodes).

When the host-based mobility management protocol was applied to a wireless mobile communication infrastructure, telecommunication service providers and standards development organizations found that existing solutions were unsuitable for mobility services. In particular, an MN must perform IP stack operations, but when mobility is required, the stack point must be upgraded or modified. This increases the complexity and operations cost of the MN. Moreover, a host-based mobility management protocol is also needed because the MN itself can lead service providers to a lack of control of mobility management resources. Therefore, the Third Generation Partnership Project (3GPP) standard development organization proposed that a new approach was needed to support mobile services in the IETF.

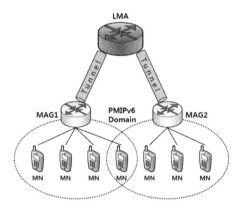

Fig. 2. Operating structure of PMIPv6 (LMA: local mobility anchor; MAG: mobile access gateway).

PMIPv6 is a network-based mobility management protocol that can be processed without having to change the connection point separately from the MN [10]; Fig. 2 shows the operating structure. The MAG (mobile access gateway) for supplying mobility services, introduced in PMIPv6, and the LMA (local mobility anchor) are the two entities that are responsible for detecting and registering the movement of the MN in the PMIPv6 domain. The MAG senses the movement of the MN and sends a proxy binding-update message to the LMA. It specifies that the LMA includes additional features and operates as the HA (home agent) [5]. The LMA notifies the MN that the proxy binding-update message has connected to a MAG, creating or updating the binding cache. The MAG receives a proxy binding-acknowledgment message containing the HNP (home network prefix) and sends an RA (router advertisement) message including the HNP to the MN. The MN assembles a pHoA (proxy home of address) based on the HNP included in the RA message.

The LMA obtains the same HNP for the movement, just as the MN in the PMIPv6 domain uses a pHoA. Because of the network-based mobility services provided by the mobile service management entity permission settings, the MN's entire PMIPv6 domain is aware of the position of a single link [10]. However, because the PMIPv6 performs MIPv6 mobility management each time an MN enters a PMIPv6 domain, the

MAG that detects the entry of the MN must go through the handover process to send its own position information to the LMA. As the MN moves through the handover, information and communication with the MN can be cut, resulting in a packet loss [11]. FPMIPv6 [12] is a protocol, developed as an extension to PMIPv6, which is intended to prevent packet loss to reduce the handover latency, so as to improve the handover performance.

3   Cross-Layer Based Adaptive Vertical Handover

This section is composed as follows. First, it introduces the features of the vertical handover in the heterogeneous wireless network, and defines the network model. Second, it explains the problems of a vertical handover. Third, it defines the number of vertical handovers, the probability of a connection interruption, and the network utilization, in order to analyze the performance in diverse environments.

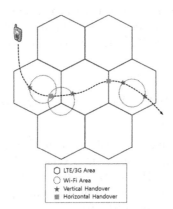

Fig. 3. MN handover methods for different network types.

A heterogeneous wireless network is defined as a combination of LTE/3G and Wi-Fi networks. Figure 3 shows two types of bonds in a heterogeneous wireless network, solid and loose. The first uses a gigabit interface and the other is a backbone network using IP. The solid bond type provides advantages such as better management, verification, and certification. The proposed vertical handover method is suitable for both types of formats.

If it is assumed that a heterogeneous wireless network, composed of N WLAN and LTE/3G networks, is responsible for a region which is L long and W wide, and $n(0 \leq n \leq N)$ is the network index, the WLAN is expressed as *WLANn*. Further, it is assumed that the LTE/3G network can cover the entire region. Each WLAN has an equal transmission radius of R_{11}, and the N WLANs distribute to random regions. RSS_{11} is the strength of the signal which the MN receives from *WLANn*. The MN contains two interfaces that can access LTE/3Gs and WLANs, respectively, and it selects the appropriate interface as needed.

When the MN moves from network i to network j according to the node mobility, if the two network types are different, the MN executes a vertical handover for the sake of permanence of data transmission; if the network types are the same, it executes a horizontal handover. They can be explained as a network-based handover or a host-based handover. If it is a network-based handover, the network decides how to execute the handover, and the network controller is provided with the RSS from the MN to decide it. The host-based handover has more accurate network information than the network-based handover, so it helps decide the handover. Thus, this study focuses on the problem of the network support handover.

When the proposed method carries out a vertical handover, it uses diverse performance criteria, including the SWGoS (sum of weighted grade of service) and the network utilization numerical value, and deduces the differences from the two. First, when the MN moves to an overlapping region between two networks, if the ping-pong phenomenon occurs, unnecessary handovers occur, and the overhead increases. Here, the fewer the handovers are, the better the handover algorithms are.

Second, when the available network bandwidth is not sufficient, a handover connection is interrupted. The damage which occurs at a connection interruption is worse than at a connection failure. In order to reduce this damage and to maximize the profit of the network supplier, the SWGoS is adopted as an important metric, and is expressed as follows.

$$SWGoS = W_B \sum_{k=1}^{K} kB_k + W_D \sum_{k=1}^{K} kD_k \tag{1}$$

B_k and D_k represent the blocking and interruption for each traffic class k, respectively, and W_B and W_D represent the weighted values of blocking and interruption for each connection, respectively. This study sets the values of W_B and W_D as 1 and 10. Third, the network usage rate is adopted as a separate performance metric. A high network usage rate indicates better performance.

The TCP (transfer control protocol) connection situation worsens in the vertical handover environment of a heterogeneous wireless network. A sudden change in available bandwidth makes it difficult for the TCP sender to determine an accurate congestion window; the TCP sender will receive three duplicate ACKs (acknowledgements), if a timeout occurs; if not, it causes a timeout. This result affects the network throughput considerably. Consequently, the proposed cross-layer mechanism is evaluated on the foundation of the average throughput.

4 Performance Evaluation

4.1 Network Modeling

Consider that a network model is configured to perform horizontal and vertical handovers between one of two different network terminals, as shown in Fig. 4.

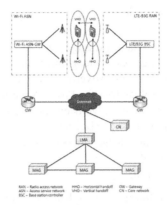

Fig. 4. Performance evaluation of other types of network topology.

Defined below are the number of hops which describe the distance at a certain point between the communicating entities.

- h_{c-H}: Average hop between the NC (Correspondent Node) and the HA (Home Agent).
- h_{c-G}: Average hop between the CN and the gateway.
- h_{H-G}: Average hop between the HA and the gateway.
- h_{L-A}: Average hop between the LMA and AR (Access Router).
- h_{A-A}: Average hop between neighboring ARs.
- h_{A-M}: Average hop between the AR and the MN; in a wireless network, h_{A-M} is assumed to be 1.

In the proposed network model, data/control packets, which are exchanged between the MN and the CN, must be transmitted through the GW. For example, if the route optimization (RO) of MIPv6 has been set, the data packet transmitted from the MN to the CN is made through $h_{c-G} + h_{L-A} + h_{A-M} + h_{A-M}$; is connected to the air of the MN and the AR. In addition, h_{A-A} is expressed by $\sqrt{h_{L-A}}$ [13, 14].

Various messages relating to IPv6 mobility support are used in the mobility management protocol. The message sizes to consider when analyzing the mobility model are as follows:

- L_{RS}: RS (Router Solicitation) message size, 52.
- L_{RA}: RA (Router Advertisement) message size, 80.
- L_{BU-HA}: BU message sent from the HA to the MN, message size, 56.
- $L_{BAck-HA}$: MN BAck in response to the HA, message size, 56.
- L_{BU-CN}: BU message transmitted from the MN to the CN, size 66.
- $L_{LBU-MAP}$: When the MN transmits the MAP LBU, message size, 56.
- $L_{LBAck-MAP}$: MAP in response to the MN LBU, message size, 56.
- $L_{PBU-LMA}$: A transfer from the LMA MAG PBU, message size, 76.
- $L_{PBAck-LMA}$: The LMA sends a PBAck to the MAG, message size, 76.
- L_{HoT}: HoTI (Home Test Init) message size, 64.
- L_{CoTI}: CoTI (Care-of Test Init) message size, 64.

- L_{HoT}: HoT (Home Test) message size, 74.
- L_{CoT}: CoT (Care-Of Test) message size, 74.
- L_{FBU}: FBU (Fast Binding Update) message size, 56.
- L_{FBAck}: FBAck (Fast Binding Acknowledgement) message size, 56.
- L_{UNA}: UNA (Unsolicited Neighbor Advertisement) message size, 52.
- $L_{RtSol\ Pr}$: RtSolPr (Router Solicitation for Proxy Advertisement) message size, 52.
- $L_{Pr\ RtAdv}$: PrRtAdv (Proxy Router Advertisement) message size, 80.
- L_{HI}: HI (Handover Initiate) message size, 52.
- L_{HAck}: HAck (Handover acknowledge) message size, 52.
- L_T: Tunneling header size, 40.
- L_D: User data packet size, 120.

The delay time required for the actual handover, excluding the authentication process, can be expressed as follows.

- $T_{L2}^{Wi-Fi\rightarrow Wi-Fi}$: Wi-Fi horizontal handover time to a Wi-Fi network. Weight 100.
- $T_{L2}^{LTE/3G\rightarrow Wi-Fi}$: LTE/3G vertical handover time to a Wi-Fi network on the network. Weight 150.
- $T_{L2}^{LTE/3G\rightarrow LTE/3G}$: LTE/3G horizontal handover time to an LTE/3G network in the network. Weight 200.
- $T_{L2}^{Wi-Fi\rightarrow LTE/3G}$: Time during a vertical handover to an LTE/3G from a Wi-Fi. Weight 300.

Like HMIPv6, PMIPv6 locally manages the movement of the MN. However, the mobility service for the MN is supported by the mobility service provision [15, 16]. As the MN is connected to the new access network, it is detected by the MAG and registers with the new access network, to obtain the same HNP included in the RA message that was sent from the MAG. If the MN performs a handover in a PMIPv6 domain [15], the address is not necessary when setting up the DAD (Disciplined Agile Delivery) process.

We define the PMIPv6 handover latency as L_{HO}^{PMIPv6}, which is expressed as follows.

$$L_{HO}^{PMIPv6} = T_{L2} + T_{LMA} \tag{2}$$

T_{LMA}[A] includes the amount of time required for the RS message transfer, an exchange of PBU/PBAck messages between the MAG and the LMA, and possibly the first data packet sent from the LMA. In this paper, when the MAG receives the RS message from the MN, it is assumed that it detects the movement of the MN. Therefore, T_{LMA} is expressed as follows.

$$T_{LMA} = d_{wl}(L_{RS}) + d_{wd}(L_{PBU}, h_{G-A}) + d_{lma-packet} \tag{3}$$

$d_{lma-packet}$ represents the time (delay) before the first data packets transmitted by the MN arrive at the LMA. A bi-directional tunnel between the LMA and the MAG can be implemented as a static tunnel between the two; tunnels do not require additional delay. This is considered static tunneling in PMIPv6. The LMA sends a data packet destined for the MN, with an instant PBAck message for receiving a valid PBU message transmitted from the MAG. Therefore, $d_{lma-packet}$ is expressed as follows.

$$d_{lma-packet} = d_{wl}(L_D) + d_{wd}(L_D + L_T, h_{G-A}) \tag{4}$$

L_T only considers $d_{wd}(L_p, h)$. This data packet for the MN is different from the tunneling method used in HMIPv6 between the LMA and MAG. If PMIPv6 and HMIPv6 manage the MN locally, PMIPv6 reduces the packet transmission overhead over a wireless link [15].

To analyze the handover failure of each mobility management protocol, [17–19] use the proposed handover blocking probability. It may fail for various reasons, such as when the handover latency is caused by a large noise in the signal which thus cannot accommodate the MN, and no radio channels are available. For example, if the MN stays in the network for less than the handover completion time, due to the loss of the link information and the radio channel, the MN handover fails.

$L_{HO}^{(\cdot)}$ is assumed to represent the handover delay in the specific mobility management protocol that was developed in the previous subsection. Note that . is used as the protocol indicator. $E\left[L_{HO}^{(\cdot)}\right]$ refers to the value of $L_{HO}^{(\cdot)}$. T_R is assumed to represent the network residence time using the probability density function $f_R(t)$. For simplicity, $L_{HO}^{(\cdot)}$ is considered using the cumulative distribution function $F_T^{(\cdot)}(t)$ with an exponential distribution. In addition, $L_{HO}^{(\cdot)}$ is the only barrier element handover; the probability of a handover block p_b is expressed as follows.

$$p_b = \Pr(L_{HO}^{(\cdot)} > T_R) = \int_0^\infty \left(1 - F_T^{(\cdot)}(u)\right) F_R(u)du = \frac{\mu_C E\left[L_{HO}^{(\cdot)}\right]}{1 + \mu_C E\left[L_{HO}^{(\cdot)}\right]} \tag{5}$$

Mobility of the MN is crossing the boundary. Assuming the coverage of the AR is in a circle, the phenomenon of μ_C is discussed in [13, 16, 20], and is expressed as follows:

$$\mu_C = \frac{2v}{\pi R} \tag{6}$$

v is the average moving velocity of the MN in the above formula; R is the radius of the coverage of the AR.

4.2   Numerical Results

In this section, the results show the performance of the mobility management protocols. For numerical analysis, we refer to [21–23] and use the following values for variables.

$h_{C-H} = 4, h_{C-L} = 6, h_{H-L} = 4, h_{L-A} = 4, h_{A-M} = 1, E(S) = 10, \tau = 20 \text{ ms}, n = 3, L_f = 19 \text{ bytes},$

$D_{wl} = [10, 40] \text{ ms}, D_{wired} = 0.5 \text{ ms}, BW_{wired} = 100 \text{ Mbps}, T_{DAD} = 1000 \text{ ms},$

$T_{L2}^{Wi-Fi \rightarrow Wi-Fi} = 45.35 \text{ ms}, T_{L2}^{LTE/3G \rightarrow Wi-Fi} = 68.03 \text{ ms}, T_{L2}^{LTE/3G \rightarrow LTE/3G} = 90.7 \text{ ms}, T_{L2}^{Wi-Fi \rightarrow LTE/3G} = 136.05 \text{ ms}$

The frame error rate (P_f) ranges from 0 to 0.7 and increases by increments of 0.05. Figure 5 show the relationship between P_f and the handover latency. The probability of transmitting the wrong packet is increased over the wireless link when P_f has a high value. Therefore, as the number of retransmissions increases in the mobility signal, the handover latency increases.

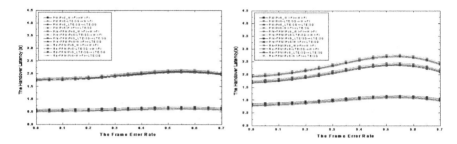

Fig. 5. $D_{wl} = 10 \text{ ms}$ (left) and $D_{wl} = 40 \text{ ms}$ (right), handover latency due to changes in P_f.

As shown in Fig. 5, P_f is a factor in each handover delay in the mobility management protocol. In addition, the value of D_{wl} also affects the handover latency. For example, the handover delay time increases sharply close to P_f, and increases with a higher value of D_{wl}. Second, the analysis proceeds to PMIPv6. In the PMIPv6 network, the MN is the local management. Further, the LMA and the MAG exchangea mobility signal. It does not have the mobility on the line of travel of the wireless signal generation, P_f and D_{wl}, to minimize the PMIPv6 performance.

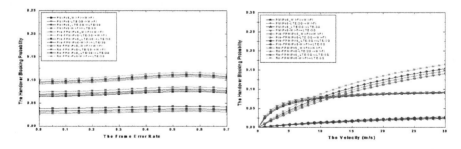

Fig. 6. Handover blocking probability due to changes in P_f (left) and v (right).

Figure 6 shows the handover blocking probability for each mobility management protocol. Handover blocking probability analysis only considers the handover latency of the barrier element. Similar to the results obtained in Fig. 6, as the value of P_f

increases, the probability of a blocked handover increases. The handover blocking probability of Pre-FMIPv6 and FPMIPv6 is low relative to the other protocols, but the blocking handover probability in MIPv6 is high.

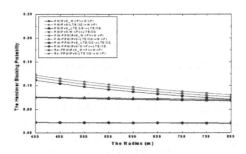

Fig. 7. Handover blocking probability due to changes in R.

In Fig. 7, the handover delays in Pre-FMIPv6 and FPMIPv6—caused by P_f, v, R, respectively—are small enough to avoid blocking issues caused by the handover. The reason the fast predictive performance is superior to other protocols is because they prepare the MN using the L2 information before performing the handover. PBU and PBAck do not affect the PMIPv6 performance value. PMIPv6 mobility uses the same message signal to move the radio link. Further, a PMIPv6 mobility signal is exchanged between the MAG and the LMA over a wired link. As with MIPv6, FPMIPv6 applies fast-handover technology to the PMIPv6 handover.

5   Conclusion

This study analyzed the existing IPv6 mobility management protocols which were developed by the IETF, and compared them in terms of handover delay time, handover interruption probability, and packet loss. The following results were confirmed through performance analysis. First, in order to improve the handover performance, the L2 (data-link layer) information must be utilized. Pre-FPMIPv6 is a high-performance mobility management protocol because it prepares in advance before the handover is carried out to a new access network. If the handover delay time is reduced, the handover blocking probability also decreases. The results of the performance analysis of this study can be used to identify the features and result index of each mobility management protocol. In addition, it can be used to facilitate the decision-making process for developing a new mobility management protocol.

Acknowledgement. This work was supported by The Components & Materials Technology Development Program (10043800, Development of Micro Smart Environmental Sensor Measurement Module, Control Chip, and Application Program) funded by The Ministry of Trade, Industry & Energy (MI, Korea).

This research was supported by the Ministry of Trade, Industry and Energy (MOTIE), Korea, through the Education Support program for Creative and Industrial Convergence (Grant Number S-2016-0117-000).

References

1. Lee, S.H., Shin, D.R., Jeong, J.P.: SePH: seamless proxy-based handoff scheme in IP-based heterogeneous mobile networks. KIPS Trans. Part C **19**(1), 71–82 (2012)
2. Kim, K.T., Jeong, J.P.: Cost analysis of mobility management schemes for IP-based next generation mobile networks. J. Internet Comput. Serv. **13**(3), 1–16 (2012)
3. Johnson, D., Perkins, C., Arkko, J.: Mobility support in IPv6, IETF RFC 3775 (2004)
4. Pack, S., Choi, J., Kwon, T., Choi, Y.: Fast-handoff support in IEEE 802.11 wireless networks. IEEE Commun. Surv. Tutorials **9**(1), 2–12 (2007)
5. Perkins, C., Johnson, D., Arkko, J.: Mobility support in IPv6, IETF RFC 6275 (2011)
6. Koodli, R.: Fast handovers for mobile IPv6, IETF RFC 5568, June, 2008
7. Soliman, H., Bellier, L., Malki, K.E.: Hierarchical mobile IPv6 mobility management (HMIPv6), IETF RFC 4140 (2005)
8. Koodli, R.: Mobile IPv6 fast handovers, IETF RFC 5568 (2009)
9. Soliman, H., Bellier, L., Elmalki, K., Castelluccia, C.: Hierarchical mobile IPv6 (HMIPv6) mobility management, IETF RFC 5380 (2008)
10. Gundavelli, S., Leung, K., Devarapalli, V., Chowdhury, K., Patil, B.: Proxy mobile IPv6, IETF RFC 5213 (2008)
11. Chai, H.S., Jeong, J.P.: Security analysis and implementation of fast Inter-LMA domain handover scheme in proxy mobile IPv6 networks. KIPS Trans. Part C **19**(2), 99–118 (2012)
12. Yokota, H., Chowdhury, K., Koodli, R., Patil, B., Xia, F.: Fast handovers for proxy mobile IPv6, IETF RFC 5949 (2010)
13. Zhang, X., Castellanos, J.G., Campbell, A.T.: P-MIP: paging extensions for mobile IP. Mob. Netw. Appl. **7**(2), 127–141 (2002)
14. Lee, J.H., Pack, S., You, I., Chung, T.M.: Enabling a paging mechanism in network-based localized mobility management networks. J. Internet Technol. **10**(5), 463–472 (2009)
15. Lee, J.H., Han, Y.H., Gundavelli, S., Chung, T.M.: A comparative performance analysis on hierarchical mobile IPv6 and proxy mobile IPv6. Telecommun. Syst. **41**(4), 279–292 (2009)
16. Lee, J.H., Chung, T.M.: How much do we gain by introducing route optimization in Proxy Mobile IPv6 networks? annals of telecommunications-annales des télécommunications **65**(5–6), 233–246 (2010)
17. Yang, S., Zhou, H., Qin, Y., Zhang, H.: SHIP: cross-layer mobility management scheme based on Session Initiation Protocol and Host Identity Protocol. Telecommun. Syst. **42**(1–2), 5–15 (2009)
18. Song, M., Cho, J.D., Jeong, J.: Global mobility management scheme for seamless mobile multicasting service support in PMIPv6 networks. TIIS **9**(2), 637–658 (2015)
19. Choi, J.Y., Cho, Y., Lee, T., Cho, J.D., Jeong, J., Roh, S.S., Kim, H.T.: Cost analysis of integrated HIP-PMIPv6 handoff scheme in multicasting-based mobile networks. Adv. Sci. Lett. **21**(3), 321–327 (2015)
20. Cho, C., Kang, J.J., Jeong, J.: Design and performance analysis of a dynamic–paging AAA mobility management scheme for PMIPv6–based wireless networks. Int. J. Sens. Netw. **15**(4), 214–222 (2014)

21. Jeong, J., Cho, C., Kang, J.J.: Hashing–based lookup service in mobile ad–hoc networks with multi–hop stretchable clustering. Int. J. Sens. Netw. **15**(3), 183–197 (2014)
22. Im, I., Jeong, J.: Cost-effective and fast handoff scheme in Proxy Mobile IPv6 networks with multicasting support. Mob. Inf. Syst. **10**(3), 287–305 (2014)
23. Song, M., Jeong, J.: Performance analysis of a novel inter-networking architecture for cost-effective mobility management support. TIIS **8**(4), 1344–1367 (2014)

On Multicasting-Based Fast Inter-Domain Handover Scheme in Proxy Mobile IPv6 Networks

Jongpil Jeong[1], Jun-Dong Cho[1], Younseok Choi[2],
Youngmin Kwon[3], and Byunghun Song[3(✉)]

[1] Department of Human ICT Convergence, Sungkyunkwan University,
2066 Seobu-ro Jangan-gu, Suwon, Kyunggi-do 440-746, South Korea
{jpjeong,jdcho}@skku.edu
[2] Graduate School of Entrepreneurship, Sungkyunkwan University,
2066 Seobu-ro Jangan-gu, Suwon, Kyunggi-do 440-746, South Korea
cys5004@naver.com
[3] IoT Convergence Research Center,
Korea Electronics Technology Institute (KETI),
Seongnam 463-816, South Korea
kwon.youngmin@gmail.com, bhsong@gmail.com

Abstract. In this paper, we review the current status of PMIPv6 (Proxy Mobile IPv6) multicast listener support being standardized in the IETF and point out limitations of the current approach and we proposed a fast multicast handover procedure in inter-domain PMIPv6 network for network-based mobility management. The proposed fast multicast handover procedure in inter domain optimizes multicast management by using the context of the MNs. We evaluate the proposed fast multicast handover procedure compared to the one based through the developed analytical models and confirm that introduced fast multicast handover procedure provides reduced service interruption time and total network overhead compared to the one based during handovers.

Keywords: Multicasting · PMIPv6 · FPMIPv6(Fast PMIPv6) · Handover

1 Introduction

Recently, IETF proposed a mobility management technique based on the standardized network through PMIPv6 protocol. PMIPv6 is similar to host-centric mobility management protocol [1, 2]. However, it also has the important difference that it does not require MN for any processing as to mobility. It provides MN IP mobility by introducing new mobile agents such as MAG and LMA. MAG within the domain in PMIPv6 recognizes the motion of MN and recognizes L2 connection notification. Moreover, it also performs the role of sending and registering PBU at LMA.

J.-D. Cho—Distinguished Visiting Professor, North University of China (Shanxi 100 people plan).

D.-S. Huang et al. (Eds.): ICIC 2016, Part III, LNAI 9773, pp. 494–505, 2016.
DOI: 10.1007/978-3-319-42297-8_46

LMA performs the role of HA to manage all MNs registered in PMIPv6 domain that was given by the receiving PBU message. LMA resends PBA including HNP for MN to MAG. MN that accessed MAG forms its own address and pHoA (Proxy Home Address) based on HNP received from MAG. When MN runs handover through new network, MN reacquires HNP and HNP acquired from the previously accessed network will be acquired from MAG of newly accessed network. PMIPv6 shows better performance than the conventional host-centric mobility management protocol [3].

However, the previously developed multicast technique cannot be applied to PMIPv6. Furthermore, it is even more difficult to apply the multicast technique between domains. On this account, this paper proposes two PMIPv6 multicast handover techniques for an MN that moves in high speed between LMA domains when multicast service is running. The first proposed PMIPv6 multicast handover technique is to reduce latency through buffering by forming a tunneling between MAGs of LMA through which MAG of LMA has MN and MN's attempt to move. The second proposed PMIPv6 multicast handover technique is to reduce latency through buffering by forming a tunneling between LMA having MN and LMA that is willing to move.

This paper implemented mathematical modeling of the PMIPv6 multicast network proposed by IETF [4, 9, 10], the latency of multicast handover procedure of the proposed two techniques and the entire network overhead and performed a performance evaluation of these. In the case of latency, IETF's proposed PMIPv6 multicast network increases latency in the linear form due to an increase in the number of hops. However, the proposed two multicast networks maintain fixed value of 85.6(ms) based on the analysis of results. That is, it shows that the number of hops does not affect them at all. Moreover, IETF's proposed PMIPv6 multicast network was found to have a difference of at least two times in the case of the entire network overhead.

This paper is organized as follows. In Sect. 2, we will explore related researches. Section 3 will describe a proposed hierarchical modeling system. Section 4 presents performance evaluation of the proposed system modeling. Finally, we will describe the conclusions in Sect. 5.

2 Related Work

With the expansion of PMIPv6, FPMIPv6(Fast PMIPv6) [5] was recently announced as an enhancement of the handover performance of PMIPv6. However, this is designed to optimize unicast communication, and not multicast communication. In other words, because the lack of contribution to the multicast group management, benefits of FPMIPv6 cannot affect to the multicast communication.

FPMIPv6 multicast handover procedure optimizes the multicast group management by using MN roaming context. Context passed from pMAG to nMAG before MN is connected with nMAG contains identification of MN, LMA address of MN, multicast driving information of MN and so on. After nMAG obtains the context of MN, and if it is necessary, it updates multicast forwarding state as well as MLD Proxy Membership Database in advance. Multicast communication is delivered to nMAG in pMAG in the same way as FPMIPv6. By doing so, as soon as MN is connected to the new network, MN is able to receive the multicast communication directly from nMAG.

FPMIPv6 multicast handover procedure does not require a new or added presence in the given PMIPv6 domain. Basic multicast handover procedure requires no changes to MN. However, the function of MAG and LMA needs to be improved [6]. In FPMIPv6 multicast handover procedure, MAG is required the following improvements. Context transfer function is necessary to send and receive the context of MN from nMAG. Context transfer is executed by L2 trigger. Therefore, it is reasonable to assume that the transmission of context to nMAG in pMAG is done before MN is actually connected to nMAG. The multicast subscription information of MN to be transmitted as part of the context updates the multicast forwarding state by nMAG and if it is necessary, it is used to perform the initial MLD Member Report. The function of multicast communication transmission and buffering is required to prevent multicast communication loss during the handover of MN. Information of nMAG such as network identification information or address is held. In order to improve LMA, the following is required. Multicast subscription information of MN transmitted from MAG is used to update the multicast forwarding state. In this case, LMA also requires multicast communication buffering for MN.

As shown in Fig. 1, in FPMIPv6 multicast handover procedure, when L2 trigger is transferred to pMAG, MN prepares the handover to nMAG in pMAG. Using L2 trigger, MN gets network connection information of nMAG such as network identification information. Then, MN provides the network identification information with the identification information of its own to pMAG. pMAG provides service to the current MN by sending a L2 report message.

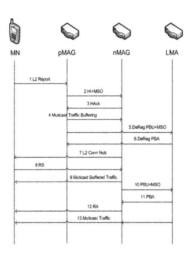

Fig. 1. FPMIPv6 multicasting handover procedures.

MLS of FPMIPv6 multicast handover procedure begins with Handover Initiate(HI) message including identification information of MN, HNP of MN, LMA address of MN, Multicast Support Option(MSO). Multicast subscription information of MN included in MSO is transmitted from pMAG to nMAG, and it is used as a parameter for

MLS. If nMAG acquires HI message and the state of nMAG can provide service to MLS for MN, it will check the message. nMAG sends back to pMAG Hack(Handover Acknowledge) message containing the value to accept or reject. If it is accepted, nMAG updates the multicast forwarding state for MN roaming. You should note that nMAG's MLS decision is determined based on the required multicast service, available buffer size, capacity of MN's number and so on. When receiving HAck message with the accepted values, pMAG delivers the multicast communication for MN to nMAG. Multicast communication transmission is performed until pMAG receives multicast communication for MN from LMA.

3 Proposed Scheme

In the case of PMIPv6 multicast and FPMIPv6 multicast handover, the mobile multicast approach is only for the case in which MN in LMA moves to nMAG in pMAG. However, network size is becoming larger and usage scope of network is expanding. Thus, this paper proposes a better technique through the performance analysis of the aforementioned two techniques by proposing the two handover techniques for the case in which MN between LMAs moves by expanding it.

Procedure between LMA domains optimizes multicast group management by utilizing MN roaming context such as FPMIPv6 multicast handover. The context to be sent from pMAG to nMAG before connecting MN to nMAG includes MN identification, LMA address of MN and multicast operation information of MN. nMAG acquires MN context. If necessary, it renews multicast transmission status in advance as well as MLD proxy membership database. Similar to FPMIPv6, multicast communication transmits pMAG in nMAG. In this way, MN becomes able to receive multicast communication directly from nMAG as soon as MN accesses new network.

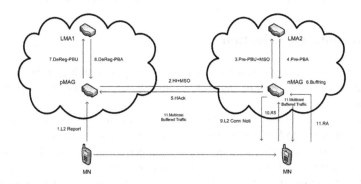

Fig. 2. FPMIPv6 multicasting communication between pMAG and nMAG.

As for FPMIPv6 multicast handover procedure between LMA domains (Fig. 2), MAG needs the following outline just like FPMIPv6 multicast handover procedure. Context transmission function needs to send and receive MN context from/to nMAG.

Context transmission is to be conducted by L2 trigger. As for the proposed multicast handover, it requires outline of nMAG unlike FPMIPv6 multicast handover procedure within domain. nMAG conducts proxy binding procedure in advance before sending HAck message to pMAG while pMAG and LMA2 exchange MSO message with pre-PBU (previously-PBU) and MSO message with pre-PBA (previously-PBA). Hereupon, multicast communication loss will be reduced substantially during MN handover.

PMIPv6 multicast handover procedure between LMA domains based on MAG transmits multicast subscription information of MN included in MSO from pMAG to nMAG by utilizing communication between pMAG and nMAG in order to conduct FPMIPv6 multicast handover procedure between LMA domains (Fig. 3). The proposed handover procedure communicates with LMA2 by sending multicast subscription information of MN included in MSO when sending DeReg.PBU message from pMAG to LMA1. Thus, it reduces multicast handover procedure. Thereupon, the loss associated with multicast communication during MN handover is reduced substantially.

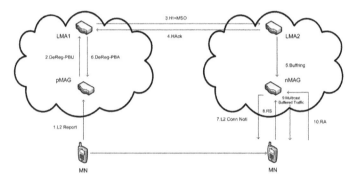

Fig. 3. Multicasting communication between LMA1 and LMA2.

4   Performance Evaluation

In this section, by mathematical modeling of the proposed multicast handover procedure's delay time and overall network overhead, we will perform a performance evaluation.

In the mathematical modeling of the handover delay time, MN performs a handover procedure between LMA domains. The communication path between LMA and LMA, LMA and MAG is a wired connection, the communication path between MAG and MN is a wireless connection. Also, we assumed that the processing time of each component can be ignored, and a message is transmitted through wired / wireless link without the error.

In the mathematical modeling of the entire network overhead, we assumed that Session Arrival Rate is same in each MN and average file size in Session is also same. In addition, all average numbers of LMA (in case of PMIPv6 network) or HA (in case of HMIPv6 network) are equal, and the standard IP switching cost is ignored. MN in the network is distributed uniformly, and finally, the location DB is searched by the binary search (Table 1).

Table 1. Parameter for defining the network overhead costs [6, 7]

Parameter	Description
N_m	The number of MN
N_c	The average number of CN per HN
N_h	The number of HA
B_q	The transmission costs of a query message per Hop
B_{dp}	The transmission costs of an average data packet per Hop
B_{da}	The transmission costs of data ACK(Acknowledgement) packet per Hop
B_{rr}	The transmission costs of RR(Return Routablitiy) message per Hop
B_{rc}	The transmission costs of RCoA(Regional Care Of Address) accept /reject message per Hop
B_{lc}	The transmission costs of LCoA(on-Link Care Of Address) accept /reject message per Hop
Φ_{hc}	Hop count between MAP(Mobility Anchor Point) and HA(Home Agent)
Φ_{mm}	Hop count between MN and MAP
Φ_{mh}	Hop count between HA and CN(Corresponding Node)
Φ_{mc}	Hop count between MN and CN
σ	Proportional constant of switching from the radio to a radio link
η	Linear coefficients of the look-up costs
T_r	Subnet duration
λ_s	AverageSession Arrival Rate of each MN
x, y	The number of AR (Access Router) in rows and columns
k	The number of AR per MAP
m	The number of MAP at m = xy/k
K	The maximum transmission unit
α	The average session size
δ_{rc}	Each RCoA registration processing costs in MAP
δ_{lc}	Each LCoA registration processing costs in MAP
δ_h	LU(Location Update) processing costs in HA
χ	Each data (ACK) average number of retransmissions
ζ	IP routing table look-up linear coefficient
ξ	Encapsulation costs
N_{LMA}	The number of LMA
β_{pc}	The transmission costs of PC(Proxy Care Of Address) accept / reject message per Hop
$\beta_{m-query}$	Transmission costs of MLD membership query message
$\beta_{m-report}$	Transmission costs of MLD membership report message
β_{m-mso}	The transmission costs of PC(Proxy Care Of Address) accept / reject message including MSO per Hop
$\Phi_{MAG-LMA}$	Hop count between MAG and LMA
Φ_{LMA-CN}	Hop count between LMA and CN

(*Continued*)

Table 1. (*Continued*)

Parameter	Description
$\Phi_{LMA-LMA}$	Hop count between LMA and LMA
$\Phi_{MAG-MAG}$	Hop count between MAG and MAG
δ_{pc}	PBU processing costs in the LMA
δ_{pc-mso}	PBU processing costs, including the MSO in LMA

4.1   Handover Delay Time Modeling

We define the handover delay time $TDT_{(HO)}^{(\cdot)}$ as the time that MN does not receive a multicast packet during the handover execution. That is, the handover delay time means that MN receives the first multicast packet since MN has moved to LMA2 in LMA1 and pMAG located in LMA1 has received DeReg.PBA from LMA1. $TDT_{(HO)}^{(BASE)}$ is defined as a handover delay time of a basic handover procedure between LMA domains.

$$TDT_{(HO)}^{(BASE)} = (DT_{L2} + DT_{RS} + DT_{LU} + DT_{MLD} + DT_{FWD} + DT_{LMA})$$

In here, DT_{L2} is a link switching time, DT_{RS} is a delay time of RS message sent to nMAG of LMA2 from MN, that is DT_{MAG-MN}. DT_{LU} is PBU and PBA procedure that is location update delay, $2DT_{LMA-MAG}$. DT_{MLD} is the procedure which register the multicast service of MN to Proxy Membership Database, $2DT_{MAG-MN} + DT_{LMA-MAG}$. DT_{FWD} is the arrival delay time of the first multicast packet of the multicast communication from LMA2 to MN, $DT_{LMA-MAG} + DT_{MAG-MN}$. DT_{LMA} is the arrival delay time between LMA domains, $nDT_{LMA-LMA}$.

We will call the handover delay time of first proposed multicast handover procedure as $TDT_{(HO)}^{(PRO_1)}$.

$$TDT_{(HO)}^{(PRO_1)} = DT_{L2} + DT_{RS} + \overline{DT_{FWD}}$$

In here, $\overline{DT_{FWD}}$ is the delay time of multicast communication's first packet buffered from nMAG to MN. It is DT_{MAG-MN}.

We will call the handover delay time of second proposed multicast handover procedure as $TDT_{(HO)}^{(PRO_2)}$.

$$TDT_{(HO)}^{(PRO_2)} = DT_{L2} + DT_{RS} + \overline{DT_{FWD}}$$

The handover delay time of the second proposed multicast handover procedure is equal to the handover delay time of the first proposed multicast handover procedure.

4.2   The Entre Overhead Modeling in the Network

The entire overhead in the network can be defined as the sum of the query message, registration message and packet tunneling costs [7].

(1) Query Message. The costs of query message in proposed FMIPv6(MAG) multicast network are similar to the costs of query message in PMIPv6 network. The costs of query message from the entire CN in the network are as follows:

$$\Lambda^{Query} = N_m N_c \lambda_s (2\beta_c \Phi_{LMA-CN} + \eta N_{LMA}(\log N_m - \log N_{LMA}))$$

(2) Registration Message. The registration message of proposed FPMIPv6(MAG) network does not need the performing procedure for MLD Membership Report of PMIPv6 multicast. Therefore, the costs of the registration message are as follows:

$$\Lambda^{Registration} = N_m \frac{2\beta_{pc}(\Phi_{MAG-LMA} - 1 + \sigma) + \delta_{pc-mso}}{MT_r}$$

(3) The Costs of Packet Tunneling. The costs of multicast network tunneling of proposed FPMIPv6(MAG) are simpler than the costs of the multicast network of PMIPv6. Packet can be transmitted between MN and cnMAG, and that cost is $(\beta_{dp} + \beta_{da})\sigma$. Packet is transmitted between cnMAG and nmMAG through the encapsulation. That costs is $(\beta_{dp} + \beta_{da})(\sigma + \Phi_{cnMAG-mnMA}) + 2\xi$. Finally, the sum of the costs of packet transmission between mnMAG and mnLMA and the costs of packet transmission between MN and mnMAG are $(\beta_{dp} + \beta_{da})$ $(\Phi_{mnMAG-mnLMA} + \sigma) + 2\xi$, which are the costs of the entire packet tunneling. Therefore, the costs of packet tunneling are as follows:

$$\Lambda^{Packet\_T} = N_m N_c \lambda_s \left\lceil \frac{\alpha}{k} \right\rceil \left[\begin{array}{l} (x+1)(\beta_{dp} + \beta_{da})(\sigma + \Phi_{cnMAG-mnMA}) + 2\xi \\ + (\beta_{dp} + \beta_{da})(\Phi_{mnMAG-mnLMA} + \sigma) + 2\xi \end{array} \right]$$

(4) The Entire Overhead in the Network. The entire overhead in the network is defined as the sum of the costs of query message, registration message and packet tunneling, then it is as follows:

$$\Lambda_{Network} = \Lambda^{Query} + \Lambda_{LC}^{Registration} + \Lambda^{Packet\_T}$$

4.3   Numerical Results

We analyzed the multicast handover delay time and the entire overhead by a mathematical model, and tried to evaluate the performance in two different ways. The handover delay time is by reference to the parameter values of [6], the entire overhead is by reference to the parameter values of [7], it is the same as in Table 2.

Table 2. Parameter vales used performance evaluation [7].

Parameter	Description	Parameter	Description	Parameter	Description
N_m	40000	σ	10	ζ	0.3
N_c	1	η	0.3	ξ	0.5
N_h	10	T_r	70	N_{LMA}	40000
B_q	0.6	λ_s	0.01	β_{pc}	0.6
B_{dp}	5.72	x, y	51, 34	$\beta_{m-query}$	0.6
B_{da}	0.60	k	12	$\beta_{m-report}$	0.6
B_{rr}	0.6	m	144.5	β_{m-mso}	0.6
B_{rc}	0.6	K	512	$\Phi_{MAG-LMA}$	35
B_{lc}	0.6	α	10240	Φ_{LMA-CN}	35
Φ_{hc}	35	δ_{rc}	5	$\Phi_{LMA-LMA}$	35
Φ_{mm}	35	δ_{lc}	30	$\Phi_{MAG-MAG}$	35
Φ_{mh}	35	δ_h	30	δ_{pc}	30
Φ_{mc}	35	χ	3	δ_{pc-mso}	30

(1) Analysis of Delay Time. This section provides performance evaluation results according to the delay time. Numerical analysis has been used with the following parameters. In wired connection, transmission delay for one Hop count is $T_{tr} = 20$ ms and $T_{tr} = DT_{MAG-MN}$, $nT_{tr} = DT_{MAG-MN}$ [1]. Figure 4 showed a change in the delay time according to n Hop count between LMA domains, and the comparison of Fig. 4 presented by changing Hop count which MN reached between LMA and MAG to 3 and 7. As shown in Fig. 4 in basic multicast procedure, when MN moves between LMA domains the handover delay time is influenced by Hop count between LMA domains. Also, if we compare Fig. 4, the basic multicast procedure is the value of Hop count's comparison between LMA and MAG in LMA domain which MN reached. That is, the delay time is influenced by two ways; one is Hop count between LMA domains, and the other is Hop count between LMA and MAG in LMA domain. However, FPMIPv6 multicast handover procedure between LMA domains by using proposed MAG and FPMIPv6 multicast handover procedure between LMA domains by using

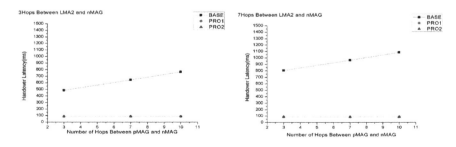

Fig. 4. n Hop Count delaytime between LMA domains.

LMA are not influenced by Hop count between LMA domains and Hop count between LMA and MAG in LMA domain which MN reached.

(2) The Entire Overhead Analysis. This section provides performance evaluation results according to the entire overhead costs. As explained in Sect. 4.2, the costs of the entire overhead is the sum of query message, registration costs, and costs of packet tunneling. The explanation and the value of the parameter which has been applied are by reference to [2]. A formula for calculating the network overhead costs of HMIPv6 and PMIPv6 is a data for a mathematical modeling of each multicast network costs, and it is excluded from the performance evaluation.

Figure 5 is the overhead of entire network in accordance with an increase in MN per MAG. With the increase of the number of MN, both PMIPv6 multicast handover procedure and two proposed multicast handover procedure linearly increased. Although multicast network between LMA by second proposed LMA tunneling is lower than PMIPv6 multicast network proposed from IETF, we can find that the entire overhead of multicast network between LMA through first proposed MAG tunneling is lowest.

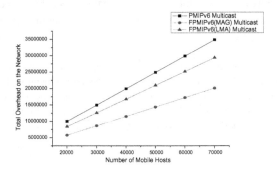

Fig. 5. The overhead of entire network in accordance with an increase in MN per MAG.

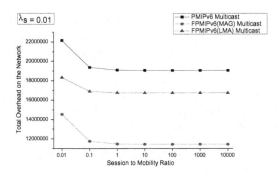

Fig. 6. The overhead of entire network in accordance with an increase in SMR (Session to Mobility Ratio). (λ_s = 0.01).

Figure 6 is the overhead of entire network in accordance with an increase in SMR. SMR is defined as $T \times_r \lambda_s$, and we fix λ_s as 0.01 and increase T_r to calculate SMR in Fig. 6. We can find that the overhead of entire network decreases according to increasing of SMR. Although multicast network between LMA by second proposed LMA tunneling is lower than PMIPv6 multicast network proposed from IETF, we can find that the entire overhead of multicast network between LMA through first proposed MAG tunneling is lowest.

5   Conclusion

This paper conducted mathematical modeling of latency and entire network overhead in accordance with multicast network handover procedure. Also, it performed performance evaluation based thereon. As shown in the graph of performance evaluation, the two proposed multicast networks were found to have substantially lower latency and overhead than PMIPv6 multicast network proposed by IETF. This paper also confirmed that multicast network between LMAs through MAG tunneling among the two proposed multicast networks was more efficient. Moreover, this paper showed more reliable information because it analyzed performance evaluation through the changes of several parameters.

Acknowledgement. This work was supported by Institute for Information & communications Technology Promotion(IITP) grant funded by the Korea government(MSIP) (No. 10047049, Development of smart service infrastructure for vulnerable group safety based on IoT).

This research was supported by the Ministry of Trade, Industry and Energy (MOTIE), Korea, through the Education Support program for Creative and Industrial Convergence (Grant Number S-2016-0117-000).

References

1. Romdhani, I., Kellil, M., Lach, H.Y., Bouabdallah, A., Bettahar, H.: IP mobile multicast: Challenges and solutions. IEEE Commun. Surv. Tutorials 6(1), 18–41 (2004)
2. Gundavelli, S., Leung, K., Devarapalli, V., Chowdhury, K., Patil, B.: Proxy mobile IPv6. IETF RFC 5213, August 2008
3. Li, Y., Su, H., Su, L., Jin, D., Zeng, L.: A comprehensive performance evaluation of PMIPv6 over IP-based cellular networks. In: 2009 IEEE 69th Vehicular Technology Conference, VTC Spring 2009, pp. 1–6. IEEE, April 2009
4. IETF MultiMob working group (2010). https://datatracker.ietf.org/wg/multimob/charter/. Accessed May 2010
5. Yokota, H., Chowdhury, K., Koodli, R., Patil, B., Xia, F.: Fast handovers for proxy mobile IPv6. IETF RFC 5949, September 2010
6. Lee, J.H., Ernst, T.: Fast PMIPv6 multicast handover procedure for mobility-unaware mobile nodes. In: 2011 IEEE 73rd Vehicular Technology Conference (VTC Spring), pp. 1–5. IEEE, May 2011

7. Hossain, M.S., Atiquzzaman, M., Ivancic, W.: Cost analysis of mobility entities of hierarchical mobile IPv6. In: 2010-MILCOM Military Communications Conference, pp. 2280–2285. IEEE, October 2010
8. Soliman, H., Bellier, L., Malki, K.E.: Hierarchical mobile IPv6 mobility management (HMIPv6). IETF RFC 5380 (2008)
9. Song, M., Cho, J.D., Jeong, J.: Global mobility management scheme for seamless mobile multicasting service support in PMIPv6 networks. TIIS 9(2), 637–658 (2015)
10. Jeong, J., Cho, C., Kang, J.J.: Hashing–based lookup service in mobile ad–hoc net-works with multi–hop stretchable clustering. Int. J. Sens. Netw. 15(3), 183–197 (2014)

Intelligent Control and Automation

The Design of SUV Anti Rollover Controller Based on Driver-in-the-Loop Real-Time Simulations

Lixin Song[1(✉)] and Yuping He[2]

[1] College of Physics and Electronic Engineering,
Hubei University of Arts and Science, Xiangyang 441053, Hubei, China
songlixin66@qq.com
[2] Faculty of Engineering and Applied Science,
University of Ontario Institute of Technology, Oshawa, ON L1H 7K4, Canada
yuping.he@uoit.ca

Abstract. This paper presents the design and validation of a differential braking (DB) controller for Sport Utility Vehicles (SUVs) using driver-in-the-loop real-time simulations. According to the driver's logic, the desired vehicle states will be decided, while actual vehicle states deviate desired values, the control system is applied to improve the lateral stability of SUVs. To derive the controller design, driver-in-the-loop real-time simulations are conducted on the UOIT (University of Ontario Institute of Technology) vehicle simulator,the Fishhook maneuver and Double Lane Change test scenarios are simulated to examine the performance of the controller. The driver-in-the-loop real-time simulation results demonstrate the effectiveness of the proposed differential braking controller in the lateral rollover stability and maneuverability improvement of the SUV.

Keywords: Differential braking control · Lateral stability · Driver-in-the-loop · Real-time simulations

1 Introduction

The Sport Utility Vehicles (SUVs) have been rapidly developed and widely promoted since 1990 s because they are able to adapt to various road conditions. However, the fact that SUVs usually have much higher center of gravities and shorter wheel bases leads to the lateral imbalance which may cause the greater rollover accident rate than other vehicles. Therefore, the lateral stability control of SUV has been paid a lot of attention by researchers and car manufacturing giants all over the world [1–3]. SUV stability control by applying corrective yaw moment can reduce the deviation of vehicle behaviors. The control of yaw moment can be achieved by a variety of approaches, including torque distribution systems, active suspensions, and differential braking control systems [4–6]. To date, differential braking control systems have been applied to road vehicles due to their cost-effective implementations based on the existing advanced braking techniques, such as anti-lock braking systems [7, 8]. With the differential braking control technique, the required yaw moment can be achieved through manipulating the braking effects at the four wheels, differentiating braking pressures in the left/right and front/rear wheel

© Springer International Publishing Switzerland 2016
D.-S. Huang et al. (Eds.): ICIC 2016, Part III, LNAI 9773, pp. 509–519, 2016.
DOI: 10.1007/978-3-319-42297-8_47

cylinders [9]. This paper presents a design and validation method for differential braking controllers of SUVs using driver-in-the-loop real-time simulations reconstructed in LabView and CarSim software packages. A linear yaw plane vehicle model with 4 degrees of freedom (DOF) is generated to derive the differential braking controller.

2   Vehicle System Model

Figure 1 shows the schematic diagram with 4 degrees of freedom (DOF) model to represent a SUV, including longitudinal motion, lateral motion, yaw motion and rollover motion. It is assumed that the SUV is front-wheel driving and the left and right steering angle is the same, the vehicle dynamics character is symmetric about X axis, the front wheel steering angle δ_f, vehicle body roll angle φ and sideslip angle of Vehicle body β are regarded as small, using small angle approximation, the motion equation of this SUV model is obtained as follow:

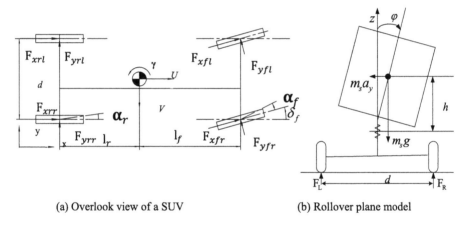

(a) Overlook view of a SUV   (b) Rollover plane model

Fig. 1. The rollover dynamic model.

The force balance equation in lateral direction (Y axis) is given by:

$$m(\dot{V} + U_\gamma) - m_s h \ddot{\varphi} = F_{yfl} + F_{yfr} + F_{yrl} + F_{yrr} \tag{1}$$

The torque balance equation in yaw direction (circling Z axis) is given by:

$$I_z \dot{\gamma} = l_f (F_{yfl} + F_{yfr}) - l_r (F_{yrl} + F_{yrr}) + \Delta M \tag{2}$$

The torque balance equation in circling X axis is given by:

$$I_x \ddot{\varphi} = m_s g h \varphi - c_\varphi \dot{\varphi} - k_\varphi \varphi - m_s h^2 \ddot{\varphi} + m_s h a_y \tag{3}$$

In (1), (2) and (3), m is Mass of the vehicle, m_s is Sprung mass, U/V is Longitudinal velocity/Lateral velocity of Central Gravity (CG), $\gamma/\dot{\gamma}$ is Yaw rate/Yaw angular acceleration, c_f / c_r is Front cornering stiffness/Rear cornering stiffness, F_{xfl} / F_{xfr} is Longitudinal force of front-left tire/Longitudinal force of front-right tire, F_{yfl} / F_{yfr} is Lateral force of front-left tire/Lateral force of front-right tire, F_{xrl} / F_{xrr} is Longitudinal force of rear-left tire/Longitudinal force of rear-right tire, F_{yrl} / F_{yrr} is Lateral force of rear-left tire/Lateral force of rear-right tire, I_x / I_z is Moment of inertia about X axis/ Z axis, l_f / l_r is Distance from CG to front axle/Distance from CG to rear axle, d is vehicle Axle tread, $\dot{\varphi}/\ddot{\varphi}$ is Vehicle rollover angle velocity/vehicle rollover angle acceleration, k_φ is Suspension roll stiffness, c_φ is Suspension roll damp, α_f is Front tire slip angle, h is Roll arm length, F_L/F_R is Left vertical load/Right vertical load. The vehicle lateral acceleration a_y can be written on the form:

$$a_y = \dot{V} + U\gamma \qquad (4)$$

As shown in Fig. 1, while a_y is bigger and the vehicle is about to roll, the front right wheel will be differentially braked, ΔM is the required compensated moment.

3 The Design of Differential Braking Controller

This paper presents a design method for differential braking controllers of SUVs by controlling the target wheels, the controller outputs the required compensated yaw moment ΔM and the braking is implemented by imposing braking force to the target wheels. The two state variables of the SUV, γ (yaw rate) and β (Sideslip angle of Vehicle body) are feedback controlled the lateral acceleration a_y to become bigger and rollover. Figure 2 shows a schematic diagram of the rollover controller.

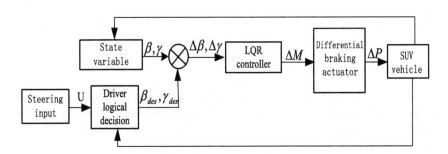

Fig. 2. Control scheme for differential braking controller design

As shown in Fig. 2, the interrelated driver, controller and the SUV (represented by the vehicle state variables) form a closed-loop control system. According to the current vehicle velocity and steering angle, the driver logically decides the desired vehicle sideslip angle β_{des} and the desired vehicle yaw rate γ_{des}, $\Delta\beta$ and $\Delta\gamma$ is the difference between the measured state variables and their desired counterparts. The controller output is target compensated yaw moment ΔM calculated by the control algorithm, ΔP is target braking pressure for control, the controller input includes vehicle state variables and driver input.

The controller parameters are derived according to the SUV model provided in Sect. 2. While the tires sideslip angles are small and load is vertical, the lateral forces and sideslip angles satisfy linear relation, their linear relation can be written as [10]:

The front tires lateral forces:

$$F_{yf} = c_f \alpha_f \tag{5}$$

The rear tires lateral forces:

$$F_{yr} = c_r \alpha_r \tag{6}$$

where c_f / c_r is front/rear axis cornering stiffness, α_f / α_r is front/rear tire sideslip angle, they satisfy following equation:

$$\alpha_f = \delta_f - \beta - \frac{l_f \gamma}{U} \tag{7}$$

$$\alpha_r = \frac{l_r \gamma}{U} - \beta \tag{8}$$

Under smaller sideslip angles and constant longitudinal velocity [11], $\beta = \frac{V}{U}$, it can be obtained:

$$\dot{\beta} = \frac{\dot{V}}{U} \tag{9}$$

From Fig. 1,

$$\Delta M = \Delta F_{xfr} \left(\frac{d}{2} \cos\delta_f + l_f \sin\delta_f \right) \tag{10}$$

Meanwhile, the brake force is simply approximated as a linear function of the corresponding brake pressure [12], including the saturation effect and the hydraulic lag. The brake pressure increment of front right wheel can be obtained:

$$\Delta P_{xfr} = \frac{0.5 T_t}{K_{bf}} - \frac{R_e \Delta M}{K_{bf}(\frac{d}{2}\cos\delta_f + l_f\sin\delta_f)} \tag{11}$$

In (11), T_t is Driving axle shaft torque, K_{bf} is Brake gain of the front tires, R_e is Effective tire radius.

Replacing the Eqs. (1) ∼ (3) with the Eqs. (4) ∼ (9) and assuming that vehicle mass equals sprung mass, that is, the wheel mass and suspension mass is ignored, the equations of motion can be expressed as follows:

$$mU\dot{\beta} - mh\ddot{\varphi} = (-c_f - c_r)\beta + \left(\frac{c_r l_r - c_f l_f}{U} - mU\right)\gamma + c_f\delta_f \qquad (12)$$

$$I_z\dot{\gamma} = (c_r l_r - c_f l_f)\beta - \frac{c_f l_f^2 + c_r l_r^2}{U}\gamma + c_f l_f\delta_f + \Delta M \qquad (13)$$

$$-mhU\dot{\beta} + \left(I_x + mh^2\right)\ddot{\varphi} + c_\varphi\dot{\varphi} = mhU\gamma + (mgh - k_\varphi)\varphi \qquad (14)$$

The Eqs. (12) $\sim$ (14) can be rearranged in a state space form as:

$$M\dot{x} = Ax + Bu + E\delta_f \qquad (15)$$

where $x = [\beta \quad \gamma \quad \dot{\varphi} \quad \varphi]^T$ is defined as system state variables, $u = \Delta M$ is control variables, δ_f is the driver input. In (15),

$$M = \begin{bmatrix} mU & 0 & -mh & 0 \\ 0 & I_z & 0 & 0 \\ -mhU & 0 & (I_x + mh^2) & c_\varphi \\ 0 & 0 & 0 & 1 \end{bmatrix}$$

$$A = \begin{bmatrix} -c_f - c_r & \frac{c_r l_r - c_f l_f}{U} - mU & 0 & 0 \\ c_r l_r - c_f l_f & -\frac{c_f l_f^2 + c_r l_r^2}{U} & 0 & 0 \\ 0 & mhU & 0 & mgh - k_\varphi \\ 0 & 0 & 1 & 0 \end{bmatrix}$$

$$B = [\, 0 \quad 1 \quad 0 \quad 0 \,]^T, \quad E = [\, c_f \quad c_f l_f \quad 0 \quad 0 \,]^T$$

The controller is designed to make the vehicle state variables approach the desired state variables as much as possible, the desired variables are defined to enable the vehicle to follow the designated path and to avoid inappropriate understeer and oversteer handling characteristics when negotiating a curved path. Previous studies have suggested that, in order to realize anticipated vehicle stability control, the desired state variables should be determined by the driver according to current vehicle velocity and steering angle. Based on Paper [13], the desired sideslip angle βdes and yaw rate γdes can be determined as follows:

$$\beta_{des} = \frac{\{2c_f c_r l_r(l_f + l_r) - c_f l_f mU^2\}\delta_f}{2c_f c_r(l_f + l_r)^2 - mU^2(c_f l_f - c_r l_r)} \qquad (16)$$

$$\gamma_{des} = \frac{2c_f c_r l_r(l_f + l_r)\delta_f}{2c_f c_r(l_f + l_r)^2 - mU^2(c_f l_f - c_r l_r)} \qquad (17)$$

The output errors between the vehicle state variables and those from the desired model need to be suppressed, the function of differential braking will be implemented to form a feedback compensator using the Linear Quadratic Regulator (LQR) technology. The LQR algorithm is used to obtain optimal feedback control gain k by minimizing the following cost function J [14]

$$J = \frac{1}{2} \int_0^\infty (e^T Q e + u^T R u) dt \qquad (18)$$

where $e = x_{des} - x$, $x = [\,\beta \quad \gamma \quad \dot{\phi} \quad \phi\,]^T$, $x_{des} = [\,\beta_{des} \quad \gamma_{des} \quad 0 \quad 0\,]^T$, e is the difference between the measured and the desired values of the state variables. Q and R are the weighting factors of state and control variables, respectively. The solution of the optimization is the feedback controller in the following form:

$$u = -k \cdot x \qquad (19)$$

with optimal gain matrix k:

$$k = R^{-1} B^T P \qquad (20)$$

where P is a symmetric, positive semi-definite symmetric matrix which satisfies the Riccati Equation:

$$A^T P + PA - PBR^{-1} BP + Q = 0 \qquad (21)$$

In order to achieve an acceptable and steady state solution in (21), the selection of the appropriate weighting factors Q and R is critical, the values of weighting factors are carefully tuned with acceptable performance measure under various vehicle maneuvers.

4   Simulation Results

In this paper, the controller design is examined with real-time simulations on the vehicle simulator, the vehicle simulator integrates the driver, vehicle dynamic model, controller and visual display in a real-time working mode. The simulator allows the interactions between the driver and the controller such that the virtual vehicle is driven under the specified testing maneuvers [15].

One of the key points in setting up the virtual testing environment is to construct the real-time version of the controller and the vehicle model [16, 17]. The integration of the controller constructed in LabView and the vehicle model developed in CarSim is implemented on the vehicle simulator. All the measured state variables can be obtained from real-time CarSim vehicle models, the LQR controller presented in Sect. 3 is designed in LabView. With the integration and synchronization of the vehicle model and the controller in real-time on the vehicle simulator, the interactions among the human driver, the controller, the virtual vehicle and road can be fully investigated for the design and validation of the differential braking controller. The driver-in-the-loop real-time simulations have been conducted in two different cases, i.e., with and without involving the differential braking controller, respectively. The driver is involved to evaluate the controller by operating the vehicle simulator. The simulations use double lane change test and fishhook test to adjust and validate the controller performance, the important variables effecting the SUV stability such as sideslip angle and yaw rate, i.e. are analyzed and compared.

4.1   Double Lane Change Simulation (The Coefficient of Friction μ = 0.75, 0.25)

Figures 3 and 4 shows the simulation results in the double lane change scenario. In the tests, the forward speed of the controlled vehicle at the 350 m mark was 90 kph compared to the baselines 100 kph (or less in the cases of spin out). It can be seen that the differential braking system does not offer drastic or substantially noticeable improvements for most performance metrics on dry road conditions providing a high or normal friction road surface ($\mu = 0.75$). A small decrease in yaw, yaw rate, and lateral acceleration is observed. A noticeable reduction in sideslip angle, an increase in

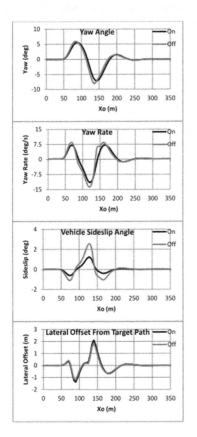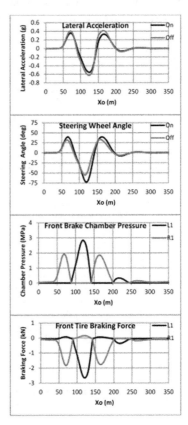

Fig. 3. Double Lane Change ($\mu = 0.75$)

required steer angle, and a corresponding minimal increase in lateral offset from the path, is observed when the system is implemented. This is indicative of a preference towards understeer and a general reluctance to changes in direction, making the vehicle appear 'rigid' and less responsive in the yaw plane, with preference to minimizing heading changes and minimizing sideslip rather than increasing turn-in responsiveness and aggressiveness while this is not entirely desired on dry roads (or performance-oriented vehicles) as sharp and immediate yaw characteristics in response to driver steering input may be necessary to avoid collisions or increase performance, it is clear that the system offers exceptional stability on deteriorating road conditions, as evidenced by the 0.25 mu tests. The system provides a tremendous increase in safety during the rainy/icy surface tests. The performance measures for both of these tests using differential braking are impressive. Sideslip angle is kept below 2 degrees, indicating a very non-dramatic and well-stabilized maneuver. Lateral acceleration is smooth and well mitigated throughout the maneuver. Yaw rate spikes up to a plateau value, which is held for a brief moment in a very smooth manner, creating the yaw angle curves that are smooth and straight. Steering wheel angle is indicative of inputs that do not require the removal of the operators hands from the wheel, as the peak angle is slightly over one quarter turn of the wheel.

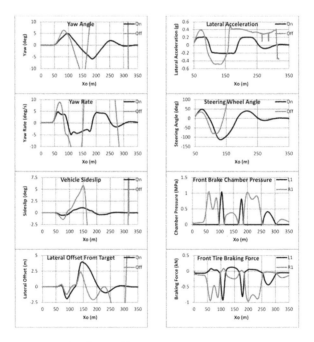

Fig. 4. Double Lane Change ($\mu = 0.25$)

4.2 Fishhook Maneuver Simulation (High μ Surface)

The initial vehicle velocity is 100 km/h. And the coefficient of friction is 0.85.

As shown in Fig. 5, the fishhook test results display a reduction in lateral acceleration, roll rate, and roll angle. As with the double lane change testing, a substantial reduction in sideslip angle is observed. The important metric here is the lateral load transfer ratio (LTR). It is clear that in both cases the LTR for the rear axle is larger than the front axle. This is the result of various passive vehicle parameters, such as front/rear weight distribution and the relative stiffness of the front and rear axles. Since the rear LTR is greater than the front, we can expect the rear inside tire to lift off the ground before the front inside tire. The baseline vehicle is approaching rear tire lift with a rear LTR of nearly 90 % at the peak.

However, the differential braking system reduces this value by approximately 10 % at the peak. Similar reductions can be seen in the front axle LTR. It is expected that a reduction in the risk of rollover would be apparent when utilizing the differential braking system.

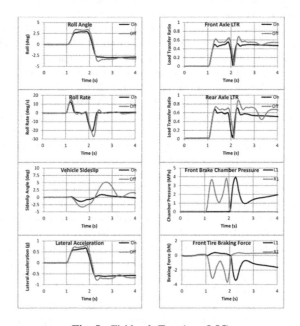

Fig. 5. Fishhook Test (μ = 0.85)

5 Conclusions

This paper presents a design and validation method for differential braking controllers of SUVs using driver-in-the-loop real-time simulations. The simulation results show that braking controller can effectively manipulate the yaw moment to improve the lateral stability and maneuverability of the SUV, lower the sideslip angle when steering and LTR peak value and enhance the anti-rollover performance of the vehicle. In the tests, the interactions among the driver, differential braking controller, and SUV model are well exposed and can be fully investigated for improving designs while provides lower cost control system assessment prior to in vehicle road testing. The next research will further validate and perfect the anti-rollover controller by the worst case evaluation method and hardware in the loop simulations.

Acknowledgement. Financial support for this research is provided by The Natural Science Foundation of Hubei Province (No. 2014CFB477) and Hubei Key Laboratory of Low Dimensional Optoelectronic Material and Devices (No. HLOM142011).

References

1. NHTSA: Traffic Safety Facts 2010: A Compilation of Motor Vehicle Crash Data from the Fatality Analysis Reporting System and General Estimates System (2012)
2. Sun, T., He, Y.: Lateral stability improvement of car-trailer systems using active trailer braking control. J. Mech. Eng. Autom. 2(9), 555–562 (2012)
3. Zong, C., Zhu, T., Wang, C., Liu, H.: Multi-object stability control algorithm of heavy tractor semi-trailer based on differential braking. Chin. J. Mech. Eng. 25(1), 1–10 (2012)
4. Zhang, S., Zhang, T., Zhou, S.: Vehicle stability control strategy based on active torque distribution and differential braking. In: International Conference on Measuring Technology and Mechatronics Automation, vol, 1, pp. 922–925 (2009)
5. Barbarisi, O., Palmieri, G., Scala, S., Glielmo, L.: LTV-MPC for raw rate control and side slip control with dynamically constrained differential braking. Eur. J. Control 15(3), 468–479 (2009)
6. Esmailzadeh, E., Goodarzi, A.: Optimal raw moment control law for improved vehicle handling. Mechatronics 13(5), 659–675 (2003)
7. Xu, Y., Weng, J., Jin, Z., Feng, T.: The vehicle rollover control based on active steering and differential braking. Comput. Simul. 28(6), 330–334 (2011)
8. Friedman, K., Hutchinson, J.: Review of existing repeatable vehicle rollover dynamic physical testing methods. In: ASME International Mechanical Engineering Congress and Exposition, Lake Buena Vista, vol. 11, pp. 51–59 (2009)
9. Guo, J., Gao, Z., Guan, X.: Improved ABS and its influence for vehicle lateral stability. J. Jilin Univ. (Engineering and Technology Edition) 39(3), 39–42 (2009)
10. Huang, A., Chen, C.: A low-cost driving simulator for full vehicle dynamics simulation. IEEE Trans. Veh. Technol. 52(1), 162–172 (2003)
11. Yu, Z.: Automobile Theory, 4th edn, pp. 114–118. China Machine Press, Beijing (2007)
12. Short, M., Pont, M.: Assessment of high-integrity embedded automotive control systems using hardware in the loop simulation. J. Syst. Softw. 81(7), 1163–1183 (2008)
13. Chung, T., Kyongsu, Y.: Design and evaluation of side slip angle-based vehicle stability control scheme on a virtual test track. IEEE Trans. Control Syst. Technol. 14(2), 224–234 (2006)
14. He, Y., Mcphees, J.: Multidisciplinary design optimization of mechatronic vehicles with active suspensions. J. Sound Vib. 283(2), 217–241 (2005)

15. Ding, X., He, Y.: Application of driver-in-the-loop real-time simulations to the design of SUV differential braking controllers. In: Su, C.-Y., Rakheja, S., Liu, H. (eds.) ICIRA 2012, Part I. LNCS(LNAI), vol. 7506, pp. 121–131. Springer, Heidelberg (2012)
16. Valasek, M., Vaculin, O., Kejval, J.: Global chassis control: integration synergy of brake and suspension control for active safety. In: Proceedings of the International Symposium on Advanced Vehicle Control (2004)
17. Lee, H.: Virtual test track. IEEE Trans. Veh. Technol. **53**(4), 1818–1826 (2004)

Efficient Computation of Continuous Range Skyline Queries in Road Networks

Shunqing Jiang[1], Jiping Zheng[1,2(✉)], Jialiang Chen[1], and Wei Yu[1]

[1] College of Computer Science and Technology,
Nanjing University of Aeronautics and Astronautics, Nanjing, China
{jiangshunqing, jzh, chenjialiang, echoyu}@nuaa.edu.cn
[2] School of Computer Science and Engineering,
University of New South Wales, Sydney, Australia

Abstract. Skyline query processing in road networks has been investigated extensively in recent years. Skyline points for road network applications may be large while the query point may only interest the ones within a certain range. In this paper, we address the issue of efficient evaluation of Continuous Range Skyline Queries (CRSQ) in road networks. Due to the computation of network distance between objects in road networks is expensive and suffers the limitation of memory resources, we propose a novel method named Dynamic Split Points Setting (DSPS) dividing a given path in road networks into several segments. For each segment, we use Network Voronoi Diagrams (NVDs) based technique to calculate the candidate skyline interest points at the starting point of the segment. After that, when the query point moves, we dynamically set the spilt points by DSPS strategy to ensure that when the query point moves within a segment, skyline points remain unchanged and only need to be updated while moving across the split points. Extensive experiments show that our DSPS strategy is efficient compared with previous approaches.

Keywords: Continuous range skyline queries · Road networks · Network voronoi diagram · Dynamic split points setting

1 Introduction

Inspired by the advance in mobile information systems and positioning technology, the research of skyline queries in road networks has received considerable attention. A traditional skyline query retrieves a set of data points that are not dominated by any other points only considering static dimensions. However, skyline queries in road networks not only take the inherent static attributes of targets into consideration, but also consider spatial attribute (network distances between query point and targets). Figure 1 shows an example where 5 points (p_1-p_5) represent hotels with two inherent static attributes: price and ranking and a query point q represents a user's location in a road network (shown in Table 1). When issuing a skyline query: *find cheaper, higher ranking and also closer hotels*, the network distance from each hotel to q becomes one of the dimensions. For detailed hotel information as shown in Table 1, the skyline query results are (p_3, p_5) for they are cheaper, higher ranking, and closer than others (p_1, p_2, p_4).

© Springer International Publishing Switzerland 2016
D.-S. Huang et al. (Eds.): ICIC 2016, Part III, LNAI 9773, pp. 520–532, 2016.
DOI: 10.1007/978-3-319-42297-8_48

In our hotel-finding example, if the query point q is a tourist in a moving car from n_1 to n_2 in the segment $n_1 n_2$. The tourist may only interest in the hotels which network distance to him/her are within a certain range, e.g. 6 km. We can see that hotel p_3 will not be a skyline point. This is because the distance between p_3 and q is greater than 6. In this paper, we address the problem of efficient Continuous Range Skyline Queries (CRSQ) processing in road network. The points retrieved by CRSQ are termed Range Skyline Points RSQ. CRSQ is defined as: given a set of spatial interest points, a distance range value, and a query point moves on a given path, retrieve RSQ to the query point. The RSQ is updated according the location of the query point on the path. And at any time, the network distance between each interest point of RSQ and query point is less than or equal to the range distance. To evaluate CRSQ, we first determine global candidate skyline points [6] by using PINE* (improved version for range queries of Progressive Incremental Network Expansion (PINE) first proposed by Safar et al. [13]) technique based on Network Voronoi Diagrams (NVD). Then we determine the split points using our DSPS strategy and update skyline points according to the location of the query point as well as locations of split points. In general, the contributions of this paper are summarized as follows:

- We introduce improved Progressive Incremental Network Expansion (PINE*) technique for efficient network distance computation to answer range skyline queries in road networks.
- To evaluate continuous range skyline queries, we propose a Dynamic Split Points Setting (DSPS) strategy which can return skyline points immediately according to the locations of the query point and split points.
- We give experimental evidence that the solutions we proposed in this paper to address CRSQ problem is effectively and clearly.

The rest of the paper is organized as follows. In Sect. 2, we discuss related work. Some important pieces of knowledge are discussed in Sect. 3. In Sect. 4 we present how our CRSQ method and DSPS strategy can be used to answer continuous range skyline queries in road networks. In Sect. 5, extensive experiments are conducted to show the efficiency of the proposed method. Finally, we conclude our work in Sect. 6.

2 Related Work

The skyline operator has been introduced to the database community by Borzsony et al. [2] in 2001, and subsequent research has focused on efficient skyline query processing. Chomicki et al. [3] developed Sort-Filter-Skyline algorithm (SFS) which improved Block-Nested-Loops (BNL) algorithm by pre-sorting the data set. Several optimizations to SFS algorithm, e.g. [5], increase its efficiency. Papadias et al. [12] proposed a progressive algorithm called branch-and-bound skyline (BBS), based on a nearest neighbor search technique supported by R-trees.

Deng et al. [4] first investigated spatial skyline problems in road networks and presented the multi-source skyline query which considers several query points at the same time. Jang et al. [7] first addressed continuous skyline queries in road networks. The method is to pre-compute a range R for each interest point p to ensure that if the

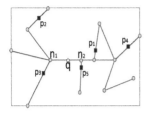

Fig. 1. Skyline query in road networks

Table 1. Attributes of interest points

Hotels	Price	Ranking	Distance
P_1	7	**	10
P_2	3	***	11
P_3	3	******	7
P_4	5	****	15
P_5	4	*****	4

query point moves in R, then p must be a skyline point. Son et al. [15] considered the road network distance as Manhattan distance and proposed Manhattan spatial skyline queries. Some other studies in road network skyline query focus on finding the skyline paths [16]. Multiple path attributes are considered and different paths would have different values at each attribute. Thus the goal is to find the set of skyline paths to allow the user to choose the most preferable paths [10, 11]. Recently, Saad Aljubayrin et al. [1] also considered the skyline trips of multiple POIs categories query problem and pre-computes and stores the distances between POIs and some geographical regions to produce near optimal results. The most related work to continuous range skyline queries is Cd_ε-SQ + based on grid index proposed by Huang et al. [6]. Given a path, Cd_ε-SQ+ broke the path to several segments without intersections. Cd_ε-SQ+ to answer continuous range skyline queries on a segment includes two phase: (1) a global skyline points determination phase; (2) a result turning point determination phase. In the first phase, Cd_ε-SQ+ was associated with the grid index to determine the skyline points on the end point of a segment. Connect these determined skyline points with the interest points on this segment as the global skyline points of this segment. In the second phase, Cd_ε-SQ+ determined result turning points by comparing the static and dynamic attributes of the global skyline points. However, Cd_ε-SQ+ is not efficient due to the time-consuming network distance computation and the determination of split points (landmarks called by Huang et al. [6]).

Kolahdouzan et al. [8] first used Network Voronoi Diagrams (NVD) to solve kNN queries in road networks. They proposed a novel approach called VN^3. VN^3 partitions a large network into smaller network Voronoi polygons (NVPs), then pre-compute distances across each NVP. Safar et al. [13] proposed a novel approach, termed Progressive Incremental Network Expansion (PINE). This approach is also based NVD and pre-computes distances across each NVP. In intra-distance, a computation similar to the Incremental Network Expansion (INE) was adopted. PINE has less disk access time and CPU time than VN^3, and PINE is applicable for all kinds of the density and distribution of the interest points. Also, Safar et al. [14] addressed some spatial queries in road networks based on PINE, include kNN and CkNN queries to show the effectiveness of PINE technique. In this paper, we take full virtues of PINE and NVD to calculate the network distance to answer continuous range skyline queries.

3 Preliminaries

NVD and the Preprocessing of PINE*. Figure 2(a) depicts a road network model G (N, P, E), which $N = \{n_1, n_2, ..., n_{13}\}$ are the intersections of a road, $P = \{p_1, p_2, p_3\}$ are the interest points, the edges set $E = \{n_1n_2, n_1n_3, ..., p_3n_{13}\}$ represented the segment of the road. Figure 2(b) shows the NVD of the road network in Fig. 2(a) which divides the network into network Voronoi polygons, namely $NVP(p_i)$. Interest points p_1, p_2 and p_3 are generators. The set $B = \{b_1, b_2, ..., b_7\}$ is called border points set. A border point is the midpoint of a shortest path from one generator to another one, e.g. b_1, $d_N(b_1, p_1) = d_N(b_1, p_2)$.

For each network Voronoi polygons, PINE* pre-computes the inter distances between border points and the network distance from each border point to the generator interest point. After that, the information about border points, all of NVP's adjacent NVP and the pre-computed distances will be stored.

Four Kinds of Split Points. For CRSQ, the split point specifies the location along with the path where the skyline results of the query point will change. That is, the skyline results set remains unchanged until q arrives at the location of the split point. Huang et al. [6] introduced four kinds of split points:

$exit_p$. Interest point $p \in RSP$, when the query point passes through $exit_p$, $d_N(p, q) > d_r$. So, p is not a skyline point at this time.

in_p. Interest point $p \notin RSP$, when the query point passes through in_p, $d_N(p, q) < d_r$. At this time, if $\nexists p^* \in RSP^{up}$, $p^* < p$ we should add point p into RSP.

$exit_{pj/pi}$. Interest point p_i, $p_j \in RSP$, $p_j \prec_{static} p_i$ but $d_N(p_j, q) > d_N(p_i, q)$. When the query point passes through $exit_{pj/pi}$, $d_N(p_j, q) < d_N(p_i, q)$, so, $p_j \prec p_i$, at this time, we should delete point p_i from RSP.

$in_{pi/pj}$. Interest point $p_i \in RSP$, $d_N(p_j, q) < d_r$ but $p_i \prec p_j$ $(p_j \notin RSP)$. If the query point passes through $in_{pi/pj}$, $d_N(p_i, q) > d_N(p_j, q)$, $p_i \nprec p_j$. At this time, if $\nexists p^* \in RSP^{up}$, $p^* < p$, we should add point p_j into RSP.

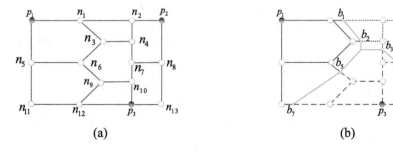

(a)   (b)

Fig. 2. Network Voronoi Diagrams

For CRSQ on a segment, the approach that introduced by Huang et al. [6] computed all split points of the segment at the beginning, as the query point moves, just to update the *RSP* according the split point that it arrives at. This process needs amount of time to compute split points and stores them. Our approach adopts a novel approach dynamically setting the split points and updating the *RSP* as the query point moves.

4 Proposed CRSQ

In this paper, the result of CRSQ for a query point q includes the skyline results set and a split point. The split point specifies the location along with the path where the skyline results set of the query point will change. Algorithm 1 shows the pseudo codes of D-CRSQ. At the beginning, we divide path P (the given path in CRSQ) into several segments not containing any interest points or road network intersections (line 1). For example, we split the path AE (dashed line in Fig. 3) into six segments, (A, B), (B, p_1), (p_1, C), (C, D), (D, p_2) and (p_2, E). For each segment, we first acquire the candidate skyline points (*CSP*) by PINE*, then use our dynamic split points setting strategy to implement CRSQ on this segment (line 2-3). Lemma 1 shows the basic idea of acquire the *CSP* of a segment.

Lemma 1. *Given a segment $n_i n_j$ of a road network G, the length of $n_i n_j$ is L, a query point q moves from n_i to n_j, distance range value d_r, the set CSP_{ninj} is a collection of interest points, and $d_N(p, n_i) < L + d_r$, $p \in CSP_{ninj}$. Wherever the query point q is located on $n_i n_j$, if $p' \in P \cap p' \notin CSP_{ninj}$, then, $d_N(p', q) > d_r$, and P is the interest points set of G.*

Proof. Because $p' \notin CSP_{ninj}$, so, $d_N(p', n_i) > L + d_r$. We proof $d_N(p', q) > d_r$. There are two situations: (1) If $d_N(p', n_i) \leq d_N(p', n_j)$, we have $d_N(p', q) = d_N(p', n_i) + d_N(n_i, q)$, because $d_N(p', n_i) > L + d_r$, so, $d_N(p', q) > d_r$. (2) If $d_N(p', n_i) > d_N(p', n_j)$, $d_N(p', n_i) = L + d_N(p', n_j)$. $d_N(p', n_i) > L + d_r$, so, $d_N(p', n_j) > d_r$. $d_N(p', q) = d_N(p', n_j) + d_N(n_j, q)$. That is, $d_N(p', q) > d_r$.

According to Lemma 1, when acquiring the *CSP* of segment $n_i n_j$, we just to find all the interest points p ($p \in P$) that, $d_N(p, n_i) < L + d_r$. For the example shown in Fig. 4, the interest points set $\{p_1, p_2, ..., p_8\}$ in the area surrounded by the larger dotted line is $CSPn_i n_j$, the distances between them to n_i are less than $L + d_r$. The *RSQ* of a query point q on the segment $n_i n_j$ will only be generated from $\{p_1, p_2, ..., p_8\}$. $d_N(p_9, n_i) > L + d_r$ and p_9 will never become a range skyline point wherever the query point q is located on the segment $n_i n_j$.

Now we use Fig. 4 to show the process of CRSQ on segment $n_i n_j$. Processing on the other segments is similar to $n_i n_j$. In Sect. 4.1, we show how to efficiently acquire all the interest points that $d_N(p, n_i) < L + d_r$ (the *CSP* of segment $n_i n_j$) and calculate the network distance $d_N(p, n_i)$ by using PINE* technique. Section 4.2 introduces our proposed method to process dynamic continuous range queries.

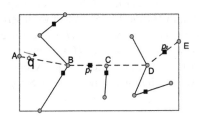

Fig. 3. Example of CRSQ

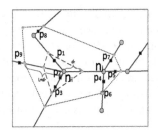

Fig. 4. Example of dynamic CRSQ

Algorithm 1. DSPS strategy of CRSQ queries (D-CRSQ)

---

1 Divide P into several segments each of which does not contain any interest
  point or road network intersections;
2 for each segment P_{segi} of P do
3   Find candidate skyline points of P_{segi} at the starting node of P_{segi};
4   Dynamicaly acquire split points and update the RSQ according to the
    candidate skyline points of P_{segi} as the query point q moving;

4.1 Calculation of Candidate Skyline Points Set of a Segment

Our PINE* approach to acquire candidate skyline points exploits the properties of the
Network Voronoi Diagrams (NVDs) and the pre-computes the distance for each cell of
the NVD. For more information about NVDs, please refer to [8] for details. Each
candidate skyline interest point p_i of segment $n_i n_j$ is in the form of $(p_i, d_N(n_i, p_i))$ stored
incrementally in CSP_{ninj}, and sorted according to $d_N(n_i, p_i)$ in ascending order. Algorithm 2 shows the pseudo codes of acquiring the CSP_{ninj} of the segment $n_i n_j$. The first
part of Algorithm 2 is the initializing process (line 1-6). For the query point, we first
use R-tree index to locate the Voronoi polygon $NVP(p_i)$ that contains node n_i. The
generator of $NVP(p_i)$ is the interest point p_i, then let the border points of $NVP(p_i)$ be
$\{b_{i1}, b_{i2}, \ldots, b_{ik}\}$ (line 2-3). After that, Dijkstra algorithm is adopted to calculate the
road network distance $d_N(n_i, p_i)$ between n_i and p_i and insert $(p_i, d_N(n_i, p_i))$ into CSP_{ninj}
(p_i is the nearest neighbor of n_i, if $d_N(n_i, p_i) > L + d_r$, there is no interest point located
in the distance range). At the same time, we also calculate the road network distances
between n_i and $\{b_{i1}, b_{i2}, \ldots, b_{ik}\}$: $d_N(n_i, b_{i1}), d_N(n_i, b_{i2}), \ldots, d_N(n_i, b_{im})$ and insert $(b_{i1},
d_N(n_i, b_{i1})), (b_{i2}, d_N(n_i, b_{i2})), \ldots, (b_{im}, d_N(n_i, b_{im}))$ into minimum heap H with the weight
$(b, d_N(n_i, b))$ (line 4-6).
    The next part is the main calculating process (line 7-17). Line 7 remove first entry
from H: $(b, d_N(n_i, b))$, while $d_N(n_i, b) < L + d_r$, then, for the other generator point p_k of
b, we use formula $d_N(n_i, p_k) = d_N(n_i, b) + d_N(b, p_k)$ to compute the distance between n_i
and p_k directly for $d_N(b, p_k)$ is pre-computed. If $d_N(n_i, p_k) < L + d_r$, insert $(p_k, d_N(n_i,
p_k))$ into CSP_{ninj} (line 9-12). For each other border point b_j of $NVP(p_k)$, we also get the

distance $d_N(n_i, b_j)$ directly by the formula: $d_N(n_i, b_j) = d_N(n_i, b) + d_N(b, b_j)$ according to pre-computed distances, and insert $(b_j, d_N(n_i, b_j))$ into H (line 13-15). When we insert a new point into CSP_{ninj} or H, if the point already belongs to CSP_{ninj} or H, we just need to update the value of road network distance. If $dN(n_i, b) > L + d_r$ or H is empty, then, there is no interest point that its road network distance to n_i is under $L + d_r$ except the points existing in CSP_{ninj}, the algorithm ends. At last, the interest points set CSP_{ninj} is the candidate skyline points set of $n_i n_j$.

Algorithm 2. Acquiring CSP_{ninj} based on PINE*

Input: n_i, d_r, length of segment $n_i n_j$ L.
Output: CSP_{ninj}
1   $CSP_{ninj} = \varnothing, H = \varnothing$;
2   $NVP(p_i)$ is the cell containing n_i in NVD;
3   $b_{i1}, b_{i2}, ..., b_{ik}$ is the borders point of $NVP(p_i)$;
4   calculate the road network distance from n_i to p_i and $b_{i1}, b_{i2}, ..., b_{ik}$;
5   insert $(p_i, d_N(n_i, p_i))$ into CSP_{ninj};
6   insert $(b_{i1}, d_N(n_i, b_{i1})), (b_{i2}, d_N(n_i, b_{i2})), ..., (b_{in}, d_N(n_i, b_{in}))$ into H;
7   remove first entry $(b, d_N(n_i, b))$ from H;
8   while $d_N(n_i, b) < d_r + L$ do
9   p_k is the other generator point of border point b;
10      $d_N(n_i, p_k) = d_N(n_i, b) + d_N(b, p_k)$;
11   if $d_N(n_i, p_k) < d_r + L$ then
12   insert $(p_k, d_N(n_i, p_k))$ into CSP_{ninj};
13   for each border point b_j of $NVP(p_k)$ do
14          $d_N(n_i, b_j) = d_N(n_i, b) + d_N(b, b_j)$;
15   insert $(b_j, d_N(n_i, b_j))$ into H;
16   if H is not empty then
17   remove first entry $(b, d_N(n_i, b))$ from H;
18   return CSP_{ninj};

4.2   Processing DSPS for CRPQ on a Segment

After acquiring the CSP_{ninj} of segment $n_i n_j$, we discuss dynamic split points setting (DSPS) for CRSQ queries (D-CRSQ) in segment $n_i n_j$. Table 2 summarizes the mathematical notations frequently used in the rest paper.

The processing steps of D-CRSQ in segment $n_i n_j$ are as follows:

Step 1: To monitor the changing trend of the network distance between the interest point in CSP_{ninj} and query point q, the first step is to divide the interest points in CSP_{ninj} into increasing and decreasing groups(represented as CSP_{ninj}^{up} and CSP_{ninj}^{down}) [9]. When we get CSP_{ninj} utilizing the approach that introduced in Sect. 4.1, we can easily

Table 2. Summary of notations

Notation	Definition
$p_i \prec p_j$	p_i dominates p_j in terms of static and dynamic attributes
$p_i \prec_{static} p_j$	p_i dominates p_j only in terms of static attributes
$p_i \not\prec p_j$	p_i cannot dominates p_j
RSP	range skyline points set q
RSP^{up}	increasing group of RSP
RSP^{down}	decreasing group of RSP
CSP_{ninj}	the candidate skyline points of segment $n_i n_j$
CSP_{ninj}^{up}	increasing group of CSP_{ninj}
CSP_{ninj}^{down}	decreasing group of CSP_{ninj}
CSP_{ninj}^{R}	interest points group within a distance range d_r of CSP_{ninj}
SPQ	split points set

find the shortest path between n_i and each points in CSP_{ninj}. As the query point q moves from n_i to n_j, if the shortest path from p to n_i passes through n_i, then, the network distance $d_N(p, q)$ is increasing, we insert it into CSP_{ninj}^{up}. Otherwise, we insert it into CSP_{ninj}^{down}.

Step 2: When obtaining the changing trend of each interest point from CSP_{ninj}, this step is to further delete the interest points from CSP_{ninj} that will not be skyline points wherever the query point is located on the segment, we introduce two Lemmas:

Lemma 2. *If $p_i \in CSP_{ninj}^{up}$ and $d_N(p_i, n_i) > d_r$, p_i will never become skyline point.*

Proof. As the query point q moves from n_i, $d_N(p_i, q) = d_N(p_i, n_i) + d_N(n_i, q)$. For $d_N(p_i, n_i) > d_r$, $d_N(p_i, q) > d_r$ is held all the time. Hence, p_i will not become skyline point.

Lemma 3. *The points in CSP_{ninj}^{up} and CSP_{ninj}^{down} being dominated will never become skyline points.*

Proof. We first prove the points in CSP_{ninj}^{up}. For two points $p_i, p_j \in CSP_{ninj}^{up}$, suppose $p_j \prec p_i$, so $d_N(p_i, q) > d_N(p_j, q)$. As the query point q moves, the distances between p_i, p_j and q have the same changing trend, and the static attributes of them keep the same, so p_i is dominated by p_j all the time. Thus, p_i will never become skyline point. Proof for the points in CSP_{ninj}^{down} is similar.

Based on the above two Lemmas, we delete corresponding interesting points from CSP_{ninj}^{up} and CSP_{ninj}^{down}. This step could improve the efficiency of program execution.

Step 3: Now, we could acquire the initial skyline points (*ISP*) of q (starting moving at the node n_i). We take skyline domination test in CSP_{ninj}^{R}, where $CSP_{ninj}^{R} \subset CSP_{ninj}$, for each $p \in CSP_{ninj}^{R}$, $dN(p, q) < d_r$, and if $p' \in (CSP_{ninj} - CSP_{ninj}^{R})$, $d_N(p', q) > d_r$. The non-dominated interesting points of CSP_{ninj}^{R} is *ISP* and return $RSP = ISP$.

Step 4: After acquiring the *ISP* of q, we will try to find the location of split point that q will arrive. For dynamically acquire split points, we introduce the generation rules of four types split points (introduced in Sect. 3). The split point is in the form of: (sp, $d_N(sp, q)$) where sp denotes the type of split points with their generating points (e.g. $in_{pi/pj}$, the type is in and the generating points are p_i, p_j) and $d_N(sp, q)$ is the distance between sp and the query point q.

$exit_p$. Select interest point p_{max}, $p_{max} \in RSP^{up}$ and $d_N(p_{max}, q) > d_N(p', q)$. $\forall p'$, $p' \in (RSP^{up} - \{p_{max}\})$, if $d_r - d_N(p_{max}, q) < L$, p_{max} generates split point: ($exit_{pmax}$, $d_r - d_N(p_{max}, q)$).

in_p. Select interest point p_{min}, $p_{min} \in (CSP_{ninj} - CSP_{ninj}^R) \cap CSP_{ninj}^{down}$ and $d_N(p_{min}, q) < d_N(p', q)$. $\forall p'$, $p' \in ((CSP_{ninj} - CSP_{ninj}^R) \cap CSP_{ninj}^{down} - \{p_{min}\})$, if $d_N(p_{min}, q) - d_r < L$, p_{min} generates split point: (in_{pmin}, $d_N(p_{min}, q) - d_r$).

$exit_{pj/pi}$. For each point p_i in RSP^{up} and p_j in RSP^{down}, if $p_j \prec_{static} p_i$, $(d_N(p_j, q) - d_N(p_i, q))/2 < L$. Then, p_i, p_j generate split point: ($exit_{pj/pi}$, $(d_N(p_j, q) - d_N(p_i, q))/2$).

$in_{pi/pj}$. For each point p_i in RSP^{up} and p_j in $CSP_{ninj}^R \cap CSP_{ninj}^{down}$ but not in RSP, if $p_i \prec p_j$, $(d_N(p_j, q) - d_N(p_i, q))/2 < L$, then, p_i, p_j generate split point: ($in_{pi/pj}$, $(d_N(p_j, q) - d_N(p_i, q))/2$).

After the split points are obtained, we use a priority queue *SPQ* to store and sort them according the network distances between them and query point q.

Step 5: Remove the top entry sp from *SPQ*. Then, when the distance query point q moves is less then $sp.dis$ (the distance between sp to start point of q), the *RSQ* keeps unchanged. When the query point moves $sp.dis$ forward arriving at a split point: add $sp.dis$ to pairs of CSP_{ninj}^{up} and subtract $sp.dis$ from pairs of CSP_{ninj}^{down}; subtract $sp.dis$ from each split point of *SPQ*. At this time, we should notice that: the distance of each member from CSP_{ninj} and *SPQ* is the network distance between them to query point q). Next, we carry out the following operations:

- If sp is $exit_p$:
  - delete the interest point p from *RSP* and CSP_{ninj};
  - delete sp' which generated by p from *SPQ*;
  - determine $exit_{p'}$ and insert it into *SPQ*;
- If sp is in_p:
  - if $\exists p' \in RSP^{up}$, $p' \prec p$, then, determine and insert split point ($in_{p'/p}$, $(d_N(p, q) - d_N(p', q)/2)$ into *SPQ* if $d_N(p', q)/2 < L$;
  - if $\nexists p* \in RSP$, $p* \prec p$, then, insert p into *RSP*. And if $\exists p' \in RSP^{up}$, $p \prec_{static} p'$, determine and insert split point ($exit_{p/p'}$, $d_N(p, q) - d_N(p', q)/2$) into *SPQ* if $d_N(p, q) - d_N(p', q)/2 < L$.
  - determine and insert split point in_{p*} into *SPQ*;
- If sp is $exit_{pj/pi}$:
  - delete the interest point p_i from *RSP* and CSP_{ninj};
  - delete sp' which generated by p_i from *SPQ*;

- If sp is $in_{pi/pj}$:
  - if $\not\exists in_{pk/pj} \in SPQ$, then, insert interest point p_j into RSP;
  - when p_j is been inserted into RSP, if $\exists p \in RSP^{up}$, $p_j \prec_{static} p$, then, determine and insert split point $(exit_{pj/p}, d_N(p_j, q) - d_N(p, q)/2)$ into SPQ if $(d_N(p_j, q) - d_N(p, q)/2) < L$;

Step 6: Repeat step 5 until query point q arrives at n_j.

5 Experiments

5.1 Experimental Setup

Experiments are implemented in Visual C++ 2010, and all algorithms are executed on a Windows 7 Service Pack 1 PC with 3.3 GHz Intel Core i5-4590 CPU and 4 GB memory. We use the road maps of California Road Network1 which consists of 21,048 nodes and 21,693 edges. The moving query point follows a random path, e.g. when a query point moves on an edge with a given speed to the end node, it randomly selects a neighbor edge and continuously moves on it.

Table 3 shows the system parameters used in our experiments: interest points (varying from 0.5 K to 10 K) are generated from the road network edges randomly, the query range set varies from 1 % to 10 % of the entire space. Each interest point has several static attributes (varying from 2 to 6), whose values satisfy Gaussian distribution ($\mu = 5$, $\sigma^2 = 3$). The numbers of segments of each path are from 1 to 20.

Table 3. Parameters and values

Parameters	Values
Number of interest points(K)	0.5, 1, 2, 5, 10
Number of query range	1 %, 2 %, 3 %, 5 %, 10 %
Number of static attributes	2, 3, 4, 5, 6
Segments of every query path	1, 5, 10, 15, 20

5.2 Results

We first evaluate the efficiency of three network distance computation approaches to acquire the candidate skyline interest points of a segment: our PINE* approach that introduced in Sect. 4.1, Grid Index based (GI) approach that proposed in [6] for network distance computation of Cd_ε-SQ and the further improved version of GI named GI + of Cd_ε-SQ +. Figures 5 and 6 show the average CPU time under the different number of interest points (varying from 0.5 k to 10 k) and different query range (varying from 1 % to 10 %) in road networks, the experiment result shows that our PINE* approach significantly outperforms Grid index based GI and GI + approach.

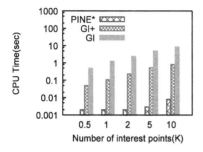
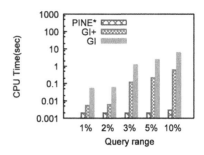

Fig. 5. *CSP* computation effect vs. interest points

Fig. 6. *CSP* computation effect vs. query range

Effect of the Number of Interest Points. Figure 7 shows the average processing time under different number of interest points (varying from 0.5 K to 10 K). We can see that as the interest points increase, the processing time of D–CRSQ increases slightly. When there are a large number of interest points, D–CRSQ is much better than Cd_ε-SQ and Cd_ε-SQ+ algorithms. This is because our D–CRSQ method reduces expensive shortest path computations greatly. As the increasing number of interest points, our DSPS strategy also decreases the number of calculation of split points to a large extent.

Effect of Query Range. Figure 8 shows the average processing time under different ranges (varying from 1 % to 10 %). As the query range increases, our PINE* technique to answer D–CRSQ queries could acquire CSR efficiently. Thus, the query range has little influence on our DSPS strategy.

Effect of Static Attributes. Figure 9 shows the average CPU time under static attributes (varying from 2 to 6). More attributes of each interest point would result in more skyline points so that more dominance tests are performed. D–CRSQ method can reduce the number of skyline domination test and the settings of split points. So the efficiency of D–CRSQ is better than Cd_ε-SQ and Cd_ε-SQ+.

Effect of Query Length. Figure 10 shows that the average processing time under different query paths. It is obvious that as the query path number increases, the

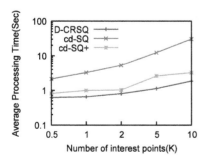
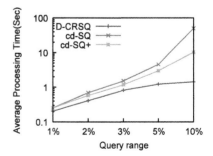

Fig. 7. Effect of interest points

Fig. 8. Effect of query range

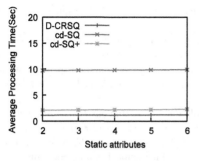

Fig. 9. Effect of static attributes

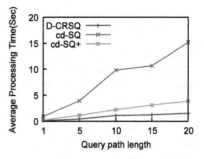

Fig. 10. Effect of query path

processing time will increase. For the same reason mentioned above, our DSPS strategy for CRSQ queries is also much better than grid index based methods.

6 Conclusion

In this paper, we address continuous range skyline queries in road networks. Due to the network distance computation is expensive, we utilize PINE* technique based on NVDs to calculate candidate skyline interest points. Further, we propose DSPS strategy to set the split points for efficient processing continuous range skyline queries. Experiments show that our proposed method to answer CRSQ queries is efficient to calculate and maintain skyline in road networks.

Acknowledgement. This work is partially supported by Natural Science Foundation of Jiangsu Province of China under grant No. BK20140826, the Fundamental Research Funds for the Central Universities under grant No. NS2015095, the Funding of Graduate Innovation Center in NUAA under grant No. KFJJ20151604.

References

1. Aljubayrin, S., He, Z., Zhang, R.: Skyline trips of multiple POIs categories. In: Renz, M., Shahabi, C., Zhou, X., Chemma, M.Aamir (eds.) DASFAA 2015. LNCS, vol. 9050, pp. 189–206. Springer, Heidelberg (2015)
2. Borzsony, S., Kossmann, D., Stocker, K.: The skyline operator. In: Proceedings of the 17th International Conference on Data Engineering (ICDE), Heidelberg, Germany (2001)
3. Chomicki, J., Godfrey, P., Gryz, J.: Skyline with presorting. In: Proceedings of the 19th International Conference on Data Engineering (ICDE), Bangalore, India (2003)
4. Deng, K., Zhou, X., Shen, H.T.: Multi-source skyline query processing in road networks. In: Proceedings of the 19th International Conference on Data Engineering (ICDE), Istanbul, Turkey (2007)
5. Godfrey, P., Shipley, R., Gryz, J.: Maximal vector computation in large data sets. In: Proceedings of the 31st International Conference on Very Large Data Bases (VLDB), Trondheim, Norway (2005)

6. Huang, Y.H., Chang, C.H., Lee, C.: Continuous distance-based skyline queries in road networks. Inf. Syst. **37**(7), 611–633 (2012)
7. Jang, S., Yoo, J.: Processing continuous skyline queries in road networks. In: International Symposium on Computer Science and its Applications (CSA) (2008)
8. Kolahdouzan, M., Shahabi, C.: Voronoi-based K nearest neighbor search for spatial network databases. In: Proceedings of the 30th International Conference on Very Large Data Bases (VLDB), Toronto, Ontario, Canada (2004)
9. Kolahdouzan, M.R., Shahabi, C.: Alternative solutions for continuous K nearest neighbor queries in spatial network databases. Geoinformatica **9**(4), 321–341 (2005)
10. Kriegel, H.P., Renz, M., Schubert, M.: Route skyline queries: a multi-preference path planning approach. In: Proceedings of the 26th International Conference on Data Engineering (ICDE), Long Beach, California, USA (2010)
11. Mouratidis, K., Lin, Y., Yiu, M.L.: Preference queries in large multi-cost transportation networks. In: Proceedings of the 26th International Conference on Data Engineering (ICDE), Long Beach, California, USA (2010)
12. Papadias, D., Tao, Y., Fu, G.: Progressive skyline computation in database systems. TODS **30**(1), 41–82 (2005)
13. Safar, M.: K nearest neighbor search in navigation systems. Mob. Inf. Syst. **1**(3), 207–224 (2005)
14. Safar, M., El-Amin, D., Taniar, D.: Optimized skyline queries on road networks using nearest neighbors. Pers. Ubiquit. Comput. **15**(8), 845–856 (2011)
15. Son, W., Hwang, S.W., Ahn, H.K.: MSSQ: manhattan spatial skyline queries. Inf. Syst. **40**, 67–83 (2014)
16. Tian, Y., Lee, K.C.K., Lee, W.-C.: Finding skyline paths in road networks. In: Proceedings of the 17th ACM SIGSPATIAL International Symposium on Advances in Geographic Information Systems (GIS), Seattle, Washington, USA (2009)

Beidou Software Receiver Based on Intermediate Frequency Signal Collector with INS-Aided

Xuwei Cheng$^{(\boxtimes)}$, Xiaqing Tang, Meng Wu, and Junqiang Gao

Department of Control Engineer, Academy of Armored Force Engineering,
Beijing, China
chengxuwei0872@126.com

Abstract. For the sake of easily developing and improving performances of BDS receiver and BDS (Beidou navigation satellite system)/INS (inertial navigation system) integration system, the INS-aided SDR (software defined radio) is investigated. The performance of the SDR with an intermediate frequency signal collector for B1-band with and without INS-aided are analysed and experimentally validated, respectively. From the optimized acquiring scheme-PMF (partial matched filter)-FFT (fast Fourier transform) with INS aiding to the steadily tracking method, all important components of the system are introduced and analysed in detail. Then, based on the principle of minimizing the tracking error, an optimal adaptive bandwidth designing method is adopted. The results show that the presented PMF-FFT based scheme is capable of realizing effective BDS B1 signal acquisition and tracking with the INS information aiding. This SDR will be a useful tool for develop the BDS and others GNSS soft- or hardware receivers and integration system.

Keywords: BDS · Software receiver · Signal acquisition · Tracking loop

1 Introduction

Development and application of GNSS (global navigation satellite system) are of great importance for the increasing services demands in the civil, military areas of both system development and technique application [1]. China's BDS (Beidou navigation satellite system, also known as COMPASS) is experiencing a rapid developing period in recent 5 years. It provides positioning, timing and a short messaging communication services to the Asian-Pacific area with 16 satellites in orbit from December, 2012.

Aimed to pursue more adaptive ability to positioning with higher precision and robustness, inspired by GPS (global positioning system), the reprogrammed BDS SR (software receiver) or SDR (software-defined radios) are greatly needed. Actually, the first SDR was introduced in 1996 [2]. Since then, several research groups have presented their research achievements [3, 4]. SDR is proving an inevitable trend of the GNSS techniques, for which it is mainly concentrating in both the laboratory and engineering field, its open architecture and flexible design make building reconfigurable, dynamic choosing parameters for simulation and test more easy and effective.

© Springer International Publishing Switzerland 2016
D.-S. Huang et al. (Eds.): ICIC 2016, Part III, LNAI 9773, pp. 533–540, 2016.
DOI: 10.1007/978-3-319-42297-8_49

In order to develop and test the software algorithms of BDS/INS, a complete B1-band antenna and RF (radio frequency) front-end has been designed.

On the other hand, the crucial challenge of current receiver design and application lie in how to improve system robustness, but accuracy [5]. To improve receiver's performance for tracking weak signals, the 2 issues are involved: adopting more advanced antenna design and digital signal processing, on the other hand, eliminating Doppler affection with multi-sensor information fusion, such as GNSS/INS (inertial navigation system) deep integration [6].

The major discussion about GPS SDR focus on structure and signal acquisition, so this paper analyzes improved acquiring scheme with new FFT algorithm an INS aiding, and compare the affection of the phase lock loop noise error models with external aiding [7], which decreases the tracking loop error and provides a guidance for the design of GNSS/INS deep integration after analyzing the tracking performance affection of different accuracy grade of the core components of INS.

The remainder of the paper is organized in the following way. Section 2 designs the SDR based on RF front-end with INS-aiding. Section 3 details a kind of fast acquiring method based on PMF-FFT. Section 4 analyzes loop tracking and its performances, and optimal bandwidth design method for INS aiding PLL. Test results and conclusion are provided in Sects. 5 and 6, respectively.

2 BDS SDR Structure

The SDR is of great necessity to develop software or hardware real-time receivers. Figure 1 shows the architecture of the BDS software receiver based on RF front-end [8]. The RF front end provides digital signals to the software receiver, which is digitized by an ADC (analog-to-digital converter) as close to the antenna as possible, after then, the host computer or PC can easily process and obtain necessary information of the vehicle. Thus, we can control the receiver through flexible changing the function from amplitude to frequency modulation.

The SDR describes the idea and function of signal acquisition tracking and positioning calculation more clearly than the hardware. For some of the processes are given by calculating with several algorithms and programs, the new acquiring and tracking methods with parameters changed can be obtained easily. Similar to Borre's SDR framework [9], the real-time software BDS receiver based on C ++ has been established with high PC operating speed and coding technology.

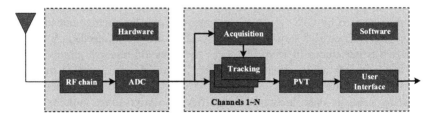

Fig. 1. Architecture of the BDS software receiver based on RF front-end

3 Signal Acquisition Method with Aiding

The BDS B1 is consisted of the carrier frequency, ranging code and NAV message. The received signal can be expressed as follows:

$$S(t) = AD(t)C(t)\sin[\omega_c(t) + \theta_0] + n(t) \tag{1}$$

Where A is signal amplitude, $D(t)$ is data modulated on ranging code, $C(t)$ is pseduo-random ranging code, $\omega_c(t)$ is Doppler carrier frequency, $n(t)$ is white Guass noise. To make intimate frequency signal's model more accuracy, the INS aiding is imported, which makes the correlactor error of local carrier and received signal decrease, and get lower error carrier frquency and code phase.

Figure 2 shows the basic structure of signal acquisition scheme with the INS aiding based on FFT. INS send position and velocity in less than 10 Hz, which keeps synchronization with satellite signal. After calculating the relative speed between vehicle and satellite in the LOS (line of sight), Doppler frequency and its rate of change can be obtained, then, the central of frequency search range can be set, which is decided by the uncertainty of INS information, and drives receiver's local NCO (numerically controlled oscillator) generate codes and carriers.

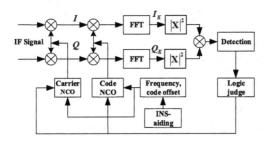

Fig. 2. Implementation of signal acquisition scheme with the INS aiding

Without consideration of multipath, ionosphere and troposphere factors, the carrier Doppler frequency shift of BDS signal between satellite and received signal can be expressed as follows:

$$f_d = f_{rec} + f_s + \Delta f_{rec} + \Delta f_s \tag{2}$$

Where f_{rec} and f_s are Doppler frequency caused by vehicle and satellites motion, respectively; Δf_{rec} and Δf_s are frequency error caused by receiver and satellite oscillator error, both can be ignored for satellite oscillator drift is very slight, while Δf_{rec} is eliminated by pre-calibrating. Then, Eq. (2) is simplified as follows:

$$f_d = f_{rec} + f_s = \frac{1}{\lambda}(\vec{v}_{rec} - \vec{v}_s) \cdot \vec{e} \tag{3}$$

Where λ is the carrier wavelength; $\vec{v}_{rec}$ and $\vec{v}_s$ are vehicle and satellite speed, respectively. Thus, and the Doppler frequency cause by vehicle and satellite motion can be calculated for given motion; $\vec{e}$ is the unit vector from satellite to vehicle. Once we discuss about Doppler drift caused by vehicle speed, the effect of satellite's speed can be ignored, and the Doppler variance is obtained from Eq. (3).

$$\sigma_{f_d} = \frac{\vec{e} \cdot E(\vec{v}_{rec}(\vec{v}_{rec})^{T}) \cdot \vec{e}}{\lambda^2} \tag{4}$$

Obviously, Doppler estimation error of external aiding is decided by speed error of INS. INS's output accuracy should be counted so that reduce the computation and improve the acquisition speed.

For satellites signals, acquisition is a key problem to accomplish for posterior solve. Different from the hardware, the software functions has more advanced and flexible configuration to estimate the carrier frequency and the code phase. Here, the parallel PMF-FFT technique is adopted for decrease acquiring time and improving characteristics. Figure 3 shows its basic structure for acquiring.

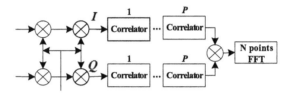

Fig. 3. Signal acquisition principle based on PMF-FFT

For PMF-FFT acquiring structure, the modulated periodic signal is divided into P correlators (each length is L) in time domain. Similarly, after synchronization, the normalized frequency response of PMF-FFT can be computed as follows:

$$G_{FFT}(\Delta f, k) = \frac{1}{M} \left| \frac{\sin(\omega_c \frac{M}{P}) \cdot \sin(\omega_c M - \pi \frac{P}{N} k)}{\sin(\omega_c) \cdot \sin(\omega_c \frac{M}{P} - \pi \frac{k}{N})} \right| \tag{5}$$

Where $\omega_c = \pi f_d T_c$, $k = 1, 2, \ldots, N-1$. If the correlator's peak values output above the given detection threshold, the signal synchronization is finished, so $f_d = k_{max}/NXT_c$, and the IFFT operation can directly provide the result sequence.

4   Design of Tracking Loop

The traditional integrated INS/GNSS system only involves INS's error compensation and lacking the modifying of the receiver. While the deep integration controls NCO with the corrected INS information and completes the local signal replica by, thus the code and carrier tracking loop is finished, which solves the serious problem that the weak signals or under interference and high dynamic environment, and enhance receiver's acquiring and tracking performance [10].

The code and frequency loop coupling mutually are simultaneously working during signal tracking and replica the local signal. The FLL is more sensitive to dynamic stress than DLL, and easily loss lock for its longer wavelength, under dynamic applying environment, the adaptively adjusting method of PLL bandwidth is adopted [11]. For the INS information for integration system can be easily got, which is utilized to tracking PLL to improve tracking performance, with the C/N_0 of received signal and measured dynamic stress, the tracking ability dramatically can be enhanced. Figure 4 shows the INS-aiding PLL structure model.

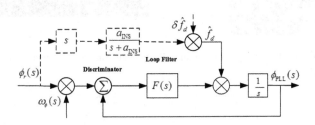

Fig. 4. The PLL model with INS aiding

Where, the $\phi_r(s)$ is input, $\omega_\phi(s)$ is external phase noise, $\hat{f}_d$ and $\delta\hat{f}_d$ are Doppler estimator error and clock frequency error, respectively, provided by INS.

$$\phi_{PLL}(s) = H_1(s)\phi_r(s) + H_2(s)\omega_\phi(s) \tag{6}$$

Where, $H_1(s) = \frac{\frac{a_{INS}}{s+a_{INS}}s + F(s)}{s+F(s)}$, a_{INS} is equivalent bandwidth of INS, For a traditional 2nd PLL, if $F(s) = K(s+a)/s$, then, $H_2(s) = \frac{2\zeta\omega_n s + \omega_n^2}{s^2 + 2\zeta\omega_n s + \omega_n^2}$, ω_n is the natural frequency, ζ is the damping factor. The performance of PLL mainly depend on the loop filter, and the loop bandwidth can be obtained as follows:

$$B_n = \frac{1}{2\pi|H(0)|}\int_0^\infty |H(j\omega)|^2 d\omega = \frac{\omega_n}{2}\left(\zeta + \frac{1}{4\zeta}\right) \tag{7}$$

As mentioned above, the loop bandwidth is the most important parameter which affect the GNSS receiver's tracking performance, while the dynamic and controlling noises ability contradicts with each other. For smaller bandwidth can control the loop noise under certain value, thus the loop works with smaller tracking error, but if the bandwidth is too small, the dynamic ability can't be meet.

5 Test Results

The experiment was implemented with intermediate frequency signal collect platform (hardware and software), which can both collect GPS and BDS signals with 2 independent channels, meanwhile the data format is assigned as *.bin for Matlab and *.dat

Fig. 5. The intermediate frequency signal collector hardware and software collecting process

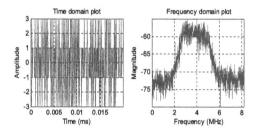

Fig. 6. Raw signal in the time and frequency domain

for Visual Studio 2013. In order to compatible with both BDS and GPS signals, the SAW filter cannot be added to RF collector, so the high performance antenna is used. Figure 5 shows the RF antenna and soft UI in PC used in experiment. Figure 6 illustrates the raw signal that is used for acquisition calculation in time & frequency domain. The parameters for the signal processing are set as: sampling frequency is 16.368 MHz, The intermediate frequency is 4.092 MHz, and data with 8 bits was stored (data transfer rate is almost 8 M/s).

By using the algorithms method above, acquisition for BDS B1 signal is calculated in the developed software receiver. There were 8 satellites identified. Figure 7 shows the acquiring, tracking, positioning results and the C/N_0 of the 8 captured satellites' signal in the experiment, which also proves the acquisition capability.

It can be seen that the proposed acquisition method is feasible for acquiring BDS signals, while the antenna was placed beside the window, so a signal transmitter is employed, and its antenna is displaced on the laboratory building roof where the satellites in space can be directly received. The C/N_0 of each satellite signal are all above 40 dB-Hz, which make acquiring and tracking more easily. For the smaller VDOP and HDOP, the positioning results (116.1465454, 39.844975) are trustworthy, while the altitude needs to be corrected with external information source, such as the altimeter.

Actually, for high dynamic application, once the signal is covered by noises, the PLL cannot make tracking error below the tracking threshold (15o) only by adjusting

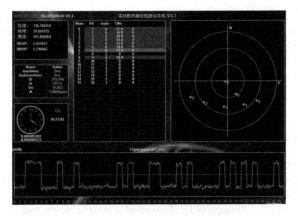

Fig. 7. Experimental results of BDS SDR

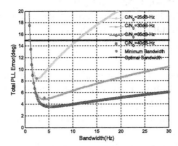

Fig. 8. The relationship between bandwidth and tracking error (Color figure online)

bandwidth parameter. Figure 8 indicates that the dynamic stress error is compensated effectively with INS aiding, the optimal bandwidth and the minimal bandwidth of PLL are respectively decreased by INS aiding. It thus clear that INS-aiding PLL tracking performance is superior to pure PLL, the tracking ability of weak signal is dramatically improved.

6   Conclusion

The SDR base on RF front end and C ++ platform is introduced, which involves signal acquisition, tracking and positioning calculation and other designs. The PMF-FFT algorithm is deployed to signal acquisition to decrease computing time with INS aiding, and the INS information is deployed and an optimal adaptive bandwidth designing method is adopted.

In order to improve the loop tracking performance, based on the estimated C/N_0 the tracking error can keep minimum with the proposed method. Actually, tactical IMU (inertial measurement unit) will be enough for improve signal tracking performance of BDS/INS deep integration, which is helpful for integrated system design and pursue high accuracy and reliability. Moreover, with the improvement of electronic technique and BDS capability, more advanced items will be discussed and special engineering application platforms may be serious focused.

References

1. Kaplan, E.D., Hegarty, C.J.: Understanding GPS: principle and applications, 2nd edn. Artech House, Boston (2006)
2. Akos, D.M., Braasch, M.S.: A software radio approach to global navigation satellite system receiver design. In: Proceedings of the 52nd Annual Meeting of the Institute of Navigation, Cambridge, MA, USA (1996)
3. Ding, J.C., Zhao, L., Huang, W.Q.: Design of a general single frequency GPS receiver research platform for high dynamic and weak signal. J. Comput. 4(12), 1195–1201 (2009)
4. Thompson, E., Clem, N., Renninger, I.: Software-defined GPS receiver on USRP-platform. J. Netw. Comput. Appl. 35(4), 1352–1360 (2012)
5. Alireza, R., Demoz, G.E., Dennis, M.A.: Carrier loop architectures for tracking weak GPS signals. IEEE Trans. AES 44(2), 697–710 (2008)
6. Soloviev, A.: Tight coupling of GPS and INS for urban navigation. IEEE Trans. AES 46(4), 1731–1746 (2010)
7. Ye, P., Zhan, X.Q., Fan, C.M.: Novel optimal bandwidth design in INS-assisted GNSS phase lock loop. IEICE Electron. Express 8(9), 650–656 (2011)
8. Liu, J., Cai, B.G., Wang, J.: B1 signal acquisition method for BDS software receiver. In: 9th International Conference on Intelligent Computing Theories and Technology, Nanning, China (2013)
9. Borre, K., Akos, D.: A software-defined GPS and galileo receiver: single-frequency approach. In: ION 18th International Technical Meeting of the Satellite Division. Long Beach, CA, USA (2005)
10. Lashley, M., Bevly, D.M.: Performance comparison of deep integration and tight coupling. Navigation 60(3), 159–178 (2013)
11. Tang, X.Q., Cheng, X.W., Meng, W.M.: Optimal bandwidth design and performance analysis of INS aided PLL. J Acad. Armored Forced Eng. 29(3), 164–171 (2015)

Fuzzy Neural Sliding Mode Control
for Robot Manipulator

Duy-Tang Hoang[1] and Hee-Jun Kang[2](✉)

[1] Graduate School of Electrical Engineering,
University of Ulsan, Ulsan 680-749, South Korea
hoang.duy.tang@gmail.com
[2] School of Electrical Engineering, University of Ulsan,
Ulsan 680-749, South Korea
hjkang@ulsan.ac.kr

Abstract. A fuzzy neural sliding mode controller (FNNSMC) is proposed for robot manipulators. Sliding mode controller is implemented based on two radial basic function neural networks and a fuzzy system. The first neural network is used to estimate the robot dynamic function. The second neural network combines with a fuzzy system to present the switching control term of sliding mode control. This combination resolves the chattering phenomenon. The stability of proposed controller is proven. Finally, simulation is done on a 2-link serial robot manipulator to verify the effectiveness.

Keywords: Fuzzy control · Neural network · Radial basic function neural network · Robot manipulator · Sliding mode control

1 Introduction

Robot manipulators are complex, high nonlinear and high coupled. It's almost impossible to obtain an exact dynamic model of a robot. Neural network with universal approximation characteristic can approximate any continuous function with arbitrary accuracy [1, 2], include robot manipulators [3]. However, additional uncertainties come from the estimation error of neural network. To solve the problem of uncertainties, a lot of control methods have been proposed, such as adaptive control [5, 6], robust control [6, 7], intelligent control [8–10], sliding mode control (SMC) [12, 13]. SMC may be the most effective method due to its robustness and fast transient response [12, 13]. But SMC causes chattering phenomenon in the control system. To deal with this problem, boundary layer neighbor [14] and saturation approximation [15] may be simplest methods and easy to implement, but as a tradeoff, while control signals are continuous, the convergence of tracking error is not guaranteed. High order SMC [16] can suppress chattering, but often require higher order derivative of sliding surface which is difficult to obtain in real application.

In this paper, we proposed a new method to deal the chattering phenomenon by combining a radial basic function neural network (RBFN) a fuzzy logic system (FLS).

The remainder of this paper is organized as follow. Section 2 describes the dynamic equation of serial rigid link robot manipulator. Section 3 is process of designing

© Springer International Publishing Switzerland 2016
D.-S. Huang et al. (Eds.): ICIC 2016, Part III, LNAI 9773, pp. 541–550, 2016.
DOI: 10.1007/978-3-319-42297-8_50

proposed control scheme. Section 4 is the simulation on a 2-link robot manipulator in Matlab. Section 5 presents some conclusions.

2   Dynamic of Robot Manipulator

The dynamic equation of a serial n-link robot manipulator is expressed as the following equation

$$M(\theta)\ddot{\theta} + C\left(\theta, \dot{\theta}\right)\dot{\theta} + G(\theta) + F\left(\dot{\theta}\right) + \tau_d = \tau \tag{1}$$

where $\theta, \dot{\theta}, \ddot{\theta} \in R^{n \times 1}$ are respectively position, velocity and acceleration of robot joints, $M(\theta) \in R^{n \times n}$ is the inertia, $C(\theta, \dot{\theta}) \in R^{n \times 1}$ is the centripetal and Coriolis, $G(\theta) \in R^{n \times 1}$ is the gravitational term, $F(\dot{\theta}) \in R^{n \times 1}$ is the Coulomb friction, $\tau_d \in R^{n \times 1}$ is the unknown disturbance bounded by a known positive constant τ_{dM}, and $\tau \in R^{n \times 1}$ is the input torque control.

Some useful properties of robot dynamic:

Property 1.   The inertia matrix $M(\theta)$ is symmetric and positive definite.

Property 2. $\dot{M}(\theta) - 2C\left(\theta, \dot{\theta}\right)$ is a skew symmetric matrix.

3   Controller Design

Firstly, position and velocity tracking error are defined as

$$e = \theta_d - \theta \tag{2}$$

$$\dot{e} = \dot{\theta}_d - \dot{\theta} \tag{3}$$

where $\theta_d, \dot{\theta}_d$ are desired position and velocity, θ and $\dot{\theta}$ are real position and velocity of robot joints obtained from position and velocity sensors.

The sliding surface is define as

$$s = \dot{e} + \alpha e \tag{4}$$

where $\alpha = diag(\alpha_1, \alpha_2, \ldots, \alpha_n)$, α_i is a positive constant.

Take derivative of sliding surface s in (4) and multiply with inertial matrix

$$M\dot{s} = M\ddot{\theta}_d - M\ddot{\theta} + M\alpha\dot{e} = M\left(\ddot{\theta}_d + \alpha\dot{e}\right) - M\ddot{\theta} \tag{5}$$

Putting (1) into (5) to substitute $M\ddot{\theta}$, we obtain

$$M\dot{s} = M\left(\ddot{\theta}_d + \alpha\dot{e}\right) + C\left(\dot{\theta}_d + \alpha e\right) + (F + G) - Cs + \tau_d - \tau$$
$$= f(X) - Cs - \tau + \tau_d \tag{6}$$

where $f(X) = M\left(\ddot{\theta}_d + \alpha\dot{e}\right) + C\left(\dot{\theta}_d + \alpha e\right) + (F + G)$ and $X = \left(e, \dot{e}, \theta, \dot{\theta}, \ddot{\theta}\right)^T$.

RBFNs have advantages of easy design, good generalization, strong tolerance to input noise and online learning ability [5]; that makes it very suitable to design flexible control systems and estimation of functions. The estimation of $f(X)$ by the first RBFN is expressed as

$$\hat{f}(X) = \hat{W}^T \phi(X) \tag{7}$$

where $\hat{W}$ and $\phi(X)$ are weight and basic function of RBFN respectively.

Define W^* is optimal weight of the first RBFN estimates $f(X)$, which means that

$$f(X) = W^{*T} \phi(X) + \varepsilon \tag{8}$$

where ε is the estimation error and bounded by a positive constant $\|\varepsilon\| \leq \varepsilon_M$.

A neural sliding mode controller (NSMC) according to [13] is design as

$$\tau = \hat{W}^T \phi(X) + K_D s + K_v sign(s) \tag{9}$$

where K_D is a symmetric positive matrix, K_v is a constant satisfies $K_v \geq \varepsilon_M + \tau_{dM}$.

Instead of using a fixed gain K_v and a sign function as in (9), we proposed a switching term as following

$$\tau_{swa} = \hat{K}_v u_{fs} \tag{10}$$

The variable $\hat{K}_v$ is adapted by the second RBFN

$$\hat{K}_v = \hat{V}^T \phi(X_k) \tag{11}$$

where $X_k = (e, \dot{e})^T$, and $\hat{V}$ and $\phi(X_k)$ are the weight and basic function of the second RBFN.

The u_{fs} term in (10) is output of a FLS designed as following [17, 18]

$$u_{fs} = FLS(s) \tag{12}$$

The input of FLS is sliding surface s, the membership functions of input and output linguistic variable are shown in Fig. 1. The fuzzy rules are described as

Rule i:If s is F_i^s then u_{fs} is R_i

F_i^s represents fuzzy set of input variable, includes NB (negative big), NM (negative medium), NS (negative small), ZE (zero), PS (positive small), PM (positive medium),

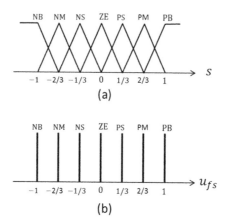

Fig. 1. Fuzzy sets of input variable (a) and output variable (b)

PB (positive big). R_i is singleton output of FLS, includes NB, NM, NS, ZE, PS, PM, PB. The rule base of FLS can be expressed as in Table 1 (Fig. 2).

Table 1. Rule base of FLS

s	NB	NM	NS	ZE	PS	PM	PB
u_{fs}	PB	PM	PS	ZE	NS	NM	NB

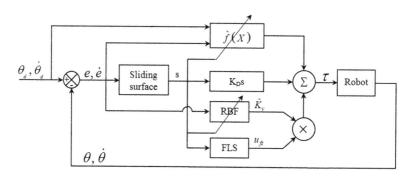

Fig. 2. Block diagram of FNNSMC

Theorem 1. Consider the robot manipulator (1), if the torque command is designed as

$$\tau = \hat{W}^T \phi(X) + K_D s + \hat{V}^T \phi(X_k) u_{fs} \qquad (13)$$

with update laws of the two RBFN are as following

$$\dot{\hat{W}} = F\phi(X)s^T \tag{14}$$

$$\dot{\hat{V}} = P\phi(X_K)\|s\| \tag{15}$$

where F, P are symmetric positive define matrices, then the close loop system is stable.

Proof. Define V^* and is optimal weight of the second RBFN computes $\hat{K}_v$, which means that

$$K_v = V^{*T}\phi(X_k) + \tilde{K}_v \tag{16}$$

where

$$\tilde{K}_v = \hat{K}_v - K_v \tag{17}$$

Consider a Lyapunov function

$$L = \frac{1}{2}s^T M s + \frac{1}{2}tr\left\{\tilde{W}^T F^{-1}\tilde{W}\right\} + \frac{1}{2}tr\left\{\tilde{V}^T P^{-1}\tilde{V}\right\} \tag{18}$$

where $\tilde{W} = W^* - \hat{W}$ and $\tilde{V} = V^* - \hat{V}$.

From the definitions of $\tilde{W}$ and $\tilde{V}$ we have

$$\dot{\tilde{W}} = -\dot{\hat{W}} = -F\phi(X)s^T \tag{19}$$

$$\dot{\tilde{V}} = -\dot{\hat{V}} = -P\phi(X_K)\|s\| \tag{20}$$

Take derivative both sides of (20) and using the fact that F, P are symmetric matrices, we get

$$\dot{L} = \frac{1}{2}\dot{s}^T M s + \frac{1}{2}s^T M \dot{s} + \frac{1}{2}s^T \dot{M} s + tr\left\{\tilde{W}^T F^{-1}\dot{\tilde{W}}\right\} + tr\left\{\tilde{V}^T F^{-1}\dot{\tilde{V}}\right\} \tag{21}$$

The Property 1 of the robot dynamic gives

$$\dot{s}^T M s = s^T M \dot{s} \tag{22}$$

Substitute (22) into (21)

$$\dot{L} = \frac{1}{2}s^T \dot{M} s + s^T M \dot{s} + tr\left\{\tilde{W}^T F^{-1}\dot{\tilde{W}}\right\} + tr\left\{\tilde{V}^T F^{-1}\dot{\tilde{V}}\right\} \tag{23}$$

Substitute (6) into (23)

$$\dot{L} = \frac{1}{2} s^T \dot{M} s + s^T \left(f(X) - Cs - \tau + \tau_d \right) + tr\left\{ \tilde{W}^T F^{-1} \dot{\tilde{W}} \right\} + tr\left\{ \tilde{V}^T F^{-1} \dot{\tilde{V}} \right\} \qquad (24)$$

The Property 2 of the robot dynamic gives

$$\frac{1}{2} s^T \dot{M} s = s^T C s \qquad (25)$$

Substitute (13), (19), (20) and (25) into (24)

$$
\begin{aligned}
\dot{L} &= -s^T K_D s + s^T (\varepsilon + \tau_d) + s^T \tilde{W}^T \phi(X) - tr\left\{ \tilde{W}^T F^{-1} F \phi(X) s^T \right\} \\
&\quad - \hat{V}^T \phi(X_K) u_{fs} - tr\left\{ \tilde{V}^T P^{-1} P \phi(X_K) \|s\| \right\} \\
&= -s^T K_D s + s^T (\varepsilon + \tau_d) - s^T \hat{V}^T \phi(X_K) u_{fs} - \tilde{V}^T \phi(X_K) \|s\|
\end{aligned}
\qquad (26)
$$

Substitute (11) and (17) into (26)

$$
\begin{aligned}
\dot{L} &= -s^T K_D s + s^T (\varepsilon + \tau_d) - \hat{K}_v s^T u_{fs} - \tilde{K}_v \|s\| \\
&\leq -s^T K_D s + s^T (\varepsilon + \tau_d) + \hat{K}_v \|s\| - \tilde{K}_v \|s\|
\end{aligned}
\qquad (27)
$$

Substitute (17) into (27) we obtain

$$\dot{L} \leq -s^T K s + s^T (\varepsilon + \tau_d) - K_v \|s\| \qquad (28)$$

Because $K_v \geq \|\varepsilon_M\| + \|\tau_{dM}\|$, we obtain the following

$$s^T (\varepsilon + \tau_d) - K_v \|s\| \leq 0 \qquad (29)$$

Finally, base on (29), the Eq. (28) infers to

$$\dot{L} \leq -s^T K_D s \leq 0 \qquad (30)$$

Since $L \geq 0$ and $\dot{L} \leq 0$, the close loop system is asymptotically stable respect to the sliding surface s. That means

$$\lim_{t \to \infty} (\dot{e} + \alpha e) = \lim_{t \to \infty} s = 0 \rightarrow \lim_{t \to \infty} e = \lim_{t \to \infty} \dot{e} = 0 \qquad (31)$$

So the tracking errors of close loop system converge to zero.

4   Simulation

In order to verify the proposed scheme control, this section considers simulation on a 2-link robot manipulator as show in Fig. 4. Its dynamic parameters are considered to be $m_1 = m_2 = 1\,(\text{kg})$, $l_1 = l_2 = 1\,(\text{m})$, $l_{c1} = l_{c2} = 0.5\,(\text{m})$ and link inertia $I_1 = I_2 = 0.8\,(\text{kgNm}^2)$. The dynamic of the robot is described as in (1), where

Inertia term

$$M(1,1) = m_1 l_{c1}^2 + m_2 (l_1^2 + l_2^2 + 2l_1 l_{c2} \cos(\theta_2)) + I_1 + I_2$$
$$M(1,2) = M(2,1) = m_2 l_{c2}^2 + m_2 l_{c2} l_1 \cos(\theta_2) + I_2$$
$$M(2,2) = m_2 l_1 l_{c2} \sin(\theta_2)$$

Centripetal and Coriolis term

$$C(1) = -2m_2 l_1 l_{c2} \sin(\theta_2)\dot{\theta}_1 \dot{\theta}_2 - m_2 l_1 l_{c2} \sin(\theta_2)\dot{\theta}_2^2$$
$$C(2) = m_2 l_1 l_{c2} \sin(\theta_2)\dot{\theta}_1^2$$

Gravitational term

$$G(1) = m_1 g l_{c1} \cos(\theta_1) + m_2 g (l_1 \cos(\theta_1) + l_{c2} \cos(\theta_1 + \theta_2))$$
$$G(2) = m_2 l_{c2} g \cos(\theta_1 + \theta_2)$$

The Coulumb friction $F\left(\dot{\theta}\right)$ and disturbance τ_d are assumed as

$$F(1) = 2\dot{\theta}_1 + 3sign\left(\dot{\theta}_1\right)$$
$$F(2) = 1.5\dot{\theta}_2 + 1.5sign\left(\dot{\theta}_2\right)$$

$$\tau_d(1) = 3\sin t + 5\sin 2t$$
$$\tau_d(1) = 1.5\sin 3t + 4\sin 2t$$

The desired trajectory is given by

$$\theta_d = [\sin t \quad \sin t]^T$$

In order to show the effectiveness, the proposed controller is compared with the neural sliding mode controller (NSMC) (9) which was presented in [13].

$$\tau = \hat{W}^T \phi(X) + K_D s + K_v sign(s) \tag{32}$$

where the parameters are chosen as $K_D = diag(55,55)$, $K_v = diag(10,10)$, the sliding surface parameter is chosen as $\alpha = diag(10,10)$. The weight of the RBFN is initialed with zero value.

Our proposed controller scheme (FNNSMC) is simulated with parameters are chosen as $K_D = diag(55,55)$, $\alpha = diag(10,10)$, weights of two RBFNs are initialed with zero values.

The result of simulation is shown in Figs. 3, 4 and 5. The Figs. 3 and 4 show the tracking performance of joint 1 and 2 respectively. Obviously, in comparing with

NSMC, the FNNSMC give better tracking performance for this robot with very small error of position.

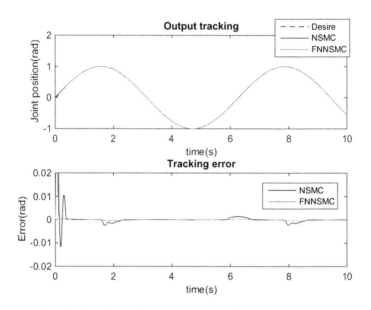

Fig. 3. Tracking performance at joint 1 (Color figure online)

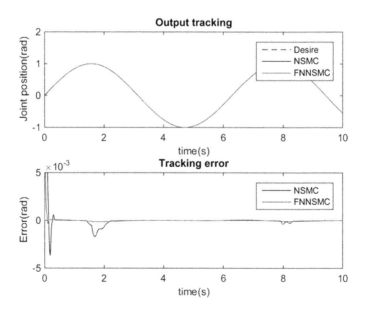

Fig. 4. Tracking performance at joint 2 (Color figure online)

Figure 5 shows the torque commands for two joints of robot. The NSMC has very big chattering phenomenon. On the contrary, our proposed controller FNNSMC gives control signal very smooth without chattering.

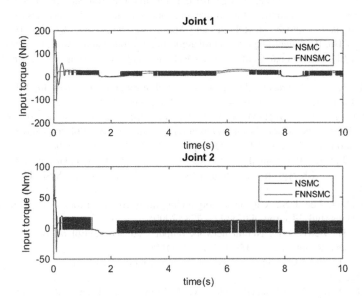

Fig. 5. Torque control command (Color figure online)

5   Conclusion

In this paper, a fuzzy neural sliding mode controller for robot manipulator is proposed. The stability of close loop system is proven. Simulations on a 2-link robot verified the performance of the proposed controller. By combining FLC, RBF and SMC, this control scheme has main advances as following.

- Guarantee the robustness of system without any knowledge about upper bounds of external disturbances and system uncertainties.
- Solve the chattering phenomenon of conventional sliding mode control.

Acknowledgments.   Following are results of a study on the "Leaders Industry-university Cooperation" Project, supported by the Ministry of Education (MOE).

References

1. Park, J., Sanberg, I.W.: Universal approximation using radial-basis-function networks. Neural Comput. **2**, 246–257 (1991)
2. Wu, Y., Wang, H., Zhang, B., Du, K.-L.: Using radial basis function networks for function approximation and classification. ISRN Appl. Math. **2012**, 34 pages (2012). Article ID 324194

3. Lewis, F.L., Liu, K., Yesildirek, A.: Neural net robot controller with guaranteed tracking performance. IEEE Trans. Neural Netw. **6**(3), 703–715 (1995)
4. Yu, H., Xie, T., Paszczynski, S., Wilamowski, B.M.: Advantages of radial basis function networks for dynamic system design. IEEE Trans. Ind. Electron. **58**(12), 5438–5450 (2011)
5. Hsia, T.C.: Adaptive control of robot manipulators - a review. In: Proceedings of 1986 IEEE International Conference on Robotics and Automation, vol. 3, pp. 183–189 (1986)
6. Feng, Z., Hu, B.: A new adaptive control algorithm of robot manipulators. In: Proceedings of 1988 IEEE International Conference on Robotics and Automation, vol. 2, pp. 867–872, 24–29 April 1988
7. Pan, H., Xin, M.: Indirect nonlinear robust control of robot manipulators. In: 2014 American Control Conference (ACC), pp. 4286–4291, 4–6 June 2014
8. Chen, C.-Y., Cheng, M.H.-M., Yang, C.-F., Chen, J.-S.: Robust adaptive control for robot manipulators with friction. In: 3rd International Conference on 2008 Innovative Computing Information and Control, ICICIC 2008, pp. 422–422, 18–20 June 2008
9. Wai, R.-J., Chen, P.-C.: Intelligent tracking control for robot manipulator including actuator dynamic via TSK-type fuzzy neural network. IEEE Trans. Fuzzy Syst. **12**(4), 552–560 (2004)
10. Huang, S.-J., Lee, J.-S.: A stable self-organizing fuzzy controller for robotic motion control. IEEE Trans. Ind. Electron. **47**(2), 421–428 (2000)
11. Kim, Y.H., Lewis, F.L.: Neural network output feedback control of robot manipulators. IEEE Trans. Robot. Autom. **15**(2), 301–309 (1999)
12. Utkin, V.I.: Variable structure systems with sliding modes. IEEE Trans. Autom. Control **22**, 212–222 (1977)
13. Liu, J., Wang, X.: Advanced Sliding Mode Control for Mechanical Systems. Tsinghua University Press, Springer, Beijing, Heidelberg (2011)
14. Slotine, J.J.E., Li, W.: Applied Nonlinear Control, pp. 41–190. Prentice Hall, EngleWood Cliffs (1991)
15. Li, T.H.S., Huang, Y.C.: MIMO adaptive fuzzy terminal sliding mode controller for robotic manipulators. Inf. Sci. **180**(23), 4641–4660 (2010)
16. Levant, A.: High-order sliding modes, differentiation and output feedback control. Int. J. Control **76**(9/10), 924–941 (2003)
17. X.-T. Tran, H.-J. Kang: Arbitrary finite-time tracking control for magnetic levitation systems. Int. J. Adv. Robot Syst. **11**(157) (2014)
18. Kuo, C.-L., Li, T.-H., Guo, N.: Design of a novel fuzzy sliding-mode control magnetic ball levitation system. J. Intell. Robot. Syst. **42**(3), 295–316 (2005)

Measure of Compatibility Based Angle Computing for Balanced Posture Control on Self-balancing Vehicles

Siyuan Ma[1(✉)], Man Wang[2(✉)], Chunye Du[1(✉)], and Yang Zhao[1(✉)]

[1] School of Electrical Engineering,
Xi'an Jiaotong University, Xi'an 710049, China
{masy91, zhaoyangofficial}@163.com,
duchunye1991@stu.xjtu.edu.cn
[2] Intelligent Transportation System Research Center,
Southeast University, Nanjing 210096, China
wangmanseu@126.com

Abstract. In this paper a method is proposed called data consistency measure computing of compatibility based multi-sensor Kalman fusion system, so that the measured data is obtained from several sensors constructed at a certain position rather than traditional methods of combining only one gyroscope and one accelerometer applied to current self-balancing vehicle, in order to attain a more reliable system for real time online measure and also prolong the lifetime of product. According to the measure computing of compatibility based outcome of measured angle and angular velocity, data is fused to gain the input of a proportional differential (PD) control part, which is adopted to reach the balancing stability of a self-balancing vehicle.

Keywords: Data fusion · Data measure of compatibility · Kalman filter · Dynamic measure

1 Introduction

The rapid development of miniature electro-mechanical devices (MEMS), or ever more perfect image processing algorithms, cause the interaction between users and their devices to be no longer limited to pushing a button or pulling a lever [1]. The demand of the level of automation increases rapidly, one of the concomitant product is self-balancing vehicles. Two-wheeled production based on self-balancing technology, including model cars and comparable scooters and other vehicles are increasingly prosperous in the field, and its relevant control algorithms become the crucial segment. Its property are more outstanding especially in those areas where there are cramped spaces, in which the requirement for flexibility is much more indispensable. Moreover, in some certain circumstances, human beings are unable to be exposed to some unsatisfactory environments, such as toxic gases for instance, the necessity of self-controlled production is considerably great. Also, as the degree of the intelligence of this type of product increases, it gradually begins to become within the spectrum of

© Springer International Publishing Switzerland 2016
D.-S. Huang et al. (Eds.): ICIC 2016, Part III, LNAI 9773, pp. 551–562, 2016.
DOI: 10.1007/978-3-319-42297-8_51

intelligence transportation system. The different aspects of algorithms for self-balancing vehicles, therefore, become a hot button during the past few years.

Self-balancing vehicles are characterized by short turning radius, low power consumption and widely applied ability [2]. In most cases, the vehicle has three degrees of mechanical freedom: translation of the vehicle, balancing and turning [3]. This paper mainly focuses on the translation aspect. Self-balancing vehicle system controls the motors on each side according to the measure of the angle of vehicle in order to keep the vehicle in balance. The real time online estimation and the algorithm to control the motor, consequently, plays an important role in the performance of the vehicle.

With the guarantee of the precision of the angle's measure, this paper analyze a concise version of model for increasing the capacity of online real time estimation, and propose an algorithm suited to the application of Kalman filtering on a microcontroller. Instead of the traditionally prevalent algorithm that coalesces only one gyroscope and one accelerometer, a muti-sensor system based on data consistency measure computing of compatibility is adopted, in which the measured data is the combination of several sensors rather than one before the fusion step, for we cannot guarantee the precision and reliability of any single sensor.

2   The Model for Self-balancing Vehicle

The structure of Self-balancing vehicle can be regarded as an inverted pendulum (see Fig. 1) with a variety of disturbances. The destination of controlling the inverted pendulum as a single pendulum and reach a stable state at the vertical position can be attained by two approaches: the first one is to change the direction of gravity, the second is push an additional force so that the force for recovering is opposite to the direction of displacement. The first approach is unpractical to implement, so we adopt the second approach.

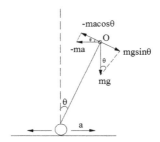

Fig. 1. The model of two-wheeled vehicle

Controlling the wheels at the bottom of the vehicle to make it under an accelerated movement while at the same time observing the inverted pendulum on the vehicle (non-inertial system), it turns out that the vehicle is under an additional stress (inertia force), the direction of which is opposite to the direction of the accelerated velocity of the wheel and proportional to it. Then the restoring force of the inverted pendulum is as following (see formula (1)).

$$F = mg\theta - ma\cos\theta \approx mg\theta - mp_1\theta \tag{1}$$

When the value of θ is small, $\sin\theta \approx \theta$, $\cos\theta \approx \theta$. Presumably, the accelerated velocity of wheels is proportional to the drift angle, and the proportional coefficient is p_1, then we get the approximate expression of F in the formula (1). When p_1 is greater than g, the direction of restoring force is contrary to that of the displacement. Besides, it needs an additional damping force so that the vehicle can gain the equilibrium. Damping force is proportional to the velocity of drift angle, namely the angle velocity of swaying of the vehicle measured by gyroscope while in opposite direction. Then we can get the formula (2) as the deduction of formula (1).

$$F = mg\theta - mp_1\theta - mp_2\theta \tag{2}$$

Then we get the relationship of accelerated velocity and drift angle at formula (3) (cutting m on the both side of the equation).

$$a\cos\theta = p_1\theta + p_2\theta' \tag{3}$$

Thus, as long as we can correctly measure the vehicle's drift angle and the accelerated velocity, we can make the vehicle at a self-balancing state and further control the speed and direction (in this situation, we only lay stress on the forward and backward direction which has no relation to the difference of velocity on two wheels for accomplishing the direction control).

3 Measure of the Drift Angle

The controlling of self-balancing vehicle is a dynamic equilibrium process, so it is necessary to consider both static properties and dynamic properties.

3.1 Measure of Drift Angle by Accelerometer

Accelerometer utilizes inertial deformation based on MEMS theory to detect accelerated velocity, which reflect the impact of resultant external force on the target's accelerated velocity [4]. Using different components on three separate axes, the drift angle can be calculated. When the component of gravity conducts an accelerated velocity on one of the axis of the sensor, the voltage corresponded to these axis changes [5].

$$\Delta u = Kg\sin\theta \tag{4}$$

Δu is the variation of voltage; K is the coefficient indicating the sensor's sensitivity; g means the acceleration of gravity; θ is drift angle. The variation of voltage and drift angle has a sine relationship. Despite its precision under the circumstances of equilibrium, in other cases, however, the vehicle is in non-equilibrium state, then discomposed disturbance by vehicle's accelerated velocity lead to data distortion.

3.2   Measure of Drift Angle by Gyroscope

Gyroscope can be applied to the measure of object's angular velocity of rotation, this kind of sensor harnesses the theory that item in rotating coordinate system suffers from Coriolis force, and make force unit by piezoelectric ceramics. The vibration of the sensor variate along with the object's rotating, and it then reflects the angular velocity. Sensors attached to the vehicle measures the angular velocity. After integration to the angular velocity of vehicle's inclination, the outcome is the angle.

$$\theta(t) = \int_0^t w(t)dt \tag{5}$$

If deviation and drift exists in the signal of angular velocity, accumulative error will be engendered after integration, because signals from the sensor need to be integrated. The deviation and drift will ultimately results in circuit saturating during the process of accumulation of time. The measuring result of the gyroscope is shown in Fig. 2.

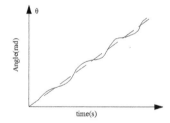

Fig. 2. Phenomenon of integration drift

As shown in Fig. 2, as factors, such as its intrinsic characteristics, temperatures, time of integration etc., influence the performance of this kind of sensors, integration drift cannot be overlooked.

3.3   Evaluation Depending on Measure of Compatibility

In order to prolong the lifetime of product and attain a more reliable data before the fusion step, we dispose of the traditional algorithm and propose a multi-sensor structure system, which utilizes measure computing of compatibility to insure data consistency. From the perspective of information error-tolerance theory, when a certain sensor's value is ambivalent to 2/3 of the other counterparts, the value is unacceptable. Therefore, for the sake of calculation, we use 6 sensors of gyroscope and accelerometer separately.

In general, the distribution of data reflects the measure of compatibility, so failure caused by disturbance or invalidity of sensor itself can be eliminated by the process of computing the measure of compatibility. It is acceptable to analyzing the measure of compatibility and consistency by statistic methods. This section, consequently, proposes an operator for measure computing based on the thought of Chauvenet statistical [6] decision suited to the control of self-balancing vehicle.

The measure of compatibility in this paper's algorithm consists of mutual-support degree and self-support degree. Mutual-support degree is calculated by

$$
\psi_{ij} = \begin{cases} 1 - \frac{|x_i(t) - x_j(t)|}{3\varsigma(t)}, & \text{if } x_i(t) \neq x_j(t) \text{ and } |x_i(t) - x_j(t)| < 3\varsigma(t) \\ 1 - \frac{|S_i - V_{t-1}| - |S_j - V_{t-1}|}{3\varsigma(t)}, & \text{if } x_i(t) = x_j(t) \text{ and } |S_i(t) - S_j(t)| < 3\varsigma(t) \\ 0, & \text{else} \end{cases}
\tag{6}
$$

In formula (6), ψ_{ij} is mutual support of two same kind of sensors, namely I and J; x_i and x_j separately represent the result of the sensor at time t; V_t means the ultimately outcome after evaluation of measure of compatibility, which is defined in formula (17), and the reason we define it in this form is that we only consider the absolute degree of deviation. S_i and S_j is the root of S_i^2 and S_j^2, which are unbiased estimation of variance of sensors I and J separately and define by

$$
\begin{cases} S_i = \frac{1}{n-1} \sum_{k=1}^{n} (x_{i_k} - \bar{x}_i)^2, & \bar{x}_i = \frac{1}{n} \sum_{k=1}^{n} x_{i_k} \\ S_j = \frac{1}{n-1} \sum_{k=1}^{n} (x_{j_k} - \bar{x}_j)^2, & \bar{x}_j = \frac{1}{n} \sum_{k=1}^{n} x_{j_k} \end{cases}
\tag{7}
$$

Self-support degree is defined by

$$
\psi_i = \begin{cases} 1 - \frac{|x_i(t) - x_{median}(t)|}{3\varsigma(t)}, & \text{if } |x_i(t) - x_{median}(t)| < 3\varsigma(t) \\ 0, & \text{else} \end{cases}
\tag{8}
$$

In formula (8), $x_{median}(t)$ means the median of all of the same kind of sensors' measured value at time t. In formula (6) and formula (8), $\varsigma(t)$ is the sensor's support factor, which is calculated by the robust standard deviation based on mediate absolute deviation (MAD) estimation:

$$
\varsigma(t) = G \cdot median_i |x_i(t) - median_j(x_j(t))|
\tag{9}
$$

The value of G depends on the distribution of population of sensors' value. In most cases, normal distribution is applied well in this model, so the value of G can be calculated by

$$P(|X - \mu| \leq MAD) = P\left(\left|\frac{X - \mu}{\sigma}\right| \leq \frac{MAD}{\sigma}\right) = P\left(|Z| \leq \frac{MAD}{\sigma}\right) = \frac{1}{2} \qquad (10)$$

Therefore, we must have that

$$\Phi(MAD/\sigma) - \Phi(MAD/\sigma) = \frac{1}{2} \qquad (11)$$

Since the sum of $\Phi(MAD/\sigma)$ and $\Phi(-MAD/\sigma)$ equals to 1, we have that

$$MAD/\sigma = \Phi^{-1}(3/4) \qquad (12)$$

from which we obtain the factor

$$G = 1/\left(\Phi^{-1}(3/4)\right) \qquad (13)$$

Hence, $G \approx 1.4826$.

In other words, the expectation of 1.4826 times the MAD for large samples of normally distributed X_i is approximately equal to the population standard deviation [7, 8]. Can also be calculated by Q_n estimation [9]. Q_n estimation use distance between pairing observed data instead of degree of dispersion that the observed data from the center of all data to obtain robust standard deviation. This method has a better calculation efficiency.

$$\varsigma_{Q_n}(t) = v_{Q_n} \cdot G_{Q_n} \cdot \left\{ |x_i(t) - x_j(t)| ; i < j \right\}_z \qquad (14)$$

In formula (14), v_{S_n} and v_{Q_n} are adjustment factors based observation. For Q_n estimation, $G_Q = 2.2219$; $z = \binom{N}{2}/4$ [10]. We define the measure of compatibility as

$$\omega_{ij}(t) = \frac{\kappa_i \cdot \psi_i \cdot \psi_j \cdot \psi_{ij}}{\eta} \qquad (15)$$

In formula (15), η is an adjustment coefficient for practice purpose; is a factor computing the compatibility in a historical perspective defined in formula (23). The advantage of the definition proposed is that all of the data is under consideration, no matter the dimension of all data at time t from all same kind of sensors, or data from one sensor gathered from every time in the past. Through an analysis we can find out following features of the algorithm:

(1) When $\omega_{ij} = 1$, it indicates that the measured value of two sensors is absolutely the same, or i and j indicates the same sensor;
(2) When $0 < \omega_{ij} < 1$, it indicates that the degree of the two sensors is relevant, the more the value of is, the more relevancy between the two sensors are;
(3) When $\omega_{ij} = 0$, it indicates that two sensors have no relevancy at all.

From all the analysis stated above, it comes to a conclusion that satisfy the requirement to evaluate the measure of compatibility, with no need to the use of integration, which shows the advantage in the implementation of microcontroller practically. We now get a matrix of the value of ω_{ij}, whose size is $n \times n$.

According to the algorithm proposed, matrix W is a symmetric matrix, so the sum of either each column or each row reaches to the same outcome which indicates the measure of compatibility of one certain sensor against to all other sensors as a whole. The bigger the value is, the more relevant the sensor to other is, which also shows that how reliable of the measure is. We define the indicator of reliability of sensor i as

$$r_i(t) = a \cdot \frac{\sum\limits_{j=1, j \neq i}^{n} \omega_{ij}(t)}{n-1} \tag{16}$$

In formula (16), a is an adjustment factor. On the basis of the value of n numbers, we sort them from small to large, and disposes of numbers if $r_i(t) < T$, where T is a constant threshold. The next step is to calculate the final measure, shown in following.

$$V_t = \sum_{i=1}^{m} \left(\left(r_i / \sum_{j=1}^{m} r_j \right) \cdot x_i(t) \right) \tag{17}$$

In formula (17), m is the number of sensors whose corresponding $R_i(t)$ is greater than T. If $m = 0$, then give up the value at t, controlled quantity at time t harnesses the value equals to or slightly less than that in $t-1$.

After calculating the ultimate value, there is an additional step used to help calculate ω_{ij} defined previously in formula (15). We firstly define a matrix based on the reliability of all same kind of sensors as following:

$$R = \begin{bmatrix} r_{1|t} & r_{1|t-1} & \cdots & r_{1|t-m} \\ r_{2|t} & r_{2|t-1} & \cdots & r_{2|t-m} \\ \cdots & \cdots & \cdots & \cdots \\ r_{n|t} & r_{n|t-1} & \cdots & r_{n|t-m} \end{bmatrix} \tag{18}$$

Every sensor's history behavior is defined as

$$\kappa_i = \frac{\left| \prod\limits_{k=1}^{m} \exp(-k \cdot R_{ik}) - median_j \left\{ \prod\limits_{k=1}^{m} \exp(-k \cdot R_{jk}) \right\} \right|}{median_j \left\{ \prod\limits_{k=1}^{m} \exp(-k \cdot R_{jk}) \right\}} \tag{19}$$

The reason of defining factor κ_i is that the genuine reliability of a certain sensor is not perfectly proportional to either the computed self-support degree or mutual-support degree. Some circumstances in which some external disturbance, such as jolting or vibration between two wheels after a sharp turn exists from time to time, will lead to sensors' suffering from distortion. If there is a sensor with poor qualification that coincidentally gets a high value of ψ_{ij} at this moment, this specific sensor will be considered as a credible one without the consideration of κ_i so that the precision of V_t will be influenced.

Also, it is noteworthy that when data sequence is not equal length caused by certain reasons, normally it can be resolved by mean generation with consecutive neighbors or by mean vanishing with consecutive neighborhood, in order to make the data sequence at the same length [11], then to use the measure computing of compatibility we propose.

3.4   Data Fusion Based on Kalman Filtering

After estimation on measure of compatibility stated in previous session, difference between gyroscope and accelerometer is not eliminated yet. Accelerometer has a reliable property of measuring the angle in static circumstances but the performance in motion state is undesirably. However, gyroscope has a totally opposite property at these two environments. This sector we use Kalman filtering data fusion algorithm to exterminate this kind of deviation in ways applicable to this situation.

Kalman filtering is a recursion data processing algorithm, regarding the minimum mean square error (MMSE) as the criteria [12], and it has a wide range of applications. The time complexity of this algorithm is $O(n^3)$, n is the dimension of the state matrix. With the guarantee of the precision as a prerequisite, in order to enhance the performance of real time estimation in a microcontroller system with regard to angel measuring, when designing the algorithm of Kalman filtering, we adopt two-dimensional state vector, which is composed of the vehicle's angle of inclination V_θ and angular velocity of the vehicle's inclination V_ω after estimation of measure of compatibility. The detailed computing algorithm is as shown following. The state space is described as

$$\begin{bmatrix} V_{\theta_t} \\ V_{\omega_t} \end{bmatrix} = \begin{bmatrix} 1 & T \\ 0 & 1 \end{bmatrix} \begin{bmatrix} V_{\theta_t} \\ V_{\omega_t} \end{bmatrix} + [0 \quad 1] w_{t-1} \tag{20}$$

And the simplified version of formula (20) is described as

$$X_t = \Phi S_{t-1} + B w_{t-1} \tag{21}$$

In formula (21), Φ is system equation; S_{t-1} is prediction matrix at time t; B is control input matrix; w_{t-1} is system noise at time $t - 1$; T is prediction period, or updating period; V_{θ_t} and V_{ω_t} are the ultimately computed values after evaluating the measure of compatibility, which is more stable than any particular value of θ_{t_i} and ω_{t_i}.

$$\begin{bmatrix} \hat{V}_{\theta_t} \\ \hat{V}_{\omega_t} \end{bmatrix} = \begin{bmatrix} 1 & 0 \\ 0 & 1 \end{bmatrix} \begin{bmatrix} V_{\theta_t} \\ V_{\omega_t} \end{bmatrix} + \begin{bmatrix} n_{tV_\theta} \\ n_{tV_\omega} \end{bmatrix} \tag{22}$$

Formula (22) is the definition of observation equation and the simplified version is defined as

$$\hat{X}_t = HS_t + N \tag{23}$$

In formula (23), X_t is the observation matrix; H is measurement matrix; N means the observation noise at time t. One-step forecasting mean square error matrix is described as

$$M_{t|t-1} = \Phi M_{t-1}\Phi^T + BC_{w_{t-1}}B^T \tag{24}$$

In formula (24), $M_{t|t-1}$ is state one-step forecasting mean square error matrix; M_{t-1} is the covariance of state estimation, which reflects the precision of state filtering; is system noise's covariance at time t.

$$G_t = \frac{M_{t-1} \cdot H^T}{H \cdot M_{t-1} \cdot H^T + C_{N_t}} \tag{25}$$

In formula (25), is observation vector N's mean square error matrix at time t; is the gain matrix of state filtering, whose value is relevant to initial state matrix of covariance, mean square error of turbulent noise, and observation vector N's mean square error. Besides, state filtering equation is

$$S_t = \Phi S_{t-1} + G_t(X_t - H\Phi S_{t-1}) \tag{26}$$

In formula (26), the adjustment of predicted value is achieved by Kalman gain G_t's dynamic variation.

Filtering mean square deviation matrix is calculated by

$$M_t = (I - G_t H)M_{t-1} \tag{27}$$

In formula (27), I is an identity matrix. The updating of Kalman gain G_t is obtained by the updating of M_t. Establishing the model of the system composed of gyroscopes and accelerometers, then consider the output of the system as output real value and white noise. With regard to the model we then determine C_{N_t} as

$$C_{Nt} = \begin{bmatrix} C_{ntV_\theta} & 0 \\ 0 & C_{ntV_\omega} \end{bmatrix} \tag{28}$$

In formula (28), C_{ntV_θ} and C_{ntV_ω} is separately the covariance of process noise of vehicle's angle of inclination measured by accelerometers and gyroscopes, which means different value represents different reliability degree. For instance, if the data from accelerometers after measure of compatibility is more credible than that from gyroscopes, then it is suitable to set a relatively small value to C_{ntV_θ}. The value of C_{N_t} is attained by further experimentation depending on the performance of vehicle itself.

4 Process of Experiment and Analysis of Corresponding Results

In the experiment, we design that every 5 *ms* the system conduct a data collection from all 12 sensors, then all of the data is transmitted to the computer by Bluetooth. Then we process all of the data by simulation software. In the simulation, all of the data collected at time t is processed by computer instead of microcontroller for the sake of analysis.

To show the effect of the algorithm proposed in this paper, comparison between final results to the initial curve is needed. However, since there are many sensors of various type, which means the non-uniqueness of initial curve, and they are in different quality, a further analysis is needed. Firstly, data collected from accelerometer is more reliable than that from gyroscope comparatively, so the curve drawn from accelerometer is regarded as a better choice. Secondly, among 6 accelerometers, according to the performance of each sensor, Sensor 2 is the most stable one. Consequently, in order to make the comparison meaningful, we decide to assign Sensor 2 to indicate the standard original signal curve.

In the first experiment, we compute the value of r_i for all the accelerometer and the results are shown in Fig. 3:

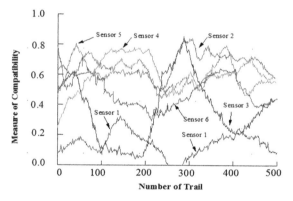

Fig. 3. The value of r_i for 500 times experiments when the vehicle is in motion (Sensor 1 to Sensor 6)

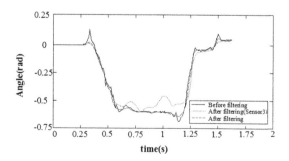

Fig. 4. Comparison between initial signal, fusion by only sensor 3 and fusion after the algorithm this paper proposes (Color figure online)

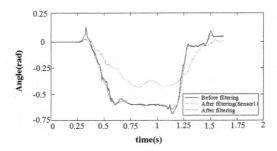

Fig. 5. Comparison between initial signal, fusion by only sensor 1 and fusion after the algorithm this paper proposes (Color figure online)

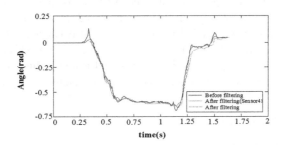

Fig. 6. Comparison between initial signal, fusion by only sensor 4 and fusion after the algorithm this paper proposes (Color figure online)

After filtering by the algorithm this paper proposes, as shown in figures above, computed curve do not contain cluster, and it has a fast response speed, especially when the system is in equilibrium, the curve after fusion is really stable.

The result of simulation is in accord with the previously presumption. As Fig. 4 shows, the precision of the curve in blue fuses well at the edges of the curve, while in the middle position, the curve distorts apparently, the largest of which is up to 0.131. In Fig. 5, the result is obvious of no precision, only the outline of the curve accord with the general trend of the initial one. When currently prevalent method of controlling self-balancing vehicle uses Sensor 1, the performance will be badly influenced. In Fig. 6, the curve of after fusion with Sensor 2 only, though the ultimate outcome is pretty well, its performance is a bit of inferior when compared with the curve using the steps this paper proposed. The largest deviation is no more than 0.04 at the point $t \approx 1.16$ s. Once the satisfied value of the angle of inclination and angle of velocity is obtained, then, according to the inverted pendulum model, it is easier to control the self-balancing vehicle by adopting proportional differential methods, in which p_1 determines whether the vehicle's center of gravity can reach a stable level in equilibrium position, and p_2 determines the speed to reach that stable level [13].

5 Conclusion

Using measure computing of compatibility resolves the problem currently prevalent methods suffered from that the vehicle is so vulnerable to the precision and reliability of specific sensor. In the computing algorithm this paper proposes, we use different sensor's value in different time where they comparatively have a better reliability. Consequently, we dispose of the indispensability of the high requirement of certain device. In addition, we consider not only the measure of compatibility mutually at time t, but also historically from the perspective of all of the past value individually, so it is much better than only using arithmetic mean and median. Then we utilize Kalman filtering to achieve the fusion of the value from accelerometers and gyroscopes, so that static characteristic and dynamic characteristic both reach an ideal result. According to the experiment, this paper proposes an ideal method to gain an accurate and fast response of the variation of angle of inclination. Based on that, a proportional differential control can easily be applied to control the self-balancing vehicle. This paper proposes the computing algorithm that can not only be used in self-balancing system, but also in other places like suspension control and gauges. This computing strategy can also be a reference in other multi-sensor system designs.

References

1. Ciężkowski, M.: Modeling the interaction between two-wheeled self-balancing vehicle and its rider. Int. J. Appl. Mech. Eng. **18**(2), 341–351 (2013)
2. Tan, M., Wu, Y., et al.: Design of two-wheeled self-balancing mobile robot guided by electromagnetism. Huazhong Univ. Sci. Technol. J. **1**(41), 249–252 (2013)
3. Grepl, R., Zouhar, F., Štěpánek, J., et al.: The development of self-balancing vehicle: a platform for education in mechatronics. Technical Comput. Prague (2011)
4. Li, Y., Zhao, Z., Gao, J.: Application of MEMS inertial sensor ADIS16355 in attitude measurement system. Data Collect. Process. **27**(4), 501–507 (2012)
5. Li, Z., Wang, D.: High-precision angle measuring device based on triaxial acceleration sensor. Instrum. Technique Sens. **8**, 30–32 (2013)
6. Zhuang, C.Q., Wu, Y.S.: An Introduction to Probability and Mathematic Statistics. South China University of Technology Press, Guangzhou (1997)
7. Ruppert, D.: Statistics and Data Analysis for Financial Engineering. Springer, New York (2011)
8. Leys, C., Ley, C., Klein, O., et al.: Detecting outliers: do not use standard deviation around the mean, use absolute deviation around the median. J. Exp. Soc. Psychol. **49**(4), 764–766 (2013)
9. Croux, C., Rousseeuw, P.J.: Time-efficient Algorithms for Two Highly Robust Estimators of Scale, pp. 411–428. Physica-Verlag, Heidelberg (1992)
10. Si, G., Zhang, Y., Zhao, W.: Confidence evaluation based on consistency measure method for multi-sensor. Xi'an Jiaotong Univ. J. **47**(8), 7–11 (2013)
11. Luo, B., Yuan, K., Chen, J., et al.: Uncertainty analysis based dynamic multi-sensor data fusion. J. Automation. **30**(03), 407–415 (2004)
12. Ye, Z., Feng, E.: Attitude stabilization based on quaternion and kalman filter for two-wheeled vehicle. J. Sens. Technol. **25**(4), 524–528 (2012)
13. Ma, S., Lu, T., Zhang, L.: Attitude measurement and control of two-wheeled self-balancing car. Measurment Control Technol. **34**(4), 71–73 (2015)

Research on Engineering Tuning Methods of PID Controller Parameters and Its Application

Shuxia Li[1] and Jiesheng Wang[1,2(✉)]

[1] School of Electronic and Information Engineering,
University of Science and Technology Liaoning, Anshan 114044, China
lishuxia9685@163.com, wang_jiesheng@126.com
[2] National Financial Security and System Equipment Engineering
Research Center, University of Science and Technology Liaoning,
Anshan 114044, China

Abstract. PID controller is widely used in the many industrial process control fields because of its simple algorithm, good robustness and high reliability. On the basis of the summary on the tuning process, the principle and the respective characteristics of Z-N frequency response method, Cohen-Coon method, Astrom-Hagglund method and CHR method, the simulation experiments are carried out on the three typical controlled objects. According to the simulation results, the advantages and disadvantages of four engineering tuning methods used for setting the PID controller parameters are analyzed and compared.

Keywords: PID controller · Parameter tuning · Cohen-Coon · Z-N · CHR

1 Introduction

The PID controller is one of the most advanced control strategies, because of its simple algorithm, good robustness and high reliability; it is widely used in the field of industrial process control. However, the parameter tuning and optimization of PID controller is an important research problem [1]. With the rapid development of the intelligent control theory, the advanced PID control method based on knowledge inference's expert PID controller, the rule based self-learning PID controller, the neural network PID controller based on the connection mechanism and the intelligent PID controller based on fuzzy logic are proposed [2, 3]. However, in the industrial field, the conventional PID controller is still widely used. According to Japan's survey, now in the existing variety of control technology, the optimized PID control technology occupies 6.8 %, the PID control technology occupies 84.5 %, the manual control accounted for 6.6 %, the modern control technology accounted for 1.6 %, and the intelligent control technology accounted for 0.6 %. All in all, in the industrial process control, the PID structure loop accounts for more than 95 % of the total control circuit.

The design and parameter tuning of PID controller has a very important effect on the performance of the control system. Those three parameters of PID controller, not only affect the control effect, but also affect the robust performance of the controller.

© Springer International Publishing Switzerland 2016
D.-S. Huang et al. (Eds.): ICIC 2016, Part III, LNAI 9773, pp. 563–570, 2016.
DOI: 10.1007/978-3-319-42297-8_52

The parameters setting methods of the PID controller can be divided into the process frequency response method, the model based method and the intelligent setting method. At present in the industrial field, the mathematical models of controlled objects are difficult to be established, so this paper focuses on the engineering tuning methods which do not depend on the mathematical model of controlled objects. The PID parameter engineering tuning method is proposed by Ziegler and Nichols in 1942. The Z-N method is generally suitable for manual calculation and controller initialized set-point, which is widely used in industrial field [4]. In 1984, Astrom proposed a new parameter tuning method of PID controller, where the relay feedback technique was applied to the PID controller in industrial process [5]. The critical point information of the process frequency response was obtained by relay control and the Z-N controller parameters were calculated. After 20 years of development, the PID parameters tuning methods based on relay feedback technique have been improved [6–8].In this paper, the research on engineering tuning methods of PID controller parameters and its application is carried out.

2 Engineering Tuning Methods of Controller Parameters

2.1 Basic Principles of PID Controller

The PID controller is a kind of effective and simple control algorithm based on the "past, present and future" in information of the deviation. The whole system is composed of the PID controller and the controlled object. It is essentially a linear controller. The control deviation is calculated according to the given value PV and the actual output value SV, that is:

$$e(\text{t}) = \text{r}(\text{t}) - \text{y}(\text{t}) \tag{1}$$

Then the object is controlled by the linear combination of the proportional, integral and differential.

$$u(t) = K_p \left[e(t) + \frac{1}{T_i} \int_0^t e(t)\mathrm{d}t + \frac{T_d \mathrm{d}e(t)}{\mathrm{d}t} \right] \tag{2}$$

where, $K_i = K_P/T_i$, $K_d = K_p T_d$, $e(t)$ is the feedback bias. The transfer function can be expressed as:

$$G(s) = \frac{U(s)}{E(s)} = K_p \left(1 + \frac{1}{T_i S} + T_d S \right) \tag{3}$$

The PID controller parameters, such as the ratio gain K_p, the integral time constant T_i and the differential time constant T_d, will affect the control effect of the controlled system.

2.2 Z-N Frequency Response Method

In 1942, Ziegler and Nichols put forward an engineering tuning method on the PID controller parameters named the critical ratio method, which does not depend on the object model. The Z-N frequency response is determined by the critical point of the process transfer function, which is located in the intersection of the Nai's curve and the negative real axis. By determining the critical gain K_u and the critical oscillation period T_u of the control object to determine the PID controller parameters according to Table 1. In order to obtain the critical point, people usually use the proportional controller to make the closed-loop system to reach the critical state. In order not to produce the critical stability, the critical amplitude oscillation can make the system to make the 4:1 attenuation oscillation, also known as the damped oscillation method.

Table 1. Tuning equations of Z-N critical ratio method.

PID controller	K_p	T_i	T_d
P controller	$0.5K_u$	–	–
PI controller	$0.4K_u$	$0.85T_u$	–
PID controller	$0.6K_u$	$0.5T_u$	$0.125T_u$

2.3 Cohen-Coon Method

The Cohen-Coon method is mainly applied to a model with the first order inertia and pure delay. Cohen-Coon's task is to configure the dominant pole so that the system's attenuation ratio reached 25 %. For the P or PD type controller, the pole assignment is to make the system get the maximum gain and also meet the decay rate, so that the system's static error reaches the minimum. For the PI or PID controller, the pole configuration will allow the controller to achieve a maximum of integral error, so that the integral error can be minimized. The PID controller parameter tuning table based on the Cohen-Coon method is shown below Table 2.

Table 2. Tuning equations of cohen-coon method.

PID controller	K_p	T_i	T_d
P controller	$\frac{1}{a}\left(1+\frac{0.35\tau}{1-\tau}\right)$	–	–
PI controller	$\frac{0.9}{a}\left(1+\frac{0.92\tau}{1-\tau}\right)$	$\frac{3.3-3.0\tau}{1+1.2\tau}L$	–
PID controller	$\frac{1.35}{a}\left(1+\frac{0.18\tau}{1-\tau}\right)$	$\frac{2.5-2.0\tau}{1-0.39\tau}L$	$\frac{0.37-0.37\tau}{1-0.81\tau}L$

Note: $a=(K_pL)/T$, $\tau=L/(L+T)$.

2.4 Astrom and Hagglund Method

In 1984, based on the relay feedback principle, Astrom and Hagglund proposed a new tuning method of PID controller parameter, whose principle is shown in Fig. 2. In the feedback loop of the process output to its input, the relay is connected to the relay in order to approximate the critical point. Most of the process will be in a limited cycle oscillation, and its critical gain and cycle is decided by the amplitude and frequency of the oscillation. This method can guarantee the stability of the closed-loop oscillation response, so it has been widely used in the tuning of industrial PID controller parameters. As shown in Fig. 2, the controlled object has at least $-\pi$ phase in high frequency, which can generate a periodic oscillation of T. This makes the intersection of the frequency phase lag $-\pi$ of the controlled object. The intersection of this frequency is the intersection of the negative real axis and the curve of the Nai, and the critical point of the critical point is described as follows:

$$\omega_n = 2\pi/T \qquad (4)$$

where μ is the amplitude of the relay. According to the Fourier series expansion of the relay characteristics, the amplitude of the first term is 4 μ/π. If the oscillation value of the output a of the controlled object is y, the amplitude of the controlled object of the point can be approximately represented as (Fig. 1):

$$K_u = \frac{\pi a}{4\mu} \qquad (5)$$

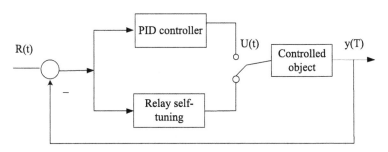

Fig. 1. Diagram of relay feedback control.

2.5 CHR Tuning Method

In the actual system, the traditional Z-N tuning method has been continuously developed. The so-called Chicn-Hrones-Reswick (CHR) algorithm is one of the Z-N tuning methods. In Table 3, the empirical formula is given, which can be divided into two parts, one is the 0 % overshoot, which allows a relatively large amount of damping to ensure that the system has no overshoot, and the other is a 20 % overshoot, which allows for a large overshoot in order to quickly respond to the changes in set-point. Table 4 gives the CHR algorithm for the PID controller with a disturbance; however, the proposed 0 % overshoot is not set up for all the objects. Compared with the Z-N tuning method, the critical parameters of the CHR tuning formula are not designed, and the time constant of the system is T, which makes the design and tuning of the controller parameters simple and easy to use.

Table 3. CHR tuning equations of set-point.

Controller	0 % Overshoot			20 % Overshoot			Object type
	K_p	T_i	T_d	K_p	T_i	T_d	$R = T/\tau$
P	$\frac{0.3}{\alpha}$			$\frac{0.7}{\alpha}$			$R > 10$
PI	$\frac{0.35}{\alpha}$	$1.2T$		$\frac{0.6}{\alpha}$	T		$7.5 < R < 10$
PID	$\frac{0.6}{\alpha}$	T	0.5τ	$\frac{0.95}{\alpha}$	$1.4T$	0.47τ	$3 < R < 7.5$

Table 4. CHR tuning equations of disturbance point.

Controller	0 % Overshoot			20 % Overshoot			Object type
	K_p	T_i	T_d	K_p	T_i	T_d	$R = T/\tau$
P	$\frac{0.3}{\alpha}$			$\frac{0.7}{\alpha}$			$R > 10$
PI	$\frac{0.6}{\alpha}$	$4T$		$\frac{0.7}{\alpha}$	$2.3T$		$7.5 < R < 10$
PID	$\frac{0.95}{\alpha}$	$2.4T$	0.5τ	$\frac{1.2}{\alpha}$	$2T$	0.42τ	$3 < R < 7.5$

Note: $\alpha = \tau k/T$.

3 Simulation Experiments and Results Analysis

In order to verify the performance of the PID controllers, the algorithm is simulated on the followed objects.

- Object 1: $G(S) = 5/(s^4 + 3s^3 + 7s^2 + 5s)$
- Object 2: $G(s) = 0.33/(134S^2 + 18.5S + 1)$
- Object 3: $G(s) = 3/(s^4 + 3s^3 + 6s^2 + 3s + 1)$

Aiming at the above three controlled objects, the Z-N frequency response method, Cohen-Coon method, and Hagglund-Astrom method and CHR method are used to tune the PID controller parameters. The obtained PID controller parameters are shown in Table 5, the simulation results on the three controlled objects are shown in Figs. 2, 3 and 4, and the performance indices of three controlled objects under different PID controllers are listed in Table 6.

It can been seen form Fig. 2 and Table 6 that for the four-order control object, the overshoot (Os) under the Z-N method is relatively large, but the rise time (Tr) and regulation time (Ts, 5 %) is smaller, and the system will be stable. But the other three methods need to experience a few cycles' oscillations to achieve stability. By Fig. 3 and Table 6 it can be seen that the four methods all experience a few cycles oscillations to achieve stability, but the oscillation of Hagglund-Astrom method is the smallest and the fastest to reach stability, while the Z-N and CHR methods are the most obvious oscillation and reach stable lastly. Through Fig. 4 and Table 6, the effects under these four kinds of methods are almost the same, that is to say they are very fast to reach stable, but the Z-N method has the largest overshoot. In conclusion, for different

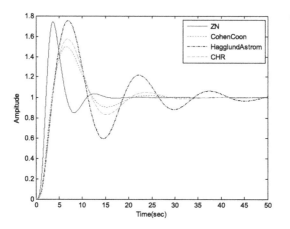

Fig. 2. Simulation results under four tuning methods (object 1).

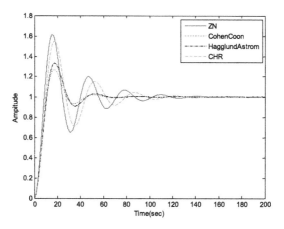

Fig. 3. Simulation results under four tuning methods (object 2).

controlled objects, four kinds of PID controller parameter tuning methods will show different control performance.

In conclusion, The PID controller based on the Z-N frequency response method is applied in the actual industrial process, which will have larger overshoot and cause the more violent oscillation. Thus the technical requirements will not be met. The CHR method is a way to get better closed loop characteristics by changing the step response. Cohen-Coon method is a comprehensive parameter tuning method based on the generalized object properties, the load disturbance and the various performance indicators. Although the way to get the characteristic parameters of the object is simple, how to determine the maximum slope of the reaction curve and how to determine the tangent remain to be solved. At the same time, the degree of approximation in this method is too larger and rough. All of these will bring out the great error to the self-tuning. Astrom-Hagglund method adopts the nonlinear part of the relay characteristics to

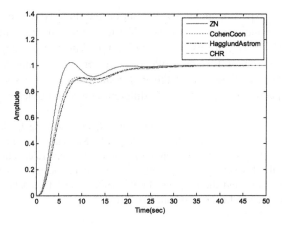

Fig. 4. Simulation results under four tuning methods (object 3).

Table 5. Parameters of PID controller under four tuning methods.

Methods	Object 1			Object 2			Object 3		
	K_p	K_i	K_d	K_p	K_i	K_d	K_p	K_i	K_d
Z-N	1.07	0.44	0.65	17.88	2.45	32.18	0.80	0.25	0.62
C-C	0.49	0.13	0.32	16.28	0.8	27.35	0.58	0.18	0.44
A-H	0.37	0.16	0.21	15.36	0.98	26.15	0.48	0.17	0.35
CHR	0.43	0.13	0.25	14.16	1.62	21.52	0.58	0.17	0.34

Table 6. Performance index of PID controller.

Performance index	Object 1				Object 2				Object 3			
	Z-N	C-C	A-U	CHR	Z-N	C-C	A-U	CHR	Z-N	C-C	A-U	CHR
Os (%)	74.26	**49.95**	75.37	57.39	61.29	**26.85**	33.18	54.92	38.89	26.31	25.41	**23.65**
Tr (s)	**1.34**	2.32	2.39	2.41	**6.58**	8.47	8.35	7.55	**22.97**	28.52	29.64	29.72
Ts (s)	**10.17**	18.38	39.14	19.11	82.12	**38.36**	41.09	74.97	146.88	121.14	124.23	**121.05**

replace the pure proportional controller in the Z-N method, which makes the system appear the limit cycle so as to obtain the required critical value. This simple self-tuning method based on relay feedback avoids the problem of long tuning time and the critical stability of the Z-N frequency response method.

4 Conclusions

Aiming at the engineering tuning of PID controller parameters, the principle and respective characteristics of the Z-N frequency response method, the Cohen-Coon method, the Hagglund-Astrom method and the CHR tuning method are summarized and the simulation experiments are carried out on the three typical controlled objects. Three performance indexes, such as the overshoot (Os), the rise time (Tr) and the regulation time (Ts, 5 %), are used to compare four kinds of the engineering setting methods on the tuning of PID controllers parameters. According to the simulation results, the proper tuning method of the PID controller parameters should be chosen based on the respective characteristics of different controlled objects.

Acknowledgement. This work was supported by the National Key Technologies R & D Program of China (Grant No. 2014BAF05B01), the Project by National Natural Science Foundation of China (Grant No. 21576127), the Program for Liaoning Excellent Talents in University (Grant No. LR2014008), the Project by Liaoning Provincial Natural Science Foundation of China (Grant No. 2014020177), and the Program for Research Special Foundation of University of Science and Technology of Liaoning (Grant No. 2015TD04).

References

1. Wang, W., Zhang, J.T., Chai, T.Y.: A survey of advanced PID parameter tuning methods. Acta Autom. Sinica **26**(3), 347–355 (2000)
2. Åström, K.J., Hägglund, T.: The futures of PID control. Control Eng. Pract. **9**(11), 1163–1175 (2001)
3. Bennett, S.: The past of PID controllers. Annu. Rev. Control **25**(1), 43–53 (2001)
4. Ziegler, J.G.: Optimum settings for automatic controllers. Trans. A.S.M.E. **115**(2), 759–768 (1942)
5. Åström, K.J., Hägglund, T.: Automatic tuning of simple regulators with specifications on phase and amplitude margins. Automatica **20**(5), 645–651 (1984)
6. Su, S.W., Jungmin, O., In-Beum, L., Lee, J., Seok-Ho, Y.: Automatic tuning of PID controller using second-order plus time delay model. J. Chem. Eng. Jpn. **29**(6), 990–999 (1996)
7. Wang, Q.G., Lee, T.H., Fung, H.W., Bi, Q., Zhang, Y.: PID tuning for improved performance. IEEE Trans. Control Syst. Technol. **7**(4), 457–465 (1999)
8. Hang, C.C., Astrom, K.J., Wang, Q.G.: Relay feedback auto-tuning of process controllers - a tutorial review. J. Process Control **12**(1), 143–162 (2002)

A Heuristic Approach to Design and Analyze the Hybrid Electric Vehicle Powertrain and Energy Transmission Systems

Khizir Mahmud[1](✉), Mohammad Habibullah[2],
Mohammad Shidujaman[3], Sayidul Morsalin[4], and Günnur Koçar[2]

[1] School of Automation, Northwestern Polytechnical University, Xi'an, China
khizir@mail.nwpu.edu.cn
[2] Solar Energy Institute, Ege University, İzmir, Turkey
habibullahiut@gmail.com, gunnur.kocar@ege.edu.tr
[3] Department of Mechanical Engineering, Tsinghua University, Beijing, China
shangtl5@mails.tsinghua.edu.cn
[4] Department of Electrical and Electronic Engineering,
Chittagong University of Engineering and Technology, Chittagong, Bangladesh
sayidul.morsalin@gmail.com

Abstract. The Hybrid Electric Vehicle (HEV) powertrain, and an efficient power conversion and energy transmission process, are significant factors to reduce conventional fuel consumption and vehicle gas emission. The scale of gas emission of an HEV depends on an efficient design process for the powertrain and an optimal energy management system. Therefore, this paper models an efficient powertrain and energy transmission process for both series and parallel HEVs, which can contribute to the emission reduction process. Different power conversion stages, energy transmission paths, emissions, and a systems response corresponding to the driver's profile are analyzed systematically. Finally, the emission of the proposed system is compared with the European standard of vehicle gas emission.

Keywords: Emission reduction · Energy management · Parallel hybrid electric vehicle · Series hybrid electric vehicle · Powertrain · Vehicle energy transmission

1 Introduction

In recent years, the automotive industry has been heading towards the trend of adopting clean energy technologies. The automobile is one of the key sources of toxic gas emission. The emission can be halved by hybridizing conventional vehicles, i.e. by using hybrid electric vehicles (HEV) which are a combination of electrical and conventional energy sources. Again, the emission of an HEV can be reduced significantly by maintaining a proper combination of electrical storage and engine, i.e. by modeling a proper powertrain, power conversion, and energy transmission process.

The performance of the HEV is highly dependent on efficient power split control, electrical storage charge-discharge control, and the optimization of the energy

© Springer International Publishing Switzerland 2016
D.-S. Huang et al. (Eds.): ICIC 2016, Part III, LNAI 9773, pp. 571–582, 2016.
DOI: 10.1007/978-3-319-42297-8_53

management systems. All of these factors are controlled by several strategies like PID control, intelligent control [1], dynamic programming, model predictive control, supervisory control, etc.

Several research articles on the HEV energy management system have been presented in [2–4]. However, this paper focuses on an efficient modeling process of the HEV powertrain, power conversion, and energy transmission. In [5, 6], the HEV powertrain and power converters have been discussed. Other research papers [7–9] focused on the placement of the motor and engine of the HEV. However, most of the vehicle powertrain modeling and analysis dealt with a single vehicle type and focused less on the power and energy transmission process on the mechanical and electrical systems for both series and parallel HEVs. This paper tries to fill the latter gaps by modeling an improved powertrain of an HEV and explaining the total energy transmission process with all possible energy paths. The emission level of the proposed system has been analyzed and compared with the European standard gas emission.

2   Parallel Hybrid Electric Vehicle Powertrain Model

The proposed architecture of the parallel hybrid powertrain is depicted in Fig. 1. A gasoline engine has been used as the conventional energy source, and a NiMH-battery pack (20 kWh) has been used as the electrical source. The diver's profile indicates the driver's commands (acceleration, brake) which will be fed to the controller, and the controller will control the electrical energy storage systems (ESS) and the mechanical systems. According to the driver's profile and controller signal, the ESS will give power to the converter (DC-DC) to drive a DC motor which is connected to the back wheel with the necessary gears and shaft. At the same time, the gasoline engine will also give power to the mechanical system of the back wheel.

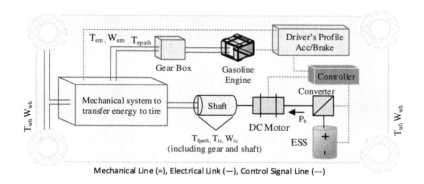

Fig. 1. PHEV powertrain and energy transmission schematic

2.1 Common Energy Path

A quasi-static approach has been used to model the system in Fig. 1. It uses a backward powertrain simulation [9]. At speed $v(t)$, the vehicle driving force is as follows.

$$F_v(t) = \frac{\rho_a v(t)^2}{2} c_D A_f + F_R(v)\cos\gamma(t) + gm_t\sin\gamma(t) \tag{1}$$

Here $F_v(t)$ = drag force, $\gamma(t)$ = road slope, F_R = rolling resistance, which is the fifth-order polynomial function of vehicle speed represented below.

$$F_R(v) = gm_t[a_0 + a_1 v(t) + a_2 v(t)^2 + a_3 v(t)^3 + a_4 v(t)^4 + a_5 v(t)^5] \tag{2}$$

The necessary torque and rotational speed of the vehicle's wheels are shown below.

$$T_{wh}(t) = r_{wh}F_y(t) + \ominus_v\frac{a(t)}{r_{wh}} \tag{3}$$

$$\Omega_{wh}(t) = \frac{v(t)}{r_{wh}} \tag{4}$$

At the wheel axle (Fig. 1), the relationship is the balance of torques represented as $T_{epath}(t) + T_{fpath}(t) = T_{wh}(t)$. The torque split factor (u) of the control variable maintains the torque distribution in parallel paths. It can be defined as $u(t) = T_{epath}(t)/T_{wh}(t)$. If $u(t) = 1$, the electrical path delivers the required energy for the torque and regenerative braking. But, if the value of $u(t)$ becomes zero, the gasoline engine will supply energy [9, 10].

2.2 Electrical Energy Path

By following the 'causality chain' [11], the electrical path calculates the available power of each block of the output and wheels. In the forward-backward calculation process, the available-power calculation in each of the block's output stages may differ from the required power because of the limited upstream path. The relationships in the forward and backward directions of the electrical path include the transmission's I/O model [11].

$$T_{em}(t)R_{egb}(n(t)) = T_{epath}(t) \tag{5}$$

$$\omega_{em}(t) = \omega_{wh}(t)R_{egb}(n(t)) \tag{6}$$

The relationship between the current I_b and the battery SOC is as follows [12].

$$P_b(t) = I_b(t)V_{b,oc}(SOC(t), P_b(t) - I_b(t)^2, R_b(SOC(t)) \tag{7}$$

The SOC variation and the corresponding current limit are as follows [12].

$$Q_{max}\left(\frac{dSOC(t)}{dt}\right) = -I_b(t) \tag{8}$$

The open-circuit voltage $V_{b,oc}$ is a tabulated function of the state of charge (SOC) of the battery. According to the system in Fig. 1, different energy path has been analyzed in Fig. 2.

Electric power only (Fig. 2(a)): the ESS provides energy for the required torque at the wheel. The energy from the ESS passes through the converter to the motor and the mechanical transmission. The energy during regenerative braking is utilized the charge ESS in the opposite direction.

Hybrid/Electric Assist (Fig. 2(b)): this is the normal operating condition on the road. In this case, both sources give power for necessary torque to the tire depending on the driver's profile.

Engine + battery charging (Fig. 2(c, d)): in this case, the battery is charged from the power through the bidirectional dc-dc converter.

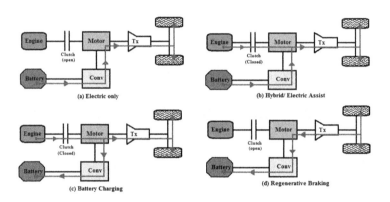

Fig. 2. Different power transmission stages and paths

3   PHEV System Simulation

3.1   Driver's Profile

A driver's profile (Fig. 3) consists of the acceleration-deceleration level and brake according to an arbitrary duration considering the practical road conditions.

At first, the vehicle driver's profile provides acceleration and continues from 0 s to 25 s. After 25 s the acceleration is reduced and at 35 s the driver applies the brake which continues till 50 s. To consider the arbitrary electric drive activation and to check the ESS and motor control according to the driver's profile (open-loop control signal), at 15 s it will be activated and keep active till 25 s. Again, the driver will accelerate the vehicle at 55 s and continue till 80 s. As in the previous strategy, at 60 s the electric drive will activate (get a signal from the controller according to the driver's profile) and continue till 80 s. Within this 60–80 s period, the vehicle power system

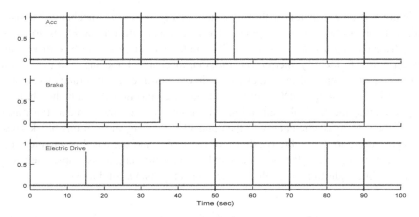

Fig. 3. Driver's profile to analyze practical road condition

will work as an HEV. At 90 s, the brake will be imposed on the system which will continue until the end of the designed simulation time.

3.2   Total System Figure

The total system of the parallel hybrid electric vehicle (PHEV) (in Fig. 4) has been modeled by Sim Power Systems software with the help of the Advanced Vehicle Simulator (ADVISOR) environment [13, 14]. In this process, both electrical and conventional energy source have been used. A driver profile has been chosen to provide an open-loop control signal to manage the vehicle power accordingly. Depending on the driver's profile and the electric controller, battery charging and both electrical and engine fuel energy transmission paths will be defined.

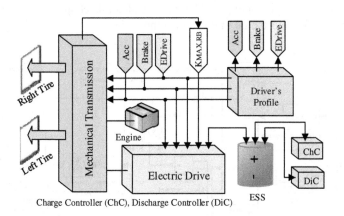

Fig. 4. Total system simulation diagram

Figure 5 is an expansion of the subsystem of Fig. 4 named 'mechanical transmission'. This block consists of the practical vehicle mechanical arrangement to transfer energy from the engine/motor to the tire. It includes a differential gear box, drive clutch, amotion sensor to identify energy transmission, mechanical housing, and the driveline environment. The input of this block is the driver's profile signal (brake, acceleration, and electric drive signal), DC motor mechanical connection through the shaft, and gasoline engine mechanical connection. The output of this mechanical transmission block is the system kinetic energy to the controller. One signal from this block goes to the electric drive to define the condition of the motor and the gasoline engine torque. Finally, the energy from the dc motor and the gasoline engine is transmitted to the vehicle left and right tires through this mechanical transmission block.

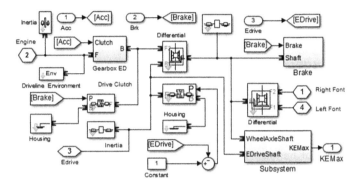

Fig. 5. Mechanical transmission block of Fig. 4

The 'electric drive' block of Fig. 4 has been elucidated in Fig. 6. It consists of a 2-quadrant dc chopper and the dc motor. The two-quadrant chopper topology includes two controllable semiconductor switches and an anti-parallel freewheeling diode with an overall half-bridge structure [15]. This two-quadrant dc chopper is connected to the battery on one side and the dc motor on another side.

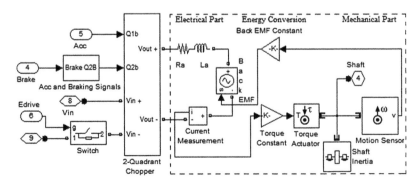

Fig. 6. Electrical drive of the PHEV

The DC motor model consists of three parts: electrical part, energy conversion, and mechanical part. The electrical part of the dc motor includes a resistive-inductive load and a constant current source with back emf generation capability, the energy conversion part includes back emf conversion and a torque conversion block, and the mechanical part includes a torque actuator, motion sensor, and shaft inertia. Open-loop control systems have been used to control the power split between mechanical and electrical systems. Different controllers such as PID, fuzzy, and neural-network-based controllers can be used to control the speed, energy, and power split between conventional vehicle, HEV, and EV. In the case of a traditional vehicle, the engine needs to 'ramp up' to provide its full torque [16–18]. However in an HEV, the electric motor can give full torque even at low speed, with lower noise and better efficiency. Moreover, this dc motor shows some excellent qualities like excellent drive control with fault tolerance, attractive 'off the line' acceleration, better tolerance of voltage fluctuations [16, 17]. The key convenience of an electromotor is the ability to work as a generator as well. In parallel HEV systems, during mechanical braking the energy is regenerated and the battery is charged, as the motor works as a generator at that time.

3.3   Gasoline Engine and Energy Storage Electrical Model

The gasoline engine and battery pack simulation model are depicted in Figs. 7(a) and (b) respectively. If the engine speed rises above the maximum allowed speed, the output torque is cut to zero. Saturation is applied to keep the throttle signal between zero and one [19]. The air-fuel combustion dynamics is not modeled in this case. The

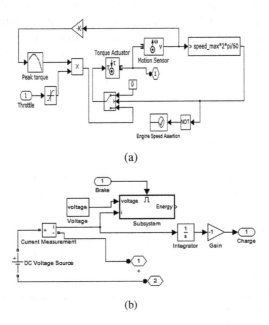

(a)

(b)

Fig. 7. (a) Gasoline engine electrical model, (b) Energy storage system electrical model

battery-pack Simulink model of the HEV includes a voltage source, gain, constant load, and charge control process according to the brake/acceleration condition. A controller attached to the battery pack is connected to the brake of the driver's profile and controls the battery charging condition during regenerative braking.

3.4   PHEV Simulation Output Analysis

The energy transmission from the engine and ESS to the tire is analyzed in this section and illustrated in Fig. 8. From the driver's profile in Fig. 3 it is shown that initially acceleration is provided for 25 s and only the gasoline engine is working from 0 to 15 s. Therefore, the output in Fig. 8 shows a linear rise in transferred energy and torque to the left/right tires. At 15 s, the electric motor also starts working and both engine and motor work together (like an HEV) till 25 s. Therefore, Fig. 8 also shows the sudden rise of transferred energy to the tire as the doubled power source acts on the vehicle. At 25 s, the driver stops accelerating the vehicle and maintains the existing acceleration till 35 s. Thus Fig. 8 shows a constant energy transfer to the tire. As no acceleration is given by the driver, therefore, the energy transferred within this time (25–35 s) is constant. According to the driver's profile, the brake works at 35 to 50 s. Therefore, Fig. 8 shows a sudden drop in transferred energy to the tires. After braking, there is no extra power given by the driver. The transferred energy output also does not change within the period 50 to 55 s. Again, at 55 s the driver will accelerate the car and continue till 80 s. Therefore the transferred energy curve also shows a sudden rise in energy transformation. Within the period 55 to 60 s, the vehicle runs in a conventional mode (only engine), therefore the energy transferred to the tire is a little bit less. However, at 60 s, both sources (engine and battery) are activated till 80 s. Therefore, the transferred energy figure has a sudden rise at 60 s which continues till 80 s. As there is no further acceleration change, the existing energy transfer will remain constant till 90 s. At 90 s, the brake has been imposed, so Fig. 8 shows a sudden drop in energy transformation.

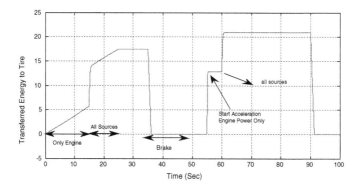

Fig. 8. Transmitted energy status to tire (energy is transmitted according to driver's profile and electric controller's decision)

4 SHEV Analysis by ADVISOR

The simulation of the Series Hybrid Electric Vehicle (SHEV) powertrain is shown in Fig. 9 [20]. The SHEV is propelled by a traction motor which can be used to charge batteries by regenerative braking. Different components' efficiencies have been listed in Table 1.

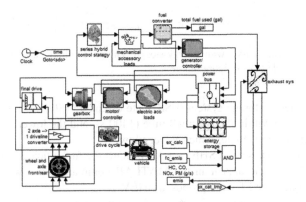

Fig. 9. SHEV simulation using ADVISOR [20]

Table 1. Different components' efficiencies for the SHEV

Name	Efficiency
Fuel economy	46.4116 mpg diesel equivalent and 40.5164 mpg gasoline equivalent
Engine average efficiency	22.6964 %
Driveline average efficiency	93.4538 %
ESS average discharge efficiency	99.0329 %

According to Table 1, the simulation time is 1369 s. The SHEV has been simulated for a distance of 7.45 miles, and to travel this specified distance the average speed of the vehicle is 19.58 mph (miles per hour) with the highest 56.7 mph. As the vehicle always considers the practical road condition, which needs both acceleration and deceleration, the simulation data shows a maximum acceleration of 4.84 ft/sec^2 (feet per second squared), where the minimum is –4.84 ft/sec^2 with an average 1.66 ft/sec^2. According to the road condition the vehicle needs acceleration/deceleration, therefore the vehicle sometimes needs a full fuel flow or sometimes less, and the vehicle sometimes goes to an idle condition. The total idle time is 259 s. The number of stops for the vehicle driving curve is taken as 17. Based on the statistical data in Table 1 and the simulation, a pictorial representation is shown in Fig. 10 [20]. In this figure, the upper plot shows the driving curve with mph versus specified time, the middle one

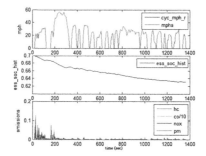

Fig. 10. SHEV driving curve with energy-storage charge condition and the corresponding emission of the vehicle (Color figure online)

represents the energy storage system (ESS) state-of-charge (SOC) condition according to the driving curve within the specified time (1400 s), and the lower curve represents the emission from the vehicle, which is very low.

After running the designed SHEV in ADVISOR, the fuel efficiency and the overall efficiency of the vehicle appear in the MATLAB command window. The following data have been achieved for the proposed SHEV.

Engine average efficiency: 22.6964 %, Driveline average efficiency: 93.4538 %, Motor/Inverter average motoring efficiency: 37.5176 %, Motor/Inverter average generating efficiency: 58.0275 %, ESS average discharge efficiency: 99.0329 %, ESS average recharge efficiency: 87.5966 %, ESS average round-trip efficiency: 86.7495 %.

Table 2. Comparison of the gas emission between the proposed SHEV and the European standard gas emission of HEV

Name of Gases	CO (g/mi)	NOX (g/mi)	HC (g/mi)
Proposed SHEV gas emission	0.13584	0.40653	0.21005
European standard of gas emission (2005)	1.611	0.4025	0.161

*CO- Carbon monoxide, NO_x- Oxides of Nitrogen, HC- Hydrocarbon, g/mi- grams per mile

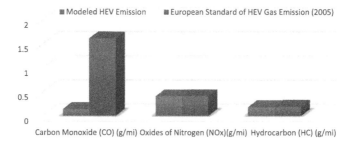

Fig. 11. Graphical representation of the designed and European standard of HEV gas emission (Color figure online)

Table 2 shows a comparison between the proposed SHEV gas emission and the European gas emission standard. The European gas emission standard of January 2005 has been taken as the standard to compare with the proposed SHEV [21] (Fig. 11).

5 Conclusion

For the PHEV, the total efficiency is higher for cruising, and in the case of long-distance driving, it uses both energy sources. It also has an additional degree of freedom to switch between electric storage and ICE power. As compared to the SHEV, a less powerful electromotor than ICE can be used, as it assists the traction. However, in the case of less rotational speed the efficiency drops, as the ICE has lower efficiency at lower RPM. As the ICE cannot be decoupled from the wheel, it also affects the overall efficiency in some cases. Considering all cases the proposed system has been designed to an optimum level with high efficiency. The simulation result justifies its better capability in the powertrain, energy transmission, and reduced emission.

References

1. Mahmud, K., Tao, L.: (2013, November). Vehicle speed control through fuzzy logic. In: 2013 IEEE Global High Tech Congress on Electronics (GHTCE), pp. 30–35. IEEE (2013)
2. Moreno, J., Ortúzar, M.E., Dixon, J.W.: Energy-management system for a hybrid electric vehicle, using ultracapacitors and neural networks. IEEE Trans. Ind. Electron. 53(2), 614–623 (2006)
3. Koot, M., Kessels, J.T., De Jager, B., Heemels, W.P.M.H., Van den Bosch, P.P.J., Steinbuch, M.: Energy management strategies for vehicular electric power systems. IEEE Trans. Veh. Technol. 54(3), 771–782 (2005)
4. Erdinc, O., Vural, B., Uzunoglu, M.: A wavelet-fuzzy logic based energy management strategy for a fuel cell/battery/ultra-capacitor hybrid vehicular power system. J. Power Sources 194(1), 369–380 (2009)
5. Wangsupphaphol, A., Idris, N.R.N., Jusoh, A., Muhamad, N.D.: Power Converter Design for Electric Vehicle Applications. J. Teknologi 67(3), 25–31 (2013)
6. Sciarretta, A., Back, M., Guzzella, L.: Optimal control of parallel hybrid electric vehicles. IEEE Trans. Control Syst. Technol. 12(3), 352–363 (2004)
7. Powell, B.K., Bailey, K.E., Cikanek, S.R.: Dynamic modeling and control of hybrid electric vehicle powertrain systems. IEEE Control Syst. 18(5), 17–33 (1998)
8. Chau, K.T., Wong, Y.S.: Overview of power management in hybrid electric vehicles. Energy Convers. Manag. 43(15), 1953–1968 (2008)
9. Sciarretta, A., Back, M., Guzzella, L.: Optimal control of parallel hybrid electric vehicles. IEEE Trans. Control Syst. Technol. 12(3), 352–363 (2004)
10. Rizzoni, G., Guzzella, L., Baumann, B.M.: Unified Modeling of hybrid electric vehicle drivetrains. IEEE/ASME Trans. Mechatron. 4(3), 246–257 (1999)
11. Kuang, M.L., Ochocinski, C.A., Mack, D., Anthony, J.W.: U.S. Patent no. 6,994,360. U.S. Patent and Trademark Office, Washington, DC (2006)
12. Ehsani, M., Gao, Y., Miller, J.M.: Hybrid electric vehicles: architecture and motor drives. Proc. IEEE 95(4), 719–728 (2007)

13. Mahmud, K., Town, G.E.: A review of computer tools for analyzing the impact of electric vehicles on power distribution. In: 2015 Australasian Universities Power Engineering Conference (AUPEC), pp. 1–6. IEEE (2015)
14. Mahmud, K., Town, G.E.: A review of computer tools for modeling electric vehicle energy requirements and their impact on power distribution networks. Appl. Energy **172**, 337–359 (2016)
15. Emadi, A., Lee, Y.J., Rajashekara, K.: Power electronics and motor drives in electric, hybrid electric, and plug-in hybrid electric vehicles. IEEE Trans. Ind. Electron. **55**(6), 2237–2245 (2008)
16. Panagiotidis, M., Delagrammatikas, G., Assanis, D.N.: Development and use of a regenerative braking model for a parallel hybrid electric vehicle', SAE Technical Paper, no. 2000-01-0995 (2000)
17. Butler, K.L., Ehsani, M., Kamath, P.: A Matlab-based modeling and simulation package for electric and hybrid electric vehicle design. IEEE Trans. Veh. Technol. **48**(6), 1770–1778 (1999)
18. Paganelli, G., Delprat, S., Guerra, T.M., Rimaux, J., Santin, J.J.: Equivalent consumption minimization strategy for parallel hybrid powertrains. In: IEEE 55th Vehicular Technology Conference, VTC Spring 2002, vol. 4, pp. 2076–2081 (2002)
19. Tremblay, O., Dessaint, L.A., Dekkiche, A.I.: A generic battery model for the dynamic simulation of hybrid electric vehicles. In: Vehicle Power and Propulsion Conference (VPPC 2007), TX, USA, 09–12 September, pp. 284–289 (2007)
20. Zhang, X., Mi, C.: Vehicle power management: modeling, control, and optimization. Springer Science & Business Media, London (2011)
21. DieselNet. Emission Standards, Cars and Light Trucks, January 1, 2016 (2015). https://www.dieselnet.com/standards/eu/ld.php

Beginning Frame and Edge Based Name Text Localization in News Interview Videos

Sanghee Lee[1], Jungil Ahn[2], Youlkyeoung Lee[1], and Kanghyun Jo[1(✉)]

[1] School of Electrical Engineering, University of Ulsan,
Daehak Rd. 93, Nam-gu, Ulsan, Korea
{shlee,yklee}@islab.ulsan.ac.kr, acejo@ulsan.ac.kr
[2] Department of Technology, Ulsan Broadcasting Corporation,
Gukyo Rd. 41, Jung-gu, Ulsan, Korea
jiahn@ubc.co.kr

Abstract. To make the automatic person indexing of interview video in the TV news program, this paper proposes the method to detect the overlay name text line among the whole overlay texts in one frame. The proposed method is based on the identification of the beginning frame and the edge using Canny edge detector. The experimental results on Korean television news videos show that the proposed method efficiently detects and localizes the overlaid name text line.

Keywords: Overlay text · Beginning frame · Edge detector · News video · Indexing

1 Introduction

In general, the texts in video sequences can be divided into the overlay text and the scene text. The overlay text, referred as graphics text or caption in other papers, is graphically generated and artificially overlaid on the image by human at the time of editing. The examples of the overlay text include the subtitles in news video, sports scores. On the other hand, the scene text naturally exists in the image being recorded in native environment. This text is found in street signs, text on trucks, and the writing on shirts in natural scenes [1–15].

Although these two type text often contain important information about the content of the video, the overlay text provides additional information and further attracts audiences' interest. Therefore, the overlay text is an essential issue on the automated content analysis systems such as the scene understanding, indexing, browsing, and retrieval. Especially, the overlay text in news videos provides concise and direct description of the content. For instance, the text annotates the names of people and places, or describes objects and the current issue. The detection and recognition of the overlay text have become a hot topic in news video analysis such as identification of person or place, name of news-worthy event, date of event, stock market, other news statistics, and news summaries [1–15].

Among these applications, the identification of the person from the overlay text raises a lot of interest in the information research community. The identification using the overlay person names (OPN) has started to be investigated [16]. Since then the

© Springer International Publishing Switzerland 2016
D.-S. Huang et al. (Eds.): ICIC 2016, Part III, LNAI 9773, pp. 583–594, 2016.
DOI: 10.1007/978-3-319-42297-8_54

research area has raised a large amount of work, especially in face clustering tasks, face naming in captioned images, and recently, automatic naming within broadcast videos [17–22]. Therefore, this paper focuses on the application to make the automatic person indexing system by the OPN in the news interview videos. The name and title information in the interview video of the TV news program are valuable for building an information retrieval and data mining system.

As shown in Fig. 1, the extraction system of overlay text information from video sequences mainly consists of four main parts: text detection, localization, segmentation, and recognition. Text detection is to find the text regions if there is text in a video frame. Text localization is to group text regions into text lines and generate a set of tight bounding boxes around all text lines. The text segmentation or tracking is to determine the temporal and spatial locations. Text recognition is segmented for text regions by binary image and performed the OCR (Optical Character Recognition) system [2, 4, 5].

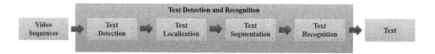

Fig. 1. Workflow of the overlay text extraction system

Although OCR is accurate, it requires a good detection of the text regions in image. Low resolution of the imagery, richness of the background and compression artifacts limit the detection accuracy that can be achieved in practice using existing text detection algorithms. Thus, the text detection step must find a maximum amount of text, but also find the exact coordinates of the boxes that contain text. Without an accurate surrounding box, the quality of text recognition is degraded, leading to poor performances. Current video text detection approaches can be classified into two categories. One is detecting text regions individual frames independently. The other is utilizing the temporality of the video sequences. The former can be divided into the connected component based methods, textures analysis base methods, and gradient edge based methods. The latter is based on the fact that the overlay texts generally last at the same position for a few seconds [3–7, 14, 15, 23].

This paper aims to achieve a good accuracy of video text detection by temporal analysis in news video sequences. And by applying the text detection on only the overlay text including frames, the accuracy of text detection can be improved compared to the methods without pre-processing. The proposed method for the pre-processing is based on the identification of the beginning frame by the edge density of the text using Canny edge detector. The detailed proposed method is presented in Sect. 2. Section 3 shows the results of the experiments. And the last section describes the brief conclusion and the future works.

2   Proposed Method

The framework of the proposed method is given in Fig. 2. The method uses the rule-based characteristics in production of the TV news program. Since the broadcasting videos are produced by professionals, many of the accepted production rules apply to the TV content. For example, text and logo are often overlaid onto the natural content in a structured manner, such as aligned text lines at the bottom or on the upper corners of the screen, to minimize the chance of covering the important content. And the common property in most news video sequences is summarized as follows: The overlay text position is fixed, generally in the range of 1/3 from the bottom of the frame. The background of the text is usually opaque or translucent, and in most case, the color of the background is eye-catching, such as white, blue, yellow, and so on. Colors of text characters are often very distinguishable from the background color. The sizes and fonts of the overlay text from the same news video generally remain unchanged for a long term [2, 12].

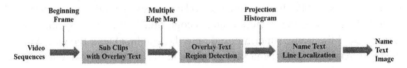

Fig. 2. The framework of the proposed method

Since the videos have the appearance and disappearance of the overlay text, the processing in every frame is unnecessary and time-wasting. First, the sub clip which contains the overlay text frames is made based on the beginning frame identification of our previous work [24]. And then, the overlay text region detection is done by the reference frames selection and the multiple-edge-map. Since the same overlay text last on the same position for a few seconds or more, the reference frames selection reduces the processing time. The multiple-edge-map image is acquired by the logical AND operation on the Canny edge map images of the selected reference frames. Based on this result, the overlay text region is detected by the number of black and white transitions. The ROI text line mask which limits the region of interest in the whole text lines is obtained by the horizontal projection analysis of the detected overlay text region. At last, the overlay name text line is detected by applying the ROI text line mask image on one of the reference frames. Details are given below.

2.1   Sub Clip Based on Beginning Frame Identification

By observing a large quantity of the TV news programs, the appearance and disappearances of the overlay text occur suddenly or slowly in most the news videos as shown in Fig. 3. By precisely locating the critical frame where each overlay text appears or disappears, the overlay text detection can be focused solely on the frame containing the overlay texts. To obtain the goal, this paper proposes the identification of the overlay text beginning frame by the edge density based on the Canny Edge detector. The beginning frame is defined as the frame which has abrupt difference at edge density of current frame and previous frame and a little difference at the edge density of current frame and next frame [24].

(a) non appearance image   (b) the transition image   (c) the apperance image

Fig. 3. Example of overlay text appearances in news video sequences

By identifying the beginning frame in the TV news videos, as shown in Fig. 4, the input video frames can be divided three periods; non-text period, transition period, and text period. Thus, since the detection and recognition is limited only the text period, which consists of the frame superimposed onto the graphical text, the whole processing time is saved. And it helps to achieve a good result of the overlay text detection in news video sequences.

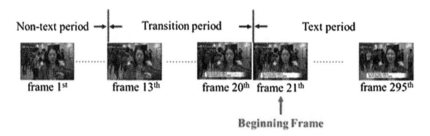

Fig. 4. Characteristics of overlay text in news video sequences

For example, as shown in Fig. 5, since the text presents many edges, the frame including the overlay text has significant changes in the edge density than the frame not containing the overly text. The edge density is defined as the number of detected edge pixels in a frame over the total number of pixels in the frame. The period between the vertical red lines from the frame 13th to the frame 20th like Fig. 5(d) has sharply different from the previous frame and the frame 21th has little different of the edge density of the frame 22th. Thus, the non-text period is from the frame 1st to the frame 12th, the transition period is from the frame 13th to the frame 20th, and the text period is from the frame 21th. As a result, the beginning frame becomes the frame 21th.

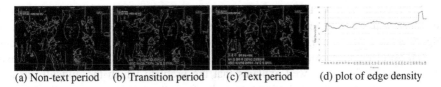

(a) Non-text period   (b) Transition period   (c) Text period   (d) plot of edge density

Fig. 5. Edge density using Canny edge detector

2.2   Overlay Text Region Detection

This paper aims to achieve a good accuracy of video text detection by the temporal analysis of the news videos. In the sub clip frames, the four reference frames are selected and to detect the overlay text region, the logical AND operation was executed on Canny edge maps of the four reference images [2].

According to the characteristics presented in previous paragraph, the four reference frames are selected in the sub clip. If the sub clip is played f frames per second, the overlay text stays in a fixed location for at least $2f$ consecutive frames. Let k be the nearest integer that is not less than f. This paper defines every consecutive k frames to be one round. To simplify the calculation, about only the 1st round, the four reference frames are selected on frame 1, $\lfloor k/3 \rfloor$, $2\lfloor k/3 \rfloor$, $3\lfloor k/3 \rfloor$ like Fig. 6. Because the same overlay text is fixed in the same position for every consecutive k frames [2].

(a) 1st reference frame   (b) 2nd reference frame   (c) 3rd reference frame   (d) 4th reference frame

Fig. 6. The four reference frame images.

The simple line deletion is applied on each edge images to remove long lines which are unlikely to be characters in the Canny edge result image. When the edge image is scanned from left to right and top to bottom, a horizontal and vertical line is removed if its length exceeds the presumed width w and height h of a character. As a result, the edge map images of four reference frames are obtained like the Fig. 7.

(a) 1st reference frame   (b) 2nd reference frame   (c) 3rd reference frame   (d) 4th reference frame

Fig. 7. The edge images of four reference frames.

Next, the logical AND operation is executed on the edge map images of the four reference frames. The result image is called the Multiple-Edge-Map, as shown in Fig. 8. Since a position (i, j) becomes an edge pixel if all four edge images are edge at (i, j), most of the background edge pixels are removed, whereas the static overlay texts are remained. The result well explains that the problem which the difficulty to distinguish whether the detected edges are really from the overlay texts is alleviated by multiple frame integration method. Because the same overlay text appears in the same location for many successive frames, while the location of background edge pixels may differ in a few pixels.

Fig. 8. The Multiple-Edge-Map image.

And then, the overlay text candidate region is detected by utilizing the number of the black and white transition. As shown in Eq. (1), the value of N_{trans} can be obtained that a window of the presumed character size $w \times h$ slides from left to right and top to bottom on the Multiple-Edge-Map image.

$$N_{trans} = \sum_{i=0}^{h-1} \left(\sum_{j=1}^{w-1} |b(i,j) - b(i,j-1)| \right) + \sum_{j=0}^{w-1} \left(\sum_{i=1}^{h-1} |b(i,j) - b(i-1,j)| \right) \quad (1)$$

where w and h are the width and height of window, and b($\cdot$) is binary image. If N_{trans} is larger than threshold T_{trans}, this window is masked. The union of all masked windows is the overlay text candidate region. The threshold T_{trans} depends on the character size and is obtained by $T_{trans} = \beta(w \times h)$ with β a constant which is empirically measured [2].

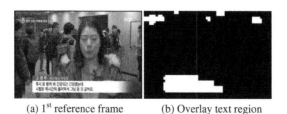

(a) 1st reference frame   (b) Overlay text region

Fig. 9. The overlay text region detection image.

Finally, to resolve the problem that characters lose some pixels in the AND operation, a morphological closing is applied first and then dilation is followed. A closing with a horizontal structuring element of size $\lceil w/3 \rceil$ is used to fill holes. A dilation with a structuring element of size $\lceil w/4 \rceil \times \lceil h/4 \rceil$ is applied to connect the characters [2]. The result image is the overlay text region like Fig. 9.

2.3   Name Text Line Localization

In general, many overlay texts can exist in the one frame of the video. To detect a name text line, it is necessary not analyzing the whole overlay texts in the one frame. Since the TV content is produced by professionals, this paper constrains the detection region based on the news program production rules. In the news interview video sequences, over a few lines, the story of interviewees is positioned at the bottom of the frame like the characteristics remarked in the previous. And the interviewee's name and the title is fixed on the top line among the interviewee's first story lines and appear in the same position for a few seconds. Therefore, only the top of the first story lines becomes the interest of region (ROI).

To detect the text line, at first, the horizontal projection histogram must be obtained by applying on the Multiple-Edge-Map image of the previous paragraph. To scan the result image from top to bottom, and count the number of edge in a row by Eq. (2) can be gotten the horizontal projection histogram image like Fig. 10(b).

$$H_{hor}(i) = \sum_{j=0}^{w-1} b(i,j), \quad i = 0, 1, \ldots, h - 1 \tag{2}$$

This projection image is analyzed in the range of 1/2 from the bottom of the frame. The text line is defined as a consecutive horizontal projection histogram, if the histogram values are more than threshold. The first top of the text line in the half bottom region is selected as the region of interest. The start point and end point of ROI along the height (vertical) axis are applied to the overlay text region image. The result is the ROI text line mask image like Fig. 10(c). At last, to apply on the one frame of four reference frame images yields the overly name text line image like the Fig. 10(d).

(a) Multiple-Edge-Map (b) Horizontal projection (c) ROI text line mask (d) name text line

Fig. 10.   The overlaid name text line localization.

3 Experimental Results

Since there was no standard dataset for the proposed method, the test videos used in the experiment were captured from the video sequences in TV news program in Korea. The resolution of the videos was 720X480, and frame rate was 29.97 frames per second. Each video clip may last less than 2 s. In order to show that the proposed method is effective, the name text localization is evaluated in terms of Recall(R) and Precision(P). And the comparison experiment is executed in case of using the beginning frame, and not.

3.1 Identification of Beginning Frame

Figure 11 and Table 1 present the results of the overlay text beginning frame identification using Canny edge detector. The two thresholds of Canny edge detector, *i.e.* T_{high} or T_{low} was decided based on our empirical studies [2]. The reference value of abrupt difference among the frames was decided 0.03 by our empirical studies. By the identification of the beginning frame, the input videos were divided three sub clips; non-text clip, transition clip, and overlay text clip.

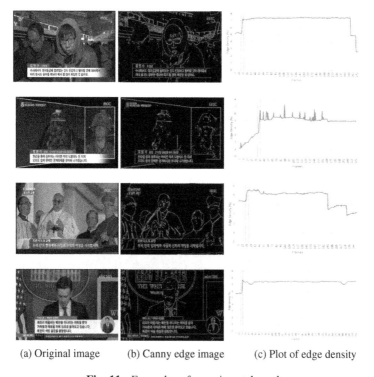

(a) Original image   (b) Canny edge image   (c) Plot of edge density

Fig. 11. Examples of experimental result.

Table 1. The results of overlay text beginning frame identification

	Ground truth		Canny edge detector	
	Transition frame	*Beginning frame*	*Transition frame*	*Beginning frame*
Test 1	16 ~ 24	25	16 ~ 27	28
Test 2	38 ~ 47	48	30 ~ 42	43
Test 3	18 ~ 25	26	14 ~ 24	25
Test 4	18 ~ 24	25	18 ~ 22	23

3.2   Name Text Localization

The presumed character size $w \times h$ was 20X20 pixels. The threshold of the N_{trans} was set to be 0.15 and the threshold of the horizontal projection histogram analysis was the presumed character width 20.

Figure 12 and Table 2 show the results of the overlaid name text line detection. To evaluate the performance of the proposed algorithm, this paper shows the block level accuracy of the results of overlaid name text line detection, and uses precision and recall as performance measures. The measures are calculated Eqs. (3) and (4).

$$R_{pre} = \frac{TP}{TP + FP} \tag{3}$$

$$R_{pre} = \frac{TP}{TP + FN} \tag{4}$$

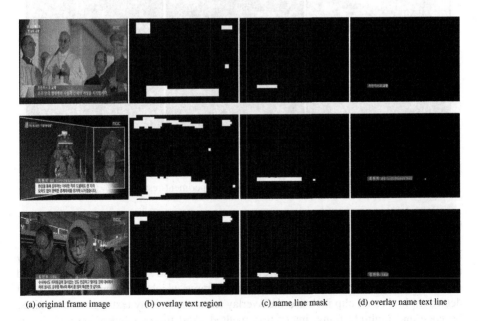

(a) original frame image   (b) overlay text region   (c) name line mask   (d) overlay name text line

Fig. 12. The example images of the experimental results.

Table 2. Experimental results of name text line detection.

Method	Precision rate (%)	Recall rate (%)
Proposed method	66.67	100

In Fig. 12, (b) column is the results of the overlaid text region, (c) column is the results of the name line mask based on horizontal projection analysis, (d) column is the results of the detected overlay name text line. The proposed method accurately detects the position of the overlaid name text line in many examples.

3.3   The Comparative Experiment

To prove that the beginning frame identification is effective and useful, this paper was experiment in two cases of using the beginning frame, and not. As shown in Fig. 13(b), using the beginning frame properly detects the name text line. In contrast to, Fig. 13(c) shows that not using the beginning frame identification method fail to detect the name text line. Therefore, for pre-processing of the overlay text detection, the usage of the beginning frame identification method helps to accurately detect the overlay text.

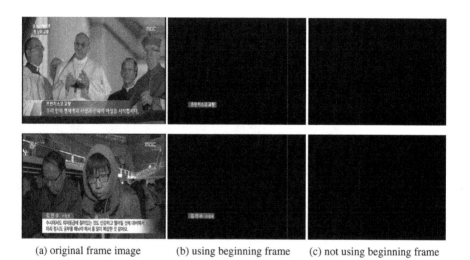

(a) original frame image   (b) using beginning frame   (c) not using beginning frame

Fig. 13. The comparative experiment result.

4   Conclusion

This paper proposes the detection method of the name text line among many overlay texts in one frame in the news interview videos. The video sequences become divided the sub clips by the identification of beginning frame which is pre-processing for text detection. In the sub clip included the overlay texts, the overlay text region is obtained by edge and multiple frame integration method. And the overlaid name text line is localized by the production rules and the horizontal projection histogram analysis.

In general, the name text line contains not only name but also age, degree, job and address. This kind of information plays an important role to develop an automatic person indexing system. Thus, the method to extract effectively the information will be developed based on learning algorithm such as neural network.

References

1. Hua, X.-S., Liu, W., Zhang, H.-J.: An automatic performance evaluation protocol for video text detection algorithms. IEEE Trans. Circ. Syst. Video Technol. **14**(4), 498–507 (2004)
2. Lee, S.H., Ahn, J.I., Jo, K.H.: Automatic name line detection for person indexing based on overlay text. J. Multimedia Inf. Syst. **2**(1), 163–170 (2015)
3. Ye, Q., Doermann, D.: Text detection and recognition in imagery: a survey. IEEE Trans. Pattern Anal. Mach. Intell. **37**(7), 1480–1500 (2015)
4. Wang, Z., Wu, X., Yang, L., Zhang, Y.: A survey on video caption extraction technology. In: The 4th International Conference on Multimedia Information Networking and Security, pp. 713–716 (2012)
5. Zhang, J., Kasturi, R.: Extraction of text objects in video documents: recent progress. In: The 8th IAPR Workshop on Document Analysis Systems, pp. 5–17 (2008)
6. Xu, J., Shivakumara, P., Lu, T., Phan, T.Q., Tan, C.L.: Graphics and scene text classification in video. In: The 22nd International Conference on Pattern Recognition, pp. 4714–4719 (2014)
7. Aradhye, H.B., Myers, G.K.: Exploiting videotext "Events" for improved videotext detection. In: The 9th International Conference on Document Analysis and Recognition, pp. 894–898 (2007)
8. Shivakumara, P., Phan, T.Q., Tan Hong, C.L., Lim, K.J.: A gradient difference based technique for video text detection. In: ICDAR, pp. 156–160 (2009)
9. Gargi, U., Antani, S., Woods, R.E.: Indexing text events in digital video database. Pattern Recogn. **1**, 1481–1483 (1998)
10. Shivakumara, P., Huang, W., Tan, C.L.: An efficient edge based technique for text detection in video frames. In: DAS, pp. 307–314 (2008)
11. Fu, X., Gao, H.: Gray-based news video text extraction approach. In: 5th International Conference on Computer Science and Convergence Information Technology (2010)
12. Yang, Z., Shi, P.: Caption detection and text recognition in news video. In: 5th International Congress on Image and Signal Processing (2012)
13. Yen, S.-H., Chang, H.-W., Wang, C.-J., Wang, C.-W.: Robust news video text detection based on edges and line-deletion. WSEAS Trans. Sig. Process. **6**(4), 186–195 (2010)
14. Sato, T., Kanade, T., Huges, E.K., Smith, M.A., Sato, S.: Video OCR: Indexing digital news libraries by recognition of superimposed caption. Multimedia Syst. Issue **5**(7), 385–395 (1999)
15. Poignant, J., Besacier, L., Quenot, G., Thollard, F.: From text detection in videos to person identification. In: International Conference on Multimedia and Expo (2012)
16. Satoh, S., Nakamura, Y., Kande, T.: Name-It: naming and detecting faces in news videos. In: Proceedings of IEEE Multimedia (1999)
17. Gay, P., Dupuy, G., Lailler, C., Odobez, J.-M., Meignier, S., Deleglise, P.: Comparison of two methods for unsupervised person identification in TV shows. In: 12th International Workshop on Content Based Multimedia Indexing (2014)

18. Pham, P.T., Tuytelaars, T., Mones, M.-F.: Naming people in news videos with label propagation. In: Proceedings of ICME (2010)
19. Jou, B., Li, H., Ellis, G., Morozoff-Abegauz, D., Chang, S.-F.: Structured exploration of who, what, when, and where in heterogeneous multimedia news source. In: Proceedings of ACM Multimedia (2013)
20. Poignant, J., Besacier, L., Le, V.B., Rosset, S., Quenot, G.: Unsupervised speaker identification in TV broadcast based on written names. In: Proceedings of Interspeech (2013)
21. Poignant, J., Bredin, H., Le, V.B., Besacier, L., Barras, C., Quenot, G.: Unsupervised speaker identification using overlay texts in TV broadcast. In: Proceedings of Interspeech (2012)
22. Bendris, M., Favre, B., Charlet, D., Damnati, G., Senay, G., Auguste, R., Martinet, J.: Unsupervised face identification in TV content using audio-visual sources. In: Proceedings of CBMI (2013)
23. Lee, C.-C., Chiang, Y.-C., Huang, H.-M., Tsai, C.-L.: A fast caption localization and detection for news videos. In: The 2nd International Conference on Innovative Computing Information and Control, pp. 226–229 (2007)
24. Lee, S.H., Ahn, J.I., Jo, K.H.: Comparison of text beginning frame detection methods for robust overlay text recognition. In: IWAIT 2016 (2016)

Intelligent Data Analysis and Prediction

Improved Collaborative Filtering Algorithm (ICF)

Xue Li, Xiaolei Zhang, and Zhixin Sun[✉]

Key Laboratory of Broadband Wireless Communication
and Sensor Network Technology, Nanjing University of Posts
and Telecommunications, Nanjing 210003, China
sunzx@njupt.edu.cn

Abstract. Collaborative filtering recommendation currently serves as a key method when it comes to how to rapidly and effectively make proper personalized recommendation for customers. For the sake of above, it becomes a research hot pot in e-commerce system. By analyzing the defects of existing collaborative filtering algorithm, we proposed an improved collaborative filtering algorithm (ICF). In our construction, we firstly establish "users-items" interest degree matrix, and introduce the mechanism of singularity degree to generate the set of similar users, and then optimizes the method of neighbor set generation by recommend importance mechanism. Finally, our experiment proves that ICF has higher performance of accuracy and coverage than other existing algorithms.

Keywords: Personalized recommendation · Collaborative filtering · Interest degree matrix · Singularity impact degree · Recommend importance

1 Introduction

By using e-commerce platform, people can enjoy the convenience of shopping without having to go out. However, due to increasing number of deals on e-commerce platform, it is impossible of people to browse all the items in short time using Internet browser. To solve this problem, personalized recommendation system arises at the historic moment. To the best of our knowledge, collaborative filtering(CF) algorithm now has been a hot spot and the most successful algorithm applied in the field of personalized recommendation. In this paper, we present an improved collaborative filtering algorithm - ICF. Innovations of ICF algorithm mainly include three aspects:

(i) ICF establishes the matrix of "user - project" interest degree, instead of traditional data exhibition form in former collaborative filtering algorithm;

(ii) By introducing the mechanism of singularity degree to calculate the similar users set, ICF provides effective improvements in solving two major problems of recommender systems: "cold start" and neighbor misjudgment;

(iii) According to diversity of recommend importance, ICF optimizes the method of neighbor set generation so that recommended results become more reliable.

© Springer International Publishing Switzerland 2016
D.-S. Huang et al. (Eds.): ICIC 2016, Part III, LNAI 9773, pp. 597–606, 2016.
DOI: 10.1007/978-3-319-42297-8_55

2 Related Research

Nowadays collaborative filtering algorithm is widely used in different websites, and becomes a vital technique personalized recommendation. In order to rate the unrated items, Badaro et al. [1] propose a new hybrid CF algorithm using one kind of weighted combination matrix based on both user-based collaborative filtering and item-based collaborative filtering. It provides effective improvement in solving the problem of data sparsity in collaborative filtering algorithm. Meanwhile, Ba et al. [2] proposes another improved CF algorithm, which combines the clustering algorithm and SVD algorithm. The algorithm judges an item whether it deserves to be recommended only depending on the recent generated recommendation neighbors.

In order to capture the sequential behaviors of users and items, a recommended system based on CF is proposed in [3]. It helps to find the set of neighbors that are the most influential to given users (items). In the solution of reference [4], item similarity matrix is not only established by nearest neighborhood of each single item, but also by the detection for underlying item neighborhood with propagated neighborhoods. Differently, influences of users and the topology of their social network are considered in [5]. It analyzes relationships among users from their scores on the same item and how influential users impact others. In addition, extension of CF algorithm based on user is realized in [6], which focuses on achieving correlation of interests among different users via exploring potential attributes of users and other data mining techniques. To sum up, [3–6] optimize the neighbor set generation method in further, making the recommended results more reliable. But they fail to make full use of contextual information, which made system facing a performance bottleneck to some extent.

Ying et al. [7] propose a personalized web page recommendation model based on user context and CF. Its improved Collaborative Filtering (CF) algorithm discovers similar users' interested web page sets of the target user, based on which Collaborative Filtering web Page Set (CFPS) of one user is filtered. Based on user attributes, a new CF recommendation algorithm is proposed in [8]. This algorithm effectively improves the accuracy, quality and user satisfaction of recommendation system in social networks. At meantime, a new method of user-based CF is proposed in [9] based on predictive value padding. It predicts the empty values in user-item matrix by integrating content-based recommendation algorithm and user activity level before calculating user similarity.

Through the above analysis we can conclude that the current collaborative filtering algorithm study focuses on three aspects: sparseness of data, computing similar users set and neighbor set generation. In this paper, we propose an improved collaborative filtering algorithm ICF. On the one hand, ICF uniquely established "users-items" interest degree matrix, on the other, singularity impact degree mechanism is introduced to compute the impact of a single component of user vector on similarity measure from the whole matrix of "users-items" interest degree and to produce weighted influence on similarity measure, which effectively solved the problems of similarity measure in collaborative filtering, including false judgment of neighbors and helplessness for new users or items and so on. What's more, recommendation importance mechanism is introduced to compute the recommendation importance of a user for an item and filter the result of similarity measure so as to optimize the neighbor set generated by traditional collaborative filtering.

3 Details of Algorithm

The whole thought of ICF is roughly the same as the traditional collaborative filtering algorithm, but in specific implementation details, ICF aims at establishing "users-items" interest degree matrix, introducing the mechanism of singularity degree to generate the similar users set, and optimizing the method of neighbor set generation by recommend importance mechanism. The following table lists frequently used notations (Table 1).

Table 1. Frequently used notations.

Notation	Description
$segTime_{start \sim end}$	Time window variable, from start to end
daygap	Time interval variable, the unit of the variable is the day
U	The user set extracted from the session file
u_t	User $u_t \in U \cap t \in \{1, 2, \cdots, n\}$
$sessionNum^j_{u_t,k}$	The number of the user u_t browses the k item in the first j day
$scanNum^j_{u_t,k}$	The number of the user u_t browses the k item in the first j day
$density^j_{u_t,k}$	The density of the user u_t browses the k item in the first j day
$stayDegree^j_{u_t,k}$	The residence density of user u_t browses the k item in the first j day
$stayTime^j_{u_t,k}$	The sum of residence time of the user u_t in the user session file browses the k item in the first j day
$I^{segTime_{start \sim end}}_{u_t,k}$	Interest degree that the user u_t to the kth item within $segTime_{start \sim end}$
IT^P	positive interest degree, IT^P : $ID_{u_t,k} \in (\frac{3}{4}, 1]$
IT^M	medium interest degree, IT^M : $ID_{u_t,k} \in [\frac{1}{4}, \frac{3}{4}]$
IT^N	negative interest degree, IT^N : $ID_{u_t,k} \in [0, \frac{1}{4})$
P^k	It represents a user set of interest degree value within the IT^P range from the n users to item K, $\#P^k$ represents the total number of elements
M^k	A user set of interest degree value within the IT^M range from the n users to item K
N^k	A user set of interest degree value within the IT^N range from the n users to item K
$SI^P_{u_t,k}$	Positive interest singularity impact factor, $SI^P_{u_t,k} = 1 - \frac{\#P^k}{\#U} \Rightarrow SI^P_{u_t,k} \in [0,1]$
$SI^M_{u_t,k}$	Medium interest singularity impact factor, $SI^M_{u_t,k} = 1 - \frac{\#M^k}{\#U} \Rightarrow SI^M_{u_t,k} \in [0,1]$

(Continued)

Table 1. (*Continued*)

Notation	Description
$SI^N_{u_t,k}$	Negative interest singularity impact factor, $SI^N_{u_t,k} = 1 - \frac{\#N^k}{\#U} \Rightarrow SI^N_{u_t,k} \in [0,1]$
L_k	$L_k = \{u_t \in U \mid ID_{u_t,k} \neq \bullet \cap t \in \{1,2,\cdots,n\}\}$
Item	Item set, $\#Item$ represents the total number of elements
M_{u_t}	$M_{u_t} = \{k \in Item \mid ID_{u_t,k} \in IT^P\}$
N_{u_t}	$N_{u_t} = \{k \in Item \mid ID_{u_t,k} \in IT^M \cup ID_{u_t,k} \in IT^N\}$

ICF-Building "Users-Items" Interest Degree Matrix. In order to build "users-items" interest degree matrix, we need to know the interest degree value at some intervals. Usually interest degree values are determined as the specific formula follows:

$$density^j_{u_t,k} = \frac{scanNum^j_{u_t,k}}{\sum\limits_{item=1}^{m} sessionNum^j_{u_t,item}} \Rightarrow density^j_{u_t,k} \in [0,1] \tag{1}$$

$$scanNum^j_{u_t,k} = \frac{1}{daygap} \sum\limits_{i=j}^{j+daygap-1} sessionNum^i_{u_t,k} \Rightarrow j \in [segTime_{start}, segTime_{end}], \tag{2}$$

where m is the number of items that are extracted from the user's session file.

This paper refers to $stayDegree^j_{u_t,k}$ so as to display length of access time. The specific formula is as follows:

$$stayDegree^j_{u_t,k} = \frac{stayTime^j_{u_t,k}}{\sum\limits_{item=1}^{m} stayTime^j_{u_t,item}} \Rightarrow stayDegree^j_{u_t,k} \in [0,1] \tag{3}$$

The calculation of degree of users' interest in items is as follows:

$$I^{segTime_{start} \sim end}_{u_t,k} = \frac{1}{\#segTime} \sum\limits_{j=start}^{\#segTime} w^j_{u_t,k} \bullet density^j_{u_t,k} \Rightarrow I^{segTime_{start} \sim end}_{u_t,k} \in [0,1], \tag{4}$$

where $w^j_{u_t,k}$ is an important factor that determines whether $density^j_{u_t,k}$ is able to affect the calculation of interest degree value. The specific calculation method of $w^j_{u_t,k}$ is given as follows:

$$w^j_{u_t,k} = \begin{cases} 1 & stayDegree^j_{u_t,k} \geq Th \\ 0 & stayDegree^j_{u_t,k} < Th \end{cases} \tag{5}$$

According to $I_{u_t,k}^{segTime_{start \sim end}}$, we can establish "users - items" interest degree matrix. As shown in Fig. 1 ($ID_{u_t,k}$ is on behalf of $I_{u_t,k}^{segTime_{start \sim end}}$):

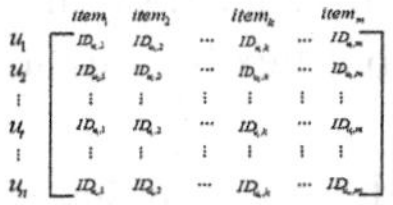

Fig. 1. "Users - Items" interest degree matrix

ICF- Similar User Set Calculation. Definition 3.1 singularity impact degree factor: the factor expresses the impact from the individual. The specific quantitative process of the singularity impact degree factor is as follows (this paper divides user's interest into three levels: positive interest, medium interest and negative interest):

Based on MSD [10] similarity calculation method, we take the product of singularity impact degree factor of different users as the weight value. Combinations are shown in the following Table 2:

From the above table we can get the formula of the similarity calculation method based on the singularity impact degree is as follows:

Table 2. 6 kinds of combinations of singularity impact degree factor product.

	Combinations	Singularity impact degree factors plot
1	(IT^P, IT^P)	$(SI_k^P)^2 (A = \{k \in Item \mid ID_{u_t,k} \in IT^P \cap ID_{u_s,k} \in IT^P\})$
2	(IT^M, IT^M)	$(SI_k^M)^2 (B = \{k \in Item \mid ID_{u_t,k} \in IT^M \cap ID_{u_s,k} \in IT^M\})$
3	(IT^N, IT^N)	$(SI_k^N)^2 (C = \{k \in Item \mid ID_{u_t,k} \in IT^N \cap ID_{u_s,k} \in IT^N\})$
4	(IT^P, IT^M) or (IT^M, IT^P)	$SI_k^P SI_k^M$ $(D = \{k \in Item \mid (ID_{u_t,k} \in IT^P \cap ID_{u_s,k} \in IT^M) \cup (ID_{u_t,k} \in IT^M \cap ID_{u_s,k} \in IT^P)\})$
5	(IT^P, IT^N) or (IT^N, IT^P)	$SI_k^P SI_k^N (E = \{k \in Item \mid (ID_{u_t,k} \in IT^P \cap ID_{u_s,k} \in IT^N) \cup (ID_{u_t,k} \in IT^N \cap ID_{u_s,k} \in IT^P)\})$
6	(IT^M, IT^N) or (IT^N, IT^M)	$SI_k^M SI_k^N (F = \{k \in Item \mid (ID_{u_t,k} \in IT^M \cap ID_{u_s,k} \in IT^N) \cup (ID_{u_t,k} \in IT^N \cap ID_{u_s,k} \in IT^M)\})$

$$MSD^{SI}(u_t, u_s) = \frac{1}{6}\left[\frac{1}{\#A}\sum_{k \in A}[1 - (ID_{u_t,k} - ID_{u_s,k})^2] \bullet (SI_k^P)^2 + \frac{1}{\#B}\sum_{k \in B}[1 - (ID_{u_t,k} - ID_{u_s,k})^2] \bullet (SI_k^M)^2\right.$$

$$+ \frac{1}{\#C}\sum_{k \in C}[1 - (ID_{u_t,k} - ID_{u_s,k})^2] \bullet (SI_k^N)^2 + \frac{1}{\#D}\sum_{k \in D}[1 - (ID_{u_t,k} - ID_{u_s,k})^2] \bullet SI_k^P SI_k^M$$

$$+ \frac{1}{\#E}\sum_{k \in E}[1 - (ID_{u_t,k} - ID_{u_s,k})^2] \bullet SI_k^P SI_k^N + \frac{1}{\#F}\sum_{k \in F}[1 - (ID_{u_t,k} - ID_{u_s,k})^2] \bullet SI_k^M SI_k^N]$$

$$\Leftrightarrow A \neq \emptyset \cup B \neq \emptyset \cup C \neq \emptyset \cup D \neq \emptyset \cup E \neq \emptyset \cup F \neq \emptyset \qquad (6)$$

$$MSD^{SI}(u_t, u_s) = \bullet \quad \Leftrightarrow \quad A = \emptyset \cap B = \emptyset \cap C = \emptyset \cap D = \emptyset \cap E = \emptyset \cap F = \emptyset \quad (7)$$

The advantages of the similarity calculation method based on singularity impact degree are as follows:

(1) It is realized on the basis of MSD similarity calculation method, so it has the ability to solve the problem of "cold start" like MSD.
(2) Interest degree division is more flexible due to its scalability, which is free from fixed hierarchy number.
(3) Focusing on the partial in the view of the whole, which means we take into account the effect of the individual components, in process of similarity calculation, based on overall distribution characteristics investigated from whole level of interest in an item to measure the similarity among the users.

ICF- Neighbor Set Generation. Definition 3.2 recommended Importance of an item: it refers to a function that the item can be recommended by the system, and the calculation formula is as follows:

$$RI_k^{item} = \begin{cases} \dfrac{\sum\limits_{u \in L_k} ID_{u,k}}{\#U} & L_k \neq \emptyset \\ \bullet & L_k = \emptyset \end{cases}, \tag{8}$$

if $L_k \neq \emptyset$, $RI_k^{item} \in [0,1]$; if $L_k = \emptyset$, RI_k^{item} has no solution.

Definition 3.3 recommended Importance of a user: it refers to a function that the user makes the system give priority to his recommendation, and its calculation formula is as follows:

$$RI_{u_t}^{user} = F_1 F_1 = \begin{cases} \left(\dfrac{\#M_{u_t} + \#N_{u_t}}{\#Item}\right)\left(1 - \dfrac{\#M_{u_t}}{\#M_{u_t} + \#N_{u_t}}\right) & M_{u_t} \neq \emptyset \cup N_{u_t} \neq \emptyset \\ \bullet & M_{u_t} \neq \emptyset \cap N_{u_t} \neq \emptyset \end{cases} \tag{9}$$

where $F_1 = \dfrac{\#M_{u_t} + \#N_{u_t}}{\#Item}$ represents the proportion of items that have a degree of interest, $F_2 = 1 - \dfrac{\#M_{u_t}}{\#M_{u_t} + \#N_{u_t}}$ represents the proportion of positive interest degree items in items that have a degree of interest. If $M_{u_t} \neq \emptyset$ or $N_{u_t} \neq \emptyset$, $RI_{u_t}^{user} \in [0,1]$; if $M_{u_t} \neq \emptyset$ and $N_{u_t} \neq \emptyset$, $RI_{u_t}^{user}$ has no solution.

Definition 3.4 recommendation importance of a user for an item: it refers to a function of the user makes the system recommend an item, and its calculation formula is as follows:

$$RI_{u_t,k} = \begin{cases} ID_{u_t,k} * RI_k^{item} * RI_{u_t}^{user} & RI_k^{item} \neq \bullet \cap RI_{u_t}^{user} \neq \bullet \cap ID_{u_t,k} \neq \bullet; \\ RI_{u_t,k} = RI_k^{item} * RI_{u_t}^{user} * \dfrac{1}{\mu} \sum\limits_{itemN \in Q_k'}[RI_{itemN}^{item} \cdot ID_{u_t,itemN} \cdot sim(k,itemN) & \\ & RI_k^{item} \neq \bullet \cap RI_{u_t}^{user} \neq \bullet \cap ID_{u_t,k} = \bullet \cap Q_k' \neq \emptyset; \\ \bullet & others; \end{cases} \tag{10}$$

where $\mu = \sum\limits_{itemN \in Q_k'} sim(k, itemN)$.

Using the similarity calculation method based on the singularity impact degree, we sort MSD^{SI} by the order from big to small. So we can get L users who are the most similar to the target user u_t: $simU_{u_t}(u_{SI_1}, u_{SI_2}, \cdots, u_{SI_L})$. Then we respectively measure diversity between L users and the target user u_t in recommendation importance of a user for an item. The less diversity between a certain user and target user u_t, the more similar ability in recommend an item such as preferences in behavior patterns, and interest. The calculation method of diversity in recommendation importance of a user for an item between the user and the target user u_t is as follows:

We assume that the vector of target user's recommended importance of m items is as follows: $RI_{u_t}^{user-item}(RI_{u_t,1}, RI_{u_t,2}, \cdots, RI_{u_t,k} \cdots, RI_{u_t,m})$. The vector represents recommended importance of m items of the user $u_{SI_i}(i = 1, 2, \cdots, L)$ in the target user's similar user set $simU_{u_t}(\neq \emptyset)$ is as follows: $RI_{u_{SI_i}}^{user-item}(RI_{u_{SI_i},1}, RI_{u_{SI_i},2}, \cdots, RI_{u_{SI_i},k} \cdots, RI_{u_{SI_i},m})$. The difference between the target user u_t and the similar user u_{SI_i} in factors of recommended importance of items is as follows:

$$RI_{u_t,u_{SI_i}}^{dif} = \begin{cases} \dfrac{\sum\limits_{itemD \in RID} \left| RI_{u_t,itemD} - RI_{u_{SI_i},itemD} \right|}{\#RID} & simU_{u_t} \neq \emptyset \cap RID \neq \emptyset; \\ \bullet & others; \end{cases} \quad (11)$$

where $RID = \{K \in Item \mid RI_{u_t,k} \neq \bullet \cap RI_{u_{SI_i},k} \neq \bullet\}$, $\#RID$ represents the total element number of the set RID. Obviously, the smaller value of $RI_{u_t,u_{SI_i}}^{dif}$, the smaller difference between the target user u_t and the similar user u_{SI_i} in recommended importance of items. So u_{SI_i} may become one of the candidates of the optimal neighbor set of u_t, which is close to u_t.

We calculate the difference between each user in $simU_{u_t}$ with the target user u_t in factors of recommended importance of m items according to the following formula. The user in $simU_{u_t}$ is sorted by the order from big to small according to the $RI_{u_t,u_{SI_i}}^{dif}$ value. We select the former KN ($0 \leq KN \leq L$) users to form target user's best neighbor set: $OPN = \{u_{opn_i} \in simU_{u_t} \mid i = 1, 2, \cdots, KN\}$. Namely, $(simU_{u_t} \neq \emptyset \cap OPN \neq \emptyset) \cap (\forall u_{any_i} \in OPN) \cap (\forall u_{any_j} \in simU_{u_t} \cap u_{any_j} \notin OPN) \cap (RI_{u_t,u_{any_i}}^{dif} \neq \bullet \cap RI_{u_t,u_{any_j}}^{dif} \neq \bullet) \Rightarrow RI_{u_t,u_{any_i}}^{dif} \leq RI_{u_t,u_{any_j}}^{dif}$.

In the "users-items" interest degree matrix, based on the target user's optimal neighbor set OPN, we predict the target user's item interest degree value that has no value of it. The prediction uses the method of modified weighted prediction, the formula is as follows:

$$if\ ID_{u_t,k} = \bullet$$

$$P_{u_t,k}^{ID} = \begin{cases} \bullet & OPN = \emptyset \\ \overline{ID_{u_t}} + \mu_{u_t,k} \sum\limits_{knu \in OPN} MSD^{SI}(u_t, knu)(ID_{knu,k} - \overline{ID_{knu}}) & OPN \neq \emptyset \end{cases}, \quad (12)$$

where $\overline{ID_{u_t}}$ is the average interest degree value of the target user's interest in an item; $\overline{ID_{knu}}$ is the average interest degree value of the optimal neighbor user knu of the target user u_t interested in the item: $\mu_{u_t,k} = \dfrac{1}{\sum\limits_{knu \in OPN} sim(u_t,knu)}$.

According to the above prediction of target user's item interest degree value, we can use TOP-N way to push $P_{u_t,k}^{ID}$ in the top N (from big to small) corresponding to the item set $ItemRec$ to target user u_t.

We define the target user's item set with no interest degree value is as follows: $ItemL = \{iteml \in Item \mid ID_{u_t,iteml} = \bullet\}, 0 \le \#ItemL \le m$. Where $\#ItemL$ represents the total element number of $ItemL$, so $ItemRec = \{itemR_i \in ItemL \mid i = 1,2,\cdots, N \cap 0 \le N \le \#ItemL\}$. Namely, $(\forall item_{any_i} \in ItemRec) \cap (\forall item_{any_j} \in ItemL \cap item_{any_j} \notin ItemRec) \cap (P_{u_t,item_{any_i}}^{ID} \ne \bullet \cap P_{u_t,item_{any_j}}^{ID} \ne \bullet) \Rightarrow P_{u_t,item_{any_i}}^{ID} \ge P_{u_t,item_{any_j}}^{ID}$. And we stipulate $ItemRec$ is the ordered set, so $P_{u_t,itemRe_i}^{ID} \ge P_{u_t,itemRe_{i+1}}^{ID}$ Where item $itemRe_i \in ItemRec$, $i = 1,2,3,\cdots,N-1$.

4   Experiments and Results

Experimental Data and Log Preprocessing. We extracted log of 15 days from June 6th, 2015 to June 20th, 2015 in server of official website of Nanjing University of Posts and Telecommunications as our experimental data. In order to establish the "user-item" interest degree matrix, we did a survey on which items users browse, the number of access of items, dwell time and other essential information. We extracted 75 % of the users in the user set, while extracting 80 % of the items in the project set as training set, and the rest were used as test set. Finally we compared ICF, in terms of mean absolute error (MAE) and coverage, based on [1, 5] at the same running environment.

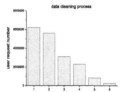

Fig. 2. Data cleaning.

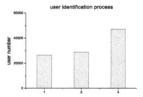

Fig. 3. User identification.

Before the user session identification, we performed data cleaning on the log, and Fig. 2 is the change chart of user request number in the process of data cleaning. As can be seen from Fig. 2, after data cleaning, the number of user requests has decreased from 6277820 to 268220. Figure 3 is the change chart of user number in the user identification process. As can be seen from Fig. 3, after user identification, the number of user has increased from 6277820 to 268220.

After user identification, we used the time window model to identify sessions on the log. We chose 25.5 min as a measure of the session timeout threshold, and identified

53110 users' sessions. After the above steps of log preprocessing, we can obtain that the number of user request log entries is 268220, the number of users is 47440 and the number of sessions is 53110. On this basis, we constructed the "user – project" interest matrix with the information of items users browse, the number of access of items, dwell time and other essential information.

Experimental Results and Analysis. According to the "user-item" interest degree matrix, we compared ICF, in terms of mean absolute error (MAE) and coverage, with traditional collaborative filtering algorithm and other improved collaborative filtering algorithm based on [1, 5] at the same running environment (In the following figure, K represents the size of the neighbor user set).

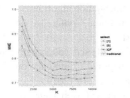   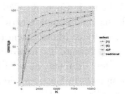

Fig. 4. MAE. (Color figure online)   **Fig. 5.** Coverage. (Color figure online)

As is shown in Fig. 4: The value of MAE with ICF is always less than others no matter what values K takes. It is apparent to see that recommendation results became more accurate. As is shown in Fig. 5: Coverage of ICF is always higher than others no matter what value K takes. Viewing the above, ICF has better coverage, and shows better personalized recommendation.

5 Conclusions

In this paper, we propose an improved collaborative filtering algorithm ICF. In this paper, we propose an improved collaborative filtering algorithm ICF. In our construction, improvement can be accounted for by following three aspects: data representation, calculation of similarity degree and generation of neighbor set. Finally, we verify its high accuracy and coverage by experiments compared with existing algorithms.

Acknowledgments. This work is supported by the National Natural Science Foundation of China (60973140, 61170276, 61373135), The research project of Jiangsu Province (BY2013011), The Jiangsu provincial science and technology enterprises innovation fund project (BC2013027), the High-level Personnel Project Funding of Jiangsu Province Six Talents Peak and Jiangsu Province Blue Engineering Project. The major project of Jiangsu Province University Natural Science (12KJA520003).

References

1. Badaro, G., Hajj, H., El-Hajj, W., et al.: A hybrid approach with collaborative filtering for recommender systems. In: 2013 9th International Conference on Wireless Communications and Mobile Computing Conference (IWCMC), pp. 349–354 (2013)
2. Ba, Q., Li, X., Bai, Z.: Clustering collaborative filtering recommendation system based on SVD algorithm. In: 2013 4th International Conference on Software Engineering and Service Science (ICSESS), pp. 963–967 (2013)
3. Sun, G.F., Wu, L., Liu, Q.: Recommendations based on collaborative filtering by exploiting sequential behaviors. J. Softw. **24**(11), 2721–2733 (2013)
4. Ji, H., Chen, X., He, M., et al.: Improved recommendation system via propagated neighborhoods based collaborative filtering. In: 2014 International Conference on Service Operations and Logistics, and Informatics (SOLI), pp. 119–122 (2014)
5. Yao, Z., Feng, J., Chen, X.: Collaborative filtering recommendation algorithms research based on influence and complex network. In: 2014 11th International Conference on Service Systems and Service Management (ICSSSM), pp. 1–6 (2014)
6. Chang, N., Terano, T.: Improving the performance of user-based collaborative filtering by mining latent attributes of neighborhood. In: 2014 International Conference on Mathematics and Computers in Sciences and in Industry (MCSI), pp. 272–276 (2014)
7. Ying, Z., Zhou, Z., Han, F., et al.: Research on personalized web page recommendation algorithm based on user context and collaborative filtering. In: 2013 4th International Conference on Software Engineering and Service Science (ICSESS), pp. 220–224 (2013)
8. Huigui, R., Shengxu, H., Chunhua, H.: User similarity-based collaborative filtering recommendation algorithm. J. Commun. **2**, 16–24 (2014)
9. Fan, J., Pan, W., Jiang, L.: An improved collaborative filtering algorithm combining content-based algorithm and user activity. In: 2014 International Conference on Big Data and Smart Computing (BIGCOMP). IEEE Computer Society, pp. 88–91 (2014)

Methods of Machine Learning for Linear TV Recommendations

Mikhail A. Baklanov[1(✉)] and Olga E. Baklanova[2(✉)]

[1] Tomsk State University, Tomsk, Russia
baklanov.ma@gmail.com
[2] D.Serikbayev East-Kazakhstan State Technical University,
Ust-Kamenogorsk, Kazakhstan
OEBaklanova@mail.ru

Abstract. This paper describes methods of improving TV-watching experience using Machine Learning for Linear TV recommendations. There is an overview of existing methods for video content recommendations and an attempt of developing new method that focused only on linear TV recommendations and takes into account all specifics around it. Recommendation system based on this approach was implemented in Russian pay TV provider ZOOM TV, and demonstrated two times churn rate reduction in comparison with same service without recommendation system. Existing methods and new method effectiveness compared with offered approach by analyzing real people content consumption during 1 year.

Keywords: Recommenders · Linear TV · SMART TV · Collaborative filtering · Content filtering · Machine learning

1 Introduction

There are 900 million pay TV customers all over the world. These people pay for hundreds of TV-channels but watch only 10−15 of them and spend about 25 % of TV watching time on channel surfing. The content discovery on TV did not changed since it was invented and anyone who watch linear TV need to zap channel one by another to find something interesting. That worked good while there were only 10 channels but it is not very pleasant user experience when TV viewer needs to check 100−200 channels manually. Popular content discovery techniques like text search and recommendations from friends are not efficient enough on TV because of limited input means.

It must be admitted that for video-on-demand (VOD) and for linear TV this problem has different important traits. VOD is rather young way of content consumption, and in terms of content discovery, there are several different use cases. The most popular use case that many researchers are trying to solve is "to find something new and interesting". In theory, if an algorithm knows how strong person enjoyed other movies, it is possible to evaluate how strong he will appreciate the new one.

For linear TV most popular case is different. Usually TV-watcher turns on the TV set and want to switch to the most interesting or less annoying TV-show that is on air now. The usage of ordinary machine learning algorithms is complicated by various

© Springer International Publishing Switzerland 2016
D.-S. Huang et al. (Eds.): ICIC 2016, Part III, LNAI 9773, pp. 607–615, 2016.
DOI: 10.1007/978-3-319-42297-8_56

factors such as poor input devices, the novelty of TV-shows, patterns of content consumption and the fact that TV-set is not a personal device, it used by a whole family or household. But from the other hand the content on TV-channels already preselected by professional producers and editors, for most cases it is relevant to its target audience.

This paper aims to analyze linear TV content specifics and develop the most effective method for linear TV recommendations.

2 Problem Formulation and Motivation

Recommendation system aims could be different and depends on business needs of the organization that is trying to integrate it. Anyway, usually it aims to improve some of the KPIs, which could make product more profitable or sustainable. That is why it is important to measure the effectiveness of recommendation system not only by some statistical measures like standard deviation but also use some business KPI to make sure that developed system is close enough to real life and demanded.

One of the first papers on linear TV recommendation systems is [1], which describes a personalized Electronic Program Guide. Many works present collaborative and content filtering based approaches [2].

In [4] authors proposed a mechanism that models, records and analyzes users' traces. It allows evaluating competencies for recommending people with more expertise on a certain subject. R. Turrin et al. in [3] are trying to extend common recommendation methods by integrating the current watching context into the user viewing model. In [5] P. Cremonesi et al. are focused on Implicit Feedback and emphasize the specifics of linear TV watching that makes it different from Video-on-Demand such as:

- Available items change over time: many TV programs are often broadcast once and then not anymore for a long time.
- Time-constrained catalog of items: all programs are transmitted in a predefined schedule. Therefore, the recommendations must consider only the items on air at moment in which they are requested.
- The user feedback is usually implicit, provided in the form of watched/not watched shows.

D. Zibriczky et al. [6] also specify different features of linear content paying special attention that the most common device for consuming linear TV-content is TV-set. Their paper is focusing on the following recommendation problems:

- Noisy input: as TV watching is a very, very laid-back environment, users do not provide explicit feedback. The taste is expressed implicitly by channel zapping, recording, recording playback events. One key challenge is how to pre-process and identify the relevant user actions that can be efficiently is used for user profiling.
- Live TV programs are always new: by nature, live TV programs are always new. Recommender algorithms that require user feedback for preference modeling cannot be used, since such data is not available before the program start, and when it becomes available, it is too late, the program is over. The metadata associated with live programs is also much less detailed and relevant to on-demand programs.

About 60 % of live TV programs belong to news, sports, reality, talk shows, and many of them are broadcasted in live. Either the program descriptions are the same or very similar that makes the recommendation problem very challenging.

- Time-based recommendation: consumption patterns changes on daily and weekly bases. Users are not interesting in the same type of content in the morning and in the evenings, and it is also true for weekends.
- Shared, multi-user device: TV is a shared, multi-user device. The algorithms should be aware to this problem and if there is no explicit indication, who is in front of the device, it should also predict.

This paper analyzes recommenders from the viewpoint of customer loyalty, which can improve such important business metrics like ARPU and churn rate. So the problem is how to make a recommender that will be efficient enough to make it users more satisfied with a product.

Let U - number of users and S - the set of all objects that can be recommended. Let h - a function that measures user satisfaction, i.e. the extent to which a subject s like user u. Thus, for each user $u \in U$ we want to select an object $s' \in S$ that would maximize user satisfaction. To put it more formally:

$$\forall c \in C, s'_c = \arg \max_{u \in U} h(u, s). \tag{1}$$

3   Common Machine Learning Methods

The basic idea of recommender is to predict what user will prefer and how strong will he or she satisfied by an object. All existing methods are start from the attempt to learn from user's actions and understand their habits and tastes. The strategy should be based on the information which recommender can access and objective function.

3.1   Item-Based Methods

Content Filtering (item-based) involves the creation of user profiles and objects. In order to take into account relevant recommendations of the object parameters corresponding user preferences. The recommendation system objects are described using keywords (tags), and create a profile that characterizes its attitude to certain objects.

In the item-based systems, the function of satisfaction $h(u, s)$ user u of a particular content item s is determined on the basis of information on user satisfaction with the content item $s_i \in S$, which are "similar" to s. For example, the recommendation system for films to recommend movies to the user u, the system tries to understand what is common between the films having the user previously praised highly. After that, the user will be recommended movies as much as possible similar to the ones that the user has given high marks.

In fact, content that the user consumes forms a profile in the form of a plurality of parameters characterizing the object s. As a parameter often, protrude keywords corresponding weights for each object. Thus, the problem arises of how to weigh these parameters. One of the most prominent ways to determine the weights of keywords in information retrieval is TF-IDF [7] measure.

3.2   User-Based Methods

Collaborative filtering uses a well-known group of users' preferences to predict the unknown preference of another user. The basic assumption is that those who are equally evaluated objects of any kind in the past tend to give a similar assessment of other subjects in the future.

In contrast to the Item-based methods, User-based methods try to predict user satisfaction with an object on the basis of how other users rate this item. Those users's u satisfaction of objects s $h(u, s)$ is calculated on the basis of users u_j which for any signs are "similar" to u. An important feature here is that the characteristics, by which the user can be considered similar to the other users, may go beyond the system. For example, social networking profiles or demographic information obtained explicitly, can be the basis to assess the similarity of the new primary user of existing ones. This approach is often used to solve "cold" start problem.

3.3   Hybrid Recommendation Systems

Hybrid recommendation systems are a combination of content filtering and collaborative filtering. Some of recommendation systems use hybrid methods combining user-based and item-based methods that partially or completely neutralize some of the disadvantages of each approach. You can select the following ways of combining different approaches in hybrid systems of reference:

1. The introduction of item-based and user-based methods separately and combining them received from assessment;
2. The introduction of some item-based settings in the user-based approach;
3. The introduction of some user-based settings in the item-based approach;
4. Creation of a general model, which uses parameters of both approaches.

One way to build hybrid recommender system is the implementation of two independent item-based and user-based systems. Then there are two possible scenarios. In the first, we can combine the ratings received from each system using a linear combination of ratings [8] or the scheme described in [9]. The second scenario is to use only one of the systems depending on what herein will work better. For example, the system «DailyLearner» [10] selects recommender systems, which can give a recommendation to the lowest level of error.

The most popular approach to this category of hybrid recommender systems is a decrease of dimension for the group of content-based user profiles. For example, in [11] uses a latent semantic analysis (LSA), to create user profiles Collaborative representation, where the user profile is represented as a set of vectors, which leads to a significant improvement in performance compared to pure item-based approach.

To construct the general advisory model is most commonly used user-based and item-based parameters (for example, the age of the users and the genre of the film) in a single advisory system.

4   Recommender Based on Personal Choice Strategies

In contrast to the classical problem of the recommendations of video on demand, which is available at any time, the aim of this work, in fact, are recommendations for changing the current time on any channel of an arbitrary limited list of channels. This clarification is so significant that even with the use of electronic program recommendation algorithms that work well in video on demand, here is difficult.

The main idea of this approach is to take into account how person choose what to watch. Analyzing the history of channel switching it is possible to make for every device for each timeslot it has the vector with specific coefficients that models how important for a household this or that tag. For example, for somebody it could be important to watch sports live and for another person the most important is to watch English Premier League. For first person the biggest impact will be made by the tag "live", for the second person the tag "league".

The final rating of transmission (Rbr_s) for any user (u) is different and, in general, is calculated as follows:

$$Rbr_{su} = Q_{su} + P_{su}, \tag{2}$$

$$Q_{su} = Q_{su-1} + a_{su}l_{su} + \sum\nolimits_{j=1}^{n} (b_{su}t_{isu} + c_{su}w_{isu}), \tag{3}$$

$$P_{su} = \sum\nolimits_{j-1}^{m} k_{jsu}tag_{jsu}, \tag{4}$$

where:

$l_s \in \{-1, 0, 1\}$ and indicates the presence of a mark by the user that this transfer him like or dislike;

$t_i, w_i \in \{0, 1, 2, \cdots, n \cdots, \infty\}$ and denote, respectively, how many times were used pause and rewind within it (timeshift), and how many issues the user looked through this transfer to the end;

a, b, c, k_j – coefficients are calculated for each individual user on the basis of how great value to the user preferences;

i – number of an episode;

j – number of a parameter;

tag_j – value of parameters for TV-show s;

s – number of TV-show;

n – the number of views made by user for all episodes of one TV-show;

m – the number of parameters that are characteristic of this transmission;

Q_S – the number of parameters specific to a given TV-show;

P_S – the sum of the weights of all the parameters of the TV-show multiplied by

Unlike most popular approaches related to the clustering of users and the issuance of recommendations based on the user's membership in a particular group, this approach presume absolute individual tastes user and provides this calculation of the importance of particular criteria on which man rests in the choice of transmission. The importance of a fixed coefficient, the larger the ratio is the higher the importance of this coefficient.

5   Results and Discussion

In this part will be analyzed and compared the effectiveness of the "classical" approach and an approach based on filtering and ordering. As the representatives of the "classical" approach using collaborative filtering algorithms based on users and object-based.

The study will take place on the data collected from more than 300,000 users of the Pay TV for 1 year of work. These data provide a history of decision-making by the user on which TV-show to watch, the order in which the channels have been submitted to it and user actions according these TV-shows.

As a measure of efficiency is considered the number of actions that are necessary for the user to find the most interesting programs for each method on average.

The most interesting programs here and further we consider the transfer, the user ultimately chose. We believe that the user has chosen the transfer, if it is:

- watched it for 5 min and looked through to the end (± 2 min);
- noted the transfer as a favorite;
- use a pause or rewind.

To implement cloud platform will be used Microsoft Azure Machine Learning Studio, which allows users to run in the cloud and scale the infrastructure for big data (Big Data). Users at any given time have access to 80 channels.

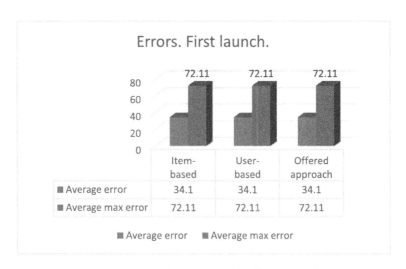

Fig. 1.   Errors. First launch (Color figure online)

For the analysis, how to lead the above methods of the recommendations used a sample of users who have not specified before using the categories that they are not interesting.

At first launch shown on Fig. 1, due to the lack of any history all systems equally ineffective.

Data accumulated over the week, it enough that each of methods significantly reduces errors compared with random TV-shows ordering. As can be seen from the chart, the offered method provides delivery much closer to what the user wants to select, but the maximum error for users is still very large. It is shown on Fig. 2.

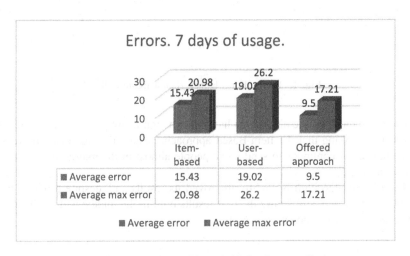

Fig. 2. Errors. Second launch (Color figure online)

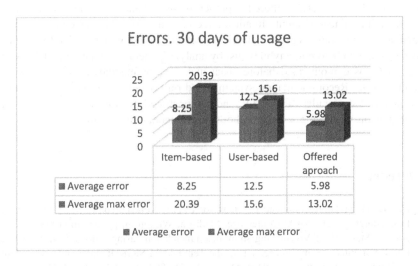

Fig. 3. Errors. Third launch (Color figure online)

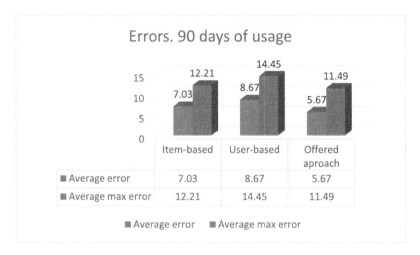

Fig. 4. Errors. Forth launch (Color figure online)

For thirty days, all the algorithms have shown significant progress, compared with 7 days. It is striking that the Item-Based approach, despite the significant decrease in the average error continues to send some people almost in the middle of the list. It is shown on Fig. 3.

The offered algorithm is slightly improved its results and, despite progress on the part of the other two, still shows the best results both in the middle and on the average maximum error. It is shown on Fig. 4.

6   Conclusion

The results showed that the offered approach is much better suits for specific tasks of recommending linear content. It allows recommendations be personal in terms of understanding persons tastes and allows to work effectively with premier TV-show that was never shown before to anybody just by analyzing meta-information and TV-viewer tastes that was demonstrated before. That makes recommendation logics work more like human do. Recommendation system based on this approach was implemented in Russian pay TV provider ZOOM TV, and demonstrated two times churn rate reduction in comparison with same service without recommendation system.

References

1. Das, D., Horst, H.: Recommender systems for TV. In: Workshop on Recommender Systems, Proceedings of 15th AAAI Conference, Madison, Wisconsin, pp. 35–36 (1998)
2. Ali, K., Stam, V.: TiVo: making show recommendations using a distributed collaborative filtering architecture. In: Proceedings of the Tenth ACM SIGKDD International Conference on Knowledge Discovery and Data Mining, KDD 2004, pp. 394–401 (2004)

3. Turrin, R., Condorelli, A, Cremonesi, P, Pagano, R.: Time-based TV programs prediction. In: 1st Workshop on Recommender Systems for Television and Online Video at ACM RecSys (2014)

4. Wang, N., Abel, M.-H., Barthes, J.-P., Negre, E.: A recommender system from semantic traces based on bayes classiffier. In: Proceedings of the 2nd International Conference on Knowledge Management, Information and Knowledge Systems (KMIKS), Hammamet, Tunisia, pp.49–60, April 2015

5. Cremonesi, P., Modica, P., Pagano, R., Rabosio, E., Tanca, L.: Personalized and context-aware TV program recommendations based on implicit feedback. In: Stuckenschmidt, H., et al. (eds.) EC Web 2015. LNBIP, vol. 239, pp. 57–68. Springer, Heidelberg (2015). doi:10.1007/978-3-319-27729-5_5

6. Zibriczky, D., Hidasi, B., Petres, Z., Tikk, D.: Personalized recommendation of linear content on interactive TV platforms: beating the cold start and noisy implicit user feedback. In: TVMMP: Proceedings of the International Workshop on TV and Multimedia Personalization (TVMMP) held UMAP 2012, Montreal, Canada (2012)

7. Salton, G.: Automatic Text Processing. Addison-Wesle, Reading (1989)

8. Claypool, M., Gokhale, A., Miranda, T., Murnikov, P., Netes, D., Sartin, M.: Combining content-based and collaborative filters in an online newspaper. In: Proceedings ACM SIGIR 1999 Workshop Recommender Systems: Algorithms and Evaluation, August 1999

9. Pazzani, M.: Framework for collaborative, content-based, and demographic filtering. Artif. Intell. Rev. 13, 393–408 (1999)

10. Billsus, D., Pazzani, M.: User modeling for adaptive news access. User Modeling User-Adap. Interact. 10(2-3), 147–180 (2000)

11. Soboroff, I., Nicholas, C.: Combining content and collaboration in text filtering. In: Proceedings International Joint Conference on Artificial Intelligence Workshop: Machine Learning for Information Filtering (1999)

UC-PSkyline: Probabilistic Skyline Queries Over Uncertain Contexts

Zhiming Zhang[1], Jiping Zheng[1,2(✉)], and Yongge Wang[1]

[1] College of Computer Science & Technology,
Nanjing University of Aeronautics and Astronautics, Nanjing, China
{Zhang-Zhiming-24,jzh,sywyg}@nuaa.edu.cn
[2] School of Computer Science and Engineering,
University of New South Wales, Sydney, Australia

Abstract. Probabilistic skyline queries as an aspect of queries on uncertain data have become an important issue. Previous work on uncertainty modeling for probabilistic skyline queries only lies within the data. However, attribute values of uncertain data are influenced by contexts in real applications while uncertainty is also along with contexts. Further, previous work on probabilistic skyline queries only retrieves those points whose skyline probabilities are higher than a given probabilistic threshold. In this paper, we develop a novel probabilistic skyline query on uncertain data over uncertain contexts called UC-PSkyline, where possible world semantics model is utilized to model uncertain contexts. To avoid unnecessary pair-wise dominance tests, we devise an in-memory tree structure ZB*-tree to process UC-PSkyline queries efficiently. We also develop preprocessing and pruning techniques that can efficiently improve performance of UC-PSkyline. Experiments show the effectiveness and efficiency of the proposed techniques on real and synthetic data sets.

Keywords: Probabilistic skyline queries · Uncertain contexts · ZB*-tree

1 Introduction

The skyline query [1] is a useful operation for many important applications including multicriteria optimal decision making. Given two certain and multi-dimensional tuples u and v, u dominates v if u is no worse than v in all dimensions, and strictly better than v in at least one dimension. However, uncertain data is inherent in many practical applications, such as location-based services, sensor networks, market analysis and environmental surveillance. Pei et al. [2] first proposes probabilistic skyline queries on uncertain data with the discrete probability distribution model. They assume that each instance of any uncertain object has explicit attribute preferences. However, attribute preferences are uncertain in many applications including dynamic skyline queries [3–5] and usually influenced by contexts [5]. Context [6–8] is a general term used to capture the current query situations. In general, contexts are from sensing devices and uncertainty is along with them. Probabilistic skyline queries proposed by [2] cannot capture the information over uncertain contexts. For instance, in the skyline analysis of NBA

© Springer International Publishing Switzerland 2016
D.-S. Huang et al. (Eds.): ICIC 2016, Part III, LNAI 9773, pp. 616–628, 2016.
DOI: 10.1007/978-3-319-42297-8_57

players, we want to find better players who can achieve better performance in some aspects. A player can be represented by his game records (i.e., instances). Records of different players are uncertain due to many uncertain contexts such as career stages, location of games and physical condition of players. Thus, skyline probabilities of records are computed only over the same contexts. Apparently, existing techniques of probabilistic skyline computations cannot directly be applied to queries over uncertain contexts. Therefore, a novel model about probabilistic skyline on uncertain data over uncertain contexts is required. In this paper, for the first time, we propose a novel probabilistic skyline query on uncertain data over uncertain contexts called UC-PSkyline.

Meanwhile, users must choose a suitable threshold in each query for interesting result sets if a given threshold is used to prune uncertain data that take skyline probabilities of at least p. However, selecting a suitable threshold is as difficult as probabilistic skyline queries themselves [9–11]. To tackle this limitation, UC-PSkyline computes exact skyline probabilities of uncertain data. Although [9] and [10] both propose exact skyline probabilities for all uncertain objects, they do not consider uncertain contexts in probabilistic skyline computation. Moreover, algorithms proposed by [9] are used only for 2 dimensional data space due to the factor exponential in the data dimensionality, and techniques presented by [10] are based on an internal memory that cannot be applied to large data sets. In this paper, we develop a novel probabilistic skyline query on uncertain data over uncertain contexts called UC-PSkyline, where possible world semantics model is utilized to model uncertain contexts.

The rest of the paper is organized as follows. The related work is reviewed in Sect. 2. Section 3 introduces uncertain contexts and defines the concept of UC-PSkyline. Section 4 analyses the two defects of existing tree structures and proposes ZB*-tree. Section 5 investigates UC-PSkyline with ZB*-tree and discusses preprocessing and pruning techniques. Extensive experiments are conducted to evaluate the effectiveness and efficiency of our techniques in Sect. 6. Finally, we conclude our paper in Sect. 7.

2 Related Work

Pei et al. [2] first investigate skyline computation problems on uncetain data where the concept of the skyline probability is introduced. They present bottom-up and top-down methods for p-skyline queries. Consequent researches focus on efficient probabilistic skyline query processing and its variations [5, 12, 13]. However, choosing a suitable value p is as difficult as probabilistic skyline queries themselves. To overcome this limitation, Atallah and Qi [9] first compute exact skyline probabilities of uncertain objects. They combine the grid algorithm and the weighted dominance counting algorithm, and the algorithm proposed by [9] is used only for 2-dimensional data space due to the factor exponential in data dimensionality. To overcome this problem, Kim et al. [10] propose PSkyline based on Z-tree which is based on ZB-tree. ZB-tree is used in the traditional skyline algorithm Z-Search proposed by [14] and employs Z-addresses of points to search potential skyline points in Z-order. PSkyline with Z-tree is based on an internal memory and does not consider uncertain contexts. Moreover, PSkyline with Z-tree have serious defects (see in Sect. 4) that cannot be applied to the problems proposed in this paper.

To the best of our knowledge, we first study introducing probabilistic skyline queries on uncertain data over uncertain contexts which are very common in real applications [11]. Although dynamic skyline queries [3–5] are studied extensively, there is only one paper formally defining skyline queries on certain data over contexts. Sacharidis et al. [4] proposed skyline queries on certain data over similar contexts called P-CSQ. However, the contexts proposed in this paper are uncertain and skyline queries introduced in this paper are based on uncertain data. Since probabilistic skyline queries on uncertain data over uncertain contexts are inherently different from existing work, previous methods cannot directly be applied to the problem proposed in this paper.

3   UC-PSkyline

In this section, we formally introduce probabilistic skyline queries on uncertain data over uncertain contexts called UC-PSkyline. Due to space limitation, some basic concepts about dominance, probabilistic skylines and uncertain contexts modeling are omitted here, please refer to [2, 11] for details.

3.1   Probabilistic Skyline Queries Over Uncertain Contexts

Traditional probabilistic skyline queries on uncertain data are performed without uncertain contexts. However, the attribute preferences of instances of uncertain objects are influenced by uncertain contexts. Each instance of uncertain objects only occurs over one certain context, and dominance tests between instances of different objects are only performed over the same certain contexts. By default, we consider tuples in a d-dimensional data space $D = (D_1, D_2, ..., D_d)$. Let UD, UC be an uncertain data set and an uncertain context set respectively, U, $V \in UD$. For possible world set W_{UC} of uncertain context UC, $W \in W_{UC}$, we write $u^W \in U$ means that u^W is an instance of U over W, and $PrSky(u^W)$ is called the skyline probability of u^W over W. $PrSky(u^W)$ can be calculated as follows:

$$PrSky(u^W) = \prod_{V \in D, V \neq U} \left(1 - \sum_{v^W \in V, v^W \prec u^W} Pr(v^W) \right) \tag{1}$$

$PrSkyW(U)$ is called the skyline probability of U over W. $PrSkyW$ (U) can be calculated as follows:

$$PrSky_C(U) = \sum_{W \in \mathbf{W}} (Pr(W) * PrSky_W(U))$$

$$= \sum_{W \in \mathbf{W}} \left(Pr(W) * \sum_{u^W \in U} Pr(u^W) * PrSky(u^W) \right) \tag{2}$$

Then, the skyline probability of U over UC is the expectation of $PrSkyW(U)$ for every possible world $W \in W_{UC}$, denoted by $PrSkyUC(U)$. $PrSkyUC(U)$ can be calculated as follows:

$$PrSky_C(U) = \sum_{W \in \mathbf{W}} (Pr(W) * PrSky_W(U))$$

$$= \sum_{W \in \mathbf{W}} \left(Pr(W) * \sum_{u^W \in U} Pr(u^W) * PrSky(u^W) \right) \tag{3}$$

Definition 1. *Given an uncertain data set UD over the uncertain context UC, probabilistic skyline queries on uncertain data over UC are to compute the exact skyline probability of each uncertain object $U \in UD$ over UC (UC-PSkyline for short).*

3.2 Dominance Tests Over the Same Certain Context

Equation (3) for calculating $PrSkyUC$ (U) indicates that we must perform dominance tests between instances of U and instances of all other objects over each possible world of UC. Apparently, pair-wise dominance tests over each possible world between instances of all different objects are too costly. However, many unnecessary pair-wise dominance tests can be avoided. Thus, we propose the novel techniques [2, 10] to reduce the number of unnecessary dominance tests over each possible world in this section. We assume that UD^W is a subset of UD over the possible world W. Let $U_{min} = (\min^{|U|} I = 1\{u_i.D_1\}, \ldots, \min^{|U|} I = 1\{u_i.D_d\})$ and $U_{max} = (\max^{|U|} I = 1\{u_i.D_1\}, \ldots, \max^{|U|} I = 1\{u_i.D_d\})$ be the minimum and maximum corners, $UD^{W1}.MBB_{min}$ and $UD^{W1}.MBB_{max}$ of the minimum bounding box (*MBB* for short) of UD^W, respectively. Apparently, for $\forall O^W \in UD_1^W$, $\forall o^W \in O^W$, $UD_{1.min}^W \leq o^W \leq UD_{1.max}^W$. For $UD^{W2} \subseteq UD$, $UD_{.min}^{W2}$ and $UD_{.max}^{W2}$ are $UD_{.MBBmin}^W$ and $UD_{.MBBmax}^W$, respectively. We can obtain the following lemma.

Lemma 1 *(Avoiding Unnecessary Dominance Tests).*

1. *If $UD_1^w.max \prec UD_2^w.min$, then $UD_1^w \prec UD_2^w$.*
2. *If $UD_1^w.min \nprec UD_2^w.max$, then $UD_1^w \nprec UD_2^w$.*
3. *If $UD_2^w.max \prec o^W$, then $\forall U^W \in UD_2^w, U^W \prec o^W$.*
4. *If $UD_2^w.min \nprec o^W$, then $\forall U^W \in UD_2^w, U^W \nprec o^W$.*

Proof sketch. *The proofs of all above items are similar. Limited by space, we only prove that the third item holds. For $\forall U^W$ UD^W, $\forall u^W \in U^W$, $u^W.Di < D_2^w.$ $max.Di(1 \leq I \leq d)$, then $u^W \prec D_2^w.max$. If $UD_2^w.max \prec o^W$, we can obtain $u^W \prec o^W$ based on the transitivity of dominance relations ($\prec$) between instances, then $U^W \prec o^W$, i.e., the third item holds.*

Lemma 1 provides a method to reduce the number of unnecessary dominance tests between a subset of uncertain objects and instances of individual uncertain object.

4   ZB*-tree

In this section, we first discuss Z-addresses [10, 12] on which all tree structures in our paper are based. Then, we discuss Z+tree which improves Z-tree based on ZB-tree in inner memory for UC-PSkyline and its two defects. Finally, motivated by defects of Z +tree, we propose a novel inner memory tree structure ZB*-tree specifically for UC-PSkyline.

4.1   Z-addresses

For $u^W = (u^W.D_1, ..., u^W.D_d)$, $\forall u^W.Di \in [0, 1)(1 \leq i \leq$ d), we assume that u^W is represented by $(\lfloor 2z * u^W.D_1 \rfloor, ..., \lfloor 2z*u^W.D_d \rfloor)$. Then, each attribute value of u^W can be represented by a z-bit binary number. We can obtain that the Z-address of u^W is a z*d-bit binary number by interleaving bits of the binary representations of all attributes, denoted by $u^W.zadd$. $u^W.zadd$ can be regarded as z*d-bit groups, which the j-th bit of $u^W.zadd$ can be obtained by the (j/d)th bit of the binary representation of the (j%d)-th attribute. Here '/' is divider operator and '%' is modulus operator. $u^W.zadd$ is inherently hierarchical which it can partition d-dimensional data space over W into 2zd equal-sized regions. Thus, Z-addresses have the following properties [10, 14].

Property 1. Z-address Properties.

1. *If $x^W \prec y^W$, then $x^W.zadd \leq y^W.zadd$.*
2. *A prefix (its length is kd and $k \leq z$) of $x^W.zadd$ is another $x^W.zadd$ which is kd-bit groups.*
3. *d-dimensional data space can be divided into 2zd equal-sized regions which each region has the same Z-address.*

ZB-tree and Z-tree are similar tree structures. ZB-tree is in an external memory and is used in the traditional skyline queries while Z-tree is in an internal memory and is used in probabilistic skyline queries on uncertain data. They are both based on Property 1. According to this, we build Z+tree which improves Z-tree based on ZB-tree in the inner memory for UC-PSkyline to find the defects of ZB-tree and Z-tree.

Z+tree is a variant of B+-tree to index instances of uncertain objects according to ordered Z-address of instances. While Z+tree exists a very prominent defect which z +tree leads to much repeated and unnecessary dominance tests or traversals from the root node of Z+tree, which increase I/O costs and elapsing time.

4.2   A New Tree Structure ZB*-tree

ZB*-tree which is based on Z+tree is a new variant of B*-tree to organize instances of uncertain data according to ordered Z-addresses of instances. ZB*-tree strategically links all adjacent sibling nodes which can form a linked list at each depth. In each linked list, the direction of a pointer is from right to left, right nodes point to left nodes, as shown in Fig. 1.

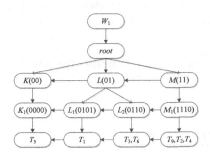

Fig. 1. The ZB*-tree structure

In ZB*-tree, an internal node $IN = \{Child[], Min, Max, Pori, Pflag, flag\}$. $IN.$ $Child[]$ is an array of child nodes of IN and child nodes are ordered ascending by their Z-addresses. $IN.Min$ and $IN.Max$ are $IN.MBBmin$ and $IN.MBBmax$, respectively. $IN.$ $Pori$ and $IN.Pflag$ are both pointers. The former pointers to the left adjacent nodes at the same depth and the nodes at each depth form a linked list through $IN.Pori$, the latter points to NULL or a node that cannot be determined whether the node dominates the goal node or not. $IN.flag$ is TRUE if $IN.Pflag$ is altered. Otherwise, $IN.flag$ is FALSE. A leaf node $LN = \{Instance[], Min, Max, Pori, Pfla\}$. Representations of $LN.Min$, $LN.$ Max, $LN.Pori$, $LN.Pflag$ and $LN.flag$ are the same as those of IN. $LN.Instance[]$ is an array of instances each of which is denoted by $\{Tuple, Zadd, Pr, Object, PrSky\}$. $Tuple$ is a d-dimensional number. $Zadd$ is the Z-address of Tuple. Pr is the probability that Tuple happens. $PrSky$ is the probability that Tupple is in the skyline set. For a node N (an internal node or leaf node) and every instance accessed in N, $N.Min < u$ and $u < N.Max$.

5   UC-PSkyline with ZB*-tree

In this section, we first introduce UC-PSkyline with ZB*-tree over W. Then, we present preprocessing techniques which can improve performance of UC-PSkyline. Finally, we discuss performance of UC-PSkyline with ZB*-tree. Since dominance tests between instances of different objects are performed only over certain contexts (i.e., the possible worlds), UC-PSkyline with ZB*-tree discussed in the following is built over the some possible world W, denoted by UC-PSkyline with ZB*-tree over W. Uncertain data over other possible worlds are built in the same way.

5.1   UC-PSkyline with ZB*-tree over W

In UC-PSkyline with ZB*-tree over W, we need two ZB*-trees over W and one arrays, denoted by EZB*, IZB* and $UO[]$, respectively. EZB* stores all instances of uncertain data over W and IZB* maintains those instances which dominate the goal nodes and is initialized to NULL. IZB* changes with UC-PSkyline processing. $UO[]$ stores skyline probabilities of all uncertain objects and all values of $UO[]$ are initialized to 0. The length of $UO[]$ is the size of UD (i.e., the size of all uncertain objects). UC-PSkyline traverses EZB* in preorder traversal for computing skyline probabilities of all uncertain objects. Algorithm 1 shows the pseudo-codes of dominance process.

Algorithm 1: Dominance_Innode (Innode M, Innode N)

---
1 if $M < N$
2 if $M.flag ==$ true then
3    $N.Pflag = M.Pflag$; $N.flag =$ true
4 else
5    $N.Pflag = M.Pori$; $N.Flag =$ true
6 Insert M to IZB*
7 else if M do not dominate N
8 if $M.flag ==$ true
9    $N.Pflag = M.Pflag$; $N.flag =$ true
10 else
11    $N.Pflag = M.Pori$; $N.Flag =$ true

When UC-PSkyline reaches a node N (i.e., the goal node) at depth i ($0 \leq i \leq H - 1$) in preorder traversal (shown in Fig. 2(a) where the dotted lines represent omitted nodes), there are three situations as shown in Algorithm 2:

1. N is an internal node. If $N.flag$ is FALSE, UC-PSkyline checks if all left sibling nodes in the linked list which $N.Pori$ points to dominate N. Otherwise, UC-PSkyline checks if all left sibling nodes in the linked list which $N.Pflag$ points to dominate N. If there is a left sibling node L that dominates N in accordance with item 1 of Lemma 1, UC-PSkyline alters $Pflag$ of the first child node M_1 of the right sibling node M adjacent to L if M has child nodes, and $M_1.Pflag$ points to the last child node K_k of left sibling node K adjacent to L if K has child nodes and $M_1.flag$ is TRUE, shown in Fig. 2(a). Meanwhile, IZB* maintains the sub-tree whose root node is L (lines 9–12). If there is a left sibling node F that does not dominates N in accordance with item 2 of Lemma 1. UC-PSkyline alters $Pflag$ of the first child node G_1 of the right sibling node G adjacent to F if G has child nodes, and $G_1.Pflag$ points to the last child node Ee of the left sibling node E adjacent to F if E has child nodes and $G.flag$ is TRUE, shown in Fig. 2(b).
2. N is a leaf node (lines 23–30). First, UC-PSkyline checks if left sibling nodes in the linked list which $N.Pflag$ or $N.Pori$ (if $N.flag$ is FALSE) points to dominate all instances of N in accordance with item 3 and 4 of Lemma 1. Then, UC-PSkyline checks if each instance of left sibling nodes (cannot be sure if the nodes dominate goal instances) in the linked list which $N.Pflag$ or $N.Pori$ (if $N.flag$ is FALSE) points

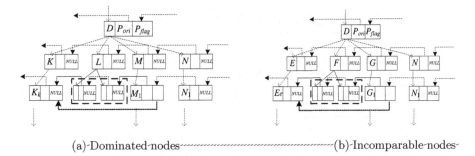

(a) Dominated nodes --- (b) Incomparable nodes

Fig. 2. UC-PSkyline with ZB*-tree over W

to dominates all instances of N.*Next*, all nodes of IZB* dominate N. UC-PSkyline calculates skyline probabilities of all instances of N through the two above steps.

3. UC-PSkyline ends till all leaf nodes of EZB* are traversed. UC-PSkyline uses Eq. (2) for calculating skyline probabilities of all uncertain objects over W (lines 31–33). Finally, UC-PSkyline calculates all uncertain objects over UC according to Eq. (3).

Which are not dominated by any other instance of K_1 are skyline points, i.e., *PrSky* $(t_5) = 1 \times 1 = 1$ (t_1 belongs to A and t_1 is not dominated by any instance of B or C). According to Eq. (3), UC-PSkyline updates the skyline probability over context S of object A (omitted next). Next, UC-PSkyline traverses the internal node L. UC-PSkyline cannot determine the dominance relation between L and K (a node that $L.Pflag$ points to). Next, UC-PSkyline will traverse L_1 and check dominance relation between L_1 and the nodes that $L.Pflag$ points to. L_1 is not dominated by only K_1. Thus, *PrSky* $(t_1) = 1 \times 1 = 1$. Next, UC-PSkyline traverses L_2, UC-PSkyline check the dominance relation among L_2, L_1 and $K(1)$, then UC-PSkyline cannot determine the dominance relation between $L(2)$ and $L(1)$, while L_2 is dominated by $K(1)$. Thus, *PrSky* $(t_3) = 1 \times (1 - 1/4) = 3/4$, $PrSky(t_6) = 1 \times (1-1/4) = 3/4$. Next, UC-PSkyline traverses the internal node M. UC-PSkyline can determine that K dominate M, while UC-PSkyline cannot determine the dominance relation between M and L. Thus, UC-PSkyline set $L_1.flag$ and $L_1.Pflag$ to *TRUE* and *NULL*, respectively. Maintains the one sub-tree, the one whose root node is K, all instances of IZB* dominate M, shown in Fig. 2(b)). Next, UC-PSkyline traverses the leaf node M_1. L_2 dominate M_1 while L_1 does not dominate M_1. And all instances of IZB* dominate M_1, then $PrSky(t_9) = (1-1/4) \times (1-1/4) \times (1-1/4) = 27/64$, $PrSky(t_2) = (1-1/4) \times (1-1/4) \times (1-1/4) = 27/64$, *PrSky* $(t_4) = (1-1/4) \times (1-1/4) \times (1-1/4) = 27/64$. Similarly, UC-PSkyline can compute the skyline probabilities of instances over the context O. Finally, UC-PSkyline calculates the skyline probabilities of all uncertain objects over uncertain contexts according to Eq. (3).

Algorithm 2: `UC-PSkyline with ZB* Tree over W`

Input: Uncertain dataset UD over W
Output: $UO[\ ]$
1 Initialize $UO[\ ]$; EZB*← ZB*-tree built from UD over W;
2 Traverse EZB* in Preorder; Reach N at depth I;
3 if N is an internal node
4   for each leftsib node of N
5   if $N.flag == true$
6             $M = N.Pflag$
7   Dominance_Innode(M, N)
8   if the relation of dominating between M and N is decided
9   if $M.flag = true$
10                    $M = M.Pflag$
11   else
12                    $M = M.Pori$
13   else
14                $M = N.Pori$;
15   Dominance_Innode(M, N)
16   if the relation of dominating between M and N is decided
17   if $M.flag = true$
18                        $M = M.Pflag$
19   else
20                        $M = M.Pori$
21 if N is a leafnode
22   for each leftsib M of N
23   if $N.flag == true$
24             $M = N.Pflag$
25   for each Instance I of N Do
26   for each Instance J of M Do
27   Calculate $PrSkyw(I)$ according to J and Instances of $IZB*$
28 for each U of UD Do
29   Calculate $PrSkyw(U)$, update $UO[\ ]$
30 return $UO[\ ]$

To further improve performance of UC-PSkyline, the next section investigates preprocessing techniques. The former attempts to calculate the skyline probabilities of instances and uncertain objects whose skyline probabilities are 0 or 1 before UC-PSkyline is performed, the latter aims at avoiding unnecessary dominance tests between nodes and instances.

5.2   Preprocessing Techniques

For $U,V \in UD$, $V \neq U$, let U_{min}^w, U_{max}^w, V_{max}^w, V_{max}^w be $U^W \cdot MBBmin$, $U \cdot MBB_{max}$, $V^w \cdot MBB_{min}$ and $V^W \cdot MBB_{max}$, respectively.

Preprocessing. The intuition way to do preprocessing is in accordance with Lemma 1: skyline probabilities of some instances or uncertain objects can be calculated beforehand. Algorithm 3 shows the pseudo-codes of this preprocessing technique. Mathematically, the preprocessing technique can calculate the skyline probability of an instance or uncertain object through the following Lemma.

Algorithm 3: Preprocessing procedure

Input: uncertain data set UD over W
Output: UD'
1: for each $U \in UD$ do
2:   calculate U^W_{min} and U^W_{max};
3:   for another $V \in UD$ do
4:     if $U^W_{max} \prec V^W_{min}$ then /* item 3 of Lemma 2*/
5:       $PrSky_W(V) \leftarrow 0$;
6:     else if $V^W_{min} \prec U^W_{max}$ then /* item 4 of Lemma 2*/
7:       $PrSky_W(U) \leftarrow 1$;
8:     else
9:       for each instance x^W of V do
10:        if $U^W_{max} \prec x^W$ then/* item 1of Lemma 2*/
11:          $PrSky(x^W) \leftarrow 0$;
12:        else if $x^W \prec U^W_{min}$ then/* item 2 of Lemma 2*/
13:          $PrSky(x^W) \leftarrow 1$;
14: UD'.insert(U^W_{min} and U^W_{max});
15: return UD';

Lemma 2 (Preprocessing). *For* $\forall V \in UD$, $V! = U$, $\forall x^W \in U$

1. *If* $V^w_{max} \prec x^W$, *then* $PrSky(x^W) = 0$.
2. *If* $x^W \prec V^w_{min}$, *then* $PrSky(x^W) = 1$.
3. *If* $U^w_{max} \prec V^w_{min}$, *then* $PrSky_W(V) = 0$.
4. *If* $V^w_{min} \nprec U^w_{max}$, *then* $PrSky_W(U) = 1$.

Proof sketch. *The proofs of the above items are similar. Limited by space, we only prove the third item. For any instance* x^W *of U, we have* $x^W \prec U^W$, *we have* $x^W \prec U^w_{max}$. *If* $U^w_{max} \prec V^w_{min}$, *then* $x^W \prec V^w$ *min based on the transitivity of dominance relations* $(\prec)$. *Meanwhile, for any instance* v^W *of V, we have* $V^w_{min} \prec v^W$ *and* $x^W \prec V^w_{min}$, *then* $x^W \prec v^W$. *Applying the above observation to Eq. (2), we can obtain*

$$\Pr Sky_W(V) = \sum_{v^W \in V} Pr(v^W) * \Pr Sky(v^W)$$

$$= \sum_{v^W \in V} \left(Pr(v^W) * (1-1) * \prod_{V_1 \in UD, V_1 \neq U} \left(1 - \sum_{v^W_1 \in V_1, v^W_1 \prec v^W} Pr(v^W_1) \right) \right) \quad (4)$$

$$= 0$$

The proofs of other items in Lemma 2 are similar.

6 Performance Evaluation

In this section, we evaluate the efficiency and effectiveness of proposed algorithms based on synthetically generated data sets.

Algorithms for Comparison. For comparison, we implement four algorithms, UC-PSkyline with ZBtree, UC-PSkyline with ZBtree with preprocessing technique, UC-PSkyline with ZB*-tree, UC-PSkyline with ZB*-tree with preprocessing technique (ZB-tree, ZB+-tree, ZB*-tree, ZB*+-tree for short, respectively). We do not compare UC-PSkyline with BBS (the most efficient skyline queries on certain data) or the algorithm AQ proposed by [2] because ZB-tree and Z-tree are obviously better than BBS and AQ for adopting in-memory indexes.

Synthetic Data Sets. We use a publicly available generator included in PTKlib available at http://www.cs.sfu.ca/~jpei/. to construct synthetic data sets and modified as follows:

1. **Uncertain Context Data Set:** The dimensionality of uncertain contexts varies from 2 up to 4. The number of tuples involved in each dimensionality follows the normal distribution N (20, 10), varying from 10 up to 30. The probabilities of independent tuples involved in each dimensionality follow the normal distribution N (0.5, 0.5).
2. **Uncertain Objects Data Set:** the data set is generated based on anti-correlated, correlated and independent distribution. The cardinality of objects varies from 100 up to 10 K objects. The dimensionality of objects varies from 2 up to 10. We generate instances in the data space [0, 1] and the cardinality of instances varies from 150 K up to 500 K. The number of instances involved in each object varies from 100 up to 10 K. The probabilities of instances involved in each object follow the normal distribution N (0.5, 0.5).

ALL algorithms are implemented in C++, compiled with Visual Studio 2010 and executed on Windows 7 Service Pack 1 with a 3.3 Ghz Intel Core i3-3220 M CPU and 4 Gigabytes of main memory. The disk page size is set to 4096 bytes. We use CPU time as the performance metric which represents the duration from the time an algorithm starts to the time the result is completely returned. In each experiment, we only vary a single parameter and set the remaining to their default values. Table 1 displays parameters under investigation and their corresponding ranges, where default values are the bold ones.

Table 1. Parameters and values

Parameter	Range
Object Cardinality	100, **500**, 1000
Instance Cardinality	100 K, 150 K, **250 K**, 500 K
Instance Dimensionality	4, **6**, 8

6.1   Evaluate Effect of Instance Dimensionality

Figure 3 depicts CPU time against instance dimensionality from 4 up to 8 for synthetically generated data sets. We can see that UC-PSkyline with ZB*-tree incurs least I/O time (except ZB*+-tree) by avoiding repeated unnecessary dominance tests between the common ancestor nodes and avoiding repeated traversals for anti-correlated and independent distributions. UC-PSkyline with ZB*+-tree incurs less CPU time than UC-PSkyline with ZB*-tree. This is because preprocessing technique can calculate skyline probabilities of some instances and uncertain objects before UC-PSkyline is performed and avoid unnecessary dominance tests between nodes. Similarly, UC-PSkyline with ZB+-tree outperforms UC-PSkyline with ZB-tree. UC-PSkyline with ZB-tree need scanning the entire data set, resulting in the higher CPU time.

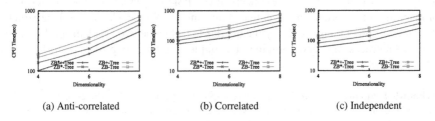

(a) Anti-correlated   (b) Correlated   (c) Independent

Fig. 3.   Instance dimensionality vs. CPU time

6.2   Evaluate Effect of Instance Cardinality

Figure 4 show CPU time against the instance cardinality from 100 K up to 500 K for synthetically generated data sets, respectively. In the two figures, UC-PSkyline with ZB*+-tree outperforms the best for anti-correlated and independent distributions.

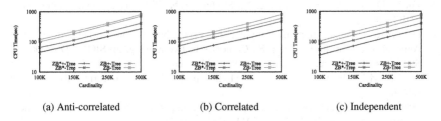

(a) Anti-correlated   (b) Correlated   (c) Independent

Fig. 4.   Instance cardinality vs. CPU time

6.3   Evaluate Effect of Object Cardinality

Figure 5 depicts CPU time against object cardinality from 100 up to 1000 for synthetically generated data sets, respectively. As observed in the figures, UC-PSkyline with ZB*-tree outperforms the best (except ZB*+-tree). UC-PSkyline with ZB*+-tree outperforms better than UC-PSkyline with ZB*-tree.

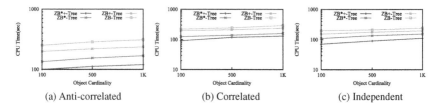

(a) Anti-correlated   (b) Correlated   (c) Independent

Fig. 5. Object Cardinality vs. CPU time

7 Conclusion

To the best of our knowledge, we first study introducing probabilistic skyline queries on uncertain data over uncertain contexts, which uncertainty lies within the data and the query. We formally introduce the concept of UC-PSkyline which computes exact skyline probabilities of uncertain data over uncertain contexts. We develop a new Tree Structure ZB*-tree for processing UC-PSkyline queries to tackle the two problems of the existing work. Preprocessing and pruning techniques further improve performance of UC-PSkyline.

Acknowledgement. This work is partially supported by the Natural Science Foundation of Jiangsu Province of China under grant No. BK20140826, Foundation of Graduate Innovation Center in NUAA No. KFJJ20161607.

References

1. Borzsony, S., Kossmann, D., Stocker, K.: The skyline operator. In: ICDE (2001)
2. Pei, J., Jiang, B., Lin, X., Yuan, Y.: Probabilistic skylines on uncertain data. In: VLDB (2007)
3. Dellis, E., Seeger, B.: Efficient computation of reverse skyline queries. In: VLDB (2007)
4. Sacharidis, D., Arvanitis, A., Sellis, T.: Probabilistic contextual skylines. In: ICDE (2010)
5. Zhang, Q., Ye, P., Lin, X., Zhang, Y.: Skyline probability over uncertain preferences. In: EDBT (2013)
6. Dey, A.K.: Understanding and using context. Pers. Ubiquit. Comput. **5**(1), 4–7 (2001)
7. Agrawal, R., Rantzau, R., Terzi, E.: Context-sensitive ranking. In: SIGMOD (2006)
8. Stefanidis, K., Pitoura, E., Vassiliadis, P.: Adding context to preferences. In: ICDE (2007)
9. Atallah, M.J., Qi, Y.: Computing all skyline probabilities for uncertain data. In: PODS (2009)
10. Dongwon, K., Hyeonseung, I., Sungwoo, P.: Computing exact skyline probabilities for uncertain databases. TKDE **24**(12), 2113–2126 (2012)
11. Zheng, J., Wang, Y., Wang, H., Yu, W.: Asymptotic-efficient algorithms for skyline query processing over uncertain contexts. In: IDEAS, pp. 106–115 (2015)
12. Atallah, M.J., Qi, Y., Yuan, H.: Asymptotically efficient algorithms for skyline probabilities of uncertain data. TODS **36**(2), 12:1–12:28 (2011)
13. Böhm, C., Fiedler, F., Oswald, A., Plant, C., Wackersreuther, B.: Probabilistic skyline queries. In: CIKM (2009)
14. Lee, K.C.K., Zheng, B., Li, H., Lee, W.C.: Approaching the skyline in z order. In: VLDB (2007)

A Multi-direction Prediction Approach for Dynamic Multi-objective Optimization

Miao Rong[1($\boxtimes$)], Dun-wei Gong[1,2], and Yong Zhang[1]

[1] School of Information and Electrical Engineering, China University of Mining and Technology, Xuzhou 221116, China
Rongmiao307@163.com
[2] School of Electrical Engineering and Information Engineering,
Lanzhou University of Technology, Lanzhou 730050, China

Abstract. In the real word, many multi-objective optimization problems are subject to dynamic changing conditions, which may occur in objectives, constraints and parameters. This paper provides a prediction strategy, called multi-direction prediction strategy (MDP), to enhance the performance of multi-objective evolutionary optimization algorithms in dealing with dynamic environments. Besides, the proposed prediction method makes use of multiple directions determined by several representative individuals. Our experimental results indicate that MDP can well tackle dynamic multi-objective problems.

Keywords: Dynamic · Multi-objective optimization · Prediction

1 Introduction

Dynamic multi-objective optimization problems (DMOPs) widely exist in our real world. Practical examples of such situations are design [1], planning [2], scheduling [3], etc. For handling these DMOPs, the optimizer should be capable of tracking the optima whose locations change with time.

Evolutionary algorithms (EAs) have been recognized as one of the most powerful optimizers [8]. The first attempt that utilizes EAs to solve DMOPs was conducted by Fogel and his colleagues in 1966 [9, 11, 12]. Since dynamic multi-objective evolutionary algorithm (DMOEA) is built on the static MOEA, under dynamic environments, converging tendency of a conventional static EA imposes severe limitations on performance of EA. In order to efficiently solve DMOPs, it is better for DMOEA to build its own framework. Moreover, some operators have been added for tackling environmental changes, such as the diversity maintenance technique and the prediction [10].

In this paper, we focus on the prediction method which aims to formulate a new population close to or even cover the PS under the new environment based on the historic information. To achieve this goal, we propose a multiple-direction prediction method for DMOEA. In this approach, several representative individuals are first sought based on the distribution information of the PS in the previous environment; next, several evolutionary directions are estimated by the representative individuals of the previous two PSs; finally, each individual evolves following the direction determined by its corresponding representative individual.

© Springer International Publishing Switzerland 2016
D.-S. Huang et al. (Eds.): ICIC 2016, Part III, LNAI 9773, pp. 629–636, 2016.
DOI: 10.1007/978-3-319-42297-8_58

2 Proposed Method

In this section, we address how to deal with the changes of the environment and the weaknesses of those available prediction approaches utilizing the proposed multiple direction prediction method (MDP), which aims to generate a population close enough to the true Pareto front. The basic idea of MDP is to estimate a set of individuals which have the ability to cover the true PS of the new environment. To achieve this goal, we first record and store several representative individuals which are able to describe the shape and diversity of the Pareto set obtained by the evolutionary algorithm at each time; subsequently, when an environmental change is detected, estimate the evolution directions in terms of information concerning representative individuals at the nearest two former time points (i.e., t and $t-1$). These evolutionary directions are employed to achieve the new locations of the representative individuals and predict the changes of the Pareto front; finally, a certain number of evolutionary individuals are generated around those new locations so as to improve the response of the population concerning the environmental change.

2.1 Determination of Representative Individuals

Due to the fact that representative individuals are designed to describe properties of the Pareto set, it is of great necessity for the selection strategy not only to choose those who are able to have both outperformed convergence and diversity. Therefore, M PF end-points are selected as representative individuals for a DMOP who has M objectives. The calculation equation is as follows.

$$p(m) = \arg \min_{x_j \in s} \left(f_m(x_j) \right) \tag{1}$$

Following that, the PS center point is calculated as one of the representative individuals due to its capability in helping the optimization algorithm converge to the PS under the new environment. In the studies [26–28], researchers utilize PS center points in the current and previous time, to predict the new locations of individuals. The definition of the abstracted PS center point is as follows.

$$C_i^t = \frac{1}{|PS^t|} \sum_{i=1}^{|PS^t|} x_i^t \tag{2}$$

where PS^t is the PS obtained at t, $|PS^t|$ is the number of individuals in PS^t, and x_i^t is the i-th solution in PS^t.

Herein, C^t is utilized to store these $(M + 1)$ representative individuals, where $C_i^t(i = 1, \ldots, M)$ are the M end-points of the PF, and $C_i^t(i = M + 1)$ is the center point at t.

2.2   Multiple Direction Estimation

In order to estimate the variation tendency of the Pareto front when an environmental change is detected, a multiple direction estimation strategy is proposed in this section. This estimation strategy utilizes the time series model, connects the two nearest historical information of representative individuals, and predicts multiple directions concerning the representative individuals of the Pareto front. Therefore, the variation trend of the Pareto front is able to be described completely by those multiple directions.

Suppose that $C^t = \{c_1^t, c_2^t \cdots, c_{K'}^t\}$ and $C^{t-1} = \{c_1^{t-1}, c_2^{t-1} \cdots, c_{K''}^{t-1}\}$ are two representative individual sets in accordance with time t and $t-1$, respectively. For one individual c_i^t in C^t, the proposed strategy will first seek for the nearest representative individual to it in the set C^{t-1}, denoted as c_j^{t-1}, which is viewed as the parent individual of c_i^t. In that way, the evolutionary direction of the individual c_i^t can be estimated by the change from the parent individual c_j^{t-1} and the current individual c_i^t as follows.

$$\Delta c_i^t = c_i^t - c_j^{t-1} \tag{3}$$

2.3   Generation of Individuals

For sake of improving the responding speed to the environmental changes, we plan to generate a certain number of evolutionary individuals around the predicted Pareto front. Hopefully, a set of individuals which have the capability to get close to or even cover the true Pareto front with a uniform distribution is likely to be achieved. When a change has been detected, we first employ simple linear model to add the evolutionary directions obtained by means of multiple direction estimation measure in 3.2 to the representative individuals in the nearest historical time. Therefore, the location and shape of the Pareto front under the new environment is able to be predicted. Subsequently, the new individuals will be generated.

Suppose that an individual x^t belongs to the i-th category, which representative individual is c^t. The evolutionary direction of c^t has been calculated by means of Sect. 2.2, denoted as Δc^t. Therefore, a new individual x^{t+1} is generated as follows.

$$x^{t+1} = x^t + \Delta c^t \tag{4}$$

Similarly, several new locations of those representative individuals are generated as follows.

$$c^{t+1} = c^t + \Delta c^t \tag{5}$$

To improve the diversity of the population, a number of individuals are generated randomly in the search space. Therefore, the new population consists of two parts, i.e., a number of predicted individuals and randomly-generated individuals in the decision space.

2.4   Framework of the Prediction Method

The framework of the proposed prediction strategy is incorporated into the particle swarm optimization algorithm. The pseudo code of the dynamic multi-objective particle swarm optimization algorithm based on the multi-direction prediction strategy is depicted as follows (Fig. 1).

Function MDP-DMOPSO : MDP-DMOPSO (N_p , N_s , $g_{\max}$, $\tilde{\delta}$)
//*Input: N_p (size of the swarm), N_s (size of the archive), g_{max} (maximum number of iterations), $\tilde{\delta}$ (an environment threshold)*//
//*output: Op (the Pareto optimal set)*//
1: $g = 0$ For i=1 to N_p
initialize $\left(x_i(0) \right)$

    $pb_i(0) \leftarrow x_i(0)$

Endfor
1.2 $F(pop_0) \leftarrow evaluate(pop_0)$
1.3 $Ar_0 \leftarrow non\_dominated(pop_0)$
2: **While** $g \le g_{max}$, do   //*Main cycle *//
2.1   **If environment changes**
    $C^t \leftarrow Select\_RPoins(ps^t, K)$
    $\Delta C \leftarrow Decide(C^t, C^{t-1}, K)$
    $Reset(Pop_g)$
    Endif
2.2   **Update the position of each particle in** pop_g **by PSO**
    2.2.1 **For** $i = 1$ to N_p
        $pbest_i(g) \leftarrow GET\_pbest()$
        $gbest_i(g) \leftarrow GET\_gbest()$
        $x_i(g+1) \leftarrow UP\_PARTICLE(gbest_i(g), pbest_i(g), x_i(g))$
        Endfor
    2.2.2 $F(pop_{+1}) \leftarrow EVALUATE(pop_{+1})$
    2.2.3 **Update the archive**
        $ps_{g+1} \leftarrow NON\_DOMINATED(pop_{g+1} \cup Ar_0)$
        If $\left| ps_{g+1} \right| > N_s$
        Using the crowding-distance to prune the archive.
        Endif
    2.3 $g \leftarrow g+1$
    Endwhile
3: $Op \leftarrow ps^t$ **and stop the algorithm**

Fig. 1. The Pseudo Code of MDP+DMOPSO

3   Results

This section evaluates the performance of the proposed algorithm, MDP-DMOPSO, on several well-known dynamic multi-objective test problems, compared with the particle swarm optimization algorithm with initialize strategy, namely, RIS.

3.1   Experiment Preparations

In this paper seven benchmark problems, FDA1, FDA2, FDA3, DMOP1, DMOP2, DMOP3, are selected. Table 1 shows these functions and their types. The inverted generational distance (IGD) metric has been widely employed to test the performance of stationary MOPs, which can measure both diversity and convergence of non-dominated optimums to true Pareto front.

Let P^{t*} be a set of uniformly distributed Pareto optimal points in the PF^t, and P^t be an approximation of PF^t. The IGD metrics is defined as follows [13].

$$IGD(P^{t*}, P^t) = \frac{\sum_{v \in P^{t*}} d(v, P^t)}{|P^{t*}|} \tag{6}$$

where $d(v, P^t) = \min_{u \in P^t} \|F(v) - F(u)\|$ is the distance between v and P^t and $|P^{t*}|$ is the cardinality of P^{t*}. If P^t gets close enough to PF^t and can cover all parts of the whole PF^t, the IGD value will be low.

For these test functions, the severity of changes is set to be $n_T = 10$. The frequency of the changes is set to be $\tau_T = 20$. The dimensions of the test problems are $n = 10$.

The population size is set to be $N = 150$ for test problems. The archive size (the Pareto set size) is set to be $V = 100$. It is assigned to be 4500 evaluation times for these functions, i.e., the environment changes every 30 generations for all the two algorithms.

3.2   Performance Analysis

Table 2 depicts the IGD values of MDP and RIS on the test instances over 20 runs. For comparing the difference between those two sets of results, t-test is employed, where + (-) supports (fails to support) the hypothesis that there is variance between the two sets of results.

Table 2 suggests that MDP performs better than RIS in almost all cases, especially on those DMOPs whose PS changes. For FDA2, even though $31 \leq t < 40$, MDP and RIS can obtain the same mean value of the IGD metric, they have great variance as the t-test result shows. Besides, on DMOP1, although $31 \leq t < 40$, the mean value of the IGD metric of MDP is 0.2008, slightly higher than RIS (0.1252), the t-test result fail to support these two sets of IGD values have difference with each other.

Furthermore, Fig. 2 shows the Pareto fronts obtained by the two strategies at t = 8, 25, 35, 40, and 50 with lowest IGD values on FDA2. We can see that MDP has good performance in tracking the changes of the environment, and improving diversity of the algorithm.

Table 1. Test instances

Test instances	Definition	Type
FDA1	$f_1(X_{\phi_{II}}) = x_1$ $f_2 = g \cdot h$ $g(X_{\phi b}) = 1 + \sum_{x_i \in X_{\phi b}} (x_i - G(t))^2$ $h(f_1, g) = 1 - \sqrt{\frac{f_1}{g}}$ $G(t) = \sin(0.5\pi t), t = \frac{1}{n_t}\lfloor \frac{\tau}{\tau_i} \rfloor$ $where: m = 10, X_1 = (x_1) \in [0,1]; X_{II} = (x_2, \cdots, x_m) \in [-1,1]$	Type I
FDA2	$f_1(X_{\phi_{II}}) = x_1$ $f_2 = g \cdot h$ $g(X_{\phi b}) = 1 + \sum_{x_i \in X_{\phi b}} x_i^2$ $h(f_1, g) = 1 - \left(\frac{f_1}{g}\right)^{H(t)^{-1}}$ $H(t) = 0.75 + 0.7\sin(0.5\pi t), t = \frac{1}{n_t}\lfloor \frac{\tau}{\tau_i} \rfloor$ $where: X_{\phi_{II}} = (x_1) \in [0,1]; X_{\phi b}, X_{\phi b} \in [-1,1], |X_{\phi b}| = |X_{\phi b}| = 15$	Type III
FDA3	$f_1(X_{\phi_{II}}) = \sum_{x_i \in X_{\phi_{II}}}^{n} x_i^{F(t)}$ $f_2 = g \cdot h$ $g(X_{\phi b}) = 1 + G(t) + \sum_{x_i \in X_{\phi b}} (x_i - G(t))^2$ $h(f_1, g) = 1 - \sqrt{\frac{f_1}{g}}$ $G(t) = |\sin(0.5\pi t)|$ $F(t) = 10^{2\sin(0.5\pi t)}, t = \frac{1}{n_t}\lfloor \frac{\tau}{\tau_i} \rfloor$ $where: |X_{\phi_{II}}| = 1, |X_{\phi b}| = 9, X_{\phi_{II}} \in [0,1]; X_{\phi b} \in [-1,1]$	Type II
DMOP1	$f_1(x_1) = x_1$ $f_2 = g \cdot h$ $g(x_2, \cdots, x_m, t) = 1 + 9\sum_{i=2}^{m} x_i^2$ $h(f_1, g) = 1 - \left(\frac{f_1}{g}\right)^{H(t)}$ $H(t) = 0.75\sin(0.5\pi t) + 1.25, t = \frac{1}{n_t}\lfloor \frac{\tau}{\tau_i} \rfloor$ $where: m = 10, x_i \in [0,1], \forall i = 1, 2, \cdots, m$	Type II
DMOP2	$f_1(x_1) = x_1$ $f_2 = g \cdot h$ $g(x_2, \cdots, x_m, t) = 1 + \sum_{i=2}^{m} (x_i - G(t))^2$ $h(f_1, g) = 1 - \left(\frac{f_1}{g}\right)^{H(t)}$ $H(t) = 0.75\sin(0.5\pi t) + 1.25, t = \frac{1}{n_t}\lfloor \frac{\tau}{\tau_i} \rfloor$ $where: m = 10, x_i \in [0,1], \forall i = 1, 2, \cdots, m$	Type II
DMOP3	$f_1(x_1) = x_1$ $f_2 = g \cdot h$ $g(x_2, \cdots, x_m, t) = 1 + \sum_{i=2}^{m} (x_i - G(t))^2$ $h(f_1, g) = 1 - \left(\frac{f_1}{g}\right)^{H(t)}$ $H(t) = 0.75\sin(0.5\pi t) + 1.25, t = \frac{1}{n_t}\lfloor \frac{\tau}{\tau_i} \rfloor$ $where: m = 10, x_i \in [0,1], \forall i = 1, 2, \cdots, m$	Type II

Table 2. The IGD values on 20 runs

		$1 \leq t < 10$	$11 \leq t < 20$	$21 \leq t < 30$	$31 \leq t < 40$	$41 \leq t \leq 50$
		Mean/std.	Mean/std.	Mean/std.	Mean/std.	Mean/std.
FDA1	MDP	**0.0120/0.0060(+)**	**0.0074/0.0032(+)**	**0.0146/0.0033(+)**	**0.0158/0.0063(-)**	**0.0126/0.0073(+)**
	RIS	0.0311/0.0020	0.0252/0.0021	0.0279/0.0056	0.0230/0.0047	0.0218/0.0027
FDA2	MDP	**0.0036/2.9575e-05 (-)**	**0.00369/1.4205e-05 (+)**	0.0036/2.0682e-05 (-)	**0.0036/1.1720e-05 (+)**	**0.0036/3.4290e-05 (+)**
	RIS	0.0047/0.0030	0.0037/7.53e-05	0.0036/3.8e-05	0.0036/3.35e-05	0.0037/6.35e-05
FDA3	MDP	**0.0036/6.896e-05 (+)**	**0.0036/3.0546e-05 (+)**	**0.0068/0.0048(+)**	**0.0101/0.0088(+)**	**0.0036/4.5897e-05 (+)**
	RIS	0.2707/0.0518	0.2337/0.0305	0.2861/0.0121	0.2874/0.0271	0.2388/0.0237
DMOP1	MDP	**0.0041/0.0001(+)**	**0.0042/0.0001(+)**	0.1760/0.1466(+)	0.2008/0.1488(-)	**0.0041/0.0002(+)**
	RIS	0.0755/0.0283	0.0414/0.0255	**0.0863/0.1067**	0.1252/0.1754	0.0102/0.0008
DMOP2	MDP	**0.0037/ 5.1830e-05(+)**	**0.0037/ 2.8288e-05(+)**	**0.0036/ 1.5351e-05(+)**	**0.0037/ 2.8446e-05(+)**	**0.0037/ 1.4142e-05(+)**
	RIS	0.0316/0.0253	0.0167/0.0095	0.0205/0.0105	0.0150/0.0080	0.0166/0.0067
DMOP3	MDP	**0.0037/ 9.1116e-05(+)**	**0.0037/ 3.4155e-05(+)**	**0.0037/ 3.4269e-05(+)**	**0.0037/ 3.4497e-05(+)**	**0.0037/ 3.2664e-05(+)**
	RIS	0.2622/ 0.2573	0.0617/ 0.0284	0.0758/ 0.0041	0.0419/ 0.0183	0.0466/ 0.0132

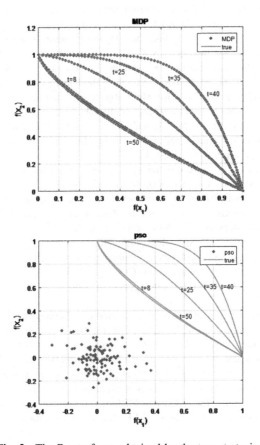

Fig. 2. The Pareto fronts obtained by the two strategies

4   Conclusion

In this paper, a new prediction strategy, MDP is proposed, which employs multiple directions based on representative individuals to provide a guide for the evolution of the population. By comparing with the strategy which generates population randomly when an environmental change is detected, MDP shows great capability in tracing the movement of PF (PS).

Acknowledgement. This research was supported by the National Natural Science Funds of China (No. 61473299), the Natural Science Foundation of Jiangsu province (No. BK20130207), and the China Postdoctoral Science Foundation funded project (No. 2014T70557, 2012M521142).

References

1. Barba, P.D.: Dynamic multi-objective optimization: a way to the shape design with transient magnetic fields. IEEE Trans. Magn. **44**(6), 962–965 (2008)
2. Bui, L.T., Michalewicz, Z.: An evolutionary multi-objective approach for dynamic mission planning. In: Proceedings of the IEEE CEC, pp. 1–8 (2010)
3. Deb, K., Rao N., U.V., Karthik, S.: Dynamic multi-objective optimization and decision-making using modified NSGA-II: a case study on hydro-thermal power scheduling. In: Obayashi, S., Deb, K., Poloni, C., Hiroyasu, T., Murata, T. (eds.) EMO 2007. LNCS, vol. 4403, pp. 803–817. Springer, Heidelberg (2007)
4. Kim, K., Mckay, R.B., Moon, B.R.: Multi-objective evolutionary algorithms for dynamic social network clustering. In: Proceedings of the GECCO, pp. 1179–1186 (2010)
5. A.-Alducin, M.Y., Efren, M.M., Nicandro, C.R.: Differential evolution with combined variants for dynamic constrained optimization. In: IEEE Congress on Evolutionary Computation (CEC), pp. 975–82 (2014)
6. Linnala, M., Madetoja, E., Ruotsalainen, H., et al.: Bi-level optimization for a dynamic multi-objective problem. Eng. Optimiz. **44**(2), 195–207 (2012)
7. Jin, Yaochu, Sendhoff, Bernhard: Constructing dynamic optimization test problems using the multi-objective optimization concept. In: Raidl, Günther R., et al. (eds.) EvoWorkshops 2004. LNCS, vol. 3005, pp. 525–536. Springer, Heidelberg (2004)
8. Azzouz, R., Bechikh, S., Said, L.B.: A multiple reference point-based evolutionary algorithm for dynamic multi-objective optimization with undetectable changes. In: IEEE Congress on Evolutionary Computation (CEC), pp. 3168–3175 (2014)
9. Fogel, L.J., Walsh, M.J.: Artificial Intelligence Through Simulation Evolution. John Wiley, New York (1966)
10. Zhou, A.M., Jin, Y.C., et al.: A population prediction strategy for evolutionary dynamic multi-objective optimization. IEEE Trans. Cybern. **44**(1), 40–53 (2014)
11. Zhang, Y., Gong, D.W., Ding, Z.H.: A bare-bones multi-objective particle swarm optimization algorithm for environmental/economic dispatch. Inf. Sci. **192**, 213–227 (2012)
12. Zhang, Y., Gong, D.W., Zhang, J.H.: Robot path planning in uncertain environment using multi-objective particle swarm optimization. Neurocomputing **103**, 172–185 (2013)
13. Zhang, Q., Zhou, A.M., Jin, Y.C.: RM-MEDA: a regularity model based multi-objective estimation of distribution algorithm. IEEE Trans. on Evolutionary Computation. **12**(1), 41–64 (2008)

A Model Predictive Current Control Method for Voltage Source Inverters to Reduce Common-Mode Voltage with Improved Load Current Performance

Huu-Cong Vu and Hong-Hee Lee$^{(\boxtimes)}$

School of Electrical Engineering, University of Ulsan, Ulsan, South Korea
vuhuucong90@gmail.com, hhlee@mail.ulsan.ac.kr

Abstract. This paper presents a new model predictive current control (MPCC) method in order to reduce common-mode voltage (CMV) for a three-phase voltage source inveter (VSI). By utilizing twenty-one vectors instead of eight vectors used in the conventional MPCC method,the proposed MPCC method can not only reduce the CMV but also improve load current performance. Simulation results are given to verify the effectiveness of the proposed MPCC method.

Keywords: Common-mode voltage · Model predictive control · Voltage source inverters

1 Introduction

Three-phase voltage source inverters (VSIs) have been widely employed in AC motor drive systems [1]. However, the common-mode voltage (CMV) generated by fast switching operation of the VSI is a main source of early motor-winding failure and bearing deterioration. Thus, the reduction of the CMV is important to increase the motor life time. In general, there are two solotions to solve the CMV problems for VSI: hardware and software solutions. The hardware solution is achieved with the aid of the aditional devices such as the isolation transformers, passive filters, and active-cancerlers [2, 3]. However, this solution resulted in increasing in inverter volume, weight, and cost. Whereas, the software solution such as PWM methods and predictive control methods, which does not need to use the additional power devices. Several PWM methods have been proposed to reduce the CMV of the VSI by avoiding the use of zero vectors because the zero vectors produce the highest CMV [4, 5]. Due to the lack of using zero vectors, the performance of these PWM methods become deteriorated.

The predictive control method is a computer control algorithm, which consists of model predictive control and dedbeat control. Recently, the model predictive current control (MPCC) method is prefer to use due to its advanced characteristics such as simplicity, control flexibility and good performance. Therefore, the MPCC method has been received many considerations for the power converters and the drive applications [6–12].

© Springer International Publishing Switzerland 2016
D.-S. Huang et al. (Eds.): ICIC 2016, Part III, LNAI 9773, pp. 637–647, 2016.
DOI: 10.1007/978-3-319-42297-8_59

In the MPCC method, the cost function is designed with a CMV constraint of the CMV reduction. The amount of CMV reduction is decided by an appropriate weighting factor [13]. However, the selection of the weighting factor is not easy due to the lack of theoretical design procedure. In order to solve this problem, two reduced CMV MPCC methods without the selection of the weighting factor are presented in [14]. These methods utilize two active vectors in one sampling period that can reduce the CMV as much as 66.7 % compared to the conventional method. However, they result in complicated and burden computation for switching times.

To overcome this drawback, this paper proposes a new MPCC method to reduce the CMV by generating twenty-one virtual vectors with the fixed dwell-time from six original active vectors. Thus, the proposed method can reduce the CMV without the switching time calculation. Moreover, the proposed method also can improve the load current quality by selecting the nearest vectors to synthesize the reference current. Simulation results are provided to validate the proposed MPCC method.

2 Conventional Model Predictive Current Control

Figure 1 shows a common three-phase VSI topology which includes six insulated gate bipolar transistor (IGBT) fed by a DC-link voltage.

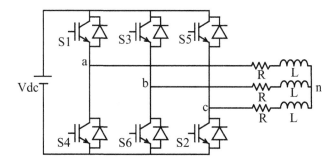

Fig. 1. Three-phase voltage source inverter.

The VSI output phase voltages with a general three-phase inductive load can be written as:

$$
\begin{aligned}
v_{an} &= Ri_a + L\frac{di_a}{dt} \\
v_{bn} &= Ri_b + L\frac{di_b}{dt} \\
v_{cn} &= Ri_c + L\frac{di_c}{dt}
\end{aligned}
\tag{1}
$$

where R and L are the load resistance and inductance, respectively.

The space vector of the output voltages is defined as follows:

$$\mathbf{v} = \frac{2}{3}\left(v_{an} + v_{bn}e^{j(2\pi/3)} + v_{cn}e^{j(4\pi/3)}\right) \tag{2}$$

The output voltage vector $\mathbf{v}$ is determined by the switching states of the VSI and DC-link voltage V_{dc}, as given in Eq. (3):

$$\mathbf{v} = \frac{2}{3}V_{dc}\left(S_a + S_b e^{j(2\pi/3)} + S_c e^{j(4\pi/3)}\right), \tag{3}$$

where the switching states S_a, S_b, and S_c are defined as follows:

$$S_a = \begin{cases} 1 & \text{if } S_4 \text{ off and } S_1 \text{ on} \\ 0 & \text{if } S_1 \text{ off and } S_4 \text{ on} \end{cases} \tag{4}$$

$$S_b = \begin{cases} 1 & \text{if } S_6 \text{ off and } S_3 \text{ on} \\ 0 & \text{if } S_3 \text{ off and } S_6 \text{ on} \end{cases} \tag{5}$$

$$S_c = \begin{cases} 1 & \text{if } S_2 \text{ off and } S_5 \text{ on} \\ 0 & \text{if } S_5 \text{ off and } S_2 \text{ on} \end{cases} \tag{6}$$

In general, the three-phase VSI can produce eight voltage vectors, including six active vectors and two zero vectors as shown in Fig. 2, to control the load current [6].

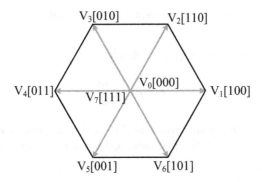

Fig. 2. Space vector diagram of voltage source inverter.

The space vector of the load current is defined as follows:

$$\mathbf{i} = \frac{2}{3}\left(i_a + i_b e^{j(2\pi/3)} + i_c e^{j(4\pi/3)}\right) \tag{7}$$

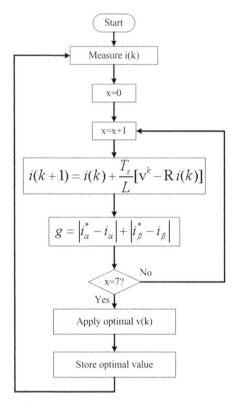

Fig. 3. The algorithm of MPCC method.

Substituting (2) and (7) into (1), we obtain:

$$\mathbf{v} = R\mathbf{i} + L\frac{d\mathbf{i}}{dt} \tag{8}$$

The derivative of the load current in the continuous-time model in (8) can be approximated by the basis of the first-order approximation with a sampling period T_s as:

$$\frac{di}{dt} = \frac{i(k+1) - i(k)}{T_s} \tag{9}$$

By substituting Eq. (9) into Eq. (8), the load current dynamics can then be expressed in the discrete-time domain as:

$$i(k+1) = i(k) + \frac{T_s}{L}[v(k) - Ri(k)], \tag{10}$$

where v(k) and $i(k)$ are the load voltage and current vectors at instant k, respectively.

Table 1. Peak value of CMV

Type	Switching states			Voltage vectors	$|v_{cm}|$
	S_a	S_b	S_c		
Zero vectors	0	0	0	V_0	$V_{dc}/2$
	1	1	1	V_7	$V_{dc}/2$
Active vectors	1	0	0	V_1	$V_{dc}/6$
	1	1	0	V_2	$V_{dc}/6$
	0	1	0	V_3	$V_{dc}/6$
	0	1	1	V_4	$V_{dc}/6$
	0	0	1	V_5	$V_{dc}/6$
	1	0	1	V_6	$V_{dc}/6$

According to (10), the seven future load currents $i(k+1)$ can be predicted by seven voltage vectors of the VSI, $v(k)$. These seven future load currents, $i(k+1)$, are used to determine the optimal voltage vector that minimizes following cost function:

$$g = \left| i_\alpha(k+1) - i_\alpha^{ref} \right| + \left| i_\beta(k+1) - i_\beta^{ref} \right| \tag{11}$$

The algorithm of MPCC method is shown in Fig. 3 and can be summarized as follows [6]:

(1) The value of the reference current i^{ref} is obtained and the load current $i(k)$ is measured.
(2) The model of the VSI is used to predict the value of the load current at the next sampling instant (k + 1).
(3) The cost function is used to evaluate the error between the reference and the predicted current in the next sampling interval for each voltage vector.
(4) The optimal voltage vector, which minimizes the current error, is determined among seven voltage vectors and the corresponding switching states are generated.

3   Proposed MPCC to Reduce CMV and Load Current Harmonics

In general, the CMV of the three-phase VSI is defined as follows:

$$v_{cm} = \frac{(v_{a0} + v_{b0} + v_{c0})}{3} \tag{12}$$

where v_{a0}, v_{b0}, v_{c0} are the output phase voltages with respect to the neutral point (Fig. 1).

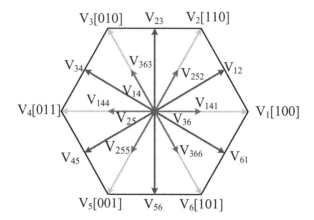

Fig. 4. Virtual space vectors.

Table 2. Construction of virtual small vectors

Virtual small vectors	Combined active states		$|v_{cm}|$
V_{141}	[100]	[011]	$V_{dc}/6$
V_{144}	[100]	[011]	$V_{dc}/6$
V_{252}	[110]	[001]	$V_{dc}/6$
V_{255}	[110]	[001]	$V_{dc}/6$
V_{363}	[010]	[101]	$V_{dc}/6$
V_{366}	[010]	[101]	$V_{dc}/6$

As can be seen, the zero vectors produce higher CMV that corresponds to $\pm V_{dc}/2$, whereas the active vectors produce smaller CMV that corresponds to $\pm V_{dc}/6$ as shown in Table 1.

In order to reduce the CMV, only six active vectors are considered to implement the predictive current control method. However, the current ripple is increased due to the absence of the zero vectors. In order overcome this problem, the proposed MPCC utilizes twenty-one vectors that include six active vectors and fifteen virtual vectors constructed from six active vectors which produce the smaller CMV that corresponds to $\pm V_{dc}/6$. By using twenty-one vectors instead of six active vectors, the distance between the predicted currents and the reference current is decreased, and the reference tracking error is reduced. Fifteen virtual vectors can be divided into three groups as below:

(1) Group I consists of virtual small vectors
(2) Group II consists of virtual medium vectors
(3) Group III consists of virtual zero vectors

The methods, used to produce these virtual vectors, are described as follow:

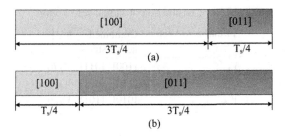

Fig. 5. Implementation of (a) virtual small vector V_{141}. (b) virtual small vector V_{144}.

The virtual small vectors are constructed by two opposite active vectors with three over four and one over four of the sampling interval, respectively as shown in Table 2. For example, V_{141} vector is constructed by applying V_1 three over four of the sampling interval, V_4 one over four of the sampling interval as shown in Fig. 5(a). V_{144} vector is constructed by applying V_1 one over four of the sampling interval, V_4 three over four of the sampling interval as shown in Fig. 5(b).

The virtual medium vectors are constructed by two adjacent active vectors with the same dwell time, as shown in Table 3. For example, V_{12} vector is constructed by applying V_1 in half of the sampling interval, and V_2 is applied in half of the sampling interval as shown in Fig. 6.

Table 3. Construction of virtual medium vectors

Virtual medium vectors	Combined active states		$|v_{cm}|$
V_{12}	[100]	[110]	$V_{dc}/6$
V_{23}	[110]	[010]	$V_{dc}/6$
V_{34}	[010]	[011]	$V_{dc}/6$
V_{45}	[011]	[001]	$V_{dc}/6$
V_{56}	[001]	[101]	$V_{dc}/6$
V_{61}	[101]	[100]	$V_{dc}/6$

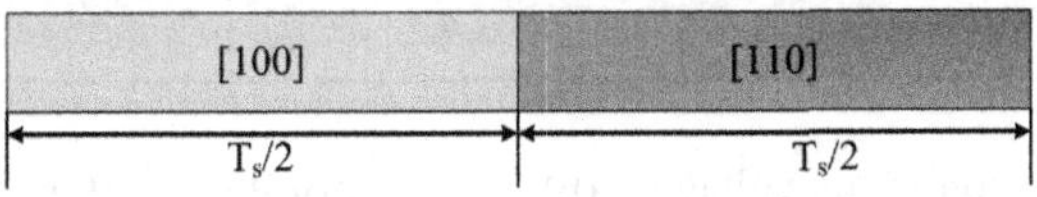

[100]	[110]
$T_s/2$	$T_s/2$

Fig. 6. Implementation of virtual medium vector V_{12}.

The virtual zero vectors are constructed by two opposite active vectors with the same dwell time, as shown in Table 4. For example, V_{14} vector is constructed by applying V_1 in half of the sampling interval, and V_4 is applied in half of the sampling interval as shown in Fig. 7.

Table 4. Construction of virtual zero vectors

Virtual zero vectors	Combined active states		$|v_{cm}|$
V_{14}	[100]	[011]	$V_{dc}/6$
V_{25}	[110]	[001]	$V_{dc}/6$
V_{36}	[010]	[101]	$V_{dc}/6$

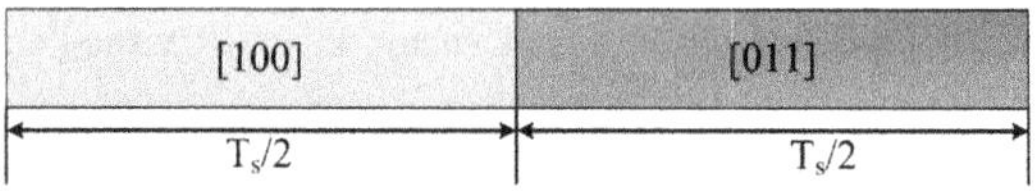

[100]	[011]
$T_s/2$	$T_s/2$

Fig. 7. Implementation of virtual zero vector V_{14}.

4   Simulation Results

In order to verify the effectiveness of the proposed predictive current control, some numerical simulations have been carried out by using PSIM 9.0. The simulation parameters are given as follows:

- DC-link Voltage: 200 V
- Sampling time: T_s = 50 μs
- Three-phase R-L load: R = 10 Ω, L = 20 mH
- Output frequency: f_0 = 50 Hz
- Amplitude of reference current: I^{ref} = 6 A

Figure 8(a) shows the CMV waveform of the VSI when the conventional MPCC method is applied. The peak value of the CMV is achieved 100 V corresponding with

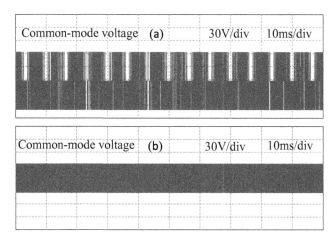

Fig. 8. Common-mode voltage of: (a) Conventional MPCC method. (b) Proposed MPCC method.

$V_{dc} / 2$. Figure 8(b) shows the CMV waveform of the VSI when the proposed MPCC method is utilized. From Fig. 8, the proposed MPCC method reduces the peak value of CMV: the peak of the CMV decreases from 100 V to 33.33 V. As can be seen, the proposed MPCC method reduces the peak value of the CMV to 66.7 % compare with conventional MPCC method.

The waveforms of the load currents of conventional MPCC method and the proposed MPCC method are shown in Figs. 9(a) and (b), respectively. Total harmonic distortion (THD) of the load current with the conventional MPCC is 1.61 %, while that with the proposed MPCC is 1.2 %. It is obvious that the load currents performance of the proposed MPCC method is better than that of the conventional MPCC method.

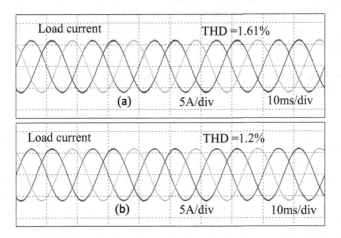

Fig. 9. Three-phase load currents of: (a) Conventional MPCC method. (b) Proposed MPCC method.

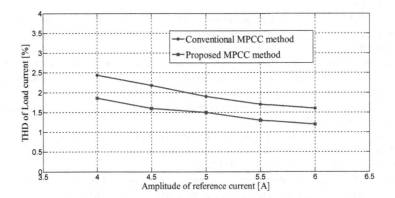

Fig. 10. Comparative results in THD of load currents.

Figure 10 shows the THDs of the conventional and proposed MPCC corresponding to the amplitude of the reference current. From Figs. 9 and 10, we can say that the proposed MPCC method also improves the load current performance compare to the conventional MPCC method.

5 Conclusion

This paper has presented a MPCC method that can reduce the CMV as well as improve the load current performance for the VSI. The proposed MPCC method is generated twenty-one virtual vectors, which restrict the peak value of the CMV to 66.7 % compared to the conventional MPCC method with eight original vectors. The proposed MPCC method also can reduce the load current distortion due to the reference current tracking error is reduced by using the virtual vectors. Simulated results are provided to confirm the performance of the proposed MPCC method.

Acknowledgment. This work was supported by the National Research Foundation of Korea Grant funded by the Korean Government (NRF- 2015R1D1A1A09058166).

References

1. Kwon, Y.-C., Kim, S., Sul, S.-K.: Voltage feedback current control scheme for improved transient performance of permanent magnet synchronous machine drives. IEEE Trans. Ind. Electron. **59**(9), 3373–3382 (2012)
2. Yuen, K.K.F., Chung, H.S.H., Cheung, V.S.P.: An active low-loss motor terminal filter for overvoltage suppression and common-mode current reduction. IEEE Trans. Power Electron. **27**(7), 3158–3172 (2012)
3. Akagi, H., Doumoto, T.: An approach to eliminating high-frequency shaft voltage and ground leakage current from an inverter-driven motor. IEEE Trans. Ind. Appl. **40**(4), 1162–1169 (2004)
4. Hava, A.M., Ün, E.: A high-performance PWM algorithm for common-mode voltage reduction in three-phase voltage source inverters. IEEE Trans. Power Electron. **26**(7), 1998–2008 (2011)
5. Tian, K., Wang, J., Wu, B., Cheng, Z., Zargari, N.R.: A virtual space vector modulation technique for the reduction of common-mode voltages in both magnitude and third-order component. IEEE Trans. Power Electron. **31**(1), 839–848 (2016)
6. Rodriguez, J., et al.: Predictive current control of a voltage source inverter. IEEE Trans. Ind. Electron. **54**(1), 495–503 (2007)
7. Rodriguez, J., Kennel, R.M., Espinoza, J.R., Trincado, M., Silva, C.A., Rojas, C.A.: High-performance control strategies for electrical drives: an experimental assessment. IEEE Trans. Ind. Electron. **59**(2), 812–820 (2012)
8. Fuentes, E., Kalise, D., Rodríguez, J., Kennel, R.M.: Cascade-free predictive speed control for electrical drives. IEEE Trans. Ind. Electron. **61**(5), 2176–2184 (2014)
9. Xie, W., et al.: Finite-control-set model predictive torque control with a deadbeat solution for PMSM drives. IEEE Trans. Ind. Electron. **62**(9), 5402–5410 (2015)

10. Cortes, P., Wilson, A., Kouro, S., Rodriguez, J., Abu-Rub, H.: Model predictive control of multilevel cascaded h-bridge inverters. IEEE Trans. Ind. Electron. **57**(8), 2691–2699 (2010)
11. Vargas, R., Rodriguez, J., Rojas, C.A., Rivera, M.: Predictive control of an induction machine fed by a matrix converter with increased efficiency and reduced common-mode voltage. IEEE Trans. Energy Convers. **29**(2), 473–485 (2014)
12. Kouro, S., Cortes, P., Vargas, R., Ammann, U., Rodriguez, J.: Model predictive control—a simple and powerful method to control power converters. IEEE Trans. Ind. Electron. **56**(6), 1826–1838 (2009)
13. Vargas, R., Rodriguez, J., Rojas, C.A., Rivera, M.: Predictive control of an induction machine fed by a matrix converter with increased efficiency and reduced common-mode voltage. IEEE Trans. Energy Convers. **29**(2), 473–485 (2014)
14. Kwak, S.: Mun, S.k.: Model predictive control methods to reduce common-mode voltage for three-phase voltage source inverters. IEEE Trans. Power Electron. **30**(9), 5019–5035 (2015)

A Fault Diagnostic Approach
for the OSGi-Based Cloud Platform

Zeng-Guang Ji[1(✉)], Rui-Chun Hou[1], and Zhi-Ming Zhou[2]

[1] College of Information Science and Engineering, Ocean University of China,
238 Songling Road, Laoshan District, Qingdao, Shandong, China
jizengguang0204@163.com
[2] Newstar Computer Engineering Center, Ocean University of China,
No. 23 Eastern Hongkong Road, Qingdao, Shandong, China

Abstract. OSGi-based cloud platform has the advantages of dynamic update, real-time monitoring. It brings convenience to cloud service. But the research of OSGi-based cloud platform doesn't get the attention it deserves now and the service diagnosis and fault recovery for the OSGi-based Cloud Platform haven't been researched systematically, especially in the resource conflicts among services, fault diagnosis and handling. These issues generally require users' interaction, increase the users' learning curve, which may turn out to be a waste of time. This paper divides those faults on the OSGi-based cloud platform into several models, and presents an approach to resolve them. Our method can detect service conflicts, handle exceptions, and diagnose service errors. And this method is proved by our experiment that can diagnose faults, repair faults, and send notifications to users, even at running time.

Keywords: OSGi · Cloud platform · Diagnostic · Fault

1 Introduction

In recent years, service computing, cloud computing and cloud platform have a quick development that provides effective utilization of service resources. There are some features in service-oriented OSGi framework, such as dynamics, lightweight, modular and great for remote deployment, etc. These features accord with the standards of basic architecture as a cloud computing platform, which means they provide support for the architecture. Therefore, there are many researchers attempt to prove the possibility of using OSGi service frameworks as the basis of cloud computing platform [1].

Building high-reliable cloud platforms within complex cloud computing environments, the point challenge of it is to locate, handle and resolve faults on the cloud environments [2]. The highly dynamic environment has a higher risk of the malfunction occurred, we cannot to avoid all the faults. Therefore, the fault diagnosis and handling methods on the cloud environment become significant. Some suitable fault diagnosis and handling methods for services composition in the cloud computing were emerge in endlessly in the recent years, such as [3–8]. Due to the unique characteristics of OSGi framework, those methods cannot apply to the OSGi-based cloud platform completely. Hence the fault diagnosis issues become a problem that need to be solved urgently.

© Springer International Publishing Switzerland 2016
D.-S. Huang et al. (Eds.): ICIC 2016, Part III, LNAI 9773, pp. 648–657, 2016.
DOI: 10.1007/978-3-319-42297-8_60

This paper proposed a fault diagnostic approach for the OSGi-based cloud platform, this approach can detect service conflicts, handle exceptions and diagnosis faults, even on-demand diagnosis on occasion. By the end of this paper, we assessed the performance of the fault diagnosis by an experiment, which has testified the validity of this approach.

2 Fault Classification

The related works on cloud platform which use OSGi framework as the architectural foundation [1, 9, 10] has researched its viability and proved it scientific. Based on the present study on the cloud platform faults, we divided various faults that might appear into several fault diagnosis models.

2.1 Fault Classification

Resource competition was the major reason for service conflict. In general, the usual solution is to set priority [11]. While on the OSGi-based cloud platform, we designed a general method that through setting and testing priority to avoid such service conflicts. And on the OSGi-based cloud platform, we can design of the OSGi framework layer, use the detection mechanism to realize automatic service conflict detection, and take the appropriate action.

2.2 Fault Diagnosis

The application of the cloud platform has its own fault tolerance mechanism. Different applications, equipment for fault tolerance is not the same. We cannot require all applications or equipment use the same fault tolerant strategy. In this case, the occurrence of fault contains a variety of possibilities, the fault diagnosis is to establish the topological, diagnosis mechanism can use topological complete diagnosis and troubleshooting, but the application of internal fault cannot be found.

2.3 Exception Handling

All the software and hardware on a cloud platform have chances to fail, the normal operations of any software are also likely to lead to exception, and the exception may contain a variety of errors. Exception may occur at any point in running time, the event will affect the normal running of the program, and may cause the entire platform is not available.

3 Diagnosis Framework Implementation

First of all, this paper studied [9, 10] the architecture of OSGi-based cloud platform, we think [9] that the management will become more scientific and effective after using D-OSGi framework for cloud platform The architecture of the OSGi-based cloud platform is shown in Fig. 1.

Fig. 1. The architecture of the OSGi-based Cloud

In the cloud, all kinds of applications, bundles are deployed in the servers and other equipment or virtual environment. The remote OSGi management server manages OSGi bundle by integration management tools. The remote OSGi framework is connected to the local OSGi framework, makes it can get the information at the local framework, which provides the possibility for the fault diagnosis in the cloud. To this end, the architecture of the terminal, VM and Hosting Server should have remote and local Bundles at the same time.

In order to make the OSGi-based cloud platform has the reliability and availability, the capability of fault diagnosis and reporting is indispensable. According to this demand, this paper designed the bottom bundle of the OSGi framework in the cloud platform to make it have the ability to diagnose and request at runtime. To this, we proposed our conflict detection methods and service mediation mechanism to achieve service recovery and exception handling.

Figure 2 is our modified framework. The Designed bundle we called Cloud Diagnostician, which contains native interface, runtime diagnosis, on-demand diagnosis and diagnosis management. To make the JVM can commune with Cloud Diagnostician, we found an appropriate Automated Logging Framework and embed Modified Automated Logging Framework in the JVM and the OSGi framework. In Automated Logging Framework we deploy Service Conflict Detector to monitor the OSGi service activities. When detects the service conflict, Automated Logging Framework informs Diagnosis Management to operate properly for the OSGi Framework. Diagnosis Management receives the message and transmits it to the Server, then the show bundle display to the user.

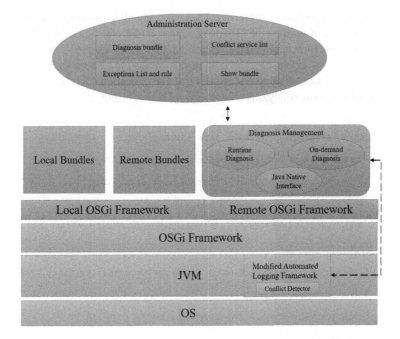

Fig. 2. The architecture of the proposed Cloud Diagnostician

Diagnosis Management we designed contains the Runtime Diagnosis and On-demand Diagnosis two sub-components. The Runtime Diagnosis supports exception detection, and even automatic detection at runtime. On-demand Diagnosis is used for the scenario in which users ask for detect faults, only according to the operational command by a user, and controlled by Diagnosis Management. Java Native Interface is an interface that Diagnosis Management communicates with Service Conflict Detector, which has the advantage of need not pass the OSGi framework communicate directly [11].

Diagnosis Management to manage all fault conditions we have defined in the list, and monitor bundles state. Management contains the recovery mechanism, can stop the fault bundle, send the state information, etc. and it manages two diagnostic components, adjusts the detection scheme. Runtime diagnosis has two fault handlers—the one is Exception Handler and the other one is Service Conflict Handler. Scheme container can remotely update and report to the administrator server.

3.1   Runtime Diagnosis

When detected the exception at running time, the Exception Handler will start to analyze detail information in the first place. In fact, The Exception Handler handles the runtime exception that is stored in files of predefined list.

The exception is stored in the hash table, and The Exception Handler operational principle is to read the predefined list. The predefined list can be updated through OSGi bundles. The error scenarios and related solutions have already predefined in the system. When an exception occurs, the Exception Handler traverse predefined error scenarios. And then it verifies the stack trace information matches or not. If the scene matching, it will perform a scheduled operation. The errors scenarios are divided into five aspects: (1) types (2) results (3) subtypes (4) reason (5) The missing file [11]. In view of these five kinds of scenarios, we designed of exception analysis algorithm. The OSGi Cloud Exception Analyzing Algorithm is shown in Fig. 3.

```
PROCEDURE Exception_Detection (trace_message)

  IF trace_message matches cloudexception_type THEN

    IF trace_message matches cloudexception_result THEN

      IF trace_message matches cloudexception_subtype THEN

      Get the corresponding predefined solution

      END IF

      ELSE IF trace_message matches cloudexception_missing_files THEN

      FOR every missing package

        Download and Install the missing file

      END IF

    ELSE

      Return that the exception handler cannot handle

    END IF

  END PROCEDUCE
```

Fig. 3. The OSGi Cloud Exception Analyzing Algorithm

Conflict Handler is used to solve service conflict problems, such as several services using the same device or resources at the same time, which is called the race condition. In order to avoid the race condition in the cloud platform, it is necessary to ensure that at the same time there is only one service can use the device or resources.

The Service Conflict Detector send information to the Conflict Handler, and the feedback's information including the method name of lead to failure, the bundle ID and the recovery operations. The Conflict Handler responsible for transmit this information to Diagnosis Management. At the same time, the Diagnosis Management delivers information to the Show Bundle of server, then it is displayed to the user, notify the Diagnosis Management performs the related operation. The Detection Algorithm is shown in Fig. 4. In the predefined conflict list, we using a for loop traverse all of the methods, If detected current_method matches the predefined, the state Flag changes, and internal for loop to find the correlative solution. The Conflict Handler gets the bundle ID first, and then takes corresponding actions on the basis of the number in the predefined list of scenarios.

```
PROCEDURE Serviceconflict_Detection (current_method)↵

  FOR every rule indexed x in conflict service list↵

  IF current_method matches EnteringMethod THEN↵

    flag[x]:=true;↵

  ELSE IF current_method matches LeavingMethod THEN↵

    flag[x]:=false;↵

  ELSE↵

    FOR every method in the parentheses↵

    IF pattern matches AND flag[x]==true THEN↵

    Get bundleID from this method object↵

    CALL handle_conflict (bundleID,Solution_NO) ↵

    END IF↵

    END IF↵

END PROCEDURE↵
```

Fig. 4. The service conflict detection algorithm

3.2   On-demand Diagnosis

On-demand Diagnosis is used in the user think that some services have wrong. The coordination among each part is shown in Fig. 5. When users triggered the diagnosis, the testing bundle will be downloaded from the bundle server. The testing bundle has some testing procedures, which is used to test service faults and diagnosis problems.

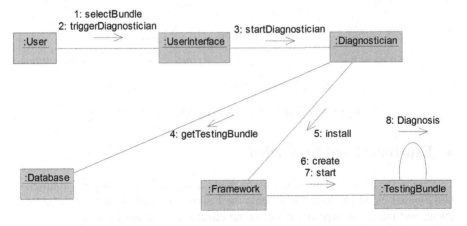

Fig. 5. Communication diagram for executing On-demand Diagnosis

The Cloud Diagnostician will request user to intervene in necessary, through the user interface. This can greatly reduce the time and may provide more accurate solutions in some occasions. The service provider can constantly adjust solutions, based on the problems that would arise, through the administration server. Of course, they can update the testing bundle with the cloud.

3.3   Modified Automated Logging Framework

The Java Native Interface is used to comminute between OSGi Framework and Automated Logging Framework. When some bundle is installed, the Automated Logging Framework is responsible for getting the bundle ID of the method from OSGi framework, meanwhile requests the Diagnosis Management to diagnose or stop by bundle ID, or take other actions.

In the Modified Automated Logging Framework, we find an object can store information of a class. This object includes all of the information about class, for example, the method ID, class name and etc. When we install a new bundle, the information of all the classes in the new bundle will be loaded into JVM. In the OSGi framework, the bundle listener finds that a new bundle is installed, then the Automated Logging Framework gets the new bundle ID.

If the users take some actions through remote control interface or the bundle start, then the conflict detection will start. The detection process is shown in Fig. 6. If this process detected conflicts, the bundle ID will be passed to the Conflict Hander, and then the Diagnosis Management will be triggered.

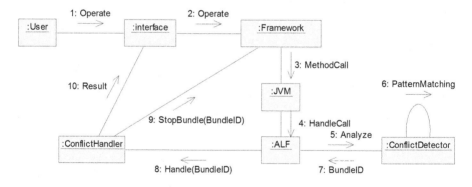

Fig. 6. Communication diagram when detecting service conflict

4   Experiment and Evaluation

To test the feasibility and efficiency of the method we proposed. This paper through the OSGi-based cloud platform implements the healthcare cloud platform and the health client, and makes an experiment to test the effectiveness of the method.

Service conflict is assumed to occur in health client by the remote control interface requires using weight detection equipment, at the same time, when the device is in used by other applications, the service will be detected conflict. The Diagnosis Management will stop the current request, and sends the message to the user, as shown in Fig. 7.

Fig. 7. Diagnosis information message

As the exception occurrence is various. Here we exemplify the functionality of exception handling by real situation. We set a lost service bundle situation at the run time and then the Exception Handler found the exception and processed it automatically, after that the Exception Handler will restart the bundle, the system running normally as well, as shown in Fig. 8.

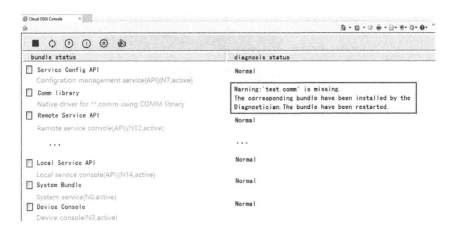

Fig. 8. A snapshot of the system after diagnosing the exception

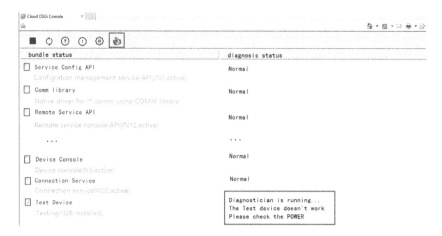

Fig. 9. A snapshot of the after using the On-demand Diagnostician

Direct trigger abnormal situation, if the users found exception, they only need to click the On-demand Diagnostician button, and then the diagnosis will start, the diagnosis information shown on the right side of the page. The users will clearly understand the location of the problem, as shown in Fig. 9.

We have made a comparative assessment of the time of exception handling. The plain OSGi means the OSGi framework is without the Cloud Diagnostician, and the time means that the cloud platform stopped. After we installed the Cloud Diagnostician, an exception be caught and handled. As shown in Fig. 10. The Automated Logging Framework analysis filtering information consumed time, but the time can be accepted because the diagnosis can help to find the reason and solution. The OSGi-based cloud

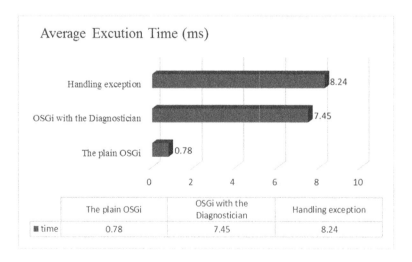

Fig. 10. The evaluation of the Diagnostician

platform is more effective to deal with the fault than without the diagnosis Bundle. If there is no diagnosis method that the user need to spend more time to find the exception and to solve it.

5 Summary

This paper studied the failure problem of the OSGi-based cloud platform and divided the model failure problem, proposed the detection method, troubleshooting, handling exception in the OSGi-based cloud platform. In the OSGi-based cloud platform add the Bundle diagnosis, detect service conflict, solve the fault, alter the user, and the user can also be independent diagnosis and get the operation suggestions. The diagnosis spends little time, the effect is obvious. We verified the method of handle fault is improved in OSGi-based cloud platform.

Acknowledgement. This work was supported by the grants of the National Science & Technology Pillar Program (No. 2015BAF04B02), and supported by the grants of the Major Projects in Shandong Province (No. 2015ZDXX0101G07).

References

1. Kim, D., Lee, C., Helal, S.: Enabling elastic services for OSGi-based cloud platforms. In: 7th International Conference on Ubiquitous and Future Networks, pp. 407–409 (2015)
2. Jia, Z., Xing, X.: Log-based service diagnosis method in cloud. In: International Conference on Mechatronics and Control, pp. 1299–1303 (2014)
3. Zhang, Y., Zheng, Z., Lyu, M.R.: BFTCloud: a byzantine fault tolerance framework for voluntary-resource cloud computing. In: 2011 IEEE 4th International Conference on Cloud Computing, pp. 444–451 (2011)
4. Kopp, O., Leymann, F., Wutke, D.: Fault handling in the web service stack. In: 8th International Conference on Service Oriented Computing, pp. 303–317 (2010)
5. Ardissono, L., Console, L., Goy, A., et al.: Enhancing web services with diagnostic capabilities. In: 3rd European Conference on Web Services, pp. 182–191 (2005)
6. Friedrich, G., Fugini, M., Mussi, E., et al.: Exception handling for repair in service-based processes. IEEE Trans. Softw. Eng. **36**, 198–215 (2010)
7. Hamadi, R., Benatallah, B., Medjahed, B.: Self-adapting recovery nets for policy-driven exception handling in business processes. Distrib. Parallel Databases **23**, 1–44 (2008)
8. Dai, Y., Yang, L., Zhang, B., et al.: Exception diagnosis for composite service based on error propagation degree. In: 2011 IEEE International Conference on Services Computing, pp. 160–167 (2011)
9. Hang, C., Can, C.: Research and application of distributed OSGi for cloud computing. In: 2010 International Conference on Computational Intelligence and Software Engineering. IEEE Wuhan Section (2010)
10. OSGi Alliance, RFP 133 - Cloud Computing Technical Report. http://citeseerx.ist.psu.edu/viewdoc/download?doi=10.1.1.370.9597&rep=rep1&type=pdf
11. Wang, P.C., Lin, C.L., Hou, T.W.: A service-layer diagnostic approach for the OSGi framework. IEEE Trans. Consum. Electron. **55**, 1973–1981 (2009)

Study of Self-adaptive Strategy Based Incentive Mechanism in Structured P2P System

Kun Lu, Shiyu Wang, Ling Xie, and Mingchu Li[(⊠)]

School of Software, Dalian University of Technology,
Dalian 16621, Liaoning, China
minchul@dlut.edu.cn

Abstract. P2P systems provide peers a dynamic and distributed environment to share resource. Only if peers are voluntarily share with each other can system stably exist. However, peers in such systems are selfish and never want to share even with tiny cost. This can lead to serious free-riding problems. Incentive mechanisms based on evolutionary game aim at designing new strategies to distinguish defective peers from cooperative peers and induce them to cooperate more. Nevertheless, the behavior patterns of peers are versatile. Using only one certain strategy to depict peers' behaviors is incomplete. In this paper, we propose an adaptive strategy which integrates advantages of 3 classic strategies. These 3 strategies form a knowledge base. Each time a peer with this strategy can select one adjusting to system status according to the adaptive function. Through experiments, we find that in structured system, this strategy can not only promote cooperation but also the system performance.

Keywords: Adaptive function · Incentive mechanism · Evolutionary game

1 Introduction

Autonomous systems, such as P2P system have been widely used due to their openness and anonymity. The stability of such systems severely relies on the selflessness of peers and cooperation among peers. But peers participate in the system are rational. These selfish peers in P2P system tend to deny service request in order to maximize their own profit. Without any incentive, free-riding problem arises and performance of system declined. So promoting cooperation in P2P system becomes significant.

To encourage peers to voluntarily share in the system, incentive mechanisms are brought in. The essence of incentive is that through providing transaction history to help peers know the service requester and have a better decision on service granting. In this way, peers with poor history can hardly get service from others. To get better service in the following transactions, peers have to voluntarily make contributions, so that cooperation among peers promoted. Incentive mechanism can be divided into different types: micro-payment based [1], reputation based [2–4], genetic algorithm based [5], global trust based [6], market mechanism based [7], social norm based [8], etc.

Normally, peers in P2P system are treated as rational and strategic. They try to maximize their utility by participating the system but never want to cost anything. So evolutionary game [9] is a suitable tool to model the peers and interactions among

© Springer International Publishing Switzerland 2016
D.-S. Huang et al. (Eds.): ICIC 2016, Part III, LNAI 9773, pp. 658–670, 2016.
DOI: 10.1007/978-3-319-42297-8_61

them. It can reveal the pattern of evolution of the system. So many studies propose new incentive strategies to distinguish cooperative peers and promote cooperation among peers. By far, incentive mechanism based on evolutionary game mainly focus on designing reciprocative strategy. Peers with reciprocative strategy make decisions according to transaction history. Although these strategies can promote cooperation in certain scenarios, there're still limits of them. Wang Yuf. proves through theoretical and experimental analysis that without considering cost of getting transaction history, defecting is the only evolutionary stable strategy (ESS) [10].

Nevertheless, peers may behave differently and versatilely when confront same opponent in different environment. So, only one strategy to depict peers behaviors is not suitable. Reference [11] provides us a novel thinking of designing a new strategy. Inspired by Ye, in this paper, we propose a new adaptive strategy. We integrate classic incentive strategies into a knowledge base. Each time a peer interacts with another one, this peer can select one suitable strategy according the surroundings. Through simulation, we find that with adaptive strategy in the system, cooperation is promoted in different network structures as well as system performance.

2   Model P2P System with Evolutionary Game

2.1   Basic Assumptions

Most incentive mechanisms based on evolutionary game consider peers to be bounded-rational. Peers in the system want to maximize their utilities, so to get a better utility they tend to learn from peers with better utility than their own. However, peers may make mistakes for emotional, moral or other unknown reasons. That is why we call it bounded-rational. Besides, peers in the system are strategic. Strategies describe the decision making under various scenarios.

2.2   P2P Evolution Model

P2P evolution model as described in Algorithm 1, each round of transactions consists of three phases: transactions, payoff calculation and strategy update.

(1)   Initialization

Before simulation, initialize network structure and all attributes of peers.

(2)   Processing transactions

In this phase, each peer randomly sends a service request to their neighbors. Peers who receive requests decide whether to grant the service or not according to their own strategy and transaction history. The transaction history is provided by the system. Each peer will perform m requests and peer behaviors are recorded by the system.

(3) Payoff calculation

Calculate each individual's payoff and the average system performance.

(4) Strategy update

We carry a synchronized update method. At the end of each round, each peer decides whether to change their strategy or not by the same function.

Algorithm 1. Evolution Framework

```
1: Initialization;
2: for each round do
3:      for each peer i do
4:          for k=0 to m do
5:              Randomly select neighbor j;
6:              Peer i send a request to j;
7:              Peer j responses the request;
8:              if j is adaptive then
9:                  Peer j implements adaptive func-
tion;
10:             end if
11:         end for
12:     end for
13:     Strategy update;
14: end for
```

2.3   Network Structure

In many previous studies, the P2P system is depicted as a well-mixed population. However, in real network environment, such a well-mixed population doesn't exist. So, besides a well-mixed population, we have chosen 3 common network structure square lattice, scale-free network and small-world network to better simulate P2P system.

Well-mixed Population. In a P2P system, we consider there to be no central server. Each peer can perform as a server as well as a client. So, in this structure each peer can interact with any other peer in the system.

Square Lattice. Square Lattice is a simple network structure. Peers are located on the vertices with von Neumann neighborhood and periodic boundary conditions. This ensures each peer has the same degree ($<k> = 4$).

Small-World Network. Small-world network was proposed by Watts and Strogatz in 1998 based on the structure of human society [12]. The formulation of small-world network starts from a circle. Then each node connects other K nodes with probability p. So the average degree of each node is K + 2. In this paper we consider a small-world network with average degree $<k> = 4$.

Scale-Free Network. In real networks, few nodes have many connections, and most nodes have only a few connections. To describe this attribute, scale-free network [13] is widely used. It's always used to describe the Internet, social networks, etc. In this paper, we use a scale-free network with average degree <k> = 4.

2.4 Strategy Description

Nodes can be typically divided into 3 types:

ALLC: Always Cooperate. Provides services to other peers unconditionally.
ALLD: Always Defect. Denies requests from other peers unconditionally.
Reciprocator: Incentive strategy. Takes transaction history into consideration when responding to service requests.

2.4.1 Incentive Strategy

In transactions, a server decides whether to grant a service or not according to its strategy. Traditionally, two strategies are considered. ALLC (Always Cooperate), always grants a service unconditionally and neglects the requesters' transaction history; ALLD (Always Defect) never grants a service. However, to distinguish defectors from cooperators and implement incentive mechanism, a third strategy, reciprocity, is needed. Reciprocators, agents with a reciprocal strategy, will grant a service according to the requesters' transaction history. However, in a real system a reciprocator may follow different rules (reciprocal policy) to decide on a response. In this paper, we mainly consider 3 different classic reciprocal policies.

Mirror Policy. The mirror reciprocal policy is described as ones serving the requester with the same probability that the requester serves others. The probability is the ratio of the number of a requester serving others (N_{j_serve}) to the number being served by others ($N_{j\_get\_request}$).

$$p_{mirror} = \frac{Nj\_serve}{Nj\_get\_request} \qquad (1)$$

Proportion Policy. Proportion policy can be described as one serving others according to the requesters' history of receiving service [14]. The probability is the ratio of the number of sending service to others (send_service) to the number of getting service from others (get_service).

$$p_{prop} = \min\left(\frac{send\_service}{get\_service}, 1\right) \qquad (2)$$

Upstream Indirect Reciprocity Policy. This policy is proposed by Nowak in 2006 [15]. Peers with this policy may consider the last transaction when it was a service requester. If it got service last time, it will provide service to the other peer this time.

The calculation of serving probability relies on transaction history. But there exists a bootstrapping problem. When no transaction occurs, a reciprocator does not know to grant the service or not. To solve this problem, we adopt a tolerant policy. A reciprocator always serves when confronted with a requester with no transaction history.

2.4.2   Adaptive Strategy

Peers using an adaptive strategy have a knowledge base with three aforementioned incentive strategies. Each time they perform a transaction, peers will select one strategy with a certain probability, and adjust the probabilities through an adaptive function [11]. The adaptive function is described as Algorithm 2. The parameter $Q(S_i)$ is the intermediate value when determining the value of $\pi(S_i)$. $\pi(S_i)$ stands for the probability that strategy S_i is chosen for its following strategy. And $S_i \in S\{\text{Mirror}, \text{Prop}, \text{Indirect}\}$.

Payoff of peer j is denoted by p_j and $\bar{P}_j$ is the average payoff of peer j's neighbors (except itself). The strategy for choosing is influenced by its own payoff (effected by parameter α), neighbors average payoff (effected by parameter γ) and payoff gap (effected by parameter β). This function means the strategy with a lower payoff has a lower probability of being selected next time.

Algorithm 2. Adaptive function

For each peer j with adaptive strategy

$Q(S_i) \leftarrow (1 - \alpha) \cdot Q(S_i) + \alpha \cdot p_j$

$\pi(S_i) \leftarrow \pi(S_i) + \beta \cdot (Q(S_i) - \gamma \cdot \bar{p}_j)$

Normalise$(\pi(S_i))$

2.5   Payoff Calculation

In this paper, we consider the average payoff for each peer. While receiving a service, a peer will have a B bonus and a C cost while providing a service. The total payoff is calculated as (3). As for a well-mixed population, we consider that each peer with same strategy share a common payoff.

$$payoff_i = B * \frac{get\_service}{send\_request} - C * \frac{send\_service}{get\_request} \tag{3}$$

2.6   Learning Model

2.6.1   Learning Best Neighbor (LBN)

As described in Algorithm 3, after each round, each node will consider whether to change its strategy with the probability γa, which is called the adaptive rate. A peer will change to the neighbors' strategy whose payoff is best in its neighborhood (strategy with highest average payoff in well-mixed population) with probability as the sensitivity γs to payoff gap as shown in (4).

$$p_{i \to best} = \gamma s * (payoff_{best} - payoff_i) \tag{4}$$

Algorithm 3. LBN
```
For each peer i

  If randf() < γₐ
```
$$p_{i \to best} \leftarrow \gamma_s * (payoff_{best} - payoff_i)$$
```
    If randf() <  pᵢ→best
        Peer I change strategy of best
    else
        Maintain its strategy
  else
        Maintain its strategy
```

2.6.2 Opportunistic Learning Model (OLM)

As shown in Algorithm 4, after each round, each node will randomly choose a teacher from its neighborhood with probability as the sensitivity γs.

Algorithm 4. OLM
```
For each peer i
    Randomly select a teacher peer j

    If i.strategy != j.strategy and i.payoff<j.payoff

      If randf() < γₛ
          Change strategy to j.strategy
      Else
          Maintain its own strategy
```

---

3 Simulation

3.1 Experiment Settings

We carry out our experiments in a well-mixed population, square lattice, scale-free network and small-world network, respectively. All these structures contain 2500 peers and the same average degree $<k> = 4$ (except well-mixed population). Initially, the fraction of cooperative peers (ALLC and R) and defective peers (ALLD) are equal, and the specific fraction is (ALLC, R, ALLD) = (0.4,0.1,0.5). Without special emphasis, we use the parameter listed in Table 1.

Table 1. Parameters

Symbol	Definition	Value
N	Size of population	2500
$<k>$	Average degree of network	4
B	benefit when getting a service	3
C	cost when providing service	1
α	how individual payoff influences strategy choosing	0.2
β	how payoff gap influences strategy choosing	0.3
γ	how neighbors' average payoff strategy choosing	0.85
steps	steps of evolution	5000
m	transactions per node	100
γ_s	sensitivity	0.04
γa	adaptive rate	0.1

3.2 Influence of Network Structure

The well-mixed population is the most widely used structure when studying incentive mechanisms in a P2P system. However, without considering cost when checking transaction history, ALLD is the only ESS. In a real network, due to the network structure, there're always some cooperators that survive. In this group of experiments, we consider the influence of network structure on cooperation and compare the results of three classic strategies in 4 different network structures.

Under the CBLM model in a well-mixed population (Figs. 1 and 5), the only surviving strategy is ALLD, whatever the incentive strategy is. But the situation changes when we add a network structure to it. For Mirror and prop strategies, with the influence of network structure the proportion of cooperators and reciprocators promoted and defectors are well restricted (see Figs. 2, 3, 4, 6, 7 and 8). However, the indirect strategy is a weak incentive strategy. When there are only defectors and indirect reciprocators, they never provide service to each other. It may seem like all peers are defectors in the system so they can co-exist in the system, but the system performance remains very low. We can get similar results under OLM.

To test the effectiveness of the adaptive strategy, we carry out simulation in three network structures and compare the results with other three strategies. Figures 9, 10, 11, 12, 13, 14, 15 and 16 show that an adaptive strategy can promote cooperation under both the CBLM and OLM learning models. Such cooperation is not only the promotion of the total proportion of cooperators and reciprocators, but also the fraction of cooperation. That means with an adaptive strategy, more and more peers are willing to share unconditionally so that the system performance can be promoted and the system can be continuously stable.

In addition to the evolution of cooperation, we also consider the system average payoff at ESS as a measurement of system performance. As shown in Fig. 17, with an adaptive strategy, the system performance cannot reach the highest possible level among all these strategies. The main reason for that is due to the probability of a reciprocator getting service from others and serving others that may not be equal to 1. When an adaptive strategy is brought in, the more cooperators are in the system, and a

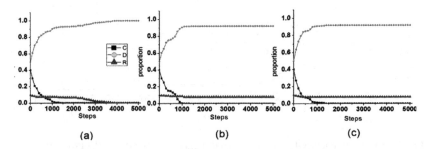

Fig. 1. Evolution under CBLM in well-mixed population; (a) Mirror (b) Prop (c) Indirect.

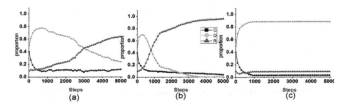

Fig. 2. Evolution under CBLM on square lattice; (a) Mirror (b) Prop (c) Indirect.

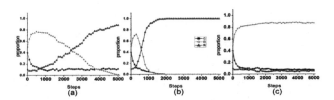

Fig. 3. Evolution under CBLM in scale-free network; (a) Mirror (b) Prop (c) Indirect.

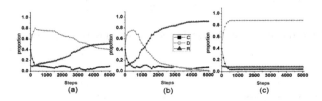

Fig. 4. Evolution under CBLM in small world network; (a) Mirror (b) Prop (c) Indirect.

relatively large cost is incurred. Therefore the system with an adaptive strategy doesn't yield the highest payoff. However, the performance of an adaptive strategy can reach high levels in different networks. This attribute helps an autonomous network maintain better stability. So, the adaptive strategy is still meaningful to the autonomous system design. Also, we can see that the system performance when using an indirect strategy is fairly low. When there are only peers with indirect strategy in the system, once a peer

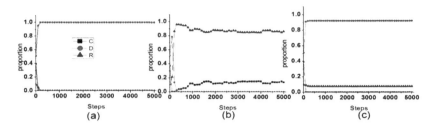

Fig. 5. Evolution under OLM in well-mixed population; (a) Mirror (b) Prop (c) Indirect.

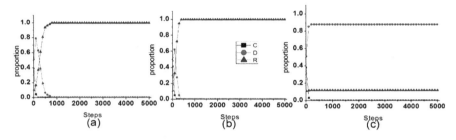

Fig. 6. Evolution under OLM on a square lattice; (a) Mirror (b) Prop (c) Indirect.

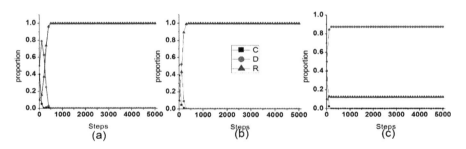

Fig. 7. Evolution under OLM in scale-free network; (a) Mirror (b) Prop (c) Indirect.

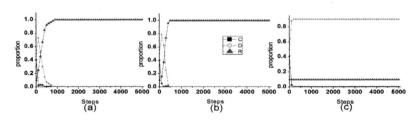

Fig. 8. Evolution under OLM in small world network; (a) Mirror (b) Prop (c) Indirect.

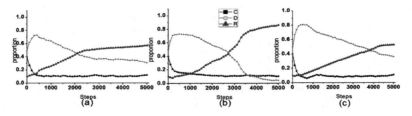

Fig. 9. Evolution under CBLM with mirror strategy; (a) Lattice (b) Scale-free (c) Small World.

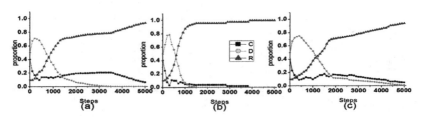

Fig. 10. Evolution under CBLM with prop strategy; (a) Lattice (b) Scale-free (c)Small World.

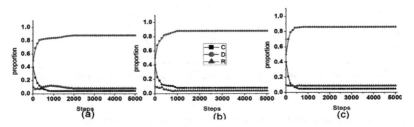

Fig. 11. Evolution under CBLM with indirect strategy; (a) Lattice (b) Scale-free (c) Small World.

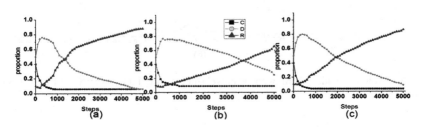

Fig. 12. Evolution under CBLM with adaptive strategy; (a) Lattice (b) Scale-free (c) Small World.

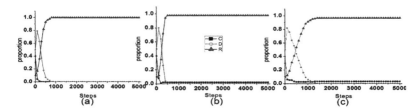

Fig. 13. Evolution under OLM with mirror strategy; (a) Lattice (b) Scale-free (c) Small World.

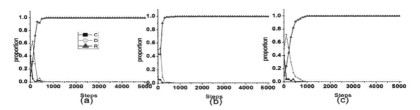

Fig. 14. Evolution under OLM with prop strategy; (a) Lattice (b) Scale-free (c) Small World.

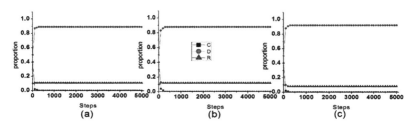

Fig. 15. Evolution under OLM with indirect strategy; (a) Lattice (b) Scale-free (c) Small World.

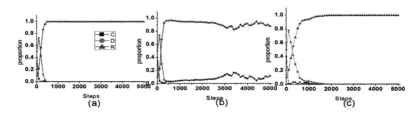

Fig. 16. Evolution under OLM with adaptive strategy; (a) Lattice (b) Scale-free (c) Small World.

defects, then the defecting peer won't cooperate next time. So the cooperation rate is quite low in such system, and the system performance is low. This also proves that indirect strategy is a weak incentive strategy.

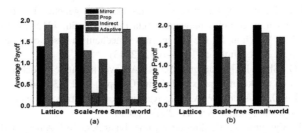

Fig. 17. Average payoff at ESS; (a) CBLM (b) OLM (Color figure online)

4 Conclusions

In this paper we mainly bring in an adaptive strategy into the P2P system. Through simulation under evolutionary game framework, we have reached conclusions as follows:

(1) Cooperation promoted through network structure

We carried out experiments in well-mixed population, square lattice, scale-free network and small-world networks. The experimental results show that with the influence of network structure, the proportion of cooperators and reciprocators rises. It shows that network structure can promote cooperation.

(2) Cooperation promoted through adaptive strategy

To test the effectiveness of adaptive strategy, we carried out experiments in three different networks and compared the results with other three incentive strategies. The results show that in a system with an adaptive strategy, cooperation emerges. Besides, no matter what the network structure is, a system with an adaptive strategy has better robustness and higher performance. So, in summary, the introduction of adaptive strategy has significance as an incentive mechanism to promote not only cooperation but also improve system performance.

Acknowledgments. This paper is supported by the National Science Foundation of China under grant No. 61272173, 61403059.

References

1. Saroiu, S., Gummadi, P.K., Gribble, S.D.: Measurement study of peer-to-peer file sharing systems, San Jose, CA (2002)
2. Kamvar, S.D., Schlosser, M.T., Garcia-Molina, H.: The eigentrust algorithm for reputation management in P2P networks, New York, NY (2003)
3. Xiong, L., Liu, L.: Peertrust: supporting reputation-based trust for peer-to-peer electronic communities. IEEE Trans. Knowl. Data Eng. **16**, 843–857 (2004)
4. Zhou, R., Hwang, K.: Powertrust: a robust and scalable reputation system for trusted peer-to-peer computing. IEEE Trans. Parall Distr. **18**, 460–473 (2007)

5. Hai, L., Chen-Xu, W.: Evolution of strategies based on genetic algorithm in the iterated prisoners dilemma on complex networks. Acta Phys. Sin. **56**, 4313–4318 (2007)

6. Ouyang, J., Lin, Y., Zhou, S., Li, W.: Incentive mechanism based on global trust values in P2P networks. J. Syst. Simul. **5**, 1046–1052 (2013)

7. Dong, G.U.O., Shan, L.U., Bao-qun, Y.I.N.: A novel incentive model for P2P file sharing system based on market mechanism. J. Chin. Comput. Syst. **33**, 1–6 (2012)

8. Xin-kao, L.I.A.O., Li-sheng, W.A.N.G.: Research on incentive mechanism based on social norms and boycott. Comput. Sci. **41**, 28–30 (2014)

9. Hofbauer, J., Sigmund, K.: Evolutionary game dynamics. Bull. Amer. Math. Soc. **40**, 479–519 (2003)

10. Wang, Y., Nakao, A., Vasilakos, A.V., Ma, J.: P2P soft security: on evolutionary dynamics of P2P incentive mechanism. Comput. Commun. **34**, 241–249 (2011)

11. Ye, D., Zhang, M.: A self-adaptive strategy for evolution of cooperation in distributed networks. IEEE Trans. Comput. **64**, 899–911 (2015)

12. Watts, D.J., Strogatz, S.H.: Collective dynamics of small-world networks. Nature **393**, 440–442 (1998)

13. Barabsi, A.-L., Albert, R., Jeong, H.: Scale-free characteristics of random networks: the topology of the world-wide web. Phys. A **281**, 6 (2000)

14. Feldman, M., Lai, K., Stoica, I., Chuang, J.: Robust incentive techniques for peer-to-peer networks, New York, NY (2004)

15. Nowak, M.A., Sigmund, K.: Evolution of indirect reciprocity. Nature **437**, 1291–1298 (2005)

Hybrid Short-term Load Forecasting Using Principal Component Analysis and MEA-Elman Network

Guangqing Bao[1], Qilin Lin[1(✉)], Dunwei Gong[2], and Huixing Shao[3]

[1] College of Electrical Engineering and Information Engineering,
Lanzhou University of Technology, Lanzhou 730050, China
{Guangqing.Bao,Qilin.Lin,15293187780}@163.com
[2] School of Information and Electrical Engineering,
China University of Mining and Technology, Xuzhou 221008, China
Dunwei.Gong@163.com
[3] State Grid Huangshan Power Supply Company, Huangshan 245000, China
Huixing.Shao@163.com

Abstract. Meteorological factors, the main causes that impact the power load, have become a research focus on load forecasting in recent years. In order to represent the influence of weather factors on the power load comprehensively and succinctly, this paper uses PCA to reduce the dimension of multi-weather factors and get comprehensive variables. Besides, in view of a relatively low dynamic performance of BP network, a model for short-term load forecasting based on Elman network is presented. When adopting the BP algorithm, Elman network has such problems as being apt to fall into local optima, many iterations and low efficiency. To overcome these drawbacks, this paper improves the active function, optimizes its weights and thresholds using MEA, and formulates a MEA-Elman model to forecast the power load. An example of load forecasting is provided, and the results indicate that the proposed method can improve the accuracy and the efficiency.

Keywords: Meteorological factor · PCA · Mind Evolutionary Algorithm · Optimization · The Elman network · Short-term load forecasting

1 Introduction

Electric load forecasting is the precondition and important information for a grid to dispatch, control, and make an operation plan or market orientation reasonably. According to the time length, electrical load forecasting can be divided into the following four types: long term, medium term, short term and very short term [1]. This paper regards the daily load forecasting as the research object, which belongs to the category of short-term load forecasting. Short-term load forecasting is the prediction of electrical load from 1 h to 1 week. It can help electric companies to make decisions on energy production and purchasing, infrastructure development and load switching [2]. Hobbs's report [3] has showed that electric power companies can gain 760 ten thousands of USD if the mean absolute percentage error of load forecasting is reduced by 1.5 %.

© Springer International Publishing Switzerland 2016
D.-S. Huang et al. (Eds.): ICIC 2016, Part III, LNAI 9773, pp. 671–683, 2016.
DOI: 10.1007/978-3-319-42297-8_62

So it is important to forecast load correctly and efficiently for electricity providers in a competitive market.

Up to present, scholars have proposed various mathematical methods to improve the accuracy and the efficiency in prediction, although several techniques aim to one-dimensional time series, e.g., regression analysis, time series analysis. These methods are mainly based on historical load data, and cannot meet the requirements from electric companies, since some important information is lack and the load forecasting precision is low. Many studies have demonstrated that load forecasting is not only related to historical load, but also is affected by meteorological factors. Therefore, it is necessary to take various meteorological factors into account for short-time load forecasting. However, it may have a great computation complexity, long time on prediction, and a low forecasting accuracy if meteorological data are directly employed. The reason is that relationship among various meteorological factors exists, resulting in redundant information. Therefore, it is necessary to preprocess meteorological factors. Principal component analysis is a powerful tool to tackle this problem by using a small number of variables [4]. Principal component analysis is employed in this paper to tackle six meteorological factors, with the purpose of getting comprehensive indexes. Then, in view of a relatively low dynamic performance of BP network, a model for short-term load forecasting based on the Elman network is proposed. When adopting the BP algorithm, the Elman network has such problems as being apt to fall into local optima, many iterations and low efficiency. To overcome these drawbacks, this paper improves the active function of the Elman network, optimizes its weights and thresholds using the Mind Evolutionary Algorithm, and formulates a MEA-Elman network model to forecast the power load. Compared with the genetic algorithm, MEA can remember information during the evolution, has an inherent parallelism in structure and improves the efficiency of the evolution by using similartaxis and dissimilation. A good convergence of MEA is proved by the martingale convergence theorem in [5].

This paper integrates meteorological factors by using principal component analysis and proposes a MEA-Elman model for short-term load forecasting. Then a forecasting and analysis for an actual power load system in a region of Eastern Europe is conducted, and the simulation results demonstrate that the proposed method can improve the accuracy and the efficiency in prediction.

2   Meteorological Factors Analysis

One of the most important factors is meteorological change for short-term load forecasting. There are various variables, such as temperature, humidity, wind speed, cloudiness, precipitation and light intensity. These factors have different influences on the load of a power system, and temperature and relative humidity have the maximal influence. For example, when changing air temperature, the demands from consumers will vary. The reduction in temperature in winter causes the use of heaters. On the contrary, the increase in temperature in summer results in the use of air conditioners. In order to improve the prediction accuracy, the forecasting model should not consider only one major factor, and ignore other factors. However, considering too many factors will increase the dimensions and raise the computational complexity. To solve these problems, this paper employs principal component analysis to tackle meteorological factors. Principal component analysis is a dimensionality reduction algorithm, which replaces more original variables with a small number of new variables to obtain comprehensive information.

Assuming that there is n samples, with each having p variables. These samples can form the following $n \times p$ matrix

$$X = \begin{bmatrix} x_{11} & x_{12} & \cdots & x_{1p} \\ x_{21} & x_{22} & \cdots & x_{2p} \\ \cdots & \cdots & \cdots & \cdots \\ x_{n1} & x_{n2} & \cdots & x_{np} \end{bmatrix} \tag{1}$$

The detail process of principal component analysis is provided as follows:

Step1: normalize raw data in the range of [0, 1].

Step2: compute the following correlation coefficient matrix:

$$R = \begin{bmatrix} r_{11} & r_{12} & \cdots & r_{1p} \\ r_{21} & r_{22} & \cdots & r_{2p} \\ \cdots & \cdots & \cdots & \cdots \\ r_{n1} & r_{n2} & \cdots & r_{np} \end{bmatrix} \tag{2}$$

where $r_{ij}(i, j = 1, 2 \ldots p)$ denotes the correlation coefficient between X_i and X_j, with its expression being as follows:

$$r_{ij} = \frac{\sum\limits_{k=1}^{n} (x_{ki} - \overline{x_i})(x_{kj} - \overline{x_j})}{\sqrt{\sum\limits_{k=1}^{n} (x_{ki} - \overline{x_i})^2 \sum\limits_{k=1}^{n} (x_{kj} - \overline{x_j})^2}} \tag{3}$$

Step3: calculate the characteristic value of the correlation coefficient matrix. According to the formula, $|\lambda I - R| = 0$, compute all the eigenvalues and rank them in the descending order.

Step4: calculate the principal component contribution rate and the cumulative contribution rate according to the following formula:

$$\beta_i = \frac{\lambda_i}{\sum\limits_{k=1}^{p} \lambda_k} (i = 1, 2 \ldots p) \tag{4}$$

$$T = \frac{\sum\limits_{k=1}^{i} \lambda_k}{\sum\limits_{k=1}^{p} \lambda_k} (i = 1, 2 \ldots p) \tag{5}$$

where β_i is the principal component contribution rate, T is the cumulative contribution rate. In general, take eigenvalues $\lambda_1, \lambda_2 \ldots \lambda_m$, to analysis when their cumulative contribution rate reaches to 85 % – 95 %.

Step5: calculate the eigenvectors corresponding to the eigenvalues, $\lambda_1, \lambda_2 \ldots \lambda_m$, and the m principal components can be expressed as:

$$
\begin{cases}
z_1 = a_{11}x_1 + a_{12}x_2 + \ldots + a_{1p}x_p \\
z_2 = a_{21}x_1 + a_{22}x_2 + \ldots + a_{2p}x_p \\
\ldots \ldots \\
z_m = a_{m1}x_1 + a_{m2}x_2 + \ldots + a_{mp}x_p
\end{cases}
\tag{6}
$$

where m is the number of principal components required, and satisfies that $m \le p$. In this way, p meteorological factors can be reduced to m comprehensive indexes.

In the following, principal components analysis is conducted by six factors, they are temperature X_1, precipitation X_2, wind speed X_3, cloudiness X_4, humidity X_5 and light intensity X_6. The process is shown in Figs. 1 to 2.

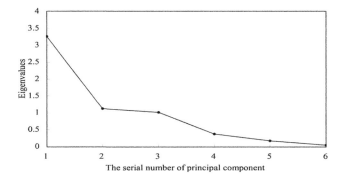

Fig. 1. Eigenvalues of principal component analysis

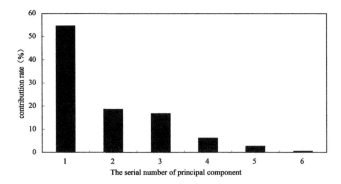

Fig. 2. Contribution rates of principal component analysis

In Fig. 1, six eigenvalues are shown in the descending order. In Fig. 2, the first 3 principal components have the main contribution, where the first, the second, and the third have 54.48 %, 18.82 %, and 16.83 % contribution rates, respectively. In total, the first 3 principal components reach to 90.13 % cumulative contribution rate. Since the cumulative contribution rate of principal components is required to reach over 85 %, the first 3 eigenvalues are selected, and their corresponding eigenvectors are calculated as follows:

$$\begin{cases} z_1 = -0.103x_1 - 0.492x_2 - 0.477x_3 - 0.484x_4 - 0.127x_5 - 0.520x_6 \\ z_1 = 0.844x_1 + 0.278x_2 - 0.079x_3 - 0.126x_4 - 0.411x_5 - 0.140x_6 \\ z_3 = 0.358x_1 + 0.031x_2 + 0.030x_3 - 0.359x_4 + 0.861x_5 - 0.003x_6 \end{cases} \quad (7)$$

From formula (7), 3 new comprehensive meteorological indexes are generated, and employed to replace six original factors, so as to reduce the number of input variables.

3 THE MEA-Elman Model

3.1 The Mind Evolution Algorithm

The mind evolution algorithm was proposed by Sun Chengyi et al., to solve the problem of premature convergence and low convergence rate in evolutionary algorithm [6, 7]. The basic diagram of MEA is shown in Fig. 3.

Here are some important concepts about MEA:

Groups and sub-groups: the sum of all individuals is called a group, and a group can be divided into a number of sub-groups, while sub-groups can be divided into superior sub-groups and temporary sub-groups.

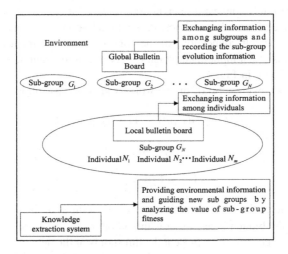

Fig. 3. The basic diagram of MEA

Bulletin board: the global bulletin board is used for exchanging information among sub-groups, and the local bulletin board is used for individuals.

Similartaxis and dissimilation: similartaxis is carried out within the sub-groups to obtain superior individuals which have a high fitness value through the competition. Dissimilation is conducted in the whole solution space, by comparing the fitness value of temporary sub-groups with superior sub-groups to obtain the higher one and abandon the lower one.

The basic idea of the mind evolution algorithm is as follows:

Step1: set the parameters, POP_{size}-the size of groups, $Best_{size}$-the number of superior sub-groups, Tem_{size}-the number of temporary sub-groups, S_G-the size of sub-groups, and S_G is shown as follows:

$$S_G = POP_{size} \, / \, (Best_{size} + Tem_{size}) \tag{8}$$

Step2: generate initial groups randomly, then compute the fitness value of each sub-group and get the superior sub-groups and temporary sub-groups according to their values.

Step3: conduct similartaxis and dissimilation operations on the sub-groups and the whole solution space, respectively.

Step4: decode the optimal individual, and export the individual if it satisfies the optimal conditions; otherwise, go to the similartaxis and dissimilation operations.

3.2   Elman Network

Elman network is a kind of dynamic recurrent network, which consists of four layers: input layer, hidden layer, context layer and output layer [8]. There are adjustable weights and thresholds connecting neighboring layers. Generally, it is considered as a special kind of feed-forward neural networks with additional memory neurons and the local feedback. The self-connections of the context nodes in Elman network make it sensitive to the historical input data, which is very useful in dynamic system modeling. Its structure is shown in Fig. 4.

The state space expression of the network is provided as follows:

$$y_h(k) = G(w_1 u(k-1) + w_2 y_c(k) + b_1) \tag{9}$$

$$y_c(k) = y_h(k-1) \tag{10}$$

$$y_0(k) = F(w_3 y_h(k) + b_2) \tag{11}$$

where $u_{(k-1)}$ denotes the input variables, $y_c(k)$ is the output of the context layer, $y_h(k)$ is the output of the hidden layer, $y_0(k)$ is the output of the network; w_1, w_2, w_3 denote the connection weights; b_1, b_2 represent the thresholds of the network; $F(k)$ is a purelin expression which denotes the transformation function of the output layer, while $G(k)$

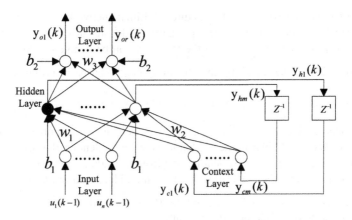

Fig. 4. The structure of Elman network

represents the active function of the hidden layer, and it is usually a sigmoid function, shown as follows:

$$G(k) = 1/(1+e^{-k}) \tag{12}$$

3.3 The Improvement of the Active Function

Because the sigmoid function has a tendency of making the convergence rate of a network slow [9], it needs to be improved. Adding constant-a, gain parameter-b, independent variable factor-c, adjustable parameter-d to the above formula, one can obtain an improved active function shown as follows:

$$G(k) = a + b \big/ (1 + e^{-c(k+d)}) \tag{13}$$

and the derivative of the formula (13) is provided as follows:

$$G'(k) = -c/b(G - (2a+b)/2)^2 + cb/4 \tag{14}$$

The learning speed of the network is related to $G'(\mathrm{k})$. When the value of $G(\mathrm{k})$ is close to $(2a+b)/2$, the derivative value becomes large, and the convergence rate becomes fast. Therefore, the convergence rate can be improved by adjusting the parameters. In this study, the values of the parameters are chosen as follows: $a = -1, b = 2, c = 0.5, d = 2$.

3.4 The Steps of Optimization

In this paper, Mind Evolution Algorithm is used to optimize the weights and thresholds of the network, and the MEA-Elman network model is established. The steps are shown as follows:

Step1: determine the topological structure of Elman network, and assume that S_1 denotes the number of neurons in the input layer, S_2 denotes that in the hidden layer, S_3 denotes that in the output layer, then the topology can be denoted as $S_1 - S_2 - S_3$.

Step2: according to the topology, the encoding length can be determined as follows:

$$s = s_1 \times s_2 + s_2 \times s_2 + s_2 \times s_3 + s_2 + s_3 \tag{15}$$

where S denotes the encoding length, $S_1 \times S_2$ denotes the number of connecting weights from the input layer to the hidden layer, $S_2 \times S_2$ denotes that from the context layer to the hidden layer, $S_2 \times S_3$ denotes that from the hidden layer to the output layer, S_2 and S_3 denotes the number of thresholds in the hidden and output layer.

Step3: the parameters of MEA are set to $POP_{size} = 100$, $Best_{size} = 1$, $Tem_{size} = 4$, $S_G = 20$ on the basis of the encoding length.

Step4: the fitness function is designed as follows:

$$Fitness = 1 \left/ \frac{(T_{sim} - T_{rel})^2}{n} \right. \tag{16}$$

where T_{sim} denotes the forecasting value, T_{rel} denotes the real value of load, n denotes the number of elements of the test set. From the expression, the higher the fitness value, the better the result.

Step5: generate an initial population, a superior population, a temporary population, and then conduct similartaxis and dissimilation.

Step6: decode the best individual according to the fitness value.

Step7: apply the optimal individual to the initial weights and the thresholds of Elman network, then train the network.

4 A Case Study

4.1 The Pretreatment of Historical Load Data

The actual data of power system load lying in East Europe is used as the samples to set the prediction model, test results and verify the accuracy of the model. In order to improve the precision of load forecasting, effective cleaning needs to be done to the historical loads [10]. In this procedure, the saturation phenomenon needs to be avoided and the dimensions of different input samples need to be unified. As a result, formula (17) can be used to complete the normalized operation for data in the input layer, and formula (18) can be used to complete the reverse normalized operation for data in the output layer.

$$y = \frac{(y_{max} - y_{min}) \times (x_{max} - x_{min})}{(x_{max} - x_{min})} + y_{min} \tag{17}$$

$$x = \frac{(x_{max} - x_{min}) \times (y_{max} - y_{min})}{(y_{max} - y_{min})} + x_{min} \tag{18}$$

where y_{min} and y_{max} denote the normalized minimum and maximum, x_{min} and x_{max} are the minimum and maximum of sample data.

4.2 The Selection of Input Variables

The meteorological factors are considered earlier in this article, but the date attribute cannot be ignored during the load forecasting. Generally, load on Sunday or Saturday are lower than those on weekday (from Monday to Friday), and the holiday is close to Saturday. The quantification for each date type is shown in Table 1.

Table 1. The quantification for each data type

Date type	Value before mapping	Value after mapping
Monday	1	0.1
Tuesday	2	0.2
Wednesday	3	0.3
Thursday	4	0.4
Friday	5	0.5
Saturday	6	0.6
Sunday	7	0.7
holiday	6	0.6

Calculate the similarity coefficient of date factors by the following formula:

$$c_i = 1 - |f(x_i) - f(x_0)| \tag{19}$$

where c_i denotes the correlation coefficient, x_i denotes the ith date type, x_0 denotes the type of the predict day, $f(x_i)$ and $f(x_0)$ are x_i and x_0 mapping value.

Above all, this paper takes the meteorological factors, date attribute, and historical load as the inputs. The input variables are shown in Fig. 5.

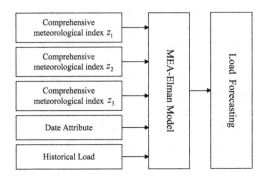

Fig. 5. The inputs variables of forecasting model

4.3   The Results and Discussion

This paper proposes the principal component analysis and the MEA-Elman model for short-term load forecasting using actual data. Choose a day in the sample data as the test day, and the historical data of the whole month before the test day are regarded as the training samples. Then three comprehensive meteorological index, data attribute and historical load are taken into account. When the forecast is complete, the model and the forecast accuracy should be measured using some statistical functions. There are a variety of statistic methods for calculating the forecasting error. In our study,

Table 2. Forecasting results of different forecasting models

Time	Actual (MW)	MEA-Elman (considering meteorological and date)		Elman(considering meteorological and date)		MEA-Elman(without considering related factors)	
		Forecast value (MW)	APE	Forecast value (MW)	APE	Forecast value (MW)	APE
1:00	720	730.354	1.44 %	709.654	1.44 %	701.356	2.59 %
2:00	698	705.482	1.07 %	678.653	2.77 %	673.452	3.52 %
3:00	648	640.347	1.18 %	633.578	2.23 %	629.587	2.84 %
4:00	656	646.243	1.49 %	644.842	1.70 %	624.843	4.75 %
5:00	651	641.342	1.48 %	643.543	1.15 %	637.861	2.02 %
6:00	604	613.473	1.57 %	589.834	2.35 %	615.584	1.92 %
7:00	578	569.342	1.50 %	601.382	4.05 %	601.246	4.02 %
8:30	598	591.42	1.10 %	603.643	0.94 %	588.542	1.58 %
9:00	644	630.572	2.09 %	622.942	3.27 %	630.258	2.13 %
10:00	666	658.427	1.14 %	657.348	1.30 %	674.359	1.26 %
11:00	700	710.642	1.52 %	672.834	3.88 %	689.452	1.51 %
12:00	700	695.326	0.67 %	686.284	1.96 %	689.254	1.54 %
13:00	714	710.567	0.48 %	698.617	2.15 %	704.387	1.35 %
14:00	702	711.537	1.36 %	690.427	1.65 %	694.842	1.02 %
15:00	691	686.352	0.67 %	683.542	1.08 %	682.632	1.21 %
16:00	677	670.834	0.91 %	666.364	1.57 %	654.254	3.36 %
17:00	688	683.761	0.62 %	680.843	1.04 %	678.542	1.37 %
18:00	713	720.831	1.10 %	702.567	1.46 %	680.543	4.55 %
19:00	735	732.452	0.35 %	730.542	0.61 %	742.658	1.04 %
20:00	743	738.452	0.61 %	735.943	0.95 %	732.418	1.42 %
21:00	717	705.643	1.58 %	721.457	0.62 %	700.546	2.29 %
22:00	698	687.513	1.50 %	675.761	3.19 %	654.238	6.27 %
23:00	691	694.243	0.47 %	677.932	1.89 %	670.846	2.92 %
24:00	704	706.428	0.34 %	718.249	2.02 %	721.861	2.54 %
MAPE		1.09 %		1.89 %		2.46 %	
RMSE		7.946		14.069		18.971	

absolute percentage error (APE), mean absolute percentage error (MAPE), root mean square error (RMSE), which are widely used in the literature [11, 12], are selected to validate the proposed models as follows:

$$APE = \left| \frac{RE - F}{RE} \right| * 100\% \tag{20}$$

$$MAPE = \frac{1}{N} \left(\sum_{i=1}^{N} APE_i \right) \tag{21}$$

$$RMSE = \sqrt{\frac{\sum_{i=1}^{N} (F - RE)^2}{N}} \tag{22}$$

where RE refers the actual value, F refers the forecasting value, N is the number of forecast samples.

The forecast values for all 24 h of a day are considered for a general assessment. Some forecasting results are presented in Table 2 and Figs. 6 and 7.

As can be seen from Table 2, under the condition of considering the meteorological factors and date attributes, MAPE and RMSE of the MEA-Elman model are 1.09 % and 7.946, respectively, while those of the Elman model are 1.89 % and 14.069, which indicates the optimization algorithm is effective; without considering related factors, MAPE and RMSE of the MEA-Elman model are 2.46 % and 18.971, and increase by 1.37 % and 11.025, showing that taking meteorological factors and date attributes into account is necessary. In a word, the MEA-Elman model has a higher forecasting accuracy than the Elman model under the same conditions, and can further improve the accuracy of the prediction and the stability of the predicted value when considering the meteorological factors and date attributes. From the curves in Figs. 6 and 7, the advantages are strongly illustrated.

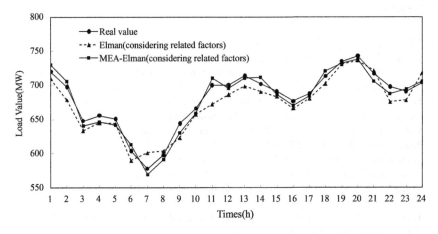

Fig. 6. The prediction curves drawn by different forecasting model

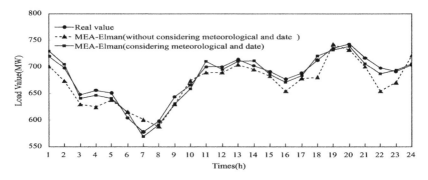

Fig. 7. The meteorological and date factors effect on the forecast accuracy

5 Conclusion

This paper proposes an electric load forecasting model based on the principal component analysis and the MEA-Elman network model. Comprehensive meteorological indexes are obtained by principal component analysis. In addition, the active function is improved and the weights and thresholds of Elman network are optimized using Mind Evolution Algorithm. Then the actual data of power system load lying in East Europe is used as the samples to test results and verify the accuracy of the model. The simulation results indicate that the proposed method can improve the accuracy and the efficiency in prediction and has an important practical significance for electric power dispatch and market orientation.

Acknowledgments. The work presented in this paper is result of the research project National Natural Science Foundation of China(51267011), partly financed by Ministry of Human Resources and Social Security of the People's Republic of China(1202ZBB136).

References

1. He, Y.Y., Xu, Q.F., Yang, S.L., Yu, B.G.: A power load probability density forecasting method based on RBF Neural Network. Proc. CSEE **33**, 93–98 (2013)
2. Sousa, J.C., Jorge, H.M., Neves, L.P.: Short-term load forecasting based on support vector regression and load profiling. Int. J. Energy Res. **38**, 350–362 (2013)
3. Hobbs, B.F., Helman, U., Jitprapaikulsarn, S., et al.: Artificial neural networks for short-term energy forecasting. Accuracy Econ. Value Neurocomputing **23**, 71–84 (1998)
4. Choi, J.Y., Lee, J.M.: Dynamic PCA algorithm for detecting types of electric poles. Trans. Korean Inst. Electr. Eng. **59**, 651–656 (2010)
5. Guo, H.G.: Migration probabilities analysis and almost sure convergence proof of mind evolutionary algorithm. Control Decis. **39**, 2201–2206 (2014)
6. Wang, F., Xie, K.M., Liu, J.X.: Swarm intelligence based MEA design. Control Decis. **25**, 145–148 (2010)
7. Li, G., Li, W.H.: Facial feature tracking based on mind evolutionary algorithm. J. Jilin Univ. (Eng. Technol. Ed.) **45**, 606–612 (2015)

8. Pham, D.T., Liu, X.D.: Training of Elman networks and dynamic system modelling. Int. J. Syst. Sci. **27**, 221–226 (1996)
9. Liu, R., Fang, G.F.: Short-term load forecasting with comprehensive weather factors based on improved Elman neural network. Power Syst. Prot. Control **40**, 113–117 (2012)
10. Kebriaei, H., Araabi, B.N., Rahimi, K.: A Short-term load forecasting with a new nonsymmetric penalty function. IEEE Trans. Power Syst. **26**, 1817–1825 (2011)
11. Hasan, H.Ç., Mehmet, Ç.: Short-term load forecasting using fuzzy logic and ANFIS. Neural Comput. Appl. **26**, 1355–1367 (2015)
12. Ding, N., Benoit, C., Foggia, G.: Neural network-based model design for short-term load forecast in distribution systems. IEEE Trans. Power Syst. **31**, 72–81 (2015)

RNG k-ε Pump Turbine Working Condition of Numerical Simulation and Optimization of the Model

Guoying Yang[✉]

LanZhou Petrochemical College of Vocational Technology,
Lanzhou 730060, China
79546245@qq.com

Abstract. The runner design uses the design principle of low specific speed Francis turbine's blade, then hydraulic design, 3D modeling and ICEM meshing were carried on a pump-turbine with given parameters. Taking into account the factors of high strain rate and curvature over the stream surface, RNG k- model was used to solve N-S equations. The finite volume method was used to discrete and SIMPLEC algorithm was used to solve pressure - speed coupling equations. By simulating the model with the software of Fluent and correcting repeatedly, the efficiency of pump-turbine under design condition was predicted. Besides, the changes of the pressure field and velocity field in volute, runner and draft tube were analyzed during different operating conditions. This design method has provided a new reference for pump-turbine runner's hydraulic design and made up for the lack of empirical data by using CFD technology to give a direct-viewing reflection of hydraulic performances simultaneously, Numerical experiment results show that the optimal efficiency of turbine's operating condition can reach 91 % and the optimal efficiency of pump's can reach 82 %. There will be a further improvement if modify the model repeatedly.

Keywords: Pump turbine · Numerical simulation · Structure optimization

1 Introduction

Pumped storage units in the power grid peak shaving unique ways of working, as an effective energy storage device, pumped storage units with electrical equipment start-stop quickly and adjust the advantages of flexible, can be effective to cope with the change of load [1], widely used in foreign countries for many years. China is also beginning to vigorously construction of pumped storage power station, according to the low cycle efficiency of pumped storage unit, in this paper, using the theory of dual hydraulic design, Pro/E modeling, ICEM tetrahedron grid, three-dimensional numerical simulation of full port and Fluent, a preliminary analysis of the internal flow field and flow mechanism of reversible turbine, on the basis of the model optimization improvement, improve the efficiency of reversible turbine.

© Springer International Publishing Switzerland 2016
D.-S. Huang et al. (Eds.): ICIC 2016, Part III, LNAI 9773, pp. 684–693, 2016.
DOI: 10.1007/978-3-319-42297-8_63

2   Hydraulic Design and Three Dimensional Modelling

Reversible machine wheel is usually based on the theory of pump design, working condition of hydraulic turbine, often called top-down design theory. In this paper, considering the actual situation, using the theory of water turbine dual design wheel, water pump working condition checking, and then using the theory of three dimensional flow in numerical calculation, and then according to the fixed hydraulic design calculation, the design process is called reverse design theory. A reversible design parameters of the turbine is: the head H = 204 m, rotational speed n = 500 r/min, flow qv = 2.35 m^3 / s, working condition of hydraulic turbine efficiency.

2.1   Nominal Diameter

Nominal diameter:

$$D_1 = \sqrt{\frac{P}{9.81 q_{V,11} H^{3/2} \eta}} \tag{1}$$

Type: P for reversible machine shaft power, kW; qv, 11units of flow, m^3 / s, H is working head, m; According to the design parameters calculated through the literature D1 = 450 mm.

2.2   Runner Exit Diameter

Runner exit diameter:

$$D_2 = K \sqrt[3]{q_V / n} \tag{2}$$

Type: K is constant, calculate the D2 = 300 mm.

2.3   Guide Vane is Relatively High

On the basis of the wheel diameter, the guide blade relative height increase will make wheel import flow area increases, discharge increases, increase the output of generating unit, the runner height increases at the same time, can make the runner blade inlet edge extension, the blade itself strength and stiffness are affected. Therefore the choice of the $\overline{b_0}$ [2] also must adapt to and n_2, this paper take $\overline{b_0}$ = 66.72 mm.

2.4   The Crown and the Ring Line

Type curve on the champions league can increase the runner exit area, the curve type ring can make water a smooth curve, this paper on the champions league, the ring adopts curve type.

2.5   Number of Runner Blades

Runner blade of the number of how many will affect the capability of the reversible hydraulic and structural strength, with an average of 6 ~ 9 reversible turbine runner blade, comprehensive consideration to determine the reversible machine for Z = 6 blade wheel [3].

2.6   The Water Area of Inspection

The working parameters according to the reversible machine, has determined the specific speed of relative height $\overline{b_0}$, guide vane, runner on the champions league, the ring line, wheel nominal diameter and mapped the reversible machine outlet diameter wheel type axial surface flow line, and the painted flow checked the distribution of the water area of along the line, as shown in Figs. 1 and 2.

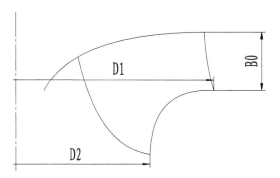

Fig. 1. Meridian projection of runner

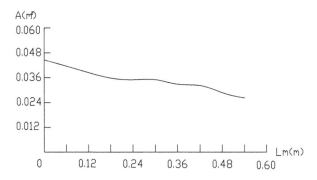

Fig. 2. Flowcross-section changes along streamline

2.7   Axial Plane Inside Draw, Airfoil Thickening and Blade Bone Line Forming

Reversible machine wheel flow axial plane inside the calculation precision of drawing and blade bone line will affect the design of the runner various performance, due to the limitation of space, not in detail, its middle plane, shaft surface cutting line as shown in Fig. 3, structural design as shown in Figs. 4 and 5.

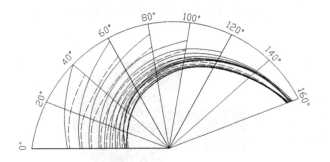

(a) Leaf blade plane cutting line

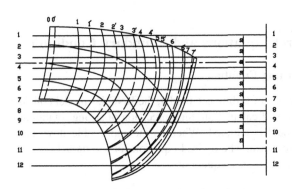

(b) Vane axial plane cutting line

Fig. 3. Blade plane shear map and meridian projection

3   Meshing

This article USES the ICEM tetrahedron structured grid generation technology [4–8], As shown in Fig. 5, Volute, runner and draft tube grid number, respectively 657372, 180532, 255703.

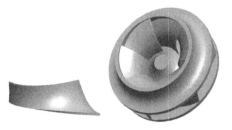

Fig. 4. The 3-D model of blade and runner

Fig. 5. The mesh diagram on calculation domain of pump-turbine

4   The Control Equations and Boundary Conditions

4.1   The Governing Equation

Among various kinds of turbulence model, the RNG k - model considering the rotation effect in the process of calculation, high precision of strong rotational flow calculation, The RNG k-ε model is adopted. When as incompressible flow, and not considering user-defined source term, The revised equation for k-ε model:

Turbulent kinetic energy equation:

$$\rho \frac{Dk}{Dt} = \frac{\delta}{\delta x_i} \left[\left(u + \frac{u_t}{\sigma_k} \right) \frac{\delta k}{\delta x_i} \right] + G_k - \rho\varepsilon \tag{3}$$

The turbulent kinetic energy dissipation rate equations:

$$\rho \frac{D\varepsilon}{Dt} = \frac{\partial}{\partial x_i} \left[\left(u + \frac{u_t}{\sigma_k} \right) \frac{\delta \varepsilon}{\delta x_j} \right] + C_{\varepsilon 1} \frac{\varepsilon}{k} G_\kappa - C_{\varepsilon 2}^* \rho \frac{\varepsilon^2}{k} \tag{4}$$

$$C_{\varepsilon 2}^* = C_{\varepsilon 2} + \frac{c_u \eta^3 (1 - \eta/\eta_0)}{1 + \beta \eta^3} \tag{5}$$

The turbulence model constants are:

$$c_u = 0.0845, \ c_{\varepsilon 1} = 1.42, \ c_{\varepsilon 2} = 1.68, \ \sigma_k = 0.72, \ \sigma_\varepsilon = 0.75.$$

4.2 The Boundary Conditions

Reversible turbine need two-way operation, condition of two different calculation need given the different boundary conditions [9–12].

(1) turbine inlet boundary using imported to ensure that the flow rate, outlet boundary using free discharge;
(2) water pump operating mode: imported boundary based on the speed of flow, use free discharge outlet boundary;
(3) wall condition: no slip wall meet the solid wall conditions, so close to the wall with a standard wall function method.

5   The Calculation Results and Analysis

5.1   Working Condition of the Turbine Blade Pressure Distribution is Analyzed

Blade surface pressure distribution and largely determine the performance of the reversible machine. Can be seen in Fig. 6, working condition of the turbine blade pressure imports to exports is regular, in line with the work principle of the blade, in contrast, lack of pump turbine in blade edge and blade pressure contour parallel import and export and streamline is not absolute, vertical pump turbine blade Angle is big, long blade was the cause of this phenomenon [13–15]. But on the whole, the reversible machine in working condition of the turbine can completely satisfy the requirement of the energy conversion.

5.2   Water Pump Working Condition of Blade Pressure Distribution is Analyzed

Pressure distribution on the surface of the blade for pump working condition of energy conversion has a decisive role. From Fig. 7 (note: vertical and horizontal wheel axis as the starting point 0) pump working condition of wheel pressure distribution can be seen that, in turn the entire field similar to the distribution of pressure distribution and the common centrifugal pump pressure, and pressure gradient increase is relatively clear, relative to the working condition of pump energy conversion relations in terms of ideal.

5.3   The Experiment Results Analysis

To directly reflect the results of numerical experiment is scientific, the Fig. 8 (note: abscissa to wheel axis as the starting point of 0, the same as in Fig. 9), you can see that

Pressure (KPa)

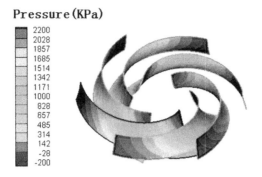

(a) Leaf blade working face

Pressure (KPa)

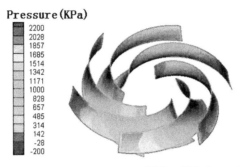

(b) Leaf blade on the back

Fig. 6. Pressure distribution on working surface and the reverse side of turbine operating (Color figure online)

working condition of the turbine wheel axis point corresponds to different distance of blade working pressure are higher than the back pressure, and uniform distribution, working condition of the pump pressure distribution curve bending is opposite bigger. Is apart from the central axis in Fig. 9 −0.05 − m, blade pressure mutations in the design process according to the working condition of hydraulic turbine, working condition of pump checking of reverse design theory has a certain relationship. Working condition of the turbine or pump working condition, is apart from the central axis of 0.10 m at the working face, the pressure distribution on the back of the blade have been mutations, this may be related to inadequate design.

5.4   Fluent Efficiency Calculation

Calculating formula for the working condition of hydraulic turbine efficiency

$$\eta = \frac{p_{\text{out}}}{p_{\text{in}}} = \frac{M\omega}{\Delta p q_V} \tag{6}$$

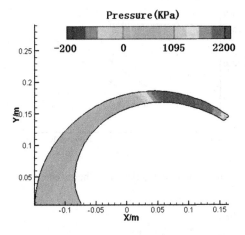

(a) Leaf blade working face

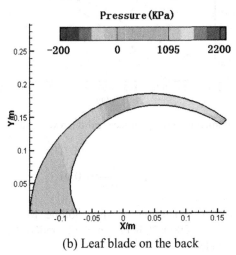

(b) Leaf blade on the back

Fig. 7. Pressure distribution on working surface and the reverse side of pump operating (Color figure online)

Water pump working condition of the calculation formula for efficiency

$$\eta = \frac{\Delta p q_V}{M \omega} \qquad (7)$$

Type the relevant parameters can be read Report feature of using Fluent software.

(1) Working condition of the turbine shaft torque M = 81.79 kN·m, traffic qV = 2.35 m³/s, $\Delta P = 2000, 312 k\,Pa$, $\omega = 52.3$ rad/s. The efficiency of the turbine operating mode:

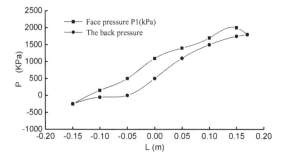

Fig. 8. Pressure profile on working surface and the reverse side of turbine operating

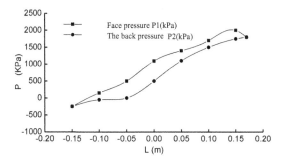

Fig. 9. Pressure profile on working surface and the reverse side of pump operating

$$\eta = \frac{M\omega}{\Delta pq_V} = 0.91$$

(2) Water pump working condition of shaft torque M = 107.765 kN·m, traffic qV = 2.31 m³/s, $\Delta P = 2000,703 k\,Pa$, $\omega = 52.3$ rad/s. The efficiency of water pump operating mode:

$$\eta = \frac{\Delta pq_V}{M\omega} = 0.82$$

6 Conclusion

(1) Reversible turbine two point does not overlap, so whether the runner design or selection requires certain compromise, on the basis of the basic parameters from a condition, using the existing hydraulic design method to calculate, to use another condition to check again. Using the theory of reverse design modeling, the model of water pump operating mode is shown slightly insufficient, working condition of hydraulic turbine hydraulic performance is good. This kind of design method is suitable for the long time running in the turbine operating mode of the pumped storage power station.

(2) Numerical calculation shows that the reversible machine working condition of the optimal efficiency of the turbine reached 91 %, water pump working condition of the optimal efficiency reached 82 %, repeated correction model further room for improvement.

References

1. Mei, Z.: Pumped Storage Power Generation Technology, pp. 105–135. Mechanical Industry Press, Beijing (2001)
2. Cao, K., Yao, Z.: Principle and Hydraulic Turbine Design. Tsinghua University Press, Beijing (1991)
3. Li, R., Su, F.: Abrasion protection design and experimental study of wicket gate of hydraulic turbines. J. Hydrodyn. **4**, 446–449 (2008)
4. Li, Q., Li, R., Han, W., et al.: Francis turbine water diversion, water inside the solid liquid two phase flow numerical analysis. J. Lanzhou Inst. Technol. **34**(6), 47–50 (2008)
5. Liu, S., Wu, Y., Zhang, L. et al.: Vorticity analysis of a cavitating two-phase flow in rotating. Int. J. Nonlinear Sci. Numerical Simul. **10**(5), 601–616 (2009)
6. Hellstroem, J.G.I., Marjavaara, B.D., Lundstroem, T.S.: Parallel CFD simulations of an original and redesigned hydraulic turbine draft tube. Adv. Eng. Softw. **38**(5), 338–344 (2007)
7. Escaler, X.R., Egusquiza, E.D., Farhat, M.D., et al.: Detection of cavitation in hydraulic turbines. Mech. Syst. Signal Proc. **20**(4), 983–1007 (2006)
8. Wang, R.-J., Zhang, K., Wang, G.: FLUENT Software Technology Foundation and Application Tutorial, pp. 134–152. Tsinghua University Press, Beijing (2007)
9. Zhou, L.: Turn the Turbine Rotating Field Calculation and Performance Prediction. Dissertation China Agricultural University (1999)
10. Hellstroem, G.I., Marjavaara, B.D., Lundstroem, T.S.: Parallel CFD simulations of an original and redesigned hydraulic turbine draft tube. Adv. Eng. Softw. **38**(5), 338–344 (2007)
11. Escaler, X., Egusquiza, E., Farhat, M., Avellan, F., Coussirat, M.: Detection of cavitation in hydraulic turbines. Mech. Syst. Signal Process. **20**(4), 983–1007 (2006)
12. Izena, A., Kihara, H., Shimojo, T., Hirayama, K., Furukawa, N., Kageyama, T., Goto, T., Okamura, C.: Practical hydraulic turbine model. In: 2006 Power Engineering Society General Meeting (PES 2006), vol. 8, pp. 3837–3843 (2006)
13. Guan, R.: Mixed-flow reversible spiral case and draft tube of the comparison experiment. Hydraulic machinery laboratory report, Tsinghua University (1976)
14. Hou, Y., Qi, X., Chang, Y., et al.: Three-dimensional modeling of the turbine runner of based on PRO/E. J. Drainage Irrig. Mach. Eng. **24**(1), 14–16 (2006)
15. Wang, L., Liu, J., Liang, C., et al.: Flow characteristics of pump-turbine at turbine braking mode. J. Drainage Irrig. Mach. Eng. **29**(1), 1–5 (2011)

Computer Vision

Efficient Moving Objects Detection by Lidar for Rain Removal

Yao Wang, Fangfa Fu, Jinjin Shi, Weizhe Xu, and Jinxiang Wang$^{(\boxtimes)}$

Microelectronics Center, Harbin Institute of Technology, Harbin, China
jxwang@hit.edu.cn

Abstract. Rain and snow are often imaged as brighter streaks, which can not only confuse human vision but degrade efficiency of computer vision algorithm. Rain removal is very important technique in these fields such as video-surveillance and automatic driving. Most existing methods rely on optical flow algorithm to detect rain pixel and estimate motion field. However, it is extremely challenging for them to achieve real-time performance. In this paper, a LIDAR based algorithm is proposed, which is capable of achieving rain pixel robustly and efficiently from motion field. The motion objects (vehicles and human) are identified for separation by LIDAR (Sick LMS200) in this paper. Then rain pixels on moving objects are removed by bilateral filter which can preserve edge information instead of causing blurring artifacts around rain streaks. Experimental results show that our method significantly outperforms the previous methods in removing rain pixel and detecting motion objects from motion field.

Keywords: Rain removal · Real-time · LIDAR · Bilateral filter

1 Introduction

The demand for computer vision system in the fields of video-surveillance, military, and automatic driving, etc. is increasing rapidly. Due to the effect of bad weather condition, such as rain, snow, fog and mist, current systems are facing great challenges for the poor performance. According to [1], the detection of rain is a difficult task because both rain and moving objects produce sharp intensity changes in dynamic scene. We should treat these two parts separately instead of causing artifacts around other pixels. Thus methods that can remove visual effects and frequent intensity fluctuations caused by undesirable weather conditions are necessary. In order to remove the rainy effect, separating object motion from rain and removing rain pixels are very crucial. Most previous algorithms have been proposed for this purpose.

Most of the applications simply consider how to remove rain pixels from images, regardless of paying more attention to detecting moving objects efficiently. One of the excellent works in rain removal and separating moving objects from dynamic scenes was studied by Grag and Nayar [2], which shows impressive result but the direction and velocity of rain drops limited its functions.

Another work adopted an alternative approach by Barnum et al. [3, 4] based on the frequency analysis of rain streaks [5, 6]. It is assumed that rain streaks in an entire video sequence have similar shapes and orientations. Therefore they detected the rain

© Springer International Publishing Switzerland 2016
D.-S. Huang et al. (Eds.): ICIC 2016, Part III, LNAI 9773, pp. 697–706, 2016.
DOI: 10.1007/978-3-319-42297-8_64

drops by selecting repeatedly occurring frequency components through the dynamic scenes. However their method always ignored rain drops from moving objects and regard moving objects as the region of rain streaks.

Chen and Chau [7] proposed a method based on optical flow to detect and separate moving objects. Their method was suitable to few moving objects in the dynamic scene. As moving objects increase, optical flow algorithm results in low performance.

To overcome these drawbacks, this paper proposes a method to detect moving objects by LIDAR without using optical flow or physical properties of rain. First we employ LIDAR to scan moving objects according to different speed and map LIDAR coordinates to image plane. Then we set rain streaks as two parts: motion objects and static background. At last we treat two parts respectively by bilateral filter and Gaussian filter instead of causing blurring.

The rest of this paper is organized as follow. We review some related work in Sect. 2. Then we propose our algorithm: Motion segmentation by LIDAR, Rain detection and Motion Exclusion are described in Sect. 3. Experimental results of detecting and removing rain pixels are presented in Sect. 4 with some conclusions follow in Sect. 5.

2   Related Work

One of the notable works in detecting rain pixels by identifying rain pixels with high intensity changes in the scene, and then recovering rain pixels by taking the average pixel values of neighboring frames. Garg and Nayar [2] based on physical and photometric properties of rain to detect and remove rain from images. However this method fails to separate rain from moving objects and consider all the pixels as rain pixels no matter if there is a moving object in the scene. Therefore this method always causes blur artifacts around moving objects. In order to reduce blurring artifacts around rain streaks, Kim and Lee [5] detected rain streak regions using shape and orientation features, then remove rain by an adaptive nonlocal means filter to remove rain from single image. Liu el at. [6] employed the detecting function based on chromatic properties and the Kalman filter for rain removal. Ding and Chen [8] removed rain and snow via guided L0 smoothing filter from single image based on edge characteristic of rain or snow. These methods fail to performance better in removing rain when facing with dynamic scenes. Chen and Chau [7] put forward a method to remove rain in dynamic scenes. They applied Expectation Maximization (EM) on an optical flow field to separate motion map into two parts: moving objects and static background. Rain pixels are treated respectively by using different filters depending on whether they belonged to moving objects or static background. Their method performance better in distinguish moving objects because of optical flow. Shen and Xue [9] proposed a fast method by using three successive frames to detect rain based on optical flow. However optical flow is not real-time performance when moving objects increased. To overcome the limitation of optical flow, Tan and Chen [10] used aligning neighboring frames combined with optical flow to detect rain pixels. Varun and Vijayan [11] employed phase congruency features to find rain pixels and applied optical flow to get motion map from dynamic scenes. However their method works poorly when moving objects increase. In general, current methods are not suitable for practical requirements with respect to their detection accuracy and recovery performance.

All the above methods suffer the same problem that they are not applicable to detect moving objects efficiently. To address the limit of the previous approaches, we put forward simple but efficient moving objects detection and rain drops removing method in this paper. That is, we employ LIDAR to scan objects in front of vehicle and distinguish their motion situation by the speed. Then we map the position of moving objects from LIDAR coordinates to image plane. Therefore we set whole rain image as two parts: moving objects and static background. Finally we treat these two parts to remove rain pixels by bilateral filter and Gaussian filter.

3   Proposed Algorithm

The proposed algorithm consists of three parts — for each frame, we apply motion segmentation by LIDAR, detect rain pixels, and then recover each rain pixel.

3.1   Motion Segmentation by LIDAR

The rain and object motion always cause pixel intensity fluctuation of a rainy scene. The fluctuations caused by rain should be removed, and the ones caused by moving object need to be retained. Thus motion field segmentation naturally becomes a fundamental procedure of rain removal algorithm. Previous approaches employed optical flow to estimate moving object. We briefly describe the method here. That method needs two adjacent frames to get motion cue by computing optical flow filed. After thresholding the magnitude of the optical flow field, applying a Gaussian Mixture Model to the resulting binary image in order to get motion objects locations in the target frame. The motion cue is a likelihood function for motion objects based on the optical flow field. However, using optical flow for motion target segmentation has its intrinsic drawbacks, as it can only detect obvious intensity changes and areas that belong to the target without obvious intensity change cannot be effectively recognized, which causes the so called "Aperture Problem" [4]. The accuracy of optical flow always depends on Gaussian Mixture Model (GMM) in order to overcome "Aperture Problem", but its computational complexity is relatively high due to cluster motion field and it is are not capable of real-time applications. We employ LIDAR to detect moving objects efficiently [12] without using other algorithms.

First, we use LIDAR to scan object in front of LIDAR [13], the scanning range of LIDAR is 180° which is capable of scanning all the moving objects. LIDAR is set on the top of our vehicle and the height of LIDAR is 0.89 m, which is the half height of vehicle or the pedestrian. We can get location of object by LIDAR data. By employing distance data from LIDAR, we can get the speed of object which aids us to identify whether the object is static or dramatic. V_P and V_c in (1) are the speed of object and car respectively, d is the distance data between object and LIDAR, t_1 represents time period when LIDAR emit light, t_2 represents time of reflection data form LIDAR. We set D_{th} and D_{th1} in (2) are 1 m/s and 15 m/s respectively. Therefore we can identify whether the object is static or dynamic by value of B_M in (2).

$$V_p = V_C = d/(t_1 - t_2) \tag{1}$$

$$B_M = \begin{cases} 1 & V_p > D_{th} || V_C > D_{th1} \\ 0 & \text{otherwise} \end{cases} \tag{2}$$

If the LIDAR detects an object (pedestrian or vehicle), it will return $M_1(x_1,y_1,z)$ on the left side and $M_2(x_2,y_2,z)$ on the right side, so the top of the vehicle $M_3(x_1,2y_1,z)$ on the left side and $M_4(x_2,2y_2,z)$ on the right side. The bottom of vehicle $M_5(x_1,y_1-0.89,z)$ and $M_6(x_2,y_2-0.89,z)$ in the LIDAR coordinates.

LIDAR coordinates map to image plane process includes the following three steps:

(1) We map LIDAR coordinate system onto the camera coordinate system which can be expressed as follows:

$$\begin{bmatrix} x_c \\ y_c \\ z_c \end{bmatrix} = R \begin{bmatrix} x_l \\ y_l \\ z_l \end{bmatrix} + T = \begin{bmatrix} r_{11} & r_{12} & r_{13} \\ r_{21} & r_{22} & r_{23} \\ r_{31} & r_{32} & r_{33} \end{bmatrix} \begin{bmatrix} x_l \\ y_l \\ z_l \end{bmatrix} + T \tag{3}$$

In the camera coordinate system, the (x_c,y_c,z_c) coordinates identify the point which transform from LIDAR coordinates (x_l,y_l,z_l). R and T are the affine matrix respectively.

(2) Then we transform camera coordinate system into image physical coordinates which can be expressed as below:

$$x = \frac{fx_c}{z_c}, y = \frac{fy_c}{z_c} \tag{4}$$

In the image physical plane, the (x,y) coordinates identify the physical position of a pixel. f represents the focal length of the camera. So the relationship between two coordinates could be easily solved by affine transformation which is show as follows:

$$z_c \begin{bmatrix} x \\ y \\ 1 \end{bmatrix} = \begin{bmatrix} f & 0 & 0 & 0 \\ 0 & f & 0 & 0 \\ 0 & 0 & 1 & 0 \end{bmatrix} \begin{bmatrix} x_c \\ y_c \\ z_c \\ 1 \end{bmatrix} \tag{5}$$

(3) In the image plane, the (u, v) coordinates identify the position of a pixel, the optical center is given by (u_0, v_0). Let $s_x = 1/dx$ and $s_y = 1/dy$ denote the value of the focal length expressed in pixels. From the perspective camera model we know that:

$$\begin{cases} u - u_0 = x/d_x = s_x x \\ v - v_0 = y/d_y = s_y y \end{cases} \tag{6}$$

The relationship between physical cameras coordinates and image coordinates could be expressed as follows:

$$\begin{bmatrix} u \\ v \\ 1 \end{bmatrix} = \begin{bmatrix} 1/d_x & 0 & u_0 \\ 0 & 1/d_y & v_0 \\ 0 & 0 & 1 \end{bmatrix} \begin{bmatrix} x \\ y \\ 1 \end{bmatrix} \tag{7}$$

At last LIDAR coordinates map to image plane, the coordinates in the image plane are M'1, M'2, M'3, M'4, M'5 and M'6 shown in the Fig. 1.

(a)   (b)

Fig. 1. LIDAR coordinates map to image plane (a) input image (b) our result

3.2 Rain Detection

First, we get image from successive frame which is shown in Fig. 2 (a). Then we detect moving vehicle by LIDAR which is shown in (b) and employ threshold method to detect rain pixels that grey scale intensity differences between two successive frames are calculated. The grey rain image map is show in (d). The threshold value are set in order to detect all the intensity fluctuations caused by rain, then we set D_{th2} is 5. The binary map is calculated which can be expressed as follows:

$$I_{diff} = \begin{cases} 1 & I_N - I_{N-1} \geq D_{th2} \\ 0 & I_N - I_{N-1} < D_{th2} \end{cases} \tag{8}$$

Then we acquire location of moving car from grey rain map according to LIDAR data in (e) which has similar result with Chen's method by optical flow shown in (c). Finally we get grey rain map without moving objects in (f).

3.3 Motion Exclusion

Rain pixels within the motion object and static background should be treated respectively, therefore we divide rain map into two parts: one is rain pixels in the moving objects and the other is rain pixels in the static background. We set rain pixels in the motion target area is A_m and rain pixels in the static background area is A_b which can be expressed as follows.

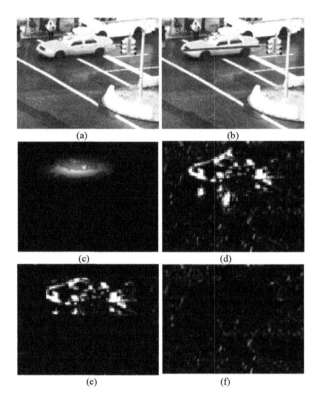

Fig. 2. The rain removal based on LIDAR (a) Our input image (b) Moving car detected by LIDAR and we map LIDAR data to image plan in order to separate moving objects. (c) Chen [6] use optical flow to detect moving objects. (d) Our result of grey images of motion object and rain map. (e) Binary image of moving object. (f) Rain map without moving object. Our results correctly distinguish moving objects from rain map, while the previous method inadequately recognizes whole region of moving object.

$$A_m = \{I(x,y) \mid I_{Rain}(x,y) = 1 \& B_M = 1\}$$
$$A_b = \{I(x,y) \mid I_{Rain}(x,y) = 1 \& B_M = 0\}$$
(9)

Under different situations, we will use different methods to remove rain pixels. Rain pixels of A_m are recovered using bilateral filter on moving objects which can preserve edge information instead of causing blurring artifacts around rain streaks. Then the rain pixels of A_b are covered by using a Gaussian filter temporally on three frame neighbors A_b.

4   Experimental Results

We first show results of our algorithm in Fig. 3. Figure 3(a) shows several successive frames from the video of the yellow car. In Fig. 3(b) LIDAR is used for detecting moving objects, then LIDAR coordinates is mapped to image plane. Figure 3(c) shows our result of binary moving objects map which can help us locate moving objects. And Fig. 3(d) is to detect moving objects from Chen's method by applying optical flow. By comparing (c) and (d) in Fig. 3, our method preserves most parts of car which are larger than the result of Chen, because the good performance of LIDAR in detecting moving objects. Then we compare the second row in Fig. 3, the moving car is in the middle of road with high speed. The optical flow can detect parts of car show in Fig. 3(d) which is similar to our result in Fig. 3(c). In the third row of Fig. 3, the moving car is

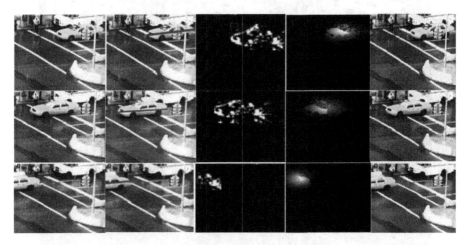

Fig. 3. Rain removal image.(a) input image. (b) moving car is detected by LIDAR. (c) separate moving car from grey rain image. (d) the result of Chen's [6] method to separate moving car. (e) rain remove image of our method. (Color figure online)

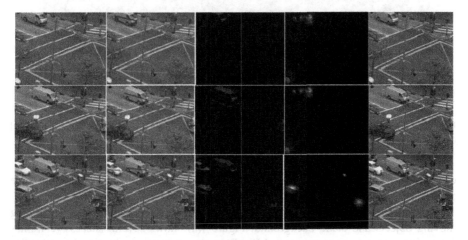

Fig. 4. increasing numbers of moving objects in dynamic scene. (a) input image. (b) moving cars detected by LIDAR. (c) map moving objects to grey image. (d) the method of Chen [6] in separating moving objects from the scene. (e) our proposed method in removing rain.

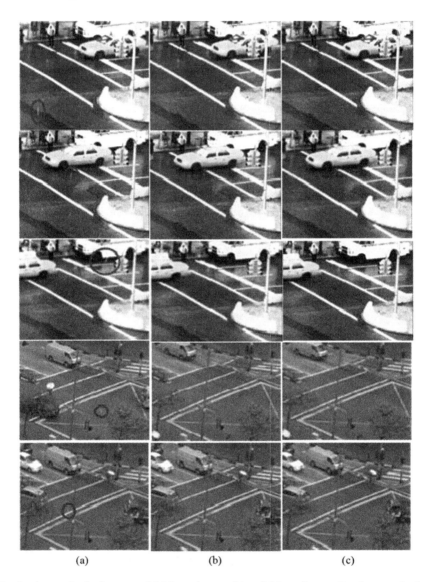

(a)   (b)   (c)

Fig. 5. the result of rain removal (a) input image. (b) and (c) are the results of our method and Chen's [6] method.

disappearing. Our result contains most parts of car while previous method lost some parts of the car.

In Fig. 4 moving objects are increasing which pose some difficulties in separating moving objects and static background. Figure 4(a) shows several successive frames from video. Our method employs LIDAR to detect moving objects. According to different speed, we can distinguish moving cars efficiently in Fig. 4(b). Due to the limit of optical flow, neither moving objects with similar speed nor with lower speed can be distinguished. Compared (c) to (d) in Fig. 4, our method detect 1,2 and 5 moving cars respectively in successive frames, however the method of Chen detect 1, 1 and 2 cars. Therefore some cars are ignored by optical flow. In Fig. 4(e) is result of our method to remove rain from image.

Next we compare the performance of two methods in removing rain streaks as shown in Fig. 5. We select five different images and rain streaks are indicated by red circle in Fig. 5(a). In Fig. 5(b) we get similar result as show in (c). Finally we compare the time of our method and previous method, Chen [6] use 300 ms to separate moving objects from motion filed by optical flow and finish rain removal every frames, however our method just costs 100 ms to detect and remove rain pixels. Our rain removal result is better than previous method especially in detecting moving objects no matter if moving objects increases or not.

5 Conclusions

In this paper, we proposed an efficient rain recovery algorithm based on LIDAR. Previous methods proposed a method which shows state of-the-art performance in removing rain in highly dynamic video scenes. However, the method is not suitable for real-time performance when moving objects increased. We put forward an efficient method to detect moving objects for rain removal. We first identified the location of moving objects by LIDAR. Then, we use several models to map LIDAR data to 2D image plan. We separate moving objects from whole images and get rain map without moving objects. Finally we employ edge-preserving bilateral filter to remove rain pixels on moving objects, and use Gaussian filter temporally on three frames to remove rain pixels in the static background.

Experiments show that the proposed method is able to remove rain from scenes taken from dynamic scenes very effectively. The performance of rain removal still has much room for improvement. A future research issue is to extend the proposed algorithm to low resolutions images.

Acknowledgement. This work was supported by a grant from National Natural Science Foundation of China (NSFC, No. 61504032).

References

1. Narasimhan, S.G., Nayar, S.K.: Vision and the atmosphere. Int. J. Comput. Vis. **48**(3), 233–254 (2002)
2. Garg, K., Nayar, S.K.: Detection and removal of rain from videos. In: Proceedings of the 2004 IEEE Computer Society Conference on Computer Vision and Pattern Recognition, CVPR 2004, vol. 1. IEEE (2004)
3. Barnum, P., Kanade, T., Narasimhan, S.: Spatio-temporal frequency analysis for removing rain and snow from videos. In: Proceedings of the First International Workshop on Photometric Analysis for Computer Vision-PACV 2007. INRIA (2007)

4. Barnum, P.C., Narasimhan, S., Kanade, T.: Analysis of rain and snow in frequency space. Int. J. Comput. Vis. **86**(2–3), 256–274 (2010)
5. Kim, J.-H., et al.: Single-image deraining using an adaptive nonlocal means filter. In: 2013 20th IEEE International Conference on Image Processing (ICIP). IEEE (2013)
6. Liu, P., Xu, J., Liu, J., et al.: A rain removal method using chromatic property for image sequence. In: 11th Joint International Conference on Information Sciences. Atlantis Press (2008)
7. Chen, J., Chau, L.P.: A rain pixel recovery algorithm for videos with highly dynamic scenes. IEEE Trans. Image Process. **23**(3), 1097–1104 (2014)
8. Ding, X., Chen, L., Zheng, X., et al.: Single image rain and snow removal via guided L0 smoothing filter. Multimedia Tools Appl. **75**, 2697–2712 (2015)
9. Shen, M., Xue, P.: A fast algorithm for rain detection and removal from videos. In: 2011 IEEE International Conference on Multimedia and Expo (ICME). IEEE (2011)
10. Tan, C.-H., Chen, J., Chau, L.-P.: Dynamic scene rain removal for moving cameras. In: 2014 19th International Conference on Digital Signal Processing (DSP). IEEE (2014)
11. Santhaseelan, V., Asari, V.K.: Utilizing local phase information to remove rain from video. Int. J. Comput. Vis. **112**(1), 71–89 (2015)
12. Arras, K.O., Mozos, ó.M., Burgard, W.: Using boosted features for the detection of people in 2D range data. In: 2007 IEEE International Conference on Robotics and Automation. IEEE (2007)
13. Kurnianggoro, L., Hernandez, D.C., Jo, K.-H.: Camera and laser range finder fusion for real-time car detection. In: 40th Annual Conference of the IEEE Industrial Electronics Society, IECON 2014. IEEE (2014)

Multiple Visual Objects Segmentation Based on Adaptive Otsu and Improved DRLSE

Yaochi Zhao, Zhuhua Hu$^{(\boxtimes)}$, Yong Bai, Xingzi Liu, and Xiyang Liu

College of Information Science and Technology, Hainan University,
Haikou 570228, Hainan, China
{yaochizi,eagler_hu}@163.com

Abstract. Aiming at the problem that contours always are fractured and not precise in the segmentation of multiple visual objects, Distance Regularized Level Set Evolution(DRLSE) is used. Considering the characteristics of background difference image, an adaptive Otsu is proposed to segment difference image. In order to take full advantage of temporal information in videos, frame difference and background difference are introduced into energy function of DRLSE. Firstly, an adaptive Otsu and asymmetric morphological filtering based method are used to obtain better initial contours. Secondly, initial contours are evolved with improved DRLSE that frame difference and background difference are integrated in as priori knowledge, which can avoid that objects contours are evolved to background edge, and reduce over segmentation. The experimental results show that the contours of multiple video targets can be obtained more precisely and rapidly with the method proposed in this paper than with the existing methods.

Keywords: Multiple visual objects segmentation · Distance regularized level set evolution · Otsu · Morphological filter

1 Introduction

Visual objects segmentation is one of the key technologies in intelligent video analysis, and it is a meaningful and greatly challenging subject to seek accurate and real-time method for video objects segmentation in areas such as computer vision and pattern recognition. Level Set Active Contour(LSAC) [1–3] has a top-down mechanism, which make it integrate prior knowledge such as motion, shape, texture and color, and accurately get the continuous and accurate contour of visual objects. Because of the precision of Active Contour Model(ACM) and the topological adaptability of Level Set, LSAC based visual objects segmentation and tacking became a hot spot in the research of computer vision.

Currently, there are some problems in the segmentation algorithms based on LSAC [4, 5]: Firstly, because contour is evolved to local minimums LSAC is sensitive to initial contour, and thus how to quickly obtain relatively accurate initial contour is a key problem, especially in multiple objects scenes. Secondly, as the traditional LSAC is driven by gradient force, contour is easily to be evolved to obvious background edge or go over weak object boundary and so error segmentation or over segmentation will appear.

© Springer International Publishing Switzerland 2016
D.-S. Huang et al. (Eds.): ICIC 2016, Part III, LNAI 9773, pp. 707–716, 2016.
DOI: 10.1007/978-3-319-42297-8_65

In this paper we present an adaptive Otsu based threshold segmentation and asymmetric morphological filtering method to rapidly get relatively accurate initial contours of multiple targets. Aiming at the precise problem caused by error or over segmentation, an improved DRLSE constrained by frame difference and background difference is proposed. With the improved DRLSE, the functional of contour can converge to weak object edge and go over background contour and reduce error or over segmentation.

2   Brief Introduction to Algorithms

The algorithms process is shown in Fig. 1. The algorithms consist of two parts: the one is to obtain initial contours, and another is to get precise contours. Firstly, background difference image is segmented with an adaptive Otsu's method. Secondly, asymmetric morphological filtering based method is used to obtain better initial contours. Finally, initial contours are evolved with improved DRLSE that frame difference and background difference are integrated in as priori knowledge.

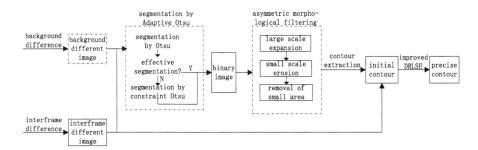

Fig. 1. Flowchart of algorithms in this paper

3   Initial Contours of Multiple Visual Objects

To obtain the initial contours of video objects, we use background subtraction, adaptive Otsu segmentation and asymmetric morphological filtering in this paper.

3.1   Adaptive Otsu Segmentation

In difference image, the distinguishing between targets and background is always by setting a threshold value, so a suitable threshold is a key. Among numerous threshold segmentation algorithms, Otsu [6, 7] is widely used because of the simple calculation and adaptivity. The idea of Otsu is to find an optimal threshold value, with which the interclass variance can get the maximum after clustering. Let t is the threshold value, and $w_1(t)$ and $w_2(t)$ are the proportion of background and objects respectively in the whole image after clustering with t, let $u_1(t)$, $u_2(t)$ and $u(t)$ are the mean value of background, objects and the whole image respectively, let t_{otsu} is the threshold value with Otsu, and the problem can be expressed as follow:

$$t_{otsu} = \arg\max\{w_1(t)(u_1(t) - u(t))^2 + w_2(t)(u_2(t) - u(t))^2\} \tag{1}$$

In the case that the number of target pixels is close to the background, the segmentation with threshold value by Otsu is effective, but when the number of two classes is not close, segmentation with t_{otsu} will tend to split the class with more pixels. In most cases, the area of background in difference image is much larger than objects, t_{otsu} is always smaller than the ideal value, and a suitable value with Otsu in the constraint range, larger than t_{otsu}, is a better option. While in the case that the areas of objects are big, t_{otsu} is a better choice. As shown in Fig. 2, the areas of objects in three scenes are not the same, the objects area in scene 1 is the smallest, and the scene 3 is the biggest. In scene 1, the area of objects is much smaller than the background, segmentation with t_{otsu} is error, and the threshold value with Otsu in constraint range is better. While, in scene 3, t_{otsu} is the idea threshold value. In this paper, we firstly segment image with t_{otsu}, and if the segmentation is invalid, we do it with constraint Otsu.

(a) a frame in videos (b) difference image (c) segmented image by Otsu's method
(d)segmented image by constraint Otsu's method

Fig. 2. Contrast of segmentation effection between Otsu and constraint Otsu

In actual application, it is a worthy question to explore: how to automatically determine the effectiveness of segmentation. We take use of connected component labeling to test the effectiveness of the Otsu segmentation and determine to whether subsequently segment with constraint Otsu or not. If the number of connected component is larger than a fixed value, the subsequent segmentation is necessary, otherwise, the subsequent is not.

3.2   Asymmetric Morphological Filtering

After differential operation and segmentation with a threshold value, in LSAC based visual objects segmentation, post-processing is necessary for continuous contours. The traditional approach is to use rectangle or fitting ellipse to stand for initial contours [8– 10] as shown in Fig. 3(a)–(c). However, there is a great deal of difference between the actual shape of object and rectangle or ellipse. What's more, multiple visual objects cannot be distinguished with those kinds of methods, which will influence the effectiveness of segmentation and cause too much time consumption. In [11], center distance of connected component was calculated to group adjacent areas, with which, the respective contours of multiple visual objects could be obtained, but initial contours were still expressed by rectangles, and most of the contours information were lost (Fig. 3(d)).

(a) initial contour in [8] and [9] (b) initial contour in [10] (c) initial contour in [5] (d) initial contours in [11]

Fig. 3. Traditional initial contour

After the adaptive Otsu segmentation, mentioned above, visual object may be broken into many connected regions. We adopt asymmetric morphological filtering proposed in this paper to get an integrate initial contour for every visual target. Firstly, multiple broken moving targets are expanded into an integrated area by using a large scale expansion coefficient. Secondly, expanded image are eroded with a small scale corrosion coefficient. Finally, noise areas are filtered out by removing the small area, and then every connected area stands for a visual object.

As shown in Fig. 4(a), the large-scale expansion has played a role of clustering connected regions in Fig. 2(d), but no need set clustering center and number in traditional cluster methods. From Fig. 4(d) we can see that small scale corrosion and removal of small area can filter noise. What's more, shape information of the visual object is reserved to a great degree. Besides, due to the asymmetry of morphological

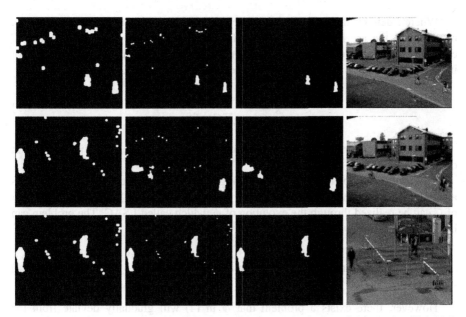

(a) large-scale expansion (b) small-scale erosion (c) removal of small area region (d) initial contour

Fig. 4. Results of asymmetric morphological filtering

operation we use, the initial contour is slightly outward expansion on the basis of the actual contour, which is beneficial to the subsequent precise contour evolution.

4   Contour Precision Evolution Based on Improved DRLSE

To avoid evolution in the whole image area, we take use of multiple Level Set Function (LSF) to express contour of every visual object. And we adopt DRLSE in their Region of Interesting(ROI) to get precise contour respectively.

4.1   DRLSE

Active contour model(ACM) proposed by Kass and Others [12] in 1988 became a hot spot in the research of computer vision, because that it integrate the upper prior knowledge and the underlying characteristics of image. The highly representative of ACM is Geodesic active contour (GAC) [13] presented by Caselles and others, which can be expressed as:

$$\min\left\{E(C) = \int_0^{L(C)} g(|\nabla I[C(s)]|)\mathrm{d}s\right\} \tag{2}$$

Where I denotes the image to be segmented, C denotes curve, $L(C)$ and $E(C)$ are, respectively, the arc length and energy of the curve C, when E is the minimal, C consists of the strongest local edges In (2), g denotes the evolution stop function of image, which is an inverse function of gradient, as shown in Eq. (3).

$$g = \frac{1}{1 + |\nabla G * I|^2} \tag{3}$$

In (3), G denotes the Gaussian kernel with standard variance, and $*$ is convolution operation.

To solve the problem of contour topology adaptivity, curve function C in (3) is always embedded in high dimensional LSF expressed by ϕ in (4). The evolution result of C can be realized through the evolution of ϕ, which called LSAC. The problem can be described as follow:

$$\min\left\{ E(\phi) = \iint_\Omega g\,(x,y)\delta(\phi)|\nabla\phi|\mathrm{d}x\mathrm{d}y \right\} \tag{4}$$

However, there exists a problem that ϕ in (4) will gradually deviate from the attribute of distance function in the process of evolution. The traditional solution is to reinitialize ϕ to one distance function regularly. In [14, 15], Distance regulation term is introduced into energy functional, called DRLSE, which can entirely avoid reinitialization in the process of evolution. The problem can be expressed as follow:

$$\min\left\{ \varepsilon(\phi) = \mu\Re_p(\phi) + \lambda \int_\Omega g\delta(\phi)|\nabla\phi|\mathrm{d}X + \alpha \int_\Omega gH(-\phi)\mathrm{d}X \right\} \tag{5}$$

Where $\Re_p(\phi)$ denotes the distance regulation term, the second term represents length term, and the third term is area term, which behaves as expansibility or contractility in the process of evolution. μ, λ, and α are the distance regulation, length, and area coefficient respectively.

4.2   Improved DRLSE Based on Frame Difference and Background Difference

The prior knowledge of image is always integrated into g in visual objects segmentation and tracking based on DRLSE. To weaken the effects of background edge on contour evolution, the temporal difference image between two frames is merged into g in [5], and g can be expressed as follow:

$$g\big(I(x,y)_n\big) = \frac{1}{1 + |\nabla I(x,y)_n|^p} + \frac{1}{1 + \left|DI_{frsub}(x,y)_{n,n-1}\right|^p}, p = 1, 2. \tag{6}$$

Where $DI_{frsub}(x,y)_{n,n-1}$ represents the difference between two frames.

To weaken the influence of background edge, at the same time reduce the excessive segmentation phenomenon, we integrate frame difference and background difference information into g, which can be written as:

$$g\left(I(x,y)_n\right) = \frac{1}{1 + \left|\nabla I(x,y)_n\right|^2} + \frac{1}{1 + \left|DI_{frsub}(x,y)_{n,n-1}\right|^2} + \frac{1}{1 + \left|DI_{bksub}(x,y)_n\right|^2} \quad (7)$$

Where $DI_{bksub}(x,y)_n$ is background difference.

Generally, the value of $DI_{bksub}(x,y)_n$ in the visual object is large, and g is small. So curve's evolutionary force in the moving object is less than in the other's from the viewpoint of dynamics.

In the existing algorithms, λ and α both are a fixed value respectively. In our algorithms, we further propose the length and area coefficients based on background difference.

Though the edges of visual object in every ROI after the self-adaption Otsu segmentation may be discontinuity, it is confirmed that they belong to the moving object area, which can be used as priori knowledge to guide contour evolution. Let $I\_obj_{n,k}(x,y)$ is the set of the kth object pixels in the nth frame, obtained by background difference and self-adaption Otsu. Let $\alpha\_obj_{n,k}(x,y)$ and $\lambda\_obj_{n,k}(x,y)$ are the length and area coefficients of the kth object's pixel (x, y) in the nth frame. We use (8) and (9) to merge the above prior knowledge into g, as described in follows.

$$\alpha\_obj_{n,k}(x,y) = \begin{cases} 0 & if(x,y) \in I\_obj_{n,k}(x,y) \\ b_1 & if(x,y) \notin I\_obj_{n,k}(x,y) \end{cases} \quad (8)$$

$$\lambda\_obj_{n,k}(x,y) = \begin{cases} 0 & if(x,y) \in I\_obj_{n,k}(x,y) \\ b_2 & if(x,y) \notin I\_obj_{n,k}(x,y) \end{cases} \quad (9)$$

Where b_1 and b_2 both are a positive integer.

From (8) and (9), we can see that the length and area coefficients both are zero at the visual object's certain pixel. In other words, the external forces to guide curve evolution at those pixels are zero, which can avoid the contour cross the weak edges of visual object, so reduce the phenomenon of over segmentation.

5   Experiment Results

Some typical videos in which there are multiple pedestrians, car, bicycle, electrombile were tested, and the partial results are shown in Fig. 5. All the above algorithms were implemented in hardware environment: Intel (R), Core (TM) i5-2500 CPU @ 3.30 GHZ, 4.00 GB memory; and software environment: Window 7 operating system, Matlab R2013a.

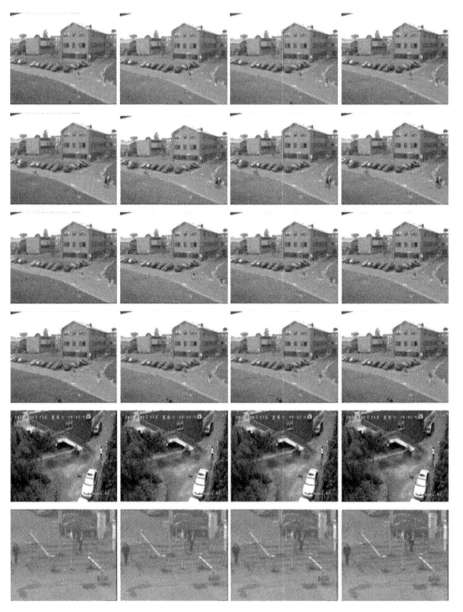

(a) a frame in videos (b) evolution results with traditional DRLSE (c) evolution results with
DRLSE in[5] (d) evolution results with improved DRLSE in this paper

Fig. 5. Contrast of results between tradition DRLSE, DRLSE in [5] and improved DRLSE in
this paper

The traditional DRLSE in [15] is influenced by background edge, for example, the first pedestrians' evolution contour on the right in frame 1 and bicycle's evolution contour in frame 2 are affected by the edge of lawn and weeds in the lawn respectively. The algorithm in [5] can eliminate the influence caused by the background edge, shown in Fig. 5(c). The improved algorithms in this article can not only eliminate this kind of influences, but also more accurately converge to the target edge, as shown in Fig. 5(d)'s corresponding figure.

For prolate objects, such as pedestrians, due to head and food's centripetal force is larger than other parts, over segmentation is likely to appear with the existing DRLSE in the case that head or foot's edges are weak. For instance, the first pedestrians head's contour on the right in frame 3 was over segmented with tradition DRLSE in [15], and DRLSE in [5], while the phenomenon can be reduced with background difference based DRLSE proposed in this paper, as shown in Fig. 5's corresponding figure.

Moreover, from the pedestrians in frame 2, the bottom pedestrian in frame 4 and the left one in the frame 6, we can know under the condition of the same evolution times (as shown in Table 1), The contour was evolved to visual object's actual boundary with our algorithm faster than the existing algorithms.

Table 1. Evolution time statistics

Scenes	Frame size	Initial contour acquisition time /s	evolution time in [15]/s	evolution time in [5] /s	evolution time in this paper /s	Iterations of target 1	Iterations of target 2	Iterations of target 3
1	384*288	2.531	11.598	11.437	11.695	4	12	12
2	384*288	2.36	16.972	16.963	17.191	16	6	17
3	384*288	2.543	10.139	10.16	10.076	13	12	–
4	384*288	2.175	14.84	14.646	13.977	11	14	12
5	352*288	2.348	12.243	12.468	12.509	21	8	–
6	352*288	1.995	19.503	19.985	20.782	27	21	–

Table 1 shows the time consume of obtaining initial contour and the contrast of evolution time between existing algorithms and the improved algorithms. From the table and Fig. 5, we can see that contour evolution takes up most of the time, and the contour obtained with our improved DRLSE is more accurate than the existing algorithms almost in the same amount of time.

It's worth mentioning that the evolution of multiple targets is in serial in this paper, and if the algorithms were optimized, the evolution speed would be greatly accelerated. In addition to, in the case that contour's accuracy is not need too high, just using the algorithm for getting initial contours can meet the requirements of the application.

6   Conclusions

In this paper, we have presented an effective segmentation method of multiple visual objects. We have the following conclusions: (1) The algorithm based on adaptive Otsu and asymmetric morphologic filter can provide more accurate multiple initial contours than tradition algorithms which express contours as regular shape and does not distinguish multiple objects. (2) Our improve DRLSE based on background difference and frame difference can evolve contour to the actual boundary more precisely than the existing algorithms in almost the same time.

Acknowledgement. The authors thank for the supports by the Doctoral Candidate Excellent Dissertation Cultivating Project of Hainan University, Hainan Province Natural Science Foundation of China (No. 20156228), National Natural Science Foundation of China (No. 61561017) and Hainan social development science and technology projects (No. 2015SF33).

References

1. Hu, W., Zhou, X., Li, W., Luo, W., Zhang, X., Maybank, S.: Active contour-based visual tracking by integrating colors, shapes, and motions. IEEE Trans. Image Process. **22**(5), 1778–1792 (2013)
2. Cremers, D., Rousson, M., Deriche, R.: A review of statistical approaches to level set segmentation integrating color, texture, motion and shape. Int. J. Comput. Vision **72**(2), 195–215 (2007)
3. Zhou, X., Hu, W.: Object contour tracking with fusion of color and incremental shape priors. ACTA Automatica Sin. **35**(11), 1395–1402 (2009). (in Chinese)
4. Wang, M., Dai, Y., Wang, Q.: A novel fast-snake object tracking approach. ACTA Automatica Sinica. **40**(6), 1108–1115 (2007). (in Chinese)
5. Hu, Z., Zhao, Y., Cheng, J., Peng, J.: Moving object segmentation method based on improved DRLSE. J. Zhejiang Univ. Eng. Sci. **48**(8), 1488–1495 (2014). (in Chinese)
6. Otsu, N.: A threshold selection method from gray-level histogram. IEEE Trans. Syst. Man Cybern. **9**(1), 62–66 (1979)
7. Hou, Z., Hu, Q., Nowinski, W.: On minimum variance thresholding. Pattern Recogn. Lett. **27**(14), 1732–1743 (2006)
8. Zhao, J., Feng, C., Shao, F., Zhang, X.: Moving object detection and segmentation based on adaptive frame difference and level set. Inf. Control **41**(2), 153–158 (2012). (in Chinese)
9. Yu, H., You, Y.: Detecting and segmenting multiple moving objects using level set method. J. Zhejiang Univ. Eng. Sci. **41**(3), 412–417 (2007). (in Chinese)
10. Li, J., Wang, J., Liang, S., Shen, W.: Method of detecting multiple moving object based on improved level set. Trans. Beijing Inst. Technol. **31**(5), 557–561 (2011). (in Chinese)
11. Zhao, H., Liu, Z., Zhang, H.: Segmentation of multiple moving targets based on modified C-V model. Chin. J. Sci. Instrum. **30**(5), 1082–1089 (2010)
12. Kass, M., Witkin, A., Terzopoulos, D.: Snakes: active contour models. In: Proceedings-First International Conference on Computer Vision London, pp. 259–268 (1987)
13. Caselles, V., Kimmel, R., Sapiro, G.: Geodesic active contours. Int. J. Comput. Vision **22**(1), 61–79 (1997)
14. Li, C., Xu, C., Gui, C., Fox, M.: Level set evolution without re-initialization: a new variational formulation. In: 2005 IEEE Computer Society Conference on Computer Vision and Pattern Recognition, CVPR 2005, pp. 430–436 (2005)
15. Li, C., Xu, C., Gui, C., Fox, M.: Distance regularized level set evolution and its application to image segmentation. IEEE Trans. Image Process. **19**(12), 3243–3254 (2010)

The Measurement of Human Height Based on Coordinate Transformation

Xiao Zhou[1], Peilin Jiang[2,3], Xuetao Zhang[1,4],
Bin Zhang[2], and Fei Wang[1(✉)]

[1] The Institute of Artificial Intelligence and Robotics, Xi'an Jiaotong University,
No.28 Xianning West Road, Xi'an 710048, China
{xuetaozh,wfx}@mail.xjtu.edu.cn
[2] The School of Software Engineering, Xi'an Jiaotong University,
No.28 Xianning West Road, Xi'an 710048, China
{pljiang,bzhang82}@mail.xjtu.edu.cn
[3] National Engineering Laboratory for Visual Information Processing
and Application, Xi'an Jiaotong University,
No.28 Xianning West Road, Xi'an 710048, China
[4] Shaanxi Digital Technology and Intelligent System Key Laboratory,
Xi'an Jiaotong University, No.28 Xianning West Road, Xi'an 710048, China

Abstract. This paper presents a new method for measuring human height from video frames. Based on coordinate transformation, the positional relationship between the coordinates of a person's feature points on the image can be transformed into the person's height with the knowledge of intrinsic parameters and rotation angle of the camera. In our method, the distance between the camera and the target is not a necessary parameter, which, in contrast, can be estimated by our algorithm. From experiments, we conclude that our method can be simply implemented to estimate a person's height from video frames in a controllable error scale.

Keywords: Visual measuring · Coordinate transformation · Camera calibration · Video

1 Introduction

Various human-recognition technology based on image processing have been presented in recent years, such as face-recognition, gait analysis, morphological analysis of human and so forth, which are mature enough to be utilized in many fields. However, the research on vision measurement plays an important role in surveillance video analysis. Specifically, the information of human height obtained from a normal surveillance camera system makes up the important characters of a target [1, 2].

This paper presents a coordinate transformation based method using a calibrated Pan Tilt Zoom (PTZ) camera to evaluate the walking object's height. Our method can work very well as long as the object is walking. Moreover, any translation or rotation of the surveillance camera will not invalidate the measurement system or reduce its precision. Certainly, to guarantee the accuracy, several parameters are indispensable for

© Springer International Publishing Switzerland 2016
D.-S. Huang et al. (Eds.): ICIC 2016, Part III, LNAI 9773, pp. 717–725, 2016.
DOI: 10.1007/978-3-319-42297-8_66

our method. Firstly, the intrinsic parameters of the camera are used to transform the image coordinate system to the camera coordinate system. The rotation angle between the camera's optical axis and the horizontal plane can help modify the camera coordinate system.

2   Related Research

Extensive research has been done in human height measurement using images. There are about two different approaches: camera-only geometry based method and multi-device based method. One of the multi-device based methods [3], considers a camera as the main hardware and a fixed laser beam is used for signal emission. Because of good identifiability of the laser beam in image, it is easy to extract distance between the laser beam projection image and the center of image, which is used to estimate the human height. This method is simple, but with a high cost due to the laser generator. Sonia Das proposed the Direct Linear Transformation method [4], to get human height variation. Specifically, the intrinsic and extrinsic parameters of the camera are needed to compute the Z (the vertical direction) coordinates of the person's sole and head using DLT method, and then the difference between them is the person's height. The method proposed by Richard Hartley [5] is one of the camera-only geometry based methods. It is necessary to extract the information told by the reference lines in a real vertical reference plane. In addition, from the whole measurement procedure, the camera cannot be moved or rotated at all. Once the camera's position is changed, the information in the reference image should be re-extracted, which in a large degree restrains the camera's vision. While our coordinate transformation based method can solve the problems mentioned above perfectly, which is one of the camera-only geometry based methods, and neither need a real reference plane nor get invalid on account of the camera's translation or rotation.

2.1   The Coordinate Transformation Based Algorithm

Our method is based on the coordinate transformation. In short, the intrinsic parameters are used to transform the image coordinate system to the camera reference frame. The angle between the camera's optical center and the ground can help modify the camera coordinate system. At last, the height of the camera is regarded as a reference to compute the height of the walking object.

Generally speaking, a person's height can be judged by the distance between the head of the person and his/her sole when the person stands upright on the floor. In other words, we can entirely establish a coordinate system matching the corner of the wall, as a result, a person's height can be estimated as the absolute value of the difference between the ordinates of the person's head and the sole. Therefore any coordinate systems without rotation compared with the corner of wall are available for us to measure a person's height. Well, as the link between the world coordinate system and the image reference frame, the camera reference frame is the best choice.

In a standard camera coordinate system (the optical axis of the camera parallels the ground and the imaging plane is perpendicular to the ground), *TB* denotes the person in the vision of the camera, *H* is the height of the person, *tb* is the image of *TB* on the virtual imaging plane, *D* shows the distance between the camera and the person; The front view of the geometric model is shown in Fig. 1.

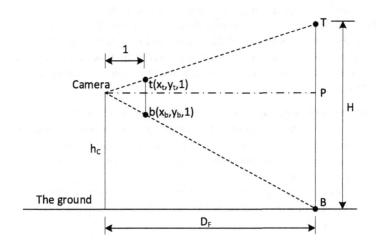

Fig. 1. The front view of the geometric model of the real scene

Supposing that the 3D coordinates, which are normalized along direction of the optical axis of the camera (z-axis), of the points t and b are $(x_t, y_t, 1)$ and $(x_b, y_b, 1)$ respectively. Then using similar triangles theorem, the height of the person denoted by *H* can be easily computed as follows:

$$H = h_c\left(1 - \frac{y_t}{y_b}\right) \tag{1}$$

Where is the height of the camera. Similarly, when the optical axis of the camera crosses through the person's body, the perpendicular distance between the lens and the person which is denoted by D_F can be computed as follows:

$$D_F = \frac{h_c}{y_b} \tag{2}$$

Nevertheless, when the optical axis of the camera doesn't crosses through the person's body, D_F is just the projection of the perpendicular distance between the lens and the person, so the real distance denoted by *D* in such circumstances should be modified by the following equation:

$$D = \frac{h_c}{y_b} \sqrt{1 + \frac{(x_t + x_b)^2}{4}} \tag{3}$$

Strictly speaking, x_t and x_b are identical when the target is a vertical segment with no width. However, the width of the walking object cannot be omitted, so we use the mean of x_t and x_b to represent the abscissa of the walking object.

So far, we have managed to estimate the height and the distance on the condition of the knowledge of the coordinates of the points t and b. However, the three-dimensional coordinates of t and b in the camera reference frame cannot be gained directly from the image. Therefore we have to use coordinate transformation to calculate these coordinates as follows.

Coordinate Transformation. There are two types of camera parameters including the intrinsic and the extrinsic. Intrinsic or internal camera parameters describe the projection of objects onto the camera image [6]. They establish the relationship between the points in the camera reference frame and the pixel coordinates of the points on the images captured from the camera.

Assuming that the pixel coordinates of t in the image reference frame is denoted by (u_t, v_t), then its coordinates denoted by $(x_1, y_1, 1)$ in the camera reference frame can be estimated as follows:

$$\begin{bmatrix} x_1 \\ y_1 \\ 1 \end{bmatrix} = A^{-1} \begin{bmatrix} u_t \\ v_t \\ 1 \end{bmatrix} \tag{4}$$

A is the intrinsic parameters matrix and given by

$$A = \begin{bmatrix} f_x & 0 & u_0 \\ 0 & f_y & v_0 \\ 0 & 0 & 1 \end{bmatrix} \tag{5}$$

where f_x, f_y are the equivalent focal length in x and y direction. u_0, v_0 are the principal point in x and y direction. Likewise, the coordinates of b denoted by $(x_2, y_2, 1)$ in the camera reference frame can be computed from the Eq. (4).

Modifying the Camera Reference Frame. The camera reference frame might not be a standard camera coordinate system because of the optical axis might not parallel the ground, in other words, there may exist a nonzero angle denoted by β between the optical axis and the ground. Thus we have to rotate the current camera reference frame to make it a standard camera coordinate system. The rotation angle of the transformation is the angle between the optical axis and the ground which can be acquired directly from a PTZ camera, and the rotation direction is the direction that can lessen the angle. The coordinates of t after the rotation transformation denoted by $(x_t, y_t, 1)$ is shown in the Eq. (6).

$$\lambda_1 \begin{bmatrix} x_t \\ y_t \\ 1 \end{bmatrix} = R \begin{bmatrix} x_1 \\ y_1 \\ 1 \end{bmatrix} \tag{6}$$

R is the rotation matrix and given by:

$$R = \begin{bmatrix} 1 & 0 & 0 \\ 0 & \cos\beta & \sin\beta \\ 0 & -\sin\beta & \cos\beta \end{bmatrix} \tag{7}$$

Where $(x_1, y_1, 1)$ is obtained by the Eq. (4). Apparently, the coordinates gained from the Eq. (6) is normalized along the z-axis. Likewise, the coordinates of b after the rotation transformation denoted by $(x_b, y_b, 1)$ can be estimated by the Eq. (6) with the change of the right side into $(x_2, y_2, 1)$.

By now, we have obtained the normalized three-dimensional coordinates of t and b in the standard camera reference frame, as long as we have the knowledge of the camera's height, we can estimated the person's height and the perpendicular distance between the lens and the person from the Eqs. (1) and (3).

As we mentioned earlier, h_c is the height of the camera, which is actually the distance between the optical center of the camera and the horizontal plane that the camera lies on. In fact, we cannot get the accurate height of the camera using a tapeline since we don't know the exact position of the optical center of the camera. But we can estimate it according to the Eq. (1): put a reference object with known height denoted by h_R in front of the camera and then h_c can be estimated as follows:

$$h_c = \frac{y_b}{y_b - y_t} h_R \tag{8}$$

3 Experiments

To testify feasibility and accuracy of our method, we designed two groups of experiments. Control group is based on Ngoc Hung Nguyen's method [7]. The first group of experiments is named feasible measurement experiment, which is implemented to estimate the human height from a series of video frames. The second group is used to testify the accuracy of our method, which considers the image of calibration plate as experimental subjects.

3.1 Human Height Measurement

The first step of our measurement system is to distinguish and extract the walking subject from a fixed background, which can be completed by using GMM [8] method. Then the ordinates of the top of the human head and his/her sole on each video frame can be easily acquired from the foreground image exported by GMM algorithm.

Finally, the measurement algorithm will be implemented to compute the height of the walking subject. In our experiments, the volunteer passed through the vision of the camera in a relatively low speed. We changed the angle between the optical axis of the camera and the horizontal plane from 7° to 13°. Correspondingly, we got 7 videos that captured the volunteer's motion.

As we mentioned before, the author's method needs six parameters including three ordinates of the reference lines on the image and their heights in the real world to evaluate the human height. Thus we have to get 7 images capturing the three reference lines on the vertical reference plane. Besides, we have to measure their real heights and extract their ordinates on each image corresponding to every different angle manually. However, our method requires only one particular parameter, which is the angle between the optical axis of the camera and the horizontal plane and can be immediately obtained from the PTZ camera. The experimental results are shown in Fig. 2.

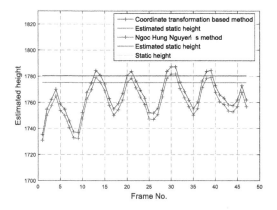

Fig. 2. The estimated heights of the volunteer when the angle between the optical axis of the camera and the horizontal plane is 7°: the star shaped points and the cross shaped points correspond to the heights computed by the author's method and our method respectively, while the static height is shown by the green horizontal straight line. The red and blue straight lines denote the estimates of the static height, which are the average of the maxima of the heights curves. (Color figure online)

Figure 2 shows the height variation of the volunteer. It can be easily observed that the results obtained by the two methods seem to match each other while the human height variation is quite significant. Figure 3 shows the relative error of the heights computed by the author's method and our method respectively in each video. Apparently, the estimated static heights computed by our method are more accurate than the author's method in the first three videos, while the others are not. More importantly, we can see the largest relative error of our method is about 1.5 % from Fig. 3, which is absolutely acceptable in measuring a person's height.

According to Ngoc Hung Nguyen's method, three reference lines are needed to compute the height of the target. One of the reference planes is arranged as shown in Fig. 4(a), and the three red circles on the calibration plate are chosen to be the reference

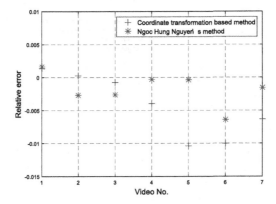

Fig. 3. The relative error of the estimated static height, which is the average of the maxima of each estimated heights' curves corresponding to every single video. The star shaped points and the cross shaped points denote the relative error pf heights computed by the author's method and our method respectively.

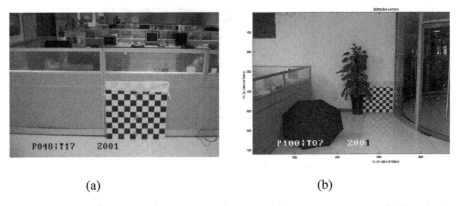

(a)   (b)

Fig. 4. (a) The reference plane with three red circles on the calibration; (b) One of test images (Color figure online)

lines. The real heights of the red corners and the ordinates of them on the image are extracted manually to compose the parameters of the author's method. One of the test images is shown in Fig. 4(b), in which we place the calibration plate about 5 meters away from the camera. We change the angle between the optical axis of the camera and the horizontal plane from 7° to 24° so that we get 18 images of the calibration plate. 48 corners are marked in red circles on each image. We compute the real height of each of the red corners on every image using both Ngoc Hung Nguyen's approach and ours. The experimental results are shown in Figs. 5 and 6.

Obviously, Fig. 5 illustrates that the average relative error of each image obtained by our approach is about 1.81 % while that of the author's method is about 3.39 %, which is nearly twice ours. Figure 6 shows the same result.

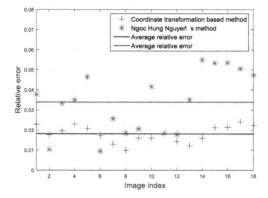

Fig. 5. The average relative error of the measurement height of 48 red corners on each calibration plate image for both methods: the blue star-like marks indicate the average relative errors of the height computed by our coordinate transformation based method, and the red marks denote Ngoc Hung Nguyen's. (Color figure online)

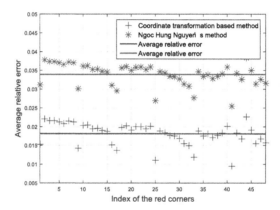

Fig. 6. The average relative error of the estimated height of each red corner on 18 calibration plate images for both methods: the blue star-like marks indicate the average relative errors of the height computed by our coordinate transformation based method, and the red marks denote Ngoc Hung Nguyen's. (Color figure online)

4   Conclusion

Our experiments show that human height can be accurately measured by using a calibrated PTZ camera with a measurement system. Compared with other methods, the arrangement of our system is very simple, including calibration of the camera and accurate measurement of the camera's height. Our height estimation algorithm can handle various situations. Especially when the camera moves or rotates instead of being fixed on the wall, our coordinate transformation based method can still work very well as long as the angle between the optical axis and the horizontal plane can be exactly obtained.

Acknowledgements. This work was supported by the National Natural Science Foundation of China under Grant Nos: 61273366 and 61231018 and the program of introducing talents of discipline to university under grant no: B13043.

References

1. Yadav, A., Patil, T.B.M.: Study of imaged based human height measurement & application. Int. J. Adv. Res. Comput. Eng. Technol. (IJARCET) **1**(6), 68–69 (2012)
2. Tsai, H.C., Wang, W.C., Wang, J.C., Wang, J.F.: Long distance person identification using height measurement and face recognition. In: IEEE Region 10 Conference, TENCON 2009. IEEE, Singapore (2009)
3. Das, S., Meher, S.: Automatic extraction of height and stride parameters for human recognition. In: 2013 Students Conference on Engineering and Systems (SCES). IEEE, Allahabad (2013)
4. Wang, C.M., Chen, W.Y.: The human-height measurement scheme by using image processing techniques. In: 2012 International Conference on Information Security and Intelligence Control (ISIC), Yunlin, Taiwan (2012)
5. Hartley, R., Zisserman, A.: Multiple View Geometry in Computer Vision. Cambridge University Press, Cambridge (2003)
6. Zhang Z.: A flexible camera calibration by viewing a plane from unknown orientations. In: Proceedings of ICCV 1999, pp.666–673, Corfu, Greece (1999)
7. Nguyen, N.H., Hartley, R.: Height measurement for humans in motion using a camera: a comparison of different methods. In: International Conference on Digital Image Computing Techniques and Applications, Fremantle, Western Australia, 3–5 December 2012
8. Zivkovic, Z.: Improved adaptive Gaussian mixture model for background subtraction. In: Proceedings of the 17th International Conference on ICPR 2004. IEEE, Cambridge (2004)

Scene Text Detection Based on Text Probability and Pruning Algorithm

Gang Zhou$^{(\boxtimes)}$, Yajun Liu, Fei Shi, and Ying Hu

The Institution of Information Science and Technology, Xinjiang University,
Shengli Road, 14, Ürümqi 830001, China
gangzhou_xju@126.com

Abstract. As the scene text detection and localization is one of the most important steps in text information extraction system, it had been widely utilized in many computer vision tasks. In this paper, we introduce a new method based on the maximally stable extremal regions (MSERs). First, a coarse-to-fine classier estimates the text probability of the ERs. Then, a pruning algorithm is introduced to filter non-text MSERs. Secondly, a hybrid method is performed to cluster connected components (CCs) as candidate text strings. Finally, a fine design classifier decides the text strings. The experimental results show our method gets a state-of-the-art performance on the ICDAR2005 dataset.

Keywords: Scene text detection · Maximum stable extreme regions · Text probability · Pruning algorithm

1 Introduction

As camera and mobile phone become more and more popular, mass images and videos need automatic analysis. Since the text information can be easily understood by computer, the research on the images and videos text information extraction system become an important research in recent years. Scene text detection and localization is the first step in a text information extraction system, and is the most important step in many vision tasks such as web images analysis, mobile phone translation and so on.

The characters in natural scene vary in size, color and font, and should be very hard to distinct from the complex background (see Fig. 1). Such characteristics of scene text detection made it a challenging research. In many comprehensive surveys of scene text detection [1], the methods can be classified into two categories: the region-based methods and the CCs-based methods. Region-based methods extract texture feature vectors by a sliding window strategy in different scales and positions of images. Candidate text regions are decided by rules or classifiers, and then are merged to generate text blocks. Although this kind of methods are robust to the influence of natural light and the image blur, it needs a large amount computation and gets a imprecise location results. In our early work [2], a cascade classifier is introduced to detect local sliding regions by three kinds of features. The results got a very high recall ratio but a very poor precision ratio. In another work [3], a conditional random field model is introduced to combine the local similarity sliding windows as a unity, and improved the precision ratio.

© Springer International Publishing Switzerland 2016
D.-S. Huang et al. (Eds.): ICIC 2016, Part III, LNAI 9773, pp. 726–735, 2016.
DOI: 10.1007/978-3-319-42297-8_67

Fig. 1. The scene text image samples

On the other hand, CCs-based methods segment candidate CCs from images, and the non-text CCs are pruned with appearance of unary components. The candidate text CCs are then clustered as CCs unites for the sequent analysis. This approach is attractive because the results can be directly used for character recognition. In our early work [4], the images were over-segmented into super-pixels, and candidate CCs were extracted by combining local contrast and color consistency. The non-text components are then pruned by a hierarchical model consisting of three stages in cascade. In another work [5], a text probability map consisting of the text position and scale information is estimated to segment CCs. To filter out the non-text CCs, a hierarchical model consisting of two classifiers in cascade is utilized to filter.

In this paper, we apply the MSER methods to extract CCs. And then, text probability is estimated by a coarse-to-fine classifier and is utilized to filtered the non-text CCs by a pruning algorithm. The candidate text CCs are clustered into strings which can be determined by a fine designed classifiers. The rest of this paper is organized as follows: (1) related work is in Sect. 2, (2) the detail of this method is introduced in Sect. 3, (3) the experiments are shown in Sect. 4, (4) the conclusions are given in Sect. 5.

2   Related Work

There are two key parts in CCs-based methods, CCs segmentation and analysis. In CCs segmentation parts, the MSER method becomes the most efficient one. This method is introduced in [6] which are widely utilized in scene text detection. In [7], the MSERs are extracted with a fixed delta and is processed with a CCs analysis part and a CCs cluster algorithm. In [8], a more common extremal region (ER) method is introduced. The ERs is extracted not only on grey channel but also in other channel. Then, incremental computable feature sets with a cascade classifier are utilized for filtering the non-text ERs. In [9], the author proposes a novel frame takes advantages of both MSERs and sliding-window based methods. The MSERs operator dramatically reduces the number of windows scanned and enhances detection of the low-quality texts. While the sliding-window with convolutional neural network is applied to correctly separate the connections of multiple characters in components. In [10], a fast and effective pruning algorithm is designed to extract MSERs as character candidates using the strategy of minimizing regularized variations.

In CCs analysis part, the researchers are focused on extraction of distinct features and CCs cluster. A stroke width transform algorithm is introduced in [11]. The stroke width map is then utilized to extract CCs and becomes a powerful feature. In [12], the approach mirrors the move from saliency detection methods to measures of objectiveness. In order to determine the characters, the author develops three novel cues that are tailored for character detection and a Bayesian method for their integration. In [13], both character appearance and structure are combined to generate representative and discriminative text descriptors.

CCs cluster part is utilized for exploring more context information. To efficiently filter out the non-text components, a conditional random field model considering unary component properties and binary contextual component relationships with supervised parameter learning is proposed in [14]. A new text model is constructed to describe text unit which includes two characters and the link connection. For every candidate text unit, they combine character and link energies to compute text unit energy which measures the likelihood that the candidate is a text object.

3 Scene Text Detection Method Introduction

In this paper, we propose a MSER-based method. The text probability is estimated by a coarse-to-fine classifier and is utilized to filter the non-text MSERs with a pruning algorithm. The candidate CCs are then linked by a rule-based algorithm and a classifier based algorithm. The linked CCs are merged into the CCs strings which are then classified as text regions and non-text regions. The flowchart of our method is shown in Fig. 2.

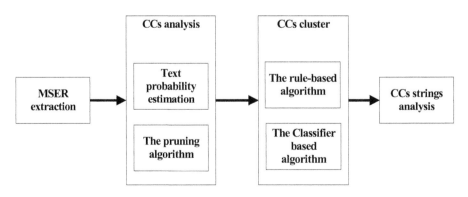

Fig. 2. The flowchart of our method

3.1   CCs Analysis

The MSER algorithm will segment repeated components which are overlap in the same region. The repeat components can be combined as a MSER tree in a parent-children relationship, see Fig. 3(a). To filter the repeat components will help us to find the proper text CCs and reduce the large amount number of non-text CCs.

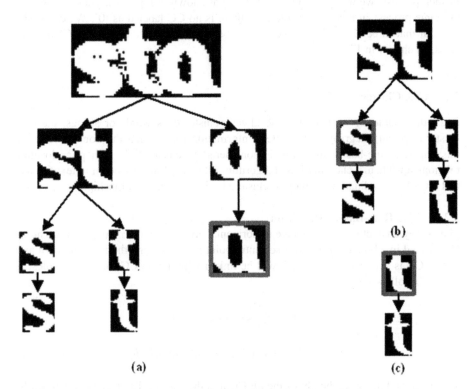

(a)

(b)

(c)

Fig. 3. The MSER tree pruning algorithm. The region with red bounding box get maximum text probability. (a) First pruning. (b) Second pruning. (c) Third pruning.

3.1.1   Text Probability Estimation

The text probability of every MSER will be estimated by a cascade classifier with two layers. In the first layer, 4 features with low computation are given with an Adaboost classifier. In the second layer, 8 features (include 4 features in the first layer) are given with a SVM classifier to estimate the text probability.

The first layer will pruning most of non-text MSERs, and 4 features are the height-width ratio, the occupation ratio, the compactness, and convex hull ratio. The second layer will add 4 more features to estimate the text probability: the boundary gradient, the stroke width occupation ratio, the stroke width difference, and the stroke width-height ratio. All the details of these features can be found in [4, 8].

3.1.2   Pruning Algorithm

In a MSER tree, all the CCs are linked as father nodes or children nodes. We first label all the CCs as '0' means not processed. Then, the CC with the maximum text probability will be selected and labeled as '1' means a text CC. As show in Fig. 3(a), the character 'a' with red bounding box is a candidate text CC. All the ancestor nodes and descendant nodes of this candidate text CC will be labeled as '2' means non-text CCs. In such process, we will get a new MSER tree as shown in Fig. 3(b). Then, we will do the same process in this MSER tree until there is no CC labeled as '0'. We can easily find out that three CCs 'a', 's', and 't' will be labeled as candidate text CCs in three steps, as shown in Fig. 3(a–c).

3.2   CCs Cluster

Since the characters in natural scene always clustered as words or strings, the CCs cluster step can integrate more context information for filtering the non-text CCs. In this paper, a rule-based algorithm is considered to cluster CCs with prior location relationship. Then, a classifier based algorithm filters the link between non-text CC and text CC. The candidate CCs will be clustered with the link relationship.

3.2.1   The Rule-Based Algorithm

The nearby characters always locate in horizontal arrangement in natural scene. The ith CC located in a bounding box whose centre point is (x_i^c, y_i^c), and the top and down axis of this CC is y_i^t and y_i^d. The jth CC will be linked with the ith CC follow two rules, as shown in following.

$$\left| x_i^c - x_j^c \right| < 4 \cdot \max(h_i, h_j) \tag{1}$$

$$\min(y_i^d, y_j^d) - \max(y_i^t, y_j^t) > 0.5 \cdot \min(h_i, h_j) \tag{2}$$

The h_i and h_j are the height of the ith CC and the jth CC. The Eq. 1 gives a nearby relationship, and the Eq. 2 gives a horizontal arrangement relationship. Following these two rules, the CCs will be linked (as shown in Fig. 4(b)).

3.2.2   The Classifier Based Algorithm

The nearby characters always shared similar color, stroke width and height. To eliminate the links between the text CCs and non-text CCs, a SVM classifier is utilized to decide the text CCs links. In the training step, the links between the text CCs are labeled as positive samples and the links between the non-text CCs and text CCs are labeled as negative samples (other links are not labeled). Then, 5 features are extracted to describe the links: the color difference, the stroke difference, the height difference, mean text probability of two CCs, and text probability difference. All the details of these features can be found in our previous work [4].The linked CCs will be clustered as candidate CCs strings.

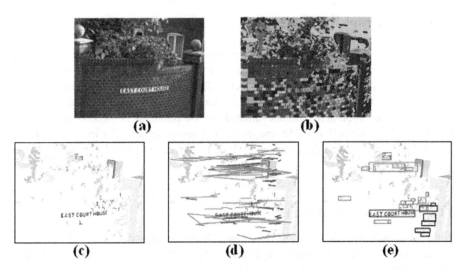

Fig. 4. CCs cluster and CCs strings analysis. (a) The original image. (b) The MSERs. (c) The text probability of the candidate CCs. (d) The CCs cluster results. All the links are produced by the rule based algorithm. The red links are decided by the classifier based algorithm. (e) The red bounding box is the text CCs, the green bounding box is the first kind of non-text CCs strings, and the blue bounding box is the second kind of non-text CCs strings. (Color figure online)

3.3   CCs Strings Analysis

Although the characters can be linked as a text string after CCs cluster step, some of non-text CCs also can be clustered as CCs strings. From the labeling step, we found out that there are two kinds of non-text CCs strings in natural scene. In first kind, some of the non-text CCs (such as leaves) are occasionally clustered as CCs strings. This kind of CCs strings always appears as low text probability, small number of CCs, and so on. In second kind, some of repeat CCs (such as barrels, bricks, windows) are clustered as CCs strings. This kind of CCs strings always appears as same shape, see Fig. 4(c).

From the observation of the labeling step, 5 features are built up for a SVM classifier. As shown in next:

- Mean text probability. In a text CCs strings, the text probability of every CC will be higher than the non-text CCs strings.
- The number of CCs. The number of CCs in a text CCs string will be higher.
- The stroke height ratio. The ratio between the mean stroke of every CCs and the height of the CCs string region can separate the text CCs stings from the repeat CCs strings (the bricks and barrels always appear high ratio).
- The occupation ratio difference. The occupation ratio in every CC of a text string will be higher than the repeat CCs strings.
- The stroke width occupation ratio difference. The stroke width occupation ratio in every CC of a text string will be higher.

As the text strings must be split into words, we use the bounding box distance to measure the distance between words [4].

4 Experimental Results

In experiments, we focus on analysis of every step and the final location results. In Subsect. 4.1, the parameters of the classifiers of three steps (CCs analysis, CCs cluster and CCs strings) are discussed. In Subsect. 4.2, the detection and localization of our method on the ICDAR2005 dataset will be compared with other methods. The performance of our approach is first evaluated on the public ICDAR2005 dataset. The ICDAR2005 dataset includes 258 images in the training set and 251 images in the test set. The text characters in these images include English and numeral characters and are horizontally aligned.

4.1 The Parameters Discussion

In the CCs analysis step, there are two stages: Adaboost classifier and SVM classifier. In Adaboost classifier, 4 features are extracted from the labeled samples. The RealAdaboost algorithm is utilized, and about 1.8 k positive samples and 32 k negative samples are training the parameters. In Fig. 5(a), the ROC curve of the Adaboost classifier obtained by cross-validation. The threshold used in the experiments is 2.2 (recall 95.85 %, precision 65.83 %).

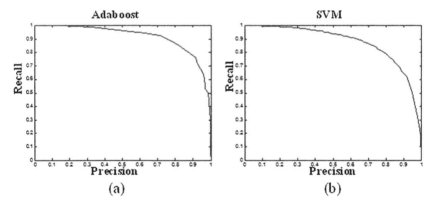

Fig. 5. The performance in CCs analysis step. (a) The Adaboost classifier performance. (b) The SVM classifier performance.

In SVM classifier, 8 features are extracted from 13 k samples (about 1.7 k positive samples and 11.3 k negative samples). In Fig. 5(b), the ROC curve of the SVM classifier obtained by cross-validation. And a logistic correction algorithm is obtained to estimate the text probability. The threshold used in the experiments is 0.05 (recall 94.83 %, precision 50.16 %).

In the CCs cluster step, a SVM classifier is obtained in classifier based algorithm. In the CCs pairs labeling, we found out there are less negative samples (about 120 negative samples and 400 positive samples). When the threshold value is 0, the classifier performance obtained 93 % recall ratio and 97 % precision ratio. And in the CCs strings step, we also use a SVM classifier to analyze the text strings. In this step about 273 positive samples and 322 negative samples are labeled. When the threshold value is 0, we got about 95 % recall ratio and 82 % precision ratio.

4.2 The Localization Results

After we obtain the text localization results, the standard precision-recall terms [15] are obtained to evaluate the word localization result. The word detection result is a set of rectangles designating bounding boxes for detected words. A set of ground truth boxes is provided in the dataset. The match value between two rectangles is defined as the area of the intersection rectangle divided by the area of the minimum containing rectangle. Hence, the best match for a rectangle in a set of rectangles can be defined as the maximum match value.

Our approach got about 69 % recall ratio, 72 % precision ratio and 70 % F-measure performance which is better than our previous work. Comparing with some other public papers, our approach is also better. All the comparison results can be seen in Table 1.

Table 1. The comparision results

Approach	Recall %	Precision %	F-measure %
Proposed methods	69	72	70
Gang et al. [4]	69	70	69
Epshtein et al. [11]	60	73	66
Pan et al. [14]	67	70	69
Gang et al. [5]	67	68	67

Fig. 6. Some of successful samples

Some of the succeed samples of localization results can be seen in Fig. 6. Our approach can get precise results, and the results can be recognized directly.

Some of the failed samples of localization results can be seen in Fig. 7. Our approach failed on some nature images due to two categories. In first category, some of characters can not be extracted by the MSER algorithm. As the MSER algorithm processed on the intensity channel, some characters in scene do not appear contrast in gray value, see Fig. 7(a). In second category, character is isolated in images. In this situation, the text CC can not be clustered, and can not be properly processed in sequential step, see Fig. 7(b).

(a) low contrast

(b) isolated character

Fig. 7. Some of failed samples

5   Conclusion

In this paper, we propose a scene text detection method based on estimating text probability and pruning algorithm. Three steps including CCs analysis, CCs cluster, and CCs strings analysis are obtained in our approach. In CCs analysis step, a coarse-to-fine classier estimates the text probability of the MSERs. Then, a pruning algorithm based on select maximum text probability is introduced. In CCs cluster step, a rule based algorithm and a classifier based algorithm are proposed to link CCs pairs. In CCs strings analysis step, a SVM classifier decides the text strings. The experimental results discussed the parameters in each step and analyzed the localization results on ICDAR2005 dataset. Our method got better results than other 4 methods. However, some low contrast and isolated character samples are failed by our method which needs to improve in our future work.

Acknowledgements. The authors thank the anonymous reviewers and editor who provided many valuable comments and suggestions. This work is funded by Scientific Research Programs of the Higher Education Institution of XinJiang (grant nos. XJEDU2014S006 and XJEDU2014I004).

References

1. Ye, Q.X., Doermann, D.: Text detection and recognition in imagery: a survey. IEEE Trans. Pattern Anal. Mach. Intell. **37**(7), 1480–1500 (2015)
2. Zhou, G., et al.: Detecting multilingual text in natural scene. In: Proceedings of the 2011 1st International Symposium on Access Spaces (ISAS), pp. 116–120 (2011)
3. Zhou, G., et al.: A new hybrid method to detect text in natural scene. In: 2011 18th IEEE International Conference on Image Processing (ICIP 2011), pp. 2605–2608 (2011)
4. Zhou, G., et al.: Scene text detection method based on the hierarchical model. IET Comput. Vision **9**(4), 500–510 (2015)
5. Zhou, G., Liu, Y.H.: Scene text detection based on probability map and hierarchical model. Opt. Eng. **51**(6), 067204-1–067204-9 (2012)
6. Matas, J., et al.: Robust wide-baseline stereo from maximally stable extremal regions. Image Vis. Comput. **22**(10), 761–767 (2004)
7. Koo, H.I., Kim, D.H.: Scene text detection via connected component clustering and nontext filtering. IEEE Trans. Image Process. **22**(6), 2296–2305 (2013)
8. Neumann, L., Matas, J.: Real-time scene text localization and recognition, In: 2012 IEEE Conference on Computer Vision and Pattern Recognition, pp. 3538–3545. IEEE, New York (2012)
9. Huang, W., Qiao, Y., Tang, X.: Robust scene text detection with convolution neural network induced MSER trees. In: Fleet, D., Pajdla, T., Schiele, B., Tuytelaars, T. (eds.) ECCV 2014, Part IV. LNCS, vol. 8692, pp. 497–511. Springer, Heidelberg (2014)
10. Yin, X.C., et al.: Multi-orientation scene text detection with adaptive clustering. IEEE Trans. Pattern Anal. Mach. Intell. **37**(9), 1930–1937 (2015)
11. Epshtein, B., et al.: Detecting text in natural scenes with stroke width transform. In: IEEE Conference on Computer Vision and Pattern Recognition 2010, pp. 2963–2970. IEEE Computer Society, Los Alamitos (2010)
12. Li, Y., et al.: Characterness: an indicator of text in the wild. IEEE Trans. Image Process. **23**(4), 1666–1677 (2014)
13. Yi, C.C., Tian, Y.L.: Text extraction from scene images by character appearance and structure modeling. Comput. Vis. Image Underst. **117**(2), 182–194 (2013)
14. Pan, Y.F., Hou, X.W., Liu, C.L.: A hybrid approach to detect and localize texts in natural scene images. IEEE Trans. Image Process. **20**(3), 800–813 (2011)
15. Lucas, S.M., et al.: ICDAR 2003 robust reading competitions: entries, results, and future directions. Int. J. Doc. Anal. Recogn. **7**(2–3), 105–122 (2005)

A Novel Feature Point Detection Algorithm of Unstructured 3D Point Cloud

Bei Tian[1], Peilin Jiang[2,3], Xuetao Zhang[1,4],
Yulong Zhang[2], and Fei Wang[1(✉)]

[1] The Institute of Artificial Intelligence and Robotics, Xi'an Jiaotong University,
No.28 Xianning West Road, Xi'an 710048, China
{xuetaozh,wfx}@mail.xjtu.edu.cn
[2] The School of Software Engineering, Xi'an Jiaotong University,
No.28 Xianning West Road, Xi'an 710048, China
pljiang@mail.xjtu.edu.cn
[3] National Engineering Laboratory for Visual Information Processing
and Application, Xi'an Jiaotong University,
No.28 Xianning West Road, Xi'an 710048, China
[4] Shaanxi Digital Technology and Intelligent System Key Laboratory,
Xian Jiaotong University, No.28 Xianning West Road, Xi'an 710048, China

Abstract. Compared with 3D mesh data, unstructured point cloud data lack adjacency relationship between points, which only contain geometric coordinates and little information. This paper focuses on the research of characteristics of unstructured point cloud detection algorithm. We put forward the multiscale 3D Harris feature point detection algorithm, which uses iteration strategy to select the optimal Harris response value in multiple scales. Compared with the classical 3D Harris feature point detection algorithm for mesh data, our algorithm can fully use the local information of point cloud models to detect feature point on point cloud models. It is very robust to rotation transformation of point clouds and noise.

Keywords: Feature point detection · 3D point cloud · Unstructured · Multiscale · Harris

1 Introduction

Along with the popularization of 3D modeling techniques, feature point extraction from point cloud is a research hotspot, which is the key technology in the processing of point clouds. However, the existing feature extraction algorithm for 3D models is mainly in mainly aiming at the mesh data, which is unable to meet the requirements of feature extraction for unstructured point clouds. Unstructured point cloud data which contain much scan noise lack adjacency relationship between points and surface information of the original model. So it is hard to detect the feature points of unstructured 3D point cloud directly. While feature point detection algorithm based on mesh point cloud is relatively diversified. This paper draws lessons from the existing feature point detection algorithm of mesh point cloud, and puts forward the feature

© Springer International Publishing Switzerland 2016
D.-S. Huang et al. (Eds.): ICIC 2016, Part III, LNAI 9773, pp. 736–744, 2016.
DOI: 10.1007/978-3-319-42297-8_68

point detection algorithm of unstructured 3D point cloud. Extending the two-dimensional image feature detection algorithm to 3D point cloud is an important research idea among the existing mesh point cloud feature detection algorithms. One typical method is the 3D Harris feature point detection algorithm for mesh data [7]. The algorithm chooses the points with local maximum of Harris response value as candidate feature points. Before calculating the Harris response value, this algorithm uses surface fitting technology and determines the final feature points after choosing the candidate feature points using two feature point selection optimization method; one is determining the first part of candidate feature points with max Harris response value as the final feature points; the other Optimization method is clustering with the candidate feature points according to the distance between them.

Figure 1 shows the result of using feature point detection algorithm on mesh point cloud 3D Harris feature point detection algorithm for unstructured 3D point cloud. Figure 1. (a) shows the result of 3D Harris feature point detection. We can see that the feature points are mainly distributed over the edge or projection area of the point cloud model. While Fig. 1. (b) shows the result of using this algorithm directly to unstructured point cloud data directly, where the feature points detected are in haphazard order. Thus the 3D Harris feature point detection algorithm for mesh point cloud cannot be used to unstructured 3D point cloud directly.

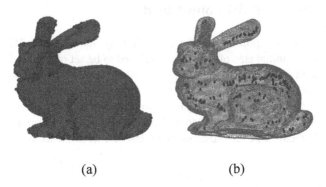

(a)   (b)

Fig. 1. D Harris feature point detection results by method used in [7]. In (a) is point cloud feature point detected result; in (b) is the unstructured point cloud detection.

2   Related Research

Feature point detection means to find a point cloud set in which the point do not changed following the subject's surface. The kind of feature point detection is divided into mesh data and unstructured data method. In first kind of detection the data firstly meshed and the linkage is obtained. Yamanny [1] made the feature point by high curvature value. Gal [2] clustered points of local curvature max value into a feature region and proposed geometric feature region detection method. Hildebrand [3] presented a method to detect feature lines by filtering for higher order surface derivatives and while [4] used topological analysis to detect feature lines. In scale space, by continuously sampling in scale space, [5] provided DoG to detect feature points, and similar to it, Zou [6] proposed to a DoG operator in Geodesic scale space. Sipiran [7] presented a method with 3D Harris operator. [5–7] extended 2D feature points detection operator into 3D surface. All these methods depend on mesh data with prior knowledge of linkage and are sensitive to noise. Otherwise, Hu and Hua [8] used Laplace-Beltrami operator and Sun [9] considered thermal diffusion theory to describe with the thermonuclear descriptor and detect the feature point with its local optimal value.

For the unstructured point cloud data, the feature point detection is basically based on geometric character of 3D object. The existing algorithms primarily focus on "convex feature" such as boundary, corner. Gal [10] used covariance analysis to cluster the point cloud and compute the minimum spanning graph (MSG). This method is not flexible in precise control. Pauly et al. [11] extended this approach to multiscale analysis. Wu [12] proposed to use local entropy to cluster the point cloud. A point is detected to be a feature point when the sampling neighborhood expanded and the local entropy obviously changing. Weber [13] projected sample points on Gaussian sphere and clustered to detect feature points. It performs poorly on complex structural surface. Park [14] proposed an improvement to use tensor voting analysis on multiscale.

3   Multiscale 3D Harris Feature Point Detection for Unstructured 3D Point Cloud

For 3D Harris feature point detection algorithm for unstructured 3D point cloud, parameters such as initial size of neighborhood need to be set manually. Since different point clouds need different parameters, this algorithm has bad adaption abilities. For the same point cloud model, different size of neighborhood cannot reflect sampling points' neighborhood structure. Thus scale space is introduced to solve the problems below. This paper uses neighborhood set to indicate scale space of sampling points and allocate a scale factor for every scale space to describe its influence on Harris response value calculating.

3.1   Multiscale 3D Harris Feature Point Detection Algorithm

The basic idea of multiscale 3D Harris feature point detection algorithm is calculating Harris response value of a sampling point under multiscale. Then determine the optimal value by comparing and selecting the value under the optimal scale. The innovation lies in taking advantage of multiscale neighborhood information so that the calculating of Harris response value is more robust. Meanwhile the self-adaptionability is improved. The difference of 3D Harris feature point detection algorithm between multiscale and single scale lies in the calculating of Harris response value, optimal scale selection and candidate feature point selection.

(1)   Calculating of multiscale Harris response value
   Since different scale space influences the Gaussian weight of sampling points' partial derivatives f_x and f_y directly, in order to establish connections between

scale factor and weight of sampling point, this algorithm chooses the average distance between sampling points and their neighborhood points under different scales as the scale factor σ_k:

$$\sigma_k = \frac{1}{\left|N_r^k(p_i)\right|} \sum_{j=1}^{\left|N_r^k(p_i)\right|} \left\| p_i - p_i^j \right\|_2 \qquad (1)$$

Where $N_r^k(p_i)$——the neighborhood point set of p_i under the kth scale.

(2) Selection of optimal scale

Local information of sampling point is used effectively under small scale. However, local features will be filtered with much too bigger scale and non-sampling point neighborhood information will be added which may result in error calculating of feature value. Figure 2 reflects the change of Harris response value under different scales.

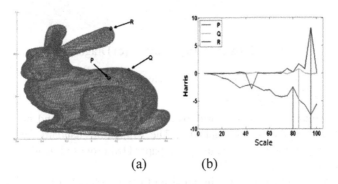

(a)   (b)

Fig. 2. Optimal scale selection (Color figure online)

In Fig. 2, point P,Q and R are three different points on the rabbit point cloud model of Fig. 2. (a)(b) reflects their change tendency of Harris response value under different scales. The mostly used optimal scale selection method now is setting the scale whose feature measure value significantly changes with the increase of scale factor as the optimal scale [15]. This paper define the optimal scale as the one whose Harris response value significantly increases with the increase of size of sampling points' neighborhood. In Fig. 2 (b) the three lines with color of red, green and blue represent the optimal scales of calculating Harris response value at point of P, Q and R.

(3) Multiscale selecting candidate feature points

Harris response value of feature points in 3D Harris feature point detective algorithm only needs to be the local maximum in feature point geometrical neighborhood; while the value in multiscale algorithm should be the local maximum in both geometrical neighborhood and scale neighborhood. Thus the constraints for feature point response values are stricter which makes the selected feature points with more invariance for transformation and noise.

4 Experiment and Analysis

In order to test effectiveness of multiscale 3D Harris feature point detection algorithm proposed for unstructured 3D point cloud, we test with point cloud data AIM@SHAPE project [16] and Stanford 3D Scanning Repository [17] provided using multiscale 3D Harris feature point detection algorithm and Sipiran feature point detection algorithm in paper [7]. Then we contrasted and analyze the feature point detection results.

Trim-Star point cloud, Bunny point cloud and Dragon point cloud are used in the experiments. Figure 3 shows the 3D picture of the cloud data.

(a) Trim-star   (b)Bunny   (c) Dragon

Fig. 3. 3D point cloud examples.

The experiments test two algorithms'robustness to rotation and noise by rotating with cloud and adding noise manually. The evaluation standard for effect of two algorithms is repetition rate of feature point paper [18] proposed. Higher repetition rate means better robustness.

In the experiments, we set the initial size of neighborhood of Harris response value to 20 and the size for comparing to 8. The final feature point's selected using Sipiran algorithm are the ones in the top 15 % of the candidates and the feature points using multiscale 3D Harris feature point detection algorithm are obtained by MST algorithm.

4.1 Analysis of Effect on Point Cloud Rotation

In order to test the influence of rotation on feature point detection results, 5 groups of rotation experiments are made with three point cloud models, whose rotation angle are 10°, 20°, 30°, 40° and 50° around z axis. Figure 4 shows the test results of two feature point detection algorithms with unstructured 3D point cloud.

In Fig. 4, a(1)~(5), b(1)~(5) and c(1)~(5) are the results of three point cloud models using Sipiran algorithm after rotating. Fig. 4 a'(1)~(5), fig b'(1)~b'(5), fig c'(1)~c'(5) are the results using our multiscale 3D Harris feature point detection algorithm with the same data. It will be seen that the results using our algorithm are better than the ones using Sipiran algorithm, which are not much influenced by point clouds. Figure 5 shows the change tendency of feature point repetition rate with rotation of point cloud model. Higher repetition rate means better robustness.

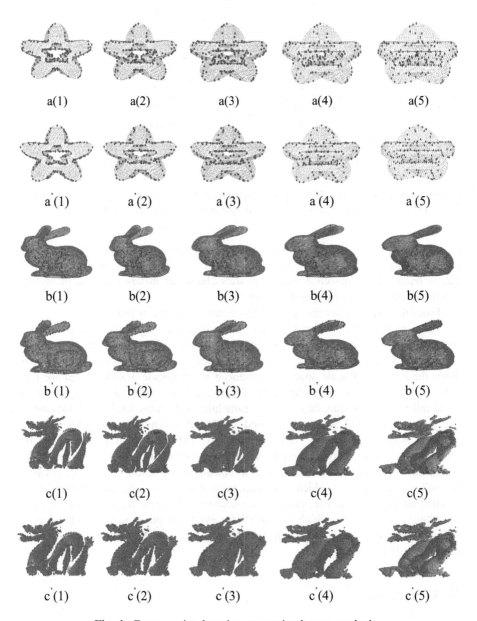

a(1)   a(2)   a(3)   a(4)   a(5)

a'(1)   a'(2)   a'(3)   a'(4)   a'(5)

b(1)   b(2)   b(3)   b(4)   b(5)

b'(1)   b'(2)   b'(3)   b'(4)   b'(5)

c(1)   c(2)   c(3)   c(4)   c(5)

c'(1)   c'(2)   c'(3)   c'(4)   c'(5)

Fig. 4. Feature point detection on rotation by two methods.

From Fig. 5 we can see that, with the increase of rotation angle, repetition rate using our multiscale 3D Harris feature point detection algorithm is high and steady, while the repetition rate using Sipiran feature point detection algorithm shows a downward trend. For Trim-Star data, the downward trend is slow, while the downward trend of Bunny data and Dragon data, whose structure is complicated is faster. The

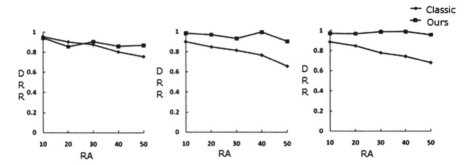

Fig. 5. Comparison of two feature point detection methods. The x axis indicates the rotation angle (RA), the Y axis indicates feature point detection repetition rate (DRR).

reason is that the algorithm did not solve the problem of arbitrary in feature point's normal vector as well as the error in surfaces fitting. In contrast, our multiscale 3D Harris feature point detection algorithm uses Hausdorff distance instead of surfaces fitting, which makes it rotation invariant.

From the experiments above, conclusion can be drawn as follows:

First, multiscale 3D Harris feature point detection algorithm can be used to unstructured 3D point cloud. Building local coordinate with invariance as well as using Hausdorff distance instead of calculating partial derivatives makes the algorithm robust. Then the concept of scale space is introduced to multiscale 3D Harris feature point detection algorithm thus the calculation of Harris response value can make good use of local information of sampling point. Meanwhile, a dynamic method of looking for every point's optimal scale is proposed so that this algorithm has some adaption abilities. Anyway the weak point of multiscale 3D Harris feature point detection algorithm lies in not solving the problem of being effected by noise completely. The establishment of local coordinate as well as calculation of Hausdorff distance is easy to be influenced by noise, which may account for this. Last, the time complexity of calculating Harris response value in multiscale 3D Harris feature point detection algorithm is rather high, which is $O(MN^2)$ for every sampling point. M represents the size of sampling point's neighborhood. N represents the amount of set points for calculating Hausdorff distance.

5   Conclusion

This paper proposes a multiscale 3D Harris feature point detection algorithm for unstructured 3D point cloud. The innovation lies in applying 3D Harris feature point detection algorithm to unstructured 3D point cloud, which is rotation invariant by setting up local coordinate and using Hausdorff distance instead of partial derivative to overcome the arbitrary when calculating sampling point's normal vector. In addition, the concept of scale space is introduced to optimize the Harris feature point detection algorithm for unstructured 3D point cloud and the innovation lies in introducing scale information to algorithm and proposing the selection strategy of optimal scale, which makes the algorithm self-adaptive. At last, contrasting the mesh cloud 3D Harris feature point detection algorithm with our multiscale 3D Harris feature point detection algorithm, we can draw a conclusion that our algorithm is robust to rotation when detecting feature points in unstructured 3D point cloud.

Acknowledgements. This work was supported by the National Natural Science Foundation of China under Grant Nos: 61273366 and 61231018, National Key Technology Support Program under Grant No: 2015BAH31F01, and the program of introducing talents of discipline to university under grant no: B13043.

References

1. Yamany, S.M., Farag, A.A.: Surface signatures: an orientation independent free-form surface representation scheme for the purpose of objects registration and matching. IEEE Trans. Pattern Anal. Mach. Intell. **24**(8), 1105–1120 (2002)
2. Gal, R., Cohen-Or, D.: Salient geometric features for partial shape matching and similarity. ACM Trans. Graph. **25**(1), 130–150 (2006)
3. Hildebrandt, K., Polthier, K., Wardetzky, M.: Smooth feature lines on surface meshes. In: Proceedings of the Third Eurographics Symposium on Geometry Processing. Eurographics Association, Vienna (2005)
4. Weinkauf, T., Gunther, D.: Separatrix persistence: extraction of salient edges on surfaces using topological methods. Comput. Graph. Forum **28**(5), 1519–1528 (2009)
5. Castellani, U., Cristani, M., Fantoni, S., et al.: Sparse points matching by combining 3D mesh saliency with statistical descriptors. Comput. Graph. Forum **27**(2), 643–652 (2008)
6. Zou, G., Hua, J., Dong, M., et al.: Surface matching with salient keypoints in geodesic scale space. Comput. Animat. Virtual Worlds **19**(3–4), 399–410 (2008)
7. Sipiran, I., Bustos, B.: A robust 3D interest points detector based on harris operator. In: Proceedings of the 3rd Eurographics Conference on 3D Object Retrieval, pp. 7–14. Eurographics Association, Norrköping (2010)
8. Hu, J., Hua, J.: Salient spectral geometric features for shape matching and retrieval. Vis. Comput. **25**(5–7), 667–675 (2009)
9. Sun, J., Ovsjanikov, M., Guibas, L.: A concise and provably informative multi-scale signature based on heat diffusion. Comput. Graph. Forum **28**(5), 1383–1392 (2009)
10. Gumhold, S., Wang, X., McLeod, R.: Feature extraction from point clouds. In: Proceedings of 10th International Meshing Roundtable, Newport Beach, CA, pp. 293–305 (2001)
11. Pauly, M., Keiser, R., Gross, M.: Multi-scale feature extraction on point-sampled surfaces. Comput. Graph. Forum **22**(3), 281–289 (2003)
12. Wu, J., Wang, Q.: Feature point detection from point cloud based on repeatability rate and local entropy. In: Proceedings of SPIE 6786, MIPPR 2007: Automatic Target Recognition and Image Analysis; and Multispectral Image Acquisition, Wuhan, China (2007)
13. Weber, C., Hahmann, S., Hagen, H.: Sharp feature detection in point clouds. In: Shape Modeling International Conference (SMI), pp. 175–186 (2010)
14. Park, M.K., Lee, S.J., Lee, K.H.: Multi-scale tensor voting for feature extraction from unstructured point clouds. Graph. Models **74**(4), 197–208 (2012)
15. Lindeberg, T.: Feature detection with automatic scale selection. Int. J. Comput. Vis. **30**(2), 79–116 (1998)

16. Tangelder, J.H., Veltkamp, R.: A survey of content based 3D shape retrieval methods. Multimedia Tools Appl. **39**(3), 441–471 (2008)
17. Tombari, F., Salti, S., Stefano, L.: Unique signatures of histograms for local surface description. In: Daniilidis, K., Maragos, P., Paragios, N. (eds.) Computer Vision – ECCV 2010, pp. 356–369. Springer, Heidelberg (2010)
18. Boyer, E., Bronstein, A.M., Bronstein, M.M., et al.: SHREC 2011: Robust feature detection and description benchmark. In: Proceedings of the 4th Eurographics Conference on 3D Object Retrieval. Eurographics Association, Llandudno (2011)

Online Programming Design of Distributed System Based on Multi-level Storage

Yang Yu, Laksono Kurnianggoro, Wahyono, and Kang-Hyun Jo[✉]

Intelligent Systems Lab, Graduate School of Electrical Engineering,
University of Ulsan, Ulsan 680749, Korea
{yuyang,laksono,wahyono}@islab.ulsan.ac.kr,
acejo@ulsan.ac.kr

Abstract. For the difficult situation of the node application upgrade in the distributed control system, this paper designs a remote online programming method. This method uses STM32F407 microcontroller and In-Application Programming (IAP) techniques. The upper network is based on Ethernet and http protocol, the lower network is based on UART and CAN. This paper plans the STM32 flash memory based on the IAP characteristics, and introduces the design principle of the distributed IAP. This paper also designs the IAP program and PC user interface program. This method has easy operation, high reliability and good stability, it has a very good value in the distributed system.

Keywords: IAP · Vehicle control system · Distributed system · Online programming · Multi-level storage

1 Introduction

After the hardware of embedded systems installs into the product, it is difficult for the upgrade of the embedded system program. With the technology development, the IAP (In-Application Programming) approach is increasingly used to update user programs, it improves the scalability and maintainability of the system [1]. The automatic vehicle system is a distributed control system; many node controllers connect to the main controller through UART or CAN. These nodes are often in narrow places or seal places; it is difficult to update the program by the downloader. So it is necessary to study the online programming technology. The main control computer connects to the upper controller through Ethernet, the upper controller connects to the lower controller through UART or CAN [2].

The IAP approach can be used in different realms: the data acquisition distributed system of river water quality [5], the monitor of mobile communications repeater equipment [3], the remote meter reading system [7], the vehicle wireless terminal in port special operations vehicle [4], the management terminal of power load [6]. This paper uses the STM32 IAP method in the automatic vehicle control system.

The update methods of STM32 program have ISP (In-System Programming), ICP (In-Circuit Programming) and IAP (In-Application Programming). ICP uses the emulator to update program; ISP needs configure the STM32 BOOT pin, it updates the program by Bootloader program and the ISP software. ICP and ISP all need a

D.-S. Huang et al. (Eds.): ICIC 2016, Part III, LNAI 9773, pp. 745–752, 2016.
DOI: 10.1007/978-3-319-42297-8_69

mechanical operation. If the circuit board has sealed in the box, it is difficult to update the program [8–12].

IAP is a means of upgrading firmware in the field using the MCU communication interfaces such as UART, USB, CAN, Ethernet and a wireless network. It can be used to remote or wireless update program. When you boot the microcontroller, you can choose to put it in either:

(1) IAP mode in order to execute the IAP code.

(2) Normal mode in order to execute the application code.

Both the IAP code and the application code are in the embedded Flash memory of the microcontroller. The flash has two memory blocks, one is the Bootloader area, another is the application area. The IAP code is usually stored in the Bootloader area, it in the first pages of the MCU Flash, and the user application code stored in the application area [13–17].

Some interference can cause the update to fail during the IAP process, some reliable mechanism can avoid such situations [18–24].

The IAP operation flow diagram is shown in Fig. 1.

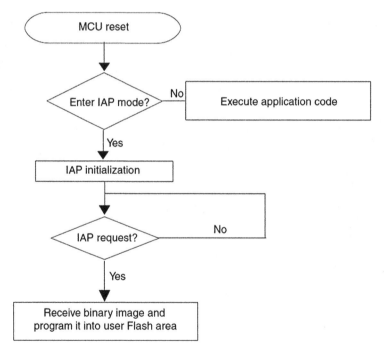

Fig. 1. IAP operation flow diagram.

2 Vehicle Control System

The automatic vehicle control system is a distributed control system. It uses four UARTs to acquire the magnetic sensor information, and uses a CAN network to control all CAN devices. It has three layers. The upper layer connects to the middle layer by Ethernet. The middle layer connects to the lower layer by UART or CAN. The block diagram of the automatic vehicle control system is shown in Fig. 2.

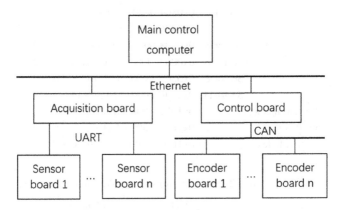

Fig. 2. The automatic vehicle control system.

The sensor board reads 16 magnetic sensor data, and sends the data to the acquisition board via UART 2. The AD22151 is a linear magnetic field transducer. It can detect the pre-set road magnetic information.

The acquisition board receives the sensor data from 4 sensor boards, and sends the data to PC by Ethernet. The UART 2 of 4 sensor boards connect to the UART 2-5 of the acquisition board. UART baud rate is 450 Kbps.

The control board can read digital and analog input and can set digital and analog output. It can communicate with the encoder board by CAN2.0 J1939 protocol. CAN baud rate is 250 Kbps.

The linear solution encoder connects to the encoder board. It can measure the vehicle wheel speed. The encoder board reads the encoder data and sends it to CAN bus. The control board reads the encoder CAN data and sends it to PC by Ethernet.

3 Flash Memory Organization

STM32F4 flash memory has a main memory, system memory, OTP area and an option byte. The main memory has 4 sectors of 16 Kbytes, 1 sector of 64 Kbytes, and 7 sectors of 128 Kbytes. The organization of STM32F407 flash module is shown in Table 1.

The main memory allocation of STM32F4 flash are shown in Table 2. The IAP area has the IAP program which can update the program of current board or sub-board. The APP area has the user-defined program of current board. The sub-board APP area

Table 1. STM32F407 flash module organization

Block	Name	Block base addresses	Kb
Main memory	Sector 0	0x0800 0000 - 0x0800 3FFF	16
	Sector 1	0x0800 4000 - 0x0800 7FFF	16
	Sector 2	0x0800 8000 - 0x0800 BFFF	16
	Sector 3	0x0800 C000 - 0x0800 FFFF	16
	Sector 4	0x0801 0000 - 0x0801 FFFF	64
	Sector 5	0x0802 0000 - 0x0803 FFFF	128
	Sector 6	0x0804 0000 - 0x0805 FFFF	128
	Sector 7	0x0806 0000 - 0x0807 FFFF	128
	Sector 8	0x0808 0000 - 0x0809 FFFF	128
	Sector 8	0x080A 0000 - 0x080B FFFF	128
	Sector 8	0x080C 0000 - 0x080D FFFF	128
	Sector11	0x080E 0000 - 0x080F FFFF	128
System memory		0x1FFF 0000 - 0x1FFF 77FF	30
OTP area		0x1FFF 7800 - 0x1FFF 7A0F	528
Option bytes		0x1FFF C000 - 0x1FFF C00F	16

Table 2. Flash main memory allocation

Board	Area	Address
Acquisition board and Control board	IAP	0x0800 0000 - 0x0801 FFFF
	APP	0x0802 0000 - 0x0803 FFFF
	Sub-board APP	0x0806 0000 - 0x0807 FFFF
	Backup APP	0x0808 0000 - 0x0809 FFFF
	Data	0x080E 0000 - 0x080F FFFF
Sensor board and Encoder board	IAP	0x0800 0000 - 0x0801 FFFF
	APP	0x0802 0000 - 0x0803 FFFF
	Backup APP	0x0808 0000 - 0x0809 FFFF
	Data	0x080E 0000 - 0x080F FFFF

has the user-defined program of sub-board. The data area has the update sub-board number, APP update flag and the board network information. The backup APP area has the backup program which can restore the APP program of current board.

4   IAP Update Process

The IAP update process is based on three layer distributed system. It uses Ethernet, UART and CAN protocol to transfer the command and data.

In the main board, after STM32F4 reset, the system jumps to the IAP main function and reads APP update flag in the data area of the flash memory.

If the flag means no updating, the system jumps to the APP main function. When receiving the IAPdownloader software command by Ethernet, the APP update flag and the update board number is saved, and the system restarts to enter the IAP program.

If the flag means updating, the system resets APP update flag and reads the update board number to perform different IAP download program. The IAP flowchart of the main board is shown in Fig. 3.

In the IAP program, the binary file can be download to the board by http software, it can be written to the corresponding flash memory area. If the board number is the current board number, the system finishes the download process and restarts. If the

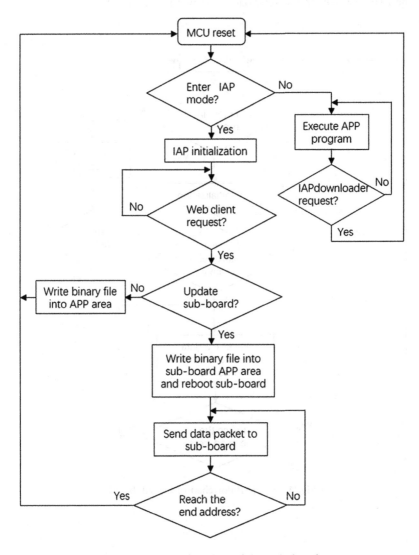

Fig. 3. The IAP flowchart of the main board.

board number is the sub-board number, the system sends command to make the sub-board reboot into the IAP program. Then the system sends a data packet to the sub-board, and writes it to the sub-board flash. When receiving a command, the system sends the next data packet. When reaching the end address of the flash area, the main board and sub-board all reboot into the APP program.

In the sub-board, after the APP program receives a command by UART or CAN, it reboots into the IAP program. In the IAP program, after receiving Start_Send command, the system continuously receives the data packet and writes it into the sub-board flash, until reaching the end address of this flash area. Then the sub-board reboots into the APP program.

The IAP flowchart of the sub-board is shown in Fig. 4.

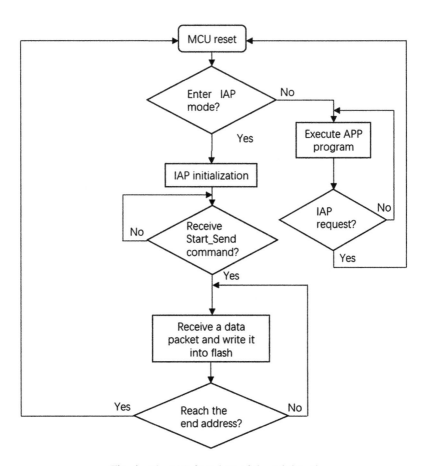

Fig. 4. The IAP flowchart of the sub-board.

5 Experiments

These IAP code size are very small (Flash in the main board: 32 KB, Flash in sub-board: 28 KB), thus it can be applied in many embedded products. When the APP object code is 80 kB Hex file, the IAP download time with different interface and baud rate is shown in Table 3. So the time is mainly determined by the maximum baud rate of sub-board. The update time of main board is about 4 s. The whole transmission and update time of sub-board is about 13 s in the UART of 450 Kbps baud rate, and is about 17 s in the CAN of 250 Kbps baud rate.

Table 3. The IAP download time with 80kB Hex file

Board	Interface	Baud Rate	Time
Acquisition board	IAP using Ethernet	10 Mbps	4 s
Control board	IAP using Ethernet	10 Mbps	4 s
Sensor board	IAP using UART	450 Kbps	13 s
		250 Kbps	16 s
		125 Kbps	21 s
		56.25 Kbps	27 s
		37.5 Kbps	31 s
Encoder board	IAP using CAN	250 Kbps	17 s
		125 Kbps	23 s
		56.25 Kbps	30 s
		37.5 Kbps	35 s

6 Conclusion

This paper describes how to design an online programming method of the distributed system by combining Ethernet, UART, CAN bus and STM32 online programming techniques. This method unified the external interfaces of the whole distributed system. It has achieved the perfect operating results for the practical automatic vehicle control system. It has the features such as simple, flexible, fast operation, so that it effectively reduces labor costs and improves the system scalability and maintainability. Experimentally it shows a very good value in the distributed practical system.

References

1. Zhang, W.J., Nan, Y.M.: Design and implementation of IAP techniques based on STM32F103VB. J. Comput. Appl. **29**(10), 2820–2822 (2009)
2. Nan, Y.M.: Programming for STM32F103xxx standard peripherals based on STM32 standard peripherals library. J. Changsha Aeronaut. Vocat. Tech. Coll. **10**(4), 41–45 (2010)

3. Xu, Y.L., Luo, D.Y., Zhang, H.: Remote online upgrade of multi-point embedded system based on GIS. Comput. Meas. Control **14**(3), 383–385 (2006)
4. Dai, Z.C., Xiang, Y.: Design of remote upgrade of equipment monitoring system software. In: 2010 Second International Conference on Information Technology and Computer Science, pp. 462–465 (2010)
5. Ni, C.Q.: Application of IAP in GPRS remote meter-reading embedded system. CD Technol. **2008**(8), 37–38 (2008)
6. Que, F.B.: Remote program upgrade based on stm32. Electron. Instrum. Customers **20**(5), 90–92 (2013)
7. Zhao, H.B., Tian, Q.C.: Realization of program remote updating through IAP function of LPC2214. Radio Eng. China **36**(7), 53–55 (2006)
8. Zhang, Y., Bao, K.J.: Design and implementation of bootloader for vehicle control unit. Comput. Eng. **37**(12), 233–235 (2011)
9. Xu, Q.Q.: A Design and implement of IAP based on HTTP. In: 2011 International Conference on Computer Science and Service System, vol. 27(29), pp. 1918–1922 (2011)
10. Xu, Y., Ma, Y.: IAP in STM32F103Series MCU. Microcontrollers Embed. Syst. **13**(8), 35–37 (2013)
11. Zhu, F.L., Yang, M.: Design of remote update system based on the SCM with IAP function. J. Mech. Electr. Eng. **27**(9), 76–79 (2010)
12. Jiang, J.C., Wang, Z.S., Feng, H.Z., Liu, T.: Design and implementation of IAP on-line upgrading technology based on software trigger. J. Comput. Appl. **32**(6), 1721–1723 (2012)
13. Tong, G.X., Fu, L., Liu, H.: Realization of STM32 in application programming based on CAN bus. Inf. Technol. **10**, 49–52 (2015)
14. Chen, X., Tang, K.: The application of IAP technology in Multi-CPU power system stability control device. Jiangsu Electr. Eng. **29**(2), 38–40 (2010)
15. Jiang, X.M., Li, X.H., Ren, Z.R., Yao, M.: IAP online upgrade and teleupgrade resolvent based on ARM. Comput. Appl. **28**(2), 519–521 (2009)
16. Yin, H., Yan, H.: Storage solution to security of embedded remote upgrade. J. Comput. Appl. **31**(4), 942–944 (2011)
17. Chao, Y.X., Han, D., Xu, J.B.: A scheme of firmware upgrade of STM32 based on CPUID and AES algorithm. Appl. Electron. Tech. **41**(3), 28–30 (2015)
18. Pan, Z., Polden, J., Larkin, N., Duin, S.V., Norrish, J.: Recent progress on programming methods for industrial robots. Robot. Comput.-Integr. Manufact. **28**(2), 87–94 (2012)
19. Gefen, D., Carmel, E.: Is the world really flat? a look at offshoring at an online programming marketplace. MIS Q. **32**(2), 367–384 (2008)
20. Yang, W.R., Jiang, C., Hu, Y.L.: Online-programming IP design and implementation based on SoC. Open Autom. Control Syst. J. **5**, 59–66 (2013)
21. Banerjee, S., Gupta, T.: Efficient online RTL debugging methodology for logic emulation systems. In: 2012 25th International Conference on VLSI Design, pp. 298–303 (2012)
22. Park, H., Xu, J., Park, J., Ji, J.: Design of on-chip debug system for embedded processor. In: IEEE International SoC Design Conference (ISOCC), ISOCC 2008, vol. 3, pp. 11–12 (2008)
23. Sang, S.J.: Design and implementation of cyclic redundancy check algorithm. In: Proceedings of the 2015 Computing, Control, Information and Education Engineering, pp. 405–409 (2015)
24. Pancholi, V.R., Patel, B.P.: Enhancement of cloud computing security with secure data storage using AES. Int. J. Innovative Res. Sci. Technol. **2**(9), 18–21 (2016)

Depth-Sensitive Mean-Shift Method
for Head Tracking

Ning Zhang[1], Yang Yang[1(⊠)], and Yun-Xia Liu[2]

[1] School of Information Science and Engineering,
Shandong University, Jinan 250100, China
yyang@sdu.edu.cn
[2] School of Control Science and Engineering,
Shandong University, Jinan 250014, China

Abstract. Target tracking is one of the most basic application in computer vision and it has attracted wide concern in recent years. Until now, to our best knowledge, most research focused on the tracking research with 2D images, including the Tracking-Learning-Detection (TLD), particle filter, Mean-shift algorithm, etc. While with the advanced technology and lower cost of sensors, 3D information can be used for target tracking problems in many researches and the data can be obtained by laser scanner, Kinect sensor and etc. As a new type of data description, depth information can not only obtain the spatial position information of target but also can protect privacy and avoid the influence of illumination changes. In this paper, a depth-sensitive Mean-shift method for tracking is proposed, which use the depth information to estimate the range of people's movement and improve the tracking efficiency and accuracy effectively. What's more, it can adjust kernel bandwidth to adapt to the target size according to the distance between target and the depth camera. In the designed system, Kinect2.0 sensor is not only used to get the depth data and track the target but also can be mobilized by steering gear flexibly when tracking. Experimental results show that these improvements make Mean-shift algorithm more robust and accurate for handling illumination problems during tracking and it can achieve the purpose of real-time tracking.

Keywords: Depth image · Mean-shift · Tracking · Kinect2.0

1 Introduction

Target tracking is one of the key technologies in computer vision and it has been widely used in various industries while creating great economic and social value in recent years, such as video surveillance, human-computer interface, and etc. However, the vast majority of researches focus on using 2D data [1, 2] in the field of target tracking. Moving target tracking is often influenced by the position of the moving target, the change of illumination and the complex background. To overcome these problems existing in tracking mechanism, the device and the algorithm has been improved. Applying depth image to pattern recognition research is a new technology in recent years [3]. Though sonar sensors [4], Infrared (IR) [5, 6] and laser range finders [7] are low in cost, they have poor angular resolution and are susceptible to false

© Springer International Publishing Switzerland 2016
D.-S. Huang et al. (Eds.): ICIC 2016, Part III, LNAI 9773, pp. 753–764, 2016.
DOI: 10.1007/978-3-319-42297-8_70

echoes. Compared with those sensors, Kinect has the advantages of high resolution and low price. The depth data of the camera can eliminate the influence by the illumination, shadow and texture of the object surface. In addition, stability of the Kinect sensor can guarantee the robustness of the system, while reducing the complexity of the study greatly.

Among the tracking algorithms, Mean-shift [8, 9] tracking algorithm has received wide attention because of its rigorous theory, simple implementation and good tracking performance. It can quickly locate the target without exhaustive search and applied in the real-time target tracking in the field of [10–13] successfully. The size of the kernel function plays an important role in the Mean-shift tracking algorithm. It not only determines the number of samples among iteration but also reflects the size of the tracking window. However, the size of windows is fixed and the positioning accuracy is low in traditional Mean-shift algorithm. As viewed from these problems, Bradski et al. [12] presented a method based on moment invariants, but the calculation of moment features would affect the real time performance of the algorithm seriously. Comaniciu et al. [13] proposed a metric derived from the Bhattacharyya coefficient as the similarity measure, and the proper bandwidth of kernel is adapted by checking a +10 % larger and a -10 % smaller ellipse and choosing the best one. The method can get good results only when the target is smaller gradually. Collins et al. [14] proposed obtaining the optimal kernel bandwidth by adding an additional scale kernel, which carried out in discrete scale space defined by Mean-shift iteration. However, almost all of the improvements referred above are not used depth information effectively.

It is well known that the velocity of human moving is limited, typically around 1 m/s. So, the human moving range in image coordinate based on the distance between human and camera can be calculated easily, which can efficiently adjust bandwidth of kernel, improve the tracking efficiency and quality. Fortunately, the depth sensors, such as laser scanning camera and Kinect, will offer the distance information of camera. In this paper, we utilize Kinect to get the distance information, and then apply it to Mean-shift algorithm, which kernel function width can be adjusted adaptively according to the change of depth value and distance. The results show that the depth-sensitive algorithm can track the target more robustly and the performance is relatively good.

The rest of the paper is organized as follows. In Sect. 2, the basic Mean-shift theory and the principle of depth image acquisition are introduced. In Sect. 3, a depth-sensitive Mean-shift algorithm is described in detail. The system design and the analysis of the results are given in Sect. 4, and the conclusion is presented in Sect. 5.

2 Basic Theory and Depth Image Acquisition

2.1 The Mean-Shift Algorithm Theory

The following Fig. 1 illustrates the basic principle of the Mean-shift tracking algorithm. As shown below: x_i^0, x_i^1 and x_i^n represent the central points of each iteration and x_i^0 stands for the initial region's central point when tracking, which the superscript 1, 2 and n indicates the number of iterations, the black spots stand for the window sample points which moving continuously, meanwhile, the black and red circles represent the density estimation window size of each iteration in traditional Mean-shift algorithm and the improved algorithm respectively. The arrow indicates the drift vector of the sample point relative to the center point of the kernel function and the Mean-shift vector will point to the densest direction of the sample point which is the gradient direction. The Mean-shift algorithm is convergent and according to iterative search for the densest area of the sample points in the feature space of the current frame, the search points can drift along the direction of the sample point density increase to the local density maxima point x_i^n, which is considered as the center point of target region, so as to it can achieve the purpose of tracking.

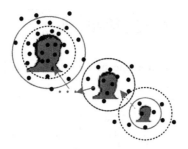

Fig. 1. Mean-shift algorithm principle

Assuming that x_0 is the center of the object region. There are n pixels in target region represented by $\{x_i\}_{i=1...n}$. There are m bins in the color histogram of target object. The probability distribution of bins is represented as formula (1), where C_1 is normalization coefficient. We can get C_1 according to the formula:

$$\hat{q}_u = C_1 \sum_{i=1}^{n} k\left(\left\|\frac{x_0 - x_i}{h}\right\|^2\right) \delta[b(x_i) - u] \tag{1}$$

$$C_1 = \frac{1}{\sum_{i=1}^{n} k\left(\left\|\frac{x_0-x_i}{h}\right\|^2\right)} \tag{2}$$

Similar to the formula (1), the model of the candidate target in the current search window can be expressed as formula (3):

$$\hat{p}_u = C_h \sum_{i=1}^{n_h} k\left(\left\|\frac{y - x_i}{h}\right\|^2\right) \delta[b(x_i) - u] \tag{3}$$

where the normalization constant is

$$C_h = \frac{1}{\sum_{i=1}^{n_h} k\left(\left\|\frac{y-x_i}{h}\right\|^2\right)} \tag{4}$$

Then the object tracking problem can be simplified to search for the optimal y to make the similarity between $\hat{q}_u$ and $\hat{p}_u$ reach the highest. Bhattacharyya coefficient [12] can be used to measure the similarity, which is defined as follow:

$$\hat{\rho}(y) \equiv \rho[p(y), q] = \sum_{u=1}^{m} \sqrt{p_u(y)\hat{q}_u} \tag{5}$$

$$\rho[p(y), q] \approx \frac{1}{2} \sum_{u=1}^{m} \sqrt{p(y_0)q_u} + \frac{C_h}{2} \sum_{i=1}^{n} w_i k\left(\left\|\frac{y - x_i}{h}\right\|^2\right) \tag{6}$$

$$w_i = \sum_{u=1}^{m} \delta[b(x_i) - u] \sqrt{\frac{q_u}{p_u(y_0)}} \tag{7}$$

Base on the Mean-shift vector, the new location of object is obtained according Eq. (8).

$$y_1 = \frac{\sum_{i=1}^{n_h} x_i w_i g\left(\left\|\frac{y - x_i}{h}\right\|^2\right)}{\sum_{i=1}^{n_h} w_i g\left(\left\|\frac{y - x_i}{h}\right\|^2\right)} \tag{8}$$

2.2   Depth Image Acquisition by Kinect2.0

In our depth-sensitive target tracking system, 3D data is obtained by the latest Kinect2.0 sensor, which is launched by Microsoft in Oct. 2014. It can realize capture the target's motion, image detection, human tracking and etc. Compared to the kinect1.0 launched in Nov. 2010, the depth of information obtained is more abundant and the field of view (FOV) is enlarged 60 %.

The video capture system of Kinect2.0 is composed of infrared emitter and VGA camera. The object's 3D image can be put into the screen by the depth camera and ordinary RGB camera of Kinect. The core of depth tracking process is an infrared sensor, which can make kinect2.0 sensor to obtain the target's position regardless of the ambient light conditions. Each pixel of the depth frame obtained by Kinect2.0 sensor represents the physical distance between the object and the camera. In addition, it can generate depth image with the speed of 30 frames per second and the resolution of the depth image is 512 × 424 generally. Microsoft uses the flight time calculation method to replace the existing light measurement method to obtain the depth value, making the new Kinect faster and more accurate than the original one. The method is based on the most accurate time calculation principle for a single photon to bounce back from an object or person. Microsoft suggests that the effective physical distance is 800 mm ~ 8100 mm. The principle is interpreted in Fig. 2. The green line indicates the effective recognition range, while the red line represents the invalid distance. The concrete method for calculating the

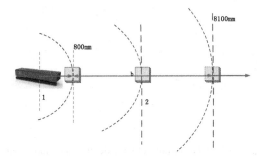

Fig. 2. Valid recognition distance and flight time calculation

flight time can be interpreted: The half time of a single photon to bounce back from an object (position 2) to camera (position 1) multiplied by light speed.

3 Depth-Sensitive Mean-Shift Algorithm

The concrete steps of the depth-sensitive Mean-shift tracking algorithm based on Kinect2.0 depth image are as follows:

(1) Obtain the depth image of the target by the depth sensor;
(2) Achieve the relationship between the distance of the target and the depth camera and the depth value of the target by the depth camera calibration, and the detailed steps include:
   a. To keep the target in the center position of the depth camera's view, and move the Kinect forward with same distance every time (each moving distance is n), then the depth image in the range of effective recognition of the depth camera can be acquired. It should be noted that the value of n can be select with the range from 1 mm to 20 mm, then the depth value of target center points of all depth frame are obtained;
   b. After getting N sets of data, the relationship between the distance and the depth value of the target's center point can be obtained according to the straight line fitting of the N sets of data based on the least square method, as shown in formula (9):

$$Dist = \alpha \cdot Depth + \beta \tag{9}$$

In formula (9), *Depth* represents the depth value of the center point of the target, and *Dist* represents the distance between the target and the depth camera. In this paper, the value of α, β can be obtained, $\alpha = 1.003$, $\beta = 126.2$; The relationship between *Dist* and *Depth* is shown as Fig. 3. To simplify the calculation, the relationship between the two variables can be evaluated into a linear model. It should be noted that the reference range of α is from 0.998 to 1.005, β is from 118.7 to 132.4. Under normal circumstances, if Kinect2.0 depth camera is used, both values of α and β should fluctuate within the range that referred above.

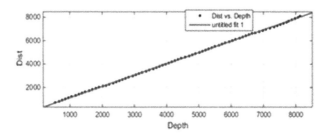

Fig. 3. The relationship between depth value and distance

(3) Using a depth camera for target tracking: update the size of the kernel bandwidth continuously according to the relationship to adapt to the changes of the target size in the camera's view, the concrete steps as follows:

a. In the $(i\text{-}1)$-*th* frame depth image, the relationship between the depth value of the target's center point and the distance between the target and the depth camera is shown in the formula (10):

$$Dist_{(i-1)} = \alpha \cdot Depth_{(i-1)} + \beta \qquad (10)$$

In the *i-th* depth frame, the relationship is shown in the formula (11):

$$Dist_i = \alpha \cdot Depth_i + \beta \qquad (11)$$

When the formulas (10) and (9) are subtraction, the (12) can be obtained:

$$\Delta Dist = \alpha \cdot \Delta Depth \qquad (12)$$

In the formula (12), *ΔDepth* represents the difference of the actual distance between the two adjacent depth frames; *ΔDist* represents the change of the depth value of target's center point between the two adjacent depth frames.

b. The change of kernel bandwidth when tracking is shown as formula (13):

$$r_i = \begin{cases} R & |\Delta Depth| \leq \Delta Depth_{opt} \\ \left(\frac{\Delta Depth_{opt}}{|\Delta Depth|}\right)^{sgn(\Delta Depth)} & |\Delta Depth| > \Delta Depth_{opt} \end{cases} \qquad (13)$$

In the formula above, $\Delta Depth_{opt} = 1000v/\alpha k$, $\Delta Depth_{opt}$ is a threshold. V (m/s) represents the maximum speed of the target when tracking; K stands for the frame rate of the depth camera; r_i represents the search radius of the *i-th* depth frame. R means the initial search radius. sgn(*ΔDepth*) is a sign function, if*ΔDepth*≥0, sgn(*ΔDepth*) = 1; otherwise, sgn(*ΔDepth*) = -1;

c. Tracking the target based on the depth images according to the kernel bandwidth r_i obtained above; if $i = n$, end of the algorithm; otherwise, $i = i+1$, jump to step a. of (3).

The principle of the improved algorithm are as shown in Fig. 4, where (a) stands for the initial depth frame with a kernel window in certain size. (e), (f) and (g) stand for the local amplification display of target field in (b), (c) and (d). The black dotted rectangles in (e), (f) and (g) are the initial kernel windows while the red ones mean that it can adapt to the size of the target object accurately.

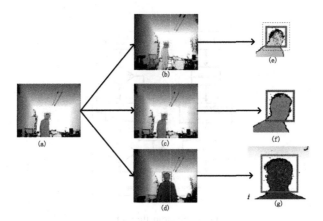

Fig. 4. The principle of depth-sensitive Mean-shift algorithm

4 Experiment and Analysis

4.1 Head Tracking System Design

In order to verify the stability and accuracy of our depth-sensitive Mean-shift algorithm, a head tracking system is designed and as shown in Fig. 5. The proposed system consists several parts including Kinect2.0, PC, steering engine and single chip microcomputer (SCM). The role of the steering gear is to control the rotation angle of kinect2.0 precisely and automatically, making the head is always in the center of camera's view, which can improve the robustness of the system effectively when tracking with our depth-sensitive method.

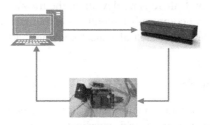

Fig. 5. System parts connection diagram

This steering gear control method based on the SCM has the advantages of simple, high precision and low cost. In this paper, the model of steering engine is HD1501, which has two degrees of freedom in horizontal and vertical direction and can be very convenient to install the camera. The Kinect2.0 can achieve around 150 degrees rotation in both two direction under the control of the steering engine in our tracking system. It expands the scope of tracking greatly while controlling the target object always in the center field of the camera's view flexibly, so it can avoid the loss problems of target effectively when tracking.

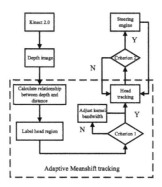

Fig. 6. System flow chart

The depth-sensitive tracking method based on the depth information proposed in this paper can be divided into several steps as follow. The overall working process of this method is showed in Fig. 6. Firstly, the kinect2.0 should be initialization and used to get the depth frame sequence. The principle of obtaining the depth image by kinect2.0 was described in detail in Sect. 3. Secondly, the relationship between depth value and real distance can be carried out according to camera calibration. Then label the target region and start tracking. The criteration1 is that whether the kernel bandwidth change or not. Suppose the depth value changes of head center point between the two adjacent frames achieve the certain threshold, the kernel bandwidth would be larger or smaller, otherwise, its value remain unchanged. So it can realize the real-time adjustment of the kernel bandwidth in the tracking process according to the change of the depth value of target. The criteration2 is that whether the human deviate from the center of the camera view. If there is no deviation, the tracking system will be keeping on, otherwise, PC will sent commands to control steering gear to rotate a certain angle to make target stay in the center area of Kinect's vision field and always be tracked.

4.2   Results and Analysis

In order to calculate the rotation angle of the steering engine, the image coordinate system should be mapped to the world coordinate system to obtain the relationship between the pixel and the real distance firstly [15]. The experiment can be set as shown in Fig. 7, which use these three pictures to explain the process of the design simply. It should be place a strip of cards with the same distance on a flat background firstly. It must be noted that each bar card should be exactly of the same size and shape and the center position of all cards should be in a horizontal line. Then mark the spot every hundred millimeters on the ground, ranging from 800 mm to 8100 mm (effective distance). The kinect2.0 should be moved vertically from far to near and the data should be collect at the same time, so that each distance corresponding to a color frame and the number of intervals of each picture and the center pixel coordinates of the card which locate in left and right edges of pictures can be received. Then the relationship between distance and the number of pixels can be obtained according to data statistics. With calculating the number of pixels corresponding to each of hundred millimeters, the results can be gotten as formula (14).

$$N = \frac{3.607 \times 10^4}{Dist + 31.34} \tag{14}$$

In the formula (14) above, N represents number of pixel per hundred millimeters, and $Dist$ represents the distance between Kinect2.0 sensor and target object.

Fig. 7. Calibration process. (a) The maximum effective recognition position of Kinect2.0 (initial position), (b) The position of Kinect in the effective range of recognition, and (c) The minimum effective recognition position of Kinect2.0 (final position)

The mathematical connection among rotate angle of steering engine, target's center coordinate and depth value can be deduced, as shown in Fig. 8. From mathematical knowledge, formula (15) can be gotten easily. The value of S can be calculated by the formula (14). Then the rotation angle will be calculated according to formulas (16) and (17). The conversion between (16) and (17) here used the formula (9) mentioned above in Sect. 2. x represents the pixel coordinate of target center, it can be obtained by the depth image when tracking. The steering gear can mobilize Kinect2.0 sensor to rotate a certain angle flexible based on the relationship obtained above, always make the head in the center field of the camera's view.

$$tan\ \alpha = \frac{s}{Dist} \tag{15}$$

$$tan\ \alpha = \frac{|x - 256| \times (Dist + 31.34)}{360 \times Dist} \tag{16}$$

$$\alpha = \arctan(\frac{|x - 256| \times (\text{Depth} + 157.54)}{360 \times (\text{Depth} + 126)}) \tag{17}$$

Our project can be run on PC with a database consisting of depth images obtained by Microsoft Kinect2.0 sensor. Three groups of result are selected in depth-sensitive head tracking process which shown in Fig. 9. They represent different conditions whether the distance between the person and the Kinect2.0 change or not. As can be seen, when people translate and rotate under a fixed vertical distance, the kernel window size is not change. When people getting closer to the Kinect2.0, the bandwidth of target becomes larger, while the size of kernel window will be smaller if people away from the Kinect2.0 sensor.

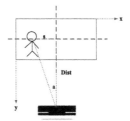

Fig. 8. Rotation angle calculation

It can be seen also that no matter how the human move, the head is remain in the central area of camera's view. As shown in Table 1, we statistics and analyses the result of the experiment. Acc means the accuracy of algorithm. It can be known that the mean kernel bandwidth in depth-sensitive Mean-shift algorithm is smaller than the traditional Mean-shift algorithm, so that the iterative calculation can be reduced effectively. In addition, the accuracy of algorithm is also be improved.

Table 1. Data statistics

Algorithm	Depth-sensitive Mean-shift				Mean-shift [10]	
Statitics / video	Max.	Min.	Avg.	Acc (%)	R	Acc (%)
a	36	32	34.5	98.0	35	91.8
b	54	35	43.5	98.2	45	92.2
c	88	59	70.8	97.4	76	94.3

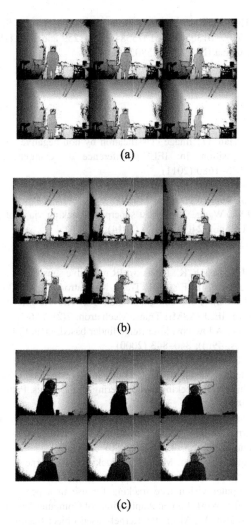

(a)

(b)

(c)

Fig. 9. The tracking result under different conditions. (a) represents the tracking results when the vertical distance has no change in translation and rotation conditions; (b) stands for the tracking results when target getting closer to the Kinect2.0 sensor and (c) is the tracking results when target away from the Kinect2.0 sensor.

5   Conclusion

This paper proposed a depth-sensitive target tracking algorithm that based on the depth information, which can adjust the kernel bandwidth to fit the size of target according to the relationship between distance and depth value. The experiment results show that the head tracking algorithm can improve the tracking accuracy and flexible control the target in the center area of the camera's view, at the same time, it can avoid the influence of illumination and complex background effectively.

Acknowledgement. This work is supported by the National Natural Science Foundation of China (Grant No. 61203269, 61305015, 61375084, 61401259 and 11474185), and Postdoctoral Science Foundation of China (No. 2015M580591).

References

1. Zhang, C., Liu, J., Tian, Q.: Image classification by non-negative sparse coding, low-rank and sparse decomposition. In: IEEE Conference on Computer Vision and Pattern Recognition, pp. 1673–1680 (2011)
2. Li, Y., Lou, J.: Mean shift tracking algorithm with hybrid target model. IEEE Trans. Comput. Eng. Appl. **46**(16), 200–203 (2010)
3. Yuan, Y., Fang, J.W., Wang, Q.: Robust superpixel tracking via depth fusion. IEEE Trans. Circ. Syst. video Technol. **24**(1), 15–26 (2014)
4. Rendas, M., Moura, J.: Ambiguity in Radar and Sonar. IEEE Trans. Signal Process. **46**(2), 294–305 (1998)
5. Shan, Y., Speich, J.E., Leang, K.: Low-cost IR reflective sensors for submicrolevel position measurement and control. IEEE/ASME Trans. Mechatron. **13**(6), 700–709 (2008)
6. Bouzit, M., Burdea, G., Popescu, G., Boian, R.: The Rutgers Master II-new design force-feedback glove. IEEE/ASME Trans. Mechatron. **7**(2), 256–263 (2002)
7. Journet, B., Bazin, G.: A low-cost laser range finder based on an FMCW-like method. IEEE Trans. Instrum. Meas. **49**(4), 840–843 (2000)
8. Fukanaga, K., Hostetler, L.D.: The estimation of the gradient of a density function, with applications in pattern recognition. IEEE Trans. Inf. Theory **21**(1), 32–40 (1975)
9. Cheng, Y.: Meanshift, mode seeking and clustering. IEEE Trans. Pattern Anal. Mach. Intell. **17**(8), 790–799 (1995)
10. Comaniciu, D., Ramesh, V., Meer, P.: Real-Time tracking of non-rigid objects using Meanshift. In: Werner, B. (ed.) IEEE International Proceedings of the Computer Vision and Pattern Recognition, vol. 2, pp.142−149 (2000)
11. Yilmaz, A., Shafique, K., Shah, M.: Target tracking in airborne forward looking infrared imagery. Int. J. Image Vis. Comput. **21**(7), 623–635 (2003)
12. Bradski G.R.: Computer vision face tracking for use in a perceptual user interface. In: Sipple, R.S. (ed.) IEEE Workshop on Applications of Computer Vision, pp. 214−219 (1998)
13. Comaniciu, D., Ramesh, V., Meer, P.: Kernel-based object tracking. IEEE Trans. Pattern Anal. Mach. Intell. **25**(5), 564–575 (2003)
14. Collins R.T.: Mean-shift blob tracking through scale space. In: Danielle, M. (ed.) IEEE International Conference on Computer Vision and Pattern Recognition, vol. 2, pp. 234−240. Victor Graphics, Baltimore (2003)
15. Li, M., Yang, Y., Liu Y.X.: Robust 3D human tracking based on kinect. In: Proceedings of the 34th Chinese Control Conference, 28–30 July 2015. IEEE (2015)

Knowledge Representation
and Expert System

Knowledge Graph Completion by Embedding with Bi-directional Projections

Wenbing Luo[(✉)], Jiali Zuo, and Zhengxia Gao

School of Computer and Information Engineering,
Jiangxi Normal University, Nanchang 330022, China
122719053@qq.com

Abstract. Knowledge graph (*KG*) completion aims at predicting the unknown links between entities and relations. In this paper, we focus on this task through embedding a KG into a latent space. Existing embedding based approaches such as TransH usually perform the same operation on head and tail entities in a triple. Such way could ignore the different roles of head and tail entities in a relation. To resolve this problem, this paper proposes a novel method for KGs embedding by preforming bi-directional projections on head and tail entities. In this way, the different information of an entity could be elaborately captured when it plays different roles for a relation. The experimental results on multiple benchmark datasets demonstrate that our method significantly outperforms state-of-the-art methods.

Keywords: Knowledge graph completion · Knowledge graph embedding · Knowledge reasoning

1 Introduction

Knowledge graphs (KGs) describe knowledge about entities and their relations with inter-linked fact triples (a triple (head entity, relation, tail entity) indicates the two entities hold the relation, denoted as (h, r, t)). The rich structured information of KGs become useful resources to support many intelligent applications such as question answering [1]. However, the low coverage is an urgent issue which hampers the wide utilization of KGs, e.g. even the largest KG of Freebase is still far from complete [2]. Knowledge graph completion (KGC) aims at predicting the links between relations and entities with the supervision of the existing KGs.

Traditional approaches for KGC usually employ logic inference rules for knowledge reasoning [3], which lack ability for supporting numerical computation in continuous spaces, and cannot be effectively extended to large-scale KGs. To address this problem, a new approach based on representation learning was recently proposed by attempting to embed a KG into a low-dimensional continuous space. In this way, it admits that the original logic inference could be completed through numerical computation [4]. Therefore, the embedding based approaches have more expansibility and are more suitable for large-scale KGs.

The promising methods usually represent the entities as point vectors in a low-dimensional spaces and represent relations as operations between two points in the

© Springer International Publishing Switzerland 2016
D.-S. Huang et al. (Eds.): ICIC 2016, Part III, LNAI 9773, pp. 767–779, 2016.
DOI: 10.1007/978-3-319-42297-8_71

entity vector spaces. Among these methods, TransE [5] and its variants are simple and effective. TransE represents a relation as a vector r indicating the semantic translation from the head entity h to the tail entity t, which aims to satisfy the equation $h + r \approx t$ when triple (h, r, t) holds. TransE effectively handles 1-1 relations but has issues in handling one-to-many, many-to-one and many-to-many relations. To address such problem, TransH and TransR are proposed to enable an entity to have distinct representations when involved in different relations. TransH first projects a head/tail entity vector (h/t) into T a relation-dependent hyper-plane by the following formulas: $h_\perp = h - w_r^T h w_r$ and $t_\perp = t - w_r^T t w_r$, where wr is the vector that spans the hyper-plane. It then aims to satisfy the equation $h_\perp + r \approx t_\perp$ in the hyper-plane of the relation r. TransR transforms a head/tail entity vector into a relation-dependent sub-space instead of a hyper-plane in TransH.

Based on the above analysis, we notice that both TransH and TransR perform the same operation (projecting into a hyper-plane or transforming into a subspace) on head and tail entities for predicting a relation, and assume that the head and tail entities own the same role for a specific relation (event) to form triples. In fact, the entities with different positions for a relation play different roles, the head and tail entities in a triple (h, r, t) indicate the agent and object of the relation (event) respectively, which makes the previous methods (sharing the same operation for head and tail entities) insufficient for modeling multiply relations. To address the aforementioned problem, this paper proposes a novel KGs embedding methods, TransBiH, which performs bi-directional projections on the head and tail for a relation into two different but related hyper-planes. The basic idea of TransBiH is illustrated in Fig. 1, which ensures the two projections of head and tail entities on two related hyper-planes are closer through a relation-dependent translation.

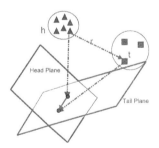

Fig. 1. Simple illustration of TransBiH

We have conducted extensive experiments on link prediction and triple classification with multiple benchmark datasets such as WordNet and Freebase. The experimental results demonstrate the effectiveness of our method. In particular, the proposed method can effectively model reflexive relations, and significantly outperforms state-of-the-art methods.

2 Related Work

Currently, the proposed methods mainly represent a KG in a low-dimensional latent space. We briefly summarize the most relevant work in Table 1. We mainly compare the models' scoring functions $f_r(h,t)$ and their complexities (the numbers of parameters in model and multiplication in learning). n_e and n_r are the number of unique entities and relations, respectively. n_t is the number of triples in a KG. k_e and k_r are the dimensions of entity and relation in the latent embedding space, h, t $\in \mathbb{R}^{k_e}$. s is the number of slices of a tensor used in NTN. g(x) is a non-linear function used in neural networks such as tanh. These methods embed entities into a vector space and define a scoring function to measure the compatibility of (h, r, t). The differences between these models are the scoring functions $f_r(h,t)$. We highlight TransE [5] and its variants in Sect. 1 already. In this section, we put more emphasis on the other methods.

Table 1. Overview of representative models for representing a KG in a latent space.

Model	Score function $f_r(h,t)$	#Parameters
LRE (2001)	$\exp(-\|W_r h - t\|^2), W_r \in \mathbb{R}^{k_e \times k_e}$	$k_e n_e + k_r^2 n_r$
SE(2011)	$\|W_{rh}h - W_{rt}t\|\ell_{1/2}, W_{rh}, W_{rt} \in \mathbb{R}^{k_r \times k_e}$	$k_e n_e + 2k_r^2 n_r$
LTM (2012)	$h^T W_r t, W_R \in \mathbb{R}^{k_r \times k_e}$	$k_e n_e + k_r^2 n_r$
UM (2012)	$\|h - t\|_2^2$	$k_e n_e$
TransE (2013)	$\|h + r - t\|\ell_{1/2}, r \in \mathbb{R}_r^k$	$k_e(n_e + n_r)$
SLM (2013)	$u_r^T g(W_{rh}h + W_{rt}t + b_r), u_r \in \mathbb{R}^s, W_{rh}, W_{rt} \in \mathbb{R}^{s \times k_e}$	$k_e n_e + (2sk_r + s)n_r$
NTN (2013)	$u_r^T g(h^T W_r t + W_{rh}h + W_{rt}t + b_r), W_r \in \mathbb{R}^{k_e \times k_e \times s}$	$k_e n_e + (sk_e^2 + 2sk_e + s)n_r$
TransH (2014)	$\|(h - W_r^T h W_r) + r - (t - W_r^T t W_r)\|\ell_{1/2}, W_r, r \in \mathbb{R}_r^k$	$k_e n_e + 2k_r n_r$
TransR (2015)	$\|(h M_r + r - t M_r\|\ell_{1/2}, r \in \mathbb{R}_r^k, M_r \in \mathbb{R}^{k_r \times k_r}$	$k_e n_e + (k_r + k_r^2)n_r$
TransBiH	$\|(h - W_{rh}^T h W_{rh}) + d_r - (t - W_{rh}^T t W_{rh})\|\ell_{1/2}$	$k_e n_e + 3k_r n_r$

To our knowledge, Linear Relational Embedding (LRE) [6] is the pioneer work in performing embedding of multi-relational data. Subsequently, many approaches have followed this line. The unstructured model (UM) [5] was proposed as the simplified version of TransE by assigning all translation vectors r = 0. Structured embedding (SE) [14] adopts two different relation-specific matrices for head and tail entities but cannot capture precious semantics of relations because the two matrices are separated in optimization. The latent factor model (LFM) [7] considers the second-order correlations between entities embedding with a quadratic form. The single layer model (SLM) and neural tensor network (NTN) were proposed by Socher [8]. To date, the NTN is the most expressive model. However, the NTN is not sufficiently simple to handle the large-scale KGs with numerous relations.

In addition to these methods based on the ranking loss framework, there have another line of related work which focus on learning the latent representations for KGs by tensor (matrix) decomposition and completion [9].

3 Knowledge Graphs Embedding by Bi-Directional Projections

We first describe some common notations: h, r and t denote the head entity, relation and tail entity for a triple fact (h, r, t). The bold letters h, t denote the corresponding embedding of h and t. Γ is the set of (positive) golden triples which are observed in a KG, and Γ' denotes the set of (negative) incorrect triples (mostly corrupted ones). ε and $\Re$ are the sets of entities and relations in a KG, respectively.

3.1 Different Roles of Two Positions in Triples

In the multi-relational knowledge graphs, the roles between the head position and the tail position in a triple are not identical. With regard to a triple (h, r, t), the head entity h represent the agent, and the tail entity t represent the object. In most cases, the agent and object of a fact (event) represent different meanings.

Firstly, as illustrated in Fig. 2, the variability of the number of unique entities occurring in head and tail positions are pervade in KGs. Secondly, most relations such as (person, place of birth, location) own different types for head and tail entities. Finally, the densities of one type's entities are very different with others, as illustrated in Fig. 3 (we use the t-SNE model [10] for compressing the embeddings of FB15k learned by TransE [5] into two dimensions.), the identical spectrum for transformation different types' entities are insufficient for modeling multiple relations. Based on the above observations, we propose to use different operations on head and tail entities for predicting a relation.

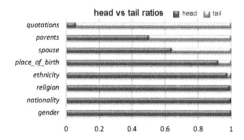

Fig. 2. The ratios of left and right entities with some sampling relations in Freebase. (Color figure online)

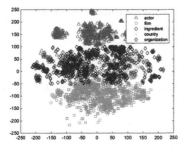

Fig. 3. The densities of entities with some sampling types.

3.2   Embedding by Bi-Directional Projections

To overcome the problems of previous methods that project (transform) head and tail entities with the identical operation, we propose TransBiH, which perform bi-directional projections on two different but related hyper-planes for the two positions' entities in one triple. As illustrated in Fig. 1, for a relation r, we span two hyper-planes with two relation-specific vectors, and the head and tail entities are projected into the two different hyper-planes rather than the same operation performed in previous methods [11, 12]. Specifically, for a triple (h, r, t), the embedding h and t are first projected to the head hyper-plane w_{rh} and the tail hyper-plane w_{rt}, and the corresponding projections are denoted as $h_\perp$ and $t_\perp$ respectively. We ensure $h_\perp$ and $t_\perp$ are more closer through the translation vector d_r in golden triples than in incorrect ones. We name this model TransBiH. Thus we define a scoring function $-\|h_\perp + d_r - t_\perp\|_{\ell_{1/2}}$ to measure the validity that the triple is correct. By restricting w_{rh} and w_{rt} are unit vectors, it is easy to obtain

$$h_\perp = h - w_{rh}^T h w_{rh}, t_\perp = t - w_{rh}^T h w_{rh}$$

Then the score function for triple is

$$f_r(h, t) = \left\| \left(h - w_{rh}^T h w_{rh} \right) + d_r - \left(t - w_{rh}^T h w_{rt} \right) \right\|_{\ell_{1/2}} \tag{1}$$

The score (distance measure) is expected to be lower for a golden triple and higher for an incorrect triple. The parameters of the model are: all the entities' embeddings, $\{e_i\}_{i=1}^{\mathcal{E}}$; all the relations' translation vectors, $\{d_i\}_{r=1}^{|\Re|}$; and the head and tail hyper-planes of all relations, $\{(w_{rh}, w_{rt})\}_{r=1}^{|\Re|}$.

Considering the relevance between head hyper-plane w_{rh} and tail hyper plane w_{rt}, we assume the head projection of a relation owning: (1) the equivalent semantic with the tail projection and (2) the reversible semantic with the head projection, of the corresponding inverse relation. For example, the semantic of the head position entities for children equate with the semantic of the tail position entities for parents.

Therefore, we make the following constraint for the two span vectors in a joint model for optimizing the two projections[1]:

$$\left(I - w_{rh}w_{rh}^T\right)\left(I - w_{rt}w_{rt}^T\right) = I \tag{2}$$

which ensures the two operations are reversible with each other.

3.3   Training

We define the following margin-based ranking loss for effective discrimination between observed (positive) triples and incorrect (negative) triples:

$$\mathcal{L} = \sum_{(h,r,t)\in\Gamma} \sum_{(h',r',t')\in\Gamma'_{(h,r,t)}} \left[f_r(h,t) + \gamma - f'_r(h',t')\right] +$$

where $[x]_+ \triangleq \max(0, x)$ aims to obtain the maximums between 0 and x, γ is the margin separating positive and negative triples, fr (h, t) indicates the score function (formula 1), Γ indicates the set of positive triples observed in the KGs, and $\Gamma'_{(h,r,t)}$ denotes the set of negative triples corresponding to (h, r, t), which will be introduced below.

Existing KGs contain only correct triples. The routine method for constructing a negative triple *(h', r', t')* is to replace the head or tail entity randomly, To obtain practical corrupted triples, we follow [11] and assign different probabilities for head/tail entity replacement. We will denote the traditional sampling method as "unif" and the new method as "bern".

To avoid over fitting, we add some regularization when we optimize the loss $\mathcal{L}$:

$$\forall e \in E, \|e\|_2 \le 1, \tag{3}$$

$$\forall r \in R, \forall p \in \{h,t\}, \left|w_{rp}^T d_r\right|/\|d_r\|_2 \le \epsilon, \tag{4}$$

$$\forall r \in R, \forall p \in \{h,t\}, \left\|w_{rp}\right\|_2 = 1, \tag{5}$$

Combined with the formula 2, we soft constraints and convert the loss $\mathcal{L}$ to the following unconstrained loss:

$$\mathcal{L} = \sum_{(h,r,t)\in\Gamma} \sum_{(h',r',t')\in\Gamma'_{(h,r,t)}} \left[f_r(h,t) + \gamma - f'_r\left(h',t'\right)\right]_+ + C\left\{\sum_{e\in\mathcal{E}} \|e\|_2^2 - 1\right\}_+$$

$$+ \sum_{p\in\{h,t\}} \sum_{r\in\Re} \left[\frac{\left(w_{rp}^T d_r\right)^2}{\|d_r\|_2^2} - \epsilon^2\right]_+ + \sum_{r\in\Re} \left\|\left(I - w_{rh}w_{rh}^T\right)\left(I - w_{rt}w_{rt}^T\right) - I\right\|_F$$

---

[1] We use the fact that the results of the formula $w^T hw$ equal with the formula $ww^T h$.

where C is the weighting hyper-parameter of the soft constraints, $\|M\|_F$ indicates the Frobenius norm of the matrix M, and we set ε to 10-3 in the experiments.

4 Experiments

4.1 Data Sets

In this work, we empirically study and evaluate the related methods for knowledge graph completion with the following two tasks: link prediction and triple classification. We use the datasets commonly used in previous methods, which are built from two typical KGs: WordNet and Freebase. We adopt two datasets from WordNet, WN18, used in [5] for link prediction, and WN11, used in [8] for triple classification. We adopt two datasets from Freebase, FB13, used in [8] for triple classification, and FB15k, used in [5] for link predication and triple classification. The statistics of these datasets are listed in Table 2.

Table 2. Datasets used in the experiments.

Dataset	# ℜ	# ε	#Triple (Train/Valid/Test)		
WN18	18	40,943	141,442	5,000	5,000
FB15k	1,345	14,951	483,142	50,000	59,071
WN11	11	38,696	112,581	2,609	10,544
FB13	13	75,043	316,232	5,908	23,733

4.2 Link Prediction

Link prediction aims to predict the missing h or t for a relation fact triple (h, r, t). Instead of obtaining one best answer, this task puts more emphasis on ranking a set of candidate entities from a KG. We conduct initial experiments using the datasets WN18 and FB15k.

Evaluation protocol. We follow the same protocol as in TransE: In the testing phrase, for each test triple (h, r, t), we replace the tail entity by all entities e in a KG and rank these entities in descending order of similarity scores, measured by the scoring function $f_r(h,e)$. A similar process for the head entity by measuring $f_r(e,t)$. Based on these entity ranking lists, we use two evaluation metrics by aggregating over all the testing triples: (1) the average rank of correct entities (denoted as *Mean Rank*) and (2) the proportion of correct entities in the top 10 ranked entities (denoted as *Hits@10*). A good method should obtain lower *Mean Rank* and higher *Hits@10*. Considering the fact that a corrupted triple for (h, r, t) may also existed in a KG, such a prediction should also be deemed correct. To eliminate this factor, we remove those corrupted triples that already appeared in training, valid or testing sets before obtaining the rank entity list of each testing triple. We term the former evaluation setting as "Raw" and the latter as "Filter".

Implementation. Because the testing datasets are the same, we directly compare our models with several baselines reported in [12]. In learning TransBiH, we select the learning rate α for SGD among {0.001, 0.01, 0.05}, the margin γ among {1, 2, 4}, the dimensions of entity and relation sharing embedding k among{20, 50, 100}, the batch size B among {20, 120, 1440, 4800}, and the hyper-parameter C among {0.0625, 0.25}. The optimal configuration is determined by the *Hits@10* in the validation set. As the strategy of constructing negative labels can greatly influence the evaluations, we use different parameters for "unif" and "bern". The default configuration for all experiments is as follows: α = 0.001, γ = 1, k = 50, B = 120, C = 0.0625, and use ℓ_1 as dissimilarity. Below, we list only the non-default parameters. Under the "unif" setting, the optimal configuration is as follows: B = 120, and C = 0.25 on WN18 and k = 100, and B = 4800 on FB15k. Under the "bern" setting, the optimal configuration is as follows: B = 120, and C = 0.25 on WN18 and B = 4800 on FB15k. For both datasets, we traverse all the training triples for 500 rounds.

Results. The results are reported in Table 3. On WN18, TransE, TransH, TransR, TransBiH and even the naive baseline unstructured model outperform other approaches in terms of the *Mean Rank* metric, but most models are poor in terms of the *Hits@10* metric. On FB15k, TransBiH consistently outperforms the other baseline models in both *Mean Rank* and *Hits@10*. As the variability of relations' semantics in FB15k is greater than that in WN18, we hypothesize that the improvements are because the proposed method can effectively handle the two positions' semantics of relations. Table 4 shows the evaluation results with separated types of relation properties. Following [5], we divide relations into four types: 1-1, 1-M, M-1 and M-N, for which the proportions in FB15k (1345 relations in total) are 24 %, 23 %, 29 % and 24 %, respectively, based on the measure used in [11]. TransBiH significantly outperform TransE, TransH, TransR and other baseline methods in 1-1, 1-M and M-1 relations. However, the proposed method presents no much advantage for M-N relations, possibly because there are various fine-grained types within an M-N relation that cannot be effectively translation by one vector.

Table 3. Experimental results on link prediction.

Data sets	WN18				FB15K			
Metric	Mean Rank Raw Filter		Hits@10 Raw Filter		Mean Rank Raw Filter		Hits@10 Raw Filter	
Unstructured [19]	315	304	35.3	38.2	1,074	979	4.5	6.3
SE [4]	1,011	985	68.5	80.5	273	162	28.8	39.8
SME (linear) [19]	545	533	65.1	74.1	274	154	30.7	40.8
SME (bilinear) [19]	526	509	54.7	61.3	284	158	31.3	41.3
LFM [7]	469	456	71.4	81.6	283	164	26.0	33.1
TransE [5]	263	251	75.4	89.2	243	125	34.9	47.1

(*Continued*)

Table 3. (*Continued*)

Data sets	WN18				FB15K			
Metric	Mean Rank Raw Filter		Hits@10 Raw Filter		Mean Rank Raw Filter		Hits@10 Raw Filter	
TransH (unif) [11]	318	303	75.4	86.7	211	84	42.5	58.5
TransH (bern) [11]	401	388	73.0	82.3	212	87	45.7	64.4
TransR (unif) [12]	232	219	78.3	91.7	226	78	43.8	65.5
TransR (bern) [12]	238	225	**79.8**	**92.0**	198	77	48.2	68.7
TransBiH (unif)	283	271	75.7	88.3	204	69	45.4	69.2
TransBiH (bern)	279	265	75.8	89.2	**190**	**66**	**48.4**	**71.0**

Table 4. Experimental results on FB15k by mapping properties of relations (%).

Tasks	Prediction Head (Hits@10)				Prediction Tail (Hits@10)			
Relation Category	1-1	1-M	M-1	M-N	1-1	1-M	M-1	M-N
Unstructured [19]	34.5	2.5	6.1	6.6	34.3	4.2	1.9	6.6
SE [4]	35.6	62.6	17.2	37.5	34.9	14.6	68.3	41.3
SME (linear) [19]	35.1	53.7	19.0	40.3	32.7	14.9	61.6	43.3
SME (bilinear) [19]	30.9	69.6	19.9	38.6	28.2	13.1	76.0	41.8
TransE [5]	43.7	65.7	18.2	47.2	43.7	19.7	66.7	50.0
TransH (unif) [16]	66.7	81.7	30.2	57.4	63.7	30.1	83.2	60.8
TransH (bern) [16]	66.8	87.6	28.7	64.5	65.5	39.8	83.3	67.2
TransR (unif) [12]	76.9	77.9	38.1	66.9	76.2	38.4	76.2	69.1
TransR (bern) [12]	78.8	89.2	34.1	**69.2**	79.2	37.4	90.4	**72.1**
TransBiH (unif)	**83.3**	86.4	48.2	65.3	**84.2**	50.4	85.3	67.0
TransBiH (bern)	83.2	**90.1**	**56.7**	68.7	83.5	**63.2**	**90.8**	69.2

4.3   Triple Classification

This task seeks to judge whether a given triple (h, r, t) is correct or not. In this task, we use three datasets: WN11, FB13 and FB15k.

Evaluation protocol. We follow the same protocol as in [8]. The evaluation of binary classification requires negative triples. WN11 and FB13 released by NTN [8] already contain negative triples, which are built by corrupting the corresponding positive (observed) triples. As FB15k has not released negative triples in previous works, we construct negative triples following the same procedure used in [8] for FB13. The setting for triple classification is very simple: for each triple *(h, r, t)*, if the dissimilarity score obtained by the score function f_r(h,t) is below a relation-specific threshold δ_r, then the triple will be classified as positive. Otherwise, it will be classified as negative. The relation-specific threshold δr is optimized by maximizing the classification accuracy on the validation set.

Implementation. Considering that the same datasets (negative triples) are used in WN11 and FB13, we directly compare our models with the baseline methods reported in [12]. For evaluation on FB15k, we implement TransE and TransH, and use the code released by Lin[2] [12] (running TransR) and Socher[3] [8] (running NTN). For TransE, TransH and TransR, we select the learning rate α for SGD among $\{0.001, 0.01, 0.05\}$, the margin γ among $\{1, 1.5, 2\}$, the dimensions of entity and relation sharing embedding k among $\{20, 50, 100\}$, the batch size B among $\{20, 120, 1440, 4800\}$, and the hyper-parameter C among$\{0.0625, 0.25\}$. For the NTN, we did not change the settings: dimension $k = 100$, and the number of slices equals 3. For TransBiH, For TransBiH, the search space of parameters is identical to the link prediction. The default configuration for all experiments are as follows: $\alpha = 0.001$, $\gamma = 1$, $k = 50$, $B = 120$, and $C = 0.0625$. Below, we list only the non-default parameters. Under the "unif" setting, the optimal configuration is as follows: $\gamma = 2$, $C = 0.25$, and use ℓ_2 as dissimilarity on WN11; $\gamma = 1$, $k = 100$, and $C = 0.25$ on FB13 and $\gamma = 1.5$, $k = 100$, and $B = 4800$ on FB15k; Under the "bern" setting, the optimal configuration is as follows: $\gamma = 2$, $C = 0.25$, and use ℓ_2 as dissimilarity on WN11; $\gamma = 2$, $k = 100$, and $C = 0.25$ on FB13 and $k = 100$, and $B = 4800$ on FB15k.

Results. The accuracies of triple classification on the three datasets are shown in Table 5. On WN11, TransBiH and TransR outperform all the other models. FB13 is sampled from the type */people/person*, it does not contains rich links among entities (e.g., comparing with the average 39.6 triples of one entity contained in FB15k, it only contains 4.6 triples), therefore, the classification model with the most parameters for relations such as SLM and NTN can outperforms the other approaches. However, on FB15k, a more practical KG with large-scale relations, the proposed method TransBiH performs much better than the other baseline models.

Table 5. Experimental results of Triple Classification (%).

Data sets	WN11	FB13	FB15k
SE [4]	53.0	75.2	-
SME [19]	70.0	63.7	-
SLM [18]	69.9	85.3	-
LFM [7]	73.8	84.3	-
NTN [18]	70.4	87.1	68.2
TransE (unif) [5])	75.9	70.9	79.2
TransE (bern) [5]	75.9	81.5	81.4
TransH (unif) [11]	77.7	76.5	85.4
TransH (bern) [11]	78.8	83.3	85.8
TransR (unif) [12]	85.5	74.7	83.7
TransR (bern) [12]	85.9	82.5	86.5
TransBiH (unif)	85.1	79.4	89.1
TransBiH (bern)	85.6	83.6	89.5

---

[2] https://github.com/mrlyk423/relation_extraction.
[3] www.socher.org.

4.4 Modeling Reflexive Relations

We assume that TransBiH can model reflexive relations better than previous methods because it use different representations for head and tail positions when scoring triples. To inspect the effectiveness of the proposed method for modeling reflexive relations, we extract some reflexive relations in FB15k for performing link prediction and triple classification. We first use the following rules: (1) the number of triples exceeds 50, and (2) the existing ratio of reverse triple (t, r, h) for triple (h, r, t) exceeds 0.8. Then, we manually verify and extract 6 representative reflexive relations for evaluations. We use the results of the best performance on triple classification for comparing TransE, TransH, TransR and TransBiH.

Tables 6 and 7 shows the evaluation results of reflexive relations on link predication (only showing the results of "bern" negative sampling strategy) and triple classification respectively. The tables illustrate that the proposed method can effectively model reflexive relations for knowledge graph completion.

Table 6. Hits@10 (Filter) of link prediction on some sampling reflexive relations (%).

Relation	TransE/TransH/TransR/TransBiH	
	Predict Head	Predict Tail
location/adjoins	34.7/59.4/62.2/64.8	37.5/58.7/66.8/68.0
person/friend	27.3/58.2/67.3/68.2	13.6/44.5/58.2/63.6
person/sibling	26.8/58.4/63.7/57.9	20.5/42.6/58.4/52.6
person/spouse	19.3/33.3/50.0/68.5	19.3/35.2/46.3/59.3
relationship/peers	28.1/56.7/76.2/71.4	23.8/47.6/80.9/76.2
military/combatant	33.3/56.7/66.7/100	0/33.3/33.3/33.3

Table 7. Accuracies of triplet classification on some sampling reflexive relations.

Relation	TransE/TransH/TransR/TransBiH	
	bern	unif
location/adjoins	0.90/0.93/0.95/0.94	0.82/0.89/**0.93**/0.91
person/friend	0.86/0.89/0.86/0.93	0.86/0.90/0.93/**0.95**
person/sibling	0.87/0.97/0.94/0.89	0.84/0.89/**0.94**/0.89
person/spouse	0.79/0.86/0.87/0.90	0.77/0.85/0.86/**0.88**
relationship/peers	0.76/0.79/0.92/0.93	0.76/0.71/0.81/**0.88**
military/combatant	0.83/0.83/0.83/1	0.5/0.5/0.33/**0.66**

5 Conclusion

We propose a novel embedding method TransBiH for performing knowledge graph completion in the latent space. TransBiH learning KGs embedding by preforming bi-directional projections on head and tail entities for a relation. With this manner, the different roles' information of an entity could be elaborately captured when it is in different position (head or tail) of a triple. Experiments results show that our method significantly outperforms state-of-the-art methods including TransE and its improving models.

Acknowledgments. This work was supported by the National Natural Science Foundation of China (Nos. 61462043, 61272212, and 61562042), and the Science and Technology Foundation of Jiangxi Province (No. 20151BAB217014).

References

1. He, S., Liu, K., Zhang, Y., Xu, L., Zhao, J.: Question answering over linked data using first-order logic. In: Proceedings of the Conference on Empirical Methods in Natural Language Processing, pp. 1092–1103 (2014)
2. Dong, X., Gabrilovich, E., Heitz, G., Horn, W., Lao, N., Murphy, K., Strohmann, T., Sun, S., Zhang, W.: Knowledge vault: a web-scale approach to probabilistic knowledge fusion. In: Proceedings of the 20th ACM SIGKDD International Conference on Knowledge Discovery and Data Mining, pp. 601–610 (2014)
3. Lao, N., Mitchell, T., Cohen, W.W.: Random walk inference and learning in a large scale knowledge base. In: Proceedings of the Conference on Empirical Methods in Natural Language Processing, pp. 529–539. Association for Computational Linguistics (2011)
4. Bordes, A., Weston, J., Collobert, R., Bengio, Y., et al.: Learning structured embeddings of knowledge bases. In: AAAI, No. EPFL-CONF-192344 (2011)
5. Bordes, A., Usunier, N., Garcia-Duran, A., Weston, J., Yakhnenko, O.: Translating embeddings for modeling multi-relational data. In: Advances in Neural Information Processing Systems, pp. 2787–2795 (2013)
6. Paccanaro, A., Hinton, G.E.: Learning distributed representations of concepts using linear relational embedding. IEEE Trans. Knowl. Data Eng. **13**(2), 232–244 (2001)
7. Jenatton, R., Roux, N.L., Bordes, A., Obozinski, G.R.: A latent factor model for highly multi-relational data. In: Advances in Neural Information Processing Systems, pp. 3167–3175 (2012)
8. Socher, R., Chen, D., Manning, C.D., Ng, A.: Reasoning with neural tensor networks for knowledge base completion. In: Advances in Neural Information Processing Systems, pp. 926–934 (2013)
9. Nickel, M., Tresp, V., Kriegel, H.P.: A three-way model for collective learning on multi-relational data. In: Proceedings of the 28th International Conference on Machine Learning, pp. 809–816 (2011)
10. Van der Maaten, L., Hinton, G.: Visualizing data using t-SNE. J. Mach. Learn. Res. **9**(85), 2579–2605 (2008)
11. Wang, Z., Zhang, J., Feng, J., Chen, Z.: Knowledge graph embedding by translating on hyperplanes. In: Proceedings of the Twenty-Eighth AAAI Conference on Artificial Intelligence, pp. 1112–1119 (2014)
12. Lin, Y., Liu, Z., Sun, M., Liu, Y., Zhu, X.: Learning entity and relation embeddings for knowledge graph completion. In: Proceedings of the Twenty-Eighth AAAI Conference on Artificial Intelligence, pp. 2181–2187 (2015)
13. Bordes, A., Glorot, X., Weston, J., Bengio, Y.: A semantic matching energy function for learning with multi-relational data. Mach. Learn. **94**(2), 233–259 (2014)

14. Bordes, A., Weston, J., Collobert, R., Bengio, Y., et al.: Learning structured embeddings of knowledge bases. In: Proceedings of the Twenty-Five AAAI Conference on Artificial Intelligence. (2011)
15. Bordes, A., Usunier, N., Garcia-Duran, A.,Weston, J., Yakhnenko, O.: Translating embeddings for modeling multi-relational data. In: Advances in Neural Information Processing Systems, pp. 2787–2795 (2013)
16. Wang, Z., Zhang, J., Feng, J., Chen, Z.: Knowledge graph embedding by translating on hyperplanes. In: Proceedings of the Twenty-Eighth AAAI Conference on Artificial Intelligence, pp. 1112–1119 (2014)
17. Lin, Y., Liu, Z., Sun, M., Liu, Y., Zhu, X.: Learning entity and relation embeddings for knowledge graph completion. In: Proceedings of the Twenty-Nine AAAI Conference on Artificial Intelligence. (2015)
18. Socher, R., Chen, D., Manning, C.D., Ng, A.: Reasoning with neural tensor networks for knowledge base completion. In: Advances in Neural Information Processing Systems, pp. 926–934 (2013)
19. Bordes, A., Glorot, X., Weston, J., Bengio, Y.: A semantic matching energy function for learning with multi-relational data. Mach. Learn. 233–259 (2012)

Petrochemical Enterprise Safety Performance Assessment Based on Interval Number

Jianwen Guo[1(✉)], Zhenzhong Sun[1], Shouyan Zhong[1], Jiaxin He[1],
Huijiang Huang[2], and Haibin Chen[1]

[1] School of Mechanical Engineering, Dongguan University of Technology,
Dongguan 523808, China
guojw@dgut.edu.cn
[2] Guangdong Shidong Industry Group, Dongguan 523980, China

Abstract. Safety performance assessment is the measurement of a petro-chemical enterprise's achievement in safety management. In order to receive a comprehensive and objective evaluation result, it is necessary to consider all evaluation factors and the numerical uncertainty caused by fuzziness when safety performance assessment is conducted. To improve conventional safety performance evaluation, an evaluation index system is established, and the interval number is used in this study by using interval number to quantify scores and calculate the safety level of petrochemical enterprise safety. A case of petrochemical enterprise is used to illustrate the effectiveness of the method and system. This method is applied to the comprehensive evaluation of petro-chemical enterprise safety to achieve good results.

Keywords: Petrochemical industry · Safety performance assessment · Interval number

1 Introduction

Safety performance assessment becomes the key means to evaluate the status of petrochemical enterprise safety [1]. Many researchers have focused on this field [2–4]. The generic safety performance assessment method is expert grading and weighted average. Because of the ambiguity and complexity of objects and the fuzziness and limitations of the human mind, there is no absoluteness in real world. Thus we need an appropriate solution to the uncertainty of measurements caused by ambiguity so as to make it be in accordance with human thinking.

Interval number can solve such problems, with more accurate expressions and operations, and more realistic results. Zhang et al. [5] proposed a new approach to multiple attribute decision making with interval numbers. Zhang et al. [6] developed a possibility degree-based micro immune optimization approach to seek the optimal solution of nonlinear interval number programming with constraints. Zhao et al. [7] researched the fuzzy multi attribute decision making in which the fuzzy numbers are given in the form of interval numbers.

In this paper, a petrochemical enterprise safety performance assessment method based on interval number is proposed. A case of petrochemical enterprise is used to

© Springer International Publishing Switzerland 2016
D.-S. Huang et al. (Eds.): ICIC 2016, Part III, LNAI 9773, pp. 780–788, 2016.
DOI: 10.1007/978-3-319-42297-8_72

illustrate the effectiveness of the method and system. The rest of this article is organized as follows. In Sect. 2, the petrochemical enterprise safety performance assessment based on interval number theory is proposed. In Sect. 3, an application example is
present. Section 4 provides the conclusions of the study.

2 Methodology

Definition 1. R is real number set. For $\underline{a}, \bar{a} \in R$, $\underline{a} \leq \bar{a}$, if $[\bar{A}] = [\underline{a}, \bar{a}]$, then $[\bar{A}]$ is called
a interval number.

Definition 2. Supposing $[\bar{A}] = [\underline{a}, \bar{a}]$ and $[\bar{B}] = [\underline{b}, \bar{b}]$. Then:

1. $[\bar{A}] = [\bar{B}]$, if and only if $\underline{a} = \underline{b}$, $\bar{a} = \bar{b}$;
2. $[\bar{A}] + [\bar{B}] = [\underline{a} + \underline{b}, \bar{a} + \bar{b}]$;
3. $[\bar{A}] \cdot [\bar{B}] = [\underline{a} \cdot \underline{b}, \bar{a} \cdot \bar{b}]$;
4. $[\bar{A}] \div [\bar{B}] = [\underline{a} \div \bar{b}, \bar{a} \div \underline{b}]$.

The proposed method is shown in Fig. 1. The assessment process is as follows.

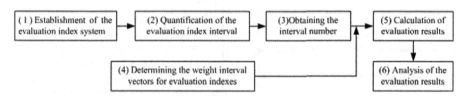

Fig. 1. Petrochemical enterprise safety assessment based on interval number

2.1 Establishment of the Evaluation Index System

The evaluation index system is shown as Fig. 2. The entire index system is composed
of 6 primary indexes and 37 secondary indexes.

2.2 Quantification of the Evaluation Index Interval

Evaluation indexes are divided into five levels numerically:

$\{[a_1, a_2], [a_3, a_4], [a_5, a_6], [a_7, a_8], [a_9, a_{10}]\}$, and $0 \leq a_1 \leq a_2 \leq a_3 \leq a_4 \leq a_5 \leq a_6 \leq$
$a_7 \leq a_8 \leq a_9 \leq a_{10} \leq 1$

The range of each level is determined by specific evaluation indexes.
Determination of the number of evaluation index intervals
Every index is scored by several experts using the interval number. The index
interval vector is as follows:

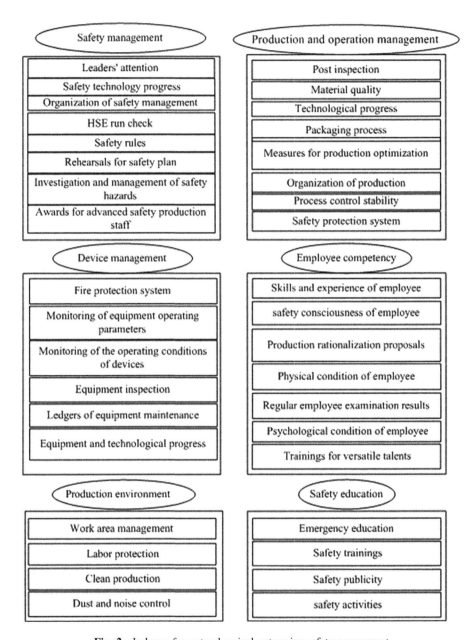

Fig. 2. Indexes for petrochemical enterprise safety assessment

$$[\vec{A}] = ([\underline{a_1}, \overline{a_1}], [\underline{a_2}, \overline{a_2}], \cdots, [\underline{a_n}, \overline{a_n}]),$$

Where n denotes the number of evaluation indexes, $[\underline{a_i}, \overline{a_i}]$ denotes the numerical value of the ith index interval.

2.3   Determination of Weight Interval Vector

The weight coefficients for the evaluation indexes are shown in Table 1.

Table 1. Weight coefficient levels of the evaluation indexes

Level	I	II	III	IV	V
Property	Particularly important	Very important	Relatively important	Important	Common
Range	(0.9,1.0)	(0.8,0.9)	(0.7,0.8)	(0.6,0.7)	(0.5,0.6)

Firstly, the numerical value of the initial weight interval is obtained:

$$\overrightarrow{[Z]} = ([\underline{Z_1},\overline{Z_1}], [\underline{Z_2},\overline{Z_2}], \cdots, [\underline{Z_n},\overline{Z_n}]),$$

where n is the number of evaluation indexes, $[\underline{Z_i},\overline{Z_i}]$ is the initial weight of the ith index, and $[\underline{Z_i},\overline{Z_i}]$ is the weight coefficient level given in Table 1.

Fuzzy mathematics is adopted to process the indexes as follows:

$$[E_i] = \frac{([Z_i] - [A])}{([B] - [A])} (1 \leq i \leq n),$$

where $[A] = [0.2, 0.5]$ and $[B] = [0.5, 0.8]$.

The indexes are normalized as follows:

$$[E_i'] = \frac{[E_i]}{\sum\limits_{i=1}^{n} [E_i]}.$$

The weighted interval vector after the processing is as follows:

$$\overrightarrow{[W]} = \left\{ [E_i'] | 1 \leq i \leq n, i \in N \right\}.$$

2.4   Calculation of the Evaluation Results

The calculations of the evaluation results are as follows:

$$[R] = \overrightarrow{[A]} \bullet \overrightarrow{[W]}^{T}.$$

2.5   Analysis of the Evaluation Results

The evaluation results levels are shown in Table 2.

Table 2. Criteria for evaluating the levels of indexes

Level		Range	Remarks
I	Excellent	[0. 08,0.1]	The enterprise safety level is extremely high, with highly coordinated indexes. It does not need any improvements.
II	Good	[0. 06,0. 08)	The enterprise safety level is quite high, with relatively coordinated indexes. It can be accepted.
III	Moderate	[0. 04,0. 06)	The enterprise safety level is common, with basically coordinated indexes. It is acceptable within a certain period
IV	Common	[0. 02,0. 04)	The enterprise safety is not very good, with poorly coordinated indexes. It needs further studies and improvements
V	Poor	[0,0. 02)	The enterprise safety is poor, with very poorly coordinated indexes. It needs improvements urgently.

3   Application

Guangdong Shidong Industry Group [27] is located at work zone of Shatian, Humen Port, Shatian Town, Dongguan City, China with an area of about 0.22 million square meters. Taking the petrochemical enterprise for example, an applied research is conducted. The quantification of evaluation indexes for the enterprise is presented from Table 2. Six experts were invited to evaluate, with the results shown from Tables 3, 4, 5, 6, 7 and 8.

Table 3. Scores of Secondary Indexes in Safety management for the enterprise

Indexes	Expert 1	Expert 2	Expert 3	Expert 4	Expert 5	Expert 6	Mean value
Importance	Excellent	Good	Moderate	Excellent	Excellent	Excellent	MV ≥ Good
	[0.8,1.0]	[0.6,0.8)	[0.4,0.6)	[0.8,1.0]	[0.8,1.0]	[0.8,1.0]	[0.7,0.9]
Safety technologies	Good	Excellent	Good	Moderate	Good	Relatively poor	Moderate≤ MV ≤ Good
	[0.6,0.8)	[0.8,1.0]	[0.6,0.8)	Table 4. [0.4,0.6)	[0.6,0.8)	[0.2,0.4)	[0.5,0.7]
Organization of safety management	Good	Good	Good	Excellent	Good	Moderate	MV = Good
	[0.6,0.8)	[0.6,0.8)	[0.6,0.8)	[0.8,1.0]	[0.6,0.8)	[0.4,0.6)	[0.6,0.8]
HSE run check	Excellent	Excellent	Good	Excellent	Moderate	Good	MV ≥ Good
	[0.8,1.0]	[0.8,1.0]	[0.6,0.8)	[0.8,1.0]	[0.4,0.6)	[0.6,0.8)	[0.7,0.9]
Safety rules	Moderate	Good	Moderate	Excellent	Good	Excellent	MV = Good
	[0.4,0.6)	[0.6,0.8)	[0.4,0.6)	[0.8,1.0]	[0.6,0.8)	[0.8,1.0]	[0.6,0.8]

(*Continued*)

Table 3. (*Continued*)

Indexes	Expert 1	Expert 2	Expert 3	Expert 4	Expert 5	Expert 6	Mean value
Rehearsals for safety plan	Excellent [0.8,1.0]	Excellent [0.8,1.0]	Good [0.6,0.8)	Excellent [0.8,1.0]	Moderate [0.4,0.6)	Excellent [0.8,1.0]	MV ≥ Good [0.7,0.9]
Investigation and management of safety hazards	Good [0.6,0.8)	Excellent [0.8,1.0]	Moderate [0.4,0.6)	Excellent [0.8,1.0]	Good [0.6,0.8)	Moderate [0.4,0.6)	MV = Good [0.6,0.8]
Awards for advanced safety production staff	Moderate [0.4,0.6)	Good [0.6,0.8)	Excellent [0.8,1.0]	Good [0.6,0.8)	Good [0.6,0.8)	Good [0.6,0.8)	MV = Good [0.6,0.8]

Table 4. Scores of Secondary Indexes in Production and operation management for the enterprise

Indexes	Expert 1	Expert 2	Expert 3	Expert 4	Expert 5	Expert 6	Mean value
Post inspection	Good [0.6,0.8)	Excellent [0.8,1.0]	Good [0.6,0.8)	Excellent [0.8,1.0]	Moderate [0.4,0.6)	Good [0.6,0.8)	MV = Good [0.6,0.8]
Material quality	Excellent [0.8,1.0]	Good [0.6,0.8)	Excellent [0.8,1.0]	Excellent [0.8,1.0]	Good [0.6,0.8)	Moderate [0.4,0.6)	MV ≥ Good [0.7,0.9]
Technological progress	Moderate [0.4,0.6)	Excellent [0.8,1.0]	Excellent [0.8,1.0]	Good [0.6,0.8)	Good [0.6,0.8)	Good [0.6,0.8)	MV = Good [0.6,0.8]
Packaging process	Excellent [0.8,1.0]	Excellent [0.8,1.0]	Good [0.6,0.8)	Moderate [0.4,0.6)	Good [0.6,0.8)	Moderate [0.4,0.6)	MV = Good [0.6,0.8]
Measures for production optimization	Excellent [0.8,1.0]	Excellent [0.8,1.0]	Good [0.6,0.8)	Moderate [0.4,0.6)	Good [0.6,0.8)	Moderate [0.4,0.6)	MV = Good [0.6,0.8]
Organization of production	Excellent [0.8,1.0]	Excellent [0.8,1.0]	Good [0.6,0.8)	Moderate [0.4,0.6)	Good [0.6,0.8)	Moderate [0.4,0.6)	MV = Good [0.6,0.8]
Process control stability	Excellent [0.8,1.0]	Moderate [0.4,0.6)	Excellent [0.8,1.0]	Good [0.6,0.8)	Good [0.6,0.8)	Excellent [0.8,1.0]	MV ≥ Good [0.7,0.9]

Table 5. Scores of Secondary Indexes in Equipment management for the enterprise

Indexes	Expert 1	Expert 2	Expert 3	Expert 4	Expert 5	Expert 6	Mean value
Security protection system	Excellent [0.8,1.0]	Moderate [0.4,0.6)	Excellent [0.8,1.0]	Good [0.6,0.8)	Good [0.6,0.8)	Excellent [0.8,1.0]	MV ≥ Good [0.7,0.9]
Fire protection system	Moderate [0.4,0.6)	Excellent [0.8,1.0]	Good [0.6,0.8)	Excellent [0.8,1.0]	Relatively poor [0.2,0.4)	Good [0.6,0.8)	MV = Good [0.6,0.8]
Monitoring of equipment operating parameters	Excellent [0.8,1.0]	Excellent [0.8,1.0]	Excellent [0.8,1.0]	Moderate [0.4,0.6)	Good [0.6,0.8)	Good [0.6,0.8)	MV ≥ Good [0.7,0.9]

(*Continued*)

Table 5. (*Continued*)

Indexes	Expert 1	Expert 2	Expert 3	Expert 4	Expert 5	Expert 6	Mean value
Monitoring of the operating conditions of devices	Moderate [0.4,0.6)	Moderate [0.4,0.6)	Good [0.6,0.8)	Excellent [0.8,1.0]	Excellent [0.8,1.0]	Excellent [0.8,1.0]	MV = Good [0.6,0.8]
Equipment inspection	Good [0.6,0.8)	Excellent [0.8,1.0]	Good [0.6,0.8)	Relatively poor [0.2,0.4)	Excellent [0.8,1.0]	Moderate [0.4,0.6)	MV = Good [0.6,0.8]
Ledgers of equipment maintenance and repair	Moderate [0.4,0.6)	Moderate [0.4,0.6)	Excellent [0.8,1.0]	Good [0.6,0.8)	Relatively poor [0.2,0.4)	Good [0.6,0.8)	Moderate ≤ MV ≤ Good [0.5,0.7]
Equipment and technological progress	Excellent [0.8,1.0]	Excellent [0.8,1.0]	Good [0.6,0.8)	Moderate [0.4,0.6)	Excellent [0.8,1.0]	Good [0.6,0.8)	MV ≥ Good [0.7,0.9]

Table 6. Scores of Secondary Indexes in Production environment for the enterprise

Indexes	Expert 1	Expert 2	Expert 3	Expert 4	Expert 5	Expert 6	Mean value
Work area management	Good [0.6,0.8)	Moderate [0.4,0.6)	Excellent [0.8,1.0]	Good [0.6,0.8)	Relatively poor [0.2,0.4)	Moderate [0.4,0.6)	Moderate ≤ MV ≤ Good [0.5,0.7]
Labor protection	Good [0.6,0.8)	Good [0.6,0.8)	Moderate [0.4,0.6)	Excellent [0.8,1.0]	Excellent [0.8,1.0]	Excellent [0.8,1.0]	MV ≥ Good [0.7,0.9]
Clean production	Moderate [0.4,0.6)	Good [0.6,0.8)	Excellent [0.8,1.0]	Good [0.6,0.8)	Moderate [0.4,0.6)	Excellent [0.8,1.0]	MV ≥ Good [0.6,0.8]
Dust and noise control	Good [0.6,0.8)	Excellent [0.8,1.0]	Moderate [0.4,0.6)	Good [0.6,0.8)	Excellent [0.8,1.0]	Moderate [0.4,0.6)	MV ≥ Good [0.6,0.8]

Table 7. Scores of Secondary Indexes in Employee competency for the enterprise

Indexes	Expert 1	Expert 2	Expert 3	Expert 4	Expert 5	Expert 6	Mean value
Skills and experience of employee	Good [0.6,0.8)	Excellent [0.8,1.0]	Moderate [0.4,0.6)	Good [0.6,0.8)	Excellent [0.8,1.0]	Moderate [0.4,0.6)	MV ≥ Good [0.6,0.8]
Safety consciousness of employee	Good [0.6,0.8)	Excellent [0.8,1.0]	Moderate [0.4,0.6)	Good [0.6,0.8)	Good [0.6,0.8)	Excellent [0.8,1.0]	MV ≥ Good [0.6,0.8]
Production rationalization proposals	Moderate [0.4,0.6)	Good [0.6,0.8)	Excellent [0.8,1.0]	Excellent [0.8,1.0]	Excellent [0.8,1.0]	Good [0.6,0.8)	MV ≥ Good [0.7,0.9]
Physical condition of employee	Excellent [0.8,1.0]	Excellent [0.8,1.0]	Moderate [0.4,0.6)	Excellent [0.8,1.0]	Good [0.6,0.8)	Excellent [0.8,1.0]	MV ≥ Good [0.7,0.9]
Regular staff examination results	Moderate [0.4,0.6)	Good [0.6,0.8)	Excellent [0.8,1.0]	Good [0.6,0.8)	Moderate [0.4,0.6)	Excellent [0.8,1.0]	MV = Good [0.6,0.8]
Psychological condition of employee	Good [0.6,0.8)	Excellent [0.8,1.0]	Good [0.6,0.8)	Excellent [0.8,1.0]	Excellent [0.8,1.0]	Moderate [0.4,0.6)	MV ≥ Good [0.7,0.9]
Training for versatile talents	Moderate [0.4,0.6)	Good [0.6,0.8)	Excellent [0.8,1.0]	Good [0.6,0.8)	Excellent [0.8,1.0]	Moderate [0.4,0.6)	MV = Good [0.6,0.8]

Table 8. Scores of Secondary Indexes in Safety education for the enterprise

Indexes	Expert 1	Expert 2	Expert 3	Expert 4	Expert 5	Expert 6	Mean value
Emergency	Excellent	Excellent	Excellent	Excellent	Good	Excellent	MV = Excellent
education	[0.8,1.0]	[0.8,1.0]	[0.8,1.0]	[0.8,1.0]	[0.6,0.8)	[0.8,1.0]	[0.8,1.0]
Safety trainings	Good	Excellent	Excellent	Excellent	Good	Moderate	MV ≥ Good
	[0.6,0.8)	[0.8,1.0]	[0.8,1.0]	[0.8,1.0]	[0.6,0.8)	[0.4,0.6)	[0.7,0.9]
Safety publicity	Moderate	Good	Excellent	Excellent	Good	Good	MV = Good
	[0.4,0.6)	[0.6,0.8)	[0.8,1.0]	[0.8,1.0]	[0.6,0.8)	[0.6,0.8)	[0.6,0.8]
Safety activities	Good	Good	Good	Relatively poor	Moderate	Excellent	Moderate ≤ MV ≤ Good
	[0.6,0.8)	[0.6,0.8)	[0.6,0.8)	[0.2,0.4)	[0.4,0.6)	[0.8,1.0]	[0.5,0.7]

The index interval vector $[\vec{A}]$ is obtained from Tables 3, 4, 5, 6, 7 and 8:

$$[\vec{A}] = ([0.7, 0.9], [0.5, 0.7], [0.6, 0.8], [0.7, 0.9], [0.6, 0.8], [0.7, 0.9], [0.6, 0.8], [0.6, 0.8], [0.6, 0.8],$$

$$[0.7, 0.9], [0.6, 0.8], [0.6, 0.8], [0.6, 0.8], [0.6, 0.8], [0.7, 0.9], [0.7, 0.9], [0.6, 0.8], [0.7, 0.9], [0.6, 0.8],$$

$$[0.6, 0.8], [0.5, 0.7], [0.7, 0.9], [0.5, 0.7], [0.7, 0.9], [0.6, 0.8], [0.6, 0.8], [0.6, 0.8], [0.6, 0.8], [0.7, 0.9],$$

$$[0.7, 0.9], [0.6, 0.8], [0.7, 0.9], [0.6, 0.8], [0.8, 1.0], [0.7, 0.9], [0.6, 0.8], [0.5, 0.7])$$

The initial weight interval vector:

$$[\vec{Z}] = ([0.8, 0.9], [0.8, 0.9], [0.7, 0.8], [0.6, 0.7], [0.7, 0.8], [0.6, 0.7], [0.6, 0.7],$$

$$[0.8, 0.9], [0.8, 0.9], [0.7, 0.8], [0.6, 0.7], [0.7, 0.8], [0.6, 0.7], [0.6, 0.7],$$

$$[0.8, 0.9], [0.8, 0.9], [0.7, 0.8], [0.6, 0.7], [0.7, 0.8], [0.6, 0.7], [0.6, 0.7],$$

$$[0.8, 0.9], [0.8, 0.9], [0.7, 0.8], [0.6, 0.7], [0.7, 0.8], [0.6, 0.7], [0.6, 0.7], [0.8, 0.9],$$

$$[0.8, 0.9], [0.7, 0.8], [0.6, 0.7], [0.7, 0.8], [0.6, 0.7], [0.6, 0.7], [0.8, 0.9], [0.8, 0.9])$$

After fuzzy mathematics process and normalized:

$$[\vec{W}] = ([\frac{2}{91}, \frac{1}{18}], [\frac{2}{91}, \frac{1}{18}], [\frac{3}{182}, \frac{5}{108}], [\frac{1}{91}, \frac{1}{27}], [\frac{3}{182}, \frac{5}{108}], [\frac{1}{91}, \frac{1}{27}], [\frac{1}{91}, \frac{1}{27}], [\frac{2}{91}, \frac{1}{18}],$$

$$[\frac{2}{91}, \frac{1}{18}], [\frac{3}{182}, \frac{5}{108}], [\frac{1}{91}, \frac{1}{27}], [\frac{3}{182}, \frac{5}{108}], [\frac{1}{91}, \frac{1}{27}], [\frac{1}{91}, \frac{1}{27}], [\frac{2}{91}, \frac{1}{18}], [\frac{2}{91}, \frac{1}{18}],$$

$$[\frac{3}{182}, \frac{5}{108}], [\frac{1}{91}, \frac{1}{27}], [\frac{3}{182}, \frac{5}{108}], [\frac{1}{91}, \frac{1}{27}], [\frac{1}{91}, \frac{1}{27}], [\frac{2}{91}, \frac{1}{18}], [\frac{2}{91}, \frac{1}{18}], [\frac{3}{182}, \frac{5}{108}], [\frac{1}{91}, \frac{1}{27}],$$

$$[\frac{3}{182}, \frac{5}{108}], [\frac{1}{91}, \frac{1}{27}], [\frac{1}{91}, \frac{1}{27}], [\frac{2}{91}, \frac{1}{18}], [\frac{2}{91}, \frac{1}{18}], [\frac{3}{182}, \frac{5}{108}], [\frac{1}{91}, \frac{1}{27}], [\frac{3}{182}, \frac{5}{108}],$$

$$[\frac{1}{91}, \frac{1}{27}], [\frac{1}{91}, \frac{1}{27}], [\frac{2}{91}, \frac{1}{18}], [\frac{2}{91}, \frac{1}{18}])$$

The evaluation results: $[\vec{R}] = [\vec{A}] \bullet [\vec{W}]^T = [\frac{14}{910}, \frac{9}{180}] = [0.02, 0.05]$.

Based on Table 2, the evaluation results are at "Level IV–III", ranging from "common to moderate". One-third of the interval spans the moderate level, which indicates that in the petrochemical enterprise, the evaluation results of safety evaluation indexes are relatively scattered, with significant disparity. The scores of some relatively important indexes are very low, while those of lesser important indexes are relatively high. This indicates that petrochemical enterprise safety is acceptable within a certain period, and will take time to improve.

4 Conclusions

In terms of the characteristics of petrochemical enterprise safety assessment, and by virtue of the advantages of interval number in solving the uncertainty of index values, a petrochemical enterprise safety assessment method based on interval number is proposed. This method is applied to the comprehensive evaluation of Guangdong Shidong Industry Group safety to achieve good results. As there are many factors that affect the safety assessment in petrochemical enterprise, only the perfect evaluation index system can get more close to the results of the facts. Furthermore, we will improve the evaluation index system according to different application scenarios.

Acknowledgments. The study was supported by the National Natural Science Foundation of China (Grant No.71201026), the Project of Department of Education of Guangdong Province (No. 2013KJCX0179, No.2014KTSCX184, and No. 2014KGJHZ014), and the Dongguan Social Science and Technology Development Project (NO.2013108101011).

References

1. Benmessaoud, T., Benazzouz, D., Benikhlef T., Mazouz L., Saadi S.: Major industrial risks assessment in chemical/petrochemical facilities. Int. J. Syst. Assur. Eng. Manag. 1–9 (2014)
2. Li, W., Liang, W., Zhang, L., Tang, Q.: Performance assessment system of health, safety and environment based on experts' weights and fuzzy comprehensive evaluation. J. Loss Prev. Process Ind. **35**, 95–103 (2015)
3. Azadeh, A., Saberi, M., Rouzbahman, M., Valianpour, F.: A neuro-fuzzy algorithm for assessment of health, safety, environment and ergonomics in a large petrochemical plant. J. Loss Prev. Process Ind. **34**, 100–114 (2015)
4. Jing S., Liu X., Cheng C., Shang X.: A HAZOP based model for safety management risk assessment in petrochemical plants. In: Proceedings of the 11th World Congress on Intelligent Control & Automation, Shenyang, China, June 2014, pp. 3551–3555. IEEE, Washington, DC (2014)
5. Zhang, Q., Gao, Q., Geng, J.: New approach to multiple attribute decision making with interval numbers. J. Syst. Eng. Electron. **19**, 304–310 (2008)
6. Zhang, Z., Tao, J.: Efficient micro immune optimization approach solving constrained nonlinear interval number programming. Appl. Intell. **43**, 276–295 (2015)
7. Zhao H.: Research and application of interval number multiple attribute decision making. In: Proceedings of the 2009 International Symposium on Intelligent Ubiquitous Computing and Education, Chengdu, China, May 2009, pp. 65–68. ACM, New York (2009)
8. GDSIG (Guang Dong ShiDong Industry Group) Website. http://www.gdshidong.com/en_asp/index.asp. Accessed 14 Feb 2016

A Non-linear Optimization Model
and ANFIS-Based Approach to Knowledge
Acquisition to Classify Service Systems

Eduyn Ramiro López-Santana[(✉)]
and Germán Andrés Méndez-Giraldo

Faculty of Engineering, Universidad Distrital Francisco José de Caldas,
Bogotá, Colombia
{erlopezs,gmendez}@udistrital.edu.co

Abstract. This paper studies the problem of knowledge acquisition to classify
service systems. We define a set of attributes and characteristics in order to
classify the service systems. To state the interactions between attributes and
characteristics we propose a non-linear optimization model and an adaptive
neuro-fuzzy inference system (ANFIS) approach. We compare both approaches
in terms of mean root square error in a data test based in International Standard
Industrial Classification. Our results present a better performance of ANFIS
approach over a set of data collected about ISIC classification in Colombia
industries.

Keywords: ANFIS · Optimization · Service classification · Service systems

1 Introduction

The purpose of the classification systems of economic activities such as ISIC (International Standard Industrial Classification) or NICE (-Nice- International Classification
of Goods and Services for the Purposes of the Registration of Marks) is to provide
some categories of activities for gathering and statistical reporting. Its aim is to normalize the economic activities of all countries in order to make measurements and
comparisons of economic and social indicators.

Usually, companies must identify their economic activity, for this is common to use
different mechanisms such as matching text, similar companies, expert advice, among
others. In some cases, the selection is simple, particularly in manufacturing systems due
to the characteristics thereof and to the large number of studies and applications that
have been developed in these systems. However, to classify services systems is not so
simple because the number of variables involved in their definition and
characterization.

The main way to classify a service activity is the use of the expertise or knowledge
to select the best category in which a type of system can enter. However not all
companies, especially small and medium enterprises (SMEs) have access to specialized
personnel because of their scarcity and high cost [1]. For this difficulty, the need for a
tool for classification of service activities is established. Possible computational tools

D.-S. Huang et al. (Eds.): ICIC 2016, Part III, LNAI 9773, pp. 789–801, 2016.
DOI: 10.1007/978-3-319-42297-8_73

are Expert Systems (ES) and Based-Rules System (BRS), which as suggested by [1], whichare a branch of Artificial Intelligence (AI) that it uses specialized knowledge to solve problems as a human specialist.

In [2] a proposal approach to classification of services systems through a BRS is presented. The authors define a set of twelve attributes to characterize an economic activity related to customer issues, process and performance measures. With these attributes, they build a BRS that explains the classification of economic activities in the three sectors defined and sections of ISIC. They propose a methodology based in three phases. The first one consist in build a knowledge base using knowledge acquisition. The second one is the engine inference based in a set of rules. The last one is a making decision process. The authors propose a waterfall process, at the beginning with the classification of the sector (Primary, Manufacturing and Services). After, the BRS obtain the Section of ISIC classification and then determine the Division. The authors use J48 algorithm to obtain the rules.

The purpose of this paper refers to state the acquisition model of the phase one of the proposed method by [2]. We define the information required and state the model to obtain the set of attributes.

The remainder of this paper is organized as follows: Sect. 2 introduces the problem statement about classification of service systems. Section 3 describes our solution approach based on non-linear optimization model and ANFIS approach. Section 4 shows the results of application of our proposal method at ISIC classification. Finally, Sect. 5 concludes this work and provides possible research directions.

2   Problem Statement

Service systems consist of activities that are offered using human or mechanical power in order to satisfy people's needs or wishes [3], also consist of a set of interacting resources provided by the customer and the provider [4]. Service can be defined as the application of competences for the benefit of another [5] and a solution and a customer's experience that satisfy her or his wishes [6]. It means that service is a kind of action, performance, or promise that is exchanged for value between provider and customer since the service is performed in close contact with a customer. Service systems is as a set of activities performed by resources (machines and people) to meet the needs or desires of people through the transformation of the initial state of any of the customer's resources [3]. Also, service systems consist of service providers and service customers working together to coproduce value in complex value networks where providers and customers might be individuals, firms, government agencies, or any organization of people and technologies [7].

Service systems are becoming a strategic area of scientific research from multidisciplinary approaches. An academic community has emerged such as *Service Science* or *Service Science Management and Engineering* [7, 8]. *Service science* is the study of complex service systems and is an inter-disciplinary integration that involves methods and theories from operations research, industrial engineering, marketing, computer science, psychology, information systems, design, and more applied to design, to manage, to control and to improve the services systems.

Figure 1 present the BRS proposed by [2]. It consists in a classification system for the International Standard Industrial Classification (ISIC) system in the Colombian context and for classification according to sector of the economy as Services, Manufacturing and Primary proposed by the authors for the section level and division level (see Fig. 2.) In addition, it is taken as instances for database-level Group, i.e. in total has 249 instances for the selection of rules. The knowledge base consists in a set of 12 attributes (see Table 1). The data was collected using interview with an expert and secondary information sources. The inference engine was made with an algorithm to generate tree J48 using the entropy information measure [9].

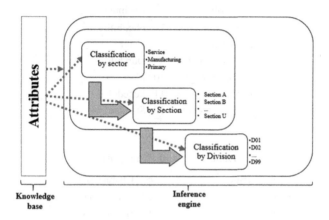

Fig. 1. A KBES to classify service systems

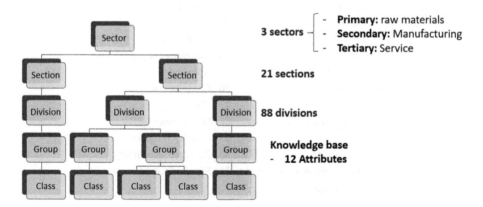

Fig. 2. Modified ISIC classification systems

ISIC classification proposed by United Nations Educational, Scientific and Cultural Organization –UNESCO- [10] aims to establish a uniform classification of productive economic activities. The term activity is understood as a process or group of operations that combine resources base such as equipment, labor, manufacturing techniques and inputs for the production of goods and services. The hierarchical structure of this classification is shown in Fig. 2., where you have 21 sections, 88 divisions, 249 groups and 495 classes, according to the adaptation made for Colombia [11].

This paper deals with the question, How to determine the set of attributes? To solve this question we define a set of 31 characteristics (see Table 2). Then, the problem consists in determine the relation between attributes and characteristics. The characteristics are collected by interviews with the experts.

Table 1. List of attributes

a	Attribute	a	Attribute
1	Coproduction	7	Flexibility
2	Dependence	8	Nature
3	Accumulation	9	Durability
4	Technology	10	Place
5	Property	11	Scheduled
6	Simultaneity	12	Standardization

Table 2. List of characteristics

r	Characteristic	r	Characteristic
1	Presence	17	Interaction
2	Behavior	18	Beneficiary
3	Continuity	19	Permission
4	Abandonment	20	Commercialization
5	Operation	21	Virtual
6	Autonomy	22	Personalization
7	Influence	23	Appearance
8	Storage	24	complementary products
9	Storage time	25	Knowledge
10	Anticipation	26	Consumption
11	Wait time	27	Place
12	Initial waiting time	28	Displacement
13	Space	29	Activity program
14	Cost of storage	30	Resource Scheduling
15	Using machines /equipment	31	Standardization
16	Using people		

3 Solution Approach

We define a set of attributes $A = \{a_1, a_2, \ldots, a_{|A|}\}$, a set of groups of ISIC classification $I = \{1, \ldots, n\}$ and a set of characteristics $R = \{r_1, r_2, \ldots, r_{|R|}\}$. Let x_{ia} as the value of each attribute $a \in A$ for each group $i \in I$ and y_{ir} as the value of each characteristic

$r \in A$ for each group $i \in I$. We establish a relation $x_{ia} = f(y_{ir})$ where f could be weighted linear combinations of structured subsets of characteristics y_{ir} as:

$$x_{ia} = \sum_{r \in R_a} w_{ar} y_{ir} \forall a \in A, i \in I \tag{1}$$

$$\sum_{r \in R_a} w_{ar} = 1 \forall a \in A \tag{2}$$

Equations (1) state the function f as weighted linear combination of structured subsets R_a of characteristics y_{ir} for each attribute $a \in A$ and each group $i \in I$ with a weight w_{ar}. Equations (2) ensure that the sum of the weights w_{ar} does not exceed one for each attribute $a \in A$.

The problem consists in determining the weights w_{ar}. We propose two approaches. The first one is a non-linear optimization model with the objective to minimize the mean square error with the assumption of x_{ia} and y_{ir} are deterministic and known. The second one is an ANFIS approach where the assumption of x_{ia} and y_{ir} are imprecise and a membership function are established.

3.1 Non-linear Optimization Model

We propose a non-linear optimization model to determine the weights w_{ar}. In the next four sections, we describe the index sets, parameters, decision variables and mathematical formulation of the model.

The indexed Sets. We define the next sets:

- A : set of attributes $\{a_1, a_2, \ldots, a_{|A|}\}$
- R: set of characteristics $\{r_1, r_2, \ldots, r_{|R|}\}$
- R_a: set of characteristics of an attribute a, $R_a \subseteq R$
- I : set of groups $\{1, \ldots, n\}$

Parameters. The parameters are described as follows:

- y_{ir}: value of each characteristic $r \in R$ for each group i
- o_{ia} = desired output of group i of an attribute a
- β_{ar} = minimum contribution of a characteristic r to an attribute a

Decision Variables. The decision variables are described as follows:

- w_{ar} = contribution of characteristic r to an attribute a
- e_{ia} = error of a group i of an attribute a
- x_{ia}: value of each attribute $a \in A$ for each group i

Model. The mathematical formulation of our MIP model is described as follows:

$$\min \sum_{i \in I} \sum_{a \in A} \frac{e_{ia}^2}{n} \tag{3}$$

s.t.,

$$e_{ia} = o_{ia} - x_{ia} \qquad \forall i \in I, a \in A \qquad (4)$$

$$\sum\nolimits_{r \in R_a} w_{ar} y_{ir} = x_{ia} \qquad \forall i \in I, a \in A \qquad (5)$$

$$\sum\nolimits_{r \in R_a} w_{ar} = 1 \qquad \forall a \in A \qquad (6)$$

$$w_{ar} \geq \beta_{ar} \qquad \forall a \in A, r \in R \qquad (7)$$

$$0 \leq w_{ar} \leq 1 \qquad \forall a \in A, r \in R \qquad (8)$$

The objective function (3) consist in minimizing the mean of square errors that is a non-linear equation. Constraints (4) compute the error e_{ia} as the difference between the desired output o_{ia} and the decision variable of the value of each attribute x_{ia}. Constraints (5) calculate the value of each attribute x_{ia} as a weighted average of characteristics $r \in R_a$. Constraints (6) ensure that the sum of contributions w_{ar} sum 1. Constraints (7) state the minimal value of each contribution w_{ar} and constraints (8) limits the lower bound as zero and the upper bound as one for each w_{ar}.

3.2 ANFIS Approach

In the literature, there are several applications of artificial neuronal networks and fuzzy inference systems to solve a non-linear optimization model. For instance, in [12] a model-free fuzzy controller is builtto of wind turbine hub assembly. In [13], a neuronal network is used to emulate a human finger for reaching a desired fingertip position in space. Moreover, in [14] that propose a fuzzy logic controller (FLC) with expert knowledge of tightening process and error detection capability in wind turbine industry. In this section, first we describe the architecture of ANFIS. Second, we show our ANFIS-based approach to knowledge acquisition to classify service system.

ANFIS Architecture. It is a flexible approach based on fuzzy logic and artificial neural networks, adaptive network-based fuzzy inference system (ANFIS) [15]. We present an architecture based in two inputs x and y and one output z based in [15, 16]. Suppose, if the rule base contains two fuzzy if-then rules such as
Rule 1: If x is A_1 and y is B_1, then $z_1 = p_1 x + q_1 y + r_1$.
Rule 2: If x is A_2 and y is B_2, then $z_2 = p_2 x + q_2 y + r_2$.
Then the membership functions and fuzzy reasoning is illustrated in Fig. 3, and the corresponding equivalent ANFIS architecture is shown in Fig. 4. The node functions in the same layer are of the same function family as described below:
Layer 1. Every node i in this layer is a square node with a node function (9)

$$O_i^1 = \mu A_i(x), \qquad (9)$$

where x is the input to node i and A_i the linguistic label (e.g., small, large, etc.) associated with this node function. $\mu A_i(x)$ is a membership functions (MF) with

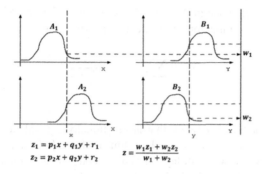

$$z_1 = p_1x + q_1y + r_1$$
$$z_2 = p_2x + q_2y + r_2$$

$$z = \frac{w_1z_1 + w_2z_2}{w_1 + w_2}$$

Fig. 3. Membership function and fuzzy reasoning (based in [15])

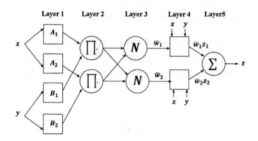

Fig. 4. Architecture of ANFIS (based in [15])

a maximum equal to 1 and minimum equal to 0. Any continuous and piecewise differentiable functions, such as commonly used trapezoidal or triangular-shaped MFs, are used for node functions in this layer. Parameters in this layer are referred to as premise parameters.

Layer 2. Every node in this layer is a circle node labeled Π, which multiplies the incoming signals and sends the product out. For instance,

$$w_i = \mu A_i(x)\mu B_i(y), i = 1, 2\ldots \tag{10}$$

Layer 3. Every node in this layer is a circle node labeled N. The i th node calculates the ratio of the i-th rule's firing strength to the sum of all rules' firing strengths:

$$\overline{w_i} = \frac{w_i}{w_1 + w_2} \tag{11}$$

For convenience, outputs of this layer will be called normalized firing strengths.
Layer 4. Every node i in this layer is a square node with a node function

$$O_i^4 = \overline{w_i}z_i = \overline{w_i}(p_ix + q_iy + r_i), \tag{12}$$

where w_i is the output of layer 3, and $\{p_i, q_i, r_i\}$ is the parameter set. Parameters in this layer will be referred to as consequent parameters.

Layer 5. The single node in this layer is a circle node labeled $\sum$ that computes the overall output as the summation of all incoming signals, i.e.,

$$O_i^5 = \sum_i \overline{w_i} z_i = \frac{\sum_i w_i z_i}{\sum_i w_i}. \tag{13}$$

Thus, it is an adaptive network that is functionally equivalent to a Sugeno fuzzy model [16]. We use a hybrid learning algorithm similar to [15], in the forward pass, the functional signals go forward till layer 4 and the consequent parameters are identified by the least squares estimate. In the backward pass, the error rates propagate backward and the premise parameters are updated by the gradient descent. The consequent parameters thus identified are optimal (in the consequent parameter space) under the condition that the premise parameters are fixed.

Our ANFIS Method. The ANFIS integrates both neural networks and fuzzy logic principle strain a Sugeno systems using neuro-adaptive learning as describe in above section. Figure 5 presents the architecture of proposed ANFIS approach. In the next

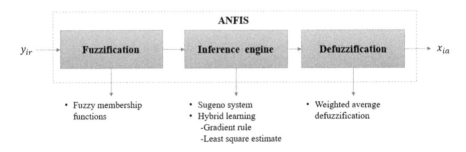

Fig. 5. Architecture of ANFIS approach

four sections, we describe the inputs, output, method, and performance measure. *Inputs.* The inputs of ANFIS are:

- A : set of attributes $\{a_1, a_2, \ldots, a_{|A|}\}$
- R: set of characteristics $\{r_1, r_2, \ldots, r_{|R|}\}$
- R_a: set of characteristics of an attribute a, $R_a \subseteq R$
- I : set of groups $\{1, \ldots, n\}$
- y_{ir}: value of each characteristic r for a group i
- m_r : number of membership functions of a characteristic r
- μ_r : type of membership functions of a characteristic r
- N : number of epochs

Output. The output of ANFIS are:

- x_{ia}: value of each attribute $a \in A$ for a group i

Method. The ANFIS is :

$$x_{ia} = ANFIS_{r \in R_a}(y_{ir}, m_r, \mu_r, N) \forall a \in A \tag{14}$$

Performance Measure. The ANFIS computes in each iteration (epoch) the $RMSE_a$: Root-mean-square error of attribute a as:

$$RMSE_a = \sqrt{\sum_{i \in I} \sum_{a \in A} \frac{e_{ia}^2}{n}}, \tag{15}$$

where e_{ia} is the error of a group i of an attribute a respect to desired output of group i of an attribute $a(o_{ia})$ compute as (4).

4 Results on ISIC System: Colombian Case

We define the subset of characteristics R_a for each attribute a in Table 3. Table 4 presents the number of MFs (m_r) and the type of MFs (μ_r) as triangular (trimf), general bell (gbelmf) or trapezoidal (trapmf). The number of epochs was 40. The experiment was developed in Matlab 2016a.

Table 3. Subset of characteristics R_a for each atribute a

a	R_a	a	R_a
1	1,2,3,4	7	7,22
2	5,6,7	8	23,24,25
3	8,9,10,11,12,13,14	9	9,26
4	15,16,17	10	27,28
5	18,19,20	11	29,3
6	1,3,21	12	7,22,31

Table 4. Inputs of ANFIS

r	m_r	μ_r	r	m_r	μ_r	r	m_r	μ_r	r	m_r	μ_r
1	3	trimf	8	2	trapmf	16	3	gbellmf	24	3	trapmf
2	3	trapmf	9	3	gbellmf	17	3	gbellmf	26	3	trimf
3	3	gbellmf	10	3	gbellmf	18	3	gbellmf	27	3	trapmf
4	3	trapmf	11	2	trapmf	19	2	trapmf	28	3	trimf
5	3	trimf	13	2	trapmf	20	2	trapmf	29	3	trapmf
6	3	gbellmf	14	3	gbellmf	22	3	trimf	30	3	trimf
7	3	gbellmf	15	3	gbellmf	23	2	trapmf	31	3	trimf

Table 5 presents the results of non-linear optimization and ANFIS of $RMSE_a$. For all attributes, the ANFIS obtain better results of *RMSE*, in average 36 % less. Figure 6 shows the graphical output of Training data, ANFIS output and Optimization output for

Table 5. Results of $RMSE_a$ for non-linear optimization and ANFIS application to ISIC system

a	Non-linear optimization	ANFIS	Improvement
1	1,223	0,718	41 %
2	1,865	1,197	36 %
3	1,456	0,525	64 %
4	1,244	1,003	19 %
5	1,375	0,987	28 %
6	1,405	1,185	16 %
7	1,608	0,955	41 %
8	0,944	0,657	30 %
9	2,571	1,038	60 %
10	1,490	1,349	9 %
11	1,709	0,922	46 %
12	1,497	0,846	44 %
Average	**1,532**	**0,948**	**36 %**

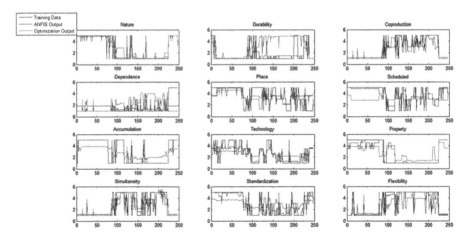

Fig. 6. Results of training data, ANFIS output and Optimization output (Color figure online)

each attribute. For example, the results of attribute Nature the ANFIS output is closer to Training Data than Optimization output. In addition, Table 5 shows in the first row the *RMSE*'s results for attribute Nature with an error of 1,223 and 0,718 for Optimization and ANFIS models, respectively. The improvement is near to 41 % with the ANFIS approach. For all attributes, the ANFIS curve is closer than Optimization curve to the Training data and is more flexible to change the values of inputs that the optimization approach.

Likewise, we can to analyze the ANFIS approach's surface of response for each attribute and their characteristics. Figure 7 shows the surface of attribute coproduction with its characteristics according with Table 3. For example, the presence and behavior characteristics influence negatively the coproduction, as the presence and behavior

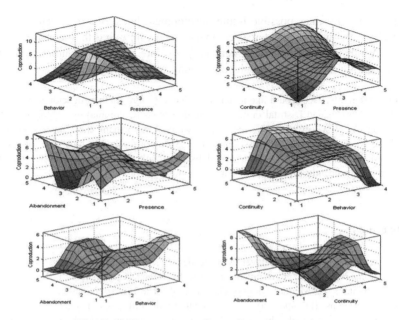

Fig. 7. Surface responses for attribute coproduction

decrease the coproduction grows up, i.e., if the customer does not been in the place of supplier (presence is low) and if it is possible that the customer's behavior affects the supplier's activity then the system has an high coproduction attribute. In addition, it is possible to explore the different combinations of these characteristics and others in order to obtain the coproduction value. The rest of characteristics could be analyzed in similar way as so the all attributes with its characteristics.

5 Conclusions

This paper review the problem of knowledge acquisition to classify service systems applied to ISIC system. The proposed scheme takes into account a set of characteristics and attributes associated with a specific system in order to determine this information under the sector, section and division to which it belongs. On the other hand, our method allows determining the attributes as input of the BRS proposed by [2] is systematic and hierarchical because it must first determine the sector to which it belongs economic activity, with this result and other attributes to determine the section, and finally with these two additional results determine the division to which it belongs. This result is useful for companies to determine their classification and for entities that consolidate information for statistical purposes.

We propose and non-linear optimization model and ANFIS-based approach. Our results show that the ANFIS is better for all attributes, but is possible to use any of both approaches. In addition, the customer could be define the weights in a customized way.

This work generates possible future development lines, one of which is the validation of results with a set of real companies and improving the database of information. Another possible improvement is to explore methods of numerical ratings based on attributes subject to rough sets, among other approaches.

Acknowledgements. We would like to acknowledge to Centro de Investigaciones y Desarrollo Científico at Universidad Distrital Francisco José de Caldas (Colombia) by supporting partially under Grant No. 2-602-468-14. The first author would like to thank the Universidad Distrital Francisco Jose de Caldas for their assistance in providing a research scholarship for his Ph.D. thesis. Last, but not least, the authors would like to thank the comments of the anonymous referees that significantly improved our paper.

References

1. Álvarez, L., Caicedo, C., Malaver, M., Méndez, G.: Design of system expert prototype to scheduling in job-shop environment (in Spanish). Revista Científica. 12 (2011)
2. López-Santana, E., Méndez-Giraldo, G.: Proposal for a rule-based classification system for service systems (in Spanish). In: Proceedings of Fifth International Conference on Computing Mexico-Colombia and XV Academic Conference on Artificial Intelligence. pp. 1–8, Cartagena (2015)
3. Pinedo, M.L.: Planning and Scheduling in Manufacturing and Services. Springer, New York (2009)
4. Böttcher, M., Fähnrich, K.-P.: Service Systems modeling: concepts, formalized meta-model and technical concretion. In: Demirkan, H., Spohrer, J.C., Krishna, V. (eds.) The Science of Service Systems. Service Science: Research and Innovations in the Service Economy, pp. 131–149. Springer, US (2011)
5. Vargo, S.L., Lusch, R.F.: Service-dominant logic: continuing the evolution. J. Acad. Mark. Sci. **36**, 1–10 (2008)
6. Spohrer, J.C., Demirkan, H., Krishna, V.: Service and Science. In: Demirkan, H., Spohrer, J. C., Krishna, V. (eds.) The Science of Service Systems. Service Science: Research and Innovations in the Service Economy, pp. 325–358. Springer, US (2011)
7. Spohrer, J., Maglio, P.P., Bailey, J., Gruhl, D.: Steps toward a science of service systems. Computer **40**, 71–77 (2007)
8. Demirkan, H., Spohrer, J.C., Krishna, V.: Introduction of the science of service systems. In: Demirkan, H., Spohrer, J.C., Krishna, V. (eds.) The Science Of Service Systems, pp. 1–11. Springer, Boston (2011)
9. Quinlan, J.R.: C4.5: Programs for Machine Learning. Morgan Kaufmann Publishers Inc, San Francisco (1993)
10. UNESCO: International Standard Industrial Classification (ISIC) of All Economic Activities, Rev.4. United Nations, New York, NY, USA (2009)
11. DANE: International Industrial Classification of all Economic Activities. Adapted for Colombia ISIC Rev. 4 A.C (in Spanish). Departamento Adminstrativo Nacional de Estadística (2005)
12. Deters, C., Secco, E.L., Wuerdemann, H.A., Lam, H.K., Seneviratne, L.D., Althoefer, K.: Model-free fuzzy tightening control for bolt/nut joint connections of wind turbine hubs. In: Proceedings - IEEE International Conference on Robotics and Automation. pp. 270–276 (2013)

13. Secco, E.L., Magenes, G.: A feedforward neural network controlling the movement of a 3-DOF finger. IEEE Trans. Syst., Man, Cybern. - Part A: Syst. Hum. **32**, 437–445 (2002)
14. Deters, C., Lam, H.K., Secco, E.L., Wurdemann, H.A., Seneviratne, L.D., Althoefer, K.: Accurate bolt tightening using model-free fuzzy control for wind turbine hub bearing assembly. IEEE Trans. Control Syst. Technol. **23**, 1–12 (2015)
15. Geethanjali, M., Raja Slochanal, S.M.: A combined adaptive network and fuzzy inference system (ANFIS) approach for overcurrent relay system. Neurocomputing **71**, 895–903 (2008)
16. Jang, J.-S.R.: ANFIS: adaptive-network-based fuzzy inference system. IEEE Trans. Syst., Man, Cybern. **23**, 665–685 (1993)

Classification of Epileptic EEG Signals with Stacked Sparse Autoencoder Based on Deep Learning

Qin Lin[1], Shu-qun Ye[1], Xiu-mei Huang[1], Si-you Li[1],
Mei-zhen Zhang[2], Yun Xue[2(✉)], and Wen-Sheng Chen[3]

[1] Guangdong Medical University Dongguan, Dongguan 523808
Guangdong, China
linqin@gdmc.edu.cn
[2] School of Physics and Telecommunication Engineering,
South China Normal University, Guangzhou 510006, Guangdong, China
xueyun@scnu.edu.cn
[3] Shenzhen Key Laboratory of Media Security, College of Mathematics
and Statistics, Shenzhen University, Shenzhen 518060, Guangdong
People's Republic of China
chenws@szu.edu.cn

Abstract. Automatic detection of epileptic seizure plays an important role in the diagnosis of epilepsy for it can obtain invisible information of epileptic electroencephalogram (EEG) signals exactly and reduce the heavy burdens of doctors efficiently. Current automatic detection technologies are almost shallow learning models that are insufficient to learn the complex and non-stationary epileptic EEG signals. Moreover, most of their feature extraction or feature selection methods are supervised and depend on domain-specific expertise. To solve these problems, we proposed a novel framework for the automatic detection of epileptic EEG by using stacked sparse autoencoder (SSAE) with a softmax classifier. The proposed framework firstly learns the sparse and high level representations from the preprocessed data via SSAE, and then send these representations into softmax classifier for training and classification. To verify the performance of this framework, we adopted the epileptic EEG datasets to conduct experiments. The simulation results with an average accuracy of 96 % illustrated the effectiveness of the proposed framework.

Keywords: Electroencephalogram (EEG) · Epilepsy · Seizures · Deep learning · Sparse autoencoder (SAE) · Softmax classifier

1 Introduction

Epilepsy is a chronic brain dysfunction syndrome caused by the hyper synchronous abnormal electrical discharge of neurons, which is one of the most common brain diseases with high morbidity and mortality. Nearly 50 million people suffer from epilepsy and the incidence of developing countries is up to 85 % [1]. Even worse, around 2.4 million new cases occur every year globally [1]. People with epilepsy are suffering

© Springer International Publishing Switzerland 2016
D.-S. Huang et al. (Eds.): ICIC 2016, Part III, LNAI 9773, pp. 802–810, 2016.
DOI: 10.1007/978-3-319-42297-8_74

from paroxysmal occurrence of seizures, resulting in dangerous and life-threatening situations. Hence, detecting and measuring the epileptic seizure is utmost important and should be a continuous process. The electroencephalogram (EEG) is the most efficient tool to diagnose the epilepsy [2], but traditional detection of epilepsy depends on artificial recognition, which is time-consuming and inefficient. To overcome these shortcomings, automatic detection technology of epileptic EEG enjoys great popularity, for it can obtain invisible information exactly and efficiently reduce the heavy burdens of doctors.

Some related works about automatic epileptic EEG signals classification are described as follows: Chandaka et al. [3] employed Cross-correlation aided SVM classifier, getting an overall accuracy of 95.96 %. Guo et al. [4] presented the line length feature and artificial neural networks with Sets A and E, obtaining an accuracy of 99.60 %. The classification accuracy shown in Guo et al. [5] is 99.20 % by using genetic programming and KNN classifier. Applying a model with permutation entropy and SVM for detection of epileptic seizures, accuracy of 93.55 % was reported in Nicolaou et al. [6]. Although above models have obtained huge achievement in theoretical analysis and practical applications, they are almost shallow learning models that are insufficient to learn and separate the complex and nonstationary epileptic EEG signals. Moreover, most of their feature extraction or feature selection methods are supervised and depend on domain-specific expertise.

To solve these problems, we proposed a novel framework to the automatic detection of epileptic EEG signals by using stacked sparse autoencoder (SSAE) based on deep learning. Compared with traditional models, this framework is capable of learning the lower-level characteristics to form more abstract and high level representations, and can better discover the significant differences between seizure and normal EEG signals from these complex and non-stationary EEG signals. Besides, it can learn the internal features without too much manual intervention. Experiments show that the proposed framework can obtain impressive results. The remainder of this paper is organized as follows: in Sect. 2, we introduce the methodology of the proposed method. Experiments and results are described in Sect. 3. Finally, conclusions are provided in Sect. 4.

2 Method

The proposed framework consists of three stages: preprocessing, SSAE networks and softmax classifier. Figure 1 explains the flow of the proposed framework.

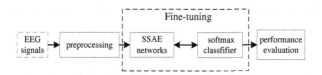

Fig. 1. The flow of the proposed framework for EEG signals classification.

In the first stage, aiming to gain more experimental data, we divided every raw EEG signals into 4 segments, then normalized the data using following formulations:

$$y_i = x_{i+1} - x_i,$$
$$z_i = (y_i - y_{min})/(y_{max} - y_{min}).$$
(1)

Secondly, SSAE networks are applied to learn the sparse and high level representations from the preprocessed data, achieving dimensionality reduction. Consequently, these representations of the preprocessed data were sent to softmax classifier for classification. The second and third stages are introduced as follows.

2.1   Stacked Sparse Autoencoder

2.1.1   Sparse Autoencoder

As an unsupervised learning method, sparse autoencoder (SAE) [7] is capable of learning the sparse representation from input data [8] to achieve dimensionality reduction. It applies back propagation to obtain the parameter gradient, setting the target values to be equal to input, formulized by $y^{(i)} = x^{(i)}$ [9]. Figure 2 is an example of autoencoder neural network.

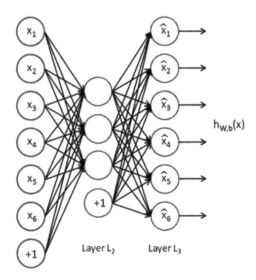

Fig. 2. One-hidden layer autoencoder: the second layer is the hidden layer.

Autoencoder network tries to learn an identical function $h_{W,b}(x) \approx x$, making the output $\hat{x}$ be approximate to the input x. The objective function of autoencoder is

$$J(W,b) = \frac{1}{m}\sum_{i=1}^{m}\left(\frac{1}{2}\left\|h_{w,b}\left(x^{(i)}\right) - y^{(i)}\right\|^2\right) + \frac{\lambda}{2}\sum_{l=1}^{2}\sum_{i=1}^{S_2}\sum_{j=1}^{n}\left(W_{ji}^{(l)}\right)^2. \tag{2}$$

where S_2 denotes the number of hidden units, the first term in Eq. (2) is an average error sum of squares; the second term is a regularization term that tends to decrease the magnitude of the weights, and it helps to prevent overfitting. The λ controls the relative importance of the two terms. To train such autoencoder, we can initialize each parameter to a small random value and apply an optimization algorithm such as gradient descent.

While the number of hidden units is large, we impose a sparsity constraint to ensure that the sparse representations will be learned. When $a_j^{(2)}(x)$ indicates the activation of hidden unit j in the autoencoder with the given input of x, the average activation of hidden unit j of training set can be

$$\hat{\rho}_j = \frac{1}{m}\sum_{i=1}^{m}[a_j^{(2)}(x^{(i)})]. \tag{3}$$

where $\hat{\rho}_j$ indirectly depends on the parameter W and b. To acquire the sparse representations of data, we approximately add a constraint

$$\hat{\rho}_j = \rho. \tag{4}$$

where ρ is a sparsity parameter which constrains the neurons to be inactive most of the time. It is generally a little value which approaches to zero, such as 0.05. Then push an extra penalty term into the objective to satisfy this constraint. Based on KL divergence, the penalty term is

$$\sum_{j=1}^{S_2} KL(\rho\|\hat{\rho}_j) = \sum_{j=1}^{S_2}\left(\rho\log\frac{\rho}{\hat{\rho}_j} + (1-\rho)\log\frac{1-\rho}{1-\hat{\rho}_j}\right). \tag{5}$$

minimizing the penalty term, $\hat{\rho}_j$ will approximate to ρ. The overall cost function is ultimately denoted as

$$J_{sparse}(W,b) = J(W,b) + \beta\sum_{j=1}^{S_2} KL(\rho\|\hat{\rho}_j). \tag{6}$$

where β means the weight of sparsity penalty term. Finally, solving above functions, the sparse representations of data will be learned.

2.1.2   Stacked Autoencoder

Consisting of multiple layers of SAE, stacked autoencoder [9] is a neural network adopted unsupervised greedy layer-wise algorithm [10] to pretrain deep neural networks, assuming the output of previous layer as the input of next layer. Formally, for a stacked autoencoder with n layers, $W^{(k,1)}$, $W^{(k,2)}$, $b^{(k,1)}$, $b^{(k,2)}$ are denoted as the corresponding parameters of $W^{(1)}$, $W^{(2)}$, $b^{(1)}$, $b^{(2)}$ of the kth autoencoder. The encoding step for stacked autoencoder is given below, which forwardly trains previous layer to a subsequent one.

$$
\begin{aligned}
a^{(l)} &= f(z^{(l)}), \\
z^{(l+1)} &= W^{(l,1)}a^{(l)} + b^{(l,1)}.
\end{aligned}
\tag{7}
$$

Similarly, the decoding step runs the encoding step in reverse order that the subsequent layers utilize the features learned from the previous layers.

$$
\begin{aligned}
a^{(n+l)} &= f(z^{(n+l)}), \\
z^{(n+l+1)} &= W^{(n-l,2)}a^{(n+l)} + b^{(n-l,2)}.
\end{aligned}
\tag{8}
$$

$a^{(n)}$ is assumed as the deepest activation of hidden units, a high-level representation of input that contains some interesting information. In short, the proposed SSAE used layer-wise unsupervised training to adjust the weights and finally adopted the BP algorithm to fine-tuning the entire network.

2.2   Softmax Classifier

As the extension of binary logistic regression, softmax regression [11] is widely used in current deep learning researches. It performs a very high classification accuracy as a nonlinear classifier while combining with the unsupervised learning parts of deep networks. Hence, we adopt softmax classifier to classify the feature vectors learned from stacked autoencoder.

2.2.1   Cost Function

If $1\{\dots\}$ is the indicative function, the evaluation criterion are: $1\{$the expression with a real value$\}$ is equal to 1 and $1\{$the expression with a false value$\}$ is equal to 0. The cost function is

$$
J(\theta) = -\frac{1}{m}\left[\sum_{i=1}^{m}\sum_{j=1}^{k} 1\{y^{(i)} = j\} \log \frac{e^{\theta_j^T x^{(i)}}}{\sum_{l=1}^{k} e^{\theta_j^T x^{(i)}}}\right].
\tag{9}
$$

The probability's value of softmax regression is

$$
p(y^{(i)} = j | x^{(i)}; \theta) = \frac{e^{\theta_j^T x^{(i)}}}{\sum_{l=1}^{k} e^{\theta_j^T x^{(i)}}}.
\tag{10}
$$

With the help of gradient descent method or L-BFGS algorithm [12], the minimum value of $J(\theta)$ would be acquired. By derivation, we get following gradient function

$$\nabla_{\theta_j} J(\theta) = -\frac{1}{m} \sum_{i=1}^{m} \left[x^{(i)} (1\{y^{(i)} = j\} - p(y^{(i)} = j | x^{(i)}; \theta)) \right] \qquad (11)$$

where $\nabla_{\theta_j} J(\theta)$ is a vector and its lth element $\frac{\partial J(\theta)}{\partial \theta_{jl}}$ is the partial derivative of $J(\theta)$ with respect to the lth component of θ_j. Aim to minimize the value of $J(\theta)$, above partial derivation is brought into gradient descent algorithm, etc.

2.2.2 Weight Decay

In order to maintain smaller parameters, we generally add a weight decay term to modify cost function, then the cost function changes to

$$J(\theta) = -\frac{1}{m} \left[\sum_{i=1}^{m} \sum_{j=1}^{k} 1\{y^{(i)} = j\} \log \frac{e^{\theta_j^T x^{(i)}}}{\sum_{l=1}^{k} e^{\theta_l^T x^{(i)}}} \right] + \frac{\lambda}{2} \sum_{i=1}^{k} \sum_{j=0}^{n} \theta_{ij}^2 . \qquad (12)$$

The decay term is able to punish greater value of parameters. It owns following functions: (1) to keep smaller value parameters; (2) for any $\lambda > 0$, the cost function would be a strict convex that ensures a unique solution. In general, optimization methods based on gradient descent or L-BFGS algorithm can guarantee a global optimal value of $J(\theta)$. Hence, the optimal cost function is

$$\nabla_{\theta_j} J(\theta) = -\frac{1}{m} \sum_{i=1}^{m} \left[x^{(i)} (1\{y^{(i)} = j\} - p(y^{(i)} = j | x^{(i)}; \theta)) \right] + \lambda \theta_j . \qquad (13)$$

3 Experimental Results

The experimental data is from the department of epileptology, University of Bonn [13]. The whole database consists of five subsets (denoted as A ∼ E), each contains 100 single-channel EEG segments of 23.6 s duration. All the segments were recorded using a 128-channel amplifier and digitized with a sampling rate of 173.61 Hz. Sets A and B obtained the EEG signals of five healthy volunteers with eyes open and closed, respectively. While Sets C, D and E were collected from the EEG signals of five epileptic patients prior to surgery. Sets C and D recorded the seizure-free intervals EEG signals from hippocampal formation of the opposite hemisphere of the brain and epileptogenic zone, respectively. And Set E described the recordings of epileptogenic zone during epilepsy seizure.

In the experiment, we adopted a 3-hidden-layer SSAE for feature extraction, with Sigmoid function as an activation function and the numbers of units in three hidden layers were 300, 250 and 150. Then sent the extracted feature to softmax classifier, whose iterated times was 150.

The performance of our method is assessed by sensitivity, specificity and classification accuracy, which is defined as:

(1) Sensitivity: the number of true positive decisions divided by the number of actual positive cases;
(2) Specificity: the number of true negative decisions divided by the number of actual negative cases;
(3) Classification accuracy: the number of correct decisions divided by the total number of cases.

The presented framework was employed to classify five different pairs of two-class EEG signals from epileptic EEG data. These pairs are: Sets A and E, Sets B and E, Sets C and E, Sets A and D, Sets D and E. In each experiment, we randomly picked up 90 % of the data for training and the rest for testing.

Table 1 shows that the highest classification accuracies by the proposed framework of Sets A and E, Sets B and E, Sets C and E, Sets A and D, Sets D and E are 100.00 %, 95.00 %, 95.00 %, 95.00 % and 95.00 %, respectively. We obtained an average classification accuracy of 96.00 %. The sensitivity and specificity are 98.67 % and 93.83 %, respectively. Through Table 1, we can see that the proposed framework provides significantly high accuracy of classification problem and is efficient to classify epileptic EEG signals. And the highest accuracy is obtained with the pair of Sets A and E.

Table 1. Performance of the proposed framework for different pairs of two-class EEG signals from the epileptic EEG data.

Different pairs	Sens(%)	Spec(%)	Accu(%)
Sets A and E	100.00	100.00	100.00
Sets B and E	100.00	87.50	95.00
Sets C and E	100.00	90.00	95.00
Sets A and D	93.33	100.00	95.00
Sets D and E	100.00	91.67	95.00
Average	98.67	93.83	96.00

Sens = sensitivity; Spec = specificity;
Accu = classification accuracy

To verify the generalization and robustness of the proposed framework, we employed the 10-time average as a further test. The classification accuracies are shown in Table 2. Almost all experiments produce good performances and the average accuracy is 91.89 %. The highest accuracy is obtained with the pair of Sets A and E while the lowest with Sets A and D.

Finally, we compared our method with other methods reported in literatures. From Table 3, the classification accuracies obtained by our method are better than the others in Sets A and E.

Table 2. Classification accuracies of 10-time average.

Different pairs	10-time average (%)
Sets A and E	95.50
Sets B and E	92.50
Sets C and E	91.67
Sets A and D	86.42
Sets D and E	93.34
Average	91.89

Table 3. Performances with our proposed method and other methods from the literatures for Sets A and E of the epileptic EEG data.

Authors	Method	Accu(%)
Chandaka et al. [3]	Cross-correlation aided SVM classifier	95.96
Guo et al. [4]	Line length feature - ANN classifier	99.60
Guo et al. [5]	GP-based feature extraction - KNN classifier	99.20
Nicolaou et al. [6]	Permutation entropy - SVM classifier	93.55
Our proposed method	SSAE networks - softmax classifier	100.00

4 Conclusion

The study proposed a framework using stacked sparse autoencoder (SSAE) based on deep learning for the automatic detection of epileptic EEG signals. Compared with other methods reported in literatures, this framework is capable of learning more abstract and high level representations, and can better discover the significant differences between seizure and normal EEG signals. Besides, it can learn the internal features without too much manual intervention. The experiment results demonstrate that the framework achieved a good performance with an average accuracy of 96 %, which suggests its bright prospects in the automatic detection of epileptic EEG signals. In the future, we will conduct more experiments with other EEG datasets by using the proposed framework and employ it into practical applications.

Acknowledgment. This study was supported by the National Natural Science Funds of China (No. 71102146), the Science and Technology Project of Guangdong Province (No.2013B010401023), the Research Funds of Guangdong Medical University (No.M2015031, No.M2015029), the Undergraduate Innovation and Entrepreneurship Training Project of School of Information Engineering of Guangdong Medical University (No.XGZD201601).

References

1. Acharya, U.R., Sree, S.V., Swapna, G., Martis, R.J., Suri, J.S.: Automated EEG analysis of epilepsy: a review. Knowl.-Based Syst. **45**, 147–165 (2013)

2. Kumar, S.P., Sriraam, N., Benakop, P.G., Jinaga, B.C.: Entropies based detection of epileptic seizures with artificial neural network classifiers. Expert Syst. Appl. **37**(4), 3284–3291 (2010)
3. Chandaka, S., Chatterjee, A., Munshi, S.: Cross-correlation aided support vector machine classifier for classification of EEG signals. Expert Syst. Appl. Int. J. **36**(2), 1329–1336 (2009)
4. Guo, L., Rivero, D.J., Rabunal, J.R., Pazos, A.: Automatic epileptic seizure detection in EEGS based on line length feature and artificial neural networks. J. Neurosci. Meth. **191**(1), 101–109 (2010)
5. Guo, L., Rivero, D., Dorado, J., Munteanu, C.R., Pazos, A.: Automatic feature extraction using genetic programming: an application to epileptic EEG classification. Expert Syst. Appl. **38**(8), 10425–10436 (2011)
6. Nicolaou, N., Georgiou, J.: Detection of epileptic electroencephalogram based on permutation entropy and support vector machines. Expert Syst. Appl. **39**(1), 202–209 (2012)
7. Hinton, G.E., Salakhutdinov, R.R.: Reducing the dimensionality of data with neural networks. Science **313**(5786), 504–507 (2015)
8. Japkowicz, N.: Jose Hanson, S., Gluck, M.A.: Nonlinear auto association is not equivalent to PCA. Neural Comput. **12**(3), 531–545 (2000)
9. Bengio, Y.: Learning deep architectures for AI. Found. Trends® Mach. Learn. **2**(1), 1–127 (2009)
10. Schölkopf, B., Platt, J., Hofmann, T.: Greedy layer-wise training of deep networks. Adv. Neural Inf. Process. Syst. **19**, 153–160 (2007)
11. Friedman, J., Hastie, T., Tibshirani, R.: Regularization paths for generalized linear models via coordinate descent. J. Stat. Softw. **33**(1), 1–22 (2010)
12. Liu, B.D.C., Nocedal, J.: On the limited memory method for large scale optimization. Math. Program. (2010)
13. EEG Time Series (Epileptic Data), November 2005, http://www.meb.unibonn.de/epileptologie/science/physik/eegdata.html.8

Employer Oriented Recruitment Recommender Service for University Students

Rui Liu[1,2], Yuanxin Ouyang[1,2(✉)], Wenge Rong[1,2], Xin Song[1,2],
Weizhu Xie[1,2], and Zhang Xiong[1,2]

[1] Engineering Research Center of Advanced Computer Application Technology,
Ministry of Education, Beihang University, Beijing 100191, China
{liurui, oyyx, w.rong, songxin,
weizhushieh, xiongz}@buaa.edu.cn
[2] School of Computer Science and Engineering, Beihang University,
Beijing 100191, China

Abstract. Currently when university students are going into job market, it is found a lot of challenges to help the students and also employers to help their efficiency in finding matching degree. However, during the job seeking peak time, for example, an event called "job fair" in China, it is found very challenging for employer to quickly filter potentially qualified applicants since an employer will probably receive huge number of resumes in a very short period. To solve this problem, in this research we proposed a student file based employer oriented job recommendation framework. In this system, a student is firstly modelled by the personal features and also academic features. Afterwards, different similarity mechanism between the fresh students with those recruited in the target employers are designed to help recommend students. Furthermore, a dynamic recruitment size aware strategy is also proposed to further polish the recommendation results. The experimental study on a Chinese university's real recruitment data has shown its potential in real applications.

Keywords: University student · Job recommendation · Employer · Personal profiling

1 Introduction

Currently the competition of employment is becoming much fiercer, particularly for fresh university graduate students since they usually do not have much work experience [19]. As such how to help students to smoothly transit from campus to the job market has then become an essential challenge [23]. To better meet the gap between students and employer, a lot of efforts have been devoted to this task. From student's perspective, they often relied on their family [14], university mentors and supervisors [23], social networks such as friends and classmates [26] to help recommend potential jobs. On the other hand, employers are also keen to recruit high quality students via different mechanisms, e.g., reference letter [5]. In the process of improving the efficiency of student employment, there is an important event called "job fairs" [18], since nearly all Chinese university has an employment office to host such event.

D.-S. Huang et al. (Eds.): ICIC 2016, Part III, LNAI 9773, pp. 811–823, 2016.
DOI: 10.1007/978-3-319-42297-8_75

This kind of job fairs does provide an important channel for students and employers to understand each other's requirements and interesting points. However, this university support has also brought some difficulties and a disadvantage is normally referred to information overload [20]. It is expected that a student has to study so many employers information and an employer will be also struggling with the overwhelming resumes, thereby making it a necessity to propose a highly efficient mechanism for the student and employer to filter useful information as quick as possible.

In fact, tackling with information overwhelming is a long challenging task in the academia and industry. One of the proven efficient solutions is personalised recommender systems. Its main idea is to evaluate matching degree of similar users and similar items [1], which have been gradually employed to deal with a wide range applications including job seeking [12, 15]. The main motivation beneath the job recommendation is to investigate the user's profile and the employer's requirement to reveal the matching possibility. To improve the employer's efficiency, an advanced method is a necessity to filter possibly matching students. Normally universities have archive of student's employment history over past years. Such data contains a lot of useful information such as individual background information and behaviour history [28]. A question is consequently raised: is it feasible to create proper recommendation for employers based on such historical data?

In this paper, we will focus on recommending qualified students to specific employers by considering employment history reports and student profiles accumulated in the universities. In the proposed framework, students are measured in terms of their similarity between those who have been recruited in the certain employer. Furthermore, based the similarity various recommendation strategies are proposed since different employer will have different recommendation requirement due to their own size and recruitment plan.

The rest of the paper is structured as follows. In Sect. 2 we will introduce the related work about job recommendation. Section 3 will present the proposed university student job recommendation framework in detail. Our experimental study based on real employment data will be described in Sects. 4 and 5 will conclude this paper and outline potential future work.

2   Related Work

Recommender systems have been proven an innovative approach for people to filter huge amount of information when making decision [24]. Afterwards a lot of technologies have been developed and successfully employed to real applications. For example, Amazon provides recommendation for books and products to different users through personalised recommendation technology [25]. Microsoft has also proven success in supplying service recommendation of product downloads and bug fixes to different users [4].

Employment recommendation is also one of the most important applications and a lot of efforts have been endowed in this area [16]. The most important point in job recommendation is to match the job description and the job seeker's personal profile to see the potentiality [2]. Therefore, it becomes an essential part to properly model the user's profile and also the job description. To this end some methods are proposed and an example is to propose a semi-structured file to describe the jobs and users. Some other researchers tried to integrate the concept of skill to the matching process to investigate the possibility between a job hunter and a specific position [3]. Though based on job hunter's own profile is easy for implementation, some scholars argued that using grouped employers and users information will be helpful [7], since similar users and similar employers might have some patterns. The technique for this idea is often referred as collaborative filtering. Collaborative filtering based job recommender systems usually create a user-job matrix which presents the scores related to employment by job hunters. Such score can be users' behaviour like visiting pattern and activity in the system [22]. A typical example is an item-based collaborative recommender system using user application records [27]. The collaborative based job recommendation has shown its potential in helping job seeking and a lot of real application have been deployed.

Some researchers further claimed that if different kinds of recommendation mechanism can be integrated, the overall performance might be able to increase to some extent [1], which is normally called hybrid recommender systems. The advantage of such approaches has been fully studied and consequently a lot of approaches have been proposed. For example, a hybrid system using description logic matchmaking and a similarity based ranking model is proposed in [11], which can utilise the advantage of logic's reasoning and collaborative filtering. Similarly, some other work makes user profile to initialise the job recommendation list and subsequently use other sophisticated mechanisms to re-ranking the recommendation list [10, 13], in which it is argued that when different methods are combined together, it is feasible to deal with dynamic features of user profiling and also job requirements.

Besides these approaches, some scholars also proposed knowledge based recommender systems for job hunting. They argued that there will be some useful hints in terms of rules or patterns from user activities in finding jobs, which can make job matching more reliable [6]. Some other researchers tried to employ data mining techniques from the archive to select potential jobs via using decision trees and association rules [8]. Furthermore, some scholars also notice that for a job recruitment system, it will be important to make both job hunter and employers satisfactory at the same time. This feature is very important and called as "reciprocal" [21]. Inspired by this argument, we think the success of giving university students a good job recommender service will be not only based on student's feedback but also the employer opinion. Therefore, in this paper, we proposed an employer oriented job recommendation framework for fresh graduate students.

3 Employer Oriented Recruitment Recommender Model

In this paper, an employer oriented recruitment recommender framework is proposed as shown in Fig. 1, where each part will be described in detail in this section. The student profile will be firstly created and then similarity calculates strategy between any two

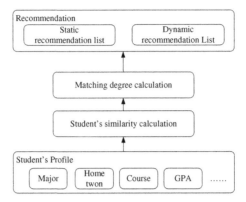

Fig. 1. Employer oriented recruitment recommender model

students will be defined. Afterwards, for a certain employer, the relation between any students with all employed graduate in the past few years will be recorded. Finally, a potential qualified students recommendation list will be provided based on such similarity and a dynamic polishing will be employed for refine the recommendation list.

3.1 Student Profiling

In this research a student profile is composed of two different types of attributes, i.e., basic attributes and achievement attributes. The basic attributes mainly consist of the student's personal background information, such as home town, gender, university, department, major and courses taken. Similarly, achievement attributes are employed to reflect this student's academic outcomes such as grades, scholarships, awards, grants and etc. Therefore, in this paper, a student profile can be formally defined as below:

$$u = \{a_1, a_2, \ldots, a_n\} \tag{1}$$

where u is the student and a_i represents the users i-th attribute.

For a graduate student who has obtained an offer from an employer in the past, we can further define that student as:

$$g = \{u, e\} = \{a_1, a_2, \ldots, a_n, e\} \tag{2}$$

where g is the graduate student with a job offer and e is his/her employer.

3.2 Student Similarity Calculation

The proper job recommendation relies on how we evaluate the similarity of students and graduates. In this research, we calculate the similarity of attributes between different students with different weights. The similarity calculation process is defined as:

$$S(u,g) = \sum_{i=1}^{n} w_i \cdot sim(u_{a_i}, g_{a_i}) \tag{3}$$

where u and g are the students who are looking for a job and a graduate who has an offer, respectively, and w_i is the i-th attribute weight in the similarity calculation, while a_i is the i-th attribute. A student's attributes can be divided into two types, i.e., bool and discrete. An example of the bool type is the home town (in this research province is employed as the granularity). If the students come from same the place, the similarity on this attribute will be 1, otherwise 0. Accordingly similarity for bool types can be defined as:

$$sim(u_{a_i}, g_{a_i}) = \begin{cases} 1, & u_{a_i} = g_{a_i} \\ 0, & u_{a_i} \neq g_{a_i} \end{cases} \tag{4}$$

For discrete attributes including the grade which ranges from 0 to 100, similarity is defined as:

$$sim(u_{a_i}, g_{a_i}) = 1 - \frac{|u_{a_i} - g_{a_i}|}{a_{i_{\max}} - a_{i_{\min}}} \tag{5}$$

where *max* and *min* define the value range. An example is the student's grade where the similarity in this perspective can be defined as the grade different divided by the whole grade range, i.e., 100. Looking at the similarity of student major subject, defining similarity as bool type is inappropriate because (1) the granularity is too coarse for two different majors since they can include similar courses to some extent, for example computer science and electronic engineering; (2) students can also select slightly different course even within the same major. To solve this problem, in this paper we use courses taken instead of major name to calculate any two student's similarity in terms of their major, which can be defined as below:

$$sim_{course}(u,g) = \frac{|C(u) \cap C(g)|}{|C(u) \cup C(g)|} \tag{6}$$

where $C(x)$ is the set of all courses that student x have taken.

3.3   Matching Degree Detection for Student and Employer

After we obtained the student's similarity calculation, we can now define matching degree of a given student u and a potential employer e. In this research, an employer e will be defined as a set of recently graduated students:

$$e = \{g_1, g_2, \ldots, g_m\} \tag{7}$$

where e is the employer and g_i is the graduate student in recent years.

Now we can further define the matching degree $M(u, e)$ between u and e as below:

$$M(u,e) = \sum_{i=1}^{m} S(u, g_i) \qquad (8)$$

where $S(u, g_i)$ is the similarity of the student u with every graduate student in the employer.

3.4   Employer Oriented Students Recommendation

After we have the matching degree between any student and any employer, along with the students similarity detection method, we proposed an employer oriented student recommendation in which the main idea is to create a list containing the most matching student with the employer. The whole sequence is listed in Algorithm 1:

Algorithm 1 Matching Based Recommendation

Input: U, e
Output: U'
1: for each student u_i in U do
2:     for each graduate g_j in e do
3:         Calculate similarity with the student $S(u_i, g_i)$
4:         $M(u_i, e) = M(u_i, e) + S(u_i, g_i)$
5:     end for
6: end for
7: Sort M(u, e) in descending order
8: Obtain top N students to form U'
9: return U'

However, Algorithm 1 will recommend student who has most similarity with all employed students. But in reality, employer will probably recruit different type students for certain positions. As such this kind of difference between different position types are also needed to be considered. Inspired by these findings and also the popular KNN algorithm, we tried an alternative voting based recommendation method as shown in Algorithm 2. Its main idea is to calculate the similarity of all students between a certain employed student in that employer and the most similar V students will get a voting ticket. After repeating this process for every employed student in the employer, we can have voting number for all students and the top N students with highest voting tickets will be recommended to the employer.

The above mentioned algorithms all have one limitation. The return list size N for all employers will be the same. However, different employer will have different recruitment plan. It is unfair and also boring if we recommend a long list to an employer if it does not need that many students, while it is also not enough for a bigger employer with more employment requirement, thereby making a balance of recommendation list length an important challenge. To this end, here we proposed a length calculation method:

$$N = \alpha \cdot avg(T(year)) \tag{9}$$

where N is the list length, α is an empirical weight and $T(year)$ is the real recruitment total number in certain year.

Algorithm 2 Voting Based Recommendation

```
Input: U, e
Output: U'
1:  for each student uᵢ in U do
2:      for each graduate gⱼ in e do
3:          Calculate similarity S_{uᵢ,gᵢ} = S(uᵢ, gᵢ)
4:      end for
5:      Sort S_{uᵢ,gᵢ} in descending order
6:      Obtain top V students
7:      for each student uₜ in V do
8:          N(uₜ) = N(uₜ)+1
9:      end for
10: end for
11: Sort N(u) in descending order
12: Obtain top N students to form U'
13: return U'
```

---

4 Experimental Study

4.1 Dataset and Evaluation Metric

In this research, a total of 129 employers that have frequently recruited postgraduate students from 2012 to 2015 are included in this experimental study. These employers include governments, state-owned companies, joint ventures, private companies, universities, institutes, and etc., which represent the employer distribution in China. The student distribution in this research is listed in Table 1, in which all students graduated from 2012 to 2014 are considered the training set and the students graduated in 2015 are used as the test set.

Inspired by the evaluation metric of $P@n$ in the field of information retrieval [9], which emphases the top n items in the return list based on the assumption that people will probably not be interested in the items beyond top k candidates [17], in this paper, we constitute a metric called $S@n$, which is defined as below:

$$S@n = \sum_{i=1}^{n} hit(e_i) \tag{10}$$

where $hit(e_i)$ represents a real employer of the student in the test set has been correctly included in the recommended list with the k employers. As such the $hit(e_i)$ is defined as:

$$hit(e_i) = \begin{cases} count(), \ u_e \in U' \\ 0, \qquad otherwise \end{cases} \tag{11}$$

where U' is the returned list defined in Algorithms 1 and 2 and $count()$ is the real number of the students obtained the offer from the employer e among the top n students recommended by the framework.

From the above equation, it is found that setting an appropriate n value is essential for the proposed framework. In this research, we will test the variance of different n value and this parameter will be set to 20, 30, and 40, respectively. Furthermore, we will mainly study the performance when n equals 40 since in the dataset the average recruitment number for every employer is 4.12.

Table 1. Student distribution

Year	Student number	Male (%)	Female (%)
2012	737	71.2	28.8
2013	731	72.0	28.0
2014	637	70.8	29.2
2015	532	72.4	27.6

4.2   Experimental Result

In this research we will studied the different attributes' influence on the recommendation to a certain employer. Due to the fact that the students involved in this experimental study are all from Beihang University, we will ignore their education background. Furthermore, since most graduates are male, we will also remove the gender information from the student profiling process. As a result, we will thoroughly study the major, course taken, home town, and grade point average (GPA) to evaluate the propose employer oriented student recommendation service.

(1) Experiment 1: Student Profiling

In this research, we empirically investigate the effect of weights of major, home town and GPA and the result is listed in Table 2. Furthermore, we also tried to replace the course with major and see if the feature of course will bring some interesting changes, accordingly the results are listed in Table 3. From this tables it is found that (1) user's profile does meet the employer's requirements to some extent; (2) Major and home town will have more important influence than GPA; and (3) Course does provide a better granularity mechanism in recommending students to the employers.

Firstly, we will empirically investigate the different weights of the attributes in student's profile on the final job recommendation. In this research, the major, home town and GPA are investigated. During the process of identifying the effect of different weight of these three variables, we firstly randomly set the weight of one variable to zero and set 50 % for the rest two variables. The results are listed in Table 2 from which it is found that the most important factor is the student's major, which is consistent to our experience. For a postgraduate student, the major is of great importance indeed. Another interesting finding is that as the weight of major increases when the rest two variables keep balanced, the overall performance does not increase accordingly. This is partly because the inappropriate usage of discrete major information. Therefore, we need to test if using course to replace major will result in improvement.

Table 2. Effect of student's personal profile in job recommendation: Major

Major	Home town	GPA	S@n		
			20	30	40
0.5	0.5	0	107	143	168
0.5	0	0.5	69	101	121
0	0.5	0.5	76	106	127
0.3	0.4	0.3	111	147	164
0.4	0.3	0.3	108	144	171
0.5	0.25	0.25	104	138	165
0.6	0.2	0.2	98	125	155

Table 3. Effect of student's personal profile in job recommendation: Course

Course	Home town	GPA	S@n		
			20	30	40
0.4	0.3	0.3	103	125	145
0.5	0.25	0.25	103	129	147
0.6	0.2	0.2	109	134	155
0.7	0.15	0.15	114	141	169
0.8	0.1	0.1	110	149	181
0.9	0.05	0.05	104	136	163

(2) Experiment 2: Voting

In this experimental study, we will empirically evaluate the matching degree based employer oriented student recommendation strategy. From the previous study, we select the values of course, home town and GPA as 0.8, 0.1, 0.1, respectively, as shown in Table 3. Afterwards, we will gradually increase the value of V to investigate its effectiveness and also the selection policy, as shown in Table 4.

(3) Experiment 3: Dynamic Recommendation List

As discussed in previous section, different employers will have different recruitment size. It is argued that using fixed length of recommendation list is not suitable for all kinds of employers. In this experimental study, we will test the effect of dynamic recommendation list size. To this end, we will use the result presented in Table 3 as baseline and empirically test α to make the total recommendation number in this experiment roughly equal to those returned by static recommendation strategy. From Table 5, it is found at the similar range of recommendation number, the recommendation result has been improved since the $S@n$ is improved while the whole recommendation number are not significantly increased, which means the efficiency is achieved.

Table 4. Matching degree in recommendation

V	S@n		
	20	30	40
20	102	130	156
30	112	142	166
40	111	143	168
50	104	133	163
60	98	128	159

Table 5. Dynamic recommendation list

α	S@L	Total recommendation number	Average recommendation number
3.74	127	2547	19.74
5.60	168	3864	29.95
7.45	193	5158	39.98

4.3   Discussion

From the experimental study, some interesting results are revealed. Firstly, student profile does show potential usefulness when employed to job recommendation. Furthermore, course and home town are most important for students and employers when they consider their matching degree. Secondly, it is interesting to find that voting based similarity recommendation does not provide satisfactory result compared with straightforward student profile based recommendation. This result deserves further study as one possible reason for this phenomena is because we only use Beihang University as study case. It is possible to have more interesting finding when more universities are involved to the experimental study. Finally, the recruitment size has essential influence indeed. This is particularly important for such recruitment recommendation system as it must balance the effectiveness and efficiency.

5   Conclusion

Student job recommendation is important for student and employer as it can reduce the effort for them in finding potential matching. In this research we proposed a student profile to describe a student and proposed different mechanism to recommend students to a certain employer. It firstly employs student profile to calculate the similarity between a fresh student and those who has been recruited in the employer. Afterwards, similar student voting strategy and dynamic recruitment size aware recommendation are also investigated. Experimental study has revealed some interesting finding and it is expected this work can shed the light for future works in this area.

Acknowledgments. This work was supported by the National High Technology Research and Development Program of China (No. 2013AA01A601), the National Natural Science Foundation of China (No. 61472021), and the Fundamental Research Funds for the Central Universities.

References

1. Adomavicius, G., Tuzhilin, A.: Toward the next generation of recommender systems: a survey of the state-of-the-art and possible extensions. IEEE Trans. Knowl. Data Eng. **17**(6), 734–749 (2005)
2. Al-Otaibi, S.T., Ykhlef, M.: Job recommendation systems for enhancing e-recruitment process. In: Proceedings of 2012 Information and Knowledge Engineering Conference (2012)
3. Almalis, N.D., Tsihrintzis, G.A., Karagiannis, N.: A content based approach for recommending personnel for job positions. In: Proceedings of 5th International Conference on Information, Intelligence, Systems and Applications, pp. 45–49 (2014)
4. Ashok, B., Joy, J., Liang, H., Rajamani, S.K., Srinivasa, G., Vangala, V.: Debugadvisor: a recommender system for debugging. In: Proceedings of the 7th Joint Meeting of the European Software Engineering Conference and the ACM SIGSOFT Symposium on the Foundations of Software Engineering, pp. 373–382 (2009)
5. Biga, K., Spott, P., Spott, E.: Smart hiring in the hospitality industry: legal and business perspectives. Am. J. Manag. **15**(4), 115 (2015)
6. Bradley, K., Rafter, R., Smyth, B.: Case-based user profiling for content personalisation. In: Brusilovsky, P., Stock, O., Strapparava, C. (eds.) AH 2000. LNCS, vol. 1892, pp. 62–72. Springer, Heidelberg (2000)
7. Cheng, Y., Xie, Y., Chen, Z., Agrawal, A., Choudhary, A., Guo, S.: Jobminer: a real-time system for mining job-related patterns from social media. In: Proceedings of the 19th ACM SIGKDD International Conference on Knowledge Discovery and Data Mining, pp. 1450–1453 (2013)
8. Chien, C.F., Chen, L.F.: Data mining to improve personnel selection and enhance human capital: a case study in high-technology industry. Expert Syst. Appl. **34**(1), 280–290 (2008)
9. Cormack, G.V., Lynam, T.R.: Statistical precision of information retrieval evaluation. In: Proceedings of the 29th Annual International ACM SIGIR Conference on Research and Development in Information Retrieval, pp. 533–540 (2006)

10. Drigas, A., Kouremenos, S., Vrettos, S., Vrettaros, J., Kouremenos, D.: An expert system for job matching of the unemployed. Expert Syst. Appl. **26**(2), 217–224 (2004)
11. Fazel-Zarandi, M., Fox, M.S.: Semantic matchmaking for job recruitment: an ontology-based hybrid approach. In: Proceedings of the 8th International Semantic Web Conference (2009)
12. Fu, Y.T., Hwang, T.K., Li, Y.M., Lin, L.F.: A social referral mechanism for job reference recommendation. In: Proceedings of 21st Americas Conference on Information Systems (2015)
13. Hong, W., Zheng, S., Wang, H.: Dynamic user profile-based job recommender system. In: Proceedings of 8th International Conference on Computer Science & Education, pp. 1499–1503 (2013)
14. Liu, D.: Parental involvement and university graduate employment in china. J. Educ. Work **29**(1), 98–113 (2016)
15. Lu, Y., Helou, S.E., Gillet, D.: A recommender system for job seeking and recruiting website. In: Proceedings of 22nd International World Wide Web Conference, Companion Volume, pp. 963–966 (2013)
16. Malinowski, J., Keim, T., Wendt, O., Weitzel, T.: Matching people and jobs: a bilateral recommendation approach. In: Proceedings of the 39th Annual Hawaii International Conference on System Sciences (2006)
17. McLaughlin, M.R., Herlocker, J.L.: A collaborative filtering algorithm and evaluation metric that accurately model the user experience. In: Proceedings of the 27th Annual International ACM SIGIR Conference on Research and Development in Information Retrieval, pp. 329–336 (2004)
18. Obukhova, E.: Motivation vs. relevance: using strong ties to find a job in urban China. Social Sci. Res. **41**(3), 570–580 (2012)
19. Paparrizos, I.K., Cambazoglu, B.B., Gionis, A.: Machine learned job recommendation. In: Proceedings of the 2011 ACM Conference on Recommender Systems, pp. 325–328 (2011)
20. Peterson, A.: On the prowl: how to hunt and score your first job. Educ. Horiz. **92**(3), 13–15 (2014)
21. Pizzato, L., Rej, T., Chung, T., Koprinska, I., Kay, J.: Recon: a reciprocal recommender for online dating. In: Proceedings of the 4th ACM Conference on Recommender Systems, pp. 207–214 (2010)
22. Rafter, R., Bradley, K., Smyth, B.: Automated collaborative filtering applications for online recruitment services. In: Brusilovsky, P., Stock, O., Strapparava, C. (eds.) AH 2000. LNCS, vol. 1892, pp. 363–368. Springer, Heidelberg (2000)
23. Renn, R.W., Steinbauer, R., Taylor, R., Detwiler, D.: School-to-work transition: mentor career support and student career planning, job search intentions, and self-defeating job search behavior. J. Vocat. Behav. **85**(3), 422–432 (2014)
24. Resnick, P., Iacovou, N., Suchak, M., Bergstrom, P., Riedl, J.: Grouplens: an open architecture for collaborative filtering of netnews. In: Proceedings of the 1994 ACM Conference on Computer Supported Cooperative Work, pp. 175–186 (1994)
25. Schafer, J.B., Konstan, J., Riedl, J.: Recommender systems in e-commerce. In: Proceedings of the 1st ACM Conference on Electronic Commerce, pp. 158–166 (1999)
26. Skeels, M.M., Grudin, J.: When social networks cross boundaries: a case study of workplace use of facebook and linkedin. In: Proceedings of the 2009 International ACM SIGGROUP Conference on Supporting Group Work, pp. 95–104 (2009)

27. Zhang, Y., Yang, C., Niu, Z.: A research of job recommendation system based on collaborative filtering. In: Proceedings of 7th International Symposium on Computational Intelligence and Design, vol. 1, pp. 533–538 (2014)
28. Zheng, S., Hong, W., Zhang, N., Yang, F.: Job recommender systems: a survey. In: Proceedings of the 7th International Conference on Computer Science & Education, pp. 920–924 (2012)

Feasibility Analysis for Fuzzy/Crisp Linear Programming Problems

Juan Carlos Figueroa-García[1]([✉]), Cesar Amilcar López-Bello[1],
and Germán Hernández-Pérez[2]

[1] Universidad Distrital Francisco José de Caldas,
Bogotá, Colombia
{jcfigueroag, clopezb}@udistrital.edu.co
[2] Universidad Nacional de Colombia, Bogotá Campus,
Bogotá, Colombia
gjhernandezp@gmail.com

Abstract. In this paper we analyze feasibility conditions for Fuzzy Linear Programming problems *(FLP)* and a special case where its constraints are composed by fuzzy numbers bounded by crisp numbers. We analyze three cases: *strong*, *weak* feasibility, and unfeasible FLPs, where strong feasibility is much more desirable than weak one since it generalizes feasible solutions.

1 Introduction and Motivation

Fuzzy Linear Programming problems (FLP) can be analyzed from different points of view, but feasibility is the property we are going to analyze here. Our analysis extends classical concepts of feasibility over LPs based on the works of Zimmermann [1], Zimmermann and Fullér [2], Fiedler et al. [3], Cerný and Hladík [4], and Hladík [5] to two main families of fuzzy LPs: problems with fuzzy parameters and fuzzy constraints (FLP), and problems with fuzzy parameters and crisp constraints.

Hladík [5] has defined basic concepts of feasibility for interval valued equations which can be extended to FLPs, so we propose similar conditions using the support of the fuzzy parameters and constraints. To do so, we define the instances of FLPs over we provide definitions of feasibility, and the case of unfeasible FLPs.

The paper is organized into 6 sections; Sect. 1 introduces the main problem; Sect. 2 presents the CCR classic model; Sect. 3 presents some basics on Interval Type-2 fuzzy sets; Sect. 4 presents the proposed Interval Type-2 fuzzy CCR model; In Sect. 5, an example is presented and solved, and finally Sect. 6 presents some concluding remarks of the study.

2 Basic Notations

In this paper, we consider X as a powerset whose elements $x \in X$ are real numbers $\mathbb{P}(X) \in \mathbb{R}$, and $\mathcal{P}(X)$ is the class of all *crisp* sets. In a crisp set $A \in X$, an element x is either a member of the set or not. The indicator function of A, χ_A is as follows:

© Springer International Publishing Switzerland 2016
D.-S. Huang et al. (Eds.): ICIC 2016, Part III, LNAI 9773, pp. 824–833, 2016.
DOI: 10.1007/978-3-319-42297-8_76

$$\chi_A(x) := \begin{cases} 1 & \text{if } x \in A \\ 0 & \text{if } x \notin A \end{cases} \tag{1}$$

A set S is called singleton $\{\tilde{S}\}$ if has a single element $x \in \mathbb{R}$ (see Fig. 1). In the real line $\mathbb{R}$, S is constant. This implies that $\chi_S(x) = 1$, and $\chi_S(\cdot) = 0 \ \forall \ x \notin S$.

$$S := \{x : x = S\} \tag{2}$$

$$S : X \rightarrow \{0, 1\} \tag{3}$$

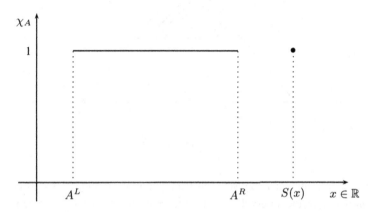

Fig. 1. Crisp set A and singleton $S(x)$

A fuzzy set $\tilde{A}$ is a generalization of a or boolean set. It is defined on an universe of discourse X and is characterized by a *Membership Function* namely $\mu_{\tilde{A}}(x)$ that takes values in the interval $[0,1]$, $\tilde{A} : X \rightarrow [0, 1]$. A fuzzy set $\tilde{A}$ may be represented as a set of ordered pairs of an element x and its membership degree, $\mu_{\tilde{A}}(x)$ i.e.,

$$\tilde{A} = \{(x, \mu_A(x)) \mid x \in X\} \tag{4}$$

where $\mathcal{F}(\mathbb{R})$ is the class of all fuzzy sets.

Now, $\tilde{A}$ is into a family of fuzzy sets $\mathcal{F} = \{\tilde{A}_1, \tilde{A}_2, \cdots, \tilde{A}_m\}$, each one with a membership function $\{\mu_{\tilde{A}_1}(x), \mu_{\tilde{A}_2}(x), \cdots, \mu_{\tilde{A}_m}(x)\}$. The *support* of $\tilde{A}$, $supp(\tilde{A})$, is composed by all the elements of X that have nonzero membership in $\tilde{A}$, this is

$$supp(\tilde{A}) = \{x \mid \mu_{\tilde{A}}(x) > 0\} \ \forall x \in X \tag{5}$$

The α-cut of $\mu_{\tilde{A}}(x)$ namely $^{\alpha}\tilde{A}$ represents the interval of all values of x which has a membership degree equal or greatest than α, this is:

$$^{\alpha}\tilde{A} = \{x \mid \mu_{\tilde{A}}(x) \geqslant \alpha\} \ \forall x \in X \tag{6}$$

where the interval of values which satisfies $^{\alpha}\tilde{A}$ is defined by

$$\tilde{A} \in \left[\inf_x {}^\alpha\mu_{\tilde{A}}(x), \ \sup_x {}^\alpha\mu_{\tilde{A}}(x) \right] = [A^L_\alpha, A^R_\alpha] \tag{7}$$

A graphical display of a triangular fuzzy set is given in Fig. 2.

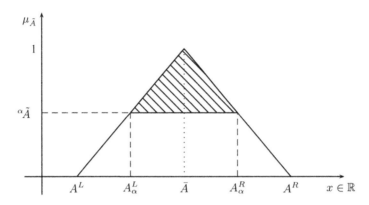

Fig. 2. Fuzzy set $\tilde{A}$

Here, $\tilde{A}$ is a triangular fuzzy set with parameters A^L, $\bar{A}$ and A^R, its poset is the set $x \in \mathbb{R}$, *supp* $(\tilde{A})$ is the interval $x \in [A^L, A^R]$. α is the degree of membership of an specific value x has regarding A. The dashed region is an α-cut done over $\tilde{A}$. A *fuzzy number* is then a convex fuzzy set defined over $\mathbb{R}^n$, defined as follows:

Definition 1 (Fuzzy Number). *Let $\tilde{A} \in \mathcal{F}(\mathbb{R})$. Then, $\tilde{A}$ is a Fuzzy Number (FN) iff there exists a closed interval $[a, b] \neq 0$ such that*

$$\mu_{\tilde{A}}(x) = \begin{cases} 1 & \text{for } x \in [a, b], \\ l(x) & \text{for } x \in [-\infty, a], \\ r(x) & \text{for } x \in [b, \infty] \end{cases} \tag{8}$$

where $l : (-\infty, a) \to [0, 1]$ is monotonic non-decreasing, continuous from the right, and $l(x) = 0$ for $x < \omega_1$, and $r : (b, \infty) \to [0, 1]$ is monotonic non-increasing, continuous from the left, and $r(x) = 0$ for $x > \omega_2$.

3 Fuzzy/Crisp LP Models

Crisp LPs i.e. $z^* = \max_x \{ z = c'x \mid Ax \leqslant b, \ x \geqslant 0 \}$ are a generalization of Fuzzy LPs (FLPs for short) in which all parameters are defined as singletons (see Eq. (1)) and the inequality $\leqslant b_i$ is a set whose indicator function is $\chi_{b_i}(x)$. In crisp LPs, the membership degree of any solution is always one, so no uncertainty is involved into the problem which means that global optimization is possible in that kind of problems.

On the other hand, FLPs involve fuzzy uncertainty $\max_x\{\tilde{z} = \tilde{c}'x \mid \tilde{A}x \lesssim \tilde{b}, x \geqslant 0\}$. In this paper, all its parameters are considered as finite-domain fuzzy numbers (this is, $supp\ (A)$ is a closed interval). The binary relation $\lesssim$ has been defined by Ramík and Rimánek [6], as follows:

Definition 2. *Let* $A, B \in \mathcal{F}(\mathbb{R})$ *be two fuzzy numbers. Then* $A \lesssim B$ *if and only if* $\sup {}^\alpha A \leqslant \sup {}^\alpha B$ *and* $\inf {}^\alpha A \leqslant \inf {}^\alpha B$ *for any* $\alpha \in [0, 1]$*, and* ${}^\alpha A := [\inf {}^\alpha A, \sup {}^\alpha A]$ *and* ${}^\alpha B := [\inf {}^\alpha B, \sup {}^\alpha B]$*. This binary relation satisfies the axioms of a partial order relation on* $\mathcal{F}(\mathbb{R})$ *and is called the fuzzy max order.*

This binary relation (fuzzy max order) has been extended to interval-valued fuzzy numbers by Figueroa-García et al. [7], and Figueroa-García [8], and it can be extended to fuzzy/crisp sets as follows:

Definition 3. *Let* $A \in \mathcal{F}(\mathbb{R})$ *be a fuzzy number, and* $b \in \mathcal{P}(\mathbb{R})$ *be a crisp set. Then* $A \lesssim b$ *if and only if* $\sup {}^\alpha A \leqslant b$ *and* $\inf {}^\alpha A \leqslant b$ *for any* $\alpha \in [0, 1]$*, where* ${}^\alpha A := [\inf {}^\alpha A, \sup {}^\alpha A] = [A_\alpha^L, A_\alpha^R]$*. This binary relation satisfies the axioms of a partial order relation on* $\mathcal{F}(\mathbb{R})$ *and is called the fuzzy/crisp max order.*

Figure 3 shows the binary relation $\lesssim$. Note that $\tilde{A} \lesssim b$ since $supp(\tilde{A}) \leqslant b$ which is equivalent to say $A^R \leqslant b$.

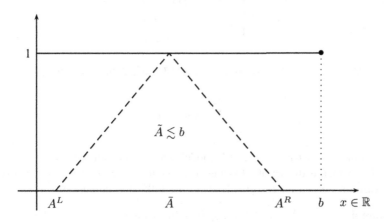

Fig. 3. Fuzzy/crisp binary relation $\lesssim$

The solution of any FLP is given by the Bellman-Zadeh extension principle:

$$\mu_{\tilde{z}}(z^*) = \sup_{z^* = \tilde{c}'x^*} \min_k \{\tilde{c}_*, \tilde{b}_1, \cdots, \tilde{b}_k, \tilde{A}_{K,*}\} \qquad (9)$$

where, $\tilde{z}$ is the fuzzy set of optimal solution in which decision making is done, $\tilde{b}_k$, $k \in K$ is the k_{th} binding constraint, $\tilde{A}_{K,*}$ is the set of fuzzy technical coefficients that compose $\tilde{A}_{K,*}x^* \lesssim \tilde{b}_k$, and $\tilde{c}_*$ is the set of fuzzy costs associated to the set of optimal variables of the problem x^*.

A crisp optimal solution x^* is the set of decision variables that solves $Ax \leqslant b$ while maximizing $z^* = c'x^*$, and x^* is a function of $A_{K,*}$ since the extreme point that locates z^* is defined by $A_{K,*}x \leqslant b_k$, this is $z^* = f(A_{K,*}x^* \leqslant b)$. In FLPs, the problem of having a global optimal solution is more complex than crisp LPs since we have infinite combinations of $\tilde{A}_{K,*}x^* \lesssim \tilde{b}$ which leads to the same value z^* and consequently NP-hard problems (see Kreinovich and Tao [9], Heindl et al. [10], Lakeyev and Kreinovich [11], and Kreinovich et al. [12]).

This way, a necessary condition for optimality in LPs is feasibility, so we focus on defining feasibility of FLPs as a necessary condition to find optimality in fuzzy constrained problems, including fuzzy/crisp problems.

4 Feasibility in FLPs

A feasible solution in a crisp LP model is any solution x' that satisfies the convex halfspace conformed by its constraints. In the fuzzy case, we define an LP with fuzzy boundaries. In this paper, we refer to a fuzzy constrained LP as the FLP proposed by Zimmermann [1] in which every set $\tilde{b}_i$ is defined as follows:

$$\tilde{b}_i := \begin{cases} 1, & f(x) \leqslant b_i^L \\ \frac{b_i^R - f(x)}{b_i^R - b_i^L}, & b_i^L \leqslant f(x) \leqslant b_i^R \\ 0, & f(x) \geqslant b_i^R \end{cases} \tag{10}$$

where $b_i^L, b_i^R \in \mathbb{R}$. Now, the constraints of an FLP are defined as follows:

$$\sum_{j=1}^{m} \tilde{A}_{ij}x_j \lesssim \tilde{b}_i \tag{11}$$

A feasible LP is basically an LP model whose constraints generate a halfspace namely $h(\cdot)$ which is the set of all values of x contained into χ_b, $x \in \chi_b$. The concept of fuzzy feasibility based on Eq. (11) leads to the following definition:

Definition 4. *Let be $\tilde{A} \in \mathcal{F}(\mathbb{R})$ a matrix of fuzzy numbers, $\tilde{b} \in \mathcal{F}(\mathbb{R})$ a set of fuzzy constraints, and $\lesssim$ a fuzzy partial order. Denote $\tilde{h}(\cdot)$ as the fuzzy polyhedron generated by $\tilde{A}x$, then an FLP is feasible only if $\tilde{h}(\cdot)$ is a non-trivial set, that is:*

$$\tilde{\mathcal{P}} = \{x \mid \tilde{h}(\cdot) \leqslant \tilde{b}\}, \tag{12}$$

where $\tilde{\mathcal{P}}$ is a non-trivial set of solutions of an FLP model.

This definition implies that $\tilde{\mathcal{P}}$ is a convex set of solutions given $\tilde{A}$ and $\tilde{b}$, which is the set of possible solutions of $\tilde{A}x \lesssim \tilde{b}$. Also, if the set x' solves $\tilde{A}x' \lesssim \tilde{b}$, then x' is sufficient for $\tilde{h}(\cdot) \leqslant \tilde{b}$, and consequently the FLP is feasible.

Computationally speaking FLPs are NP-hard problems, and only feasible ones can be optimal. The main question is: how to find at least one feasible solution x'. To do so, we propose to analyze $\mathrm{supp}(\tilde{A})$ and $\mathrm{supp}(\tilde{b})$ since they contain all possible choices of parameters $\tilde{A}$ and $\tilde{b}$. Based on concepts of feasibility introduced by Fiedler et al. [3], and extended by Cerný & Hladík [4], and Hladík [5], we present the following definition.

Definition 5. *Denote supp $(\tilde{A})$ as A_s, and supp $(\tilde{b})$ as b_s. An FLP is said to be **feasible** iff $\exists x' : \{A_s x' \leqslant b_s, x' \geqslant 0\}$.*

The following two definitions provide necessary conditions for feasibility of FLPs.

Definition 6. *Let $A_s = [A^L, A^R] \in \mathcal{P}(\mathbb{R})$ be an interval matrix, and $b_s = [0, b^L, b^R] \in \mathcal{P}(\mathbb{R})$ be a interval vector. Then the system $A_s x \leqslant b_s$ is said to be **strong feasible** iff $\exists x(A^R x \leqslant b^L)$, and its solution x^s is called **strong solution**.*

Definition 7. *Let $A_s = [A^L, A^R] \in \mathcal{P}(\mathbb{R})$ be an interval matrix, and $b_s = [0, b^L, b^R] \in \mathcal{P}(\mathbb{R})$ be a interval vector. Then the system $A_s x \leqslant b_s$ is said to be **weak feasible** iff $\exists x(A^L x \leqslant b^R)$, and its solution x^w is called **weak solution**.*

Definition 7 states that x^s is sufficient for a feasible FLP since x^s solves any combination of $A_s x^s \leqslant b_s$, $x^s \geqslant 0$, and Definition 6 states that x^w is necessary for a feasible FLP since x^w at least solves $A_s x^w \leqslant b_s$, $x^w \geqslant 0$. When no solution for $A^R x \leqslant b^L$ exists then the FLP is unfeasible, in other words:

Definition 8. *Let $A_s = [A^L, A^R] \in \mathcal{P}(\mathbb{R})$ be an interval matrix, and $b_s = [0, b^L, b^R] \in \mathcal{P}(\mathbb{R})$ be a interval vector. Then the system $A_s x \leqslant b_s$ is said to be **unfeasible** iff $\nexists x(A^L x \leqslant b^R)$.*

4.1 The Fuzzy/Crisp Problem

A less studied case is the fuzzy/crisp problem which was barely defined by Zimmermann [1] which solves the problem $\tilde{z}^* = \max_x \{z = c'x | Ax \lesssim \tilde{b}, x \geqslant 0\}$ based on the idea of having constraints in the form $\sum_j a_{ij} x_j \lesssim \tilde{b}_i$. In this problem, a crisp number e.g. $\sum_j a_{ij} x_j$ is compared to a fuzzy number $\tilde{b}$ using a partial order $\lesssim$, but the case in which a fuzzy number $\sum_j \tilde{A}_{ij} x_j$ is compared to a crisp set b has not been discussed in the literature.

The case $\sum_j \tilde{A}_{ij} x_j \lesssim b_i$ is more common in practice. Think on a problem where a company manufactures different products using human workforce and the time used to do the job has no historical information but human-like estimations. On the other hand, there is an exact estimation of the available human workforce provided by company's human resources department. In this case, manufacturing times are defined by human-like information that leads to define fuzzy sets in order to be compared to an exact (crisp) total availability of human workforce.

Again, it is enough to analyze $\text{supp}(\tilde{A})$ and $\chi_b(x)$ in order to check feasibility. This way, we propose the following two definitions:

Definition 9. *Let* $A_s = [A^L, A^R] \in \mathcal{P}(\mathbb{R})$ *be an interval matrix, and* $b = [0, b] \in \mathcal{P}(\mathbb{R})$ *be a crisp set. Then the system* $A_s x \leqslant b$ *is said to be **strong feasible** iff* $\exists x (A^R x \leqslant b)$, *and its solution* x^s *is called **strong solution**.*

Definition 10. *Let* $A_s = [A^L, A^R] \in \mathcal{P}(\mathbb{R})$ *be an interval matrix, and* $b = [0, b] \in \mathcal{P}(\mathbb{R})$ *be a crisp set. Then the system* $A_s x \leqslant b$ *is said to be **weak feasible** iff* $\exists x (A^L x \leqslant b)$, *and its solution* x^w *is called **weak solution**. If* $\nexists x (A^L x \leqslant b)$, *then the system is **unfeasible**.*

Figure 4 shows what strong, weak, and unfeasible FLPs are. Finally, if a fuzzy/crisp LP is strong feasible then it is sufficient to be optimal, and any optimal problem is necessarily weak feasible. Also note that the solution of a strong feasible problem. Also note that a strong solution x^s solves any combination of $\text{supp}(\tilde{A})$ and $\text{supp}(\tilde{b})$ while a weak solution does not guarantee to be the solution of different combinations of $\text{supp}(\tilde{A})$ and $\text{supp}(\tilde{b})$.

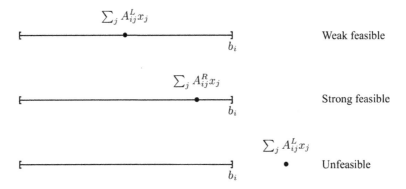

Fig. 4. Strong, weak, and unfeasible FLPs

5 Application Examples

We present three examples that illustrate the concepts presented here. A first FLP that illustrates the concept of strong and weak feasibility, a second fuzzy/crisp LP that illustrates strong and weak feasible solutions, and a third fuzzy/crisp LP that shows a unfeasible problem. We denote $T_{ij}(A^L_{ij}, \bar{A}_{ij}, A^R_{ij})$ as triangular fuzzy sets, every fuzzy constraint $Tr_i(0, b_i^L, b_i^R)$ as a trapezoidal fuzzy set, and every crisp constraint $\chi_{b_i} = [0, b_i]$ as a crisp set.

Every example is completely described and defined as presented in a mathematical programming format in order to simplify replication by the reader. All examples were intended to be simple for interpretation and computation purposes where all of them are similar with small modifications.

5.1 First Example (Feasible FLP)

This example shows a feasible FLP problem where $T_{11} = (3, 5, 6)$, $T_{12} = (2, 3, 5)$, $T_{21} = (2, 5, 7)$, $T_{22} = (5, 6, 8)$, $\tilde{b}_1 = (0, 12, 15)$, $\tilde{b}_2 = (0, 15, 18)$, $c_1 = 1$, $c_2 = 1$, and $x_j \geqslant 0$. Its description is shown as follows:

$$\max z = x_1 + x_2$$
$$\text{s.t.}$$
$$\tilde{5}x_1 + \tilde{3}x_2 \lesssim 12$$
$$\tilde{5}x_1 + \tilde{6}x_2 \lesssim 15$$

To compute strong feasibility, we solve the following problem:

$$\max z = x_1 + x_2$$
$$\text{s.t.}$$
$$6x_1 + 5x_2 \leqslant 12$$
$$7x_1 + 8x_2 \leqslant 15$$

The reader can verify that the solution of this problem $x_1^s = 1.6155$, $x_2^s = 0.4615$ is also a solution for any combination $A_{ij} \in \text{supp}(\tilde{A}_{ij})$, $b_i \in [b_i^L, b_i^R]$, so the problem is strong feasible with a strong solution.

5.2 Second Example (Feasible Fuzzy/Crisp LP)

In this example we show a feasible fuzzy/crisp problem in which $T_{11} = (2, 3, 5)$, $T_{12} = (3, 4, 5)$, $T_{21} = (4, 5, 7)$, $T_{22} = (1, 3, 4)$, $b_1 = [0, 10]$, $b_2 = [0, 12]$, $c_1 = 1$, $c_2 = 1$, and $x_j \geqslant 0$. Its description is shown as follows:

$$\max z = x_1 + x_2$$
$$\text{s.t.}$$
$$\tilde{3}x_1 + \tilde{4}x_2 \lesssim 10$$
$$\tilde{5}x_1 + \tilde{3}x_2 \lesssim 12$$

To compute strong feasibility, we solve the following problem:

$$\max z = x_1 + x_2$$
$$\text{s.t.}$$
$$5x_1 + 5x_2 \leqslant 10$$
$$7x_1 + 4x_2 \leqslant 12$$

The reader can verify that the solution of this problem $x_1^s = 1.3333$, $x_2^s = 0.6667$ is also a solution for any combination $A_{ij} \in \text{supp}(\tilde{A}_{ij})$, $b_i \in [0, b_i]$, so the problem is strong feasible with a strong solution.

5.3   Third Example (Unfeasible Fuzzy/Crisp LP)

This example shows a fuzzy/crisp unfeasible problem in which $T_{11} = (1, 2, 3)$, $T_{12} = (2, 3, 4)$, $T_{21} = (-1, 0, 1)$, $T_{22} = (3, 2, 1)$, $b_1 = [0, 8], b_2 = [0, 13]$, $c_1 = 1$, $c_2 = 1$, and $x_j \geqslant 0$. Its description is shown as follows:

$$\max z = x_1 + x_2$$
$$\text{s.t.}$$
$$\tilde{2}x_1 + \tilde{3}x_2 \lesssim 8$$
$$-\tilde{0}x_1 - \tilde{2}x_2 \lesssim -13$$

It is enough to see that this problem is unfeasible for strong conditions, this means that the problem:

$$\max z = x_1 + x_2$$
$$\text{s.t.}$$
$$x_1 + 2x_2 \leqslant 8$$
$$-\tilde{1}x_1 - \tilde{3}x_2 \leqslant -13$$

is unfeasible. The reader can verify that there is no any value $A_{ij} \in \text{supp}(\tilde{A}_{ij})$ that could lead to a feasible problem.

6   Concluding Remarks

Feasibility as a necessary condition for optimality in FLPs has been discussed and some interesting definitions have been provided and illustrated through three examples, for different cases.

The case of fuzzy coefficients with crisp boundaries in the constraints of an LP problem (fuzzy/crisp) is also discussed, and feasibility of this kind of problems is defined as well. Its applicability in hybrid contexts where human knowledge alongside crisp parameters is present in decision making and optimization problems is wide, so we provide necessary conditions for checking feasibility in fuzzy/crisp LPs. We have provided feasibility conditions for FLPs and fuzzy/crisp LP problems from a theoretical point of view. We encourage readers to work on computability of feasible and optimal solutions of this kind of problems since NP-Hardness is always present in interval-valued computations.

References

1. Zimmermann, H.J.: Fuzzy programming and linear programming with several objective functions. Fuzzy Sets Syst. 1(1), 45–55 (1978)
2. Zimmermann, H.J., Fullér, R.: Fuzzy reasoning for solving fuzzy mathematical programming problems. Fuzzy Sets Syst. 60(1), 121–133 (1993)
3. Fiedler, M., Nedoma, J., Ramík, J., Rohn, J., Zimmermann, K.: Linear Optimization Problems with Inexact Data. Springer, New York (2006)
4. Cerný, M., Hladík, M.: Optimization with uncertain, inexact or unstable data: linear programming and the interval approach. In: Nemec, R., Zapletal, F. (eds.) Proceedings of the 10th International Conference on Strategic Management and its Support by Information Systems, pp. 35–43. Technical University of Ostrava, VSB (2013)
5. Hladík, M.: Weak and strong solvability of interval linear systems of equations and inequalities. Linear Algebra Appl. 438(11), 4156–4165 (2013)
6. Ramík, J., Rimánek, J.: Inequality relation between fuzzy numbers and its use in fuzzy optimization. Fuzzy Sets Syst. 16, 123–138 (1985)
7. Figueroa-García, J.C., Chalco-Cano, Y., Román-Flores, H.: Distance measures for interval Type-2 fuzzy numbers. Discrete Appl. Math. 197(1), 93–102 (2015)
8. Figueroa-García, J.C., Hernández-Pérez, G.: On the computation of the distance between interval Type-2 fuzzy numbers using α-cuts. In: IEEE Proceedings of NAFIPS 2014, pp. 1–5. IEEE (2014)
9. Kreinovich, V., Tao, C.W.: Checking identities is computationally intractable (NP-Hard), so human provers will be always needed. Int. J. Intell. Syst. 19(1), 39–49 (2004)
10. Heindl, G., Kreinovich, V., Lakeyev, A.V.: Solving linear interval systems is NP-Hard even if we exclude overflow and underflow. Reliable Comput. 4(4), 383–388 (1998)
11. Lakeyev, A.V., Kreinovich, V.: NP-Hard classes of linear algebraic systems with uncertainties. Reliable Comput. 3(1), 51–81 (1997)
12. Kreinovich, V., Lakeyev, A.V., Noskov, S.I.: Optimal solution of interval linear systems is intractable (NP-Hard). Interval Comput. 1, 6–14 (1993)

A Simulation Study on Quasi Type-2 Fuzzy Markov Chains

Sandy Katina Salamanca-Rivera, Luis Carlos Giraldo-Arcos,
and Juan Carlos Figueroa-García$^{(\boxtimes)}$

Universidad Distrital Francisco José de Caldas, Bogotá, Colombia
{sksalamancar, lcgiraldoa}@correo.udistrital.edu.co,
jcfigueroag@udistrital.edu.co

Abstract. This paper presents a simulation study on Quasi Type-2 fuzzy Markov chains. Its main objective is to identify its stationary behavior using two methods: the *Greatest Eigen Fuzzy Set* and the *Powers of a Fuzzy Matrix* based on other reports which shows that most of fuzzy Markov chains does not have an ergodic behavior. To do so, we simulate Quasi Type-2 fuzzy Markov chains at different sizes to collect some interesting statistics.

1 Introduction and Motivation

Markov chains is an important stochastic process mainly studied over a probabilistic context (see Grimmet & Strizaker [1], Ross [2], and Stewart [3]). The fuzzy approach have been defined by Avrachenkov & Sánchez [4, 5], and Interval Type-2 fuzzy Markov chains (IT2FM) are studied by Figueroa-García [6, 7], Figueroa-García et al. [8–10]. Fuzzy Markov chains can represent human-like perceptions about the states of a discrete random process which is particularly useful in cases where statistical information of the Markov process is either absent or unreliable, so experts are asked about their perceptions about the process to obtain fuzzy sets.

While probabilistic Markov chains mostly exhibit ergodic stationary distributions (including periodicity), fuzzy Markov chains usually show non ergodic behaviors (see Figueroa-García et al. [8]) whose stationary properties have been studied by Figueroa-García et al. [8] who shown that most of fuzzy Markov chains exhibit either non ergodic or periodical distributions. Non ergodic or periodical behaviors are not desirable in decision making, so we simulate Quasi Type-2 fuzzy Markov chains (QT2FM) which are Type-2 fuzzy sets whose secondary membership functions are bounded fuzzy sets i.e. triangular or trapezoidal, to analyze its limiting properties.

This paper focuses on simulating QT2FMs and it is divided into seven sections. Section 1 is an Introductory section. Section 2 presents some basics about Markov chains; in Sect. 3, some concepts about QT2FMs and their properties are provided. Section 4 presents how to compute the stationary distribution of a QT2FM. Section 5 presents the simulation strategy and their results, Sect. 6 presents its results, and Sect. 7 shows the concluding remarks of the study.

© Springer International Publishing Switzerland 2016
D.-S. Huang et al. (Eds.): ICIC 2016, Part III, LNAI 9773, pp. 834–845, 2016.
DOI: 10.1007/978-3-319-42297-8_77

2 Basics on Fuzzy Markov Chains

Avrachenkov & Sánchez [4] defined a fuzzy Markov chain as a squared matrix which contains the possibility that a discrete state at instant t becomes into any state at next instant $t + 1$. Let $S = \{1, 2, \cdots, n\}$ be a finite fuzzy set for a fuzzy distribution on S is defined by a mapping x from S to [0,1] and a vector $x = \{x_1, x_2, \cdots, x_n\}$ where $0 \leqslant x_i \leqslant 1$, $i \in S$. x_i is the membership degree that a state i has regarding a fuzzy set S, $i \in S$ with cardinality m, $\mathcal{C}(S) = m$. The transition law of a fuzzy Markov chain is a fuzzy relational matrix P at the instant t, $T = 1, 2, \cdots, n$, as follows:

$$x_j^{(t+1)} = \max_{i \in S}\{x_j^{(t)} \wedge p_{ij}\}, j \in S \tag{1}$$

$$x^{(t+1)} = x^{(t)} \circ P \tag{2}$$

where $i, j = 1, 2, \cdots, m$ are the initial and final states of the transition and $x^{(0)}$ is their *Initial Distribution*. The powers of a fuzzy transition matrix P are computed as:

$$p_{ij}^t := \max_{k \in S}\{p_{ik} \wedge p_{kj}^{t-1}\} \tag{3}$$

Note that $p_{ij}^1 = p_{ij}$ and $p_{ij}^0 = \delta_{ij}$ where δ_{ij} is a *Kronecker Delta*. In other words:

$$P^t := P \circ P^{t-1} \tag{4}$$

The *Stationary Distribution* of a fuzzy markov states that if the powers of a fuzzy transition matrix P converge in τ steps to a non-periodic solution, then the associated fuzzy Markov chain is called *Aperiodic Fuzzy Markov Chain*, this is:

$$P^\tau = \lim_{t \to \infty} P^t \tag{5}$$

which is called *Stationary Fuzzy Transition Matrix*, otherwise it is a periodical Markov chain (see Gavalec [11–13]). Additionally, Figueroa-García et al. [8] defined two useful properties of a T1FM, which are presented below:

Definition 1 (Strong Ergodicity for Markov Chains). *A fuzzy Markov chain is called* Strong Ergodic *if it is aperiodic and its stationary transition matrix has identical rows.*

Definition 2 (Weak Ergodicity for Markov Chains). *A fuzzy Markov chain is called* Weakly Ergodic *if it is aperiodic and its stationary transition matrix is stable with no identical rows.*

3 Quasi Type-2 Fuzzy Markov Chains

A Quasi Type-2 fuzzy relational matrix $\tilde{P}$ defined over $\mathcal{C}(S) \times \mathcal{C}(S)$ has $m \times m$ elements $\{\tilde{p}_{ij}\}_{i,j=1}^m$, and n_a embedded values in the closed interval $[\underline{p}_{ij}, \overline{p}_{ij}]$ characterized by a secondary membership function $f_x(u)/u$, $J_x \subseteq [0,1]$ $\forall x \in S, j \in S$. Denote $x_{ij} = \{x^t = j | x^{t-1} = i\}$ then we have:

$$\tilde{S}_i = \sum_{x_{ij}=1}^{m} \left[\int_{u \in J_{x_{ij}}} f_{x_{ij}}(u)/u \right] \Big/ x_{ij} \ \forall \ i \in S \tag{6}$$

and consequently:

$$J_{x_{ij}} = \left[\int f_{x_{ij}}(u_{jk})/u_{jk} \right] \Big/ x_{ij}, \ i,j \in S \tag{7}$$

The set $\tilde{P}$ can be obtained from the m_{th} T2FS $\tilde{S}_i$ as follows:

$$\tilde{S}_i = \{ (x_{ij}, \mu_{\tilde{S}_i}(x_{ij})) \mid x_{ij} = i \} \quad i,j \in S \tag{8}$$

This yields the following matrix representation:

$$\tilde{P} = \begin{bmatrix} \mu_{(\tilde{S}_1)}(x_{1j}) \\ \mu_{(\tilde{S}_2)}(x_{2j}) \\ \vdots \\ \mu_{(\tilde{S}_m)}(x_{mj}) \end{bmatrix} = \begin{bmatrix} \tilde{p}_{11} & \tilde{p}_{12} & \cdots & \tilde{p}_{1m} \\ \tilde{p}_{21} & \tilde{p}_{22} & \cdots & \tilde{p}_{2m} \\ \vdots & \vdots & \ddots & \vdots \\ \tilde{p}_{m1} & \tilde{p}_{m2} & \cdots & \tilde{p}_{mm} \end{bmatrix} \tag{9}$$

$$\tilde{P} = \begin{bmatrix} \begin{bmatrix} \underline{p}_{11} & \underline{p}_{12} & \cdots & \underline{p}_{1m} \\ \underline{p}_{21} & \underline{p}_{22} & \cdots & \underline{p}_{2m} \\ \vdots & \vdots & \ddots & \vdots \\ \underline{p}_{m1} & \underline{p}_{m2} & \cdots & \underline{p}_{mm} \end{bmatrix}, \begin{bmatrix} \overline{p}_{11} & \overline{p}_{12} & \cdots & \overline{p}_{1m} \\ \overline{p}_{21} & \overline{p}_{22} & \cdots & \overline{p}_{2m} \\ \vdots & \vdots & \ddots & \vdots \\ \overline{p}_{m1} & \overline{p}_{m2} & \cdots & \overline{p}_{mm} \end{bmatrix} \end{bmatrix} \tag{10}$$

and its secondary membership function $f_{x_{ij}}(u)/u$. Here, $\tilde{p}_{ij}$ is the membership degree of any state $x^{(t)} = j$ given an initial state i, $i,j \in S$ where $\mathcal{C}(S) = m$. Thus $\tilde{P}$ is defined by two matrices $\underline{P}, \overline{P}$, and $f_{x_{ij}}(u)/u$:

$$\tilde{P} = [\underline{P}, \overline{P}], f_{x_{ij}}(u)/u \tag{11}$$

According to Mendel [14] and Hamrawi & Coupland [15], a T2FS can be also represented trough α-planes which are z-slices over $f_x(u)/u$. An α-plane for a general T2FS $\tilde{A}$, denoted $\tilde{A}_\alpha$ is the union of all primary memberships of $\tilde{A}$ whose secondary grades are greater than or equal to $\alpha, (0 \leqslant \alpha \leqslant 1)$, i.e.

$$\tilde{A}_\alpha = \bigcup_{x \in X} \{ (x,u) | \mu_{\tilde{A}}(x,u) \geqslant \alpha \} \tag{12}$$

$$\tilde{A}_\alpha = \bigcup_{x \in X} S_{\tilde{A}}(x|\alpha) \tag{13}$$

Quasi Type-2 fuzzy systems (QT2FS) have been proposed by Mendel & Liu [14] to approximate full Type-2 fuzzy systems by using closed secondary membership functions $f_x(u)/u$ and α-planes for $\alpha = \{0, 1\}$ to defuzzify the resulting system assuming the original $f_x(u)/u$. In this paper, we use triangular and trapezoidal secondary membership functions (see Mendel [14] and Hamrawi & Coupland [15]). An example of a QT2FM S_i with triangular $f_x(u)/u$ is displayed in Fig. 1.

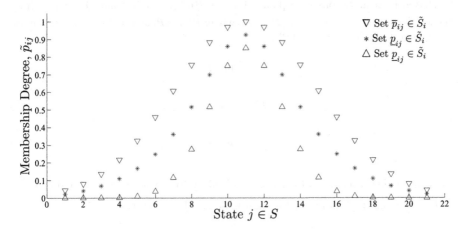

Fig. 1. Primary membership function of $\tilde{p}_{ij}$

Figure 1 shows the *Foot of Uncertainty* namely *FOU* where n_a sets $\tilde{S}^l_{e(i)}$ are bounded by three values: ∇ represents $\overline{p}_{ij}$, Δ represents $\underline{p}_{ij}$, and * represents the α-plane = 1. To visualize $f_{x_{ij}}(u)/u$ we provide Fig. 2.

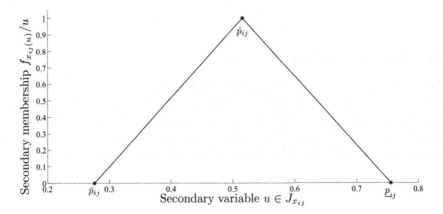

Fig. 2. Secondary membership function of $x_{ij} = 8$

Figure 2 shows the secondary membership function of $x_{ij} = 8$. For simplicity, $f_{x_{ij}=8}(u)/u$ is defined as a triangular fuzzy set in order to see it as a QT2FM.

3.1   Limiting Distribution of a QT2FM

Now, if we compute the transition law of $\tilde{P}$ of a QT2FM then we only need to compute the stationary distributions of α-planes for $\alpha = \{0, 1\}$. A triangular QT2FM is fully characterized by three matrices $\overline{P}, \underline{P}$ and $\dot{P}$ where $\overline{P} \leqslant \dot{P} \leqslant \underline{P}$, and its limiting distribution is obtained through Eq. (5) as follows

$$\underline{P}^* = \lim_{\tau \to \infty} \underline{P}^\tau \tag{14}$$

$$\dot{P}^* = \lim_{\tau \to \infty} \dot{P}^\tau \tag{15}$$

$$\overline{P}^* = \lim_{\tau \to \infty} \overline{P}^\tau \tag{16}$$

A trapezoidal QT2FM is fully characterized by four matrices $\overline{P}, \underline{P}, \vec{P}$ and $\overleftarrow{P}$ where $\overline{P} \leqslant \vec{P} \leqslant \overleftarrow{P} \leqslant \underline{P}$, and its limiting distribution is obtained through (5) as follows

$$\underline{P}^* = \lim_{\tau \to \infty} \underline{P}^\tau \tag{17}$$

$$\vec{P}^* = \lim_{\tau \to \infty} \vec{P}^* \tag{18}$$

$$\overleftarrow{P}^* = \lim_{\tau \to \infty} \overleftarrow{P}^\tau \tag{19}$$

$$\overline{P}^* = \lim_{\tau \to \infty} \overline{P}^\tau \tag{20}$$

3.2   Non Ergodic QT2FM

An ergodic QT2FM is a QT2FM whose α-planes for $\alpha = \{0, 1\}$ are strong ergodic which is equivalent to say that the powers of $\tilde{P}$ converges to $\lim_{n \to \tau} \tilde{P}^n = \tilde{P}^*$ into an idempotent matrix with equal rows/columns. A non ergodic $\tilde{P}$ can be weak ergodic or periodic (which is the most common case). This way, we provide an example of a non-ergodic triangular QT2FM defined by three transition matrices $\underline{P}, \dot{P}$, and $\overline{P}$:

$$\underline{P} = \begin{bmatrix} 0.253 & 0.201 & 0.157 & 0.154 & 0.159 \\ 0.180 & 0.008 & 0.124 & 0.201 & 0.297 \\ 0.227 & 0.226 & 0.241 & 0.267 & 0.101 \\ 0.199 & 0.292 & 0.146 & 0.024 & 0.266 \\ 0.220 & 0.112 & 0.096 & 0.198 & 0.082 \end{bmatrix} \dot{P} = \begin{bmatrix} 0.355 & 0.464 & 0.528 & 0.518 & 0.525 \\ 0.367 & 0.553 & 0.576 & 0.372 & 0.417 \\ 0.372 & 0.511 & 0.518 & 0.419 & 0.526 \\ 0.514 & 0.512 & 0.477 & 0.306 & 0.371 \\ 0.335 & 0.401 & 0.505 & 0.474 & 0.391 \end{bmatrix}$$

$$\overline{P} = \begin{bmatrix} 0.976 & 0.785 & 0.810 & 0.821 & 0.958 \\ 0.893 & 0.606 & 0.843 & 0.891 & 0.724 \\ 0.905 & 0.934 & 0.604 & 0.888 & 0.645 \\ 0.939 & 0.759 & 0.654 & 0.675 & 0.924 \\ 0.697 & 0.617 & 0.934 & 0.797 & 0.671 \end{bmatrix}$$

By using Eqs. (14), (15) and (16), the following non ergodic distributions $\underline{P}^\tau, \dot{P}^\tau$ and $\overline{P}^\tau$ are obtained:

$$\underline{P}^\tau = \begin{bmatrix} 0.253 & 0.201 & 0.157 & 0.201 & 0.201 \\ 0.220 & 0.201 & 0.157 & 0.201 & 0.201 \\ 0.227 & 0.241 & 0.241 & 0.241 & 0.241 \\ 0.220 & 0.201 & 0.157 & 0.201 & 0.201 \\ 0.220 & 0.201 & 0.157 & 0.201 & 0.201 \end{bmatrix} \dot{P}^\tau = \begin{bmatrix} 0.474 & 0.512 & 0.518 & 0.514 & 0.518 \\ 0.474 & 0.553 & 0.553 & 0.474 & 0.526 \\ 0.474 & 0.511 & 0.518 & 0.474 & 0.518 \\ 0.514 & 0.512 & 0.514 & 0.474 & 0.514 \\ 0.474 & 0.505 & 0.505 & 0.474 & 0.505 \end{bmatrix}$$

$$\overline{P}^\tau = \begin{bmatrix} 0.976 & 0.934 & 0.934 & 0.891 & 0.958 \\ 0.893 & 0.893 & 0.893 & 0.891 & 0.893 \\ 0.905 & 0.905 & 0.905 & 0.891 & 0.905 \\ 0.939 & 0.934 & 0.934 & 0.891 & 0.939 \\ 0.905 & 0.905 & 0.905 & 0.891 & 0.905 \end{bmatrix}$$

Definition 3. *Let $\tilde{P}^*$ the stationary distribution of $\tilde{P}$. It is said that $\tilde{P}$ is a strong ergodic QT2FM iff the $\alpha = \{0,1\}$ planes are strong ergodic, this is $\tilde{P}^*$ converges to strong ergodic limiting distributions $\lim_{n \to \tau} \tilde{P}^n = \tilde{P}^*$.*

4 Computation of the Fuzzy Stationary Distribution

Several methods can be used to compute the limiting distribution of the process. A first method which uses the max-min relation on $\tilde{P}$ is described below:

$$P^n = P \circ P^{n-1} = P \circ P \circ P^{n-2} = \cdots = \underbrace{P \circ P \circ \cdots \circ P}_{n times} \tag{21}$$

Now, if the stationary distribution of $\tilde{P}$ is given by $P^* = P^\tau$ where $\lim_{n \to \tau} \tilde{P}^n = \tilde{P}^*$, then $\tilde{P}$ becomes an idempotent matrix. Sánchez [5, 16, 17] provided definitions about the concept of an *Eigen Fuzzy Set* (analogous to a crisp *eigenvector* and *eigenvalue*):

Definition 4. *Let $\tilde{P}$ be a fuzzy relation in a given matrix form. Then x is called an Eigen Fuzzy set of $\tilde{P}$, iff:*

$$x \circ P = x \tag{22}$$

Definition 5. *The Fuzzy set $x \in \mathcal{F}(S)$ is contained in the fuzzy set $y \in \mathcal{F}(S)$, this is $(x \subseteq y)$, iff $x_i \leqslant y_i$ for all $i \in S$.*

Definition 6. *Let $\mathcal{X}$ be the set of eigen fuzzy sets of the fuzzy relation P namely:*

$$\mathcal{X} \triangleq \{ x \in \mathcal{F}(S) \mid x \circ P = x \} \tag{23}$$

the elements of $\mathcal{X}$ are invariants of P according to the $\circ - (max - min)$ composition. Then, if $\exists \{ x^* \in \mathcal{F}(S) \mid x \subseteq x^*, \forall x \in \mathcal{X} \}$, it is called the Greatest Eigen Fuzzy Set *of the relation P.*

Now, the idea is to find a max eigen fuzzy set, idempotent and stable:

$$x^* = \max_{i \in S} P_{ij}^n \tag{24}$$

To do so, we present the following algorithm.

Algorithm 1 (Method III, Sanchéz [5]).

i. Determine first x^1 whose elements are the greatest elements per column of P.

ii. Compute $P^2 = P \circ P$ and determine the greatest elements per column of P^2. They give x^2 where $\max_{i \in S} P_{ij}^k = (x^k \circ P^{k-1})_j = x_j^k$, $j = \overline{1, n}$, $\forall k \succ 0$ and $k = 2$, $j = \overline{1, 5}$.

iii. Compare x^2 to x^1: If they are different, compute $P^3 = P^2 \circ P$ to get x^3 where $\max_{i \in S} P_{ij}^3 = (x^3 \circ P^2)_j = x_j^3$, $j = \overline{1, 5}$.

iv. Compare x^3 to x^2: If they are different, compute $P^4 = P^3 \circ P$ to get x^4 where $\max_{i \in S} P_{ij}^4 = (x^4 \circ P^3)_j = x_j^4$, $j = \overline{1, 5}$ and so on. Stop when it is found $\vee$ such that $x^{n+1} = x^n$, that is $x^* = x^n \circ P$.

The above method finds the eigen fuzzy set x^* which is the stationary distribution of P. It is important to recall that if P is a *Strong Ergodic* fuzzy Markov chain, then its greatest eigen fuzzy set x^* converges to an idempotent matrix P^τ.

5 Methodology of Simulation

Some important aspects about the simulation process are presented next.

- **Size of the Markov Chain:** The size of $\tilde{P}$ denoted by m is the cardinality of S, $\mathcal{C}(S)$, for any $(S) \in \mathcal{F}(S)$. Four sizes of P are simulated in this paper: $m = \{5, 10, 50, 100\}$.
- **Random Number Generator:** All elements $\{\tilde{p}_{ij}\}$ of $\tilde{P}$ are obtained by using a uniform distribution, $x_n = (a_1 x_{n-1} + \cdots + a_k x_{n-k})$ mod m', $U_i = x_n/m'$ where m' is the *modulus*, and k is the order of the polynomial e.g. $\{p_{ij}\} \in [0, 1] \leftrightarrow U_{ij}[0, 1]$. For triangular QT2FMs it is mandatory that $\underline{p}_{ij} \leqslant \dot{p}_{ij} \leqslant \overline{p}_{ij}$, and for trapezoidal QT2FMs it is mandatory that $\underline{p}_{ij} \leqslant \overleftarrow{p}_{ij} \leqslant \overrightarrow{p}_{ij} \leqslant \overline{p}_{ij}$.
- **Algorithms:** Two algorithms are applied to find the steady state of a Markov Chain. The first one is presented in (21), and the second one is the Method III proposed by Sánchez [5, 16, 17] shown in Algorithm 1. While the first algorithm shows if P is either strong or weak ergodic, the second one does not identify it and only gets their greatest eigen fuzzy set $\{x_j^*\}$.
- **Number of Runs:** 1000 Runs are simulated per each size of P, a total of 4000 simulations were performed. Some interesting statistics are collected and analyzed jointly whose description is presented below.
- **Statistics of Interest:** All collected statistics are described next:

  (a) **Number of the powers of P^τ:** It is the amount of powers of P needed to obtain its steady state τ, if P is periodical then τ does not exist, otherwise the type of Markov chain must be registered.
  (b) **Type of Fuzzy Markov Chain:** If P is strong ergodic (see Definition 1) then is registered as *SE*. If it is weak ergodic (see Definition 2) then is registered as *WE*, and finally if P is periodic then is registered as such.
  (c) **Amount of iterations to obtain x_j^*:** It is the amount of iterations $\vee$ needed to obtain the greatest eigen fuzzy set x_j^n (see Method III in the Appendix).
  (d) **Computing Time:** It is time to obtain either τ or the needed time to identify a periodical behavior.

All simulations are computed by using *MatLab 2015* over an AMD Tl-64 Turion machine with 8 GB of RAM.

6 Simulation Results

Table 1 shows the amount of Markov chains which are either strong *(SE)*, weak ergodic *(WE)* or periodical per every size of triangular $\tilde{P}$.

It is clear that most of chains are periodical and the remaining ones are ergodic either in a weak or strong way.

Table 2 shows the amount of Markov chains which has either a SE, WE or periodical behavior per each size of trapezoidal chain. It is clear that most of chains have a periodic oscillation and the remaining ones are ergodic.

Table 3 (see Appendix) shows the amount of powers of $\tilde{P}$, $\vee$ needed to achieve the eigen fuzzy set of triangular QT2FMs $\tilde{P}$. $\vee_=$ is the amount of times that $\vee$ reaches a single value for all states (which can be seen as a dominance of a state). Also it can be viewed that all triangular QT2FMs achieve their greatest eigen fuzzy set around a value

Table 1. Amount of SE, WE or Periodic triangular QT2FM

Size	SE	WE	Periodic	Avg. time (sec)	Total
$\underline{P}, m = 5$	0	445	555	1.78	1000
$\dot{P}, m = 5$	0	512	488	1.74	1000
$\overline{P}, m = 5$	0	427	563	1.62	1000
$\underline{P}, m = 10$	0	281	719	5.97	1000
$\dot{P}, m = 10$	0	286	714	5.49	1000
$\overline{P}, m = 10$	0	257	743	5.41	1000
$\underline{P}, m = 50$	0	101	899	66.5	1000
$\dot{P}, m = 50$	0	98	902	69.5	1000
$\overline{P}, m = 50$	0	106	894	65.4	1000
$\underline{P}, m = 100$	0	84	916	275.5	1000
$\dot{P}, m = 100$	0	164	836	221.4	1000
$\overline{P}, m = 100$	0	87	913	271.9	1000

Table 2. Amount of SE, WE or Periodic trapezoidal QT2FM

Size	SE	WE	Periodic	Avg. time (sec)	Total
$\underline{P}, m = 5$	0	472	528	1.63	1000
$\vec{P}, m = 5$	2	460	538	1.79	1000
$\dot{P}, m = 5$	0	538	462	1.12	1000
$\overline{P}, m = 5$	0	467	533	1.97	1000
$\underline{P}, m = 10$	0	247	753	5.98	1000
$\vec{P}, m = 10$	0	321	679	5.16	1000
$\dot{P}, m = 10$	0	326	674	5.21	1000
$\overline{P}, m = 10$	0	308	692	5.78	1000
$\underline{P}, m = 50$	0	121	879	67.1	1000
$\vec{P}, m = 50$	0	97	903	61.5	1000
$\dot{P}, m = 50$	0	146	854	61.3	1000
$\overline{P}, m = 50$	0	137	863	67.9	1000

smaller than m iterations, for instance, if P is 10×10 then most part of the chains achieves x_j^* in less than 10 iterations, that is, $\lim_{n \to \infty} x_j^n < 10$.

In general, Method III obtains the greatest eigen fuzzy set faster than P^n powers to find P^τ. In contrast to, Method III does not show if the process is either periodic or not, so decision making based on this method could not be consistent. Note that while m is increased then $\tilde{P}$ has a tendency to be periodic, that is, when P is large than $m = 10$ is more probably to find a periodic behavior than a size less than $m = 10$.

7 Concluding Remarks

An important fact is: as large $\tilde{P}$ is, $\tilde{P}$ does not converge to τ, and as small $\tilde{P}$ is, as $\tilde{P}$ converges to an ergodic behavior. This reveals that the asymptotic behavior of QT2FMs is very unstable and they tend to be periodical unlike crisp Markov chains.

While the fuzzy max-min operator conduces to periodical distributions of $\tilde{P}$, it is less sensitive to perturbations than crisp operators. Also, we have observed that trapezoidal Markov chains exhibit less weak ergodic cases than triangular chains due to the amount of parameters that composes the secondary membership function: four instead of three which means less distance between parameters.

The eigen fuzzy set proposed by Sánchez is faster than the computation of $\tilde{P}^\tau$ to find the stationary distribution of the process, which implies less computations. On the other hand, it does not detect periodicity or unstable behaviors, so it is useful only in cases where a limiting distribution is needed.

QT2FMs have mostly a periodical behavior, so x_j^* can be used as its stationary distribution (for reference only). The eigen fuzzy set of P is a good crisp measure of the steady state of P, and the powers of $\tilde{P}$ are better for decision making (keeping in mind most of them are periodical). Finally, we have found that some QT2FMs have dominant states which lead to get an eigen fuzzy set with only one value.

Appendix: Iterations for Triangular QT2FM

This appendix shows how many iterations are needed to get a stationary triangular $\tilde{P}$.

Table 3. Amount of iterations ∨.

Type	#	1	2	3	4	5	6
$\underline{P},\vee,m=5$	988	0	54	476	352	91	25
$\underline{P},\vee_=,m=5$	2	---	---	---	---	1	1
$P,\vee,m=5$	1000	0	43	394	440	102	21
$P,\vee_=,m=5$	---	---	---	---	---	---	---
$\overline{P},\vee,m=5$	950	0	91	324	352	151	32
$\overline{P},\vee_=,m=5$	5	---	---	---	2	3	---

Type	#	1	2	3	4	5	6	7	8	9
$\underline{P},\vee,m=10$	1000	0	18	79	277	289	276	17	26	18
$\underline{P},\vee_=,m=10$	---	---	---	---	---	---	---	---	---	---
$P,\vee,m=10$	1000	0	0	126	327	308	117	91	19	12
$P,\vee_=,m=10$	---	---	---	---	---	---	---	---	---	---
$\overline{P},\vee,m=10$	999	0	0	112	311	203	231	109	33	0
$\overline{P},\vee_=,m=10$	1	---	---	---	---	---	---	1	---	---

Type	#	5	6	7	8	9	10	11	12	13	14	15	16	17	21
$\underline{P},\vee,m=50$	1000	22	23	145	186	228	54	112	91	63	32	9	18	8	9
$\underline{P},\vee_=,m=50$	---	---	---	---	---	---	---	---	---	---	---	---	---	---	---
$P,\vee,m=50$	1000	11	65	87	145	139	192	142	83	43	37	29	18	9	0
$P,\vee_=,m=50$	---	---	---	---	---	---	---	---	---	---	---	---	---	---	---
$\overline{P},\vee,m=50$	1000	0	82	87	164	163	168	123	99	67	18	0	9	11	9
$\overline{P},\vee_=,m=50$	---	---	---	---	---	---	---	---	---	---	---	---	---	---	---

Type	#	5	6	7	8	9	10	11	12	13	14	15	16	17	18	19	20	22	23
$\underline{P},\vee,m=100$	1000	0	21	44	176	102	119	178	53	112	63	67	19	20	17	0	9	0	0
$\underline{P},\vee_=,m=100$	---	---	---	---	---	---	---	---	---	---	---	---	---	---	---	---	---	---	---
$P,\vee,m=100$	1000	0	21	77	143	168	96	91	148	32	34	47	21	61	19	22	11	9	0
$P,\vee_=,m=100$	---	---	---	---	---	---	---	---	---	---	---	---	---	---	---	---	---	---	---
$\overline{P},\vee,m=100$	1000	0	31	39	97	114	167	92	103	83	87	62	47	39	12	12	9	0	6
$\overline{P},\vee_=,m=100$	---	---	---	---	---	---	---	---	---	---	---	---	---	---	---	---	---	---	---

References

1. Grimmet, G., Stirzaker, D.: Probability and Random Processes. Oxford Press, Oxford (2001)
2. Ross, S.M.: Stochastic Processes. Wiley, New York (1996)
3. Stewart, W.J.: Introduction to the Numerical Solution of Markov Chains. Princeton Press, Princeton (1994)
4. Avrachenkov, K.E., Sanchez, E.: Fuzzy Markov chains and decision making. Fuzzy Optim. Decis. Mak. 1(2), 143–159 (2002)
5. Avrachenkov, K.E., Sanchez, E.: Fuzzy Markov chains: specifities and properties. In IEEE (ed.) 8th IPMU 2000 Conference, Madrid, Spain, pp. 1851–1856. IEEE (2000)

6. Figueroa-García, J.C.: Interval type-2 fuzzy Markov chains. In: Sadeghian, A., Mendel, J.M., Tahayori, H. (eds.) Advances in Type-2 Fuzzy Sets and Systems. STUDFUZZ, vol. 301, pp. 49–64. Springer, New York (2013)

7. Figueroa-García, J.C.: Interval Type-2 fuzzy Markov chains: An approach. North Am. Fuzzy Inf. Process. Soc. (NAFIPS) **28**(1), 1–6 (2010)

8. Figueroa-García, J.C., Kalenatic, D., López, C.A.: A simulation study on fuzzy Markov chains. Commun. Comput. Inf. Sci. **15**(1), 109–117 (2008)

9. Kalenatic, D., Figueroa-García, J.C., Lopez, C.A.: Scalarization of type-1 fuzzy Markov chains. In: Huang, D.-S., Zhao, Z., Bevilacqua, V., Figueroa, J.C. (eds.) ICIC 2010. LNCS, vol. 6215, pp. 110–117. Springer, Heidelberg (2010)

10. Figueroa-García, J.C., Kalenatic, D., Lopez, C.A.: Interval type-2 fuzzy Markov chains: type reduction. In: Huang, D.-S., Gan, Y., Gupta, P., Gromiha, M. (eds.) ICIC 2011. LNCS, vol. 6839, pp. 211–218. Springer, Heidelberg (2012)

11. Gavalec, M.: Computing orbit period in max-min algebra. Discrete Appl. Math. **100**(1), 49–65 (2000)

12. Gavalec, M.: Periods of special fuzzy matrices. Tatra Mountains Math. Publ. **16**(1), 47–60 (1999)

13. Gavalec, M.: Reaching matrix period is NP-complete. Tatra Mountains Math. Publ. **12**(1), 81–88 (1997)

14. Mendel, J.M., Liu, F.: On new quasi type-2 fuzzy logic systems. In IEEE (ed.) Proceedings of 2008 International Conference on Fuzzy Systems (FUZZ 2008), pp. 1–7. IEEE (2008)

15. Hamrawi, H., Coupland, S.: Type-2 fuzzy arithmetic using alpha-planes. In: Proceedings of IFSA-EUSFLAT 2009, pp. 606–611. IEEE (2009)

16. Sanchez, E.: Eigen fuzzy sets and fuzzy relations. J. Math. Anal. Appl. **81**(1), 399–421 (1981)

17. Sanchez, E.: Resolution of eigen fuzzy sets equations. Fuzzy Sets Syst. **1**(1), 69–74 (1978)

Bioinformatic

ICPFP: A Novel Algorithm for Identification of Comorbidity Based on Properties and Functions of Protein

Feng He[(⊠)] and Ning Li

Institute of Machine Learning and Systems Biology, College of Electromics
and Information Engineering, Tongji University, Shanghai 201804, China
{hefeng19890620,jclining}@126.com

Abstract. The term comorbidity refers to the coexistence of multiple diseases or disorders along with a primary disease in a patient. Hence, the prediction of disease comorbidity can identify the comorbid diseases when dealing with a primary disease. Unfortunately, since the records of comorbidity in clinic are far from complete, we can't get enough knowledge to understand the reason for comorbidity. Though many researches have been done to predict disease comorbidity, the accuracy of these algorithms need to be improved. By investigating the factors underlying disease comorbidity, we found that a number of comorbidities are caused by common modules comprised by proteins. Thus, we here propose a novel algorithm to identify disease comorbidity by integrating different types of datasets ranging from properties to functions of protein. Results on real data of comorbidity display that our algorithm can perform better than previous approaches, and some of our new predictions are reported in literature, which can prove the effectiveness of our algorithm, and help deeply explain the molecular mechanism of disease comorbidity.

Keywords: Disease comorbidity · Disease similarity · Protein property · Protein function

1 Introduction

Disease comorbidity means two or more diseases occur together in a patient not by chance [1–3], which usually suggests more difficult prognosis and more serious damage due to the modern extreme specialization of physicians [4–9].

In this paper, we propose a novel algorithm named ICPFP, which is used for prediction of disease comorbidity based on integration of different types of datasets ranging from properties to functions of protein [10–15]. Compared with traditional methods, ICPFP makes use of heterogeneous information of protein depicting diseases, including disease associated proteins, amino acid frequency, physicochemical properties, protein functions, where these multiple datasets can effectively complement with each other to improve the accuracy of prediction. Besides, based on the same 20 kinds of amino acids for every protein, more comorbidities can be found by similarities of every two diseases than just discriminating the comorbidity association between two

© Springer International Publishing Switzerland 2016
D.-S. Huang et al. (Eds.): ICIC 2016, Part III, LNAI 9773, pp. 849–855, 2016.
DOI: 10.1007/978-3-319-42297-8_78

diseases by common modules based on incomplete mutant genes, PPIs or pathways, which can help deeply explain the molecular mechanism of disease comorbidity.

2   Materials and Methods

2.1   Disease Similarity Based on Proteins

With disease-associated genes, we adopted the disease-associated proteins and their corresponding amino acid sequences for every disease from Uniprot database [18]. Since for each disease, there was a corresponding protein list, the disease similarity between disease i and j based on disease associated proteins was formulated as follows.

$$sim_{pro}(d_i, d_j) = \frac{\left|D_{pro}^i \cap D_{pro}^j\right|}{\left|D_{pro}^i \cup D_{pro}^j\right|} \tag{1}$$

where D_{pro}^i represents disease-associated protein list of disease i, and $D_{pro}^i \cap D_{pro}^j$ and $D_{pro}^i \cup D_{pro}^j$ denotes the intersection and union of two disease-associated protein lists, respectively.

2.2   Disease Similarity Based on Protein Properties

Firstly, 20 kinds of amino acid frequency were calculated for every protein to form a 20 dimensional vector. For every disease, we added this vector of each protein including inside the disease-associated protein list together so that we got the amino acid frequency vector for disease as $w_i = \left(t_1^i, t_2^i, \cdots, t_{20}^i\right)^T$, where t_j^i means j-th amino acid frequency for disease i.

Secondly, ten physicochemical properties for twenty amino acids in protein-protein interface respectively was shown in Table 1. Based on these ten physicochemical properties for twenty amino acids, protein property vector for each disease could be defined as follows.

$$v_i = \left(w_i^T \cdot P\right)^T \tag{2}$$

where P means the matrix of protein property defined in Table 1. Then, the disease similarity based on protein properties was formulated as follows.

$$sim(d_i, d_j)_{pp} = \frac{v_i^T \cdot v_j}{\|v_i\| \times \|v_j\|} \tag{3}$$

Table 1. Ten physicochemical properties for twenty amino acids

Residue	NA	NE	NP	HB	HL	PP	IP	MA	EN	EI
A	5	0	2	0.25	3	−0.17	6.11	71.1	−0.22	0.0373
C	6	0	2	0.04	−1	0.43	6.31	103.1	4.66	0.0829
D	8	−1	4	−0.72	3	−0.38	5.945	115.1	−4.12	0.1263
E	9	−1	4	−0.62	3	−0.13	5.785	129.1	−3.64	0.0058
F	11	0	2	0.61	−2.5	0.82	5.755	147.2	5.27	0.0946
G	4	0	2	0.16	0	−0.07	6.065	57	−1.62	0.005
H	10	0	4	−0.4	−0.5	0.41	5.565	137.1	1.28	0.0242
I	8	0	2	0.73	−1.8	0.44	6.04	113.2	5.58	0
K	9	1	2	−1.1	3	−0.36	5.61	128.2	−4.18	0.0371
L	8	0	2	0.53	−1.8	0.4	6.035	113.2	5.01	0
M	8	0	2	0.26	−1.3	0.66	5.705	131.2	3.51	0.0823
N	8	0	4	−0.64	0.2	0.12	5.43	114.1	−2.65	0.0036
P	7	0	2	−0.07	0	−0.25	6.295	97.1	−3.03	0.0198
Q	9	0	4	−0.69	0.2	−0.11	5.65	128.1	−2.76	0.0761
R	11	1	4	−1.76	−0.5	0.27	5.405	156.2	−0.93	0.0959
S	6	0	4	−0.26	0.3	−0.33	5.7	87.1	−2.84	0.0829
T	7	0	4	−0.18	−0.4	−0.18	5.595	101.1	−1.2	0.0941
V	7	0	2	0.54	−1.5	0.27	6.015	99.1	4.45	0.0057
W	14	0	3	0.37	−3.4	0.83	5.935	186.2	5.2	0.0548
Y	12	0	3	0.02	−2.3	0.66	5.705	163.2	2.15	0.0516

2.3 Disease Similarity Based on Protein Functions

If two diseases affect the same molecular function, they tend to have similar mechanisms. Based on the Gene Ontology (GO) database where we can obtain molecular functions, the disease protein associated molecular functions were first identified. A GO term was seen to be associated with a disease if this term was enriched with the disease-associated proteins with hyper-geometric test (P-value < 0.005).

Given two diseases and their associated GO terms about molecular function, the disease similarity based on protein functions can be defined as follows.

$$sim(d_i, d_j)_{pf} = \frac{\sum_{k=1}^{m} \sum_{l=1}^{n} sim_{Lin}(r_k^i, r_l^j)}{m \times n} \qquad (4)$$

where r_k^i is the k-th term in the set of GO terms about protein associated with disease i so that m and n represents the corresponding number of GO terms associated with disease i and disease j and sim_{Lin} is the semantic similarity between two GO terms r_k^i and r_l^j, which was made by the Lin's entropy similarity theory and was implemented with the R package GOSemSim [17].

2.4   Identification of Disease Comorbidity

With various information of disease similarity said above, we defined similarity between two diseases i and j used in our proposed method ICPFP as follows.

$$csim(d_i, d_j) = \frac{1}{3} \sum_{k \in \{pro, pp, pf\}} sim_k(d_i, d_j) \tag{5}$$

To make predictions, the threshold based on $csim(d_i, d_j)$ from known pairs of disease comorbidity association was formulated as follows.

$$threshold = \frac{1}{N} \sum_{i=1}^{N} csim(d_1^i, d_2^i) \tag{6}$$

where N is number of known disease comorbid pairs, and d_1^i and d_2^i denotes disease 1 and disease 2 belonging to i-th pair of known disease comorbidity.

To evaluate the performance of our proposed algorithm, the 10-fold cross-validation was employed here with *precision, recall* and *F1 score* to test the performance of this algorithm (Fig. 1).

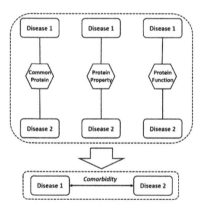

Fig. 1. Flow chart of ICPFP

3   Results

3.1   Benchmark Results on Real Data

To validate the performance of ICPFP, we applied it to the 1395 disease comorbidities collected from [12]. Besides, we also proposed three other algorithms derived from ICPFP, named ICP, ICPP, ICFP. In ICP method, two diseases were regarded in comorbidity association if their disease similarity based on disease-associated proteins above the certain threshold. The ICPP and ICFP were defined in the similar way, which was based on protein properties and protein functions, respectively. Furthermore, we also compared our algorithm with another three methods included in comoR package by Moni et al. [3, 18], named comorbidityOMIM, comorbidityPath and comorbidityDO. Since Moni et al. [3] have demonstrated that these three algorithms can perform better than methods proposed in [16], we just compared our algorithm with these three methods here to show the performance of ICPFP. Table 2 shows the performance of different algorithms and we can see that our algorithm ICPFP performs best in all metrics, which powerfully demonstrates the effectiveness of ICPFP.

Table 2. The Benchmark Results of Different Algorithms

Algorithms	Precision	Recall	F1
ICPFP	**0.687**	**0.733**	**0.698**
ICP	0.511	0.534	0.518
ICFP	0.545	0.571	0.553
ICPP	0.574	0.615	0.587
comorbidityOMIM	0.415	0.352	0.387
comorbidityPath	0.538	0.519	0.521
comorbidityDO	0.522	0.481	0.496

3.2 Prediction of Novel Comorbidities

Except for these 1395 comorbid pairs from 356 diseases, we also predict another 188 pairs of comorbidities, in which 86 pairs have been recorded in literature. For example, lymphoma (ICD-9-CM: 202.8) and breast cancer (ICD-9-CM: 174) were predicted to be in comorbidity by our method [19]. Furthermore, by searching the PubMed database, there are more examples of predicted comorbidities reported in literature can be seen in Table 3.

Table 3. The Predicted Comorbidities Validated in Literature

Disease 1 Name (ICD-9-CM)	Disease 2 Name (ICD-9-CM)	PMID
Ovarian cancer (183)	nasopharyngeal carcinoma (147.9)	24586749
Ovarian cancer (183)	epithelioma (173.9)	23633455
Prostate cancer (185)	colon cancer (153)	26376852
Lymphoma (202.8)	pilomatricoma (216.4)	25880568
Histiocytoma (216.9)	chondrosarcoma (170.9)	25493233

4 Conclusion and Discussion

In this paper, we propose a novel algorithm named ICPFP, which is used for prediction of disease comorbidity based on integration of different types of datasets ranging from properties to functions of protein. Compared with traditional methods, ICPFP makes use of heterogeneous information of protein depicting diseases, including disease associated proteins, amino acid frequency, physicochemical properties, protein functions, where these multiple datasets can effectively complement with each other to improve the accuracy of prediction. Besides, based on the same 20 kinds of amino acids for every protein, more comorbidities can be found by similarities of every two diseases than just discriminating the comorbidity association between two diseases by common modules based on incomplete mutant genes, PPIs or pathways, which can help deeply explain the molecular mechanism of disease comorbidity.

Acknowledgements. This work was supported by the grants of the National Science Foundation of China, Nos. 61133010, 61520106006, 31571364, 61532008, 61572364, 61373105, 61303111, 61411140249, 61402334, 61472282, 61472280, 61472173, 61572447, and 61373098, China Postdoctoral Science Foundation Grant, Nos. 2014M561513 and 2015M580352.

References

1. Capobianco, E., Lio, P.: Comorbidity: a multidimensional approach. Trends Mol. Med. **19** (9), 515–521 (2013)
2. Hidalgo, C.A., Blumm, N., Barabasi, A.L., Christakis, N.A.: A dynamic network approach for the study of human phenotypes. PLoS Comput. Biol. **5**(4), e1000353 (2009)
3. Moni, M.A., Lio, P.: comoR: a software for disease comorbidity risk assessment. J. Clin. Bioinf. **4**, 8 (2014)
4. Haffner, S.M., Lehto, S., Ronnemaa, T., Pyorala, K., Laakso, M.: Mortality from coronary heart disease in subjects with type 2 diabetes and in nondiabetic subjects with and without prior myocardial infarction. New Engl. J. Med. **339**(4), 229–234 (1998)
5. Zhu, L., Deng, S.P., Huang, D.S.: A two-stage geometric method for pruning unreliable links in protein-protein networks. IEEE Trans. Nanobiosci. **14**(5), 528–534 (2015)
6. Gijsen, R., Hoeymans, N., Schellevis, F.G., Ruwaard, D., Satariano, W.A., van den Bos, G. A.: Causes and consequences of comorbidity: a review. J. Clin. Epidemiol. **54**(7), 661–674 (2001)
7. Starfield, B., Lemke, K.W., Bernhardt, T., Foldes, S.S., Forrest, C.B., Weiner, J.P.: Comorbidity: implications for the importance of primary care in 'case' management. Ann. Fam. Med. **1**(1), 8–14 (2003)
8. Struijs, J.N., Baan, C.A., Schellevis, F.G., Westert, G.P., van den Bos, G.A.: Comorbidity in patients with diabetes mellitus: impact on medical health care utilization. BMC Health Serv. Res. **6**, 84 (2006)
9. Zhu, L., You, Z.H., Huang, D.S., Wang, B.: t-LSE: a novel robust geometric approach for modeling protein-protein interaction networks. PLoS ONE **8**(4), e58368 (2013)
10. Goh, K.I., Cusick, M.E., Valle, D., Childs, B., Vidal, M., Barabasi, A.L.: The human disease network. Proc. Natl. Acad. Sci. U.S.A. **104**(21), 8685–8690 (2007)
11. Zhu, L., Guo, W.L., Deng, S.P., Huang, D.S.: ChIP-PIT: enhancing the analysis of ChIP-Seq data using convex-relaxed pair-wise interaction tensor decomposition. IEEE/ACM Trans. Comput. Biol. Bioinform. / IEEE, ACM **13**(1), 55–63 (2016)
12. Park, J., Lee, D.S., Christakis, N.A., Barabasi, A.L.: The impact of cellular networks on disease comorbidity. Mol. Syst. Biol. **5**, 262 (2009)
13. Lee, D.S., Park, J., Kay, K.A., Christakis, N.A., Oltvai, Z.N., Barabasi, A.L.: The implications of human metabolic network topology for disease comorbidity. Proc. Natl. Acad. Sci. U.S.A. **105**(29), 9880–9885 (2008)

14. Rual, J.F., Venkatesan, K., Hao, T., Hirozane-Kishikawa, T., Dricot, A., Li, N., Berriz, G.F., Gibbons, F.D., Dreze, M., Ayivi-Guedehoussou, N., Klitgord, N., Simon, C., Boxem, M., Milstein, S., Rosenberg, J., Goldberg, D.S., Zhang, L.V., Wong, S.L., Franklin, G., Li, S., Albala, J.S., Lim, J., Fraughton, C., Llamosas, E., Cevik, S., Bex, C., Lamesch, P., Sikorski, R.S., Vandenhaute, J., Zoghbi, H.Y., Smolyar, A., Bosak, S., Sequerra, R., Doucette-Stamm, L., Cusick, M.E., Hill, D.E., Roth, F.P., Vidal, M.: Towards a proteome-scale map of the human protein-protein interaction network. Nature **437**(7062), 1173–1178 (2005)

15. Zheng, C.H., Zhang, L., Ng, V.T., Shiu, S.C., Huang, D.S.: Molecular pattern discovery based on penalized matrix decomposition. IEEE/ACM Trans. Comput. Biol. Bioinform. / IEEE, ACM **8**(6), 1592–1603 (2011)

16. Yu, G., Li, F., Qin, Y., Bo, X., Wu, Y., Wang, S.: GOSemSim: an R package for measuring semantic similarity among GO terms and gene products. Bioinformatics **26**(7), 976–978 (2010)

17. Chatr-Aryamontri, A., Breitkreutz, B.J., Heinicke, S., Boucher, L., Winter, A., Stark, C., Nixon, J., Ramage, L., Kolas, N., O'Donnell, L., Reguly, T., Breitkreutz, A., Sellam, A., Chen, D., Chang, C., Rust, J., Livstone, M., Oughtred, R., Dolinski, K., Tyers, M.: The BioGRID interaction database: 2013 update. Nucleic Acids Res. **41**(Database issue), D816–D823 (2013)

18. Siraj, A.K., Beg, S., Jehan, Z., Prabhakaran, S., Ahmed, M., Hussain, A.R., Al-Dayel, F., Tulbah, A., Ajarim, D., Al-Kuraya, K.S.: ALK alteration is a frequent event in aggressive breast cancers. Breast Cancer Res. BCR **17**, 127 (2015)

19. Fang, L.T., Lee, S., Choi, H., Kim, H.K., Jew, G., Kang, H.C., Chen, L., Jablons, D., Kim, I. J.: Comprehensive genomic analyses of a metastatic colon cancer to the lung by whole exome sequencing and gene expression analysis. Int. J. Oncol. **44**(1), 211–221 (2014)

Identification of HOT Regions in the Human Genome Using Differential Chromatin Modifications

Feng He[(⊠)] and Ning Li

Institute of Machine Learning and Systems Biology, College of Electronics
and Information Engineering, Tongji University, Shanghai 201804, China
{hefeng19890620,jclining}@126.com

Abstract. HOT regions, short for high occupied target regions, bound by many transcription factors (TFs) are considered to be one of the most intriguing findings of the recent large-scale sequencing studies. Recent researches have reported that HOT regions are enriched with so many biological processes and functions, which are related with promoters, enhancers and fraction of motifs. Hence, there are a lot of studies focused on the discovery of HOT regions with TFs datasets. Unfortunately, because of the limited TFs datasets from next generation sequencing (NGS) technology and huge time consuming, the HOT regions in each cell line of human genome can't be fully marked. Here, unlike the previous jobs, we have made an identification of HOT regions by means of machine learning algorithms in 14 different human cell-lines with chromatin modification datasets. The outperform results of these cell-lines can prove the effectiveness and precision of our assumption enough. In addition, we have discovered the cell-type specific HOT regions (CSHRs) of each cell line, which is used to elucidate the associations with cell-type specific regulatory functions.

Keywords: HOT regions · Histone modification · Machine learning · Cell-type specificity

1 Introduction

Transcription is regulated through functional elements in the genome such as promoters and enhancers. Identifying these genomic elements, many of which are recognized by transcription factors (TFs), is a necessary first step in their functional analysis. Toward this goal, the modENCODE and ENCODE Projects have used chromatin immuno-precipitation (ChIP) to map the binding sites for a large number of TFs in *C. elegans*, Drosophila, and humans [1–5]. In contrast, these mapping studies also identified an unusual class of sites bound by the majority of mapped TFs, termed HOT (Highly Occupied Target) regions. Consistent with this idea, Drosophila HOT regions are depleted for annotated enhancers; however, a significant fraction displays enhancer activity in transgenic analyses [6]. Drosophila and human HOT regions have features of open chromatin, such as nucleosome depletion and high turnover, and in *C. elegans* they are usually located near genes with ubiquitous expression [4, 5, 7, 8]. These previous reports suggest that HOT regions are active genomic elements but fail to

© Springer International Publishing Switzerland 2016
D.-S. Huang et al. (Eds.): ICIC 2016, Part III, LNAI 9773, pp. 856–861, 2016.
DOI: 10.1007/978-3-319-42297-8_79

provide a clear picture of their function, possibly because of differences in the definition and characterization of these regions [9–13].

Here, unlike the previous jobs, we have made an identification of HOT regions by means of machine learning algorithms in 14 different human cell-lines with chromatin modification datasets. The outperform results of these cell-lines can prove the effectiveness and precision of our assumption enough, which can be used as a new method for identification of HOT regions regardless of the limitation of ChIP-seq datasets of TFs and the issue of time consuming. In addition, we have discovered the CSHRs of each cell line, which is used to elucidate the associations with cell-type specific regulatory processes.

2   Material and Methods

2.1   Identification of HOT Regions

Genome-wide maps of 11 histone modifications, which is comprised of H3K27me3, H3K36me3, H4K20me1, H3K4me1, H3K4me2, H3K4me3, H3K27ac, H3K9ac, H2az, H3K9me3 and H3K79me2 and DNase I hypersensitive sites (DHSs) [14] from ENCODE database have been used with ChIP-seq data style here to identify the HOT regions in each of these 14 cell lines. And the HOT region datasets of these 14 cell-lines were achieved from GSE54296 in Gene Expression Omnibus (GEO) database with binary format to represent HOT regions or not.

The raw ChIP-seq data of DHSs and 11 kinds of histone modifications were preprocessed as described in [15]. Whole-genome data were divided into non-overlapping 200 bp bins. Then, the HOT regions in every cell-line were identified with chromatin modification datasets by machine learning algorithms. Firstly, for each cell type, 5000 bins of HOT regions and 5000 bins of non-HOT regions were randomly selected just like what the previously study did [7]. Secondly, chromatin modification datasets were used in 10000 bins have been used with SVM, Random Forest and Logistic Regression algorithms to identify the HOT regions in each cell-line.

To evaluate the performance of our assumption, 10-fold cross-validation was employed here with precision, recall and F1 score. This procedure of classification in each cell-line will be processed in 1000 times with random sample selection.

2.2   Identification of CSHRs

After data preprocessing, we obtained binary HOT regions profile in every cell-line. We introduce the following steps to identify CSHRs just like the research on the discovery of cell-type specific regulatory elements (CSREs) [16].

Step 1: Calculation of the differential Hamming score.
For each genomic position (a 200 bp bin) c, hamming distance was used to measure differences of the modification characteristics between each pair of cell types just as follows.

$$DHS_i(c) = \sum_{j=1}^{14} HD\big(b_i(c), b_j(c)\big) \tag{1}$$

where i represents the i-th cell type, HD means the Hamming Distance operator, b is the operator of binary value of HOT region in the certain cell type of coordinate c.
Step 2: Normalization and correction of the DHSs.
First, we normalized the sum of squares of each cell type's DHS profile to a constant. We further calculated the Z-score of DHSs for each bin among all cell types.

$$z_i(c) = \frac{DHS_i(c) - \mu(c)}{\sigma(c)} \tag{2}$$

where $\mu(c)$ and $\sigma(c)$ are the mean and the standard deviation of the DHS at genomic position c, respectively. The normalized DHS were then multiplied by the corresponding Z-scores to get the corrected scores as follows.

$$DS_i(c) = DHS_i(c) \times z_i(c) \tag{3}$$

The corrected scores reflect both the scale and relative size of the original scores.
Step 3: Wavelet smoothing.
The wavelet transform is a widely used filtering and smoothing technique, which has been comprehensively applied in computational biology [17, 18]. We applied the maximal overlap discrete wavelet transform [19], a modification of discrete wavelet transform, to the corrected DHS profiles to further reduce noises and enhance the signal-to-noise ratio just like in Fig. 1. Finally, the results with height = 1.5 and length = 12 were used to illustrate the strong biological relevance of CSHRs. As to the selection of these two parameters, we performed enrichment analysis on CSHRs identified from four other groups of parameter settings to show their robustness.

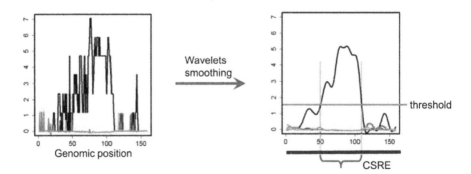

Fig. 1. The wavelets smoothing workflow in H1 cell.

3   Results

3.1   Benchmark Results of HOT Regions Identification

To validate our assumption, we performed 1000 times of experiments by random selection of 10000 200 bp bins in 14 cell types. And to identify these HOT regions with three classification algorithms Random Forest, SVM and Logistic Regression, respectively. The benchmark results of HOT region identification of 14 cell-lines have been displayed in Table 1. From Table 1, we can observed that the results of three algorithms can fully prove the correction and effectiveness of our assumption that though HOT regions in human genome are defined with TFs, they can be identified with chromatin modification datasets.

Table 1. Mean results of three algorithms on 14 cell-lines

Cell-lines	Random forest			SVM			Logistic regression		
	Precision	Recall	F1	Precision	Recall	F1	Precision	Recall	F1
A549	0.813	0.875	0.842	0.773	0.835	0.803	0.701	0.766	0.723
HepG2	0.711	0.745	0.723	0.651	0.695	0.672	0.593	0.661	0.621
K562	0.804	0.775	0.782	0.726	0.7	0.713	0.684	0.65	0.671
NHA	0.767	0.828	0.796	0.704	0.748	0.725	0.623	0.648	0.625
NHDFAd	0.703	0.744	0.733	0.665	0.689	0.677	0.574	0.632	0.618
GM12878	0.759	0.793	0.774	0.685	0.75	0.716	0.664	0.683	0.672
H1hESC	0.716	0.779	0.742	0.676	0.738	0.706	0.622	0.705	0.649
NHLF	0.756	0.809	0.781	0.702	0.747	0.724	0.581	0.596	0.583
HeLaS3	0.727	0.83	0.743	0.651	0.74	0.693	0.564	0.589	0.571
HMEC	0.731	0.811	0.789	0.697	0.778	0.736	0.688	0.746	0.723
HSMM	0.699	0.725	0.717	0.653	0.681	0.667	0.544	0.628	0.582
HSMMtube	0.762	0.815	0.778	0.711	0.755	0.732	0.624	0.686	0.655
HUVEC	0.713	0.797	0.754	0.664	0.748	0.704	0.594	0.639	0.607
NHEK	0.786	0.798	0.791	0.735	0.749	0.742	0.673	0.702	0.681

3.2   Regulatory Functions of CSHRs

As we obtained the CSHRs of 14 cell-lines with method described above, we investigated the transcriptional behavior of CSHR neighboring genes in the 14 cell types studied just like the work in [16]. Each cell type was divided into two groups: group denotes CSHR neighboring genes, and 'other' group denotes genes that are not in a proximal location to CSHRs. Only genes that could be mapped to Affymetrix probes are involved in this analysis. Expression levels were compared between these two groups in all nine cell types. All 'CSHRs neighboring' groups except that of H1hESC had significantly higher expression levels than the corresponding 'other' groups (P-value < 0.001, two-sample Wilcoxon tests) in Table 2. We also found that the neighboring genes for all cell types expect H1hESC contains significantly more highly expressed genes than expectation (P-value < 0.005).

Table 2. Expression level of neighboring genes of CSHRs

Cell-line	Expression level of neighboring genes (P-value, 10^{-4})	Cell-line	Expression level of neighboring genes (P-value, 10^{-4})
A549	3.214	NHLF	4.866
HepG2	4.441	HeLaS3	7.294
K562	1.729	HMEC	6.528
NHA	2.283	HSMM	3.343
NHDFAd	6.052	HSMMtube	3.376
GM12878	5.533	HUVEC	1.133
H1hESC	30.213	NHEK	8.852

4   Conclusion and Discussion

In this paper, we proposed an assumption that chromatin modification datasets could make identification of HOT regions in human genome unlike the previous jobs used a large number of TFs ChIP-seq datasets. We have made the identification of HOT regions by means of machine learning algorithms in 14 different human cell-lines with chromatin modification datasets. The outperform results of these cell-lines can prove the effectiveness and precision of our assumption enough, which can be used as a new method for identification of HOT regions regardless of the limitation of ChIP-seq datasets of TFs and the issue of time consuming. In addition, we have discovered the CSHRs of each cell line, which is used to elucidate the associations with cell-type specific regulatory processes.

Acknowledgements. This work was supported by the grants of the National Science Foundation of China, Nos. 61133010, 61520106006, 31571364, 61532008, 61572364, 61373105, 61303111, 61411140249, 61402334, 61472282, 61472280, 61472173, 61572447, and 61373098, China Postdoctoral Science Foundation Grant, Nos. 2014M561513 and 2015M580352.

References

1. ENCODE Project Consortium, Birney, E., Stamatoyannopoulos, J.A., Dutta, A., Guigo, R., Gingeras, T.R., et al.: Identification and analysis of functional elements in 1 % of the human genome by the ENCODE pilot project. Nature **447**(7146), 799–816 (2007)
2. ENCODE Project Consortium: A user's guide to the encyclopedia of DNA elements (ENCODE). PLoS Biol. **9**(4), e1001046 (2011)
3. ENCODE Project Consortium: An integrated encyclopedia of DNA elements in the human genome. Nature **489**(7414), 57–74 (2012)
4. Gerstein, M.B., Lu, Z.J., Van Nostrand, E.L., Cheng, C., Arshinoff, B.I., et al.: Integrative analysis of the Caenorhabditis elegans genome by the modENCODE project. Science **330** (6012), 1775–1787 (2010)

5. Gerstein, M.B., Kundaje, A., Hariharan, M., Landt, S.G., Yan, K.K., Cheng, C., Mu, X.J., et al.: Architecture of the human regulatory network derived from ENCODE data. Nature **489**(7414), 91–100 (2012)

6. Kvon, E.Z., Stampfel, G., Yanez-Cuna, J.O., Dickson, B.J., Stark, A.: HOT regions function as patterned developmental enhancers and have a distinct cis-regulatory signature. Genes Dev. **26**(9), 908–913 (2012)

7. Zhu, L., Deng, S.P., Huang, D.S.: A two-stage geometric method for pruning unreliable links in protein-protein networks. IEEE Trans. Nanobiosci. **14**(5), 528–534 (2015)

8. Zhu, L., You, Z.H., Huang, D.S., Wang, B.: t-LSE: a novel robust geometric approach for modeling protein-protein interaction networks. PLoS ONE **8**(4), e58368 (2013)

9. modENCODE Consortium, Roy, S., Ernst, J., Kharchenko, P.V., Kheradpour, P., Negre, N., Eaton, M.L., Landolin, J.M., Bristow, C.A., et al.: Identification of functional elements and regulatory circuits by Drosophila modENCODE. Science **330**(6012), 1787–1797 (2010)

10. Chen, H., Li, H., Liu, F., Zheng, X., Wang, S., Bo, X., Shu, W.: An integrative analysis of TFBS-clustered regions reveals new transcriptional regulation models on the accessible chromatin landscape. Sci. Rep. **5**, 8465 (2015)

11. Li, H., Chen, H., Liu, F., Ren, C., Wang, S., Bo, X., Shu, W.: Functional annotation of HOT regions in the human genome: implications for human disease and cancer. Sci. Rep. **5**, 11633 (2015)

12. Chen, R.A., Stempor, P., Down, T.A., Zeiser, E., Feuer, S.K., Ahringer, J.: Extreme HOT regions are CpG-dense promoters in C. elegans and humans. Genome Res. **24**(7), 1138–1146 (2014)

13. Yan, J., Enge, M., Whitington, T., Dave, K., Liu, J., Sur, I., Schmierer, B., Jolma, A., Kivioja, T., Taipale, M., Taipale, J.: Transcription factor binding in human cells occurs in dense clusters formed around cohesin anchor sites. Cell **154**(4), 801–813 (2013)

14. Gaszner, M., Felsenfeld, G.: Insulators: exploiting transcriptional and epigenetic mechanisms. Nat. Rev. Genet. **7**(9), 703–713 (2006)

15. Ernst, J., Kheradpour, P., Mikkelsen, T.S., Shoresh, N., Ward, L.D., Epstein, C.B., Zhang, X., Wang, L., Issner, R., Coyne, M., Ku, M., Durham, T., Kellis, M., Bernstein, B.E.: Mapping and analysis of chromatin state dynamics in nine human cell types. Nature **473**(7345), 43–49 (2011)

16. Chen, C., Zhang, S., Zhang, X.S.: Discovery of cell-type specific regulatory elements in the human genome using differential chromatin modification analysis. Nucleic Acids Res. **41**(20), 9230–9242 (2013)

17. Klevecz, R.R., Murray, D.B.: Genome wide oscillations in expression. Wavelet analysis of time series data from yeast expression arrays uncovers the dynamic architecture of phenotype. Mol. Biol. Rep. **28**(2), 73–82 (2001)

18. Zhu, L., Guo, W.L., Deng, S.P., Huang, D.S.: ChIP-PIT: enhancing the analysis of ChIP-Seq data using convex-relaxed pair-wise interaction tensor decomposition. IEEE/ACM Trans. Comput. Biol. Bioinform. **13**(1), 55–63 (2016)

19. Bullmore, E., Long, C., Suckling, J., Fadili, J., Calvert, G., Zelaya, F., Carpenter, T.A., Brammer, M.: Colored noise and computational inference in neurophysiological (fMRI) time series analysis: resampling methods in time and wavelet domains. Hum. Brain Mapp. **12**(2), 61–78 (2001)

Novel Algorithm for Multiple Quantitative Trait Loci Mapping by Using Bayesian Variable Selection Regression

Lin Yuan[1], Kyungsook Han[2], and De-Shuang Huang[1(✉)]

[1] College of Electromics and Information Engineering,
Tongji University, Shanghai 201804, China
yuanlindc@126.com, dshuang@tongji.edu.cn
[2] Department of Computer Science and Engineering,
Inha University, Incheon, South Korea
khan@inha.ac.kr

Abstract. Most quantitative trait loci (QTL) mapping experiments typically collect phenotypic data on single traits. However, Research complex correlated traits may provide more available information. We develop a novel algorithm for multiple traits quantitative trait loci mapping by using Bayesian Variable Selection Regression, or BVSR, that allows a new robust genetic models for different and correlated traits. We develop computationally efficient Markov chain Monte Carlo (MCMC) algorithms for performing joint analysis. Taken together, these factors put a premium on having interpretable measures of confidence for individual covariates being included in the model. We conduct extensive simulation studies to assess the performance of the proposed methods and to compare with the conventional single-trait model and existing multiple-trait model. More generally, we demonstrate that, despite the apparent computational challenges, our proposed new algorithm can provide useful inferences in quantitative trait loci mapping.

Keywords: Bayesian variable selection regression · MCMC · Gibbs sampler · Acceptance-Rejection sampling

1 Introduction

Nowadays expression quantitative trait loci (eQTL) mapping has proven to be an effective tool in the area of studying the relationship between gene expression values or phenotypic traits and genetic variants [1]. The location of single nucleotide polymorphisms (SNPs) which are connected with gene expression values are known as 'expression quantitative trait loci' (eQTL) [2]. In the past few years, numerous eQTL mapping studies dedicated to find genetic basis of complex traits. There are mainly three issues in the study of eQTL [3].

Most of these methods are applicable to mapping QTL for a single trait. However, in QTL experiments typically data on more than one trait are collected and, more often than not, they are correlated. It seems natural to jointly analyze these correlated traits [4]. Biologically, it is imperative to jointly analyze correlated traits to answer questions like pleiotropy (one gene influencing more than one trait) and/or close linkage

© Springer International Publishing Switzerland 2016
D.-S. Huang et al. (Eds.): ICIC 2016, Part III, LNAI 9773, pp. 862–868, 2016.
DOI: 10.1007/978-3-319-42297-8_80

(different QTL physically close to each other influencing the traits) [5]. Testing these hypotheses is key to understanding the underlying biochemical pathways causing complex traits, which is the ultimate goal of QTL mapping [6, 7].

The remainder of the paper is organized as follows. In Sect. 2 we describe BVSR and our choice of priors. In Sect. 3 we discuss computation and inference, including Markov chain Monte Carlo algorithms used. In Sect. 4 we examine, through simulations, the effectiveness of our new algorithm for various tasks [8].

2 Models and Priors

This section introduces notation and specifies the details of BVSR model and priors used. Our formulation is in the same vein as much previous work on BVSR, but with particular emphasis out putting priors on hyperparameters that are often considered fixed and known [9].

We consider the standard normal linear regression

$$y|u, \beta, X, \tau \sim N_n(u + X\beta, \tau^{-1}I_n) \tag{1}$$

Relating a response variable y to covariates X. Here y is an n-vector of observation n individuals, μ is an n-vector with components all equal to the same scalar μ, X is an n by p matrix of covariates, β is a p-vector of regression coefficients, τ denotes the inverse variance of the residual errors, $N_n(\cdot, \cdot)$ denotes the n-dimensional multivariate normal distribution and I_n the n by n identity matrix [10]. In many contexts, the number of covariates is very large—and, in particular, $p \gg n$—but only a small subset of the covariates are expected to be associated with the response (i.e., have nonzero β_j). Indeed, the main goal of GWAS is to identify these relevant covariates. To this end, we define a vector of binary indicators $\gamma = (\gamma_1, ..., \gamma_p) \in \{0, 1\}^P$ that indicate which elements of β are nonzero. Thus,

$$y|\gamma, u, \tau, \beta, X \sim N_n(u + X_\gamma \beta_\gamma, \tau^{-1}I_n) \tag{2}$$

Taking a Bayesian approach to inference, we put priors on the parameters:

$$\tau \sim Gamma(\lambda/2, \kappa/2) \tag{3}$$

$$u|\tau \sim N(0, \sigma_u^2/\tau) \tag{4}$$

$$\gamma_j \sim Bernoulli(\pi) \tag{5}$$

$$\beta_\gamma|\tau, \gamma \sim N_{|\gamma|}\left(0, (\sigma_a^2/\tau)I_{|\gamma|}\right) \tag{6}$$

$$\beta_{-\gamma}|\gamma \sim \delta_0 \tag{7}$$

Where $|\gamma| := \sum_j \gamma_j$, $\beta_{-\gamma}$ denotes the vector of β coefficients for which $\gamma_j = 0$, and δ_0 denotes a point mass on 0. Here π, σ_a, λ, κ, and σ_u are hyperparameters. The hyperparameters π and σ_a have important roles, with π reflecting the sparsity of the model, and σ_a reflecting the typical size of the nonzero regression coefficients. Rather than setting these hyperparameters to fixed values, we place priors on them, hence allowing their values to be informed by the data [10].

Here we suggest specifying a prior on σ_a^2 given γ by considering the induced prior on the PVE, and, in particular, by making this induced prior relatively flat in the range of $(0, 1)$. To formalize this, let $V(\beta, \tau)$ denote the empirical variance of $X\beta$ relative to the residual variance τ^{-1}:

$$V(\beta, \tau) = \frac{1}{n} \sum_{i=1}^{n} \left[(X\beta)_i \right]^2 \tau \tag{8}$$

Where this expression for the variance assumes that the covariates have been centered, and so $X\beta$ has mean 0. Then the total proportion of variance in $\mathbf{y}$ explained by X if the true values of the regression coefficients are β is given by

$$PVE(\beta, \tau) = V(\beta, \tau)/(1 + V(\beta, \tau)) \tag{9}$$

Our aim is to choose a prior on β given τ so that the induced prior on PVE (β, τ) is approximately uniform. To do this, we exploit the fact that the expected value of $V(\beta, \tau)$ (with expectation being taken over $\beta|\tau$) depends in a simple way on σ_a:

$$v(\gamma, \sigma_a) = E[V(\beta, \tau)|\gamma, \sigma_a, \tau] = \sigma_a^2 \sum_{j:\gamma_j=1} s_j \tag{10}$$

where $s_j = \frac{1}{n} \sum_{i=1}^{n} x_{ij}^2$ is the variance of covariate j. Define

$$h(\gamma, \sigma_a) = v(\gamma, \sigma_a)/(1 + v(\gamma, \sigma_a)) \tag{11}$$

To accomplish our goal of putting approximately uniform prior on PVE, we specify a uniform prior on h, independent of γ, which induces a prior on σ_a given γ via the relationship

$$\sigma_a^2(h, \gamma) = \frac{h}{1-h} \frac{1}{\sum_{j:\gamma_j=1} s_j} \tag{12}$$

3 Computation and Inference

We use Markov chain Monte Carlo to obtain samples from the posterior distribution of (h, π, γ) on the product space $(0, 1) \times (0, 1) \times \{0, 1\}p$, which is given by

$$p(h, \pi, \gamma|\mathbf{y}) \propto p(\mathbf{y}|h, \gamma)p(h)p(\gamma|\pi)p(\pi) \tag{13}$$

In particular, it explores the space of covariates included in the model, γ, by proposing to add, remove, or switch single covariates in and out of the model. To improve computational performance, we use two strategies [11, 12, 13]. First, in addition to the local proposal moves, we sometimes make a longer-range proposal by compounding randomly many local moves.

To develop MCMC an efficient strategy. We use acceptance/rejection sampling. Recall that a random variable with density f may be sampled using the accept/reject algorithm by (1) proposing X from a density g; (2) drawing $U \sim U(0, cg(X))$ where $\|f/g\|_\infty \leq c$; and (3) accepting X if $U \leq f(X)$ and rejecting X otherwise [14, 15].

4 Design and Method

With an increased complexity and sophistication of a proposed method, it is very important to compare its performance with existing methods in an objective way. We consider a backcross population with sample sizes of 200, 400, and 800 to represent very small, small, and large sample sizes. Two continuous traits (y_1 and y_2) are considered for simplicity. We simulate a genome with 19chromosomes, each of length 100 cM with 11 equally spaced markers (markers placed 10 cM apart) on each chromosome. Ten percent of the genotypes of these markers were assumed to be randomly missing in all cases. For each of the three sample sizes, we consider two correlation structures, namely, low and high with $\rho_{y1y2} = 0.5$ and $\rho_{y1y2} = 0.8$. Therefore, we have six cases with three samples sizes and two correlation structures. For each of these six cases, we simulate six QTL (E_1–E_6) that control the phenotypes: E_1 and E_2 (E_3 and E_4) are nonpleitropic QTL, influencing only the trait y_1 (y_2) with moderate-sized and weak effects, respectively; E_5 is a moderate-sized pleiotropic QTL affecting both y_1 and y_2; while E_6 is a weak pleiotropic QTL affecting both $y_1$and y_2. Table 1 presents the simulated positions of six QTL, their effect values, and their heritability (proportion of the phenotypic variation explained by a QTL) [16].

Table 1. True positions of six QTL, their effects, and heritability

	E_1	E_2	E_3	E_4	E_5	E_6
Chromosome	2	2	4	4	5	6
Position	24	56	24	65	65	45
y_1	0.8	0.6	0	0	0.8	0.6
y_2	0	0	−0.8	−0.6	0.8	0.6
y_1 (%)	8.9	4.8	0	0	8.9	4.8
y_2 (%)	0	0	9.2	5.4	9.2	5.4

For all analyses, pseudo markers were placed every 1 cM across the entire genome, resulting in a total of 1919 possible QTL positions. The prior expected number of main-effect QTL was setat $l_0 = 4$, and the upper bound on the number of QTL was then $L = 10$. To check posterior sensitivity to these fixed values, we analyzed the data with several other values of l_0 and Land obtained essentially identical results. We ran the MCMC algorithm for 12×10^4 times after discarding the first 1000 iterations as burn-in. The chain was thinned by considering one in every 40 samples, rendering

Table 2. Average correct and incorrect QTL detected for traits y_1 (first row) and y_2 (second row)

(n, ρ_{y1y2})	Correct			Extraneous		
	STA	TMV	BVMC	STA	TMV	BVMC
(200,0.5)	1.32	1.61	1.34	1.40	2.68	1.08
	1.48	1.56	1.49	0.78	2.72	1.18
(200,0.8)	0.68	2.02	2.04	0.48	3.72	1.63
	1.56	2.14	2.54	1.42	3.44	1.68
(400,0.5)	3.40	4.26	4.24	2.12	5.12	1.62
	3.52	4.4	4.32	1.26	5.12	1.64
(400,0.8)	3.02	5.21	4.67	1.26	5.98	1.48
	3.52	5.22	5.15	1.92	5.96	1.68
(800,0.5)	6.08	6.45	6.41	2.02	6.21	1.68
	7.21	7.61	7.42	1.52	6.14	1.42
(800,0.8)	7.12	7.62	7.41	2.2	6.28	2.02
	7.28	7.60	7.42	2.36	6.22	1.41

3000 samples from the joint posterior distribution. The saved posterior samples were used to make inference about the genetic architecture [17].

For all six cases we simulate 100 null (no-QTL) data sets and compute the genome wide maximum 2 logeBF (twice the natural logarithm of Bayes factors) for each trait. The 95th percentile of the max 2 logeBF empirical distribution is considered as the threshold value above which a QTL would be deemed "significant." At each replication, the number of correctly identified QTL and the number of incorrectly identified or extraneous QTL are recorded. A peak in the 2 $\log_e$BF profile is considered a QTL if it crosses the significance threshold. It is deemed correct if it is within 10 cM [9] of a true QTL. If there is more than one peak within 10 cM of the true QTL, only one is considered correct. Results: Table 2 represents the average correct and extraneous (incorrect) QTL detections for the six situations and for all three methods for y_1 and y_2, respectively [18]. The average times taken to conduct each MCMC for all six cases and three methods are presented in Table 3. TMV was the fastest in all cases followed by BVMC, STA. However, the maximum difference between the fastest and the slowest was only 0.48 min (29 s). So computationally complexity does not really pose a great threat [19].

Table 3. Average MCMC time (in minutes) for three methods

$\left(n, \rho_{y_1y_2}\right)$	STA	TMV	BVMC
(200,0.5)	2.35	1.97	2.21
(200,0.8)	2.35	1.99	2.22
(400,0.5)	4.97	3.99	4.46
(400,0.8)	4.98	4.02	4.47
(800,0.5)	10.12	9.56	10.01
(800,0.8)	10.15	9.67	10.02

In conclusion, it is evident and expected that the multivariate procedures outperform STA in the small sample size and high correlation situations. The BVMC models performed well. If one wants to detect only pleiotropic QTL, a traditional multivariate model can be used, but, in any other situations, a BVMC procedure is recommended in light of a marginal increase in computational time.

Acknowledgements. This work was supported by the grants of the National Science Foundation of China, Nos. 61133010, 61520106006, 31571364, 61532008, 61572364, 61373105, 61303111, 61411140249, 61402334, 61472282, 61472280, 61472173, 61572447, and 61373098, China Postdoctoral Science Foundation Grant, Nos. 2014M561513 and 2015M580352.

References

1. Li, Y., Zhang, M., Zhao, M.: eQTL, Quantitative Trait Loci (QTL). Springer 265–279 (2012)
2. Mah, J.T., Chia, K.S.: A gentle introduction to SNP analysis: resources and tools. J. Bioinf. Comput. Biol. **5**(5), 1123–1138 (2007)
3. Cookson, W., Liang, L., Abecasis, G., Moffatt, M., Lathrop, M.: Mapping complex disease traits with global gene expression. Nat. Rev. Genet. **10**(3), 184–194 (2009)
4. Smith, L., Kruglyak, M.: Gene-environment interaction in yeast gene expression. PLoSBiol **6**(4), e83 (2008)
5. Brem, B., Kruglyak, L.: The landscape of genetic complexity across 5,700 gene expression traits in yeast. Proc. Natl. Acad. Sci. U.S.A. **102**(5), 1572–1577 (2005)
6. Gao, C., Tignor, N.L., Strulovici-Barel, Y., Hackett, N.R., Crystal, R.G.: HEFT: eQTL analysis of many thousands of expressed genes while simultaneously controlling for hidden factors. Bioinformatics **30**(3), 369–376 (2014)
7. Higo, K., Ugawa, M., Iwamoto, Y., Korenaga, T.: Plant CIS-acting regulatory DNA elements (PLACE) database. Nucleic Acids Res. **27**(1), 297–300 (1999)
8. Mahr, S., Burmester, G.R., Hilke, D., Göbel, U., Grützkau, A., Häupl, T., Hauschild, M., Koczan, D., Krenn, V., Neidel, J.: CIS-and trans-acting gene regulation is associated with osteoarthritis. Am. J. Hum. Genet. **78**(5), 793–803 (2006)
9. Cheng, W., Shi, Y., Zhang, X., Wang, W.: Fast and robust group-wise eQTL mapping using sparse graphical models. BMC Bioinformatics **16**, 2 (2015)
10. Huang, D.S., Yu, H.J.: Normalized feature vectors: a novel alignment-free sequence comparison method based on the numbers of adjacent amino acids. IEEE/ACM Trans. Comput. Biol. Bioinf. **10**(2), 457–467 (2013)
11. Zheng, C.H., Zhang, L., Ng, V.T.-Y., Shiu, C.K., Huang, D.S.: Molecular pattern discovery based on penalized matrix decomposition. IEEE/ACM Trans. Comput. Biol. Bioinf. **8**(6), 1592–1603 (2011)
12. Zheng, C.H., Huang, D.S., Zhang, L., Kong, K.Z.: Tumor clustering using non-negative matrix factorization with gene selection. IEEE Trans. Inf Technol. Biomed. **13**(4), 599–607 (2009)
13. Deng, S.P., Zhu, L., Huang, D.S.: Predicting hub genes associated with cervical cancer through gene co-expression networks. IEEE/ACM Trans. Comput. Biol. Bioinf. **13**(1), 27–35 (2016)

14. Zhu, L., Guo, W.L., Deng, S.P., Huang, D.S.: ChIP-PIT: Enhancing the analysis of ChIP-Seq data using convex-relaxed pair-wise interaction tensor decomposition. IEEE/ACM Trans. Comput. Biol. Bioinf. **13**(1), 55–63 (2016)

15. Zhu, L., Deng, S.P., Huang, D.S.: Two-stage geometric method for pruning unreliable links in protein-protein networks. IEEE Trans. Nano Biosci. **14**(5), 528–534 (2015)

16. Deng, S.P., Zhu, L., Huang, D.S.: Mining the bladder cancer-associated genes by an integrated strategy for the construction and analysis of differential co-expression networks. BMC Genom. **16** (2015)

17. Deng, S.P., Huang, D.S.: SFAPS: an R package for structure/function analysis of protein sequences based on informational spectrum method. Methods **69**(3), 207–212 (2014)

18. Huang, D.S., Zhang, L., Han, K., Deng, S.P., Yang, K., Zhang, H.B.: Prediction of protein-protein interactions based on protein-protein correlation using least squares regression. Curr. Protein Pept. Sci. **15**(6), 553–560 (2014)

19. Zhu, L., You, Z.H., Huang, D.S., Wang, B.: t-LSE: a novel robust geometric approach for modeling protein-protein interaction networks. PLoS ONE **8**(4), e58368 (2013)

Author Index

Printed in the United States
By Bookmasters